CAMBRIDGE AERONAUTICAL SERIES

General Editors
W. A. MAIR, M.A.
ERNEST F. RELF, C.B.E., F.R.S.

II

WING THEORY

WING THEORY

BY

A. ROBINSON, Ph.D., A.F.R.Ae.S.

*Associate Professor of Mathematics at
the University of Toronto, and formerly Deputy Head of
the Department of Aerodynamics at the
College of Aeronautics, Cranfield*

AND

J. A. LAURMANN, M.A., D.C.Ae.

Research Engineer at the University of California, Berkeley

CAMBRIDGE
AT THE UNIVERSITY PRESS
1956

CAMBRIDGE UNIVERSITY PRESS

Cambridge, New York, Melbourne, Madrid, Cape Town,
Singapore, São Paulo, Delhi, Mexico City

Cambridge University Press
The Edinburgh Building, Cambridge CB2 8RU, UK

Published in the United States of America by Cambridge University Press, New York

www.cambridge.org
Information on this title: www.cambridge.org/9781107622654

First published 1956
First paperback edition 2013

A catalogue record for this publication is available from the British Library

ISBN 978-1-107-62265-4 Paperback

PREFACE

ἔργμασιν ἐν μεγάλοις
πᾶσιν ἁδεῖν χαλεπόν

The present book is to a large extent based on lectures which were given by the senior author over a number of years at the College of Aeronautics, Cranfield, and at the University of Toronto. It should therefore be suitable as a text-book for advanced courses on the subject. More generally, it is intended for all those, in industry, at the universities, or at research establishments, who are interested in aerofoil theory for either practical or theoretical reasons. To follow it the reader requires a sound knowledge of the calculus and of the theory of functions of a complex variable. Detailed references are given whenever special functions such as Legendre polynomials or Bessel functions arise in the analysis. Those parts of general hydrodynamics which are required for the development of aerofoil theory are included in the first chapter.

Mathematical wing theory is of great practical importance and at the same time of considerable theoretical depth and interest. When writing a book on a subject such as this, one is faced, inevitably, with a number of conflicting requirements. There is the desirability of presenting a comprehensive account of the theory as against the practical need for keeping the undertaking within reasonable bounds; there are the demands of mathematical rigour involving at times the introduction of subtle techniques which might be a heavy burden for those who are interested in the subject chiefly for practical reasons; there is the question whether to describe only the most efficient method for the solution of a particular problem, although an earlier and less efficient method provides a better insight into the topic under consideration; and so on. We have tried to resolve all these difficulties by compromise, inclining sometimes to one side and sometimes to the other. We shall not try to justify our choice in detail; the result constitutes the present book, and it is left to the reader to decide whether we have found a reasonable solution to the problem.

We wish to express our thanks to Mr E. F. Relf, first Principal of the College of Aeronautics, for inviting us to contribute a volume to the Cambridge Aeronautical Series, and to Professor W. J. Duncan for his friendly interest in the progress of the project. It is no exaggeration to say that without Professor Duncan's continuous encouragement this book would not have been completed. We are also aware that we owe much of our understanding of the subject to discussions with friends in Great Britain and elsewhere, and in this sense we wish to put on record particularly our indebtedness to Professor H. B. Squire and Professor A. D. Young. Our thanks are due also to the Syndics and officials of the Cambridge University Press for undertaking the publication of the present work.

A. R.

J. A. L.

University of Toronto
University of California at Berkeley

December 1953

ERRATA

CONTENTS

CHAPTER 1

FOUNDATIONS

1·1 Introduction

The object of wing theory is the investigation and calculation of the aerodynamic forces which act on a wing, or on a system of wings, following a prescribed motion in a fluid medium, usually air. The theory is based on the assumption that the medium is continuous, and it can be shown that this assumption does not lead to any appreciable errors except at very high speeds or at very low pressures (compare § 4·1 below).

The fundamental results of classical hydrodynamics will be developed in this chapter to the extent required later. For a more complete exposition of the theory the reader is referred to Lamb's standard work.‡

1·2 Fluid kinematics

We consider the motion of a given expanse of fluid as referred to a fixed right-handed rectangular system of coordinates (x, y, z). Although the medium is supposed to be continuous, we use the term 'fluid particle' to indicate an infinitesimally small volume of fluid. A fluid particle can be identified by means of its coordinates at a specified time, say (a, b, c) at time $t = 0$. Then

$$x = x(a, b, c, t), \quad y = y(a, b, c, t), \quad z = z(a, b, c, t) \qquad (1·2, 1)$$

is a set of equations for the path of the particle in question. By assumption, these equations satisfy the conditions

$$a = x(a, b, c, 0), \quad b = y(a, b, c, 0), \quad c = z(a, b, c, 0). \qquad (1·2, 2)$$

In vector notation (1·2, 1) and (1·2, 2) may be written as

$$\mathbf{r} = \mathbf{r}(\mathbf{r}_0, t), \quad \mathbf{r}_0 = \mathbf{r}(\mathbf{r}_0, 0), \qquad (1·2, 3)$$

where $\mathbf{r} = (x, y, z) = x\mathbf{i} + y\mathbf{j} + z\mathbf{k}$ is the position vector of the particle, and $\mathbf{r}_0 = (a, b, c) = a\mathbf{i} + b\mathbf{j} + c\mathbf{k}$.

The velocity of the particle is given by

$$u = \dot{x} = \frac{\partial}{\partial t} x(a, b, c, t), \quad v = \dot{y} = \frac{\partial}{\partial t} y(a, b, c, t), \quad w = \dot{z} = \frac{\partial}{\partial t} z(a, b, c, t), \quad (1·2, 4)$$

or, briefly, by

$$\mathbf{q} = \dot{\mathbf{r}} = \frac{\partial}{\partial t} \mathbf{r}(\mathbf{r}_0, t). \qquad (1·2, 5)$$

Similarly, the particle acceleration is given by $\ddot{\mathbf{r}} = (\ddot{x}, \ddot{y}, \ddot{z})$.

Now consider a point (x, y, z), fixed in space (so that x, y, z are now independent variables). At a given moment of time, (x, y, z) will be the affix

‡ H. Lamb, *Hydrodynamics* (6th ed. revised, Cambridge University Press, 1932).

of a specific particle, moving with velocity $\mathbf{q} = (u, v, w)$ say. Thus we may regard $\mathbf{q}$ as a vector function of the space variables x, y, z and of time. Given $\mathbf{q} = \mathbf{q}(x, y, z, t)$ we may obtain the paths of the particles as solutions of the system of ordinary differential equations

$$\frac{dx}{dt} = u(x, y, z, t), \quad \frac{dy}{dt} = v(x, y, z, t), \quad \frac{dz}{dt} = w(x, y, z, t).$$

On the other hand, the curves which are given by the system

$$\frac{dx}{u(x, y, z, t)} = \frac{dy}{v(x, y, z, t)} = \frac{dz}{w(x, y, z, t)}$$

are called the streamlines of the flow. In general, these are different from the paths of the particles, but the two systems coincide if the motion is steady, that is, if u, v, w are independent of time.

Let $F(x, y, z, t)$ be any scalar function defined within the given expanse of fluid. F may also be regarded as a function of time for any specific particle. We denote by DF/Dt the rate of change of F for a given particle. Now the particle which occupied the position x, y, z at time t will have moved to

$$x + \delta x \doteqdot x + u\,\delta t, \quad y + \delta y \doteqdot y + v\,\delta t, \quad z + \delta z \doteqdot z + w\,\delta t,$$

at time $t + \delta t$, where δt is a small increment. Hence, the corresponding increment of F is

$$\delta F \doteqdot F(x + u\,\delta t, y + v\,\delta t, z + w\,\delta t, t + \delta t) - F(x, y, z, t)$$

$$= \left(u \frac{\partial F}{\partial x} + v \frac{\partial F}{\partial y} + w \frac{\partial F}{\partial z} + \frac{\partial F}{\partial t} \right) \delta t + o(\delta t),$$

where $o(\delta t)$ is a quantity such that $o(\delta t)/\delta t$ tends to 0 as δt tends to 0. Dividing by δt and passing to the limit we obtain

$$\frac{DF}{Dt} = u \frac{\partial F}{\partial x} + v \frac{\partial F}{\partial y} + w \frac{\partial F}{\partial z} + \frac{\partial F}{\partial t}. \tag{1·2, 6}$$

The operation

$$\frac{D}{Dt} = u \frac{\partial}{\partial x} + v \frac{\partial}{\partial y} + w \frac{\partial}{\partial z} + \frac{\partial}{\partial t}$$

is called 'differentiation following the motion of the fluid'. In particular, if $F(x, y, z, t)$ is one of the velocity components, then DF/Dt is the rate of change of that component for a given particle. Hence

$$\frac{D}{Dt} \mathbf{q} = \left(\frac{Du}{Dt}, \frac{Dv}{Dt}, \frac{Dw}{Dt} \right) \tag{1·2, 7}$$

is the acceleration vector expressed in terms of the velocity components and their derivatives with respect to x, y, z and t.

Again, let

$$F(x, y, z, t) = 0$$

be the equation of a solid boundary. If the boundary is fixed then F will be independent of time, $\partial F / \partial t = 0$. The definition of a solid boundary implies that no particle of fluid can cross the boundary away from, or into, the fluid. Thus the rate of displacement of the boundary at one of its points in a direction normal to itself must equal the rate of displacement of the fluid particle in that direction at the point in question. In other words, the relative rate of displacement of a fluid particle at the boundary must be tangential to the boundary. To formulate this condition analytically we may assume that the solid boundary is embedded in a one-parametric family of surfaces,

$$F(x, y, z, t) = c,$$

so that the boundary corresponds to $c = 0$. Now the above condition states that a particle at the boundary is displaced along the boundary so that the rate of change of F following the motion of the particle is zero,

$$\frac{DF}{Dt} = u\frac{\partial F}{\partial x} + v\frac{\partial F}{\partial y} + w\frac{\partial F}{\partial z} + \frac{\partial F}{\partial t} = 0. \tag{1·2, 8}$$

For a fixed boundary, the condition becomes

$$u\frac{\partial F}{\partial x} + v\frac{\partial F}{\partial y} + w\frac{\partial F}{\partial z} = 0. \tag{1·2, 9}$$

Now $\dfrac{\partial F}{\partial x} : \dfrac{\partial F}{\partial y} : \dfrac{\partial F}{\partial z} = \lambda : \mu : \nu$, where λ, μ, ν are the direction cosines normal to the surface. Hence (1·2, 9) is equivalent to

$$u\lambda + v\mu + w\nu = 0, \tag{1·2, 10}$$

and in this form the condition states simply that the velocity component of the fluid particle in a direction normal to the boundary equals zero.

If the equation to the solid boundary is given in the form

$$z = f(x, y, t),$$

i.e. $z - f(x, y, t) = 0$, then (1·2, 8) and (1·2, 9) reduce to

$$\left. \begin{aligned} w &= u\frac{\partial f}{\partial x} + v\frac{\partial f}{\partial y} + \frac{\partial f}{\partial t} \\ w &= u\frac{\partial f}{\partial x} + v\frac{\partial f}{\partial y}, \end{aligned} \right\} \tag{1·2, 11}$$

and

respectively.

1·3   The equation of continuity

Let S be a closed surface drawn in an expanse of fluid, and let $\mathbf{n}$ be the normal unit vector at any point of S; $\mathbf{n}$ will always be chosen so as to point away from the interior volume V bounded by S. Let $\rho = \rho(x, y, z, t)$ be the density of the fluid; ρ may be variable in time and in space. Then

$$m(V) = \int_V \rho \, dV = \int_V \rho(x, y, z, t) \, dx \, dy \, dz$$

is the total mass of fluid occupying V. Hence the rate of change of $m(V)$ is

$$\frac{d}{dt}m(V) = \frac{d}{dt}\int_V \rho\, dV = \int_V \frac{\partial \rho}{\partial t}\, dV.$$

On the other hand, the volume of fluid which crosses the surface element $\delta \mathbf{S} = \mathbf{n}\,\delta S$ in the outward direction in a short interval of time δt is

$$\mathbf{q}\mathbf{n}\,\delta S\,\delta t = \mathbf{q}\,\delta \mathbf{S}\,\delta t,$$

and the corresponding mass of fluid is $\rho \mathbf{q}\,\delta \mathbf{S}\,\delta t$. Integrating over the surface S, dividing by δt and passing to the limit, we find that the rate at which fluid crosses S in the outward direction is

$$\int_S \rho\mathbf{q}\, d\mathbf{S}.$$

According to the law of conservation of mass, this balances the rate of change of $m(V)$ so that

$$\int_V \frac{\partial \rho}{\partial t}\, dV + \int_S \rho\mathbf{q}\, d\mathbf{S} = 0. \tag{1·3, 1}$$

Now by the divergence theorem of Green and Gauss,

$$\int_S \rho\mathbf{q}\, d\mathbf{S} = \int_V \operatorname{div}(\rho\mathbf{q})\, dV,$$

and so (1·3, 1) is equivalent to

$$\int_V \left(\frac{\partial \rho}{\partial t} + \operatorname{div}(\rho\mathbf{q})\right) dV = 0. \tag{1·3, 2}$$

Since this identity must be true for arbitrary V we conclude that the integral vanishes identically,

$$\frac{\partial \rho}{\partial t} + \operatorname{div}(\rho\mathbf{q}) = 0. \tag{1·3, 3}$$

This is the equation of continuity. It may be written in the alternative forms

$$\frac{\partial \rho}{\partial t} + u\frac{\partial \rho}{\partial x} + v\frac{\partial \rho}{\partial y} + w\frac{\partial \rho}{\partial z} + \rho\left(\frac{\partial u}{\partial x} + \frac{\partial v}{\partial y} + \frac{\partial w}{\partial z}\right) = 0, \tag{1·3, 4}$$

and

$$\frac{D\rho}{Dt} + \rho\operatorname{div}\mathbf{q} = 0. \tag{1·3, 5}$$

This relation reduces to

$$\operatorname{div}\mathbf{q} = 0, \tag{1·3, 6}$$

i.e.

$$\frac{\partial u}{\partial x} + \frac{\partial v}{\partial y} + \frac{\partial w}{\partial z} = 0 \tag{1·3, 7}$$

if $D\rho/Dt = 0$. This will be the case if ρ is constant in space and time. The fluid is then called incompressible.

1·4 Euler's equations of motion

Let S be a small plane surface drawn in a fluid. Then the force $\mathbf{f}$ exerted by the fluid on one side of S on the fluid on the other side, across S, will in general have components both normal and tangential to S. If the tangential components are negligible compared with the normal component everywhere, then the fluid is called perfect or inviscid. The fluid is called viscous if tangential forces are present. Air possesses only a small degree of viscosity under normal conditions. Nevertheless, this is sufficient to control a large part of the drag forces which act on a wing, and even the lift forces are determined by a condition which is indirectly due to viscosity (see § 1·14 below). However, the former effect can be treated separately, and the latter can be expressed ultimately in a form which does not include viscosity directly. As a result the greater part of aerofoil theory can be developed on the assumption that the air is perfect.

For a perfect fluid, the force $\mathbf{f}$ is, by definition, normal to S. By Newton's third law the forces which act across S on the volumes of fluid on either side of the surface are of equal magnitude and opposite sign. Thus $\mathbf{f} = \pm f\mathbf{n}$, where f is the modulus (magnitude) of $\mathbf{f}$ and $\mathbf{n}$ is a unit vector normal to S. Now assume that S shrinks indefinitely to a point P while maintaining its position normal to $\mathbf{n}$. Let $A = A(S)$ be the area of S. Then the limit

$$p_{\mathbf{n}} = \lim_{A \to 0} \pm \frac{f}{A}$$

will be called the pressure of the fluid at P in the direction $\mathbf{n}$. In this formula, we choose the sign so as to make $p_{\mathbf{n}}$ positive if the force exerted across S on the volume of fluid towards which $\mathbf{n}$ points, also points in that direction.

We shall now show that the pressure acting at any given point P is in fact independent of $\mathbf{n}$.

Through P draw straight lines PA, PB, PC which are parallel to the coordinate axes and choose the points A, B, C so that their distances from P are small (Fig. 1·4, 1). Let $\mathbf{n}$ be the unit vector normal to ABC and pointing away from P, $\mathbf{n} = (\lambda, \mu, \nu)$, and let Δ, Δ' be the areas of the triangles ABC and PBC. Consider the equilibrium of forces which act on the volume of fluid within the tetrahedron $PABC$. The effect of the pressure acting on the volume across ABC is

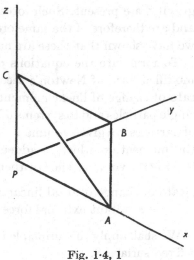

Fig. 1·4, 1

$$-p_{\mathbf{n}}\mathbf{n}\Delta = -p_{\mathbf{n}}\Delta(\lambda\mathbf{i} + \mu\mathbf{j} + \nu\mathbf{k}).$$

The x-component of this force is

$$-p_{\mathbf{n}}\Delta\lambda = -p_{\mathbf{n}}\Delta',$$

since $\Delta' = \Delta\lambda$. The pressure acting across PBC produces a force parallel to the x-axis which equals $p_{\mathbf{i}}\Delta'$, since $\mathbf{i}$ is the unit vector normal to PBC. The

x-components of the forces which are due to the pressure across PAB and PBC vanish. Hence the x-component of the resultant force due to the pressure across the faces of the tetrahedron is

$$p_i \Delta' - p_n \Delta' = (p_i - p_n) \Delta'.$$

Let V be the volume of the tetrahedron. Then, assuming for the moment that external body forces are absent,

$$\rho V \ddot{x} = (p_i - p_n) \Delta'$$

by Newton's second law of motion, or

$$p_i - p_n = \rho \ddot{x} \frac{V}{\Delta'}. \tag{1·4, 1}$$

In all these formulae, p_i, p_n, ρ, $\ddot{x}$ may be taken at P, and the formulae are correct except for terms which are of a higher order of smallness (in terms of a linear dimension, say the distance PA). Now as A approaches P indefinitely while $\mathbf{n}$ is kept constant, the ratio V/Δ', on the right-hand side of (1·4, 1), tends to 0, and so

$$p_i - p_n = 0, \quad p_i = p_n.$$

A similar argument shows that

$$p_j = p_n, \quad p_k = p_n,$$

and since the direction $\mathbf{n}$ is arbitrary we conclude that for a given point P, p_n is independent of $\mathbf{n}$. It follows that the pressure $p = p(x, y, z, t)$ is a scalar function defined for all points of a perfect fluid, at any given time. This conclusion is not modified if external body forces, such as the force of gravity, are present. Such forces are by assumption proportional to mass and are therefore of the same order of magnitude as the inertia forces, and we have shown that these are negligible in the limit.

To formulate the equations of motion of an inviscid fluid we apply a modified form of Newton's second law of motion. Expressed in terms of rate of change of linear momentum, Newton's second law applies both to single particles and to systems of particles. We may also apply it to a system of particles in a given volume V, but in that case we have to take into account the momentum which is added or abstracted by particles entering and leaving the volume. The law then reads:

'Rate of change of total linear momentum in volume V
   = resultant external force on V + rate of flow of momentum into V.'

We shall apply this principle to a volume of fluid V which is bounded by a fixed surface S.

The total linear momentum in V is

$$\mathbf{M} = \int_V \rho \mathbf{q} \, dV,$$

since $\rho \, dV$ is the element of mass, and momentum is the product of mass and velocity. Hence the rate of change of linear momentum in V equals

$$\frac{d\mathbf{M}}{dt} = \frac{d}{dt} \int_V \rho \mathbf{q} \, dV = \int_V \frac{\partial}{\partial t} (\rho \mathbf{q}) \, dV. \qquad (1\cdot4, 2)$$

The rate at which fluid leaves V across an element of surface dS is $\rho \mathbf{q} \, d\mathbf{S}$, and so $\rho \mathbf{q}(\mathbf{q} \, d\mathbf{S})$ is the rate at which linear momentum leaves V across $d\mathbf{S}$. Hence

$$\mathbf{f} = - \int_S \rho \mathbf{q}(\mathbf{q} \, d\mathbf{S}) \qquad (1\cdot4, 3)$$

is the total rate of flow of linear momentum *into* V.

The external forces consist of the pressure across S, and possibly external body forces, say $\mathbf{Q} = (X, Y, Z)$ per unit mass. In general, $\mathbf{Q}$ may be variable, but in the only case which is of interest in connexion with aerofoil theory, $\mathbf{Q}$ is the force of gravity per unit mass and may then be taken as

$$\mathbf{Q} = (0, 0, -g) = -g\mathbf{k},$$

provided the z-axis points upwards in a vertical direction.

The resultant pressure on V across S is $-\int_S p \, d\mathbf{S}$, and the total body force is given by $\int_V \rho \mathbf{Q} \, dV$ in the general case. The resultant external force on V is therefore given by

$$\mathbf{R} = - \int_S p \, d\mathbf{S} + \int_V \rho \mathbf{Q} \, dV. \qquad (1\cdot4, 4)$$

Then

$$\frac{d\mathbf{M}}{dt} = \mathbf{R} + \mathbf{f}, \qquad (1\cdot4, 5)$$

or

$$\int_V \frac{\partial}{\partial t}(\rho \mathbf{q}) \, dV + \int_S p \, d\mathbf{S} - \int_V \rho \mathbf{Q} \, dV + \int_S \rho \mathbf{q}(\mathbf{q} \, d\mathbf{S}) = 0. \qquad (1\cdot4, 6)$$

This corresponds to three equations in scalar notation, of which the first is

$$\int_V \left(\frac{\partial}{\partial t}(\rho u) - \rho X \right) dx \, dy \, dz + \int_S p \, dy \, dz + \int_S \rho u(u \, dy \, dz + v \, dz \, dx + w \, dx \, dy) = 0. \qquad (1\cdot4, 7)$$

Now by Green's theorem

$$\int_S p \, dy \, dz = \int_V \frac{\partial p}{\partial x} dx \, dy \, dz = \int_V \frac{\partial p}{\partial x} dV,$$

$$\int_S \rho u^2 \, dy \, dz = \int_V \frac{\partial}{\partial x}(\rho u^2) \, dV,$$

$$\int_S \rho uv \, dx \, dz = \int_V \frac{\partial}{\partial y}(\rho uv) \, dV,$$

$$\int_S \rho uw \, dx \, dy = \int_V \frac{\partial}{\partial z}(\rho uw) \, dV.$$

Substituting in (1·4, 7)

$$\int_V \left(\frac{\partial}{\partial t}(\rho u) - \rho X + \frac{\partial p}{\partial x} + \frac{\partial}{\partial x}(\rho u^2) + \frac{\partial}{\partial y}(\rho uv) + \frac{\partial}{\partial z}(\rho uw)\right) dV = 0.$$

But the choice of the volume V is arbitrary, and so we conclude as before that the integrand vanishes identically. This yields

$$u\frac{\partial \rho}{\partial t} + \rho\frac{\partial u}{\partial t} - \rho X + \frac{\partial p}{\partial x} + u\frac{\partial}{\partial x}(\rho u) + u\frac{\partial}{\partial y}(\rho v) + u\frac{\partial}{\partial z}(\rho w)$$

$$+ \rho\left(u\frac{\partial u}{\partial x} + v\frac{\partial u}{\partial y} + w\frac{\partial u}{\partial z}\right) = 0. \quad (1·4, 8)$$

But

$$u\frac{\partial \rho}{\partial t} + u\frac{\partial}{\partial x}(\rho u) + u\frac{\partial}{\partial y}(\rho v) + u\frac{\partial}{\partial z}(\rho w) = 0$$

by the equation of continuity, (1·3, 4). Hence

$$\left.\begin{aligned}
\rho\left(\frac{\partial u}{\partial t} + u\frac{\partial u}{\partial x} + v\frac{\partial u}{\partial y} + w\frac{\partial u}{\partial z}\right) &= \rho X - \frac{\partial p}{\partial x}, \\
\text{and similarly} \quad \rho\left(\frac{\partial v}{\partial t} + u\frac{\partial v}{\partial x} + v\frac{\partial v}{\partial y} + w\frac{\partial v}{\partial z}\right) &= \rho Y - \frac{\partial p}{\partial y}, \\
\rho\left(\frac{\partial w}{\partial t} + u\frac{\partial w}{\partial x} + v\frac{\partial w}{\partial y} + w\frac{\partial w}{\partial z}\right) &= \rho Z - \frac{\partial p}{\partial z}.
\end{aligned}\right\} \quad (1·4, 9)$$

These are Euler's equations of motion. They may be written more briefly as

$$\frac{Du}{Dt} = X - \frac{1}{\rho}\frac{\partial p}{\partial x}, \quad \frac{Dv}{Dt} = Y - \frac{1}{\rho}\frac{\partial p}{\partial y}, \quad \frac{Dw}{Dt} = Z - \frac{1}{\rho}\frac{\partial p}{\partial z}, \quad (1·4, 10)$$

or

$$\frac{D\mathbf{q}}{Dt} = \mathbf{Q} - \frac{1}{\rho}\operatorname{grad} p. \quad (1·4, 11)$$

The equations can also be obtained by applying Newton's second law to a small cube whose edges are parallel to the coordinate axes.

Now assume that the fixed surface S bounds a volume which is occupied partly by a perfect fluid and partly by a *fixed* body B. More generally, B may be part of a fixed solid body which projects into S. Let $\mathbf{T}$ be the resultant external force which acts on B excluding the force which is exerted on B by the fluid within S. Then

$$\frac{d\mathbf{M}}{dt} = \mathbf{R} + \mathbf{f} + \mathbf{T}, \quad (1·4, 12)$$

where $\mathbf{R}$ is the external force which acts on the fluid within S, excluding the force exerted on the fluid by B. $\mathbf{R}$ is still given by (1·4, 4), provided the volume integral on the right-hand side of that equation is taken over that part of the volume within S which is occupied by the fluid and, similarly, the surface integral is taken only over the area of S which is immersed in the fluid (i.e. excluding the areas where the solid body penetrates into the volume). With the same qualifications, dM/dt and $\mathbf{f}$ are still given by (1·4, 2) and (1·4, 3). Hence

$$\mathbf{T} = \int_V \left(\frac{\partial}{\partial t}(\rho \mathbf{q}) - \rho \mathbf{Q} \right) dV + \int_S p \, d\mathbf{S} + \int_S \rho \mathbf{q}(\mathbf{q} \, d\mathbf{S}). \qquad (1 \cdot 4, 13)$$

Now since the body B is at rest, by assumption, it follows that the external force $\mathbf{T}$ must be balanced by the force exerted by the fluid on B. The latter is therefore given by

$$-\mathbf{T} = -\left\{ \int_V \left(\frac{\partial}{\partial t}(\rho \mathbf{q}) - \rho \mathbf{Q} \right) dV + \int_S p \, d\mathbf{S} + \int_S \rho \mathbf{q}(\mathbf{q} \, d\mathbf{S}) \right\}, \qquad (1 \cdot 4, 14)$$

or

$$-\mathbf{T} = -\left\{ -\int_V \rho \mathbf{Q} \, dV + \int_S p \, d\mathbf{S} + \int_S \rho \mathbf{q}(\mathbf{q} \, d\mathbf{S}) \right\}, \qquad (1 \cdot 4, 15)$$

if the flow is steady.

The same formula can be deduced from the fact that the right-hand side of (1·4, 14) vanishes if no solid body is present within S. It follows that for the case now under consideration, we may replace S by the surface of B, S' say; indeed, the volume between S and S' does not contain any solid. Then (1·4, 14) reduces to $\int_{S'} p \, d\mathbf{S}$, since there is no flow of fluid across S'.

And $\int_{S'} p \, d\mathbf{S}$ clearly is the force exerted on B by the fluid within S.

If external body forces are absent or negligible, (1·4, 15) becomes

$$-\mathbf{T} = -\left\{ \int_S p \, d\mathbf{S} + \int_S \rho \mathbf{q}(\mathbf{q} \, d\mathbf{S}) \right\}. \qquad (1 \cdot 4, 16)$$

In these circumstances it is therefore possible to calculate the force exerted on a body B without a detailed knowledge of the pressure distribution over the surface of B, provided one knows the pressure and velocity distribution over some surface surrounding B. So long as the flow is steady, the same applies even if an external body force is present, since the corresponding integral in (1·4, 15) can be calculated separately. For example, if the external body force is the force of gravity, then $\mathbf{Q} = -g\mathbf{k}$, and so

$$\int_V \rho \mathbf{Q} \, dV = -\rho g \mathbf{k} \int_V dV = -\rho g (V_2 - V_1) \mathbf{k}, \qquad (1 \cdot 4, 17)$$

where V_1 is the volume of B and V_2 is the volume enclosed by S.

Just as the principle of linear momentum leads to a formula for the force which acts on a body immersed in a fluid, so the principle of angular momentum leads to an expression for the moment or couple exerted by the fluid on a body which is immersed in it.

The principle of angular momentum for a system of particles is subject to the restriction that the internal forces between any two particles act along the line connecting the two particles. However, in the present case the principle is valid (even for a viscous fluid), since the internal forces between adjacent fluid particles act at coincident points and so the resultant moment of internal forces vanishes in any case.

Let S be a fixed surface which encloses a volume occupied partly by a fixed body B as before and partly by a perfect fluid. Let $\mathbf{N}$ be the resultant moment of the external forces which act on B, about a fixed point P, and let $\mathbf{r}$ be the position vector of any point with respect to P. We then obtain in place of (1·4, 13)

$$\mathbf{N} = \int_V \left(\frac{\partial}{\partial t}(\rho \mathbf{r} \wedge \mathbf{q}) - \rho \mathbf{r} \wedge \mathbf{Q} \right) dV + \int_S p\mathbf{r} \wedge d\mathbf{S} + \int_S \rho \mathbf{r} \wedge \mathbf{q}(\mathbf{q}\, d\mathbf{S}). \quad (1·4, 18)$$

Now B is fixed, by assumption, and so we may deduce as before that $-N$ is the couple exerted by the fluid on B. In particular, under steady conditions and in the absence of body forces

$$-\mathbf{N} = -\left\{ \int_S p\mathbf{r} \wedge d\mathbf{S} + \int_S \rho \mathbf{r} \wedge \mathbf{q}(\mathbf{q}\, d\mathbf{S}) \right\}. \quad (1·4, 19)$$

The above formulae still apply if there is a number of solid bodies immersed in the fluid within S, provided that $\mathbf{T}$ and $\mathbf{M}$ are taken as the vector sums of the forces and moments which act on these bodies.

1·5 Vorticity and circulation

Let
$$\boldsymbol{\omega} = \operatorname{curl} \mathbf{q}, \quad\quad\quad (1·5, 1)$$

where $\mathbf{q}$ is the velocity vector in a given field of flow. $\boldsymbol{\omega}$ is called the vorticity of the flow. Its components are given by

$$\xi = \frac{\partial w}{\partial y} - \frac{\partial v}{\partial z}, \quad \eta = \frac{\partial u}{\partial z} - \frac{\partial w}{\partial x}, \quad \zeta = \frac{\partial v}{\partial x} - \frac{\partial u}{\partial y}.$$

It is known from vector analysis that if

$$\boldsymbol{\omega} = \operatorname{curl} \mathbf{q} = 0 \quad\quad\quad (1·5, 2)$$

identically in a given region, then $\mathbf{q}$ can be written as minus the gradient of a scalar function,

$$\mathbf{q} = -\operatorname{grad} \Phi, \quad u = -\frac{\partial \Phi}{\partial x}, \quad v = -\frac{\partial \Phi}{\partial y}, \quad w = -\frac{\partial \Phi}{\partial z}. \quad (1·5, 3)$$

Φ is called the velocity potential of the flow. It is determined by $\mathbf{q}$ except for the addition of an arbitrary constant. The introduction of the minus sign in (1·5, 3) is a convention which is not, however, accepted universally. If a velocity potential exists then (1·5, 2) must be satisfied in view of the vector identity $\operatorname{curl} \operatorname{grad} \Phi = 0$. The flow is then called irrotational. Let

$$\gamma(C) = \int_A^B \mathbf{q}\, d\mathbf{l} = \int_A^B (u\, dx + v\, dy + w\, dz)$$

be the line integral of the velocity vector along a curve C from A to B.

$\gamma(C)$ is called the flow along C, and if C is closed then $\gamma(C)$ is called the circulation around C. If $\mathbf{q}$ possesses a velocity potential, then

$$\gamma(C) = -\int_A^B \left(\frac{\partial \Phi}{\partial x} dx + \frac{\partial \Phi}{\partial y} dy + \frac{\partial \Phi}{\partial z} dz \right) = -\int_A^B d\Phi = -(\Phi_B - \Phi_A), \quad (1·5, 4)$$

where Φ_A and Φ_B denote the value of the velocity potential at the points A and B respectively. It follows that the flow is the same along any two curves C and C' which lead from a given point A to a given point B provided C can be deformed continuously into C' without leaving the region within which the flow is irrotational. Similarly, the circulation around a closed curve C must vanish for irrotational flow only if C bounds a surface which is immersed entirely in the region of irrotational flow. More generally, if the closed curve C bounds a surface S which is immersed in the fluid, then the circulation around C is given by

$$\gamma(C) = \int_C \mathbf{q}\, d\mathbf{l} = \int_S \operatorname{curl} \mathbf{q}\, d\mathbf{S} = \int_S \boldsymbol{\omega}\, d\mathbf{S} \qquad (1·5, 5)$$

by Stokes's theorem.

Consider a closed curve C which moves with the fluid. In agreement with our earlier notation we denote the rate of change of the circulation around C by $\frac{D}{Dt}\gamma(C) = \frac{D\gamma}{Dt}$. Then

$$\frac{D\gamma}{Dt} = \frac{D}{Dt}\int_C (u\,dx + v\,dy + w\,dz)$$

$$= \int_C \left(\frac{Du}{Dt} dx + \frac{Dv}{Dt} dy + \frac{Dw}{Dt} dz \right) + \int_C \left(u\frac{D}{Dt} dx + v\frac{D}{Dt} dy + w\frac{D}{Dt} dz \right). \quad (1·5, 6)$$

Assume now that the fluid is perfect. We may transform the first integral on the right-hand side of (1·5, 6) by using Euler's equations (1·4, 10). To interpret the second integral, we see, for example, that the rate of change of the x-component δx of a small segment $\delta l = (\delta x, \delta y, \delta z)$ which moves with the fluid is

$$\frac{D}{Dt}\delta x = u + \delta u - u = \delta u,$$

where u and $u + \delta u$ are the values of the x-component of velocity at the end-points of the segment. Hence

$$\frac{D\gamma}{Dt} = \int_C \left\{ \left(X - \frac{1}{\rho}\frac{\partial p}{\partial x} \right) dx + \left(Y - \frac{1}{\rho}\frac{\partial p}{\partial y} \right) dy + \left(Z - \frac{1}{\rho}\frac{\partial p}{\partial z} \right) dz \right\}$$
$$+ \int_C (u\,du + v\,dv + w\,dw).$$

Assume that $\mathbf{Q} = (X, Y, Z)$ is a conservative force, such that $\mathbf{Q} = -\operatorname{grad} W$, where W is one-valued, while ρ is a constant, or, more generally, a function of p. Then

$$X - \frac{1}{\rho}\frac{\partial p}{\partial x} = -\frac{\partial}{\partial x}\left(W + \int \frac{dp}{\rho} \right),$$

with similar expressions for $Y - \frac{1}{\rho}\frac{\partial p}{\partial y}$ and $Z - \frac{1}{\rho}\frac{\partial p}{\partial z}$. Hence

$$\frac{D\gamma}{Dt} = \int_C \left\{ -\frac{\partial}{\partial x}\left(W + \int\frac{dp}{\rho}\right)dx - \frac{\partial}{\partial y}\left(W + \int\frac{dp}{\rho}\right)dy - \frac{\partial}{\partial z}\left(W + \int\frac{dp}{\rho}\right)dz \right\}$$
$$+ \int_C (u\,du + v\,dv + w\,dw) = \left[-\left(W + \int\frac{dp}{\rho}\right) + \tfrac{1}{2}q^2 \right]_C = 0.$$

This shows that if for a perfect fluid the external body force is derived from a one-valued potential and the density is a function of the pressure only, then the circulation round a closed circuit which moves with the fluid is constant. The theorem is due to Lord Kelvin. We say that a fluid particle is irrotational at a given moment if there exists a finite volume of fluid V which includes the particle such that curl $\mathbf{q} = 0$ for all points of V. Thus, it is not sufficient in this definition that curl $\mathbf{q} = 0$ at a single point. Kelvin's theorem then implies the following important result which is due to Lagrange.

Consider a particle which belongs to a perfect fluid whose density depends on the pressure only, such that the external body forces, if any, are derived from a one-valued potential. Then if the particle is irrotational at any one time, it must be irrotational at all times. To prove this, suppose that at time $t = t_1$ the particle is enclosed in a volume V within which curl $\mathbf{q} = 0$ identically. We have to show that curl $\mathbf{q} = 0$ at any other time for all the particles which constitute V.

Assume on the contrary that curl $\mathbf{q} = \boldsymbol{\omega} \neq 0$ at time $t = t_2$ for a particle which belongs to V. Let S be a small plane area normal to $\boldsymbol{\omega}$ at the point in question, bounded by a curve C, such that all the particles of S belong to V. Then

$$\gamma(C) = \int_C \mathbf{q}\,d\mathbf{l} = \int_S \text{curl }\mathbf{q}\,dS \doteqdot \pm \omega A,$$

approximately, where A is the area of S, and so $\gamma(C) \neq 0$ at time $t = t_2$. On the other hand, at time $t = t_1$ the particles of C bound a surface over which curl $\mathbf{q}$ vanishes, and so the corresponding circulation also vanishes. This is contrary to Kelvin's theorem, and proves our assertion.

In particular, Lagrange's theorem applies if the external body force is the force of gravity. It follows that if a perfect fluid is initially at rest and is set in motion by the action of surface forces acting at a boundary only, then the resulting motion will be irrotational, provided the density depends only on the pressure and the only external body force present is gravity.

We shall now assume that the fluid is endowed with vorticity which will, in general, be distributed continuously in space. Let curl $\mathbf{q} = \boldsymbol{\omega} = (\xi, \eta, \zeta)$ as before. Then the system of differential equations

$$\frac{dx}{\xi} = \frac{dy}{\eta} = \frac{dz}{\zeta}$$

defines a family of curves whose tangents are everywhere parallel to the direction of the local vorticity vector. These curves are called 'vortex lines'. The vortex lines through a small closed curve constitute a 'vortex tube'. The volume of fluid within a vortex tube is a 'vortex filament' or briefly a 'vortex'. A surface which consists of vortex lines is called a 'vortex surface'. Thus a vortex tube is also a vortex surface. If a closed curve C on a vortex surface bounds an area S which forms part of the vortex sheet then the circulation round C vanishes. Thus

$$\gamma(C) = \int_C \mathbf{q}\, d\mathbf{l} = \int_S \operatorname{curl}\mathbf{q}\, d\mathbf{S} = 0,$$

since $\operatorname{curl}\mathbf{q}$ is everywhere perpendicular to the directed surface element $d\mathbf{S}$.

Consider two closed curves C, C', which are drawn around the same vortex tube (Fig. 1·5, 1). We wish to show that the circulation round C is equal to the circulation round C' (taken in the same sense). Indeed, we may cut the area between C and C' by the line AA' and consider the single circuit C^* which consists of C and of C' (taken in the opposite sense) joined together by the line AA' which is taken twice, once from C to C' and once from C' to C. Then C^* bounds a surface which forms part of the vortex tube. Ac-cordingly, $\gamma(C^*) = 0$. But $\gamma(C^*)$

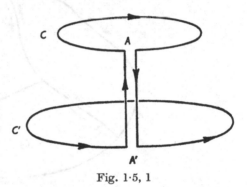

Fig. 1·5, 1

consists of the circulation round C minus the circulation round C' (when the two are taken in the same sense), while the contributions along AA' cancel. Hence

$$\gamma(C) - \gamma(C') = 0$$

as asserted. $\gamma(C)$ is therefore independent of the particular choice of C. It is called the strength of the vortex.

We conclude that a vortex cannot either begin or end in the interior of the fluid; it must either be closed or else it must begin and end on the boundaries of the fluid.

If we take the curve C as the line of intersection of a vortex tube with a plane which is normal to it then

$$\gamma(C) = \int_C \mathbf{q}\, d\mathbf{l} = \omega A, \tag{1·5, 7}$$

where A is the cross-sectional area of the vortex tube, supposed small.

Now let C be any closed curve which bounds a surface S immersed in the fluid. We subdivide S into a number of small areas bounded by curves $C_1, C_2, C_3, \ldots, C_n$. Then the circulation round C is the algebraic sum of the values of the circulation around $C_1, C_2, \ldots, C_n$, taken with appropriate sign (see Fig. 1·5, 2). Now each one of the curves $C_1, C_2, ..., C_n$ determines a vortex whose strength is precisely the circulation round the C_K in question. We therefore have the following theorem which is almost self-evident:

The circulation round a closed curve C equals the algebraic sum of the strengths of the vortex filaments embraced by C. The theorem holds both for continuous distributions of vorticity and for isolated vortices.

Now suppose that the conditions of Kelvin's theorem are satisfied. That is to say, the fluid is perfect and its density is constant or a function of the pressure, while the external body force is derived from a one-valued potential. Then we propose to show that vortex lines move with the fluid. That is to say, particles which constitute a vortex line at one time, constitute a vortex line also at any other time.

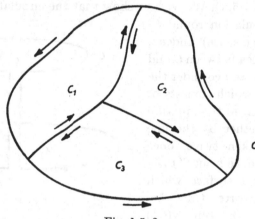

Fig. 1·5, 2

We shall establish first that particles which constitute a vortex surface at one time, constitute a vortex surface also at any other time.

It was shown earlier that for any closed curve C drawn on a vortex surface S which bounds an area which belongs entirely to the surface, the circulation round C vanishes. In particular, this must be true for all sufficiently small closed curves drawn on S. Conversely, if a surface S is such that the circulation round every curve which bounds a surface on S vanishes, then S can consist only of vortex lines or irrotational particles. But a particle which is irrotational at any time will remain so at all times. We may therefore conclude that the particles which constitute a vortex sheet at one time will constitute a vortex surface at any other time. Or briefly, vortex surfaces move with the fluid. Moreover, any vortex line may be regarded as the line of intersection of two vortex surfaces to which the line belongs. And since these sheets move with the fluid, it follows that the vortex line also moves with the fluid.

The above properties of vortices are essentially due to Helmholtz.

1·6   Bernoulli's equation

Let $\boldsymbol{\omega} = (\xi, \eta, \zeta)$ be the vorticity vector. Then

$$\frac{Du}{Dt} = \frac{\partial u}{\partial t} + u\frac{\partial u}{\partial x} + v\frac{\partial u}{\partial y} + w\frac{\partial u}{\partial z}$$

$$= \frac{\partial u}{\partial t} + u\frac{\partial u}{\partial x} + v\frac{\partial v}{\partial x} + w\frac{\partial w}{\partial x} + v\left(\frac{\partial u}{\partial y} - \frac{\partial v}{\partial x}\right) + w\left(\frac{\partial u}{\partial z} - \frac{\partial w}{\partial x}\right)$$

$$= \frac{\partial u}{\partial t} + \frac{\partial}{\partial x}\{\tfrac{1}{2}(u^2 + v^2 + w^2)\} - v\zeta + w\eta,$$

with similar expressions for Dv/Dt and Dw/Dt. In vector notation

$$\frac{D\mathbf{q}}{Dt} = \frac{\partial \mathbf{q}}{\partial t} + \tfrac{1}{2}\operatorname{grad} q^2 - \mathbf{q} \wedge \boldsymbol{\omega}. \tag{1·6, 1}$$

Accordingly, we may rewrite Euler's equations (1·4, 11) in the form

$$\frac{\partial \mathbf{q}}{\partial t} + \tfrac{1}{2}\operatorname{grad} q^2 - \mathbf{q} \wedge \boldsymbol{\omega} = \mathbf{Q} - \frac{1}{\rho}\operatorname{grad} p. \tag{1·6, 2}$$

Now assume that the flow is irrotational, that $\mathbf{Q}$ is a conservative body force, $\mathbf{Q} = -\operatorname{grad} W$, and that ρ is a function of p, $\frac{1}{\rho}\operatorname{grad} p = \operatorname{grad}\int\frac{dp}{\rho}$. Then (1·6, 2) becomes

$$\frac{\partial \mathbf{q}}{\partial t} + \tfrac{1}{2}\operatorname{grad} q^2 + \operatorname{grad} W + \operatorname{grad}\int\frac{dp}{\rho} = 0.$$

Introducing the velocity potential Φ and bearing in mind that

$$\frac{\partial \mathbf{q}}{\partial t} = -\frac{\partial}{\partial t}\operatorname{grad}\Phi = \operatorname{grad}\left(-\frac{\partial\Phi}{\partial t}\right),$$

we then obtain $\quad \operatorname{grad}\left(-\dfrac{\partial\Phi}{\partial t} + \tfrac{1}{2}q^2 + W + \displaystyle\int\frac{dp}{\rho}\right) = 0.$

That is to say, the space derivatives of the expression in the brackets vanish identically, and so

$$-\frac{\partial\Phi}{\partial t} + \tfrac{1}{2}q^2 + W + \int\frac{dp}{\rho} = \text{constant (in space)} = h(t). \tag{1·6, 3}$$

This is Bernoulli's equation. Since Φ is arbitrary to the extent of a constant in space, we might assume that $h(t)$ is absorbed in Φ. However, if Φ has been defined independently, then $h(t)$ must be included in the equation. For steady flow we may assume $\partial\Phi/\partial t = 0$ and so

$$\tfrac{1}{2}q^2 + W + \int\frac{dp}{\rho} = \text{constant} = H. \tag{1·6, 4}$$

Again, for incompressible unsteady flow, we obtain

$$-\frac{\partial\Phi}{\partial t} + \tfrac{1}{2}q^2 + W + \frac{p}{\rho} = h(t), \tag{1·6, 5}$$

and for incompressible steady flow

$$\tfrac{1}{2}q^2 + W + \frac{p}{\rho} = H. \qquad (1·6, 6)$$

If, for example, we take $\mathbf{Q}$ to be the body force of gravity, $\mathbf{Q} = (0, 0, -g)$, then $W = gz$, and Bernoulli's equation becomes, taking the case of steady incompressible flow,

$$\tfrac{1}{2}q^2 + gz + \frac{p}{\rho} = H. \qquad (1·6, 7)$$

A weaker form of Bernoulli's equation can be obtained even if the flow is not irrotational. Taking the case of steady flow, we consider a particular streamline l. At any point of l we take one of the axes of coordinates tangent to l. Using Euler's equations we then obtain

$$q\frac{\partial q}{\partial s} = Q_s - \frac{1}{\rho}\frac{\partial p}{\partial s}, \qquad (1·6, 8)$$

where s is the arc length of the streamline taken in the direction of flow, and Q_s is the component of the body force in that direction. Integration of $(1·6, 8)$ now yields

$$\tfrac{1}{2}q^2 - \int Q_s\,ds + \int \frac{dp}{\rho} = H, \qquad (1·6, 9)$$

or, for a conservative body force,

$$\tfrac{1}{2}q^2 + W + \int \frac{dp}{\rho} = H. \qquad (1·6, 10)$$

However, the constant H now depends on the particular choice of the streamline.

1·7 Source and doublet distributions

In this section we consider the irrotational flow of an incompressible and perfect fluid. Accordingly, there exists a velocity potential Φ, which satisfies the equation

$$\operatorname{div}\operatorname{grad}\Phi = \frac{\partial^2\Phi}{\partial x^2} + \frac{\partial^2\Phi}{\partial y^2} + \frac{\partial^2\Phi}{\partial z^2} = 0 \qquad (1·7, 1)$$

by $(1·5, 3)$ and $(1·3, 6)$ or $(1·3, 7)$. This is Laplace's equation.

A particular solution of $(1·7, 1)$ is given by

$$\Phi(x, y, z) = \frac{\sigma}{4\pi\sqrt{\{(x - x_0)^2 + (y - y_0)^2 + (z - z_0)^2\}}}, \qquad (1·7, 2)$$

as can be easily verified by straightforward differentiation. In this expression, σ, x_0, y_0, z_0 are arbitrary constants. Φ is said to be the potential of a source of strength σ located at (or, with origin at) the point $P_0 = (x_0, y_0, z_0)$. Note that Φ can also be written in the form

$$\Phi(x, y, z) = \frac{\sigma}{4\pi PP_0} = \frac{\sigma}{4\pi r},$$

where P is the point (x, y, z) and

$$r = PP_0 = \sqrt{\{(x - x_0)^2 + (y - y_0)^2 + (z - z_0)^2\}}.$$

If σ is negative then Φ is sometimes also called a sink. The reason for this terminology will appear presently.

Consider the rate of fluid flow ('flux') across a sphere S' of radius R with P_0 as centre. At any point of S' the normal velocity in an outward direction is

$$\left(-\frac{\partial \Phi}{\partial r}\right)_{r=R} = \frac{\sigma}{4\pi R^2}.$$

It follows that the total volume of fluid which crosses the sphere in unit time is

$$-\int_{S'} \frac{\partial \Phi}{\partial r} dS = \frac{\sigma}{4\pi R^2} 4\pi R^2 = \sigma.$$

Now consider any other closed surface S which includes P_0 in its interior. We propose to show that the rate of flow across S also is σ.

Let S' be a sphere with centre at P_0 which is wholly enclosed within S. Then the rate of flow across S' is σ. Now let V be the volume bounded by S on the outside and by S' in its interior. Then by the divergence theorem,

$$\int_V \operatorname{div} \operatorname{grad} \Phi \, dV = \int_S \operatorname{grad} \Phi \, d\mathbf{S} - \int_{S'} \operatorname{grad} \Phi \, d\mathbf{S},$$

if we take normals to both S and S' in an outward direction, i.e. pointing away from P_0. But $\operatorname{div} \operatorname{grad} \Phi = 0$ in the volume V and so

$$\int_S \mathbf{q} \, d\mathbf{S} = -\int_S \operatorname{grad} \Phi \, d\mathbf{S} = -\int_{S'} \operatorname{grad} \Phi \, d\mathbf{S} = -\int_{S'} \frac{\partial \Phi}{\partial r} dS = \sigma.$$

But $\int_S \mathbf{q} \, d\mathbf{S}$ is the rate of fluid flow across S, which is therefore equal to σ, as asserted. Thus the physical picture which corresponds to Φ is that of fluid being created (or destroyed, for $\sigma < 0$) continuously at P_0 at the rate of σ units of volume, or $\rho\sigma$ units of mass in unit time.

In addition to single sources (or sinks) or systems which consist of a finite number of sources of different strengths, located at different points, we need also to consider continuous distributions of sources on curves, surfaces or volumes. The velocity potential due to isolated sources is given by

$$\Phi(x, y, z) = \frac{1}{4\pi} \sum_{k=1}^{n} \frac{\sigma_k}{\sqrt{\{(x - x_k)^2 + (y - y_k)^2 + (z - z_k)^2\}}}, \qquad (1\cdot7, 3)$$

where σ_k, x_k, y_k, z_k are given constants. $\Phi(x, y, z)$ is defined at all points of space except at the points (x_k, y_k, z_k). It is evident that $\Phi(x, y, z)$ is again a solution of Laplace's equation. Similarly, for continuous distributions, Φ is given by one of the integrals

$$\Phi(x,y,z) = \frac{1}{4\pi} \int_C \frac{\sigma(x_0,y_0,z_0)\,dl}{\sqrt{\{(x-x_0)^2+(y-y_0)^2+(z-z_0)^2\}}}, \tag{1·7, 4}$$

$$\Phi(x,y,z) = \frac{1}{4\pi} \int_S \frac{\sigma(x_0,y_0,z_0)\,dS}{\sqrt{\{(x-x_0)^2+(y-y_0)^2+(z-z_0)^2\}}}, \tag{1·7, 5}$$

or $$\Phi(x,y,z) = \frac{1}{4\pi} \int_V \frac{\sigma(x_0,y_0,z_0)\,dV}{\sqrt{\{(x-x_0)^2+(y-y_0)^2+(z-z_0)^2\}}}, \tag{1·7, 6}$$

where C is a given curve along which the 'source density' $\sigma(x_k,y_k,z_k)$ is defined, so that the integral (1·4, 7) exists, except possibly at the curve, and S and V are a surface or a volume on which σ is similarly defined. However, for the case of a volume distribution, it is of some importance to define Φ also at points (x,y,z) of the given volume. Since the integrand of (1·7, 6) becomes infinite for $x=x_0$, $y=y_0$, $z=z_0$, this requires a passage to the limit. We surround the point (x,y,z) with a small sphere of radius ϵ and define $\Phi_\epsilon(x,y,z)$ by the right-hand side of (1·7, 6), where we have excluded the interior of the sphere from the volume of integration. Now the integrand of (1·7, 6) becomes infinite at (x,y,z) like ϵ^{-1}, while the volume of the sphere tends to zero as ϵ^3. Provided σ is finite (bounded) at the point in question, it follows therefore that the limit

$$\Phi(x,y,z) = \lim_{\epsilon \to 0} \Phi_\epsilon(x,y,z)$$

exists. In general, the function Φ which is obtained in this way, although continuous and differentiable, is not a solution of Laplace's equation as we shall see presently. On the other hand, differentiation under the sign of the integral shows $\Phi(x,y,z)$ as given by (1·7, 4)–(1·7, 6) is a solution of Laplace's equation at all points which do not belong to the curve, surface or volume of the distribution.

It was shown earlier that the rate of flow, or flux, across a surface S which contains a single source in its interior, equals the strength of the source. On the other hand, the divergence theorem shows that for a field of flow due to a single source which is external to a closed surface S, the flux across S is zero. It follows that if the field of flow is due to a number of isolated sources, then the rate of flow across a surface S equals the algebraic sum of the strengths of the sources enclosed within S.

The algebraic sum of the strengths of a number of isolated sources is called the total strength of these sources. Similarly, for a given line, surface or volume distribution of sources, the integral $\int \sigma(x_0,y_0,z_0)$ (taken over the line, surface or volume in question) is called the total strength of the distribution. We have shown that for a field of flow due to isolated sources the rate of flow across a closed surface S equals the total strength of the sources enclosed in it. Regarding a continuous distribution of sources as the limit of a distribution of isolated sources we would expect that the same statement is still true for continuous distributions. This can be verified without difficulty by straightforward calculation. Thus, for a distribution of sources along a curve C, which is enclosed within a surface S, the rate of flow equals

$$\int \mathbf{q}\,d\mathbf{S} = -\int_S \operatorname{grad} \Phi \, d\mathbf{S} = -\int_S \operatorname{grad} \left(\int_C \frac{\sigma(x_0, y_0, z_0)\, dl}{\sqrt{\{(x-x_0)^2 + (y-y_0)^2 + (z-z_0)^2\}}} \right) d\mathbf{S}$$

$$= -\int_C dl \int_S \operatorname{grad} \left(\frac{\sigma(x_0, y_0, z_0)}{\sqrt{\{(x-x_0)^2 + (y-y_0)^2 + (z-z_0)^2\}}} \right) d\mathbf{S} = \int_C \sigma \, dl.$$

This proves our assertion, since $\displaystyle\int_C \sigma \, dl$ is the total strength of the distribution along C. On the other hand, if C is outside S, then the total rate of flow across S which is due to the source distribution on C is again zero. The same applies to surface and volume distributions, and to combinations of all the types of distributions considered so far. We have therefore obtained the following result which is known as the theorem of total flux:

The outward rate of flow across a closed surface equals the total strength of the sources contained in it.

It appears from the above argument that this theorem holds wherever the field of flow can be regarded as being due to the superposition of a distribution of sources within S on a flow whose velocity potential is harmonic (i.e. is a solution of Laplace's equation with continuous second derivatives) in S.

We shall now derive two important conclusions from the theorem of total flux. First, consider a distribution of sources over a surface S and let $P_0 = (x_0, y_0, z_0)$ be any particular point on S. Surround P_0 with a small cylindrical surface S' whose bases are small but finite and are parallel to S on either side of it, and so close to S that the distance between the two bases is small compared with the dimensions of the bases themselves. Let A be the area of the two bases, then A measures approximately the area of S which is enclosed within S'. Let $\sigma = \sigma(x_0, y_0, z_0)$ be the source density of the distribution at P_0. Then the total source strength enclosed by S' is given by σA approximately. On the other hand, the rate of flow across S' is given, to the same degree of accuracy, by

$$-\left(\frac{\partial \Phi}{\partial n}\right)_+ A - \left(\frac{\partial \Phi}{\partial n}\right)_- A$$

(since we may neglect the flow across the envelope of the cylinder compared with the flow across its bases). In this formula $\partial \Phi / \partial n$ denotes the normal derivative of Φ at the point P_0 of S, taken in a direction which points away from S. The signs $+$ and $-$ have been added in order to distinguish between the values of $\partial\Phi/\partial n$ on either side of S according to an arbitrary convention. Then

$$-\left(\frac{\partial\Phi}{\partial n}\right)_{+} A - \left(\frac{\partial\Phi}{\partial n}\right)_{-} A = \sigma A$$

approximately, according to the theorem of total flux, and so in the limit exactly

$$\left(\frac{\partial\Phi}{\partial n}\right)_{+} + \left(\frac{\partial\Phi}{\partial n}\right)_{-} = -\sigma. \tag{1·7, 7}$$

If we choose the direction of differentiation as pointing from $-$ to $+$ on both sides, then (1·7, 7) is replaced by

$$\left(\frac{\partial\Phi}{\partial n}\right)_{+} - \left(\frac{\partial\Phi}{\partial n}\right)_{-} = -\sigma. \tag{1·7, 8}$$

Secondly, apply the theorem of total flux to a closed surface S drawn within a volume distribution of sources. Let V be the volume enclosed by S. Then the theorem of total flux states

$$\int_{S} \mathbf{q}\, d\mathbf{S} = -\int_{S} \operatorname{grad} \Phi\, d\mathbf{S} = \int_{V} \sigma\, dV.$$

But

$$\int_{S} \operatorname{grad} \Phi\, d\mathbf{S} = \int_{V} \operatorname{div\, grad} \Phi\, dV$$

by the divergence theorem, and so

$$\int_{V} \operatorname{div\, grad} \Phi\, dV = -\int_{V} \sigma\, dV.$$

Since this is true for any arbitrary value within the given volume distribution, it follows that

$$\operatorname{div\, grad} \Phi = -\sigma \tag{1·7, 9}$$

or

$$\frac{\partial^2\Phi}{\partial x^2} + \frac{\partial^2\Phi}{\partial y^2} + \frac{\partial^2\Phi}{\partial z^2} = -\sigma \tag{1·7, 10}$$

in scalar notation. This is Poisson's equation.

If we differentiate the potential of a source (1·7, 2) with respect to x_0, y_0 or z_0, then we obtain again solutions of Laplace's equation. A physical interpretation of these solutions is provided by the following procedure.

Let $P_0 = (x_0, y_0, z_0)$ be a fixed point, and $\mathbf{n} = (\lambda, \mu, \nu)$ a unit vector in a fixed direction. Let P_1 be a point at a small distance d from P_0 such that $P_0 P_1$ points in the direction of $\mathbf{n}$. Then the coordinates of P_1 are

$$(x_0 + \lambda d, y_0 + \mu d, z_0 + \nu d).$$

Now assume that there is a source of strength σ at P_1 and a source of strength $-\sigma$ at P_0. The resulting velocity potential is

$$\Phi(x, y, z) = \Phi_0(x_0 + \lambda d, y_0 + \mu d, z_0 + \nu d) - \Phi_0(x_0, y_0, z_0),$$

where Φ_0 is given by the right-hand side of (1·7, 2), and we have indicated the dependence of Φ_0 on its parameters. For small d we then have

$$\Phi(x, y, z) = d\left(\lambda \frac{\partial \Phi}{\partial x_0} + \mu \frac{\partial \Phi}{\partial y_0} + \nu \frac{\partial \Phi}{\partial z_0}\right)$$

approximately. We now suppose that P_1 approaches P_0 indefinitely ($d \to 0$), while the product σd is kept at a constant value, $\sigma d = \tau$. Then, in the limit

$$\Phi(x, y, z) = \frac{\tau}{\sigma}\left(\lambda \frac{\partial}{\partial x_0} + \mu \frac{\partial}{\partial y_0} + \nu \frac{\partial}{\partial z_0}\right)\Phi$$

$$= \frac{\tau}{4\pi}\left(\lambda \frac{\partial}{\partial x_0} + \mu \frac{\partial}{\partial y_0} + \nu \frac{\partial}{\partial z_0}\right)\frac{1}{\sqrt{\{(x - x_0)^2 + (y - y_0)^2 + (z - z_0)^2\}}}.$$

Now $\partial \Phi_0 / \partial x_0 = -\partial \Phi_0 / \partial x$, etc., and so the expression for Φ may be written also as

$$\Phi(x, y, z) = -\frac{\tau}{4\pi}\left(\lambda \frac{\partial}{\partial x} + \mu \frac{\partial}{\partial y} + \nu \frac{\partial}{\partial z}\right)\frac{1}{\sqrt{\{(x - x_0)^2 + (y - y_0)^2 + (z - z_0)^2\}}},$$
$$(1·7, 11)$$

or as

$$\Phi(x, y, z) = -\frac{\tau}{4\pi}\frac{\partial}{\partial n}\left(\frac{1}{r}\right), \qquad (1·7, 12)$$

where r is the distance between (x_0, y_0, z_0) and (x, y, z) and $\partial / \partial n$ indicates differentiation with respect to (x, y, z) in a direction parallel to $\mathbf{n}$. $\Phi(x, y, z)$ is said to be the velocity potential of a doublet with origin P_0, axis $\mathbf{n}$ and strength τ. We may again consider line, surface and volume distributions of doublets. Of particular importance are the surface distributions for which the axes of the doublets are normal to the surface of the distribution.

The rate of flow across a closed surface which contains isolated doublets in its interior equals zero, since a doublet may be regarded as the limit of two sources whose strengths are of equal magnitude but of opposite sign. It follows that the theorem of total flux still applies if the surface contains a distribution of doublets as well as a distribution of sources in its interior.

Consider a surface distribution of doublets whose axes are everywhere normal to the surface of the distribution, S. It is not difficult to see that in general the velocity potential Φ will be discontinuous across S. Let P be a point of S. Since Φ will be continuous at P if we remove from the distribution a small area around P, it follows that the discontinuity depends entirely on the distribution in the neighbourhood of P. To simplify the discussion we may therefore suppose that the distribution is confined to a small plane disk which includes P. Denoting the values of Φ on either side of the disk by Φ_+ and Φ_- we then obtain

$$\Phi_+ - \Phi_- = -\left(\frac{1}{4\pi}\int \tau \frac{\partial}{\partial n}\left(\frac{1}{r}\right)dS\right)_+ + \left(\frac{1}{4\pi}\int \tau \frac{\partial}{\partial n}\left(\frac{1}{r}\right)dS\right)_-.$$

In this formula, **n** is a direction which is approximately constant over the surface of integration. Hence

$$\Phi_+ - \Phi_- = -\left\{\left(\frac{\partial}{\partial n}\left(\frac{1}{4\pi}\int\frac{\tau}{r}dS\right)\right)_+ - \left(\frac{\partial}{\partial n}\left(\frac{1}{4\pi}\int\frac{\tau}{r}dS\right)\right)_-\right\}, \quad (1\cdot7, 13)$$

and so
$$\Phi_+ - \Phi_- = \tau \qquad (1\cdot7, 14)$$

on account of (1·7, 8), since we may regard the right-hand side of (1·7, 13) as the discontinuity of normal derivatives for the potential

$$\frac{1}{4\pi}\int\frac{\tau}{r}dS,$$

which is due to a source distribution of strength τ over S.

1·8 Green's formulae

In the divergence theorem

$$\int_S \mathbf{a}\,d\mathbf{S} = \int_V \operatorname{div}\mathbf{a}\,dV, \qquad (1\cdot8, 1)$$

where S is a surface bounding a volume V and $\mathbf{a}$ an arbitrary vector function, substitute
$$\mathbf{a} = \phi\operatorname{grad}\psi,$$

where ϕ and ψ are single-valued scalar functions. Then (1·8, 1) becomes

$$\int_S \phi\operatorname{grad}\psi\,d\mathbf{S} = \int_V (\operatorname{grad}\phi\operatorname{grad}\psi + \phi\operatorname{div}\operatorname{grad}\psi)\,dV. \qquad (1\cdot8, 2)$$

Similarly
$$\int_S \psi\operatorname{grad}\phi\,d\mathbf{S} = \int_V (\operatorname{grad}\phi\operatorname{grad}\psi + \psi\operatorname{div}\operatorname{grad}\phi)\,dV.$$

Subtracting

$$\int_S (\phi\operatorname{grad}\psi - \psi\operatorname{grad}\phi)\,d\mathbf{S} = \int_V (\phi\operatorname{div}\operatorname{grad}\psi - \psi\operatorname{div}\operatorname{grad}\phi)\,dV. \qquad (1\cdot8, 3)$$

This is Green's reciprocal formula. If ϕ and ψ are both solutions of Laplace's equation
$$\operatorname{div}\operatorname{grad}\phi = \operatorname{div}\operatorname{grad}\psi = 0,$$

then we obtain
$$\int_S (\phi\operatorname{grad}\psi - \psi\operatorname{grad}\phi)\,d\mathbf{S} = 0. \qquad (1\cdot8, 4)$$

(1·8, 3) may also be written as

$$\int_S \left(\phi\frac{\partial\psi}{\partial n} - \psi\frac{\partial\phi}{\partial n}\right)dS = \int_V (\phi\operatorname{div}\operatorname{grad}\psi - \psi\operatorname{div}\operatorname{grad}\phi)\,dV,$$

where $\partial/\partial n$ indicates normal differentiation across S in an outward direction. In this notation, (1·8, 4) becomes

$$\int_S \left(\phi\frac{\partial\psi}{\partial n} - \psi\frac{\partial\phi}{\partial n}\right)dS = 0. \qquad (1\cdot8, 5)$$

Now let $P_0 = (x_0, y_0, z_0)$ be any point which does not belong to the volume V, and let

$$\psi(x, y, z) = \frac{1}{r} = \frac{1}{\sqrt{\{(x - x_0)^2 + (y - y_0)^2 + (z - z_0)^2\}}}. \qquad (1\cdot8, 6)$$

Then ψ is a solution of Laplace's equation, $\operatorname{div} \operatorname{grad} \psi = 0$, so that $(1\cdot8, 3)$ becomes

$$\int_S \left(\phi \frac{\partial}{\partial n} \left(\frac{1}{r} \right) - \frac{1}{r} \frac{\partial \phi}{\partial n} \right) dS = - \int_V \frac{\operatorname{div} \operatorname{grad} \phi}{r} dV. \qquad (1\cdot8, 7)$$

Now let P_0 be a point in the interior of V. Then we cannot apply $(1\cdot8, 7)$ directly since $1/r$ becomes infinite at P_0. Surround P_0 with a small sphere S', radius ϵ, and let V' be the volume which is obtained by excluding from V the interior of S'. V' is bounded by S and S'. Then

$$\int_S \left(\phi \frac{\partial}{\partial n} \left(\frac{1}{r} \right) - \frac{1}{r} \frac{\partial \phi}{\partial n} \right) dS + \int_{S'} \left(\phi \frac{\partial}{\partial n} \left(\frac{1}{r} \right) - \frac{1}{r} \frac{\partial \phi}{\partial n} \right) dS = - \int_{V'} \frac{\operatorname{div} \operatorname{grad} \phi}{r} dV.$$
$$(1\cdot8, 8)$$

The direction of normal differentiation on S' points away from V', that is, towards the interior of the sphere. Thus

$$\left[\frac{\partial}{\partial n} \left(\frac{1}{r} \right) \right]_{r = \epsilon} = - \left[\frac{\partial}{\partial r} \left(\frac{1}{r} \right) \right]_{r = \epsilon} = \frac{1}{\epsilon^2}.$$

Hence the second integral on the left-hand side of $(1\cdot8, 8)$ becomes

$$\int_{S'} \left(\phi \frac{1}{\epsilon^2} - \frac{1}{\epsilon} \frac{\partial \phi}{\partial n} \right) dS = \frac{1}{\epsilon^2} \int_{S'} \phi \, dS - \frac{1}{\epsilon} \int_{S'} \frac{\partial \phi}{\partial n} dS. \qquad (1\cdot8, 9)$$

Now as ϵ tends to zero, the area of S' tends to zero as ϵ^2, therefore (provided only the derivatives of ϕ are continuous, as is implicit in the application of the divergence theorem) it is readily seen that the second term on the right-hand side of $(1\cdot8, 9)$ tends to zero as ϵ tends to zero. On the other hand, $\frac{1}{4\pi\epsilon^2} \int_{S'} \phi \, dS$ is the arithmetic mean value of ϕ on the sphere S', and so this integral tends to $\phi(x_0, y_0, z_0)$ as ϵ tends to zero. It follows that the right-hand side of $(1\cdot8, 9)$ becomes, in the limit, $4\pi\phi(x_0, y_0, z_0)$. Substituting in $(1\cdot8, 8)$ we obtain after a slight modification

$$\phi(x_0, y_0, z_0) = \frac{1}{4\pi} \int_S \left(\frac{\partial \phi}{\partial n} \frac{1}{r} - \phi \frac{\partial}{\partial n} \left(\frac{1}{r} \right) \right) dS - \frac{1}{4\pi} \int_V \frac{\operatorname{div} \operatorname{grad} \phi}{r} dV. \quad (1\cdot8, 10)$$

This is Green's formula. In particular, if $\operatorname{div} \operatorname{grad} \phi = 0$, as is the case for the velocity potential in irrotational incompressible flow, then

$$\phi(x_0, y_0, z_0) = \frac{1}{4\pi} \int_S \left(\frac{\partial \phi}{\partial n} \frac{1}{r} - \phi \frac{\partial}{\partial n} \left(\frac{1}{r} \right) \right) dS. \qquad (1\cdot8, 11)$$

The first part of this integral represents a source distribution on S, the second a distribution of doublets whose axes are normal to S. However, it will be shown presently that in general the values of ϕ on S alone determine its value in the interior of S, so that it is not permissible to specify both ϕ and $\partial\phi/\partial n$ in (1·8, 11) arbitrarily.

To show that the function ϕ is determined uniquely in V by its values over the boundary S, set $\psi = \phi$ in (1·8, 2), where ϕ satisfies Laplace's equation in V. The equation becomes

$$\int_S \phi \frac{\partial\phi}{\partial n} dS = \int_V (\operatorname{grad}\phi)^2 dV. \qquad (1·8, 12)$$

Suppose that $\phi = 0$ at all points of S. Then the left-hand side of (1·8, 12) vanishes. But the integrand on the right-hand side of (1·8, 12) is non-negative ($\geqslant 0$) and so this integral can vanish only if

$$\operatorname{grad}\phi = 0$$

identically in V. This implies $\phi = $ constant in V, or taking into account that $\phi = 0$ at all points of S,

$$\phi = 0 \qquad (1·8, 13)$$

everywhere in V.

Assume now that there are two solutions of Laplace's equation in V, ϕ_1 and ϕ_2, which take the same values on S. Then $\phi = \phi_1 - \phi_2$ on S, and so

$$\phi_1 - \phi_2 = 0, \quad \phi_1 = \phi_2$$

everywhere in V, by (1·8, 13). This proves that the solution of Laplace's equation in V, for given boundary values, is unique. The existence of such a solution, subject to certain conditions of regularity, is proved in standard texts on potential theory.‡

Similarly, if $\partial\phi/\partial n = 0$ on S, then (1·8, 12) shows

$$\operatorname{grad}\phi = 0, \quad \phi = \text{constant}$$

in V. It follows, as before, that if two solutions of Laplace's equation in V possess the same normal derivatives on S, then they can differ only by a constant. This constant may in fact be any real number, since the addition of a constant to any function does not change its normal derivatives. Moreover, in this case $\partial\phi/\partial n$ cannot be specified arbitrarily on S, since

$$\int_S \frac{\partial\phi}{\partial n} dS = \int_S \operatorname{grad}\phi \, d\mathbf{S} = \int_V \operatorname{div}\operatorname{grad}\phi \, dV = 0$$

by the divergence theorem. Thus the specified values of $\partial\phi/\partial n$ on S must satisfy the condition

$$\int_S \frac{\partial\phi}{\partial n} dS = 0. \qquad (1·8, 14)$$

If this condition, which is simply the condition that the total flux across S be zero, is satisfied, then the problem possesses a solution in V, again subject to certain conditions of regularity.

‡ Such as O. D. Kellogg, *Foundations of Potential Theory* (Ungar Publishing Co., 1929), ch. 11.

Assume now that the function $\phi(x, y, z)$ is a solution of Laplace's equation in the region outside a closed and bounded surface S, and suppose more particularly that its first derivatives $\partial\phi/\partial x$, $\partial\phi/\partial y$, $\partial\phi/\partial z$, vanish at infinity. This will be the case, for example, if ϕ is the velocity potential of an irrotational field of flow of an incompressible fluid which is at rest at infinity. Let F be the outward flux across S, so that F is given by

$$F = -\int_S \frac{\partial\phi}{\partial n}\,dS. \tag{1·8, 15}$$

According to the divergence theorem this is also the outward flux across any sphere S' of radius R and centre $P_0 = (x_0, y_0, z_0)$ which includes S in its interior. Thus

$$F = -\int_0^\pi \int_0^{2\pi} \left(\frac{\partial\phi}{\partial r}\right)_{r=R} R^2 \sin\theta\, d\theta\, d\chi$$

in spherical polar coordinates which are given by

$$x = x_0 + r\cos\theta, \quad y = y_0 + r\sin\theta\cos\chi, \quad z = z_0 + r\sin\theta\sin\chi.$$

This is equivalent to

$$\int_0^\pi \int_0^{2\pi} \left(\frac{\partial\phi}{\partial r}\right)_{r=R} \sin\theta\, d\theta\, d\chi = -\frac{F}{R^2}. \tag{1·8, 16}$$

Integrating (1·8, 16), we obtain

$$\int_0^\pi \int_0^{2\pi} \phi \sin\theta\, d\theta\, d\chi = \frac{F}{R} + C. \tag{1·8, 17}$$

This may also be written as

$$\frac{1}{4\pi R^2} \int_{S'} \phi\, dS = \frac{F}{4\pi R} + \frac{C}{4\pi}, \tag{1·8, 18}$$

so that the left-hand side denotes the mean value of ϕ over the sphere S' (whose area is $4\pi R^2$).

The constant of integration C which occurs in (1·8, 17) appears a priori to depend on x_0, y_0, z_0. However, if we displace the sphere S' in a direction parallel to the x-axis while keeping the radius of the sphere constant, then

$$\frac{\partial C}{\partial x_0} = \int_0^\pi \int_0^{2\pi} \frac{\partial\phi}{\partial x} \sin\theta\, d\theta\, d\chi \tag{1·8, 19}$$

by (1·8, 17). The right-hand side of (1·8, 19) equals 4π times the mean value of $\partial\phi/\partial x$ over the sphere S'. But $\partial\phi/\partial x$ vanishes at infinity, and so, by choosing the radius R sufficiently large we can make the right-hand side of (1·8, 19) as small as we please numerically. We conclude that $\partial C/\partial x_0 = 0$, and similarly $\partial C/\partial y_0 = \partial C/\partial z_0 = 0$, which shows that C is independent of the position of the centre P_0. Let us now take P_0 to be any point in the region external to S, and let S' be a circle of large radius R, as before. P_0 belongs to the finite volume enclosed between S and S' so that we may apply Green's formula, $(1\cdot8, 11)$. We obtain

$$\phi(x_0, y_0, z_0)$$

$$= -\frac{1}{4\pi}\int_S \left(\frac{\partial\phi}{\partial n}\frac{1}{r} - \phi\frac{\partial}{\partial n}\left(\frac{1}{r}\right)\right) dS + \frac{1}{4\pi}\int_{S'} \frac{\partial\phi}{\partial n}\frac{1}{r} dS - \frac{1}{4\pi}\int_{S'} \phi\frac{\partial}{\partial n}\left(\frac{1}{r}\right) dS.$$

In this formula, we take the normal differentiation across S in a direction pointing towards the exterior of S, i.e. towards the interior of the region bounded by S and S'. Accordingly, we have to prefix the sign minus to the first integral. The second integral is

$$\frac{1}{4\pi}\int_{S'} \frac{\partial\phi}{\partial n}\frac{1}{r} dS = \frac{1}{4\pi R}\int_{S'} \frac{\partial\phi}{\partial n} dS = -\frac{F}{4\pi R}$$

by $(1\cdot8, 15)$. Also, the third integral equals

$$-\frac{1}{4\pi}\int_{S'} \phi\frac{\partial}{\partial n}\left(\frac{1}{r}\right) dS = \frac{1}{4\pi R^2}\int_{S'} \phi\, dS = \frac{1}{4\pi}\left(\frac{F}{R} + C\right)$$

by $(1\cdot8, 18)$. Hence

$$\phi(x_0, y_0, z_0) = \frac{1}{4\pi}\int_S \left(\phi\frac{\partial}{\partial n}\left(\frac{1}{r}\right) - \frac{\partial\phi}{\partial n}\frac{1}{r}\right) dS + \frac{C}{4\pi}. \tag{1·8, 20}$$

In particular, if we assume $C = 0$, the formula becomes

$$\phi(x_0, y_0, z_0) = \frac{1}{4\pi}\int_S \left(\phi\frac{\partial}{\partial n}\left(\frac{1}{r}\right) - \frac{\partial\phi}{\partial n}\frac{1}{r}\right) dS. \tag{1·8, 21}$$

If the flux across S vanishes, $F = 0$, then this formula may be modified further, as follows. Since $F = 0$, there exists a function $\phi'(x, y, z)$ which is a solution of Laplace's equation in the interior of S, whose normal derivative takes the value $\partial\phi/\partial n$ on S. Now the point P_0 does not belong to S and so

$$0 = \frac{1}{4\pi}\int_S \left(\phi'\frac{\partial}{\partial n}\left(\frac{1}{r}\right) - \frac{\partial\phi'}{\partial n}\frac{1}{r}\right) dS \tag{1·8, 22}$$

by $(1\cdot8, 7)$, where r has the same meaning as in $(1\cdot8, 21)$. Subtracting $(1\cdot8, 22)$ from $(1\cdot8, 21)$ and bearing in mind that $\partial\phi/\partial n = \partial\phi'/\partial n$ on S, we obtain

$$\phi(x_0, y_0, z_0) = \frac{1}{4\pi}\int_S (\phi - \phi')\frac{\partial}{\partial n}\left(\frac{1}{r}\right) dS. \tag{1·8, 23}$$

This is a representation of ϕ in terms of a distribution of doublets only, over the surface S. As P_0 tends to infinity, $\frac{\partial}{\partial n}\left(\frac{1}{r}\right)$ varies as r^{-2} and so the derivatives of ϕ vary as r^{-3}.

Now consider an irrotational field of flow due to the uniform translational motion of a rigid body B in a fluid which is at rest at infinity. Then the field of flow possesses a velocity potential Φ, $\mathbf{q} = -\text{grad }\Phi$, which is a solution of Laplace's equation. Also $\partial\Phi/\partial x = -u$, $\partial\Phi/\partial y = -v$, $\partial\Phi/\partial z = -w$, vanish at infinity, and since B is a rigid body, the flux across any surface S which surrounds B, e.g. the surface of B itself, is equal to zero. It follows that we may assume that $\Phi = 0$ at infinity, so that (1·8, 23) applies, and hence that the velocity components u, v, w vanish at infinity as the third negative power of the distance from B.

The field of flow which is produced by the motion of the body will in general be unsteady, i.e. the velocity vector at any given point will vary with the time. Let U, V, W be the velocity components of the rigid body B. Then if we transform the velocity components of the fluid to a system of coordinates whose axes are parallel to the original axes but which is fixed relative to B, the new velocity components at any point of the fluid will be $u - U$, $v - V$, $w - W$. In particular, the velocity components at infinity now are $-U$, $-V$, $-W$, and at any point of B, which is at rest in the new system of coordinates, the normal component of the fluid velocity is equal to zero. This determines the velocity potential, except for a constant, and hence determines the field of flow uniquely. Thus the field of flow is now steady. To calculate the reaction of the fluid on the body B we may therefore apply (1·4, 15) or, neglecting body forces, (1·4, 16). Now by Bernoulli's equation (1·6, 6) the pressure p, with no body forces, is given by

$$p = \rho H - \tfrac{1}{2}\rho\{(u - U)^2 + (v - V)^2 + (w - W)^2\}.$$

Substituting this expression in (1·4, 16), where S is now a sphere of large radius which surrounds B, we obtain for the x-component of the reaction on B

$$-\left\{ \int_S (\rho H - \tfrac{1}{2}\rho[(u - U)^2 + (v - V)^2 + (w - W)^2])\, dy\, dz \right.$$
$$\left. + \int_S \rho(u - U)\,[(u - U)\, dy\, dz + (v - V)\, dz\, dx + (w - W)]\, dx\, dy \right\}. \quad (1\cdot8, 24)$$

Now the integrals $\int_S dy\, dz$, $\int_S dx\, dz$, $\int_S dx\, dy$ vanish, since they are taken over a closed surface. Expanding the expressions in the integrands of (1·8, 24) we then find that all the constant terms (i.e. ρH, $\tfrac{1}{2}\rho U^2$, etc.) can be omitted for the reason just stated while all the other terms involve either u, v or w as factors. But as we let the radius of S increase indefinitely, u, v and w tend to zero inversely as the cube of the radius, while the area of S tends to infinity only as the square of the radius. This shows that we can make all terms in (1·8, 24) as small as we please by making the radius of S sufficiently large. It follows that the x-component of the reaction on B is equal to zero, and an exactly similar reasoning shows that the same is true for the other two components. This conclusion is known as D'Alembert's paradox. It is contrary to an overwhelming mass of experimental evidence which shows that even a rigid body which moves through a fluid in uniform translational motion experiences a reaction from the medium. Accordingly, we shall have to amend our assumptions concerning the nature of the flow which is produced by the motion of a rigid body. We may add that, even on the assumptions made so far, a rigid body does experience a reaction from a fluid when the motion of the body is non-uniform, but it is now clear that even in that case we cannot expect an adequate answer on the basis of the theory so far developed.

1·9   Vortex distributions

Let θ, ξ, η, ζ be continuous functions of x, y, z which differ from zero only in a finite (bounded) volume V. We wish to determine a vector $\mathbf{q}_0$ such that

$$\operatorname{div} \mathbf{q}_0 = 0, \quad \operatorname{curl} \mathbf{q}_0 = \boldsymbol{\omega} \tag{1·9, 1}$$

in V, where $\boldsymbol{\omega} = (\xi, \eta, \zeta)$, and it is assumed that $\operatorname{div} \boldsymbol{\omega} = 0$. As a first step towards the solution of this problem we determine a scalar function Φ such that

$$\operatorname{div} \operatorname{grad} \Phi = -\theta. \tag{1·9, 2}$$

According to (1·7, 9), this condition is satisfied by the function

$$\Phi(x, y, z) = \int_V \frac{\theta}{r} dV = \int_V \frac{\theta(x_0, y_0, z_0)\, dx_0\, dy_0\, dz_0}{\sqrt{\{(x - x_0)^2 + (y - y_0)^2 + (z - z_0)^2\}}}. \tag{1·9, 3}$$

We note that at the same time

$$\operatorname{curl} \operatorname{grad} \Phi = 0,$$

since this is an identity which holds for any scalar function Φ.

Next we determine a vector function $\boldsymbol{\Psi}$ such that

$$\operatorname{curl} \operatorname{curl} \boldsymbol{\Psi} = \boldsymbol{\omega}, \tag{1·9, 4}$$

while at the same time $\operatorname{div} \boldsymbol{\Psi} = 0.$ (1·9, 5)

Now there is a vector identity

$$\operatorname{curl} \operatorname{curl} \boldsymbol{\Psi} = \operatorname{grad} \operatorname{div} \boldsymbol{\Psi} - \operatorname{div} \operatorname{grad} \boldsymbol{\Psi}. \tag{1·9, 6}$$

It follows that if (1·9, 5) is satisfied then (1·9, 4) can be replaced by

$$\operatorname{div} \operatorname{grad} \boldsymbol{\Psi} = -\boldsymbol{\omega}. \tag{1·9, 7}$$

According to (1·7, 9) this condition is satisfied by the vector $\boldsymbol{\Psi} = (F, G, H)$ whose components are given by

$$F = \frac{1}{4\pi} \int_V \frac{\xi}{r} dV, \quad G = \frac{1}{4\pi} \int_V \frac{\eta}{r} dV, \quad H = \frac{1}{4\pi} \int_V \frac{\zeta}{r} dV$$

or

$$\boldsymbol{\Psi} = \frac{1}{4\pi} \int_V \frac{\boldsymbol{\omega}}{r} dV, \tag{1·9, 8}$$

where $r = \sqrt{\{(x - x_0)^2 + (y - y_0)^2 + (z - z_0)^2\}}$ and the integration is carried out with respect to (x_0, y_0, z_0). Now

$$\frac{\partial}{\partial x}\left(\frac{1}{r}\right) = -\frac{\partial}{\partial x_0}\left(\frac{1}{r}\right),$$

and so
$$\operatorname{div}\boldsymbol{\Psi} = -\frac{1}{4\pi}\int_V\left\{\xi\frac{\partial}{\partial x_0}\left(\frac{1}{r}\right)+\eta\frac{\partial}{\partial y_0}\left(\frac{1}{r}\right)+\zeta\frac{\partial}{\partial z_0}\left(\frac{1}{r}\right)\right\}dV = -\frac{1}{4\pi}\int_V\boldsymbol{\omega}\operatorname{grad}\left(\frac{1}{r}\right)dV,$$
(1·9, 9)

where the gradient is taken with respect to (x_0, y_0, z_0). But

$$\int_V\boldsymbol{\omega}\operatorname{grad}\left(\frac{1}{r}\right)dV = \int_V\operatorname{div}\left(\frac{1}{r}\boldsymbol{\omega}\right)dV-\int_V\frac{1}{r}\operatorname{div}\boldsymbol{\omega}\,dV = \int_S\frac{1}{r}\boldsymbol{\omega}\,d\mathbf{S}, \quad (1·9, 10)$$

since $\operatorname{div}\boldsymbol{\omega} = 0$, where S is any closed surface which contains V in its interior. But $\boldsymbol{\omega} = 0$ outside V and so (1·9, 10) becomes

$$\int_V\boldsymbol{\omega}\operatorname{grad}\left(\frac{1}{r}\right)dV = 0.$$
(1·9, 11)

Hence
$$\operatorname{div}\boldsymbol{\Psi} = 0$$

by (1·9, 9), and
$$\operatorname{curl}\operatorname{curl}\boldsymbol{\Psi} = -\operatorname{div}\operatorname{grad}\boldsymbol{\Psi}$$

by (1·9, 6). It follows that $\boldsymbol{\Psi}$ satisfies not only (1·9, 7) but also (1·9, 4) and (1·9, 5).

Define the vector $\mathbf{q}_0$ by
$$\mathbf{q}_0 = -\operatorname{grad}\Phi+\operatorname{curl}\boldsymbol{\Psi}.$$
(1·9, 12)

Then
$$\operatorname{div}\mathbf{q}_0 = -\operatorname{div}\operatorname{grad}\Phi+\operatorname{div}\operatorname{curl}\boldsymbol{\Psi} = -\operatorname{div}\operatorname{grad}\Phi = \theta,$$

since
$$\operatorname{div}\operatorname{curl}\boldsymbol{\Psi} = 0,$$

according to a vector identity, and

$$\operatorname{curl}\mathbf{q}_0 = -\operatorname{curl}\operatorname{grad}\Phi+\operatorname{curl}\operatorname{curl}\boldsymbol{\Psi} = \boldsymbol{\omega},$$

so that $\mathbf{q}_0$ satisfies the conditions (1·9, 1).

Now let $\mathbf{q}$ be the velocity vector in a field of flow which is endowed with vorticity. Suppose that the fluid is incompressible, so that $\operatorname{div}\mathbf{q} = 0$. The vorticity is given by $\boldsymbol{\omega} = \operatorname{curl}\mathbf{q}$. Assuming, to begin with, that $\boldsymbol{\omega}$ is continuous and is confined to a bounded (finite) volume, we construct the vector

$$\mathbf{q}_0 = \operatorname{curl}\boldsymbol{\Psi}$$

as in (1·9, 12). We observe that $\Phi = 0$ since $\operatorname{div}\mathbf{q}_0 = \theta = 0$ identically.

Consider the vector
$$\mathbf{q}' = \mathbf{q}-\mathbf{q}_0.$$

The divergence of $\mathbf{q}'$ vanishes everywhere since

$$\operatorname{div}\mathbf{q}' = \operatorname{div}\mathbf{q}-\operatorname{div}\mathbf{q}_0 = -\operatorname{div}\operatorname{curl}\boldsymbol{\Psi} = 0, \quad (1·9, 13)$$

and the curl of the vector also vanishes throughout, since

$$\operatorname{curl}\mathbf{q}' = \operatorname{curl}\mathbf{q}-\operatorname{curl}\mathbf{q}_0 = \boldsymbol{\omega}-\boldsymbol{\omega} = 0.$$

It follows that $\mathbf{q}'$ can be derived from a potential Φ', $\mathbf{q}' = -\operatorname{grad}\Phi'$, and further than Φ' satisfies Laplace's equation

$$\operatorname{div}\operatorname{grad}\Phi' = 0$$

by (1·9, 13). We have therefore shown that $\mathbf{q}$ can be represented by the formula

$$\mathbf{q} = -\operatorname{grad}\Phi' + \operatorname{curl}\boldsymbol{\Psi}, \tag{1·9, 14}$$

where Φ' is a solution of Laplace's equation, and $\boldsymbol{\Psi} = (F, G, H)$ is given by (1·9, 8). The formula (1·9, 14) holds even if $\boldsymbol{\omega}$ is discontinuous at the surface of a vortex filament. For in this case the vorticity vector is always tangential to such a surface, and it follows that if we take S in (1·9, 10) as the surface of the vortex filament in question then the integral $\int_S \dfrac{1}{r}\boldsymbol{\omega}\,d\mathbf{S}$ still vanishes, so (1·9, 11) and the subsequent arguments which lead to (1·9, 14) still apply. On the other hand, in cases where the vorticity distribution extends to infinity, due attention must be paid to the convergence of the integrals involved.

The preceding discussion shows that

$$\mathbf{q}_0 = \operatorname{curl}\boldsymbol{\Psi}$$

may be regarded quite generally as the component of the velocity vector which is due to a given distribution of vorticity. In particular, if the given (incompressible fluid) extends over the whole of three-dimensional space and is at rest at infinity, while the vorticity is limited to a finite volume, then it can be shown that $\mathbf{q}_0 = \mathbf{q}$, so that $\mathbf{q}_0$ constitutes the entire velocity vector. To prove this we only have to show that the vector $\mathbf{q}' = \mathbf{q} - \mathbf{q}_0$ which was introduced earlier vanishes identically. Since the vorticity is limited to a finite volume, it is easy to see that the components of $\mathbf{q}_0$ vanish at infinity. It follows that the components of $\mathbf{q}'$ also vanish at infinity. Moreover, $\mathbf{q}' = -\operatorname{grad}\Phi'$, where Φ' is a solution of Laplace's equation at all points of three-dimensional space. It follows that the flux $F = -\displaystyle\int_S \dfrac{\partial\Phi'}{\partial n}\,dS$ across any closed surface is equal to zero, and hence by (1·8, 18) that the mean value of Φ over a sphere S of radius R is given by

$$\frac{1}{4\pi R^2}\int_S \Phi'\,dS = \frac{C}{4\pi}, \tag{1·9, 15}$$

where C is independent of R and independent of the coordinates of the centre of the sphere. Now apply Green's formula (1·8, 11) to the case where (x_0, y_0, z_0) is the centre of the sphere under consideration. Then

$$\Phi'(x_0, y_0, z_0) = \frac{1}{4\pi R}\int_S \frac{\partial\Phi'}{\partial n}\,dS + \frac{1}{4\pi R^2}\int_S \Phi'\,dS = \frac{C}{4\pi}.$$

But (x_0, y_0, z_0) is arbitrary and so Φ' is constant throughout and $\mathbf{q}'$ vanishes everywhere, as asserted.

Later we shall have occasion to consider also fields of flow which are due to isolated line vortices or to surface distributions of vorticity (vortex sheets). These may be regarded as limiting cases of volume distributions. A line vortex is obtained by letting the cross-section of a vortex filament tend to zero while keeping the strength of the filament constant. To arrive at the notion of a vortex sheet, consider a surface of discontinuity S in a fluid, such that the normal velocity components are continuous across S, while the tangential velocities are discontinuous across it. Let us distinguish between the velocities on either side of S by appending subscripts $+$ and $-$ to them,

$$\mathbf{q}_+ = (u_+, v_+, w_+), \quad \mathbf{q}_- = (u_-, v_-, w_-).$$

The vector $\Delta\mathbf{q} = \mathbf{q}_+ - \mathbf{q}_-$ is then by assumption tangential to S.

Consider a small rectangular circuit $ABCD$ such that the sides AB and DC are close to the surface S and parallel to it on either side of it, and the sides BC and AD are normal to S, and are small compared with AB and DC. Denoting the vector $AB = DC$ by $\delta\mathbf{l}$, and assuming that AB is on the side distinguished by the subscript $+$ and DC on the side distinguished by the subscript $-$, we then obtain for the circulation around the circuit $ABCD$

$$\gamma(ABCD) = \mathbf{q}_+ \delta\mathbf{l} - \mathbf{q}_- \delta\mathbf{l} = (\mathbf{q}_+ - \mathbf{q}_-)\,\delta\mathbf{l} = \Delta\mathbf{q}\,\delta\mathbf{l}.$$

In particular, if $\delta\mathbf{l}$ is parallel to the direction of $\Delta\mathbf{q}$ then

$$\gamma(ABCD) = \Delta q\,\delta l. \tag{1·9, 16}$$

Now replace the surface S by a thin layer of vorticity of thickness δn and such that at every point of the layer the vorticity vector $\boldsymbol{\omega}$ is tangential to the surface and normal to $\Delta\mathbf{q}$, while its magnitude is given by

$$\omega\,\delta n = |\,\mathbf{q}_+ - \mathbf{q}_-\,| = \Delta q. \tag{1·9, 17}$$

Then the area intercepted on the layer by the rectangle $ABCD$ is $\delta n\,\delta l$, and so the corresponding circulation around $ABCD$ is

$$\gamma'(ABCD) = \omega\,\delta n\,\delta l = \Delta q\,\delta l. \tag{1·9, 18}$$

(In order to obtain the positive sign on the right-hand side of (1·9, 18) we have to choose the direction to which $\boldsymbol{\omega}$ points in such a way that the circuit $ABCD$ is positive when viewed against that direction.)

Comparing (1·9, 16) and (1·9, 18) we see that $\gamma = \gamma'$. Thus we may regard the surface of discontinuity S as the limit of the vorticity layer as the thickness of the layer tends to zero. S is therefore also known as a vortex sheet. Δq is the magnitude of the 'surface vorticity', but the vorticity vector is normal to $\Delta\mathbf{q}$ in the tangent plane to S.

The field of flow due to a closed line vortex C can be obtained from (1·9, 8) as a limiting case. Assuming first that the cross-section of the vortex at any given point is the small but finite quantity A, and ω is the vorticity at that point, we have

$$\Gamma = \omega A$$

by (1·5, 7), where the constant Γ is the circulation around the vortex. To apply (1·9, 8) we then have to equate dV to $A\,dl$, where dl is the element of length of the vortex, and so

$$\omega\,dV = \omega A\,dl = \omega A\,dl = \Gamma\,dl,$$

where we have used the fact that the direction of ω is also the direction of dl at any point of the vortex. Then (1·9, 8) becomes

$$\Psi = \frac{\Gamma}{4\pi}\int_C \frac{1}{r}\,dl = \frac{\Gamma}{4\pi}\int_C \left(\frac{dx_0}{r}, \frac{dy_0}{r}, \frac{dz_0}{r}\right). \qquad (1\cdot9, 19)$$

Hence

$$u(x, y, z) = \frac{\Gamma}{4\pi}\int_C \left\{\frac{\partial}{\partial y}\left(\frac{1}{r}\right) dz_0 - \frac{\partial}{\partial z}\left(\frac{1}{r}\right) dy_0\right\}$$

$$= \frac{\Gamma}{4\pi}\int_C \frac{(z - z_0)\,dy_0 - (y - y_0)\,dz_0}{r^3},$$

and similarly

$$v(x, y, z) = \frac{\Gamma}{4\pi}\int_C \frac{(x - x_0)\,dz_0 - (z - z_0)\,dx_0}{r^3}, \qquad (1\cdot9, 20)$$

$$w(x, y, z) = \frac{\Gamma}{4\pi}\int_C \frac{(y - y_0)\,dx_0 - (x - x_0)\,dy_0}{r^3}.$$

This may be written in vector notation as

$$\mathbf{q} = -\frac{\Gamma}{4\pi}\int_C \frac{\mathbf{r}\wedge d\mathbf{l}}{r^3}, \qquad (1\cdot9, 21)$$

where $\mathbf{r} = (x - x_0, y - y_0, z - z_0)$ in agreement with the notation

$$r = \sqrt{\{(x - x_0)^2 + (y - y_0)^2 + (z - z_0)^2\}}.$$

We may interpret this formula in a simple geometrical way if we regard $\mathbf{q}$ as being made up of elementary vectors $d\mathbf{q}$ due to the elementary vortices $\Gamma\,d\mathbf{l}$ which constitutes the line vortex, so

$$d\mathbf{q} = -\frac{\Gamma}{4\pi}\frac{\mathbf{r}\wedge d\mathbf{l}}{r^3} = \frac{\Gamma}{4\pi}\frac{d\mathbf{l}\wedge\mathbf{r}}{r^3}. \qquad (1\cdot9, 22)$$

Thus the direction of $d\mathbf{q}$ is perpendicular to $d\mathbf{l}$ and $\mathbf{r}$, such that the vectors $d\mathbf{l}$, $\mathbf{r}$ and $d\mathbf{q}$, taken in that order, form a right-handed system, while its magnitude is

$$dq = \left|\frac{\Gamma\sin\theta}{4\pi r^2}\,dl\right|, \qquad (1\cdot9, 23)$$

where θ is the angle between $\mathbf{r}$ and $d\mathbf{l}$ $(0 \leqslant \theta \leqslant \pi)$. This is the 'law of Biot and Savart' which possesses a well-known counterpart in electromagnetism. It should be observed, however, that an elementary vortex as postulated in (1·9, 22) cannot exist in isolation, so that this formula is merely another and for some purposes more convenient way of writing (1·9, 21). Formulae (1·9, 20)–(1·9, 23) still apply if the line vortex, instead of being closed, extends to infinity in both directions.

We shall be interested particularly in the field of flow due to a line vortex which consists of a number of straight segments. The effect of each segment can be calculated separately, from (1·9, 22).

Let AB be such a segment, with the vorticity vector pointing from A to B, and let Q be any point in the fluid, whose normal distance from AB is $QQ' = d$ (Fig. 1·9, 1). Let $\theta_1 = \angle QAB$ and $\theta_2 = \pi - \angle QBA$, and let r be the algebraic distance of a point T on AB from Q, where the positive direction points to B. Then the velocity elements dq are all normal to the plane through A, B and Q, and such that in the case shown in Fig. 1·9, 1, for example, they all point away from the reader. Also

$$l = -d \cot \theta, \quad r = d \operatorname{cosec} \theta,$$

and so
$$dq = \frac{\Gamma}{4\pi r^2} \sin \theta \, dl = \frac{\Gamma}{4\pi d} \sin \theta \, d\theta. \tag{1·9, 24}$$

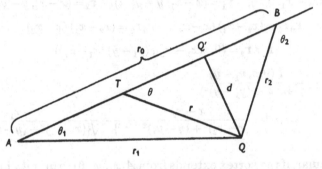

Fig. 1·9, 1

Integrating between A and B we find that the magnitude of the velocity due to AB, q_{AB}, is

$$q_{AB} = \frac{\Gamma}{4\pi d} \int_{\theta_1}^{\theta_2} \sin \theta \, d\theta = \frac{\Gamma}{4\pi d} (\cos \theta_1 - \cos \theta_2). \tag{1·9, 25}$$

In particular, if the line vortex is straight and extends to infinity in both directions, then $\theta_1 = 0$, $\theta_2 = \pi$, and so

$$q = \frac{\Gamma}{2\pi d}. \tag{1·9, 26}$$

Let $\mathbf{r}_0$, $\mathbf{r}_1$, $\mathbf{r}_2$ be the vectors AB, AQ and BQ respectively, so that $\mathbf{r}_0 = \mathbf{r}_1 - \mathbf{r}_2$. Then

$$d = |\mathbf{r}_1 \wedge \mathbf{r}_2|/r_0, \quad \cos \theta_1 = \frac{\mathbf{r}_0 \mathbf{r}_1}{r_0 r_1}, \quad \cos \theta_2 = \frac{\mathbf{r}_0 \mathbf{r}_2}{r_0 r_2},$$

and so
$$q_{AB} = \frac{\Gamma}{4\pi |\mathbf{r}_1 \wedge \mathbf{r}_2|} \mathbf{r}_0 \left(\frac{\mathbf{r}_1}{r_1} - \frac{\mathbf{r}_2}{r_2} \right). \tag{1·9, 27}$$

Also, the unit vector in the direction parallel to q_{AB} equals $\mathbf{r}_1 \wedge \mathbf{r}_2 / |\mathbf{r}_1 \wedge \mathbf{r}_2|$.
Hence

$$q_{AB} = \frac{\Gamma}{4\pi} \frac{\mathbf{r}_1 \wedge \mathbf{r}_2}{(\mathbf{r}_1 \wedge \mathbf{r}_2)^2} \mathbf{r}_0 \left(\frac{\mathbf{r}_1}{r_1} - \frac{\mathbf{r}_2}{r_2} \right). \tag{1·9, 28}$$

This formula is more convenient than $(1·9, 27)$ if it is desired to express q_{AB} in terms of the coordinates of A, B, Q, since

$$\mathbf{r}_0 = (x_2 - x_1, y_2 - y_1, z_2 - z_1),$$

$$\mathbf{r}_1 = (x - x_1, y - y_1, z - z_1),$$

$$\mathbf{r}_2 = (x - x_2, y - y_2, z - z_2),$$

where (x_1, y_1, z_1), (x_2, y_2, z_2), (x, y, z) are the coordinates of A, B and Q, respectively.

Assume that AB is a straight segment in the x, y-plane, parallel to the x-axis. Then the coordinates of A and B are $(x_1, y_1, 0)$, $(x_2, y_1, 0)$, and so

$$\mathbf{r}_0 = (x_2 - x_1, 0, 0), \quad \mathbf{r}_1 = (x - x_1, y - y_1, z), \quad \mathbf{r}_2 = (x - x_2, y - y_1, z),$$

$$\mathbf{r}_0 \mathbf{r}_1 = (x_2 - x_1)(x - x_1), \quad \mathbf{r}_0 \mathbf{r}_2 = (x_2 - x_1)(x - x_2),$$

$$\mathbf{r}_1 \wedge \mathbf{r}_2 = (0, z(x_1 - x_2), (y_1 - y)(x_1 - x_2)).$$

Hence $\quad q_{AB} = \dfrac{\Gamma(0, z, y_1 - y)}{4\pi(z^2 + (y_1 - y)^2)}$

$$\times \left(\frac{x - x_2}{\sqrt{\{(x - x_2)^2 + (y - y_1)^2 + z^2\}}} - \frac{x - x_1}{\sqrt{\{(x - x_1)^2 + (y - y_1)^2 + z^2\}}} \right). \tag{1·9, 29}$$

In particular, if the vortex extends from $A(x_1, y_1, 0)$ to infinity in a positive direction parallel to the x-axis, we obtain from $(1·9, 29)$,

$$q_{AB} = -\frac{\Gamma(0, z, y_1 - y)}{4\pi(z^2 + (y_1 - y)^2)} \left(1 + \frac{x - x_1}{\sqrt{\{(x - x_1)^2 + (y - y_1)^2 + z^2\}}} \right). \tag{1·9, 30}$$

On the other hand, if AB is a straight segment in the x, y-plane, parallel to the y-axis, where $(x_1, y_1, 0)$ and $(x_1, y_2, 0)$ are the coordinates of A and B, then

$$\mathbf{r}_0 = (0, y_2 - y_1, 0), \quad \mathbf{r}_1 = (x - x_1, y - y_1, z), \quad \mathbf{r}_2 = (x - x_1, y - y_2, z),$$

$$\mathbf{r}_0 \mathbf{r}_1 = (y_2 - y_1)(y - y_1), \quad \mathbf{r}_0 \mathbf{r}_2 = (y_2 - y_1)(y - y_2),$$

$$\mathbf{r}_1 \wedge \mathbf{r}_2 = (-z(y_1 - y_2), 0, -(x_1 - x)(y_1 - y_2)).$$

Hence $\quad q_{AB} = \dfrac{\Gamma(z, 0, x_1 - x)}{4\pi(z^2 + (x_1 - x)^2)}$

$$\times \left(\frac{y - y_1}{\sqrt{\{(x - x_1)^2 + (y - y_1)^2 + z^2\}}} - \frac{y - y_2}{\sqrt{\{(x - x_1)^2 + (y - y_2)^2 + z^2\}}} \right). \tag{1·9, 31}$$

Now consider a 'horseshoe vortex' in the x, y-plane consisting of a straight segment AB parallel to the y-axis and of the two straight segments parallel to the x-axis which extends from A and B to infinity (Fig. 1·9, 2). Let the coordinates of A and B be $(x_1, y_1, 0)$ and $(x_1, y_2, 0)$. Applying (1·9, 30) and (1·9, 31) we obtain the following formula for the induced velocity q,

$$q = \frac{\Gamma}{4\pi} \left\{ \frac{(z, 0, x_1 - x)}{z^2 + (x_1 - x)^2} \left(\frac{y - y_1}{\sqrt{\{(x - x_1)^2 + (y - y_1)^2 + z^2\}}} - \frac{y - y_2}{\sqrt{\{(x - x_1)^2 + (y - y_2)^2 + z^2\}}} \right) \right.$$

$$+ \frac{(0, z, y_1 - y)}{z^2 + (y_1 - y)^2} \left(1 + \frac{x - x_1}{\sqrt{\{(x - x_1)^2 + (y - y_1)^2 + z^2\}}} \right)$$

$$\left. - \frac{(0, z, y_2 - y)}{z^2 + (y_2 - y)^2} \left(1 + \frac{x - x_1}{\sqrt{\{(x - x_1)^2 + (y - y_2)^2 + z^2\}}} \right) \right\}, \quad (1·9, 32)$$

Fig. 1·9, 2

where the vorticity vector $\boldsymbol{\omega}$ points from A to B, i.e. the sense of rotation of the vortex is anticlockwise for an observer looking from B towards A. In particular, the expression for the z-component of the induced velocity is

$$w = \frac{\Gamma}{4\pi} \left\{ \frac{x_1 - x}{z^2 + (x_1 - x)^2} \left(\frac{y_2 - y}{\sqrt{\{(x_1 - x)^2 + (y_2 - y)^2 + z^2\}}} - \frac{y_1 - y}{\sqrt{\{(x_1 - x)^2 + (y_1 - y)^2 + z^2\}}} \right) \right.$$

$$+ \frac{y_1 - y}{z^2 + (y_1 - y)^2} \left(1 - \frac{x_1 - x}{\sqrt{\{(x_1 - x)^2 + (y_1 - y)^2 + z^2\}}} \right)$$

$$\left. - \frac{y_2 - y_1}{z^2 + (y_2 - y_1)^2} \left(1 - \frac{x_1 - x}{\sqrt{\{(x_1 - x)^2 + (y_2 - y)^2 + z^2\}}} \right) \right\}. \quad (1·9, 33)$$

Let us find the limiting form of (1·9, 32) as B approaches A indefinitely, while Γ increases in such a way that the product $\Gamma(y_2 - y_1)$ remains constant, $\Gamma(y_2 - y_1) = \Gamma'$. Regarding x, y, z and x_1 as fixed, we may write

$$q = \frac{\Gamma}{4\pi} (\mathbf{f}(y_2) - \mathbf{f}(y_1)), \quad (1·9, 34)$$

where the vector function $\mathbf{f}(\eta)$ is defined as

$$\mathbf{f}(\eta) = \frac{(z, 0, x_1 - x)}{z^2 + (x_1 - x)^2} \frac{y - \eta}{\sqrt{\{(x - x_1)^2 + (y - \eta)^2 + z^2\}}}$$

$$+ \frac{(0, z, \eta - y)}{z^2 + (\eta - y)^2} \left(1 + \frac{x - x_1}{\sqrt{\{(x - x_1)^2 + (y - \eta)^2 + z^2\}}} \right). \quad (1·9, 35)$$

Then
$$\mathbf{q} = \frac{\Gamma'}{4\pi} \frac{\mathbf{f}(y_2) - \mathbf{f}(y_1)}{y_2 - y_1},$$

and so in the limit

$$\mathbf{q} = \frac{\Gamma'}{4\pi}\left[\frac{d\mathbf{f}}{d\eta}\right]_{\eta=y_1} = \frac{\Gamma'}{4\pi}\frac{\partial}{\partial y_1}\left[\frac{(z,0,x_1-x)}{z^2+(x_1-x)^2}\frac{y-y_1}{\sqrt{\{(x-x_1)^2+(y-y_1)^2+z^2\}}}\right.$$
$$\left. + \frac{(0,z,y_1-y)}{z^2+(y_1-y)^2}\left(1+\frac{x-x_1}{\sqrt{\{(x-x_1)^2+(y-y_1)^2+z^2\}}}\right)\right], \quad (1\cdot9, 36)$$

i.e. in scalar notation

$$\left.\begin{aligned}
u &= \frac{\Gamma'}{4\pi}\frac{\partial}{\partial y_1}\left(\frac{z}{z^2+(x_1-x)^2}\frac{y-y_1}{\sqrt{\{(x-x_1)^2+(y-y_1)^2+z^2\}}}\right), \\
v &= \frac{\Gamma'}{4\pi}\frac{\partial}{\partial y_1}\left\{\frac{z}{z^2+(y_1-y)^2}\left(1+\frac{x-x_1}{\sqrt{\{(x-x_1)^2+(y-y_1)^2+z^2\}}}\right)\right\}, \\
w &= \frac{\Gamma'}{4\pi}\frac{\partial}{\partial y_1}\left\{\frac{(x_1-x)(y-y_1)}{(z^2+(x_1-x)^2)\sqrt{\{(x-x_1)^2+(y-y_1)^2+z^2\}}} + \frac{y_1-y}{z^2+(y_1-y)^2}\right. \\
&\qquad\qquad \left. \times \left(1+\frac{x-x_1}{\sqrt{\{(x-x_1)^2+(y-y_1)^2+z^2\}}}\right)\right\}.
\end{aligned}\right\} \quad (1\cdot9, 37)$$

$(1\cdot9, 37)$ may be regarded as the field of flow due to an 'infinitesimal horse-shoe vortex'.

Conversely, a horseshoe vortex of finite width can also be obtained by integrating a distribution of infinitesimal horseshoe vortices of constant strength as in $(1\cdot9, 37)$, with respect to y_1.

1·10   Two-dimensional flow

We shall now consider fields of flow in which the velocity components in a fixed direction, say the direction of the z-axis, vanish throughout, while at the same time the velocity vector is independent of z. Thus the flow is the same in all planes parallel to the x, y-plane, and we may confine our attention exclusively to the x, y-plane.

The field is now described completely by the velocity components $u = u(x, y)$, $v = v(x, y)$.

The flow along a curve, and hence the circulation around a closed curve, are given simply as special cases of the general definition for three-dimensional flow by

$$\int_C \mathbf{q}\,d\mathbf{l} = \int_C (u\,dx + v\,dy).$$

On the other hand, the concept of flow across a surface is now replaced by the flow across a curve.

Thus, let $d\mathbf{l}$ be the directed element of length of a curve C in the x, y-plane and let $\mathbf{n} = (\lambda, \mu)$ be the unit normal to $d\mathbf{l}$ such that $\mathbf{n}$ and $d\mathbf{l}$, taken in that order, form a right-handed system. Then the rate of flow or flux across C is defined by

$$\int_C \mathbf{q}\mathbf{n}\, dl = \int_C (u\lambda + v\mu)\, dl = \int_C \left(u\frac{dy}{dl} - v\frac{dx}{dl}\right) dl = \int_C (u\, dy - v\, dx).$$

The connexion between this definition and the motion of fluid in general three-dimensional flow is established by the remark that the above integral measures the three-dimensional flux across the surface which is generated by displacing C by unit distance in a direction parallel to the z-axis.

The vorticity is now given by

$$\operatorname{curl} \mathbf{q} = \operatorname{curl}(u, v, w) = (0, 0, \zeta),$$

where

$$\zeta = \frac{\partial v}{\partial x} - \frac{\partial u}{\partial y}.$$

Thus the vorticity is always parallel to the z-axis, and it is therefore customary to refer to ζ as the vorticity. If $\zeta = 0$ throughout a given region then there exists a velocity potential Φ,

$$u = -\frac{\partial \Phi}{\partial x}, \quad v = -\frac{\partial \Phi}{\partial y}, \tag{1·10, 1}$$

and for any curve C which leads from a point A to a point B within the given region, the flow along C is given by

$$\int_A^B (u\, dx + v\, dy) = -(\Phi_B - \Phi_A).$$

All this follows immediately from the facts established for three-dimensional flow. In particular, if Φ is single-valued in the given region then the circulation around a closed curve vanishes. This will be the case, for example, if the region is simply connected.

Now consider a closed curve C which bounds an area S situated within a given field of flow of an incompressible fluid. Thus the equation of continuity, which is now

$$\frac{\partial u}{\partial x} + \frac{\partial v}{\partial y} = 0, \tag{1·10, 2}$$

is satisfied within S. The flux across the curve C vanishes since

$$\int_C (u\, dy - v\, dx) = \int_S \left(\frac{\partial u}{\partial x} + \frac{\partial v}{\partial y}\right) dx\, dy \tag{1·10, 3}$$

by Green's theorem in the plane. The same can be deduced directly from physical considerations and also from the corresponding fact in three dimensions.

Let C_1 and C_2 be two curves connecting two fixed points A and B such that C_2 can be deformed into C_1 continuously within a region in which the fluid is incompressible. Then C_1 and C_2 together constitute a closed curve C, so that

$$\int_{C_1}(u\,dy-v\,dx)-\int_{C_2}(u\,dy-v\,dx)=\int_C(u\,dy-v\,dx)=0$$

by (1·10, 3). Thus, subject to the condition specified, the flux across a curve C from A to B is independent of the particular curve chosen. It follows that for fixed A, the fluid across C is a function of the coordinates of $B=(x,y)$ only. We define

$$\Psi(x,y)=-\int_A^B(u\,dy-v\,dx),$$

and call $\Psi(x,y)$ the stream function of the flow. For small displacements dy or dx only, we obtain $d\Psi=-u\,dy$ or $d\Psi=v\,dx$ respectively, and so

$$u=-\frac{\partial\Psi}{\partial y},\quad v=\frac{\partial\Psi}{\partial x}. \tag{1·10, 4}$$

Displacement of the point A will have the effect of adding a constant to Ψ, and this clearly does not effect (1·10, 4). Thus, for a given velocity field, Ψ is determinate except for the addition of an arbitrary constant.

The vorticity ζ is expressed in terms of Ψ by

$$\zeta=\frac{\partial v}{\partial x}-\frac{\partial u}{\partial y}=\frac{\partial^2\Psi}{\partial x^2}+\frac{\partial^2\Psi}{\partial y^2}.$$

It follows that if the flow is also irrotational, then Ψ satisfies Laplace's equation

$$\frac{\partial^2\Psi}{\partial x^2}+\frac{\partial^2\Psi}{\partial y^2}=0.$$

In that case there exists a velocity potential and this velocity potential also satisfies Laplace's equation

$$\frac{\partial^2\Phi}{\partial x^2}+\frac{\partial^2\Phi}{\partial y^2}=0$$

by (1·10, 2). Furthermore, by (1·10, 1) and (1·10, 4)

$$\frac{\partial\Phi}{\partial x}=\frac{\partial\Psi}{\partial y},\quad \frac{\partial\Phi}{\partial y}=-\frac{\partial\Psi}{\partial x}, \tag{1·10, 5}$$

and these are the equations of Cauchy–Riemann for the complex function

$$\Pi(z)=\Phi+i\Psi, \tag{1·10, 6}$$

where the independent variable z is given by $z=x+iy$. It follows that for irrotational incompressible fluid flow Π is an analytic function of z. Π is called the complex potential of the flow. For a given flow, Π is determinate except for the addition of an arbitrary complex constant. Conversely, we may regard every analytic function $\Pi(z)$ as the complex potential of an irrotational field of flow of an incompressible fluid.

Since the complex potential Π is an analytic function of z, its derivative also is an analytic function of z. Thus

$$-\frac{\partial \Pi}{\partial z} = -\frac{\partial}{\partial x}(\Phi + i\Psi) = -\frac{\partial \Phi}{\partial x} - i\frac{\partial \Psi}{\partial x} = u - iv$$

also is an analytic function of z. It is called the complex velocity.

To find a particular example of two-dimensional fluid flow we therefore only have to interpret the physical meaning of an arbitrary analytic function of a complex variable regarded as a complex potential. However, it is instructive to derive a first example from the field of flow of an isolated source in three-dimensional flow (compare § 1·7).

The velocity potential of a source in three-dimensional flow is (see (1·7, 2))

$$\Phi = \frac{\sigma}{4\pi} \frac{1}{\sqrt{\{(x-x_0)^2 + (y-y_0)^2 + (z-z_0)^2\}}}. \tag{1·10, 7}$$

It appears obvious on geometrical grounds that a field of flow which is due to a uniform distribution of sources along a straight line parallel to the z-axis is two-dimensional. However, if we try to integrate (1·10, 7) with respect to z_0 from $-\infty$ to $+\infty$ we find that this integral diverges. To avoid this difficulty we consider instead the velocity components which are obtained from (1·10, 7) by differentiation,

$$u = -\frac{\partial}{\partial x}\left(\frac{\sigma}{4\pi} \frac{1}{\sqrt{\{(x-x_0)^2 + (y-y_0)^2 + (z-z_0)^2\}}}\right)$$

$$= \frac{\sigma}{4\pi} \frac{x-x_0}{\{(x-x_0)^2 + (y-y_0)^2 + (z-z_0)^2\}^{\frac{3}{2}}}$$

and

$$v = \frac{\sigma}{4\pi} \frac{y-y_0}{\{(x-x_0)^2 + (y-y_0)^2 + (z-z_0)^2\}^{\frac{3}{2}}},$$

and integrate these over the range $-\infty < z_0 < \infty$. As for the third velocity component, this must vanish by symmetry although the corresponding integral is still divergent. Then

$$u^*(x, y) = \frac{\sigma}{4\pi} \int_{-\infty}^{\infty} \frac{(x-x_0)\,dz_0}{\{(x-x_0)^2 + (y-y_0)^2 + (z-z_0)^2\}^{\frac{3}{2}}}$$

$$= \frac{\sigma(x-x_0)}{4\pi\{(x-x_0)^2 + (y-y_0)^2\}} \left[\frac{z-z_0}{\{(x-x_0)^2 + (y-y_0)^2 + (z-z_0)^2\}^{\frac{1}{2}}}\right]_{-\infty}^{\infty},$$

or

$$u^*(x, y) = \frac{\sigma(x-x_0)}{2\pi\{(x-x_0)^2 + (y-y_0)^2\}}, \tag{1·10, 8}$$

and similarly

$$v^*(x, y) = \frac{\sigma(y-y_0)}{2\pi\{(x-x_0)^2 + (y-y_0)^2\}}, \tag{1·10, 9}$$

where u^* and v^* are the velocity components of the resultant two-dimensional field of flow, and σ is now the two-dimensional source strength, equal to the total strength of the three-dimensional sources in unit length z.

Since z and the corresponding velocity component w have now disappeared, we may again make use of these symbols to denote the complex variable $x+iy$ and the complex velocity $u-iv$. Besides, we shall also omit the asterisks appended to u and v in (1·10, 8) and (1·10, 9).

The corresponding velocity potential is

$$\Phi(x,y) = -\frac{\sigma}{4\pi}\log\{(x-x_0)^2+(y-y_0)^2\} = -\frac{\sigma}{2\pi}\log r, \qquad (1\cdot10, 10)$$

as can be verified easily by differentiation, where

$$r = \sqrt{\{(x-x_0)^2+(y-y_0)^2\}}.$$

This is said to be the velocity potential of a two-dimensional source of strength σ which is located at (x_0, y_0). The corresponding complex potential is

$$\Pi(z) = -\frac{\sigma}{2\pi}\log(z-z_0). \qquad (1\cdot10, 11)$$

Indeed, $z-z_0 = re^{i\theta}$,   where $\theta = \arg(z-z_0) = \tan^{-1}\dfrac{y-y_0}{x-x_0}$,

and r has the same meaning as before. Hence

$$\Pi = -\frac{\sigma}{2\pi}\log(z-z_0) = -\frac{\sigma}{2\pi}(\log r + i\theta),$$

so that $\mathscr{R}(\Pi) = -\dfrac{\sigma}{2\pi}\log r$,   as required.

Since an analytic function is determined, except for a constant, by its real or by its imaginary part, taken separately, it follows that the stream function of a two-dimensional source is given by

$$\Psi = \mathscr{I}(\Pi) = -\frac{\sigma}{2\pi}(\theta + 2k\pi),$$

where k is an arbitrary integer. Thus Ψ is many-valued.

Let w be the complex velocity of a given field of flow and let C be a closed curve in the field. Denote the circulation around C by γ and the flux across C by χ so that

$$\gamma = \int_C (u\,dx + v\,dy), \quad \chi = \int_C (u\,dy - v\,dx).$$

Then $\gamma + i\chi = \displaystyle\int_C (u\,dx + v\,dy) + i\int_C (u\,dy - v\,dx) = \int_C (u-iv)(dx+i\,dy)$,

or $\gamma + i\chi = \displaystyle\int_C w\,dz. \qquad (1\cdot10, 12)$

In particular, for a source the complex velocity is

$$w = -\frac{\partial\Pi}{\partial z} = \frac{\sigma}{2\pi(z-z_0)},$$

and so if C is a single closed curve which encloses z_0 in its interior,

$$\gamma + i\chi = \frac{\sigma}{2\pi}\int_C \frac{dz}{z-z_0} = i\sigma.$$

It follows that the circulation around C vanishes, while the flux across C equals the strength of the source. Or, in terms of the stream function, Ψ decreases by σ every time the circuit along C is described in a mathematically positive direction.

The fact that the flux across C equals the strength of the source can also be derived from the general theorem of total flux (§ 1·7).

A procedure which is very similar to that used for three-dimensional flow leads us to the concept of a 'two-dimensional doublet'. We place a source of strength σ at a point $z_0 + \delta e^{i\alpha}$ and a source of strength $-\sigma$ at z_0, where z_0 and α are constant. Letting δ tend to zero while the product $\sigma\delta = \tau$ is kept constant, we then obtain the complex velocity potential for a doublet

$$\Pi(z) = \frac{\tau e^{i\alpha}}{2\pi(z - z_0)}. \tag{1·10, 13}$$

Writing $z - z_0 = r e^{i\theta}$ as before, and separating real and imaginary parts, we obtained for the velocity potential and the stream function of a doublet

$$\Phi = \frac{\tau}{2\pi r}\cos(\alpha - \theta), \quad \Psi = \frac{\tau}{2\pi r}\sin(\alpha - \theta). \tag{1·10, 14}$$

Another important field of flow is obtained by multiplying the complex velocity potential for a source, (1·10, 11), by $-i$. Then

$$\Pi(z) = \frac{i\sigma}{2\pi}\log(z - z_0). \tag{1·10, 15}$$

The corresponding velocity potential and stream function are given by

$$\Phi = -\frac{\sigma}{2\pi}(\theta + 2k\pi), \quad \Psi = \frac{\sigma}{2\pi}\log r, \tag{1·10, 16}$$

also by (1·10, 12), $\qquad \gamma + i\chi = -\frac{i\sigma}{2\pi}\int_C \frac{dz}{z - z_0} = \sigma,$

and so $\qquad\qquad\qquad \gamma = \sigma, \quad \chi = 0. \tag{1·10, 17}$

(1·10, 16) shows that the streamlines of the flow are the circles with z_0 as centre. The magnitude of the velocity vector is inversely proportional to the distance from z_0.

(1·10, 15) is said to represent the field of flow due to a vortex of strength σ at z_0. This is in keeping with the fact that the circulation round a curve which includes z_0 equals σ. More precisely, it can be shown that the same field of flow is produced by a concentrated straight-line vortex which is perpendicular to the x, y-plane (compare §§ 1·9 and 3·8).

Yet another simple example of a type of flow which will be considered later is given by $\qquad \Pi(z) = -U e^{-i\alpha}(z - z_0). \tag{1·10, 18}$

The corresponding complex velocity is

$$w = U e^{-i\alpha},$$

so that $\qquad u = U\cos\alpha, \quad v = U\sin\alpha.$

This represents a uniform stream of fluid moving with velocity U in a direction which makes an angle α with the x-axis.

Other fields of flow are obtained by superposition from the examples listed above. Thus, we may consider the problem of finding the field of flow due to a given source of strength σ at a point z_0 in the upper half-plane ($\mathscr{I}(z_0) > 0$) in the presence of a rigid wall along the x-axis. To satisfy the boundary condition at the wall we place a source of equal strength at the point $\bar{z}_0$ which is conjugate to z_0. The resulting complex velocity potential is

$$\Pi(z) = -\frac{\sigma}{2\pi}(\log(z-z_0) + \log(z-\bar{z}_0))$$

$$= -\frac{\sigma}{2\pi}\log(z^2 - z(z_0+\bar{z}_0) + z_0\bar{z}_0).$$

The expression under the sign of the logarithm has real coefficients, $z_0 + \bar{z}_0$, $z_0\bar{z}_0$, and is therefore real positive for sufficiently large z. Moreover, the expression does not vanish anywhere on the real axis (since its roots are z_0, $\bar{z}_0$) and so it must be positive throughout. Hence

$$\Psi = \mathscr{I}(\Pi) = 0$$

for real z, so that the x-axis is a streamline, as required.

This is an example of the 'method of images'. The source at $\bar{z}_0$ is said to be the image of the source at z_0 with respect to the x-axis. The flow due to doublets and vortices in the presence of a rigid straight wall can be determined by the same method. The method can also be applied to determine the flow due to sources, doublets or vortices in the presence of a circular or of a rectangular boundary, but the solution of the latter problem requires the introduction of an infinite number of images (see § 3·11).

We consider now the field of flow which is obtained by the superposition of a uniform field of flow (1·10, 18) on the field of a doublet (1·10, 13). The resulting complex potential is

$$\Pi(z) = -U e^{-i\alpha}(z-z_0) + \frac{\tau e^{i\alpha}}{2\pi(z-z_0)}. \qquad (1\cdot10, 19)$$

The velocity potential and the stream function are given by

$$\Phi = \mathscr{R}(\Pi) = -Ur\cos(\theta-\alpha) + \frac{\tau}{2\pi r}\cos(\theta-\alpha)$$

$$= -Ur\left(1 - \frac{\tau}{2\pi Ur^2}\right)\cos(\theta-\alpha),$$

$$\Psi = \mathscr{I}(\Pi) = -Ur\sin(\theta-\alpha) - \frac{\tau}{2\pi r}\sin(\theta-\alpha)$$

$$= -Ur\left(1 + \frac{\tau}{2\pi Ur^2}\right)\sin(\theta-\alpha),$$

where $z = z_0 + r e^{i\theta}$. In particular, putting $\tau = -2\pi U a^2$, where a is a positive number we, obtain

$$\left.\begin{aligned}\Phi &= -Ur\left(1 + \frac{a^2}{r^2}\right)\cos(\theta - \alpha),\\[2mm]\Psi &= -Ur\left(1 - \frac{a^2}{r^2}\right)\sin(\theta - \alpha).\end{aligned}\right\} \qquad (1·10, 20)$$

Thus $\Psi = 0$ for $r = a$. This shows that the circle of radius a about z_0 as centre is a streamline. Or, in the three-dimensional interpretation, (1·10, 20) represents the flow of a uniform main stream round a circular cylinder of radius a.

The complex velocity is

$$\begin{aligned}w &= -\frac{d\Pi}{dz} = U\left(e^{-i\alpha} - \frac{a^2 e^{i\alpha}}{(z - z_0)^2}\right)\\[2mm]&= U\left\{\left[\cos\alpha - \frac{a^2}{r^2}\cos(\alpha - 2\theta)\right] - i\left[\sin\alpha + \frac{a^2}{r^2}\sin(\alpha - 2\theta)\right]\right\}.\end{aligned}$$

At the cylinder $(r = a)$

$$\begin{aligned}w &= U\{[\cos\alpha - \cos(\alpha - 2\theta)] - i[\sin\alpha + \sin(\alpha - 2\theta)]\}\\&= 2U\sin(\alpha - \theta)(-\sin\theta - i\cos\theta)\\&= 2Ui\,e^{-i\theta}\sin(\theta - \alpha).\end{aligned}$$

Hence $q = |w| = 2U\,|\sin(\theta - \alpha)|.$

This shows that q vanishes at the point A facing the main stream $(\theta = \alpha)$ and also at the point B which is diametrically opposed to A, the 'stagnation points'. There are no other stagnation points in the fluid. On the other hand, q reaches its maximum values for $\theta - \alpha = \pm\frac{1}{2}\pi$. At these points, q equals twice the main-stream velocity, $q = 2U$.

On this field of flow we now superimpose the motion due to a vortex at $z = z_0$. The resultant complex potential is (compare (1·10, 15) and (1·10, 19)),

$$\Pi(z) = -U\left(e^{-i\alpha}(z - z_0) + \frac{a^2 e^{i\alpha}}{z - z_0}\right) + \frac{i\Gamma}{2\pi}\log(z - z_0). \qquad (1·10, 21)$$

The complex velocity is

$$w = U\left(e^{-i\alpha} - \frac{a^2 e^{i\alpha}}{(z - z_0)^2}\right) - \frac{i\Gamma}{2\pi(z - z_0)}. \qquad (1·10, 22)$$

The formulae (1·10, 21) and (1·10, 22) represent a more general type of flow around the same circular cylinder. Indeed, the circle $r = a$ is still a streamline, since $r = a$ is a streamline for (1·10, 15) and (1·10, 19) taken separately. It can be shown that this is the most general expression for the flow of a uniform main stream around the given circular cylinder. Provided

$$\left|\frac{\Gamma}{4\pi Ua}\right| < 1,$$

there are still two stagnation points on the surface of the cylinder. The two points coalesce for $\Gamma = \pm\,4\pi Ua$, and for $|\,\Gamma/4\pi Ua\,| > 1$ there is only a single stagnation point, which is now in the interior of the fluid. However, the case $|\,\Gamma/4\pi Ua\,| \geqslant 1$ is of no importance for aerofoil theory. Application of $(1\cdot10, 12)$–$(1\cdot10, 22)$ shows that Γ is the circulation around a curve which encloses the cylinder.

The most important tool for the calculation of the flow around bodies of other shapes is provided by the theory of conformal representation. Let the complex potential $\Pi(z)$ describe a field of flow in a region R of the z-plane and let $z = f(\zeta)$ be an analytic function which maps R conformally on a region R' in the ζ-plane. Then $\Pi(f(z))$ is also the complex potential of a certain field of flow in the ζ-plane. Moreover, a streamline is characterized by the condition $\mathscr{I}(\Pi) = \text{constant}$, it follows that the relation $z = f(\zeta)$ transforms streamlines into streamlines. And since any boundary is a streamline and conversely any streamline can be taken as a rigid boundary in steady flow, it follows that the transformed curve of a boundary in the z-plane may be taken as a boundary in the ζ-plane.

Now

$$\frac{d\Pi}{d\zeta} = \frac{d\Pi}{dz}\frac{dz}{d\zeta},$$

and so if we denote the complex velocity in the ζ-plane by $\tau(\zeta)$ then the relation between τ and w is given by

$$\tau(\zeta) = w(z)\frac{dz}{d\zeta}. \tag{1·10, 23}$$

Let C be any closed curve in R, and let C' be the corresponding curve in R'. Then

$$\int_{C'} \tau\,d\zeta = \int_{C'} w\frac{dz}{d\zeta}d\zeta = \int_{C} w\,dz. \tag{1·10, 24}$$

This solution shows, according to $(1\cdot10, 12)$, that the flux across C' equals the flux across C, and the circulation round C' equals the circulation round C.

For example, consider the function

$$z = \zeta + \frac{a^2}{\zeta}. \tag{1·10, 25}$$

This function maps the entire z-plane, cut along the real axis from $x = -2a$ to $x = 2a$, on the region outside a circle of radius a about the origin in the ζ-plane. Writing $\zeta = r\,e^{i\theta}$, we obtain for $r = a$,

$$z = 2a\cos\theta.$$

As θ varies from 0 to π, z varies along the axis from $2a$ to $-2a$, to return to $2a$ as θ varies from π to 2π. The point at infinity in the z-plane corresponds to the point at infinity in the ζ-plane, and $dz/d\zeta = 1$ at that point.

We note that the segment $-2a \leqslant x \leqslant 2a$ represents a flat plate in the three-dimensional interpretation. Thus, if we wish to find the field of flow due to a main stream which is perpendicular to the plate, then $U\,e^{-\frac{1}{2}i\pi}$ is the complex velocity of the main stream both in the z-plane and in the ζ-plane. The complex potential in the ζ-plane is obtained by writing $z_0 = 0$, $\alpha = \frac{1}{2}\pi$ in (1·10, 21) and by replacing z by ζ in that formula, so

$$\Pi = Ui\left(\zeta - \frac{a^2}{\zeta}\right) + \frac{i\Gamma}{2\pi}\log\zeta. \qquad (1·10, 26)$$

Solving (1·10, 25) for ζ, we get

$$\zeta = \tfrac{1}{2}(z + \sqrt{[z^2 - 4a^2]}), \qquad (1·10, 27)$$

where the sign of the square root is taken as positive for large positive z, to make ζ tend to infinity as z tends to infinity. To obtain the complex potential of the flow about the flat plate, we substitute the right-hand side of (1·10, 27) for ζ in (1·10, 26). In particular, for $\Gamma = 0$,

$$\Pi = Ui\left(\zeta - \frac{a^2}{\zeta}\right) = Ui\left(2\zeta - \left(\zeta + \frac{a^2}{\zeta}\right)\right) = Ui\,\sqrt{(z^2 - 4a^2)},$$

and so the complex velocity is given by

$$w(z) = -\frac{d\Pi}{dz} = -Ui\,\frac{z}{\sqrt{(z^2 - 4a^2)}}.$$

At the plate, $z = 2a\cos\theta$,

$$u - iv = w = -Ui\,\frac{\cos\theta}{i\sin\theta} = -U\cot\theta.$$

Thus u becomes infinite along the edges of the plate.

Similarly, the flow without circulation around a flat plate which is at incidence α, $0 < \alpha < 2\pi$, relative to the free stream is given by

$$\Pi(z) = -U\left(e^{-i\alpha}\zeta + \frac{a^2 e^{i\alpha}}{\zeta}\right), \qquad (1·10, 28)$$

where

$$z = \zeta + \frac{a^2}{\zeta},$$

as in (1·10, 25). Writing $\zeta = a\,e^{i\theta}$ we obtain for the complex velocity at the plate

$$w = -\frac{d\Pi}{d\zeta}\bigg/\frac{dz}{d\zeta} = U(e^{i(\theta-\alpha)} - e^{-i(\theta-\alpha)})/(e^{i\theta} - e^{-i\theta}),$$

$$w = U\,\frac{\sin(\theta - \alpha)}{\sin\theta}. \qquad (1·10, 29)$$

This is of course a real quantity, since the velocity must be tangential to the plate, $w = u$, and u still becomes infinite at the edges of the plate. The stagnation points P_1, P_2 ($u = 0 = 0$) are given by $\theta_1 = \alpha$ (on the top surface of the aerofoil) and by $\theta_2 = \pi + \alpha$ (on the bottom surface of the aerofoil facing the free stream). We note that for $\theta_1 < \theta < \pi$, i.e. between P_1 and the trailing edge, and for $\theta_2 < \theta < 2\pi$, i.e. between the leading edge and P_2, the flow is in the positive direction of the real axis, while elsewhere on the plate it follows the opposite direction.

If the required transformation function is known, then the same method can be used to find the flow of a uniform main stream around any other contour—or, in the three-dimensional interpretation, any cylindrical surface—and this procedure will be discussed in greater detail in Chapter 2. It will then be shown there that the reaction of the fluid on a fixed cylindrical surface in a uniform stream—or, which is the same, the reaction on a cylindrical surface in uniform rectilinear motion through a fluid which is otherwise at rest—is different from zero only if the circulation around the body is different from zero. But according to Kelvin's theorem (§ 1·5) the circulation around a closed curve had to be zero if the circulation around the curve consisting of the same fluid particles was equal to zero at some preceding instant of time. Now in all practical cases the motion will have started from rest at some instant, and so our present theory does not explain how any circulation round the body might be produced. Moreover, even if we confine ourselves entirely to the consideration of steady motion, so that the circulation explains the existence of a fluid force on a body in two-dimensional flow, it is clear that this picture is incomplete if we cannot explain it as a limiting case of the more realistic type of flow around a finite body in three dimensions. Again, in three dimensions the force on a body which is in uniform rectilinear motion in a fluid which is otherwise at rest, must vanish by D'Alembert's paradox if the fluid is irrotational throughout (§ 1·8). And this again must be the case if the motion started from rest at some earlier instant of time, by Lagrange's theorem (§ 1·5). In order to obtain a consistent picture of the physical phenomenon involved we shall have to take into account the effects of viscosity.

1·11 Viscous fluid flow

So far we have assumed throughout that the internal forces in a fluid are always perpendicular to the surface across which they act; in other words, that the tangential or shear forces, if any, are negligible. However, such tangential forces do arise in real fluids, and this property of a fluid is known as its viscosity. The viscosity of air is relatively small, but nevertheless it controls, directly or indirectly, the forces which act on the wing of an aircraft. In particular, viscosity explains the generation of vorticity. However, it will appear that the mutual arrangement and strength of the resulting vorticity, and hence the circulation and the forces which act on a wing, can be determined to some extent without recourse to viscous fluid theory, if only one adopts a single condition which no longer refers to the viscosity explicitly (§ 1·14).

Consider first a field of flow in which the y- and z-components of the velocity vanish throughout, and in which the non-vanishing component u is constant in planes parallel to the x, z-plane, $u = u(y, t)$. Then it is an experimental fact (which, for gases, finds its explanation in the kinetic theory of gases) that the shear force which adjacent fluid volumes exert on each other depends on their relative velocity. That is to say, in the present case, assuming that the y-axis points in an upward direction, the fluid above any particular plane $y = $ constant exerts on the fluid below this plane a force per unit area τ which depends on the local gradient of the velocity $\partial u / \partial y$. More precisely it is found that τ is proportional to $\partial u / \partial y$,

$$\tau = \mu \frac{\partial u}{\partial y}, \tag{1·11, 1}$$

and μ is called the coefficient of viscosity of the fluid. τ is a force per unit area or stress, and so its dimensions are

$$[\tau] = ML^{-1}T^{-2},$$

where M, L, T denotes the dimensions of mass, length and time, as usual. Hence, by (1·11, 1), the dimensions of μ are

$$[\mu] = ML^{-1}T^{-1}.$$

It will appear presently that for a viscous fluid the field of flow depends on the quantity $\nu = \mu / \rho$, where, as before, ρ is the density of the fluid, and since $[\rho] = ML^{-3}$

$$[\nu] = L^2 T^{-1}. \tag{1·11, 2}$$

In the absence of normal stresses, the equation of motion of the type of flow now under consideration can be determined from (1·11, 1). Thus, consider the shear force which acts on a volume of fluid bounded by two parallel planes $y = $ constant which are a small distance δy apart. By (1·11, 1) this force is given by

$$\tau(y + \delta y) - \tau(y) = \mu \left\{ \left(\frac{\partial u}{\partial y} \right)_{y+\delta y} - \left(\frac{\partial u}{\partial y} \right)_y \right\} \doteq \mu \frac{\partial^2 u}{\partial y^2} \delta y$$

per unit area, where the approximation on the right-hand side becomes exact in the limit. Hence, by Newton's third law of motion,

$$\rho \, \delta y \frac{Du}{Dt} = \mu \frac{\partial^2 u}{\partial y^2} \delta y, \tag{1·11, 3}$$

where the left-hand side is the product of acceleration and of mass per unit area. Moreover, in the present case $Du/Dt = \partial u/\partial t$, and so, dividing (1·11, 3) by $\rho \, \delta y$ and passing to the limit, we obtain the equation of motion

$$\frac{\partial u}{\partial t} = \nu \frac{\partial^2 u}{\partial y^2}. \tag{1·11, 4}$$

It is found that under normal flight conditions the frictional forces which act between a solid boundary and a viscous fluid prevent any motion of the fluid relative to the solid at the boundary. This is the 'zero-slip condition'. It breaks down in regions of excessive rarefaction, but in the present book we shall scarcely be concerned with such conditions. The zero-slip condition holds for general three-dimensional flow, both when the boundary is at rest and when it is in motion.

Consider a semi-infinite expanse of viscous fluid which is bounded by the plane $y = 0$. The fluid is supposed to be initially at rest but at time $t = 0$ the plane, supposed solid, begins to move with constant velocity U in a direction parallel to the x-axis. We wish to find the subsequent motion of the fluid.

We may assume, for reasons of symmetry, that the velocity component u is independent of x and z, while the other two velocity components vanish. Thus, we may take (1·11, 4) as the differential equation of the problem. The boundary conditions for $u = u(y, t)$ are

$$\left.\begin{aligned} u(y, 0) &= 0 \quad \text{for} \quad y > 0, \\ u(0, t) &= U \quad \text{for} \quad t > 0, \end{aligned}\right\} \tag{1·11, 5}$$

where the second equation follows from the zero-slip condition.

It is clear in view of the linearity of the boundary conditions that, for given y and t, $u(y, t)$ is proportional to U. It follows that $u(y, t)/U$ is independent of U, and so can depend functionally only on y, t and ν. A non-dimensional combination of these quantities is $\omega = y^2/t\nu$, any other non-dimensional product of y, t and ν is a power of ω. Hence u/U can be written as a function of ω only,

$$\frac{u}{U} = g(\omega),$$

by dimensional analysis. Then

$$\frac{1}{U}\frac{\partial u}{\partial t} = -\frac{\omega}{t}g'(\omega), \quad \frac{1}{U}\frac{\partial^2 u}{\partial y^2} = \frac{2}{\nu t}g'(\omega) + \frac{4\omega}{\nu t}g''(\omega).$$

Substituting in (1·11, 4), we obtain the following differential equation for $g(\omega)$:

$$4\omega g''(\omega) + (2 + \omega)g'(\omega) = 0. \tag{1·11, 6}$$

This is equivalent to

$$\frac{g''(\omega)}{g'(\omega)} = -\frac{2 + \omega}{4\omega}.$$

Integrating

$$g'(\omega) = \frac{c_1}{\sqrt{\omega}}e^{-\frac{1}{4}\omega},$$

where c_1 is a constant which is yet to be determined. Integrating again, and introducing $\theta = \frac{1}{2}\sqrt{\omega}$ as variable of integration,

$$g(\omega) = 4c_1\int_0^{\frac{1}{2}\sqrt{\omega}} e^{-\theta^2}\,d\theta + c_2. \tag{1·11, 7}$$

The boundary conditions for $g(\omega)$ are

$$g(0) = 1, \quad \lim_{\omega \to \infty} g(\omega) = 0$$

by (1·11, 5). Of these, the first condition yields $c_2 = 1$. The second condition then shows that

$$4c_1 \int_0^\infty e^{-\theta^2} d\theta + 1 = 0.$$

Now $\int_0^\infty e^{-\theta^2} d\theta$ is a standard integral which is known to be equal to $\frac{1}{2}\sqrt{\pi}$.

Hence $4c_1 = -2/\sqrt{\pi}$, and so finally

$$u(y, t) = U\left(1 - \frac{2}{\sqrt{\pi}} \int_0^{\sqrt{(y^2/4\nu t)}} e^{-\theta^2} d\theta\right). \qquad (1·11, 8)$$

The integral on the right-hand side of (1·11, 8) occurs in various branches of applied mathematics. It is frequently called the error function, regarded as a function of its upper limit; sometimes the integral including the factor $2/\sqrt{\pi}$ is called the error function.

The vorticity in the fluid is given by

$$\zeta = -\frac{\partial u}{\partial y} = \frac{U}{\sqrt{(\pi \nu t)}} e^{-y^2/4\nu t}, \qquad (1·11, 9)$$

and the shear force per unit area (or 'shear stress') is

$$\tau = \mu \frac{\partial u}{\partial y} = -U\rho \sqrt{\left(\frac{\nu}{\pi t}\right)} e^{-y^2/4\nu t}. \qquad (1·11, 10)$$

In particular at the boundary, $y = 0$, and so

$$\tau_0 = \mu\left(\frac{\partial u}{\partial y}\right)_{y=0} = -U\rho \sqrt{(\nu/\pi t)} \qquad (1·11, 11)$$

is the drag on the wall per unit area.

Let us fix our attention on some definite positive y. (1·11, 8) shows that the velocity at y increases monotonically with increasing t and ultimately tends to U, the velocity of the moving plane. We note that the right-hand side of (1·11, 8) is different from zero for all positive t, so that the effect of the motion of the bounding plane $y = 0$ is felt immediately at any distance whatever. This is characteristic of a process of diffusion and is in keeping with the fact that the partial differential equation (1·11, 4) is of parabolic type. However, at points which are sufficiently far away from the boundary, the velocity is very small during the initial period after the start of the motion.

Similar remarks apply to the vorticity at a given point. As equation (1·11, 9) shows, ζ, which is initially zero, begins to increase immediately after the start of the motion, but eventually it decreases again and approaches zero ultimately, as the flow becomes uniform. Indeed,

$$\frac{\partial \zeta}{\partial t} = \frac{U}{\sqrt{(4\pi \nu t^5)}} \left(\frac{y^2}{2\nu} - t\right) e^{-y^2/4\nu t}.$$

Thus ζ increases numerically so long as $t < y^2/2\nu$, and decreases thereafter.

The amount of vorticity between the boundary $y = 0$ and any plane $y = y_0 > 0$, is measured by

$$\int_0^{y_0} \zeta \, dy = -\int_0^{y_0} \frac{\partial u}{\partial y} \, dy = -[u]_0^{y_0} = \frac{2U}{\sqrt{\pi}} \int_0^{\sqrt{(y_0^2/4\nu t)}} e^{-\theta^2} d\theta. \quad (1\cdot11, 12)$$

As y_0 tends to infinity this quantity tends to U, so that the total amount of vorticity in the fluid is constant. Now for given y_0, the upper limit of the integral on the right-hand side of ($1\cdot11$, 12), $\sqrt{(y_0^2/4\nu t)}$, becomes very large provided t is sufficiently small. This shows that in the early stages of the motion the vorticity is concentrated near the bounding plane. In other words, the vorticity is diffused into the fluid from the boundary. This is an important conclusion which helps us to understand the mechanism by which vorticity is produced in a viscous fluid.

Now assume that the semi-infinite expanse of viscous fluid, and the bounding plane $y = 0$, move with constant velocity U in the direction of the x-axis initially, and that the plane $y = 0$ is stopped suddenly at time $t = 0$. The boundary conditions now are

$$\left. \begin{array}{ll} u(y, 0) = U & \text{for} \quad y > 0, \\ u(0, t) = 0 & \text{for} \quad t > 0. \end{array} \right\} \quad (1\cdot11, 13)$$

To solve the problem, we only have to subtract the solution ($1\cdot11$, 8) from a uniform velocity U. We obtain

$$u(y, t) = \frac{2U}{\sqrt{\pi}} \int_0^{\sqrt{(y^2/4\nu t)}} e^{-\theta^2} d\theta, \quad (1\cdot11, 14)$$

which satisfies the differential equation ($1\cdot11$, 4) and the boundary conditions ($1\cdot11$, 13). The vorticity in the fluid is now given by

$$\zeta = -\frac{\partial u}{\partial y} = -\frac{U}{\sqrt{(\pi \nu t)}} e^{-y^2/4\nu t}, \quad (1\cdot11, 15)$$

while the shear stress on the wall is

$$\tau_0 = U\rho \sqrt{(\nu/\pi t)}. \quad (1\cdot11, 16)$$

The formula shows that the fluid now tends to drag the plane along with it, as might be expected.

In general three-dimensional flow, we consider the internal force $\mathbf{f_n}$ which acts across a small plane surface normal to $\mathbf{n}$ at any given point. Dividing by the area of the surface and passing to the limit we obtain a stress $\mathbf{p_n}$. In particular, if $\mathbf{n}$ is parallel to the x-axis, we denote the components of $\mathbf{p_n}$ by p_{xx}, p_{xy}, p_{xz}, and if $\mathbf{n}$ is parallel to the y- or z-axis, then we denote the components of the corresponding stresses by p_{yx}, p_{yy}, p_{yz} and p_{zx}, p_{zy}, p_{zz} respectively. It is easy to show that $p_{xy} = p_{yx}, p_{xz} = p_{zx}, p_{yz} = p_{zy}$, but this will not be required here.

Consider a small rectangular prism whose edges are parallel to the co-ordinate axes and are of lengths $\delta x, \delta y, \delta z$ respectively. The forces which act on the volume of fluid within the prism are given by the internal stresses and by the external body forces, (X, Y, Z) per unit mass, say. By considering the stresses on opposite faces of the prism we find that these contribute

$$\frac{\partial p_{xx}}{\partial x}\,\delta x\,\delta y\,\delta z + \frac{\partial p_{yx}}{\partial y}\,\delta y\,\delta z\,\delta x + \frac{\partial p_{zx}}{\partial z}\,\delta z\,\delta x\,\delta y$$

in the direction of the x-axis, except for terms of higher order of smallness. Newton's second law of motion as applied to the volume of fluid in the prism yields

$$\rho\,\delta x\,\delta y\,\delta z\,\frac{Du}{Dt} = \rho\,\delta x\,\delta y\,\delta z\,X + \left(\frac{\partial p_{xx}}{\partial x} + \frac{\partial p_{yx}}{\partial y} + \frac{\partial p_{zx}}{\partial z}\right)\delta x\,\delta y\,\delta z.$$

Hence, dividing by $\delta x\,\delta y\,\delta z$ and passing to the limit, we obtain

$$\rho\left(\frac{\partial u}{\partial t} + u\frac{\partial u}{\partial x} + v\frac{\partial u}{\partial y} + w\frac{\partial u}{\partial z}\right) = \rho X + \frac{\partial p_{xx}}{\partial x} + \frac{\partial p_{yx}}{\partial y} + \frac{\partial p_{zx}}{\partial z},$$

and similarly

$$\rho\left(\frac{\partial v}{\partial t} + u\frac{\partial v}{\partial x} + v\frac{\partial v}{\partial y} + w\frac{\partial v}{\partial z}\right) = \rho Y + \frac{\partial p_{xy}}{\partial x} + \frac{\partial p_{yy}}{\partial y} + \frac{\partial p_{zy}}{\partial z},$$

$$\rho\left(\frac{\partial w}{\partial t} + u\frac{\partial w}{\partial x} + v\frac{\partial w}{\partial y} + w\frac{\partial w}{\partial z}\right) = \rho Z + \frac{\partial p_{xz}}{\partial x} + \frac{\partial p_{yz}}{\partial y} + \frac{\partial p_{zz}}{\partial z}.$$

$$(1\cdot11, 17)$$

These equations apply to both incompressible and compressible fluids. Their further development depends on the detailed analysis of the stresses on one hand, and of the rates of strain, or rates of deformation, at any point of the fluid on the other. In particular, it can be shown that the arithmetical mean of the normal stresses across three mutually perpendicular planes at any point (e.g. $\frac{1}{3}(p_{xx} + p_{yy} + p_{zz})$) is independent of the particular orientation of these planes. This quantity is called the mean pressure. Except for the sign, it reduces to the ordinary pressure defined in § 1·4 if shear stresses are absent. On the other hand, the spatial rates of change of the velocity at a particular fluid particle can be divided into two parts, of which one corresponds to a rigid rotation of the particle as a whole while the second part indicates the rate of strain or deformation of the particle. It is then assumed that the mean pressure is independent of the rates of strain, while the remaining stresses can be expressed as linear functions of these. This leads to the equations of Navier–Stokes,‡ whose validity has been established experimentally for most cases which are of interest in aeronautics. In particular, the analysis shows that the relations between the stresses and the rates of strain do not involve any physical constants beyond the constant μ defined above. Accepting this fact alone, we may draw some important conclusions of a general nature by means of dimensional analysis.

Consider the flow of a uniform main stream of incompressible viscous

‡ See Lamb, *Hydrodynamics* (6th ed. revised, Cambridge University Press, 1932), p. 577.

fluid round a fixed body B. Let μ be the coefficient of viscosity of the fluid, ρ its density and U the main stream velocity. Also let l be a typical linear dimension of the body, for example, its maximum length in the direction of the main stream. We shall assume that there are no external body forces, or that such forces are negligible. Then the reaction F of the fluid on the body in a given direction must be expressible in terms of the constants μ, ρ, U, l by a formula　　　$F = \rho U^2 l^2 . f.$

In this formula $\rho U^2 l^2$ has the dimensions of a force, as can be verified immediately, while f is a function of the non-dimensional product or products of powers of μ, ρ, U, l. Such a product is

$$Re = \frac{\rho U l}{\mu} = \frac{U l}{\nu},$$

and all other non-dimensional products of powers of μ, ρ, U, l are powers of Re. Re is called the Reynold's number of the flow after 0. Reynolds, who was one of the founders of the modern theory of viscous fluids. Note that Re is to some extent arbitrary since its value depends on the choice of l. It is not difficult to show that any other product of powers of μ, ρ, U, l is a power of Re. Thus F can be written as

$$F = \rho U^2 l^2 f(Re) = \rho U^2 l^2 f\left(\frac{U l}{\nu}\right). \tag{1·11, 18}$$

The practical significance of this formula is as follows. Consider a field of steady flow of a viscous incompressible fluid round a body B for which the initial data are geometrically similar to the flow considered earlier. That is to say, the body B' is obtained from B by scaling it up in the ratio $l':l$, and it is disposed in the same way as B relative to the main stream. Let μ', ρ', U' be the coefficient of viscosity, the density and the main-stream velocity in the second field of flow, and assume that the Reynolds number Re' is the same as for the first flow,

$$Re' = \frac{\rho' U' l'}{\mu'} = \frac{U' l'}{\nu'} = \frac{U l}{\nu} = Re.$$

In this case, (1·11, 18) states that the force F' in the second flow which corresponds to F in the first flow is related to F by the equation

$$F' \div F = \rho' U'^2 l'^2 \div \rho U^2 l^2. \tag{1·11, 19}$$

Moreover, the velocity components at points (x', y', z') and (x, y, z) in the two fields, which are similarly situated relative to B' and B, are related to one another in the ratio $U' : U$. The argument can be extended to unsteady flow if the boundary conditions are similar at corresponding moments. That is to say, that the conditions are similar at some initial moments t_0' and t_0. and subsequently after intervals of time

$$t' \div t = \frac{l'}{U'} \div \frac{l}{U}.$$

In this relation, l, U (l', U') are typical lengths and velocities chosen at time t_0 (t_0').

The above discussion is of considerable importance in connexion with the testing of models in wind-tunnels. It is now clear that it would be desirable to achieve the same Reynolds number in the tunnel as is obtained for the free-flight conditions which are to be simulated. However, this may be difficult in practice, particularly at high speeds when another non-dimensional parameter complicates the picture (compare §4·1).

The equations of Navier–Stokes together with the equation of continuity (1·3, 7) constitute the fundamental system of differential equations for the motion of an incompressible viscous fluid. However, only a few classes of closed solutions of this system are known at present, and it is quite impossible to give an exact theoretical description of the flow round a body of arbitrary shape, or indeed of the flow round a single particular body. However, for fluids of relatively small viscosity such as air, approximate theoretical methods have been developed which, taken in conjunction with certain experimental data, permit the quantitative explanation and prediction of many of the phenomena which occur in a viscous fluid. In view of their distinct character these methods are usually not regarded as part of aerofoil theory, but they are nevertheless essential for the calculation of the drag of an aerofoil. For that reason, an introductory account of the subject has been included in the present book. The equations of Navier–Stokes are not required for the purpose and so will not be derived here. We should add that the equations of Navier–Stokes do serve as a theoretical standard with which to compare our approximate assumptions. However, no rigorous theoretical justification of these assumptions exists at present; they are justified ultimately by the agreement of their consequences with experiment.

Consider the steady flow of an incompressible viscous fluid past a semi-infinite flat plate which occupies the region $x \geqslant 0$ of the x, z-plane. Suppose that the free-stream velocity U is directed along the x-axis. In the neighbourhood of the plate the flow will be slowed down so as to satisfy the condition of zero-slip at the plate. Thus the chief effect due to the presence of the plate will be the existence of a steep gradient $\partial u/\partial y$ in its neighbourhood. The variations of the y-component of the velocity will be small by comparison and may be neglected. More precisely, $v = 0$ in the free stream and at the plate, so that we shall assume $v = 0$ throughout. Then the shear stress on a small surface normal to the y-axis is

$$\tau = \mu \frac{\partial u}{\partial y}$$

by (1·11, 1). However, τ now depends on x as well as on y.

To establish the equation of motion of the flow consider a small rectangle whose sides are parallel to the x- and y-axes, and of lengths δx and δy

respectively. Then the mass of the fluid within the rectangle (per unit width along the z-axis) is given by $\rho\,\delta x\,\delta y$ and its acceleration in the direction of the x-axis is

$$\frac{Du}{Dt}=u\frac{\partial u}{\partial x}+v\frac{\partial u}{\partial y}=u\frac{\partial u}{\partial x}, \qquad (1\cdot11,20)$$

according to our assumptions. The shear stresses on the sides parallel to the x-axis yield a force

$$\{\tau(y+\delta y)-\tau(y)\}\,\delta x=\frac{\partial\tau}{\partial y}\,\delta y\,\delta x=\mu\frac{\partial^2 u}{\partial y^2}\,\delta x\,\delta y$$

in the direction of the x-axis, and the effect of the normal stresses on the other two sides may be expected to be small by comparison. Hence, by Newton's second law,

$$\rho\,\delta x\,\delta y\,\frac{Du}{Dt}=\mu\frac{\partial^2 u}{\partial y^2}\,\delta x\,\delta y,$$

or, by (1·11, 20),

$$u\frac{\partial u}{\partial x}=\nu\frac{\partial^2 u}{\partial y^2}. \qquad (1\cdot11,21)$$

This equation is non-linear, and in order to replace it by a linear equation we reflect that, except in the vicinity of the flat plate, u is approximately equal to U. We therefore take as the differential equation of the motion

$$U\frac{\partial u}{\partial x}=\nu\frac{\partial^2 u}{\partial y^2}. \qquad (1\cdot11,22)$$

The condition at the plate is

$$u(x,0)=0 \quad\text{for}\quad x>0. \qquad (1\cdot11,23)$$

Also the primary action of the shear force is to slow down the flow on either side of the plate but not ahead of it. We conclude that

$$u(x,y)=0 \quad\text{for}\quad x\leqslant 0 \text{ and all } y. \qquad (1\cdot11,24)$$

Hence, in particular, $\quad u(0,y)=0 \quad\text{for}\quad y>0.$ $\qquad (1\cdot11,25)$

We now make the formal substitution $x=Ut$. Then (1·11, 22) becomes

$$\frac{\partial u}{\partial t}=\nu\frac{\partial^2 u}{\partial y^2},$$

which is the differential equation (1·11, 4), while the boundary conditions (1·11, 23) and (1:11, 25) are translated precisely into (1·11, 13). It follows that the formulae (1·11, 14)–(1·11, 16) apply to the present case if we replace t by x/U. Thus the solution for the velocity is

$$u(x,y)=\frac{2U}{\sqrt{\pi}}\int_0^{\sqrt{(Uy^2/4\nu x)}}e^{-\theta^2}d\theta \quad (x>0), \qquad (1\cdot11,26)$$

and this formula holds for both positive and negative y. The expression for the vorticity becomes

$$\zeta=-\frac{U^{\frac{3}{2}}}{\sqrt{(\pi\nu x)}}e^{-Uy^2/4\nu x} \quad (x>0), \qquad (1\cdot11,27)$$

while the shear stress on the plate is

$$\tau_0 = \rho U^{\frac{3}{2}} \sqrt{(\nu/\pi x)} \quad (x > 0). \tag{1·11, 28}$$

If the plate is of finite breadth in the direction of the x-axis, $0 \leqslant x \leqslant c$, say, then we may assume that the above formulae still hold upstream of the 'trailing edge' $x = c$. Hence the total reaction of the fluid on the plate (the 'skin-friction drag') is, according to the present theory,

$$2 \int_0^c \tau_0 \, dx = 4\rho U^{\frac{3}{2}} \sqrt{(\nu c/\pi)} = \rho U^2 c \frac{4}{\sqrt{\pi}} Re^{-\frac{1}{2}}, \tag{1·11, 29}$$

where $Re = Uc/\nu$ may be regarded as the Reynolds number of the flow. The factor 2 on the left-hand side of (1·11, 29) is due to the fact that we have to take into account the shear force on both sides of the plate. This formula is due to Lord Rayleigh.‡ Comparison with experiment and with more exact theory shows that the skin-friction drag given by (1·11, 29) exceeds the correct value by a factor of about 1·7. The discrepancy is not surprising if we consider how many approximations were involved in our analysis. In fact, despite its inaccuracy, the solution gives quite a good picture of the general features of the flow.

For any $x > 0$ we put $Re' = Ux/\nu$, so that Re' is a local Reynolds number. Then (1·11, 26) becomes

$$u(x, y)/U = \frac{2}{\sqrt{\pi}} \int_0^{\sqrt{Re'}\,(y/2x)} e^{-\theta^2} \, d\theta. \tag{1·11, 30}$$

The Reynolds numbers which are of interest in aerodynamics are of the order of 10^5–10^8. For example, for $Re' = 10^6$, the upper limit of the integral in (1·11, 30) is $500y/x$. Now a table of the error function§ shows that for $500y/x = 2$, the right-hand side of (1·11, 30) exceeds 0·995. That is to say, for $y/x = 0·004$, i.e. at a distance from the plate which is very small compared with the distance measured along the plate, the velocity is already very close indeed to the free-stream value U. We conclude that for the sufficiently large Reynolds numbers the effect of the viscosity is confined to a thin layer near the plate, or near any other solid boundary. This region is called the boundary layer (first introduced by Prandtl‖ in 1904). It must be understood that there is no definite line or surface which separates the boundary layer from the remainder of the fluid, but nevertheless the concept of the boundary layer is both concrete and fruitful. The fact that the boundary layer is very thin permits certain approximations within that

‡ Lord Rayleigh, 'On the motion of solid bodies through viscous liquid', *Phil. Mag.* 6th series, vol. 21 (1911), pp. 697–711.

§ For example, E. Jahnke and F. Emde, *Tables of Functions* (4th ed., Dover Publications, 1945), p. 24.

‖ L. Prandtl and A. Betz, *Vier Abhandlungen zur Hydrodynamik und Aerodynamik* (Göttingen, 1927). (Originally given in *Verhandlungen des dritten internationalen Mathematiker-Kongresses* (Heidelberg, 1904), pp. 484–91.)

region, while outside the boundary layer the viscosity is negligible, so that perfect fluid theory is applicable.

Although we have drawn a number of useful conclusions from Rayleigh's solution, it must be said that some of the assumptions involved can hardly be justified. In particular, the two steps which led to the linearization of the equation

$$\frac{Du}{Dt} = u\frac{\partial u}{\partial x} + v\frac{\partial u}{\partial y} \doteq u\frac{\partial u}{\partial x} \doteq U\frac{\partial u}{\partial x}$$

are both open to serious criticism. For although v may be presumed to be small compared with u, the gradient $\partial u/\partial y$ may well be large compared with $\partial u/\partial x$, and so $v(\partial u/\partial y)$ is not in general small compared with $u(\partial u/\partial x)$. Again, when replacing u by U as the coefficient of $\partial u/\partial x$, we neglected the fact that this approximation breaks down near the plate, i.e. precisely where the solution is of the greatest interest.

Nevertheless, the general conclusions which we derived from Rayleigh's solution are confirmed by a more exact analysis which makes no use of the approximations criticized in the preceding paragraph. The method is due to Blasius‡ and will now be described.

The equation of motion which replaces (1·11, 22) is

$$\rho\frac{Du}{Dt} = \rho\left(u\frac{\partial u}{\partial x} + v\frac{\partial u}{\partial y}\right) = \mu\frac{\partial^2 u}{\partial y^2},$$

or

$$u\frac{\partial u}{\partial x} + v\frac{\partial u}{\partial y} = \nu\frac{\partial^2 u}{\partial y^2}, \tag{1·11, 31}$$

to which we may add the equation of continuity

$$\frac{\partial u}{\partial x} + \frac{\partial v}{\partial y} = 0. \tag{1·11, 32}$$

(1·11, 32) implies the existence of a stream function ψ such that

$$u = -\frac{\partial\psi}{\partial y}, \quad v = \frac{\partial\psi}{\partial x}.$$

The boundary conditions are

$$\left.\begin{aligned} u(x, 0) = v(x, 0) = 0 \quad (x > 0), \\ u(0, y) = U, \end{aligned}\right\} \tag{1·11, 33}$$

and, moreover

$$\lim_{y\to\pm\infty} u(x, y) = U, \tag{1·11, 34}$$

since u must approach the free-stream velocity in regions which are far removed from the plate. In Rayleigh's analysis this condition is satisfied automatically.

The solution of the present problem depends on the fact that it is again possible to replace the partial differential equation (1·11, 31) by an ordinary

‡ H. Blasius, 'Grenzschichten in Flüssigkeiten mit kleiner Reibung', Z. Math. Phys. vol. 56 (1908), pp. 4–13.

differential equation. We recall that this method was also used for solving the partial differential equation (1·11, 4). The independent variable of the corresponding ordinary differential equation (1·11, 6) was $\omega = y^2/t\nu$. The corresponding variable for the solution of (1·11, 22) is obtained by the formal substitution $t = x/U$ as above, so that $\omega = Uy^2/\nu x$. Any function of ω will do equally well for the purpose of this reduction, and reference to (1·11, 26) shows that u can be expressed conveniently as a function of

$$\eta = \tfrac{1}{2}\sqrt{\omega} = y\sqrt{(U/4\nu x)}, \qquad (1·11, 35)$$

this being the upper limit of the integral for u. The idea now suggests itself that u might be a function of the single variable η, even for the present case (1·11, 31). In that case ψ, which is an integral of $-u$ with respect to y, would be the product of $\sqrt{x}$ and a function of η. To make the unknown function non-dimensional, we write accordingly

$$\psi = \sqrt{(\nu U x)}\, f(\eta).$$

Then
$$u = -\frac{U}{2} f'(\eta), \quad v = -\frac{1}{2}\sqrt{\left(\frac{\nu U}{x}\right)}\,(\eta f'(\eta) - f(\eta)),$$

$$\frac{\partial u}{\partial x} = \frac{U\eta}{4x} f''(\eta), \quad \frac{\partial u}{\partial y} = -\frac{U^{\frac{3}{2}}}{4\sqrt{(\nu x)}} f''(\eta), \quad \frac{\partial^2 u}{\partial y^2} = -\frac{U^2}{8\nu x} f'''(\eta),$$

and so (1·11, 31) becomes $\quad f'''(\eta) - f(\eta) f''(\eta) = 0.$ $\qquad (1·11, 36)$

The boundary conditions $u = v = 0$ for $y = 0$ imply

$$f(0) = f'(0) = 0. \qquad (1·11, 37)$$

Also η tends to infinity both as y tends to infinity, for given x, and as x tends to zero for given y. Hence the condition

$$\lim_{\eta \to \infty} f'(\eta) = -2 \qquad (1·11, 38)$$

implies both (1·11, 34) and the last condition in (1·11, 33). These boundary conditions are inconvenient because they are specified at two points of which one, moreover, is at infinity. If instead of (1·11, 38), we knew the value of $f''(0)$, then we should be able to determine the solution of the problem by straightforward step-by-step numerical integration. For that reason we consider first a modification of our present problem, as follows. Let $F(\chi)$ be a function which satisfies the differential equation

$$F'''(\chi) - F(\chi) F''(\chi) = 0 \qquad (1·11, 39)$$

with the initial conditions

$$F(0) = F'(0), \quad F''(0) = -1. \qquad (1·11, 40)$$

Then $F(\chi)$ is determined uniquely by these conditions and we may calculate the function by numerical integration as stated. Now put

$$\chi = \beta\eta, \quad F = \gamma f,$$

where the constants β, γ will be determined presently. Then

$$\frac{dF}{d\chi}=\frac{\gamma}{\beta}\frac{df}{d\eta}, \quad \frac{d^2F}{d\chi^2}=\frac{\gamma}{\beta^2}\frac{d^2f}{d\eta^2}, \quad \frac{d^3F}{d\chi^3}=\frac{\gamma}{\beta^3}\frac{d^3f}{d\eta^3}.$$

Substituting in (1·11, 39) we obtain

$$f'''(\eta)-\gamma\beta f(\eta) f''(\eta)=0, \tag{1·11, 41}$$

which reduces to (1·11, 36) provided $\gamma=\beta^{-1}$. With this value of γ, we obtain from (1·11, 40)

$$f(0)=f'(0)=0, \quad f''(0)=-\beta^3. \tag{1·11, 42}$$

Thus the boundary conditions (1·11, 37) are satisfied, and we have to determine β so that
$$\lim_{\eta\to\infty} f'(\eta)=\lim_{\chi\to\infty} \beta^2 F'(\chi)=-2,$$

i.e.
$$\beta=\left(-\frac{2}{\lim\limits_{\chi\to\infty} F'(\chi)}\right)^{\frac{1}{2}}. \tag{1·11, 43}$$

The numerical integration of (1·11, 39) confirms that $F'(\chi)$ does in fact tend to a finite negative value. Thus the problem is solved. The shear stress at the plate is
$$\tau_0=\mu\left(\frac{\partial u}{\partial y}\right)_{y=0}=-\mu\frac{U^{\frac{3}{2}}}{4\sqrt{(vx)}}f''(0)=-\tfrac{1}{4}\rho U^{\frac{3}{2}}\beta^3\sqrt{\frac{v}{x}},$$

and so the drag on a plate of breadth c is

$$2\int_0^c \tau_0 dx=\rho U^2 c\beta^3 Re^{-\frac{1}{2}}, \tag{1·11, 44}$$

where $Re=Uc/v$. The numerical value of β^3 is 1·328 (see *Modern Developments*, where further references are given).‡ The corresponding coefficient, according to Rayleigh's solution, (1·11, 29), is $4/\sqrt{\pi}=2\cdot256$, exceeding the value given by Blasius's theory by a factor of about 1·7 as stated earlier.

Now let $Re'=Ux/v$ be the local Reynolds number used in (1·11, 30). Then in our present notation

$$\eta=\sqrt{Re'}\,\frac{y}{2x}.$$

Numerical calculations show that the value of u/U for $\eta=2$ is 0·956 according to Blasius's solution, as compared with 0·995 in Rayleigh's method. For $\eta=2\cdot7$ Blasius's solution yields already $u/U=0\cdot996$. It follows that our general conclusions regarding the existence and extent of the boundary layer are confirmed by the present analysis. We add that the assumption that the Reynolds number Re' is of the order 10^5–10^8 evidently breaks down near the leading edge of the plate ($x=0$). Indeed, no great accuracy can be claimed for the results of boundary-layer theory in that

‡ *Modern Developments in Fluid Dynamics*, ed. S. Goldstein (Oxford University Press, 1938), pp. 135 et seq.

region. Earlier we made the statement that the theory holds for fluids of small viscosity, but this statement has no absolute meaning unless it is understood to signify that the Reynolds number of the flow is sufficiently large.

1·12   Boundary-layer theory

Consider two-dimensional flow of an incompressible fluid past a wall along the x-axis. Outside the boundary layer, for $y \geqslant \delta$, say, the effect of the viscosity is negligible, so that Euler's equations of motion apply. Hence

$$\frac{\partial u}{\partial t} + u\frac{\partial u}{\partial x} + v\frac{\partial u}{\partial y} = -\frac{1}{\rho}\frac{\partial p}{\partial x}, \qquad (1\cdot 12, 1)$$

where the pressure p is the same in all directions. Now consider the equilibrium of forces for a small rectangle in the boundary layer ($y < \delta$) whose sides are parallel to the axes of coordinates. Let $\delta x\,\delta y$ be the side lengths of the rectangle, and let p be the normal pressure on the vertical sides. The shear stresses on the sides δx are taken as $\tau = \mu(\partial u/\partial y)$ as before. Hence the resultant external force on the rectangle (per unit width along the z-axis) is

$$-(p(x+\delta x) - p(x))\,\delta y + (\tau(y+\delta y) - \tau(y))\,\delta x,$$

and this equals $$\left(-\frac{\partial p}{\partial x} + \mu\frac{\partial^2 u}{\partial y^2}\right)\delta x\,\delta y,$$

except for terms of higher order of smallness. Equating this expression to $\rho\,\delta x\,\delta y(Du/Dt)$, by Newton's second law, and dividing by $\rho\,\delta x\,\delta y$, we obtain

$$\frac{\partial u}{\partial t} + u\frac{\partial u}{\partial x} + v\frac{\partial u}{\partial y} = -\frac{1}{\rho}\frac{\partial p}{\partial x} + \nu\frac{\partial^2 u}{\partial y^2}. \qquad (1\cdot 12, 2)$$

The definition of p as the normal pressure on a vertical side within the boundary layer agrees with the definition used in (1·12, 1), since outside the boundary layer p is the same in all directions. Moreover, it can be shown that even within the boundary layer, p, as defined above, differs only very little from the mean pressure. The mean pressure does not depend directly on the viscosity (§ 1·11), and we shall therefore assume that it is constant across the boundary layer. Also while $\partial u/\partial y$ would normally be large compared to $\partial u/\partial x$ in the interior of the boundary layer, it must be small at the edge of the boundary layer where u merges smoothly into the main flow. Moreover, v vanishes at the wall and does not vary greatly across the boundary layer. It may therefore be expected to be small compared with the value of u at the edge of the layer, which we write as u_1. Accordingly, we replace (1·12, 1) by

$$\frac{\partial u_1}{\partial t} + u_1\frac{\partial u_1}{\partial x} = -\frac{1}{\rho}\frac{\partial p}{\partial x}. \qquad (1\cdot 12, 3)$$

Substituting the left-hand side of this equation for $-\dfrac{1}{\rho}\dfrac{\partial p}{\partial x}$ in (1·12, 2), we obtain

$$\frac{\partial u}{\partial t}+u\frac{\partial u}{\partial x}+v\frac{\partial u}{\partial y}=\frac{\partial u_1}{\partial t}+u_1\frac{\partial u_1}{\partial x}+\nu\frac{\partial^2 u}{\partial y^2}. \tag{1·12, 4}$$

(1·12, 4), together with the equation of continuity

$$\frac{\partial u}{\partial x}+\frac{\partial v}{\partial y}=0, \tag{1·12, 5}$$

constitutes the fundamental system of equations for the calculation of the flow in a boundary layer. The boundary conditions at the wall are

$$u=v=0, \tag{1·12, 6}$$

while
$$u=u_1 \tag{1·12, 7}$$

at the edge of the boundary layer. The velocity component u_1 is determined by the main flow and is supposed known for the purpose of the present analysis. In addition, we require the values of u across the boundary layer for some initial value of x and, if the flow is unsteady, the values of u and v throughout the boundary layer at some initial moment of time.

We note that by the approximations developed so far, the term $\partial v/\partial x$ may be neglected compared with $\partial u/\partial y$ in the expression for the vorticity,

$$\zeta=\frac{\partial v}{\partial x}-\frac{\partial u}{\partial y}.$$

Accordingly we take $\zeta=-\partial u/\partial y$. This shows that ζ is proportional to the shear stress $\tau=\mu(\partial u/\partial y)$.

According to our concept of the boundary layer, the flow inside the layer passes smoothly into the main flow at the edge of the layer. And, as already stated, the gradient $\partial u/\partial y$ in the main flow is very small compared with the order of magnitude of $\partial u/\partial y$ within the layer. Thus we might want to add the condition

$$\frac{\partial u}{\partial y}=0 \tag{1·12, 8}$$

to be satisfied at the edge of the boundary layer. This condition, together with (1·12, 7) and (1·12, 4), implies

$$\frac{\partial^2 u}{\partial y^2}=0. \tag{1·12, 9}$$

However, it is found that the condition (1·12, 8) is redundant so that in general it is impossible to satisfy this condition for finite $y=\delta$, as well as all the other boundary conditions. One can overcome this difficulty by assuming that conditions (1·12, 7) and (1·12, 8) are satisfied only in the limit, as y tends to infinity. It may appear puzzling at first sight that in a theory which is based on the fact that δ, the thickness of the boundary layer, is small, we can replace δ ultimately by infinity. However, having found a solution, we can always test whether u and $\partial u/\partial y$ do indeed approximate their ultimate values u_1 and 0 already at a short distance from the wall.

To make the concept of the thickness of the boundary layer somewhat more definite we consider the flux F across a vertical line $x = x_1$ between the wall $y = 0$ and the point $y = \delta$, where u has reached its main-stream value u_1. Then

$$F = \int_0^\delta u \, dy.$$

Now but for the presence of the boundary layer the horizontal velocity component would be $u = u_1$ throughout, and so the flux would be

$$F_1 = \int_0^\delta u_1 \, dy.$$

Thus the deficiency in the flux owing to the presence of the boundary layer is

$$F_1 - F = \int_0^\delta (u_1 - u) \, dy = u_1 \int_0^\delta \left(1 - \frac{u}{u_1}\right) dy = u_1 \delta_1,$$

where

$$\delta_1 = \int_0^\delta \left(1 - \frac{u}{u_1}\right) dy. \tag{1·12, 10}$$

The deficiency $F_1 - F$ is therefore the same as it would be if $u = u_1$ throughout, while the wall is displaced a distance δ, in the direction of the y-axis. δ_1 is called the displacement thickness of the boundary layer. The upper limit δ may be either some boundary-layer thickness introduced earlier, or $\delta = \infty$, according to the method used. A related quantity, whose importance will appear presently, is the so-called momentum thickness which is defined by

$$\vartheta_1 = \int_0^\delta \left(1 - \frac{u}{u_1}\right) \frac{u}{u_1} \, dy. \tag{1·12, 11}$$

The displacement thickness is always greater than the momentum thickness since

$$\delta_1 - \vartheta_1 = \int_0^\delta \left\{\left(1 - \frac{u}{u_1}\right) - \left(1 - \frac{u}{u_1}\right)\frac{u}{u_1}\right\} dy = \int_0^\delta \left(1 - \frac{u}{u_1}\right)^2 dy > 0.$$

For the case of steady flow with zero external pressure gradient,

$$-\frac{1}{\rho}\frac{dp}{dx} = u_1 \frac{du_1}{dx} = 0,$$

a solution of the boundary-layer equations is provided by Blasius's analysis (see § 1·11). There is an extensive body of work, sometimes rather intricate, which deals with the flow in the presence of an external pressure gradient $\partial p/\partial x \neq 0$. However, for practical purposes an approximate method has proved very successful. It is based on von Kármán's momentum equation[‡] which will now be derived.

[‡] Th. von Kármán, 'Über laminare und turbulente Reibung', *Z. angew. Math. Mech.* vol. 1 (1921), pp. 235–6.

Let us integrate (1·12, 4) with respect to y across the boundary layer, between 0 and δ. Then the integral

$$\int_0^\delta \left(u\frac{\partial u}{\partial x} + v\frac{\partial u}{\partial y} \right) dy$$

which appears on the left-hand side may be transformed as follows, using integration by parts and the equation of continuity:

$$\int_0^\delta \left(u\frac{\partial u}{\partial x} + v\frac{\partial u}{\partial y} \right) dy = \int_0^\delta u\frac{\partial u}{\partial x}\, dy + [uv]_0^\delta - \int_0^\delta u\frac{\partial v}{\partial y}\, dy$$

$$= \int_0^\delta \frac{\partial}{\partial x}(u^2)\, dy + u_1 \int_0^\delta \frac{\partial v}{\partial y}\, dy$$

$$= \frac{\partial}{\partial x}\int_0^\delta u^2\, dy - (u^2)_{y=\delta}\frac{\partial \delta}{\partial x} - u_1 \int_0^\delta \frac{\partial u}{\partial x}\, dy$$

$$= \frac{\partial}{\partial x}\int_0^\delta u^2\, dy - u_1^2\frac{\partial \delta}{\partial x} - u_1\frac{\partial}{\partial x}\int_0^\delta u\, dy + u_1^2\frac{\partial \delta}{\partial x}$$

$$= \frac{\partial}{\partial x}\int_0^\delta u^2\, dy - u_1\frac{\partial}{\partial x}\int_0^\delta u\, dy.$$

On the right-hand side of equation (1·12, 4) we obtain from the last term

$$\int_0^\delta \nu\frac{\partial^2 u}{\partial y^2}\, dy = \left[\nu\frac{\partial u}{\partial y} \right]_0^\delta = -\nu\left(\frac{\partial u}{\partial y} \right)_{y=0}.$$

Hence, the result of the integration is

$$\int_0^\delta \frac{\partial u}{\partial t}\, dy + \frac{\partial}{\partial x}\int_0^\delta u^2\, dy - u_1\frac{\partial}{\partial x}\int_0^\delta u\, dy = \left(\frac{\partial u_1}{\partial t} + u_1\frac{\partial u_1}{\partial x} \right)\delta - \nu\left(\frac{\partial u}{\partial y} \right)_{y=0}. \qquad (1\cdot12, 12)$$

This is von Kármán's momentum equation for an incompressible boundary layer. Using only the general physical notion of the boundary layer, but not the explicit equations (1·12, 4) and (1·12, 5), it can also be derived directly by considering the equilibrium of linear momentum for a thin slice of the boundary layer between two straight lines parallel to the y-axis (Fig. 1·12, 1).

The rate of flow of fluid across CD is

$$\int_0^\delta u\, dy,$$

where δ is the boundary-layer thickness regarded as a function of x. A similar ex-

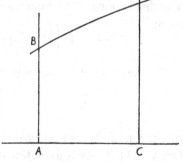

Fig. 1·12, 1

pression holds for the flow across AB, and so the total outflow from $ACDB$ through AB and CD is

$$\delta x\frac{d}{dx}\int_0^\infty u\, dy, \qquad (1\cdot12, 13)$$

where δx is the distance AC, supposed small. Now the fluid is incompressible and so (1·12, 13) is also the inflow into $ACBD$ across BD. Hence the x-compnoent of linear momentum which is carried across BD in unit time is

$$\rho u_1 \delta x \frac{d}{dx} \int_0^\delta u\, dy.$$

On the other hand, the outward flow of the x-component of linear momentum across CD and AB yields

$$\delta x \frac{d}{dx} \int_0^\delta \rho u^2\, dy,$$

and the rate of change of the x-component of linear momentum within $ACDB$ is

$$\delta x \int_0^\delta \rho \frac{\partial u}{\partial t}\, dy.$$

The external forces which act on the fluid in $ACDB$ in the direction of the x-axis are due to the normal pressures on AB and CD, and to the shear stress at the wall. The former contributes

$$-\delta x \frac{d}{dx} \int_0^\delta p\, dy = -\delta x \frac{\partial p}{\partial x} \delta = \delta x \rho \left(\frac{\partial u_1}{\partial t} + u_1 \frac{\partial u_1}{\partial x} \right) \delta,$$

according to the basic assumptions of boundary-layer theory, and the latter yields $-\tau_0 \delta x$, where τ_0 is the shear stress at the wall. Now the principle of linear momentum, as formulated in §1·4, states that the rate of change of linear momentum in $ACDB$ plus the outflow across the boundaries equals the resultant external force. Thus in the present case

$$\rho\, \delta x \left\{ \int_0^\delta \frac{\partial u}{\partial t}\, dy + \frac{\partial}{\partial x} \int_0^\delta u^2\, dy - u_1 \frac{\partial}{\partial x} \int_0^\delta u\, dy \right\} = \delta x \left\{ \rho \left(\frac{\partial u_1}{\partial t} + u_1 \frac{\partial u_1}{\partial t} \right) \delta - \tau_0 \right\}.$$

Hence

$$\rho\left\{ \int_0^\delta \frac{\partial u}{\partial t}\, dy + \frac{\partial}{\partial x} \int_0^\delta u^2\, dy - u_1 \frac{\partial}{\partial x} \int_0^\delta u\, dy \right\} = \rho \left(\frac{\partial u_1}{\partial t} + u_1 \frac{\partial u_1}{\partial x} \right) \delta - \tau_0. \quad (1\cdot12, 14)$$

Remembering that $\tau_0 = \mu(\partial u/\partial y)$, we see that (1·12, 14) agrees with (1·12, 12).

We confine ourselves to steady flow and rewrite (1·12, 12) in the form

$$\frac{1}{\rho u_1^2} \mu \left(\frac{\partial u}{\partial y} \right)_{y=0} = \frac{1}{u_1^2} \left\{ u_1 \frac{du_1}{dx} \delta - \frac{d}{dx} \int_0^\delta u^2\, dy + u_1 \frac{d}{dx} \int_0^\delta u\, dy \right\}. \quad (1\cdot12, 15)$$

Now

$$\frac{1}{u_1} \frac{d}{dx} \int_0^\delta u\, dy - \frac{1}{u_1^2} \frac{d}{dx} \int_0^\delta u^2\, dy$$

$$= \frac{d}{dx} \int_0^\delta \frac{u}{u_1}\, dy + \frac{1}{u_1^2} \frac{du_1}{dx} \int_0^\delta u\, dy - \frac{d}{dx} \int_0^\delta \frac{u^2}{u_1^2}\, dy - \frac{2}{u_1^3} \frac{du_1}{dx} \int_0^\delta u^2\, dy$$

$$= \frac{d}{dx} \int_0^\delta \left(1 - \frac{u}{u_1} \right) \frac{u}{u_1}\, dy + \frac{1}{u_1} \frac{du_1}{dx} \left(\int_0^\delta \frac{u}{u_1}\, dy - 2 \int_0^\delta \frac{u^2}{u_1^2}\, dy \right)$$

$$= \frac{d\vartheta_1}{dx} + \frac{1}{u_1} \frac{du_1}{dx} \left(2\vartheta_1 - \int_0^\delta \frac{u}{u_1}\, dy \right)$$

$$= \frac{d\vartheta_1}{dx} + \frac{1}{u_1} \frac{du_1}{dx} (2\vartheta_1 + \delta_1) - \frac{1}{u_1} \frac{du_1}{dx} \delta.$$

Substituting this expression on the right-hand side of (1·12, 15), we obtain

$$\frac{\tau_0}{\rho u_1^2} = \frac{d\vartheta_1}{dx} + \frac{\vartheta_1}{u_1}\frac{du_1}{dx}(H+2), \tag{1·12, 16}$$

where τ_0 is the shear stress at the wall, ϑ_1 is the momentum thickness, and $H = \delta_1/\vartheta_1$ is the ratio of the displacement and momentum thicknesses.

To apply the momentum equation we assume an approximate expression for the 'velocity profile' $u(y)$ across the boundary layer. Thus, let

$$u = u_1 f(\eta) \quad (0 \leqslant \eta \leqslant 1),$$

where $\eta = y/\delta$ and where f may at the same time depend on x. Then

$$\left. \begin{aligned} \delta_1 &= \delta \int_0^1 (1-f)\,d\eta, \quad \vartheta_1 = \delta \int_0^1 (1-f)f\,d\eta, \\ H &= \int_0^1 (1-f)\,d\eta \Big/ \int_0^1 (1-f)f\,d\eta, \quad \tau_0 = \mu u_1 f'(0)/\delta, \end{aligned} \right\} \tag{1·12, 17}$$

and, by (1·12, 16),

$$\frac{\nu}{u_1}f'(0) = \delta \frac{d}{dx}\left(\delta \int_0^1 (1-f)f\,d\eta\right) + \frac{\delta^2}{u_1}\frac{du_1}{dx}(H+2)\int_0^1 (1-f)f\,d\eta. \tag{1·12, 18}$$

In particular, if the external pressure gradient vanishes, then $du_1/dx = 0$, $u_1 = $ constant, and the equation becomes

$$\delta \frac{d}{dx}\left(\delta \int_0^1 (1-f)f\,d\eta\right) = \frac{\nu}{u_1}f'(0).$$

To determine a suitable function $f(\eta)$ we have the boundary conditions

$$u = v = 0 \quad \text{for} \quad y = 0, \quad u = u_1, \frac{\partial u}{\partial y} = 0 \quad \text{for} \quad y = \delta. \tag{1·12, 19}$$

To these we may add other conditions which are obtained by considering (1·12, 4) for the steady case. Then

$$u\frac{\partial u}{\partial x} + v\frac{\partial u}{\partial y} = u_1\frac{\partial u_1}{\partial x} + \nu\frac{\partial^2 u}{\partial y^2}. \tag{1·12, 20}$$

Putting $y = 0$ in this equation we get

$$\left(\frac{\partial^2 u}{\partial y^2}\right)_{y=0} = -\frac{u_1}{\nu}\frac{du_1}{dx}.$$

Again differentiating (1·12, 20) with respect to y and putting $y = 0$ afterwards

$$\frac{\partial u}{\partial y}\frac{\partial u}{\partial x} + \frac{\partial v}{\partial y}\frac{\partial u}{\partial y} = \nu\frac{\partial^3 u}{\partial y^3}.$$

The left-hand side of the equation vanishes according to the equation of continuity, and so

$$\left(\frac{\partial^3 u}{\partial y^3}\right)_{y=0} = 0. \tag{1·12, 21}$$

Additional conditions at the wall are obtained by repeating this procedure. The conditions at the wall obtained so far imply, for $f(\eta)$,

$$f(0)=0, \quad f''(0)=-\frac{du_1}{dx}\frac{\delta^2}{\nu}, \quad f'''(0)=0. \qquad (1·12, 22)$$

We may also add other conditions for the edge of the boundary layer, which require that u merges more and more smoothly into the main stream. They are

$$\frac{\partial^2 u}{\partial y^2}=0, \quad \frac{\partial^3 u}{\partial y^3}=0, \quad \dots, \quad \text{for } y=\delta. \qquad (1·12, 23)$$

Then the conditions for $f(\eta)$ are, for $\eta=1$,

$$f(1)=1, \quad f'(1)=f''(1)=\dots=0. \qquad (1·12, 24)$$

In practice, we can only satisfy the first few of these boundary conditions. For example, for zero external pressure gradient $u_1 = \text{constant} = U$, the conditions

$$f(0)=f''(0)=0, \quad f(1)=1, \quad f'(1)=0,$$

are satisfied by

$$f(\eta)=\sin\frac{\pi}{2}\eta. \qquad (1·12, 25)$$

Then the displacement thickness is

$$\delta_1=\delta\int_0^1 (1-f)\,d\eta=\delta\left(1+\left[\frac{2}{\pi}\cos\frac{\pi}{2}\eta\right]_0^1\right)=\left(1-\frac{2}{\pi}\right)\delta,$$

while

$$\vartheta_1=\delta\int_0^1 (1-f)f\,d\eta=\delta\int_0^1\left(\sin\frac{\pi}{2}\eta-\sin^2\frac{\pi}{2}\eta\right)d\eta=\left(\frac{2}{\pi}-\frac{1}{2}\right)\delta.$$

Hence (1·12, 16) becomes

$$\left(\frac{2}{\pi}-\frac{1}{2}\right)\delta\frac{d\delta}{dx}=\frac{\pi}{2}\frac{\nu}{U}.$$

Integrating this expression we obtain

$$\delta^2=\frac{2\pi^2}{4-\pi}\frac{\nu x}{U}+\text{constant}. \qquad (1·12, 26)$$

If x is measured from the leading edge of a flat plate then the constant on the right-hand side of (1·12, 26) vanishes and so

$$\delta=\frac{2\pi}{\sqrt{(8-2\pi)}}\sqrt{\frac{\nu x}{U}}.$$

The shear stress at the wall is

$$\tau_0=\mu U f'(0)/\delta=\mu U\frac{\sqrt{(8-2\pi)}}{4}\sqrt{\frac{U}{\nu x}}.$$

Hence the corresponding skin-friction drag on a flat plate of breadth c is

$$2\int_0^c \tau_0\,dx=\rho U^{\frac{3}{2}}\sqrt{\{(8-2\pi)\,\nu c\}}=\rho U^2 c\sqrt{(8-2\pi)}\,Re^{-\frac{1}{2}}, \qquad (1·12, 27)$$

where $Re=Uc/\nu$. This may be compared with (1·11, 44), in which $\beta^3=1·328$,

           R & L

while in the present formula $\sqrt{(8-2\pi)}=1\cdot311$, a very good approximation. The displacement thickness according to the present method is

$$\delta_1=\left(1-\frac{2}{\pi}\right)\delta=\frac{2\pi-4}{\sqrt{(8-2\pi)}}\sqrt{\frac{\nu x}{U}}=1\cdot741\times Re^{-\frac{1}{2}}. \qquad (1\cdot12,28)$$

The corresponding quantity obtained by evaluating the integral $(1\cdot12,10)$ between the limits 0 and ∞ for the velocity profile $u(y)$ supplied by Blasius's solution is
$$\delta_1=1\cdot721\times Re^{-\frac{1}{2}}. \qquad (1\cdot12,29)$$

The agreement between $(1\cdot12,28)$ and $(1\cdot12,29)$ is again satisfactory.

Instead of the sine function used above we may express the velocity profile as a polynomial whose coefficients are determined from the boundary conditions. This method will be described later in greater detail in connexion with its application to aerofoil theory (§ 2·6).

So far the boundary-layer equations were formulated only for flow past a straight wall. Let us now consider conditions in the neighbourhood of a curved wall. We introduce a special system of orthogonal curvilinear coordinates in the following way (Fig. $1\cdot12,2$):

We choose an arbitrary point O on the wall. This may be, for example, the forward stagnation point on a cylindrical body (compare § 1·10). Through any point P in the fluid we draw the straight line PP' normal to the wall meeting the wall at P'. Then we define the x-coordinate of P as the distance OP' measured along the wall, and the y-coordinate as the distance $P'P$. Thus the lines $x=$ constant are straight and the lines $y=$ constant, in general curved, are at a constant distance from the wall.

Consider a curvilinear quadrilateral $P'Q'QP$, where P and Q are equidistant from the wall and QQ' as well as PP' is normal to it (Fig. $1\cdot12,2$). The distances PQ and $P'Q'$ are supposed small. Then

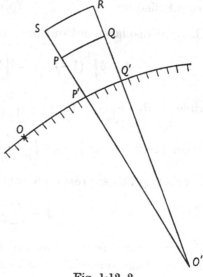

Fig. $1\cdot12,2$

the interior angles of the quadrilateral at P' and Q' are right angles, and $PP'=QQ'$. We conclude that the angles at P and Q are also right angles so that the system of coordinates is indeed orthogonal. The element of length in the direction of varying y is simply dy. Let $P'Q'=dx$. Producing PP' and QQ' beyond the wall we see that the two straight lines meet at O', the centre of curvature of the wall at P' or Q'. Then

$$\frac{PQ}{P'Q'}=\frac{PO'}{P'O'}.$$

Now $PP' = y$ and $\qquad\qquad P'O' = r = 1/\kappa$

is the local radius of curvature, while κ is the curvature. Hence

$$PQ = \frac{y+r}{r} P'Q' = (1 + \kappa y)\,dx \qquad\qquad (1 \cdot 12, 30)$$

is the element of length in the direction of varying x.

To derive the equation of continuity, we consider the rate of change of fluid mass in a small curved rectangle $PQRS$ whose sides are coordinate lines (see above, Fig. 1·12, 2). Since we confine ourselves to the case of an incompressible fluid, the total rate at which fluid crosses the boundary $PQRS$ must be equal to zero. This is, except for terms of higher order,

$$u(x+dy, y)\,dy - u(x, y)\,dy + v(x, y+dy)\,(1 + \kappa(y+dy))\,dx$$
$$- v(x, y)\,(1 + \kappa y)\,dx = 0.$$

Hence, dividing by $dx\,dy$ and passing to the limit,

$$\frac{\partial u}{\partial x} + \frac{\partial}{\partial y}((1 + \kappa y)\,v) = 0. \qquad\qquad (1 \cdot 12, 31)$$

$(1 \cdot 12, 31)$ also follows immediately from the general formula for the divergence in orthogonal curvilinear coordinates.

In the boundary layer, $(1 \cdot 12, 31)$ reduces to the ordinary equation of continuity in rectangular Cartesian coordinates if $\kappa \delta$ is small compared with unity or, which is the same, if the thickness of the boundary layer δ is small compared with the radius of curvature of the wall $r = 1/\kappa$. For $(1 \cdot 12, 31)$ may also be written as

$$\frac{\partial u}{\partial x} + (1 + \kappa y)\frac{\partial v}{\partial y} + \kappa v = 0,$$

and so, for $y \leqslant \delta$, $\qquad\qquad \dfrac{\partial u}{\partial x} + \dfrac{\partial v}{\partial y} + \kappa v \doteqdot 0. \qquad\qquad (1 \cdot 12, 32)$

But $\partial v/\partial y$ is of the order of v/δ, and this is large compared with $\kappa v = v/r$. Hence κv may be omitted from $(1 \cdot 12, 32)$, and

$$\frac{\partial u}{\partial x} + \frac{\partial v}{\partial y} = 0$$

still applies within the boundary layer.

The x-component of the acceleration of a fluid particle is given by

$$\frac{\partial u}{\partial t} + \frac{u}{1 + \kappa y}\frac{\partial u}{\partial x} + v\frac{\partial u}{\partial y} + \frac{\kappa}{1 + \kappa y}uv, \qquad\qquad (1 \cdot 12, 33)$$

where the last term is due to the curvature of the lines $x = $ constant. If $\kappa \delta$ is small compared with unity then we may replace $(1 \cdot 12, 33)$ by

$$\frac{\partial u}{\partial t} + u\frac{\partial u}{\partial x} + v\left(\frac{\partial u}{\partial y} + \kappa u\right).$$

But κu is small compared with $\partial u/\partial y$ which is of the order u/δ. Hence (1·12, 33) is reduced ultimately to the ordinary expression for the particle acceleration in rectangular Cartesian coordinates. A detailed investigation of the viscous stresses‡ shows that, provided $(d\kappa/dx)\,\delta^2$ as well as $\kappa\delta$ is small compared with unity, only the stress term $\nu(\partial^2 u/\partial y^2)$ need be taken into account, as before. Finally, the pressure gradient in the direction of varying y now has to balance also the centrifugal force which acts on the fluid in the boundary layer, but this is again small provided the product $\kappa\delta$ is small. We conclude that under the stated conditions the boundary layer equations (1·12, 4) and (1·12, 5) still apply to flow past a curved wall, and it will be seen that the boundary conditions also are unaffected.

1·13   Turbulence

The methods described so far are adequate for the analysis of the flow some distance downstream from the leading edge of a flat plate, or, more generally, of any two-dimensional body placed in a uniform air stream. However, as the thickness of the boundary layer increases towards the rear of the plate, conditions are complicated further by the fact that the flow may cease to present the smooth and steady aspect described, and to some extent postulated by the theory. Instead, there occur rapid fluctuations of the velocity, both in time and in space, on a scale which, though small in relation to the dimensions of the body, is macroscopic compared with the molecular structure of the fluid (e.g. the mean free path of the molecules if the fluid is a gas). In these circumstances the flow is said to be turbulent. By way of contradistinction, the smooth type of flow considered so far is said to be laminar. Turbulence can be detected by means of small tufts which are affixed to the surface of the body. Under conditions of laminar flow these tufts will point steadily in a fixed direction, while in turbulent flow, the scale and associated momentum of the fluctuations are sufficient to agitate the tufts so that they sway continuously to and fro. There are various more refined experimental methods for the detection and analysis of turbulence which do not concern us here. The phenomenon is one of fundamental physical importance, and an extensive mathematical theory has been developed to cope with some of its aspects. Even so, conditions in a turbulent boundary layer and more particularly the transition from laminar to turbulent flow have so far been inaccessible to a purely theoretical approach. It is known that transition to turbulence in the boundary layer depends on the Reynolds number of the boundary layer, $Re = U\delta/\nu$, on the velocity profile in the layer, on the surface roughness of the wall, and to a lesser degree on the amount of free turbulence which is

‡ See *Modern Developments in Fluid Dynamics*, ed. S. Goldstein (Oxford University Press, 1938), p. 119.

present in the surrounding fluid, and the effect of all these factors can be explained to some extent by theory. Thus, it is found that under certain conditions any small disturbance which is set up in the boundary layer, will be amplified and may eventually lead to a breakdown of laminar flow. However, the precise mechanism by which this leads to fully developed turbulence is only imperfectly understood at present.

As mentioned above, the scale of turbulence is large compared with the scale of molecular phenomena, so that the general law of viscous flow may be presumed to hold. To apply these laws to the particular circumstances of turbulent flow, we divide the velocity components and stresses into mean and fluctuating components. Thus, the mean velocities $\bar{u}, \bar{v}, \bar{w}$ are defined by

$$\bar{u}(x,y,z,t) = \frac{1}{2\tau} \int_{t-\tau}^{t+\tau} u(x,y,z,t')\,dt', \quad \bar{v}(x,y,z,t) = \frac{1}{2\tau} \int_{t-\tau}^{t+\tau} v(x,y,z,t')\,dt',$$

$$\bar{w}(x,y,z,t) = \frac{1}{2\tau} \int_{t-\tau}^{t+\tau} w(x,y,z,t')\,dt'.$$

$$(1\cdot13, 1)$$

The time interval τ in these definitions is to be chosen so that it is large compared with the time scale of these turbulent fluctuations, but small compared with the time scale of the macroscopic variations of the field of flow. The latter condition is redundant if the flow is steady. It follows from the randomness of the fluctuations that, subject to the limitations just mentioned, $\bar{u}, \bar{v}, \bar{w}$ are independent of τ. The fluctuating components of the velocities are then defined by

$$u' = u - \bar{u}, \quad v' = v - \bar{v}, \quad w' = w - \bar{w}. \qquad (1\cdot13, 2)$$

The operation by which $\bar{u}, \bar{v}, \bar{w}$ are obtained from u, v, w is called averaging with respect to time. Thus the time average of any function (x,y,z,t) is

$$\bar{f}(x,y,z,t) = \frac{1}{2\tau} \int_{t-\tau}^{t+\tau} f(x,y,z,t')\,dt'. \qquad (1\cdot13, 3)$$

This operation commutes with the operations of partial differentiation with respect to x, y, z, or t, i.e. $\dfrac{\partial \bar{f}}{\partial x} = \dfrac{\overline{\partial f}}{\partial x}$, etc. For x, y and z this follows immediately from the rule of differentiation under the sign of the integral as applied to $(1\cdot13, 3)$. Also,

$$\frac{\partial \bar{f}}{\partial t} = \frac{\partial}{\partial t}\left(\frac{1}{2\tau} \int_{t-\tau}^{t+\tau} f(x,y,z,t')\,dt'\right)$$

$$= \frac{1}{2\tau}\left(f(x,y,z,t+\tau) - f(x,y,z,t-\tau)\right) = \frac{1}{2\tau} \int_{t-\tau}^{t+\tau} \frac{\partial f}{\partial t'}\,dt',$$

which confirms the assertion for differentiation with respect to time.

It follows from the definition of the fluctuating velocity components $(1\cdot13, 2)$ that their time averages vanish. Similarly, we define the mean stresses $\overline{p}_{xx}$, $\overline{p}_{xy}$, etc., as the time averages of p_{xx}, p_{xy}, etc., respectively, and the fluctuating stresses by $p'_{xx}=p_{xx}-\overline{p}_{xx}$, $p'_{xy}=p_{xy}-\overline{p}_{xy}$. Then the time averages of the fluctuating stresses also vanish.

But although the time averages of the fluctuating velocities vanish, the same is not usually true for the time averages of the momentum carried across any given surface by the fluctuating velocities. It was first realized by Reynolds‡ that this fact is equivalent to the generation of additional internal stresses in the fluid. To see this, we first rewrite the first equation in (1·11, 17) in the form

$$\rho\frac{\partial u}{\partial t}+\rho u\frac{\partial u}{\partial x}+u\frac{\partial}{\partial x}(\rho u)+\rho v\frac{\partial u}{\partial y}+u\frac{\partial}{\partial y}(\rho v)+\rho w\frac{\partial u}{\partial z}+u\frac{\partial}{\partial z}(\rho w)$$

$$-u\left\{\frac{\partial}{\partial x}(\rho u)+\frac{\partial}{\partial y}(\rho v)+\frac{\partial}{\partial z}(\rho w)\right\}=\rho X+\frac{\partial p_{xx}}{\partial x}+\frac{\partial p_{yx}}{\partial y}+\frac{\partial p_{zx}}{\partial z}.$$

But the expression in the curly brackets on the left-hand side of this equation vanishes, by virtue of the equation of continuity, and so

$$\rho\frac{\partial u}{\partial t}+\frac{\partial}{\partial x}(\rho u^2)+\frac{\partial}{\partial y}(\rho uv)+\frac{\partial}{\partial z}(\rho uw)=\rho X+\frac{\partial p_{xx}}{\partial x}+\frac{\partial p_{yx}}{\partial y}+\frac{\partial p_{zx}}{\partial z},$$

or $\quad \rho\dfrac{\partial u}{\partial t}=\rho X+\dfrac{\partial}{\partial x}(p_{xx}-\rho u^2)+\dfrac{\partial}{\partial y}(p_{yx}-\rho uv)+\dfrac{\partial}{\partial z}(p_{zx}-\rho uw),$

and similarly

$$\rho\frac{\partial v}{\partial t}=\rho Y+\frac{\partial}{\partial x}(p_{xy}-\rho vu)+\frac{\partial}{\partial y}(p_{yy}-\rho v^2)+\frac{\partial}{\partial z}(p_{zy}-\rho vw),$$

and $\quad \rho\dfrac{\partial w}{\partial t}=\rho Z+\dfrac{\partial}{\partial x}(p_{xz}-\rho wu)+\dfrac{\partial}{\partial y}(p_{yz}-\rho wv)+\dfrac{\partial}{\partial z}(p_{zz}-\rho w^2).$

$$(1·13,4)$$

These equations hold both for incompressible and for compressible fluids. We confine ourselves to the incompressible case, and average the equations (1·13, 4) with respect to time. Bearing in mind that differentiation commutes with the operation of averaging, we then obtain, from the first equation,

$$\rho\frac{\partial\overline{u}}{\partial t}=\rho X+\frac{\partial}{\partial x}(\overline{p}_{xx}-\rho\overline{u^2})+\frac{\partial}{\partial y}(\overline{p}_{yx}-\rho\overline{uv})+\frac{\partial}{\partial z}(\overline{p}_{zx}-\rho\overline{uw}).$$

Now $\qquad \overline{u^2}=\overline{(\overline{u}+u')^2}=\overline{\overline{u}^2+2\overline{u}u'+u'^2}=\overline{u}^2+2\overline{u}\overline{u}'+\overline{u'^2},$

and so $\qquad \overline{u^2}=\overline{u}^2+\overline{u'^2},$

and similarly $\qquad \overline{uv}=\overline{u}\overline{v}+\overline{u'v'}$

and $\qquad \overline{uw}=\overline{u}\overline{w}+\overline{u'w'}.$

‡ O. Reynolds, 'On the dynamical theory of incompressible viscous fluids and the determination of the criterion', *Phil. Trans.* A, vol. 186 (1895), pp. 123–64.

Hence

$$\rho\left(\frac{\partial \bar{u}}{\partial t}+\frac{\partial}{\partial x}\,\bar{u}^2+\frac{\partial}{\partial y}\,(\overline{uv})+\frac{\partial}{\partial z}\,(\overline{uw})\right)$$

$$=\rho X+\frac{\partial}{\partial x}\,(\bar{p}_{xx}-\rho\overline{u'^2})+\frac{\partial}{\partial y}\,(\bar{p}_{yx}-\rho\overline{u'v'})+\frac{\partial}{\partial z}\,(\bar{p}_{zx}-\rho\overline{u'w'}).\quad(1\cdot13,5)$$

Now, averaging the equation of continuity for incompressible fluids we find that

$$\frac{\partial \bar{u}}{\partial x}+\frac{\partial \bar{v}}{\partial y}+\frac{\partial \bar{w}}{\partial z}=0.$$

Hence

$$\frac{\partial}{\partial x}\,\bar{u}^2+\frac{\partial}{\partial y}\,(\overline{uv})+\frac{\partial}{\partial z}\,(\overline{uw})=\frac{\partial}{\partial x}\,\bar{u}^2-\bar{u}\frac{\partial \bar{u}}{\partial x}+\frac{\partial}{\partial y}\,(\overline{uv})-\bar{u}\frac{\partial \bar{v}}{\partial y}+\frac{\partial}{\partial z}\,(\overline{uw})-\bar{u}\frac{\partial \bar{w}}{\partial z}$$

$$=\bar{u}\frac{\partial \bar{u}}{\partial x}+\bar{v}\frac{\partial \bar{u}}{\partial y}+\bar{w}\frac{\partial \bar{u}}{\partial z}.$$

Substituting this expression on the left-hand side of (1·13, 5), we obtain

$$\rho\left(\frac{\partial \bar{u}}{\partial t}+\bar{u}\frac{\partial \bar{u}}{\partial x}+\bar{v}\frac{\partial \bar{u}}{\partial y}+\bar{w}\frac{\partial \bar{u}}{\partial z}\right)$$

$$=\rho X+\frac{\partial}{\partial x}\,(\bar{p}_{xx}-\rho\overline{u'^2})+\frac{\partial}{\partial y}\,(\bar{p}_{yx}-\rho\overline{u'v'})+\frac{\partial}{\partial z}\,(\bar{p}_{zx}-\rho\overline{u'w'}),$$

and similarly

$$\rho\left(\frac{\partial \bar{v}}{\partial t}+\bar{u}\frac{\partial \bar{v}}{\partial x}+\bar{v}\frac{\partial \bar{v}}{\partial y}+\bar{w}\frac{\partial \bar{v}}{\partial z}\right)$$

$$=\rho Y+\frac{\partial}{\partial x}\,(\bar{p}_{xy}-\rho\overline{v'u'})+\frac{\partial}{\partial y}\,(\bar{p}_{yy}-\rho\overline{v'^2})+\frac{\partial}{\partial z}\,(\bar{p}_{zy}-\rho\overline{v'w'}),$$

$$\rho\left(\frac{\partial \bar{w}}{\partial t}+\bar{u}\frac{\partial \bar{w}}{\partial x}+\bar{v}\frac{\partial \bar{w}}{\partial y}+\bar{w}\frac{\partial \bar{w}}{\partial z}\right)$$

$$=\rho Z+\frac{\partial}{\partial x}\,(\bar{p}_{xz}-\rho\overline{w'u'})+\frac{\partial}{\partial y}\,(\bar{p}_{yz}-\rho\overline{w'v'})+\frac{\partial}{\partial z}\,(\bar{p}_{zz}-\rho\overline{w'^2}).$$

$$(1\cdot13,6)$$

Comparing (1·13, 6) with (1·11, 17) we see that the effect of turbulence is the addition of terms $-\rho\overline{u'^2}$, $-\rho\overline{u'v'}$, $-\rho\overline{u'w'}$, ... to the ordinary stresses p_{xx}, p_{xy}, p_{xz}, ... These additional terms, which are due to the transport of momentum by the functuating velocities, are called the Reynolds stresses. In fully developed turbulence they tend to be rather more important than the stresses due to viscosity.

While the above analysis gives us some insight into the mechanism of turbulence, it does not enable us to cope with the detailed calculation of the flow in a turbulent boundary layer. Our knowledge of the turbulent boundary layer is in fact due largely to empirical evidence combined with a certain amount of theoretical reasoning. It is found that once the boundary layer has become turbulent it tends to thicken rapidly. The flow remains laminar, however, in a very thin layer adjacent to the wall, the so-called laminar sub-layer.

Von Kármán's momentum equation is still valid in the form (1·12, 14). Averaging the equation with respect to time, we obtain

$$\rho\left\{\int_0^\delta \frac{\partial \bar{u}}{\partial t}\,dy + \frac{\partial}{\partial x}\int_0^\delta \overline{u^2}\,dy - u_1\frac{\partial}{\partial x}\int_0^\delta \bar{u}\,dy\right\} = \rho\left(\frac{\partial u_1}{\partial t} + u_1\frac{\partial u_1}{\partial x}\right) - \bar{\tau}_0, \quad (1·13,7)$$

where it has been assumed that the turbulence outside the boundary layer is negligible.

The displacement and momentum thicknesses are now defined by

$$\delta_1 = \int_0^\delta \left(1 - \frac{\bar{u}}{u_1}\right)dy, \quad \vartheta_1 = \int_0^\delta \overline{\left(1 - \frac{u}{u_1}\right)\frac{u}{u_1}}\,dy, \quad (1·13,8)$$

which reduces to the previous definition in the absence of turbulence. Proceeding as in § 1·12 (compare (1·12, 15) and (1·12, 16)) we then obtain for the case of steady mean flow

$$\frac{\bar{\tau}_0}{\rho u_1^2} = \frac{d\vartheta_1}{dx} + \frac{\vartheta_1}{u_1}\frac{du_1}{dx}(H+2), \quad (1·13,9)$$

where $H = \delta_1/\vartheta_1$.

It is convenient for some purposes to replace the definition of ϑ_1, in (1·13, 8), by

$$\vartheta_1 = \int_0^\delta \left(1 - \frac{\bar{u}}{u_1}\right)\frac{\bar{u}}{u_1}\,dy, \quad (1·13,10)$$

while still using the momentum equation in the form (1·13, 9). This involves the following approximation:

We have

$$\frac{d}{dx}\int_0^\delta \overline{u^2}\,dy = \frac{d}{dx}\int_0^\delta \bar{u}^2\,dy + \frac{d}{dx}\int_0^\delta \overline{u'^2}\,dy, \quad (1·13,11)$$

since $\overline{u'u} = 0$. Assuming that the gradient of the transport of the x-component of momentum along the boundary layer is chiefly due to the mean velocity, we may omit the second integral on the right-hand side of (1·13, 11). Then (1·13, 7) becomes

$$\rho\left\{\int_0^\delta \frac{\partial \bar{u}}{\partial t}\,dy + \frac{\partial}{\partial x}\int_0^\delta \bar{u}^2\,dy - u_1\frac{\partial}{\partial x}\int_0^\delta \bar{u}\,dy\right\} = \rho\left(\frac{\partial u_1}{\partial t} + u_1\frac{\partial u_1}{\partial x}\right) - \bar{\tau}_0. \quad (1·13,12)$$

Proceeding as before, we again obtain (1·13, 9), where ϑ_1 is now defined by (1·13, 10).

1·14   The Joukowski condition

The flow of a viscous fluid past an obstacle is exceedingly varied and its character depends both on the Reynolds number and on the geometry of the obstacle. Consider, for example, the flow past a bluff cylindrical body. At moderately high Reynolds numbers the flow is laminar. The flow is accelerated in the region of the front stagnation point, and this corresponds to a drop in pressure (a negative or 'favourable' pressure gradient) along the wall in the direction of the main flow. A point of minimum pressure, or 'maximum suction', is reached and beyond that point the pressure gradient is positive, and acts in concert with the shear stress at the wall to slow down the flow in the boundary layer. Eventually this leads to a region of reverse flow in the boundary layer (Fig. 1·14, 1). This phenomenon is known as separation, and the point P at which $\partial u/\partial y = 0$ is called the point of separation, since downstream of P the main flow is separated from the wall by a region of reverse flow. Separation occurs further downstream if the boundary layer is turbulent, for in that case the interchange of momentum due to turbulence helps to pull the fluid along against the adverse forces near the wall.

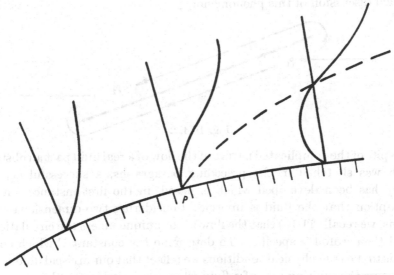

Fig. 1·14, 1

Additional complications arise further downstream owing to the appearance of concentrated vortices whose pattern depends on the Reynolds number. Separation does not occur (or occurs only very near the trailing edge) on a thin or slender body which is at small incidence relative to the free stream. It is therefore not of much importance in low-speed aerofoil theory, which is concerned with wings at moderate angles of incidence (maximum angles of 12–15°, under two-dimensional conditions, say). At higher incidences separation does take place nearer to the leading edge. This leads to a general breakdown of the flow and the wing is then said to be stalled. There is no theory to account adequately for the forces which act on the wing in these circumstances.

Consider now an aerofoil which is initially at rest, and then begins to move forward in a fixed direction, so that the aerofoil is at a small incidence relative to the direction of motion. We assume two-dimensional conditions. The case of a flat plate of unlimited span (§ 1·10) may serve as an illustration. Then the flow past the aerofoil is initially that given by potential flow (compare (1·10, 28) and (1·10, 29), where the flow is referred to a system of coordinates which is at rest relative to the aerofoil). Thus, there is a rear stagnation point P_2 on the top surface of the aerofoil near the trailing edge, and the fluid at the lower surface has to turn round the trailing edge to reach P_2 (Fig. 1·14, 2). However, very soon after the beginning of the motion, a concentrated vortex is formed at the trailing edge and cast off into the fluid, and a compensating circulation of opposite sign is set up round the aerofoil in such a way that the stagnation point P_2 moves toward the trailing edge. The reader is referred to *Modern Developments in Fluid Dynamics* for a detailed discussion of this phenomenon.‡

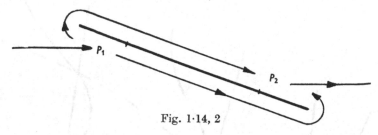

Fig. 1·14, 2

In spite of the complicated nature of the flow of a real fluid past an obstacle which was sketched in the preceding paragraphs, a successful aerofoil theory has been developed which is based, in the first instance, on the assumption that the fluid is inviscid. Considering two-dimensional conditions, we recall (§ 1·10) that the flow is not unique unless the circulation Γ round the aerofoil is specified. To determine the constant Γ which corresponds to the true physical conditions we reflect that our inviscid fluid really represents the limiting case of a fluid whose viscosity is vanishingly small. On the other hand, as we have seen, the development of the circulation depends on conditions near the trailing edge. For a given variation of the forward velocity in time the process therefore depends on a local Reynolds number which is given by a length associated with the geometry of the trailing edge. Such a length is the radius of curvature of the trailing edge, r. Thus, an appropriate Reynolds number is $Re = rU/\nu$. Now, assuming that ν can be varied at will, let it tend to 0. In order to maintain similarity (§1·11), Re must be kept constant at the same time, and this implies that the radius of curvature of the trailing edge also tends to 0. In other words the trailing edge must be assumed sharp. At the same time the velocity near the trailing edge must of course remain finite (i.e. bounded) in the actual physical case. This leads us to the following fundamental

Joukowski condition. The velocity remains finite in the neighbourhood of the trailing edge, even though the latter is assumed to be sharp.

The derivation of this principle explains why we shall accept infinities elsewhere in the fluid, in particular at a sharp leading edge (see § 2·8 below) but not at the trailing edge.

‡ Ed. S. Goldstein (Oxford University Press, 1938), pp. 40–1, 66–9.

1·15   The wake

The Joukowski condition, together with the assumption that the flow is irrotational and continuous everywhere in the fluid, is sufficient to determine the flow round a two-dimensional aerofoil (Chapter 2). This assumption neglects the layer of fluid aft of the aerofoil, which is slowed down (relative to the aerofoil) through the action of the viscous shearing forces in the boundary layer. As a result, the theory, while consistent in itself, cannot account for the drag experienced by an aerofoil even in two-dimensional motion. Thus the calculation of the drag requires the resources of boundary-layer theory (§ 2·12). The layer of fluid just mentioned may be called the 'viscous wake'.

Conditions are more complicated for the flow round a wing of finite dimensions. This is of course the most important case in practice, since conditions of two-dimensional flow can only be established artificially in a wind tunnel (and even then only approximately). Speaking purely in terms of the theory of inviscid fluids, we may say that the assumption that the velocity is bounded in the neighbourhood of a sharp trailing edge is not, in general three-dimensional flow, compatible with the assumption that the flow is irrotational and continuous everywhere in the fluid. In fact the 'classical' continuous solution of the problem of irrotational flow round a body immersed in a fluid is known, generally speaking, to produce infinite (unbounded) velocities in the neighbourhood of sharp corners. This is exemplified by the case of the flat plate (§ 1·10) and can also be established more generally. Thus, the theory of inviscid fluids alone forces us to admit the presence of discontinuities or, alternatively, the presence of regions in which the flow is no longer irrotational. However, in order to know where to look for such a discontinuity or region of vorticity, we are obliged once again to appeal to the theory of viscous fluids. We then find, as before, that the region in question must consist of particles that have passed through the boundary layer of the aerofoil, since this is the only place where vorticity can be created. As mentioned above, these particles constitute the wake, which occupies a thin layer aft of the aerofoil. Moreover, for the purposes of three-dimensional aerofoil theory it is assumed that this layer is infinitely thin, in other words, that it is a sheet which is joined to the aerofoil at its trailing edge. By assuming that the layer is infinitely thin we neglect the momentum of the fluid contained in it, and as a result the theory still does not account for some of the drag associated with the wake and this has to be calculated separately. Nevertheless, we shall still refer to this infinitely thin sheet as the wake.

Let λ, μ, ν be the direction cosines of the direction normal to the wake at a point P_1 and let u_+, v_+, w_+ and u_-, v_-, w_- be the velocity components at

P on either side of the wake, respectively. Then the velocity components normal to the wake on either side of it are

$$\lambda u_+ + \mu v_+ + \nu w_+ \quad \text{and} \quad \lambda u_- + \mu v_- + \nu w_-,$$

respectively. Now let S be a small cylindrical surface which contains P in its interior, and whose bases are parallel to the wake and adjacent to it, while the generators of S are normal to the wake (Fig. 1·15, 1). We may then neglect the area of the envelope of S compared with the area A of each one of its bases, in the first approximation.

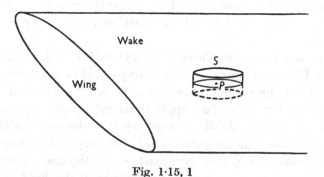

Fig. 1·15, 1

Thus the flux across S is given by

$$A((\lambda u_+ + \mu v_+ + \nu w_+) - (\lambda u_- + \mu v_- + \nu w_-))$$

in the first approximation, and this expression must vanish since no fluid is created or destroyed within S. It follows that

$$\lambda u_+ + \mu v_+ + \nu w_+ = \lambda u_- + \mu v_- + \nu w_- = q_n, \quad \text{say,}$$

so that the normal velocity component is continuous across the wake. Thus, the tangential velocity only may be discontinuous, the wake forms a vortex sheet (§ 1·9). Again, applying the principle of linear momentum (1·4, 3) we obtain, for the component of momentum parallel to the wake,

$$\rho(\mathbf{q}_t^{(+)} - \mathbf{q}_t^{(-)}) q_n A = 0, \qquad (1·15, 1)$$

approximately, where $\mathbf{q}_t^{(+)}$ and $\mathbf{q}_t^{(-)}$ are the velocity components tangential to the wake on either side of it, since no external force is acting within S. Now the wake is a sheet of discontinuity, so in general, $\mathbf{q}_t^{(+)} \neq \mathbf{q}_t^{(-)}$. Hence, (1·15, 1), implies

$$q_n = 0.$$

That is to say, the normal velocity component at the wake vanishes and the wake consists of streamlines. Furthermore, applying (1·4, 3) to the component of momentum normal to the wake, we obtain approximately

$$(p_+ - p_-) A = 0,$$

where p_+ and p_- denote the pressure on either side of the wake. Thus $p_+ = p_-$, the pressure is continuous across the wake. Bernoulli's law (1·6, 6) then yields

$$(q_l^{(+)})^2 = (q_l^{(-)})^2,$$

so that the magnitude of the velocity vector, though not its direction, is again continuous across the wake. It follows, by simple geometry, that the vector difference, $q_l^{(+)} - q_l^{(-)}$, is normal to the mean velocity vector at the wake, $\frac{1}{2}(q_l^{(+)} + q_l^{(-)})$.

We have just shown that the pressure is continuous across the wake. Also, according to the Joukowski condition in conjunction with Bernoulli's law, the pressure remains finite at the trailing edge. It follows that the pressure difference between the top and bottom surfaces in any direction which is roughly normal to the chord of the aerofoil tends to zero at the trailing edge. This condition is used frequently in place of the Joukowski condition in general three-dimensional flow (see Chapter 3).

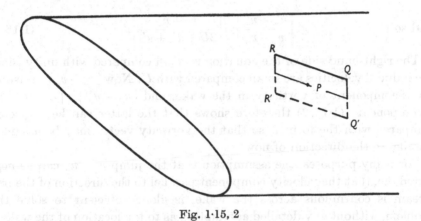

Fig. 1·15, 2

Let $C = QRR'Q'Q$ be a small rectangular circuit such that the segments QR, $Q'R'$, are parallel to $q_l^{(+)} - q_l^{(-)}$ and adjacent to the wake on either side of it, and such that the segments QQ', RR' are small by comparison (Fig. 1·15, 2) Then the circulation round C is given by $|q_l^{(+)} - q_l^{(-)}| l$, where l is the length of the segments QR, $Q'R'$. On the other hand, the circulation round a similar circuit in the plane of the vector $\frac{1}{2}(q_l^{(+)} + q_l^{(-)})$ vanishes, so that the surface distribution of vorticity in the wake is of magnitude $|q_l^{(+)} - q_l^{(-)}|$ and is normal to $q_l^{(+)} - q_l^{(-)}$ and parallel to the mean velocity vector.

The theoretical determination of the exact position of the wake is beyond our resources, and approximations must be adopted in all concrete cases. It is usually assumed that the wake consists of straight lines parallel to the direction of flow. In general, the condition that the velocity component normal to the wake should vanish is violated as a result. However, it is argued that since the velocities induced by the presence of the aerofoil are small compared with the free-stream velocity, the streamlines must be at least approximately parallel to the direction of flow. The wake consists of streamlines, as shown above.

Accepting this, we consider a wing at small incidence which is situated approximately in the x, y-plane, while the free-stream velocity is directed along the x-axis. Write $u = U + u'$, where U is the free-stream velocity, so that u' is the induced velocity component in the direction of the free stream. According to our assumption the wake occupies a portion of the x, y-plane downstream of the wing. It was shown earlier that both the magnitude of the velocity, q, and its normal component, w, are continuous across the wake. Hence

$$u_+^2 + v_+^2 = u_-^2 + v_-^2,$$

$$u_+^2 - u_-^2 = (U + u_+')^2 - (U + u_-')^2 = 2U(u_+' - u_-') + u_+'^2 - u_-'^2$$
$$= v_-^2 - v_+^2,$$

and so
$$\left| \frac{u_+' - u_-'}{v_+ - v_-} \right| = \left| \frac{v_+ + v_-}{2U + u_+' + u_-'} \right|. \qquad (1·15, 2)$$

The right-hand side of the equation is small compared with unity, since the induced velocities are small compared with U. Now $|v_+ - v_-|$ measures the x-component of vorticity in the wake, and $|u_+' - u_-'| = |u_+ - u_-|$ its y-component. (1·15, 2) therefore shows that the latter can be neglected compared with the former, so that the vorticity vector may be assumed parallel to the direction of flow.

For many purposes, the assumption that the jump $u_+ - u_-$ can be neglected, i.e. that the velocity component parallel to the direction of the free stream is continuous across the wake, is alone sufficient to solve the problem, without any detailed assumption as to the location of the wake.

It is found that some way aft of the trailing edge the wake tends to 'roll up' into isolated vortices,‡ but it appears that the effect of this phenomenon on the pressure distribution on the wing can be neglected.

The general picture of the flow round an aerofoil which has been presented in the last two sections does not take into account the influence of compressibility. This will be discussed in detail in Chapter 4. The assumption that the air is incompressible leads to satisfactory results for moderate forward speeds (up to about 200 m.p.h. under normal conditions).

When calculating the aerodynamic forces which act on an aerofoil it is customary to neglect the effect of gravity. It is evident that in incompressible potential flow, gravity does not affect the velocity distribution at all, since the latter is derived from a solution of Laplace's equation which

‡ See *Modern Developments in Fluid Dynamics*, ed. S. Goldstein (Oxford University Press, 1938), pp. 44–5.

depends only on the geometry of the boundaries. The force acting on the aerofoil is then given by

$$-\int_S p\,d\mathbf{S},$$

where S is the surface of the aerofoil. Substituting the value of p as given by Bernoulli's equation (1·6, 5), we obtain

$$-\int_S p\,d\mathbf{S} = \rho \int_S \left\{ -\frac{\partial \Phi}{\partial t} - h(t) + \tfrac{1}{2}q^2 + gz \right\} d\mathbf{S},$$

where g is the acceleration due to gravity which is supposed to act along the direction of the z-axis. Hence the contribution due to gravity is

$$\rho g \int_S z\,d\mathbf{S} = \rho g V \mathbf{k}, \qquad\qquad (1·15, 3)$$

according to the divergence theorem, where V is the volume of the aerofoil and $\mathbf{k}$ is the unit vector along the z-axis. Thus (1·15, 3) is simply the weight of the fluid displaced by the aerofoil, its 'buoyancy'. The fact that this force acts on a body immersed in a fluid under static conditions is known as the principle of Archimedes. In general, this force is quite negligible compared with the aerodynamic forces due to the variations of the velocity potential as they occur in the applications of aerofoil theory, and the same holds for the corresponding moment. Similar remarks apply to viscous fluid theory, although in that case Bernoulli's law does not apply and the problem requires the consideration of the equations of Navier–Stokes. Gravity may have some effect on turbulence, but the conditions under which this occurs are of no importance to aerofoil theory.

CHAPTER 2

AEROFOIL THEORY FOR STEADY FLOW IN TWO DIMENSIONS

2·1 General

In a spanwise plane of symmetry of an aerofoil in three dimensions, we must have $v = 0$, so that the velocity vector is everywhere in the plane of symmetry. Now according to § 1·15 the velocity vectors above and below the wake are tangential to it and of equal magnitude. Hence the two vectors are parallel and either of equal or of opposite directions. Ruling out the second possibility, we conclude that the velocity vector is actually continuous in a plane of symmetry of the aerofoil. Now if we regard two-dimensional flow as the limit of the three-dimensional flow round an aerofoil of rectangular planform whose aspect ratio becomes very large, then every plane normal to the span of the aerofoil becomes a plane of symmetry ultimately. It follows that the velocity vector must now be assumed to be continuous everywhere, in contradistinction to the conditions in general three-dimensional flow.

We have ruled out the possibility that the velocity vectors above and below the aerofoil be of equal magnitude but of opposite directions. This is an ancillary condition which was not derived from the general analysis of § 1·15. The justification for this and other assumptions stated earlier is, in the last resort, provided by experimental evidence.

In two-dimensional theory, we shall use the symbols z, w, Π to denote the complex variable $x + iy$, the complex velocity $u - iv$, and the complex potential $\Phi + i\Psi$ respectively, where Ψ is the stream function.

2·2 Blasius's formulae

We shall be concerned primarily with the flow round a single aerofoil in two dimensions. To calculate the forces acting on such an aerofoil, we may use formulae (1·4, 16) and (1·4, 19), taking opposite signs on the right-hand side of these formulae since they express the forces exerted by the *aerofoil* on the *fluid*. The directed surface element of a cylinder S whose generators are parallel to the span of the aerofoil, i.e. normal to the x, y-plane, is then given by $(dy, -dx)$ per unit span (Fig. 2·2, 1). Denoting the force acting on the aerofoil, per unit span, by $\mathbf{F} = (X, Y)$, we then obtain from (1·4, 16)

$$(X, Y) = -\left[\int_S p(dy, -dx) + \rho \int_S (u, v)(u\, dy - v\, dx) \right], \qquad (2\cdot2, 1)$$

or
$$X = -\int_S p\,dy + \rho\int_S u(v\,dx - u\,dy),$$

$$Y = \int_S p\,dx + \rho\int_S v(v\,dx - u\,dy).$$

Hence
$$X - iY = -\int_S p(dy + i\,dx) + \rho\int_S (u - iv)(v\,dx - u\,dy).$$

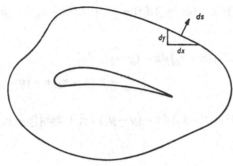

Fig. 2·2, 1

Now, by Bernoulli's theorem (1·6, 7)

$$p + \tfrac{1}{2}\rho(u^2 + v^2) = \text{constant} = H,$$

so that
$$p = H - \tfrac{1}{2}\rho(u^2 + v^2) \tag{2·2, 2}$$

and

$(X - iY)$

$$= -\int_S H(dy + i\,dx) + \tfrac{1}{2}\rho\int_S (u^2 + v^2)(dy + i\,dx) + \rho\int_S (u - iv)(v\,dx - u\,dy)$$

$$= -H\int_S (dy + i\,dx) + \tfrac{1}{2}\rho i\int_S [(u^2 - 2iuv - v^2)\,dx + (u^2 - 2iuv - v^2)\,i\,dy]$$

$$= -H\int_S (dy + i\,dx) + \tfrac{1}{2}\rho i\int_S (u - iv)^2(dx + i\,dy).$$

Now $\int_S (dy + i\,dx) = 0$, since the integral is taken round a closed contour. Hence, writing w for the complex velocity $u - iv$ and $z = x + iy$, we obtain finally

$$X - iY = \tfrac{1}{2}\rho i\int_S w^2\,dz. \tag{2·2, 3}$$

Similarly, we may apply (1·4, 19) to the calculation of the moment about a point $N = (x_0, y_0)$, again taking opposite signs on the right-hand side. We are now concerned only with the moment component normal to the x, y-plane, which will be denoted simply by M (instead of $-N$ in (1·4, 19)). Then

$$\mathbf{q}\,ds = (u, v)(dy, -dx) = u\,dy - v\,dx,$$

while the components of $\mathbf{r}\wedge d\mathbf{s}$ and $\mathbf{r}\wedge\mathbf{q}$ parallel to the span are

$$-[(x - x_0)\,dx + (y - y_0)\,dy] \quad \text{and} \quad (x - x_0)v - (y - y_0)u,$$

respectively. **r** is the radius vector from $N = (x_0, y_0)$ to a point (x, y) on S. Hence M is given by

$$M = \int_S p[(x-x_0)\,dx + (y-y_0)\,dy] + \rho \int_S [(x-x_0)\,v - (y-y_0)\,u]\,(v\,dx - u\,dy).$$

We replace p by its value according to (2·2, 2) and take into account that

$$\int_S H[(x-x_0)\,dx - (y-y_0)\,dy] = \tfrac{1}{2}H \int_S d((x-x_0)^2 + (y-y_0)^2) = 0.$$

Then

$$M = -\tfrac{1}{2}\rho \int_S [(u^2+v^2)\{(x-x_0)\,dx + (y-y_0)\,dy\}$$
$$- 2\{(x-x_0)\,v - (y-y_0)\,u\}\,(v\,dx - u\,dy)],$$

or

$$M = -\tfrac{1}{2}\rho \int_S [(u^2-v^2)\{(x-x_0)\,dx - (y-y_0)\,dy\} + 2uv\{(y-y_0)\,dx + (x-x_0)\,dy\}].$$
$$(2·2, 4)$$

Now

$$w^2(z-z_0)\,dz = (u-iv)^2\,[(x-x_0) + i(y-y_0)]\,(dx + i\,dy)$$
$$= (u^2 - v^2 - 2iuv)\,[(x-x_0)\,dx - (y-y_0)\,dy$$
$$+ i\{(y-y_0)\,dx + (x-x_0)\,dy\}]$$
$$= (u^2-v^2)\{(x-x_0)\,dx - (y-y_0)\,dy\} + 2uv\{(y-y_0)\,dx + (x-x_0)\,dy\}$$
$$+ i[2uv\{(y-y_0)\,dy - (x-x_0)\,dx\} + (u^2-v^2)\{(y-y_0)\,dx$$
$$+ (x-x_0)\,dy\}],$$

where $z_0 = x_0 + iy_0$. Comparing this expression with (2·2, 4) we see that the integrand of (2·2, 4) is the real part of $w^2(z-z_0)\,dz$, and so

$$M = -\tfrac{1}{2}\rho \mathscr{R} \int_S w^2(z-z_0)\,dz, \qquad (2·2, 5)$$

where $\mathscr{R}$ denotes the real part of a complex number, as usual. Formulae (2·2, 3) and (2·2, 5) were derived by Blasius.‡

The knowledge of the complex velocity $w = w(z)$ would enable us to calculate the force and moment acting on a given aerofoil. In actual fact, the calculation of the complex velocity for a given aerofoil normally involves the conformal mapping of the region outside the aerofoil on to another region, e.g. the region outside a circle or an approximately circular contour. According to the general theory of functions of a complex variable such a representation is made definite if we require that the point at infinity be transformed into the point at infinity, and that a specific direction through the point at infinity be transformed into a specific direction through the point at infinity in the second plane. Denoting the complex variable in the

‡ H. Blasius, 'Funktionentheoretische Methoden in der Hydrodynamik', *Z. angew. Math. Mech.* vol. 58 (1910), p. 90.

second plane by ζ, we may write the mapping function $f(\zeta)$ as a Laurent series,

$$z = f(\zeta) = C_{-1}(\zeta - \zeta_0) + C_0 + \sum_{n=1}^{\infty} \frac{C_n}{(\zeta - \zeta_0)^n}. \qquad (2\cdot2, 6)$$

The series converges for sufficiently large ζ, for any given point ζ_0 in the ζ-plane.

The derivative of $f(\zeta)$ is given by

$$\frac{dz}{d\zeta} = f'(\zeta) = C_{-1} - \frac{C_1}{(\zeta - \zeta_0)^2} - \frac{2C_2}{(\zeta - \zeta_0)^3} + \text{higher negative powers of } (\zeta - \zeta_0).$$
$$(2\cdot2, 7)$$

We shall require presently expressions for the coefficients of $d\zeta/dz$, as a Laurent series of $(\zeta - \zeta_0)$ in terms of the coefficients of $dz/d\zeta$. Let

$$\frac{d\zeta}{dz} = C_0' + \frac{C_1'}{\zeta - \zeta_0} + \frac{C_2'}{(\zeta - \zeta_0)^2} + \text{higher negative powers of } (\zeta - \zeta_0).$$

Multiplying $dz/d\zeta$ by $d\zeta/dz$ we obtain

$$1 = \frac{dz}{d\zeta}\frac{d\zeta}{dz} = \left(C_{-1} - \frac{C_1}{(\zeta - \zeta_0)^2} - \frac{2C_2}{(\zeta - \zeta_0)^3} + \dots\right)\left(C_0' + \frac{C_1'}{\zeta - \zeta_0} + \frac{C_2'}{(\zeta - \zeta_0)^2} + \dots\right)$$

$$= C_{-1}C_0' + \frac{C_{-1}C_1'}{\zeta - \zeta_0} + \frac{C_{-1}C_2' - C_1 C_0'}{(\zeta - \zeta_0)^2} + \dots.$$

Comparing coefficients, we obtain

$$1 = C_{-1}C_0', \quad 0 = C_{-1}C_1', \quad 0 = C_{-1}C_2' - C_1 C_0',$$

and so
$$C_0' = \frac{1}{C_{-1}}, \quad C_1' = 0, \quad C_2' = \frac{C_1}{C_{-1}^2}. \qquad (2\cdot2, 8)$$

Assume now that the complex velocity $w = w(z) = w(f(\zeta))$ is given as a Laurent series in the ζ-plane,

$$w = A_0 + \frac{A_1}{\zeta - \zeta_0} + \frac{A_2}{(\zeta - \zeta_0)^2} + \dots. \qquad (2\cdot2, 9)$$

Then $(2\cdot2, 3)$ yields

$$X - iY = \tfrac{1}{2}\rho i \int_S w^2 \, dz$$

$$= \tfrac{1}{2}\rho i \int_{S'} w^2 \frac{dz}{d\zeta} \, d\zeta$$

$$= \tfrac{1}{2}\rho i \int_{S'} \left(A_0 + \frac{A_1}{\zeta - \zeta_0} + \frac{A_2}{(\zeta - \zeta_0)^2} + \dots\right)^2 \left(C_{-1} - \frac{C_1}{(\zeta - \zeta_0)^2} - \dots\right) d\zeta$$

$$= \tfrac{1}{2}\rho i \int_{S'} \left(A_0^2 + \frac{2A_0 A_1}{\zeta - \zeta_0} + \frac{A_1^2 + 2A_0 A_2}{(\zeta - \zeta_0)^2} + \dots\right)\left(C_{-1} - \frac{C_1}{(\zeta - \zeta_0)^2} - \dots\right) d\zeta$$

$$= \tfrac{1}{2}\rho i \cdot 2\pi i \cdot 2 A_0 A_1 C_{-1},$$

where S' is the contour corresponding to S in the ζ-plane, and where the higher negative powers of $\zeta - \zeta_0$ have been omitted, since only the first negative power contributes to the result. Thus, finally,

$$X - iY = -2\pi\rho A_0 A_1 C_{-1}. \tag{2·2,10}$$

Similarly, (2·2, 5) becomes

$$M = -\tfrac{1}{2}\rho\mathscr{R}\int_S w^2(z - z_0)\,dz = -\tfrac{1}{2}\rho\mathscr{R}\int_{S'} w^2(z - z_0)\frac{dz}{d\zeta}d\zeta$$

$$= -\tfrac{1}{2}\rho\mathscr{R}\int_{S'} \left(A_0 + \frac{A_1}{\zeta - \zeta_0} + \frac{A_2}{(\zeta - \zeta_0)^2} + \ldots\right)^2$$

$$\times \left(C_{-1}(\zeta - \zeta_0) + C_0 - z_0 + \frac{C_1}{(\zeta - \zeta_0)} + \ldots\right)\left(C_{-1} - \frac{C_1}{(\zeta - \zeta_0)^2} - \ldots\right)d\zeta$$

$$= -\tfrac{1}{2}\rho\mathscr{R}\int_{S'} \left(A_0^2 + \frac{2A_0 A_1}{\zeta - \zeta_0} + \frac{A_1^2 + 2A_0 A_2}{(\zeta - \zeta_0)^2} + \ldots\right)$$

$$\times (C_{-1}^2(\zeta - \zeta_0) + (C_0 - z_0)C_{-1} + \ldots)\,d\zeta,$$

where only negative powers of $(\zeta - \zeta_0)$ higher than the first are omitted in the second bracket. Thus

$$M = -\tfrac{1}{2}\rho\mathscr{R}\int_{S'} \bigg(A_0^2 C_{-1}^2(\zeta - \zeta_0) + 2A_0 A_1 C_{-1}^2 + A_0^2(C_0 - z_0)C_{-1}$$

$$+ \frac{2A_0 A_1(C_0 - z_0)C_{-1} + (A_1^2 + 2A_0 A_2)C_{-1}^2}{\zeta - \zeta_0}\bigg)d\zeta$$

$$= -\tfrac{1}{2}\rho\mathscr{R}[2\pi i(2A_0 A_1(C_0 - z_0)C_{-1} + (A_1^2 + 2A_0 A_2)C_{-1}^2)],$$

or $\qquad M = \pi\rho\mathscr{I}[2A_0 A_1(C_0 - z_0)C_{-1} + (A_1^2 + 2A_0 A_2)C_{-1}^2], \tag{2·2,11}$

where $\mathscr{I}$ denotes the real coefficient of the imaginary part of a complex number, as usual.

Again, it may happen that the complex potential Π is known as a function of ζ, so that the complex velocity of the flow in the ζ-plane, τ say, is obtainable more easily than the corresponding w. We have

$$w = -\frac{\partial\Pi}{\partial z}, \quad \tau = -\frac{\partial\Pi}{\partial\zeta},$$

and so $\qquad w = -\frac{\partial\Pi}{\partial\zeta}\frac{\partial\zeta}{dz} = \tau\frac{d\zeta}{dz}.$

Let $\qquad \tau = B_0 + \frac{B_1}{\zeta - \zeta_0} + \frac{B_2}{(\zeta - \zeta_0)^2} + \ldots, \tag{2·2, 12}$

while $\qquad \frac{d\zeta}{dz} = C_0' + \frac{C_1'}{\zeta - \zeta_0} + \frac{C_2'}{(\zeta - \zeta_0)^2} + \ldots,$

where C_0', C_1', C_2' are given by (2·2, 8). Then

$$w = \tau\frac{d\zeta}{dz} = \left(B_0 + \frac{B_1}{\zeta - \zeta_0} + \frac{B_2}{(\zeta - \zeta_0)^2} + \ldots\right)\left(C_0' + \frac{C_1'}{\zeta - \zeta_0} + \frac{C_2'}{(\zeta - \zeta_0)^2} + \ldots\right)$$

$$= B_0 C_0' + \frac{B_0 C_1' + B_1 C_0'}{\zeta - \zeta_0} + \frac{B_0 C_2' + B_1 C_1' + B_2 C_0'}{(\zeta - \zeta_0)^2} + \ldots.$$

Comparison with (2·2, 9) shows that

$$A_0 = B_0 C_0' = \frac{B_0}{C_{-1}}, \quad A_1 = B_0 C_1' + B_1 C_0' = \frac{B_1}{C_{-1}},$$

$$A_2 = B_0 C_2' + B_1 C_1' + B_2 C_0' = \frac{B_0 C_{-1}}{C_{-1}^2} + \frac{B_2}{C_{-1}} = \frac{B_0 C_1 + B_2 C_{-1}}{C_{-1}^2}.$$

Substituting in (2·2, 10), we obtain

$$X - iY = -2\pi\rho \frac{B_0 B_1}{C_{-1}}, \tag{2·2, 13}$$

while the formula for the moment, (2·2, 11), now becomes

$$M = \pi\rho\mathscr{I}\left[\frac{2B_0 B_1(C_0 - z_0) + B_1^2 C_{-1} + 2B_0(B_0 C_1 + B_2 C_{-1})}{C_{-1}}\right]. \tag{2·2, 14}$$

In the particular case where the z- and ζ-planes coincide,

$$z = \zeta = (\zeta - \zeta_0) + \zeta_0,$$

we have
$$C_{-1} = 1, \quad C_0 = \zeta_0, \quad C_1 = C_2 = \dots = 0,$$

so that (2·2, 10) and (2·2, 11) become

$$X - iY = -2\pi\rho A_0 A_1, \tag{2·2, 15}$$

and
$$M = \pi\rho\mathscr{I}[2A_0 A_1(\zeta - z_0) + (A_1^2 + 2A_0 A_2)], \tag{2·2, 16}$$

respectively, results which can also be obtained directly.

In general, we have
$$\lim_{z \to \infty} w = \lim_{\zeta \to \infty} w = A_0 = \frac{B_0}{C_{-1}},$$

so that $A_0 = B_0/C_{-1}$ is the complex free-stream velocity in the z-plane. Similarly, $\lim_{\zeta \to \infty} \tau = B_0$ is the complex velocity of the corresponding flow in the ζ-plane, and in order that both shall be equal, we must choose the mapping function such that

$$\lim_{\zeta \to \infty} \frac{dz}{d\zeta} = C_{-1} = 1.$$

2·3   The law of Kutta–Joukowski

Consider the integral

$$\int_C w\,dz = \int_C (u - iv)(dx + i\,dy)$$

$$= \int_C (u\,dx + v\,dy) + i\int_S (u\,dy - v\,dx), \tag{2·3, 1}$$

taken round a contour C which encloses the aerofoil in the z-plane. It will be seen that the real part of this integral expresses the circulation round C, which will be denoted by Γ, while the imaginary part indicates the flux across C outwards. We shall assume for the remainder of this chapter that the aerofoil neither emits nor absorbs fluid, so that the flux across C vanishes. Again, using the notation of (2·2, 9) for $\zeta = z$, we have

$$\int_C w\,dz = \int_C \left(A_0 + \frac{A_1}{z-z_0} + \frac{A_2}{(z-z_0)^2} + \ldots \right) dz = 2\pi i A_1. \qquad (2\cdot3, 2)$$

Comparing (2·3, 1) and (2·3, 2) we see that $2\pi i A_1 = \Gamma$, $A_1 = -i\Gamma/2\pi$, so that A_1 is pure imaginary. Also, denoting the modulus (magnitude) of the free-stream velocity by U as before, and the angle between the positive x-axis and that velocity by α, we have for the complex free-stream velocity, $A_0 = U e^{-i\alpha}$. Substituting these values of A_0 and A_1 in (2·2, 15) we obtain

$$X - iY = -\rho U \Gamma(-i e^{-i\alpha}) = \rho U \Gamma e^{-i(\alpha - \frac{1}{2}\pi)}. \qquad (2\cdot3, 3)$$

Thus
$$X = \rho U \Gamma \sin\alpha, \quad Y = -\rho U \Gamma \cos\alpha. \qquad (2\cdot3, 4)$$

(2·3, 4) shows that the resultant aerodynamic force on the aerofoil is of magnitude $\rho U \Gamma$ and acts in a direction normal to the free stream. This is the law of Kutta–Joukowski.‡

Denoting the component of the aerodynamic force parallel to the direction of the free stream by D—drag—and the component normal to it (at an angle $+\frac{1}{2}\pi$) by L—lift—we may write this result

$$L = -\rho U \Gamma, \quad D = 0. \qquad (2\cdot3, 5)$$

Since
$$\Gamma = \int_C w\,dz = \int_C \tau \frac{d\zeta}{dz}\,dz = \int_{C'} \tau\,d\zeta,$$

we see that Γ also denotes the circulation round a corresponding contour of the corresponding velocity field in the ζ-plane, and so $\Gamma = 2\pi i B_1$, with the notation of (2·2, 12). However, in order to link Γ with the geometrical data of the aerofoil we have to study the flow near the trailing edge in more detail.

Assuming that we have mapped the region outside the aerofoil in the z-plane on the region outside a smooth curve in the ζ-plane as in §2·2, e.g. a circle. Let z_T be the affix of the trailing edge in the z-plane and ζ_T the affix of the corresponding point in the ζ-plane. Since the trailing edge is supposed to be sharp, the mapping function $f(\zeta)$ cannot be regular in the region of the trailing edge. We shall assume that $f(\zeta)$ can be expanded in the form
$$z - z_T = D_1(\zeta - \zeta_T)^{\alpha_1} + D_2(\zeta - \zeta_T)^{\alpha_2} + \ldots, \qquad (2\cdot3, 6)$$

where $\alpha_1, \alpha_2, \ldots$ are real numbers $\alpha_1 < \alpha_2 < \ldots$. Let z', z'' be two points near the trailing edge on the top and bottom surface of the aerofoil respectively, and let ζ', ζ'' be the corresponding points in the ζ-plane (Fig. 2·3, 1). Then

$$z' - z_T = D_1(\zeta' - \zeta_T)^{\alpha_1} + D_2(\zeta' - \zeta_T)^{\alpha_2} + \ldots,$$
$$z'' - z_T = D_1(\zeta'' - \zeta_T)^{\alpha_1} + D_2(\zeta'' + \zeta_T)^{\alpha_2} + \ldots,$$

‡ W. M. Kutta, 'Auftriebskräfte in strömenden Flüssigkeiten', *Illustrierte Aeronautische Mitteilungen* (July, 1902), p. 133. And N. E. Joukowski, 'Sur les Tourbillons Adjoints', *Trans. of the Physical Section of the Imperial Society of Friends of Natural Sciences, Moscow*, vol. 23, no. 2 (1906).

and so
$$\frac{z'-z_T}{z''-z_T} = \left(\frac{\zeta'-\zeta_T}{\zeta''-\zeta_T}\right)^{\alpha_1} + \cdots, \tag{2·3, 7}$$

where the dots indicate a quantity which tends to 0 as z' and z'' approach z_T indefinitely. The argument of the left-hand side of (2·3, 7) then becomes

$$\arg\frac{z'-z_T}{z''-z_T} = \arg(z'-z_T) - \arg(z''-z_T) = 2\pi - \epsilon_T,$$

where ϵ_T is the trailing edge angle, i.e. the angle between the tangents to the top and bottom surfaces of the aerofoil section at the trailing edge. On the other hand, the argument $\dfrac{\zeta'-\zeta_T}{\zeta''-\zeta_T}$ tends to π as ζ' and ζ'' approach ζ_T indefinitely, since we have assumed that the aerofoil is transformed into a smooth curve in the ζ-plane. Hence, taking arguments on both sides of (2·3, 7), we obtain

$$2\pi - \epsilon_T = \pi\alpha_1, \quad \alpha_1 = 2 - \frac{\epsilon_T}{\pi}, \quad \frac{\epsilon_T}{\pi} = 2 - \alpha_1. \tag{2·3, 8}$$

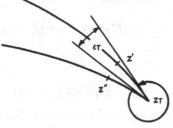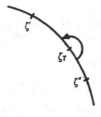

Fig. 2·3, 1

ϵ_T will normally be a small angle, certainly smaller than π, so that $\alpha_1 > 1$. Now by (2·3, 6)

$$\frac{dz}{d\zeta} = \alpha_1 D_1(\zeta-\zeta_T)^{\alpha_1-1} + \alpha_2 D_2(\zeta-\zeta_T)^{\alpha_2-1} + \cdots,$$

and since the indices of the powers of $(\zeta-\zeta_T)$ which occur in (2·3, 8) are all positive, it follows that $dz/d\zeta$ vanishes at the trailing edge; in symbols

$$\lim_{\zeta \to \zeta_T} \frac{dz}{d\zeta} = 0, \tag{2·3, 9}$$

or, more precisely,

$$\frac{dz}{d\zeta} \sim \left(2 - \frac{\epsilon_T}{\pi}\right) D_1(\zeta-\zeta_T)^{1-\epsilon_T/\pi} \quad (\zeta \to \zeta_T). \tag{2·3, 10}$$

Assume that we have transformed the region outside the aerofoil section in the z-plane into the region outside a circle S of radius a and centre ζ_0 in the ζ-plane by means of a transformation function (2·2, 6) such that $C_{-1} = 1$. Then the free-stream velocities of corresponding flows in the z- and ζ-plane coincide, $A_0 = B_0 = U e^{-i\alpha}$, say, where α is the angle between the free-stream direction and the positive x-axis as before (Fig. 2·3, 2). The general flow with circulation round the circle in the ζ-plane, with the specified free-stream velocity, is given by the complex velocity potential (equation (1·10, 21))

$$\Pi(\zeta) = - U\, e^{-i\alpha}\, (\zeta - \zeta_0) + \frac{i\Gamma}{2\pi} \log\,(\zeta - \zeta_0) - U a^2 e^{i\alpha} \frac{1}{\zeta - \zeta_0}. \qquad (2·3, 11)$$

The corresponding complex velocity in the ζ-plane is

$$\tau = -\frac{d\Pi}{d\zeta} = U\, e^{-i\alpha} - \frac{i\Gamma}{2\pi} \frac{1}{\zeta - \zeta_0} - \frac{U a^2 e^{i\alpha}}{(\zeta - \zeta_0)^2}. \qquad (2·3, 12)$$

Comparison with (2·2, 12) shows that we now have

$$B_0 = U\, e^{-i\alpha}, \quad B_1 = -\frac{i\Gamma}{2\pi}, \quad B_2 = - U a^2 e^{i\alpha}, \quad B_3 = B_4 = \ldots = 0,$$

where Γ is the circulation. The application of (2·2, 13) then leads to

$$X - iY = \rho U \Gamma\, e^{-i(\alpha - \frac{1}{2}\pi)}, \qquad (2·3, 13)$$

in agreement with (2·3, 3), while (2·2, 16) yields

$$M = \pi\rho\mathscr{I}\left[- 2U\, e^{-i\alpha} \frac{i\Gamma}{2\pi}\, (C_0 - z_0) - \frac{\Gamma^2}{4\pi^2} + 2U\, e^{-i\alpha}\, (U\, e^{-i\alpha} C_1 - U a^2 e^{i\alpha}) \right]. \qquad (2·3, 14)$$

This may also be written

$$M = - \rho U \Gamma \mathscr{I}[e^{-i(\alpha - \frac{1}{2}\pi)}\, (C_0 - z_0)] + 2\pi\rho U^2 \mathscr{I}(e^{-2i\alpha}\, C_1), \qquad (2·3, 15)$$

since the omission of real terms in the square brackets on the right-hand side of (2·3, 14) does not affect the result. The complex velocity in the z-plane is given by

$$w = \tau \frac{\partial \zeta}{\partial z} = \tau \Big/ \frac{\partial z}{\partial \zeta},$$

and since $\partial z / \partial \zeta$ vanishes at the trailing edge (see equation (2·3, 9)), while on the other hand w must remain finite at the trailing edge according to the Joukowski condition (§ 1·14), it follows that

$$\tau = 0 \quad \text{at the trailing edge.}$$

Writing
$$\zeta_T = \zeta_0 + a\, e^{-i\beta} \qquad (2·3, 16)$$

for the affix of the point corresponding to the trailing edge in the ζ-plane (see Fig. 2·3, 2) and substituting in (2·3, 12), we obtain the following equation, which may be regarded as a condition for the circulation Γ:

$$\tau(\zeta_T) = U\, e^{-i\alpha} - \frac{i\Gamma}{2\pi a}\, e^{i\beta} - U\, e^{i(\alpha + 2\beta)} = 0. \qquad (2·3, 17)$$

This yields $\Gamma = -\dfrac{2\pi a}{i}\, U(e^{i(\alpha+\beta)} - e^{-i(\alpha+\beta)}) = - 4\pi U a \sin\,(\alpha + \beta). \qquad (2·3, 18)$

Substituting this expression for the circulation in (2·3, 5) we obtain for the lift,
$$L = 4\pi\rho U^2 a \sin\,(\alpha + \beta). \qquad (2·3, 19)$$

This shows that $\alpha = -\beta$ is the angle of zero lift. $\alpha + \beta$ is called the 'effective angle of attack'.

We put
$$C_0 - z_0 = r\,e^{i\psi}, \quad C_1 = d^2\,e^{2i\gamma}. \tag{2·3, 20}$$

Then the expression for the moment (2·3, 15) becomes

$$M = 4\pi\rho U^2 ar \sin(\alpha + \beta)\,\mathscr{I}(e^{i(\frac{1}{2}\pi + \psi - \alpha)}) + 2\pi\rho U^2 d^2\,\mathscr{I}(e^{i(2\gamma - 2\alpha)})$$

or
$$M = 2\pi\rho U^2\{2ar \sin(\alpha + \beta)\cos(\psi - \alpha) + d^2 \sin 2(\gamma - \alpha)\}. \tag{2·3, 21}$$

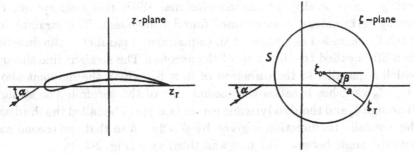

Fig. 2·3, 2

It will be seen that (2·3, 19) and (2·3, 21) express the lift and moment entirely in terms of the physical data of the flow.

The moment round z_0 for the angle of zero lift ($\alpha = -\beta$) is

$$M_0 = 2\pi\rho U^2 d^2 \sin 2(\gamma + \beta). \tag{2·3, 22}$$

Taking into account that

$$\sin 2(\gamma - \alpha) - \sin 2(\gamma + \beta) = -2\cos(2\gamma + \beta - \alpha)\sin(\alpha + \beta),$$

we then have

$$M - M_0 = 4\pi\rho U^2 \sin(\alpha + \beta)\,[ar\cos(\psi - \alpha) - d^2 \cos(2\gamma + \beta - \alpha)].$$

It follows that, if $r = d^2/a$, $\psi = 2\gamma + \beta$, i.e. if

$$z_0 = C_0 - \frac{d^2}{a}e^{i(2\gamma + \beta)} = C_0 - \frac{C_1}{\zeta_T - \zeta_0}, \tag{2·3, 23}$$

then $M - M_0 = 0$, so that the moment about z_0 is independent of incidence. z_0 is called the 'aerodynamic centre' of the aerofoil. We may say that the lift acts through this point, in the sense that the resultant force due to the pressure which is additional to the pressure at zero lift acts through it. If the position of the aerodynamic centre, and the magnitude and direction of the lift are known, then the moment about any other point is found by simple statics. However, it can also be calculated directly by the evaluation of (2·3, 21). Thus for $z_0 = C_0$, $r = 0$, so that (2·3, 21) becomes

$$M = 2\pi\rho U^2 d^2 \sin 2(\gamma - \alpha). \tag{2·3, 24}$$

Thus, the moment about $z_0 = C_0$ vanishes when $\alpha = \gamma$. It should be noted that, since

$$\gamma = \tfrac{1}{2}\arg C_1 = \tfrac{1}{2}\arg \frac{1}{2\pi i}\int_S f(\zeta)\,d\zeta,$$

it follows that C_1 and γ are in fact independent of the particular point about which $f(\zeta)$ is expanded into a Laurent series. Again, it will be seen that a modification of the value of C_0 results in pure translation of the aerofoil in the physical z-plane. We may therefore assume $C_0 = \zeta_0$ without affecting the generality of the conclusions. With this assumption, the following terminology is sometimes found convenient. The straight line through ζ_0 (regarded as a point of the z-plane) and parallel to the direction of zero lift is called the first axis of the aerofoil. The straight line through ζ_0 which is parallel to the direction of flow for which the moment about $z_0 = C_0 = \zeta_0$ vanishes is called the second axis of the aerofoil. The straight line through ζ_0 and the aerodynamic centre (a.c.) may be called the third axis of the aerofoil. Its direction is given by $\psi = 2\gamma + \beta$ so that the second axis bisects the angle between the first and third axes (Fig. 2·3, 3).

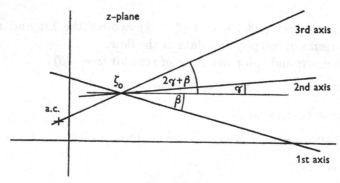

Fig. 2·3, 3

To calculate the pressure distribution round the aerofoil, we require an expression for the velocity τ on the circle S. Substituting the value of Γ as given by (2·3, 18), in equation (2·3, 12) we obtain

$$\tau = U\left\{e^{-i\alpha} + \frac{2ia\sin(\alpha+\beta)}{\zeta-\zeta_0} - \frac{a^2 e^{i\alpha}}{(\zeta-\zeta_0)^2}\right\}. \tag{2·3, 25}$$

Writing $\zeta = \zeta_0 + a\,e^{i\theta}$ for points on the circle S,

$$\tau = U e^{-i\alpha} + 2iU\sin(\alpha+\beta)e^{-i\theta} - U e^{i(\alpha-2\theta)}$$

$$= U e^{-i\theta}(e^{-i(\alpha-\theta)} - e^{i(\alpha-\theta)} + 2i\sin(\alpha+\beta))$$

$$= 2Ui\,e^{-i\theta}(\sin(\alpha+\beta) - \sin(\alpha-\theta)),$$

or $$\tau = 4Ui\,e^{-i\theta}\cos\left(a+\frac{\beta-\theta}{2}\right)\sin\frac{\beta+\theta}{2}, \tag{2·3, 26}$$

Hence $$q=|w|=|\tau|\left|\frac{d\zeta}{dz}\right|=4U\left|\cos\left(\alpha+\frac{\beta-\theta}{2}\right)\sin\frac{\beta+\theta}{2}\right|\left|\frac{d\zeta}{dz}\right| \tag{2·3, 27}$$

for points on the surface of the aerofoil.

The pressure p is then found from Bernoulli's equation which, for the present case, yields $$p+\tfrac{1}{2}\rho q^2 = p_0 + \tfrac{1}{2}\rho U^2,$$

where p_0 is the free-stream pressure, or

$$p = p_0 + \tfrac{1}{2}\rho(U^2 - q^2). \tag{2·3, 28}$$

By the choice of the circulation, $\tau_T = \tau(\zeta_T) = 0$, so that τ at a point ζ near ζ_T is given by $(\zeta - \zeta_T)\left(\dfrac{d\tau}{d\zeta}\right)_T$, or

$$\tau = U(\zeta - \zeta_T)\left\{-\frac{2ia\sin(\alpha+\beta)}{(\zeta_T - \zeta_0)^2} + \frac{2a^2 e^{i\alpha}}{(\zeta_T - \zeta_0)^3}\right\}$$
$$= \frac{U}{a}(\zeta - \zeta_T)\,e^{2i\beta}\{2e^{i(\alpha+\beta)} - 2i\sin(\alpha+\beta)\},$$

$$\tau = \frac{2U}{a}e^{2i\beta}\cos(\alpha+\beta)(\zeta - \zeta_T), \tag{2·3, 29}$$

approximately. It then follows from (2·3, 10) that

$$q=|\tau|\left|\frac{d\zeta}{dz}\right| = \frac{2\pi U\,|\cos(\alpha+\beta)|}{(2\pi - \epsilon_T)\,|D_1|\,a}\,|\zeta - \zeta_T|^{\epsilon_T/\pi}, \tag{2·3, 30}$$

so that the velocity vanishes at the trailing edge of the aerofoil (i.e. z_T is a stagnation point) except when the trailing edge is a cusp. In that case the magnitude of the velocity at the trailing edge is

$$q_T = \frac{U}{|D_1|\,a}\,|\cos(\alpha+\beta)|. \tag{2·3, 31}$$

2·4　Joukowski aerofoils

The shape of the aerofoil is defined by the mapping function $z = f(\zeta)$ and by the centre ζ_0 and radius a of the circle into which the aerofoil is transformed. The determination of the appropriate mapping function for a given aerofoil is relatively difficult. Accordingly, we shall first study a family of aerofoils which is given by a mapping function of a very simple type, viz.

$$z = \zeta + \frac{d^2}{\zeta} \quad (d > 0). \tag{2·4, 1}$$

The family of aerofoils which are obtained from (2·4, 1) for varying ζ_0 and a was in fact the first family of aerofoils to be studied theoretically, and it is known by the name of its discoverer, Joukowski.‡ Since the coefficient of the first negative power of a Laurent expansion of $f(\zeta)$ is the same for all expansions which converge for sufficiently large ζ, it follows that the definition of d in (2·4, 1) agrees with its definition in (2·3, 20) for $\gamma = 0$.

The irregularities of (2·4, 1) are given by

$$\frac{dz}{d\zeta} = 1 - \frac{d^2}{\zeta^2} = 0 \quad \text{or} \quad \infty,$$

i.e. the transformation is singular for

$$\zeta = 0, \ d, \ -d. \tag{2·4, 2}$$

Thus the circle corresponding to the aerofoil in the ζ-plane must pass through $\zeta_T = d$, which is the point corresponding to the trailing edge, and must include $\zeta = -d$, and hence $\zeta = 0$, in its interior, or in the limit, on its circumference. The affix of the trailing edge in the physical z-plane is

$$z_T = \zeta_T + \frac{d^2}{\zeta_T} = 2d. \tag{2·4, 3}$$

Now

$$z - 2d = \zeta + \frac{d^2}{\zeta} - 2d = \frac{(\zeta - d)^2}{d\left(1 + \frac{\zeta - d}{d}\right)} = \frac{1}{d}(\zeta - d)^2 \left\{ 1 - \frac{1}{d}(\zeta - d) + \frac{1}{d^2}(\zeta - d)^2 - \dots \right\},$$

and so the expansion of z about the trailing edge corresponding to (2·3, 6) is

$$z - z_T = \frac{1}{d}(\zeta - \zeta_T)^2 - \frac{1}{d^2}(\zeta - \zeta_T)^3 + \frac{1}{d^3}(\zeta - \zeta_T)^4 - \dots. \tag{2·4, 4}$$

Thus $\alpha_1 = 2$ and $\epsilon_T = 0$ by (2·3, 8); the trailing edge angle vanishes and the trailing edge is a cusp.

The aerofoil is now determined completely by the position of the centre of the circle S in the ζ-plane. We take first $\zeta_0 = 0$, so that the radius of the circle equals $a = d$, while $\beta = 0$ (see (2·3, 16)). Writing $\zeta = d\,e^{i\theta}$ for points on the circle S, we see that the corresponding points on the aerofoil are given by

$$z = d\,e^{i\theta} + \frac{d^2}{d\,e^{i\theta}} = d(e^{i\theta} + e^{-i\theta}) = 2d \cos\theta. \tag{2·4, 5}$$

Thus z is real and varies from $z = 2d$ at the trailing edge to $z = -2d$ at the leading edge, and back again, as θ varies from 0 to 2π; the aerofoil is a flat plate. The chord length, i.e. the distance from the leading edge to the trailing edge, equals $c = 4d = 4a$. The formula for the lift (2·3, 19) therefore becomes, in terms of c,

$$L = \pi\rho U^2 c \sin\alpha = 2\pi \sin\alpha \tfrac{1}{2}\rho U^2 c. \tag{2·4, 6}$$

Now the local lift coefficient C_L, is defined by

$$L = C_L \cdot \tfrac{1}{2}\rho U^2 c, \tag{2·4, 7}$$

‡ N. E. Joukowski, 'Über die Konturen der Tragflächen der Drachenflieger', *Z. Flugtech.* vol. 22, no. 2 (1910).

and so
$$C_L = 2\pi \sin \alpha, \quad \frac{dC_L}{d\alpha} = 2\pi \cos \alpha, \qquad (2\cdot4, 8)$$

or
$$\frac{dC_L}{d\alpha} = 2\pi, \quad \text{approximately.} \qquad (2\cdot4, 9)$$

The position of the aerodynamic centre, determined by substituting $C_0 = 0$, $\beta = \gamma = 0$, $a = d$ in $(2\cdot3, 23)$, is given by

$$z_0 = -d = -\tfrac{1}{4}c. \qquad (2\cdot4, 10)$$

Thus, the aerodynamic centre is situated at the quarter-chord. Since the aerofoil is symmetrical, the moment at zero lift vanishes, and this can also be seen directly from $(2\cdot3, 22)$. It follows that, for the case of a flat plate, the aerodynamic centre coincides with the centre of pressure. The moment about any other point can be found by using some statics or by substituting in the relevant formula of § 2·3. For example, the moment about the mid-chord ($z_0 = 0$) is obtained from $(2\cdot3, 24)$ as

$$M = -\tfrac{1}{8}\pi\rho U^2 c^2 \sin 2\alpha = -\frac{\pi \sin 2\alpha}{4} \tfrac{1}{2}\rho U^2 c^2. \qquad (2\cdot4, 11)$$

The corresponding local moment coefficient, defined by

$$M = C_M . \tfrac{1}{2}\rho U^2 c^2, \qquad (2\cdot4, 12)$$

is therefore given by

$$C_M = -\frac{\pi \sin 2\alpha}{4}, \quad \frac{dC_M}{d\alpha} = -\frac{\pi}{2} \cos 2\alpha,$$

or
$$\frac{dC_M}{d\alpha} = -\frac{\pi}{2} \quad \text{approximately.} \qquad (2\cdot4, 13)$$

To find the velocity distribution at the aerofoil, we calculate $dz/d\zeta$ for $\zeta = d\, e^{i\theta}$. This is

$$\frac{dz}{d\zeta} = 1 - \frac{d^2}{\zeta^2} = 1 - e^{-2i\theta} = 2i\, e^{-i\theta} \sin \theta, \qquad (2\cdot4, 14)$$

so that
$$\left| \frac{d\zeta}{dz} \right| = \frac{1}{2\,|\sin \theta|}.$$

Substituting this value of $|\, d\zeta/dz\,|$ in $(2\cdot3, 27)$ we obtain

$$q = U \left| \frac{\cos (\alpha - \tfrac{1}{2}\theta)}{\cos \tfrac{1}{2}\theta} \right|. \qquad (2\cdot4, 15)$$

This shows that the velocity becomes infinite at the leading edge for non-vanishing (say positive) α. There is a stagnation point on the lower surface of the aerofoil near the leading edge, for $\tfrac{1}{2}\theta - \alpha = \tfrac{1}{2}\pi$, $\theta = \pi + 2\alpha$. The velocity at the trailing edge does not vanish but becomes equal to $q = U \cos \alpha$ in agreement with $(2\cdot3, 31)$.

To calculate the longitudinal induced velocity $u' = u - U \cos \alpha$, we go back to $(2\cdot3, 26)$ which, combined with $(2\cdot4, 14)$, yields

$$u = \mathscr{R}(w) = \mathscr{R}\left(\tau \left/ \frac{dz}{d\zeta} \right. \right) = \mathscr{R}\left(2U \frac{\cos (\alpha - \tfrac{1}{2}\theta) \sin \tfrac{1}{2}\theta}{\sin \theta} \right) = U \frac{\cos (\alpha - \tfrac{1}{2}\theta)}{\cos \tfrac{1}{2}\theta}.$$

Hence

$$u' = u - U\cos\alpha = U\left(\frac{\cos(\alpha - \tfrac{1}{2}\theta)}{\cos\tfrac{1}{2}\theta} - \cos\alpha\right) = \frac{U}{\cos\tfrac{1}{2}\theta}\{\cos(\alpha - \tfrac{1}{2}\theta) - \cos\alpha\cos\tfrac{1}{2}\theta\},$$

or
$$u' = -U\sin\alpha\tan\tfrac{1}{2}\theta. \tag{2·4, 16}$$

We observe that the integration of the pressure over the aerofoil yields a resultant force normal to the aerofoil, whereas according to the general theory (§ 2·2) the resultant force acts in a direction normal to the main stream. The explanation of this apparent discrepancy is that the locally infinite pressure at the leading edge contributes a finite force, the so-called 'suction force', which together with the pressure integral yields a total resultant force in agreement with (2·3, 4). This suction force can be calculated directly by considering the flux of momentum across a small surface (contour) surrounding the leading edge (see § 2·8 below).

Formulae (2·4, 9) and (2·4, 10), which determine the lift curve slope and the position of the aerodynamic centre for a flat plate aerofoil, apply, roughly speaking, to all other aerofoils, and may be used as standards of reference against which the effect of various parameters of the aerofoil geometry may be assessed. We shall consider first the effect of camber or curvature of the aerofoil.

While the definition of the leading edge is obvious in the case of a sharp-nosed aerofoil, it is to some extent a matter of convention in the general case. It is possible to define the leading edge as the point of maximum curvature at the nose of the aerofoil. The chord is defined as the straight segment from leading to trailing edge. The mean camber line is defined as the locus of the midpoints of the segments which are cut out by the aerofoil on the straight lines normal to the chord. Thus if the chord is parallel to the x-axis and the equations of the upper and lower surface of the aerofoil are given by $y = y_u(x)$ and $y = y_l(x)$ respectively, the equation of the mean camber line is

$$y = y_c(x) = \tfrac{1}{2}\{y_u(x) + y_l(x)\}. \tag{2·4, 17}$$

We now take ζ_0 on the positive imaginary axis of the ζ-plane (Fig. 2·4, 1). Then
$$a = d\sec\beta, \tag{2·4, 18}$$

since the circle must pass through the singular point $\zeta_T = d$. We write $\zeta = r e^{i\theta}$, so that, taking real and imaginary parts in (2·4, 1),

$$x = \left(r + \frac{d^2}{r}\right)\cos\theta, \quad y = \left(r - \frac{d^2}{r}\right)\sin\theta, \tag{2·4, 19}$$

or
$$\frac{x^2}{\cos^2\theta} - \frac{y^2}{\sin^2\theta} = r^2 + 2d^2 + \frac{d^4}{r^2} - \left(r^2 - 2d^2 + \frac{d^4}{r^2}\right) = 4d^2. \tag{2·4, 20}$$

But, from Fig. (2·4, 1), we see that

$$a^2 = d^2\sec^2\beta = r^2 + d^2\tan^2\beta - 2rd\tan\beta\sin\theta.$$

Now $$\sec^2 \beta - \tan^2 \beta = 1,$$

and so $$r^2 - d^2 = 2rd \tan \beta \sin \theta,$$

or $$y = 2d \tan \beta \sin^2 \theta. \qquad (2·4, 21)$$

$(2·4, 21)$ shows that y is always non-negative. Also

$$\sin^2 \theta = \frac{y}{2d \tan \beta}, \quad \cos^2 \theta = \frac{2d \tan \beta - y}{2d \tan \beta}.$$

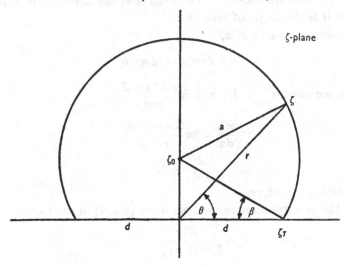

Fig. 2·4, 1

Substituting in $(2·4, 20)$, we obtain

$$\frac{x^2 2d \tan \beta}{2d \tan \beta - y} - y 2d \tan \beta = 4d^2,$$

or $$x^2 \tan \beta = (2d + y \tan \beta)(2d \tan \beta - y),$$

$$x^2 + y^2 - 2d \frac{\tan^2 \beta - 1}{\tan \beta} y = 4d^2.$$

Now $\dfrac{\tan^2 \beta - 1}{\tan \beta} = -2 \cot 2\beta$, and so the equation of the aerofoil is

$$x^2 + y^2 + 4d \cot 2\beta y - 4d^2 = 0, \qquad (2·4, 22)$$

or $$x^2 + (y + 2d \cot 2\beta)^2 = 4d^2 \operatorname{cosec}^2 2\beta \quad (y \geqslant 0). \qquad (2·4, 23)$$

The aerofoil is a circular arc, part of a circle whose centre is on the y-axis, at $y = -2d \cot 2\beta$ and whose radius is $2d \operatorname{cosec} 2\beta$. It will be seen from $(2·4, 22)$ that its leading and trailing edges are at $z = -2d$ and $z = 2d$ respectively, and from $(2·4, 23)$, that its maximum ordinate, obtained for a point on the y-axis, is

$$y = 2d(\operatorname{cosec} 2\beta - \cot 2\beta) = 2d \frac{1 - \cos 2\beta}{\sin 2\beta} = 2d \tan \beta. \qquad (2·4, 24)$$

The aerofoil coincides with its mean camber line and its chord length equals $c = 4d$. If we define the camber χ as the ratio of the maximum normal distance of the mean camber line from the chord, and of the chord length, then for the present case

$$\chi = \frac{2d \tan \beta}{4d} = \tfrac{1}{2} \tan \beta \doteqdot \tfrac{1}{2} \beta. \qquad (2\cdot4, 25)$$

Thus the interpretation of β in terms of the aerofoil geometry is that it equals about twice the camber. We recall that the aerodynamic definition of β is that it is the angle of zero lift.

Substituting the value of a,

$$a = d \sec \beta = \tfrac{1}{4} c \sec \beta,$$

in $(2\cdot3, 19)$ we obtain

$$L = \pi \rho U^2 c \frac{\sin (\alpha + \beta)}{\cos \beta}, \qquad (2\cdot4, 26)$$

so that

$$\frac{dC_L}{d\alpha} = 2\pi \frac{\cos (\alpha + \beta)}{\cos \beta},$$

or

$$\frac{dC_L}{d\alpha} = 2\pi, \qquad (2\cdot4, 27)$$

approximately, as before.

To find the aerodynamic centre, we write $(2\cdot4, 1)$ in the form

$$z = (\zeta - \zeta_0) + \zeta_0 + \frac{d^2}{\zeta_0 + (\zeta - \zeta_0)}$$

$$= (\zeta - \zeta_0) + \zeta_0 + \frac{d^2}{\zeta - \zeta_0} \left(1 - \frac{\zeta_0}{\zeta - \zeta_0} + \left(\frac{\zeta_0}{\zeta - \zeta_0} \right)^2 - \ldots \right)$$

$$= (\zeta - \zeta_0) + \zeta_0 + \frac{d^2}{\zeta - \zeta_0} - \frac{d^2 \zeta_0}{(\zeta - \zeta_0)^2} + \frac{d^2 \zeta_0^2}{(\zeta - \zeta_0)^3} - \ldots,$$

so that in the notation of $(2\cdot2, 6)$

$$C_0 = \zeta_0, \qquad (2\cdot4, 28)$$

or, in the present case, $C_0 = id \tan \beta$. By $(2\cdot3, 23)$ the position of the aerodynamic centre is then given by

$$z_0 = id \tan \beta - d \cos \beta \, e^{i\beta}$$

$$= \tfrac{1}{4} c (- \cos^2 \beta + i \tan \beta \sin^2 \beta). \qquad (2\cdot4, 29)$$

Thus, except for terms of the second order of smallness, the aerodynamic centre is still located at the quarter-chord.

The part of the moment which is independent of incidence, i.e. the moment at zero lift, is, by $(2\cdot3, 22)$,

$$M_0 = \tfrac{1}{8} \pi \rho U^2 c^2 \sin 2\beta \doteqdot \tfrac{1}{2} \pi \rho U^2 c^2 \chi, \qquad (2\cdot4, 30)$$

so that the corresponding moment coefficient is

$$C_{M_0} = \pi \chi. \qquad (2\cdot4, 31)$$

Next we consider symmetrical Joukowski aerofoils of finite thickness. It should be remarked, however, that while the overall effect of camber on the aerodynamic characteristics of an aerofoil is predicted correctly by the theory given above, the same does not in general apply to the influence of thickness. This is due to boundary-layer effects which are neglected in the present theory. Nevertheless, the pressure distribution in the forward region of the aerofoil will still in general be predicted correctly by potential theory.

We assume that the centre of the circle S is at $\zeta_0 = -d\epsilon$, where ϵ is a small positive quantity. Then the radius of S is $a = (1+\epsilon)d$, and the leading edge is naturally defined as the point z_L corresponding to $\zeta_L = -(1+2\epsilon)d$, i.e. $\theta = \pi$. Thus

$$z_L = -d(1+2\epsilon) - \frac{d^2}{d(1+2\epsilon)} \doteq -d(1+2\epsilon) - d(1-2\epsilon+4\epsilon^2) \doteq -2d(1+2\epsilon^2).$$

$$(2 \cdot 4, 32)$$

It follows that the chord length of the aerofoil equals

$$c = |z_T - z_L| = |2d + 2d(1+2\epsilon^2)| = 4d,$$

except for terms of second or higher order of smallness.

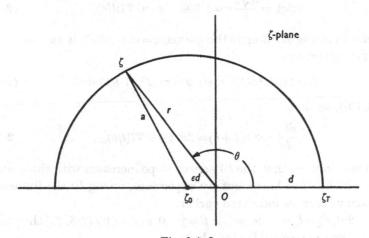

Fig. 2·4, 2

To find the shape of the aerofoil, we put $\zeta = r e^{i\theta}$, as before. From the triangle $O\zeta\zeta_0$ in Fig. 2·4, 2, we then obtain the relation

$$a^2 = d^2(1+\epsilon^2) = r^2 + d^2\epsilon^2 + 2rd\epsilon \cos\theta,$$

or, modifying a term of the order of smallness of ϵ^2,

$$d^2(1+\epsilon)^2 \doteq r^2 + 2rd\epsilon \cos\theta + d^2\epsilon^2 \cos^2\theta = (r + d\epsilon \cos\theta)^2,$$

$$r = d(1 + \epsilon(1 - \cos\theta)).$$

Hence

$$x = \left(r + \frac{d^2}{r}\right)\cos\theta = \left[d(1+\epsilon(1-\cos\theta)) + \frac{d}{1+\epsilon(1-\cos\theta)}\right]\cos\theta \doteqdot 2d\cos\theta,$$

$$y = \left(r - \frac{d^2}{r}\right)\sin\theta = 2d\epsilon(1-\cos\theta)\sin\theta = d\epsilon(2\sin\theta - \sin 2\theta).$$

$$(2\cdot4,33)$$

Now $dy/d\theta = 2d\epsilon(\cos\theta - \cos 2\theta)$, so that the condition for y to attain a maximum is

$$\cos\theta - \cos 2\theta = 0, \quad \theta = \frac{2\pi}{3} = 120°. \qquad (2\cdot4,34)$$

The corresponding value of y is

$$y_0 = \frac{3\sqrt{3}}{2}d\epsilon, \qquad (2\cdot4,35)$$

and it is easy to confirm that this is indeed a maximum, and the only one. Thus, the maximum thickness of the aerofoil is $2y_0$, and the 'maximum thickness-chord ratio' is

$$(t/c)_0 = \frac{2y_0}{c}$$

or $$(t/c)_0 = \frac{3\sqrt{3}}{4}\epsilon = 1\cdot30\epsilon, \quad \epsilon = 0\cdot77(t/c)_0. \qquad (2\cdot4,36)$$

The thickness of the aerofoil at the y-axis, $x=0$, $\theta = 90°$, is $2y = 4d\epsilon$.

The lift is given by

$$L = 4\pi\rho U^2 d(1+\epsilon)\sin\alpha = \pi\rho U^2 c(1+\epsilon)\sin\alpha \qquad (2\cdot4,37)$$

(see (2·3, 19)), so that

$$\frac{dC_L}{d\alpha} \doteqdot 2\pi(1+\epsilon) = 2\pi(1+0\cdot77(t/c)_0). \qquad (2\cdot4,38)$$

This formula implies that the lift-curve slope increases with thickness, an assertion which is not borne out by experience, owing to the influence of the boundary layer, as indicated earlier.

By (2·4, 28), $C_0 = \zeta_0 = -d\epsilon$, while $\beta = \gamma = 0$, and so by (2·3, 23) the position of the aerodynamic centre is given by

$$z_0 = -d\epsilon - \frac{d^2}{d(1+\epsilon)} \doteqdot -d\epsilon - d(1-\epsilon+\epsilon^2 - \ldots) \doteqdot -d(1+\epsilon^2).$$

Thus z_0 is at a distance $d(1+3\epsilon^2)$ aft of the leading edge (compare (2·4, 32)); the aerodynamic centre is still at the quarter chord, except for terms of second or higher order of smallness in the maximum thickness-chord ratio.

Detailed calculations of the pressure distributions round a number of Joukowski aerofoils will be found in papers by W. G. A. Perring‡ and

‡ W. G. A. Perring, 'The theoretical pressure distribution around Joukowski aerofoils', *Rep. Mem. Aero. Res. Comm., Lond.,* no. 1106 (1927).

O. Blumenthal.‡ Two examples taken from the latter work are shown in Figs. 2·4, 3 and 2·4, 4, in which the pressure coefficients $\frac{p-p_0}{\frac{1}{2}\rho U^2}$, where p_0 is the free-stream pressure, are plotted against the chordwise coordinate. Fig. 2·4, 3 illustrates the effect of camber in increasing the lift produced by the suction forces on the upper surface as compared with a flat plate at the same incidence, whilst Fig. 2·4, 4 shows how the large suction peak present for the thin aerofoil and the flat plate can be avoided by using a thick section with small curvature leading edge.

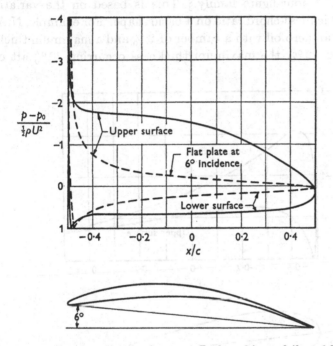

Fig. 2·4, 3. Pressure distribution on a Joukowski aerofoil on 6% thickness and 10% camber at 6° incidence. (*Courtesy N.A.C.A.*)

Generalizations of the family of Joukowski aerofoils have been obtained in various ways. Thus, instead of (2·4, 1) one may consider transformations of the more general form§

$$z = \zeta + \frac{a_1}{\zeta} + \dots + \frac{a_n}{\zeta^n}. \tag{2·4, 39}$$

All aerofoils belonging to this family have cusps at the trailing edge. To

‡ O. Blumenthal, 'Pressure distribution on Joukowski wings', *Tech. Memor. Nat. Adv. Comm. Aero., Wash.*, no. 336 (1925). (Translated from *Z. Flugtech.* May 1913.)

§ R. von Mises, 'Zur Theorie des Tragflächenauftriebs', *Z. Flugtech.* vol. 11 (1920), p. 68.

obtain aerofoils with any desired trailing edge angle, v. Kármán and Trefftz‡ chose the transformation

$$\frac{z-2d}{z+2d} = \left(\frac{\zeta-d}{\zeta+d}\right)^{\alpha_1}.$$

(2·4, 40)

The index α_1 is connected with the trailing edge angle as in (2·3, 8), $\alpha_1 = 2 - \epsilon_T/\pi$, and the Joukowski transformation is obtained for $\alpha_1 = 2$.

Other families of aerofoils have been evolved mainly through considerations of aerofoil geometry, and their characteristics established by systematic wind-tunnel testing. The classical example of such a set is provided by the N.A.C.A four-figure family.§ This is based on the variation of camber and thickness-chord ratio on a basic shape. For example, N.A.C.A. 2312 denotes an aerofoil with a camber of 2 % and a maximum thickness-chord ratio of 12 %, the maximum thickness occurring 30 % aft of the leading edge.

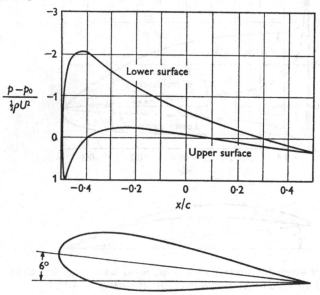

Fig. 2·4, 4. Pressure distribution on a symmetric Joukowski aerofoil of 22 % thickness at 6° incidence.

All the aerofoils mentioned in this section are now mainly of historical interest. Instead of describing their properties in detail we turn to the more general problems of calculating the pressure distribution round aerofoils of given arbitrary shape, on the one hand, and of designing aerofoils for specified pressure distributions on the other.

‡ Th. von Kármán and E. Trefftz, 'Potentialströmung um gegebene Tragflächenquerschnitte', *Z. Flugtech.* vol. 9 (1918), p. 111.

§ See E. N. Jacobs, N. Eastman, K. E. Ward and R. M. Pinkerton, 'The characteristics of 78 related airfoil sections from tests in the variable-density wind tunnel', *N.A.C.A. Rep.* no. 460 (1933).

2·5   Conjugate functions

In this section we shall derive some trigonometrical formulae which are of wide application in various branches of aerofoil theory. We shall disregard questions of convergence and refer the reader to standard textbooks‡ for a rigorous treatment.

Let $T(\theta)$ be an arbitrary trigonometrical series with real coefficients,

$$T(\theta) = \tfrac{1}{2}a_0 + a_1 \cos\theta + b_1 \sin\theta + a_2 \cos 2\theta + b_2 \sin 2\theta + \dots \qquad (2\text{·}5, 1)$$

The series $\quad \overline{T}(\theta) = b_1 \cos\theta - a_1 \sin\theta + b_2 \cos 2\theta - a_2 \sin 2\theta + \dots \qquad (2\text{·}5, 2)$

is called the conjugate series of $T(\theta)$. If $T(\theta)$ is the (convergent) Fourier series of a function $f(\theta)$, then $\overline{T}(\theta)$ converges subject to certain conditions on $f(\theta)$, and represents a function $\bar{f}(\theta)$ which is called the conjugate function of $f(\theta)$.

Let

$$g(\zeta) = C_0 + \frac{C_1}{\zeta} + \frac{C_2}{\zeta^2} + \dots \qquad (2\text{·}5, 3)$$

be an analytic function of ζ which is regular at infinity, and such that C_0 is real, $C_0 = \tfrac{1}{2}a_0$, say, while $C_1 = a_1 + ib_1$, $C_2 = a_2 + ib_2$, ... Writing $\zeta = r\,e^{i\theta}$, we then have

$$g(\zeta) = \frac{a_0}{2} + \frac{1}{r}(a_1 + ib_1)(\cos\theta - i\sin\theta) + \frac{1}{r^2}(a_2 + ib_2)(\cos 2\theta - i\sin 2\theta) + \dots$$

$$= \frac{a_0}{2} + \frac{a_1}{r}\cos\theta + \frac{b_1}{r}\sin\theta + \frac{a_2}{r^2}\cos 2\theta + \frac{b_2}{r^2}\sin 2\theta + \dots$$

$$+ i\left(\frac{b_1}{r}\cos\theta - \frac{a_1}{r}\sin\theta + \frac{b_2}{r^2}\cos 2\theta - \frac{a_2}{r^2}\sin 2\theta + \dots\right). \qquad (2\text{·}5, 4)$$

(2·5, 4) shows that for given r, the imaginary part of the expansion for $g(\zeta)$ is the conjugate series of the real part. Both series converge if r is greater than the radius of convergence of the expansion. On the circle of convergence they converge or diverge as the case may be.

To find the conjugate function $\bar{f}(\theta)$ of a given $f(\theta)$ we may expand $f(\theta)$ into a Fourier series, write down the conjugate series, and then sum it. However, it is convenient to have a compact formula which expresses $\bar{f}(\theta)$ directly in terms of $f(\theta)$. Such a formula is provided by 'Poisson's integral':

$$\bar{f}(\theta_0) = \frac{1}{2\pi}\int_0^{2\pi} f(\theta)\cos\frac{\theta - \theta_0}{2}\,d\theta. \qquad (2\text{·}5, 5)$$

The integral on the right-hand side is singular, and it is understood that Cauchy's principal value is to be taken in all cases, i.e. the integral $\displaystyle\int_0^{2\pi}$ is to be interpreted as $\qquad \displaystyle\int_0^{2\pi} = \lim_{\epsilon \to 0}\left(\int_0^{\theta_0 - \epsilon} + \int_{\theta_0 + \epsilon}^{2\pi}\right). \qquad (2\text{·}5, 6)$

‡ For example, G. H. Hardy and W. W. Rogosinski, *Fourier Series* (Cambridge University Press, 1944).

In order to derive (2·5, 5), we may assume that $f(\theta)$ is periodic with period 2π, $f(\theta+2\pi)=f(\theta)$. Substituting $t=\theta-\theta_0$ in the integral of (2·5, 5), we obtain

$$\int_0^{2\pi} f(\theta)\cot\frac{\theta-\theta_0}{2}\,d\theta = \int_0^{2\pi} f(\theta_0+t)\cot\tfrac{1}{2}t\,dt$$

$$= \int_0^{\pi} f(\theta_0+t)\cot\tfrac{1}{2}t\,dt + \int_\pi^{2\pi} f(\theta_0+t)\cot\tfrac{1}{2}t\,dt$$

$$= \int_0^{\pi} f(\theta_0+t)\cot\tfrac{1}{2}t\,dt - \int_\pi^0 f(\theta_0+2\pi-\tau)\cot\tfrac{1}{2}(2\pi-\tau)\,d\tau,$$

where we have put $\tau=2\pi-t$ in the second integral. Now

$$f(\theta_0+2\pi-\tau)=f(\theta_0-\tau) \quad \text{and} \quad \cot\tfrac{1}{2}(2\pi-\tau)=-\cot\tfrac{1}{2}\tau$$

and so $$\int_0^{2\pi} f(\theta)\cot\frac{\theta-\theta_0}{2}\,d\theta = \int_0^{\pi} (f(\theta_0+t)-f(\theta_0-t))\cot\tfrac{1}{2}t\,dt, \qquad (2·5, 7)$$

where the integral on the right-hand side is, strictly, $\lim\limits_{\epsilon\to 0}\int_\epsilon^{\pi}$. However, this precaution may now be omitted for reasonably well-behaved functions, e.g. if $f(\theta)$ is differentiable at $\theta=\theta_0$. Expanding $f(\theta_0+t)$ and $f(\theta_0-t)$, we obtain

$$f(\theta_0+t)-f(\theta_0-t)$$
$$= \sum_{n=1}^{\infty} \{a_n[\cos n(\theta_0+t)-\cos n(\theta_0-t)] + b_n[\sin n(\theta_0+t)-\sin n(\theta_0-t)]\}.$$

Now $$\cos n(\theta_0+t)-\cos n(\theta_0-t) = -2\sin n\theta_0 \sin nt,$$

$$\sin n(\theta_0+t)-\sin n(\theta_0-t) = 2\cos n\theta_0 \sin nt,$$

and so

$$(f(\theta_0+t)-f(\theta_0-t))\cot\tfrac{1}{2}t = 2\sum_{n=1}^{\infty} (b_n\cos n\theta_0 - a_n\sin n\theta_0)\sin nt\,\frac{\cos\tfrac{1}{2}t}{\sin\tfrac{1}{2}t}.$$

Accepting the admissibility of term-by-term integration

$$\int_0^{\pi} (f(\theta_0+t)-f(\theta_0-t))\cot\tfrac{1}{2}t\,dt$$
$$= 2\sum_{n=1}^{\infty} \left[(b_n\cos n\theta_0 - a_n\sin n\theta_0)\int_0^{\pi} \sin nt\,\frac{\cos\tfrac{1}{2}t}{\sin\tfrac{1}{2}t}\,dt \right].$$

Comparing the expression on the right-hand side with (2·5, 2), we find that in order to establish (2·5, 5) we only have to show

$$\int_0^{\pi} \sin nt\,\frac{\cos\tfrac{1}{2}t}{\sin\tfrac{1}{2}t}\,dt = \pi \quad (n=1, 2, \dots). \qquad (2·5, 8)$$

In fact, for $n=1$, $$\int_0^{\pi} \sin t\,\frac{\cos\tfrac{1}{2}t}{\sin\tfrac{1}{2}t}\,dt = \int_0^{\pi} 2\sin\tfrac{1}{2}t\,\frac{\cos^2\tfrac{1}{2}t}{\sin\tfrac{1}{2}t}\,dt$$

$$= \int_0^{\pi} (1+\cos t)\,dt$$

$$= \pi.$$

To prove (2·5, 8) for higher n, it is sufficient to show that

$$\int_0^\pi \sin(n+1)\,t\,\frac{\cos\frac{1}{2}t}{\sin\frac{1}{2}t}\,dt - \int_0^\pi \sin nt\,\frac{\cos\frac{1}{2}t}{\sin\frac{1}{2}t}\,dt = 0 \quad (n=1,2,\ldots). \quad (2\!\cdot\!5,9)$$

The left-hand side of (2·5, 9) equals

$$\int_0^\pi 2\cos(n+\tfrac{1}{2})\,t\sin\tfrac{1}{2}t\,\frac{\cos\frac{1}{2}t}{\sin\frac{1}{2}t}\,dt = \int_0^\pi 2\cos(n+\tfrac{1}{2})\,t\cos\tfrac{1}{2}t\,dt$$

$$= \int_0^\pi [\cos(n+1)\,t + \cos nt]\,dt = \left[\frac{\sin(n+1)t}{n+1} + \frac{\sin nt}{n}\right]_0^\pi = 0 \quad \text{as required.}$$

If $f(\theta)$ is an even function of θ, $f(\theta) = f(-\theta) = f(2\pi-\theta)$, (2·5, 5) may be replaced by an alternative formula. We then have

$$\bar{f}(\theta_0) = \frac{1}{2\pi}\int_0^{2\pi} f(\theta)\cot\frac{\theta-\theta_0}{2}\,d\theta$$

$$= \frac{1}{2\pi}\left[\int_0^\pi f(\theta)\cot\frac{\theta-\theta_0}{2}\,d\theta + \int_\pi^{2\pi} f(2\pi-\theta)\cot\frac{\theta-\theta_0}{2}\,d\theta\right].$$

Putting $\psi = 2\pi - \theta$ in the second integral,

$$\bar{f}(\theta_0) = \frac{1}{2\pi}\left[\int_0^\pi f(\theta)\cot\frac{\theta-\theta_0}{2}\,d\theta - \int_\pi^0 f(\psi)\cot\left(\pi - \frac{\psi+\theta_0}{2}\right)d\psi\right]$$

$$= \frac{1}{2\pi}\int_0^\pi f(\theta)\left(\cot\frac{\theta-\theta_0}{2} - \cot\frac{\theta+\theta_0}{2}\right)d\theta$$

$$= \frac{1}{2\pi}\int_0^\pi f(\theta)\frac{\sin\theta_0}{\sin\frac{1}{2}(\theta-\theta_0)\sin\frac{1}{2}(\theta+\theta_0)}\,d\theta.$$

Hence

$$\bar{f}(\theta_0) = -\frac{\sin\theta_0}{\pi}\int_0^\pi \frac{f(\theta)}{\cos\theta - \cos\theta_0}\,d\theta. \quad (2\!\cdot\!5,10)$$

Formula (2·5, 10) is called the 'Glauert integral'. Similarly, if $f(\theta)$ is odd, $f(\theta) = -f(-\theta) = -f(2\pi-\theta)$,

$$\bar{f}(\theta_0) = \frac{1}{2\pi}\int_0^\pi f(\theta)\left(\cot\frac{\theta-\theta_0}{2} + \cot\frac{\theta+\theta_0}{2}\right)d\theta$$

$$= \frac{1}{2\pi}\int_0^\pi f(\theta)\frac{\sin\theta}{\sin\frac{1}{2}(\theta-\theta_0)\sin\frac{1}{2}(\theta+\theta_0)}\,d\theta,$$

and so

$$\bar{f}(\theta_0) = -\frac{1}{\pi}\int_0^\pi \frac{f(\theta)\sin\theta}{\cos\theta - \cos\theta_0}\,d\theta. \quad (2\!\cdot\!5,11)$$

2·6   Theodorsen's method

Several methods have been put forward for the calculation of the pressure distribution round given aerofoils, and we proceed to describe one‡ which has found wide application.

‡ Th. Theodorsen, 'Theory of wing sections of arbitrary shape', *N.A.C.A. Rep.* no. 411 (1932).

For a given aerofoil, let T denote the trailing edge, as usual, while L is a point inside the aerofoil near its nose. In the particular case when the leading edge is sharp we let L coincide with it. In all other cases the exact choice of L is of no consequence for the following analysis; the most favourable position of L from the computational point of view will be discussed later.

We choose the x-axis in the physical plane so that it joins L and T, and we fix the origin of coordinates at the mid-point of the segment LT. Let the distance LT be denoted by $4d$ so that the coordinates of L and T on the x-axis are $-2d$, $2d$ respectively.

The transformation
$$z = z' + \frac{d^2}{z'} \qquad (2·6, 1)$$

maps the segment LT in the (physical) z-plane on a circle S' of radius d round the origin in the z'-plane. At the same time, the aerofoil section will be mapped on a curve C' in the z'-plane, which does not differ greatly from S' (Fig. 2·6, 1).

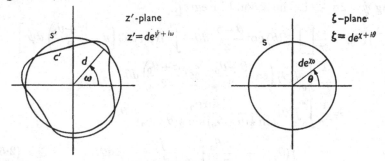

Fig. 2·6, 1

Let
$$z' = d\, e^{\psi + i\omega} \qquad (-\infty < \psi < \infty,\ 0 \leqslant \omega < 2\pi), \qquad (2·6, 2)$$

so that ψ and ω may be regarded as coordinates of the z'-plane. At the same time
$$z = d\, e^{\psi + i\omega} + d\, e^{-\psi - i\omega} = d(e^{\psi} + e^{-\psi})\cos\omega + id(e^{\psi} - e^{-\psi})\sin\omega$$
$$= 2d\cosh\psi\cos\omega + 2id\sinh\psi\sin\omega,$$

so that
$$\left. \begin{array}{l} x = 2d\cosh\psi\cos\omega, \\ y = 2d\sinh\psi\sin\omega. \end{array} \right\} \qquad (2·6, 3)$$

Thus, we may regard ψ and ω as (elliptic) coordinates of the z-plane. (The z-plane is covered twice as ψ varies from $-\infty$ to ∞ and ω varies from 0 to 2π.)

To eliminate ψ from (2·6, 3), divide the first equation by $2d\cos\omega$, the second equation by $2d\sin\omega$, square and subtract. Then
$$\left(\frac{x}{2d\cos\omega}\right)^2 - \left(\frac{y}{2d\sin\omega}\right)^2 = 1, \qquad (2·6, 4)$$

or
$$\left(\frac{x}{2d}\right)^2 \sin^2 \omega - \left(\frac{y}{2d}\right)^2 (1 - \sin^2 \omega) = \sin^2 \omega (1 - \sin^2 \omega),$$

i.e.
$$\sin^4 \omega - h \sin^2 \omega - \left(\frac{y}{2d}\right)^2 = 0,$$

where
$$h = 1 - \left(\frac{x}{2d}\right)^2 - \left(\frac{y}{2d}\right)^2.$$

It follows that
$$\sin^2 \omega = \frac{1}{2}\left[h + \sqrt{\left\{h^2 + \left(\frac{y}{d}\right)^2\right\}}\right]. \qquad (2·6, 5)$$

Similarly, we obtain, by eliminating ω from (2·6, 3),

$$\left(\frac{x}{2d \cosh \psi}\right)^2 + \left(\frac{y}{2d \sinh \psi}\right)^2 = 1, \qquad (2·6, 6)$$

or
$$\sinh^4 \psi + h \sinh^2 \psi - \left(\frac{y}{2d}\right)^2 = 0,$$

$$\sinh^2 \psi = \frac{1}{2}\left[-h + \sqrt{\left\{h^2 + \left(\frac{y}{d}\right)^2\right\}}\right]. \qquad (2·6, 7)$$

Note that (2·6, 4) and (2·6, 6) are the equations of families of confocal hyperbolae and ellipses respectively.

Since C' is approximately circular, we may assume that ψ is numerically small compared with unity for points of C', and so

$$\psi = \log(e^\psi) = \log(\sinh \psi + \cosh \psi) = \log\{\sinh \psi + \sqrt{(1 + \sinh^2 \psi)}\}$$
$$= \log(1 + \sinh \psi + \tfrac{1}{2}\sinh^2 \psi + \ldots)$$
$$= \sinh \psi + \tfrac{1}{2}\sinh^2 \psi - \tfrac{1}{2}(\sinh \psi + \tfrac{1}{2}\sinh^2 \psi)^2 + \tfrac{1}{3}(\sinh \psi + \tfrac{1}{2}\sinh^2 \psi)^3 - \ldots$$
$$= \sinh \psi - \tfrac{1}{6}\sinh^3 \psi + \ldots.$$

Taking into account (2·6, 3),

$$\psi = \frac{y}{2d \sin \omega} - \frac{1}{6}\left(\frac{y}{2d \sin \omega}\right)^3. \qquad (2·6, 8)$$

(2·6, 8) provides an approximation which is valid for small values of ψ, as required.

An expression for dz/dz' which will be used presently is obtained as follows

$$\frac{dz}{dz'} = \frac{1}{z'}\left(z' - \frac{d^2}{z'}\right) = \frac{1}{z'}(d\,e^{\psi + i\omega} - d\,e^{-\psi - i\omega})$$

$$= \frac{1}{z'}[d(e^\psi - e^{-\psi})\cos \omega + id(e^\psi + e^{-\psi})\sin \omega]$$

$$= \frac{1}{z'}[2d \sinh \psi \cos \omega + 2id \cosh \psi \sin \omega].$$

But, by (2·6, 3), $2d \sinh \psi = \dfrac{y}{\sin \omega}, \quad 2d \cosh \psi = \dfrac{x}{\cos \omega},$

and so
$$\frac{dz}{dz'} = \frac{1}{z'}(y \cot \omega + ix \tan \omega), \qquad (2·6, 9)$$

and
$$\left|\frac{dz}{dz'}\right| = \frac{1}{|z'|}\sqrt{(y^2\cot^2\omega + x^2\tan^2\omega)}. \tag{2·6, 10}$$

Assume that we have mapped the region outside the curve C' in the z'-plane on the region outside a circle S round the origin of the ζ-plane, in such a way that the point at infinity is mapped on the point at infinity and such that $dz'/d\zeta = 1$. As a consequence of Riemann's fundamental theorem there is just one such transformation, so that the radius of the circle S is determined by the other conditions imposed on the transformation.

Consider the function $\log z'/\zeta$. This is a regular function of ζ outside S which tends to 0 as ζ tends to infinity. We may therefore write

$$\log\frac{z'}{\zeta} = \sum_{n=1}^{\infty}\frac{\alpha_n + i\beta_n}{\zeta^n}, \tag{2·6, 11}$$

so that
$$z' = \zeta\exp\left(\sum_{n=1}^{\infty}\frac{\alpha_n + i\beta_n}{\zeta^n}\right).$$

Then
$$\frac{dz'}{d\zeta} = \exp\left(\sum_{n=1}^{\infty}\frac{\alpha_n + i\beta_n}{\zeta^n}\right) - \zeta\exp\left(\sum_{n=1}^{\infty}\frac{\alpha_n + i\beta_n}{\zeta^n}\right)\sum_{n=1}^{\infty} n\frac{\alpha_n + i\beta_n}{\zeta^{n+1}}$$

$$= \exp\left(\sum_{n=1}^{\infty}\frac{\alpha_n + i\beta_n}{\zeta^n}\right)\left(1 - \sum_{n=1}^{\infty} n\frac{\alpha_n + i\beta_n}{\zeta^n}\right),$$

so that $\lim_{\zeta\to\infty} dz'/d\zeta = 1$, as required.

Let
$$\zeta = d\,e^{\chi + i\theta},$$

and let the circle S be given by $\chi = \chi_0$, so that the radius of S equals $a = d\,e^{\chi_0}$ (Fig. 2·6, 1). Then

$$\log\frac{z'}{\zeta} = \log\left(\frac{d\,e^{\psi + i\omega}}{d\,e^{\chi + i\theta}}\right) = \psi - \chi + i(\omega - \theta),$$

and so, by (2·6, 11),

$$(\psi - \chi) + i(\omega - \theta) = \sum_{n=1}^{\infty}\frac{\alpha_n + i\beta_n}{\zeta^n}$$

$$= \sum_{n=1}^{\infty}\frac{\alpha_n + i\beta_n}{d^n}e^{-n\chi}(\cos n\theta - i\sin n\theta). \tag{2·6, 12}$$

Separating real and imaginary parts, we obtain for the points of S where $\chi = \chi_0$

$$\left.\begin{aligned}
\psi - \chi_0 &= \sum_{n=1}^{\infty}(\alpha_n d^{-n}e^{-n\chi_0}\cos n\theta + \beta_n d^{-n}e^{-n\chi_0}\sin n\theta),\\
\omega - \theta &= \sum_{n=1}^{\infty}(\beta_n d^{-n}e^{-n\chi_0}\cos n\theta - \alpha_n d^{-n}e^{-n\chi_0}\sin n\theta).
\end{aligned}\right\} \tag{2·6, 13}$$

The first of these two equations yields

$$\alpha_n d^{-n}e^{-n\chi_0} = \frac{1}{\pi}\int_0^{2\pi}\psi\cos n\theta\,d\theta, \quad \beta_n d^{-n}e^{-n\chi_0} = \frac{1}{\pi}\int_0^{2\pi}\psi\sin n\theta\,d\theta,$$

$$\chi_0 = \frac{1}{2\pi}\int_0^{2\pi}\psi\,d\theta. \tag{2·6, 14}$$

Also, (2·6, 13) shows that the difference $\delta = \omega - \theta$ is conjugate to $\psi - \chi_0$, and, since χ_0 is a constant, δ is also conjugate to ψ. Hence, by (2·5, 5),

$$\delta_0 = (\omega - \theta)_0 = \frac{1}{2\pi} \int_0^{2\pi} \psi \cot \frac{\theta - \theta_0}{2} d\theta, \tag{2·6, 15}$$

where the suffix 0 indicates that we are considering the difference $\delta_0 = (\omega - \theta)$ for some specific value of the argument, θ.

In the integrals (2·6, 14) and (2·6, 15), ψ may be regarded as a function of θ through the intermediary of ω, where $\psi = g(\omega)$, say, is the equation of the curve C' in the z'-plane. However, since C' does not differ greatly from S', we may assume that the difference between θ and ω is small, so that $\psi = g(\theta)$ in the first approximation. Substituting this expression for ψ on the right-hand side of (2·6, 15) we obtain a first approximation $\delta^{(1)}$ for the value of $\delta = \omega - \theta$,

$$\delta_0^{(1)} = \frac{1}{2\pi} \int_0^{2\pi} g(\theta) \cot \frac{\theta - \theta_0}{2} d\theta. \tag{2·6, 16}$$

Then $\omega = \theta + \delta^{(1)}$ is a better approximation for the argument of g, and so a second approximation $\delta^{(2)}$ for δ is

$$\delta_0^{(2)} = \frac{1}{2\pi} \int_0^{2\pi} g(\theta + \delta^{(1)}) \cot \frac{\theta - \theta_0}{2} d\theta. \tag{2·6, 17}$$

We might continue in this way, but in practice the first approximation is considered satisfactory. Similarly, to evaluate (2·6, 14) we may use $\psi = g(\omega)$ or, more exactly, $\psi = g(\theta + \delta^{(n)})$.

To find the pressure distribution, we still require an expression for $dz'/d\zeta$. Differentiating the equation $\log z' = \log d + \psi + i\omega$ with respect to ζ we obtain

$$\frac{dz'}{d\zeta} = z' \frac{d}{d\zeta} (\psi + i\omega). \tag{2·6, 18}$$

Also, differentiating along the curve C' in the z'-plane, we obtain from $\log \zeta = \log d + \chi_0 + i\theta$,

$$\frac{1}{\zeta} \frac{d\zeta}{d\omega} = \frac{1}{\zeta} \frac{d\zeta}{d\theta} \frac{d\theta}{d\omega} = i\left(\frac{d}{d\omega} (\theta - \omega) + 1 \right),$$

or

$$\frac{d\zeta}{d\omega} = i\zeta \left(1 - \frac{d\delta}{d\omega} \right).$$

Using this relation, we obtain from (2·6, 18),

$$\frac{dz'}{d\zeta} = z' \frac{d}{d\omega} (\psi + i\omega) \frac{d\omega}{d\zeta} = \frac{z'}{i\zeta} \frac{d}{d\omega} (\psi + i\omega) \Big/ \left(1 - \frac{d\delta}{d\omega} \right),$$

or, denoting the derivatives of δ and ψ with respect to ω by δ' and ψ' respectively,

$$\frac{dz'}{d\zeta} = \frac{z'}{\zeta} \frac{1 - i\psi'}{1 - \delta'}. \tag{2·6, 19}$$

Now δ' may be assumed small compared with 1, and so the modulus of $dz'/d\zeta$ is

$$\left|\frac{dz'}{d\zeta}\right| = \frac{|z'|}{|\zeta|}\frac{\sqrt{(1+\psi'^2)}}{1-\delta'} = \frac{|z'|}{d}\frac{\sqrt{(1+\psi'^2)}}{1-\delta'}e^{-\chi_0}. \tag{2·6, 20}$$

Combining this with (2·6, 10), we obtain

$$\left|\frac{dz}{d\zeta}\right| = \frac{\sqrt{(1+\psi'^2)}}{d(1-\delta')}e^{-\chi_0}\sqrt{(y^2\cot^2\omega + x^2\tan^2\omega)}$$

$$= \frac{2e^{-\chi_0}\sqrt{(1+\psi'^2)}}{1-\delta'}\sqrt{\left\{\left(\frac{y}{2d}\right)^2\cot^2\omega + \left(\frac{x}{2d}\right)^2\tan^2\omega\right\}}.$$

But, by (2·6, 3)
$$\left(\frac{y}{2d}\right)^2\cot^2\omega = \sinh^2\psi\,\cos^2\omega,$$

$$\left(\frac{x}{2d}\right)^2\tan^2\omega = \cosh^2\psi\,\sin^2\omega,$$

and so $\left(\dfrac{y}{2d}\right)^2\cot^2\omega + \left(\dfrac{x}{2d}\right)^2\tan^2\omega = \sinh^2\psi\,\cos^2\omega + \cosh^2\psi\,\sin^2\omega$

$$= \sinh^2\psi + \sin^2\omega$$

$$= \left(\frac{y}{2d\sin\omega}\right)^2 + \sin^2\omega.$$

Hence $\left|\dfrac{dz}{d\zeta}\right| = \dfrac{2e^{-\chi_0}}{1-\delta'}\sqrt{\left[(1+\psi'^2)\left\{\left(\dfrac{y}{2d\sin\omega}\right)^2 + \sin^2\omega\right\}\right]}.$ (2·6, 21)

Substituting (2·6, 21) in the expression for the modulus of the velocity (2·3, 27), we obtain

$$q = \frac{2U(1-\delta')\left|\cos\left(\alpha+\dfrac{\beta-\theta}{2}\right)\sin\dfrac{\beta+\theta}{2}\right|e^{\chi_0}}{\sqrt{\left[(1+\psi'^2)\left\{\left(\dfrac{y}{2d\sin\omega}\right)^2 + \sin^2\omega\right\}\right]}}. \tag{2·6, 22}$$

Or, substituting $\theta = w - \delta$,

$$q = \frac{2U(1-\delta')\left|\cos\left(\alpha+\dfrac{\beta+\delta-\omega}{2}\right)\sin\dfrac{\beta-\delta+\omega}{2}\right|e^{\chi_0}}{\sqrt{\left[(1+\psi'^2)\left\{\left(\dfrac{y}{2d\sin\omega}\right)^2 + \sin^2\omega\right\}\right]}}. \tag{2·6, 23}$$

By the choice of the mapping of the z-plane on the z′-plane, $\omega = 0$ corresponds to the trailing edge. Hence, according to the Joukowski condition, the corresponding velocity in the ζ-plane must vanish (although q may be positive, if the aerofoil possesses a cusped trailing edge). Now the velocity in the ζ-plane is, writing $\theta = \omega - \delta$ in (2·3, 26),

$$\tau = 4Ui\,e^{-i\theta}\cos\left(\alpha+\frac{\beta+\delta-\omega}{2}\right)\sin\frac{\beta-\delta+\omega}{2}. \tag{2·6, 24}$$

Thus, for $\omega = 0$, we must have, for all α,

$$\cos\left(\alpha + \frac{\beta + \delta}{2}\right) \sin\frac{\beta - \delta}{2} = 0,$$

and so

$$\beta = (\delta)_{\omega=0}. \qquad (2\cdot6, 25)$$

In order to calculate the velocity distribution, and hence the pressure, round a given aerofoil, for given free-stream velocity U and incidence α, we regard all the variables occurring in $(2\cdot6, 23)$ as functions of ω. Accepting the first approximation, we may use the following procedure.

Calculate ω and ψ for the points of the aerofoil by $(2\cdot6, 5)$ and $(2\cdot6, 7)$, or, alternatively, $(2\cdot6, 8)$, and plot ψ against ω, $\psi = g(\omega)$. Plot $\psi' = g'(\omega)$ and determine y as a function of ω from $(2\cdot6, 3)$. Calculate δ by $(2\cdot6, 16)$, and plot it against $\omega = \theta + \delta$; plot $\delta' = d\delta/d\omega$. Finally, determine χ_0 from $(2\cdot6, 14)$ with $\psi = g(\theta)$ and $\beta = (\delta)_{\omega=0}$. This yields all the data required for the evaluation of $(2\cdot6, 23)$. The pressure distribution is then given by $(2\cdot3, 28)$.

Equation $(2\cdot6, 23)$ was derived using the Joukowski condition. We can find a more general expression for the velocity, starting from equation $(2\cdot3, 12)$ for the velocity about a circle, with circulation Γ. On the circle $\zeta = \zeta_0 + a\,e^{i\theta}$ this velocity is

$$\tau = U\left(e^{-i\alpha} - \frac{i\Gamma}{2\pi a}e^{-i\theta} - e^{-i\alpha - 2i\theta}\right).$$

We use the law of Kutta–Joukowski, equation $(2\cdot3, 5)$, to express Γ in terms of the lift; thus

$$\tau = 2iU\,e^{-i\theta}\left(\sin(\theta - \alpha) + \frac{C_L}{2\pi}\right),$$

where $C_L = L/\frac{1}{2}\rho U^2 4a$. Hence the velocity in the z-plane is

$$q = 2U\left|\sin(\theta - \alpha) + \frac{C_L}{2\pi}\right|\left|\frac{d\zeta}{dz}\right|.$$

Assuming a lift-curve slope a_0 such that

$$C_L = a_0 \sin(\alpha + \beta),$$

and using $(2\cdot6, 21)$ we obtain, in place of $(2\cdot6, 23)$, the equation

$$q = \frac{U\left|e^{\chi_0}(1 - \delta')\right|}{\sqrt{\left[(1 + \psi'^2)\left\{\left(\frac{y}{2d\sin\omega}\right)^2 + \sin^2\omega\right\}\right]}}$$
$$\times \left|\sqrt{\left(1 - \frac{C_L^2}{a_0^2}\right)}\sin(\omega - \delta + \beta) - \frac{C_L}{a_0}\cos(\omega - \delta + \beta) + \frac{C_L}{2\pi}\right|. \qquad (2\cdot6, 26)$$

If the Joukowski condition holds, $a_0 = 2\pi$, and equation $(2\cdot6, 26)$ reduces to $(2\cdot6, 23)$. If, however, the lift is not predicted correctly by the Joukowski condition (as well may be the case, due to boundary-layer effects), equation

(2·6, 26) provides a method of correcting potential theory for viscous effects by using experimental or empirical values for a_0.

It is shown in Theodorsen's report that L is conveniently chosen as the point midway between the nose (i.e. the point of maximum curvature) and the centre of curvature of the nose. The report also contains an approximate formula for the computation of δ and a formula for the value of q at a cusped trailing edge.

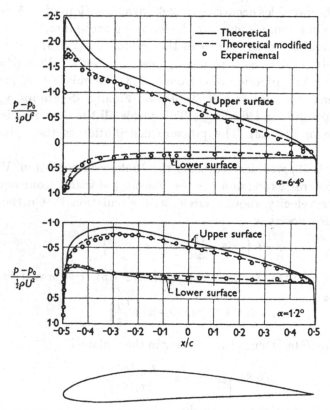

Fig. 2·6, 2. Theoretical pressure distribution about the N.A.C.A. 4412 aerofoil (calculated by Theodorsen's method) compared with experiment. (*Courtesy N.A.C.A.*)

Theodorsen's theory has been compared with experiment by Pinkerton[‡] for an N.A.C.A. 4412 aerofoil (Fig. 2·6, 2). Shown in the figure is the pressure ratio $(p-p_0)/\frac{1}{2}\rho U^2$ on the aerofoil surface, p_0 being the free-stream pressure. The solid line indicates the theoretical pressure distribution and the circles denote experimental results. The discrepancies are accounted for by boundary-layer effects, and may be corrected by using equation (2·6, 26) in place of (2·6, 23) with an experimental value for the lift coeffi-

‡ R. M. Pinkerton, 'Calculated and measured pressure distributions over the midspan section of the N.A.C.A. 4412 Airfoil', *N.A.C.A. Rep.* no. 562 (1936).

cient C_L. Following this type of procedure Pinkerton modified the theoretical curve to bring it into good agreement with experiment, as shown by the dashed line in Fig. 2·6, 2. To avoid an infinite velocity at the trailing edge (the Joukowski condition not being used, compare § 2·9), a slight alteration of the trailing edge shape was introduced in the modified theory.

In an alternative procedure to Theodorsen's due to von Kármán and Trefftz‡ the first transformation—from the z-plane to the z'-plane—is effected by means of a formula of the type of (2·4, 40). A suitable choice of the index then ensures that the trace of the aerofoil in the z'-plane is a smooth curve, but the analysis becomes more complicated in other respects.

2·7 Lighthill's method

We now come to the converse problem of designing an aerofoil for a specific pressure distribution. In order to make the problem precise, we have to detail the argument as a function of which the pressure distribution is given. In the method which will now be described§ it is assumed that the pressure, and hence the velocity distribution, are specified around the circle S into which the (hypothetical) aerofoil is transformed in the ζ-plane, by means of a transformation function whose derivative equals 1 at infinity,

$$\lim_{\zeta \to \infty} \frac{dz}{d\zeta} = 1.$$

In addition, we shall assume that $\beta = 0$, and that $U = 1$ and $a = 1$ in some specified system of units, where a is the radius of the circle S. We know from § 2·4 that this implies $c = 4$ approximately, where c is the chord of the aerofoil.

By (2·3, 27) the modulus of the velocity, q, at a point of the aerofoil, is now given by

$$q = 4 \left| \cos\left(\alpha - \tfrac{1}{2}\theta\right) \sin \tfrac{1}{2}\theta \right| \left| \frac{d\zeta}{dz} \right|. \tag{2·7, 1}$$

We write $dz = ds\, e^{i\omega}$, where ds is the element of length of the aerofoil, and ω its local incidence. Correspondingly, $d\zeta = d(e^{i\theta}) = i\, e^{i\theta}\, d\theta$, and so q becomes

$$q = 4 \left| \cos\left(\alpha - \tfrac{1}{2}\theta\right) \sin \tfrac{1}{2}\theta \frac{d\theta}{ds} \right| = 2 \left| \frac{\cos\left(\alpha - \tfrac{1}{2}\theta\right)}{\cos \tfrac{1}{2}\theta} \right| \left| \sin \theta \frac{d\theta}{ds} \right|,$$

or

$$q = 2 \left| \frac{\cos\left(\alpha - \tfrac{1}{2}\theta\right)}{\cos \tfrac{1}{2}\theta} \right| \bigg/ \left| \frac{ds}{d(\cos \theta)} \right|. \tag{2·7, 2}$$

In particular, at zero incidence ($\alpha = 0$)

$$q_0 = 2 \bigg/ \left| \frac{ds}{d(\cos \theta)} \right|, \tag{2·7, 3}$$

‡ Th. von Kármán and E. Trefftz, 'Potentialströmung um gegebene Tragflächenquerschnitte', *Z. Flugtech.* vol. 9 (1918), p. 140.

§ M. J. Lighthill, 'A new method of two-dimensional aerodynamic design', *Rep. Memor. Aero. Res. Comm., Lond.*, no. 2112 (1945).

so that the modulus of the velocity for arbitrary incidence may be written in the form

$$q = q_0 \left| \frac{\cos(\alpha - \frac{1}{2}\theta)}{\cos \frac{1}{2}\theta} \right|. \tag{2·7, 4}$$

The complex velocity, w, is a regular analytic function of z in the region outside the aerofoil, including the point at infinity. It therefore is a regular analytic function of ζ outside the unit circle. Moreover, w does not vanish in that region, and becomes equal to 1 at infinity, and so $\log w$ also is a regular function of ζ outside the unit circle. Since $\beta = 0$, $\alpha = 0$ is the angle of zero lift, and so, by (2·2, 10) with $A_0 = C_{-1} = 1$, we must have $A_1 = 0$. This implies that the expansion (2·2, 9) becomes, with $\zeta_0 = 0$, writing w_0 for the complex velocity at zero incidence,

$$w_0 = 1 + \frac{A_2}{\zeta^2} + \dots . \tag{2·7, 5}$$

To find the first term in the expansion of $\log w$ we use the formula $\log(1+t) = t - \dfrac{t^2}{2} + \dfrac{t^3}{3} - \dots .$ Then

$$\log w_0 = \frac{A_2}{\zeta^2} + \dots , \tag{2·7, 6}$$

so that the constant and the first-order term in the expansion of $\log w_0$ vanish.

We write $w = q e^{-i\omega}$, $w_0 = q_0 e^{-i\omega}$, so that ω indicates the direction of motion. This is in agreement with the definition of ω by $dz = ds\, e^{i\omega}$ at the aerofoil, since the flow must be tangential to the surface, provided we measure s in the local direction of flow. Then

$$\log w_0 = \log q_0 - i\omega .$$

Referring to (2·7, 6), we may expand $\log q_0$ and $-\omega$ in conjugate Fourier series on the unit circle, $\zeta = e^{i\theta}$:

$$\left. \begin{aligned} \log q_0 &= \alpha_2 \cos 2\theta + \beta_2 \sin 2\theta + \alpha_3 \cos 3\theta + \beta_3 \sin 3\theta + \dots, \\ -\omega &= \beta_2 \cos 2\theta - \alpha_2 \sin 2\theta + \beta_3 \cos 3\theta - \alpha_3 \sin 3\theta + \dots. \end{aligned} \right\} \tag{2·7, 7}$$

It will be observed that since q_0 may vanish at the trailing edge, $\log q_0$ may become infinite at that point, but since the infinity is only logarithmic, $\log q_0$ will still possess a Fourier series.

(2·7, 7) shows that $\log q_0$ satisfies the following three conditions:

$$\left. \begin{aligned} \int_0^{2\pi} \log q_0 \, d\theta &= 0, \\ \int_0^{2\pi} \log q_0 \cos \theta \, d\theta &= 0, \\ \int_0^{2\pi} \log q_0 \sin \theta \, d\theta &= 0. \end{aligned} \right\} \tag{2·7, 8}$$

They express the fact that the first three terms in the Fourier expansion of $\log q_0$ vanish.

Since $\log q_0$ and $-\omega$ are conjugate functions of θ, it follows that ω is linked directly to $\log q_0$ by Poisson's integral (2·5, 5). Thus for any particular $\theta = \theta_0$,

$$\omega(\theta_0) = -\frac{1}{2\pi} \int_0^{2\pi} \log q_0(\theta) \cot \frac{\theta - \theta_0}{2} d\theta. \qquad (2\cdot7, 9)$$

If the postulated velocity distribution is symmetrical, i.e. if

$$q_0(2\pi - \theta) = q_0(\theta)$$

for all θ, we may replace (2·7, 9) by Glauert's integral (2·5, 10),

$$\omega(\theta_0) = \frac{\sin \theta_0}{\pi} \int_0^\pi \frac{\log q_0(\theta)}{\cos \theta - \cos \theta_0} d\theta. \qquad (2\cdot7, 10)$$

In this case, the third condition in (2·7, 8) is satisfied automatically.

Having determined ω for given q_0, we may then find the slope of the aerofoil in the following way.

We have $dz = ds\, e^{i\omega}$, provided we measure ds in the (local) direction of the flow. Hence

$$\frac{dz}{d(\cos \theta)} = \frac{ds}{d(\cos \theta)} e^{i\omega} = \pm \left| \frac{ds}{d(\cos \theta)} \right| e^{i\omega}, \qquad (2\cdot7, 11)$$

where the sign $+$ applies if the direction of flow is in the direction of increasing $\cos \theta$, while $-$ applies in the alternative case. Now, by (2·7, 3), $\left| \dfrac{ds}{d(\cos \theta)} \right| = \dfrac{2}{q_0}$, and so

$$z = \int \pm \frac{2}{q_0} e^{i\omega} d(\cos \theta).$$

Or, writing $z = x + iy$ and $e^{i\omega} = \cos \omega + i \sin \omega$, and separating real and imaginary parts,

$$x = \int \pm \frac{2}{q_0} \cos \omega\, d(\cos \theta), \quad y = \int \pm \frac{2}{q_0} \sin \omega\, d(\cos \theta). \qquad (2\cdot7, 12)$$

Now if q_0 is symmetrical, the front stagnation point occurs for $\theta = \pi$, and so the direction of flow always coincides with the direction of increasing $\cos \theta$. In that case, the sign $+$ applies in (2·7, 12) throughout, or

$$\left. \begin{aligned} x &= -\int \frac{2}{q_0} \cos \omega \sin \theta\, d\theta, \\ y &= -\int \frac{2}{q_0} \sin \omega \sin \theta\, d\theta, \end{aligned} \right\} \qquad (2\cdot7, 13)$$

where we may take 0 as the lower limit of integration. In any case, the sign $-$ applies only in a very restricted region near the leading edge.

The pitching moment at zero incidence can be obtained from (2·2, 11) with $A_0 = C_{-1} = 1$, $A_1 = 0$, so that

$$M_0 = 2\pi\rho \mathscr{I}(A_2). \qquad (2\cdot7, 14)$$

Comparing (2·7, 6) and (2·7, 7), we see that $A_2 = \alpha_2 + i\beta_2$, and so

$$\mathscr{I}(A_2) = \beta_2 = \frac{1}{\pi} \int_0^{2\pi} \log q_0 \sin 2\theta \, d\theta,$$

and
$$M_0 = 2\rho \int_0^{2\pi} \log q_0 \sin 2\theta \, d\theta. \tag{2·7, 15}$$

The moment coefficient C_{M_0} is then given by

$$C_{M_0} = \frac{4}{c^2} \int_0^{2\pi} \log q_0 \sin 2\theta \, d\theta, \tag{2·7, 16}$$

where $c = 4$, approximately.

The lift at incidence α is found by substituting $U = a = 1$, $\beta = 0$ in (2·3, 19)

$$L = 4\pi\rho \sin \alpha. \tag{2·7, 17}$$

This yields
$$C_L = \frac{8\pi}{c} \sin \alpha, \tag{2·7, 18}$$

so that the lift curve slope is still given by $dC_L/d\alpha = 2\pi$, approximately.

The principal aim in the designing of an aerofoil is that of producing a certain desirable pressure distribution for a given lift coefficient or, which is approximately the same, for a given incidence. There are two ways in which the above analysis can be used for this purpose. The first consists in the modification of a given aerofoil of known characteristics, e.g. of a Joukowski aerofoil. Let q_{J0} and $q_{J\alpha}$ be the moduli of velocity for the given aerofoil, for incidences 0 and α respectively, and let q_0 and q_α be the corresponding quantities for the modified aerofoil. We wish to prescribe q_α for the specified design incidence, α. Writing

$$\log q_0 = \log q_{J0} + \log q_M, \tag{2·7, 19}$$

so that $\log q_M$ is the required correction, we have

$$\log q_\alpha = \log \left(q_0 \left| \frac{\cos(\alpha - \tfrac{1}{2}\theta)}{\cos \tfrac{1}{2}\theta} \right| \right)$$
$$= \log q_{J0} + \log q_M + \log \left| \frac{\cos(\alpha - \tfrac{1}{2}\theta)}{\cos \tfrac{1}{2}\theta} \right|$$
$$= \log q_{J\alpha} + \log q_M.$$

Hence $\log q_M$ is given by
$$\log q_M = \log q_\alpha - \log q_{J\alpha}, \tag{2·7, 20}$$

where $q_{J\alpha}$ and q_α denote moduli of velocities for the given aerofoil, and for the desired aerofoil respectively. Also, from (2·7, 19),

$$\omega = \omega_J + \omega_M, \tag{2·7, 21}$$

where ω_J and ω_M are the functions conjugate to $\log q_J$ and $\log q_M$. Since ω_J may be regarded as known for the given aerofoil, we only have to find ω_M to determine ω. The shape of the aerofoil is then given by (2·7, 12) and (2·7, 13).

Following the alternative method, we determine $\log q_0$ directly in terms of the postulated velocity distribution at incidence α, q_α. By (2·7, 4),

$$\log q_\alpha = \log q_0 + \log \left| \frac{\cos(\alpha - \tfrac{1}{2}\theta)}{\cos \tfrac{1}{2}\theta} \right|,$$

or

$$\log q_0 = \log q_\alpha + \log \left| \frac{\cos \tfrac{1}{2}\theta}{\cos(\alpha - \tfrac{1}{2}\theta)} \right|. \qquad (2·7, 22)$$

Hence $\log q_0$ is defined if $\log q_\alpha$ has been chosen for the particular incidence α for which the aerofoil is to be designed. In all cases, q_α must be chosen in such a way that the three conditions of (2·7, 8) are satisfied by $\log q_0$. This ensures that the aerofoil section closes up; even so, an injudicious choice may lead to the physically impossible result that top and bottom surfaces are interlaced, or to the technically undesirable feature that the nose is sharp. It is the advantage of the first method given above, which is otherwise the more complicated of the two, that it provides some guidance by the introduction of an aerofoil of known shape and characteristics. A number of worked examples and further auxiliary formulae will be found in Lighthill's paper.

The fundamental principles which govern the choice of a favourable pressure distribution belong to boundary-layer theory, and their detailed discussion is outside the scope of this book. However, the following general indications may be given.

A pressure gradient which is numerically large and positive in the direction of flow creates conditions which are favourable both to transition from laminar to turbulent flow, and to separation at high incidences. It is therefore desirable that the pressure gradient should be negative, i.e. that the flow should be accelerated, at least in the forward regions of the aerofoil. This can be achieved by placing the point of maximum suction fairly far back, e.g. at midchord (see, for example, the low drag aerofoil illustrated in Fig. 2·9, 1), whereas on a 'conventional' (e.g. Joukowski) aerofoil, the suction peak occurs near the nose, followed by a rapid rise in pressure, as seen in Fig. 2·6, 2. If the chief purpose is the prevention of transition from laminar to turbulent flow at moderate incidence, then the aerofoil is called a 'laminar flow aerofoil' or 'low drag aerofoil'. The resulting reduction of drag is confirmed by experiment, but there are considerations of surface waviness and cleanliness which impose practical restrictions on the use of the laminar flow wing.

A. A. Griffith has suggested that, instead of letting the pressure rise again gently in the rearward regions of the aerofoil, one may concentrate the entire pressure rise at a point. To avoid transition or separation at that point, suction is applied through a slot which leads into the interior of the wing. This has the effect of sucking away the boundary layer, and it is hoped that the new boundary layer aft of the slot may be laminar even for

relatively thick wings. Figs. 2·7, 1 and 2·7, 2 show the shapes and theoretical velocity distributions about two such types of aerofoil on which
extensive tests have been made.‡ The practical results fall short of the

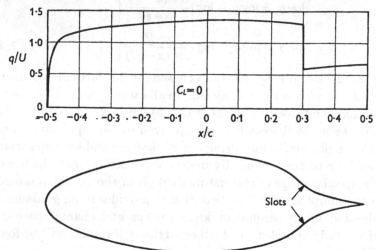

Fig. 2·7, 1. Velocity distribution about a 30 % thick
symmetrical Griffith suction aerofoil.

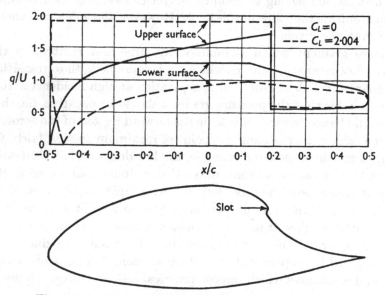

Fig. 2·7, 2. Velocity distribution about the G.L.A.S. II aerofoil.

theoretical predictions. It is found that a larger suction quantity is required
than expected, and that the retention of laminar flow past the slot positions
is often dependent critically on the slot shape.

‡ These two figures are taken from a paper by S. Goldstein, 'Low-drag and suction
airfoils', *J. Aero. Sci.* vol. 15 (1948), pp. 189–220.

On the other hand, for the production of high lift at high angles of incidence suction at or near the leading edge has been suggested, so that flow separation and the stall are delayed. Fig. 2·7, 3 gives the velocity distribution calculated by the Lighthill method for a nose suction aerofoil. ‡ The effect of suction is to reduce the rapid rise in pressure which otherwise occurs near the leading edge at high incidence. A review of some of this work may be found in a paper by E. F. Relf.§

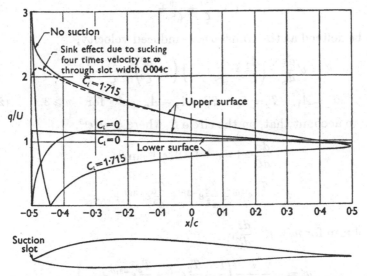

Fig. 2·7, 3. Velocity distribution on a Lighthill type high-lift nose-suction aerofoil. (*Courtesy H.M.S.O.*)

2·8   Thin aerofoil theory

There are various possible degrees of approximation to the exact theory, some of which will be discussed in § 2·9. In this section we shall describe a first approximation method for thin aerofoils, which possesses a fundamental importance of its own. The analysis is independent of §§ 2·6, 2·7.

We choose a system of coordinates so that the main chord of the aerofoil extends from $x = -\frac{1}{2}c = -2a$ at the leading edge to $x = \frac{1}{2}c = 2a$ at the trailing edge, and map the region outside the main chord on the region outside a circle S of radius a around the origin by means of the transformation

$$z = \zeta + \frac{a^2}{\zeta}, \tag{2·8, 1}$$

‡ Details are given in a report by M. J. Lighthill, 'A theoretical discussion of wings with leading edge suction', *Rep. Memor. Aero. Res. Comm., Lond.*, no. 2162 (1945).

§ E. F. Relf, 'Recent aerodynamic developments', *J. R. Aero. Soc.* vol. 50 (1946), pp. 421–48.

so that $dz/d\zeta = 1 - a^2/\zeta^2$, and in particular $dz/d\zeta = 1$ at infinity. The complex velocity in the z-plane, w, may then be written as a Laurent expansion of ζ (see (2·2, 9)),

$$w = A_0 + \frac{A_1}{\zeta} + \frac{A_2}{\zeta^2} + \dots, \qquad (2·8, 2)$$

where $A_0 = U\,e^{-i\alpha}$, and A_1 is pure imaginary (see § 2·3). We write w' for the 'induced velocity', $w = U\,e^{-i\alpha} + w'$, so that

$$w' = \frac{A_1}{\zeta} + \frac{A_2}{\zeta^2} + \dots. \qquad (2·8, 3)$$

Let τ' be defined as the transformed induced velocity,

$$\tau' = w'\frac{dz}{d\zeta} = \left(\frac{A_1}{\zeta} + \frac{A_2}{\zeta^2} + \dots\right)\left(1 - \frac{a^2}{\zeta^2}\right) = \frac{T_1}{\zeta} + \frac{T_2}{\zeta^2} + \dots,$$

where $\qquad T_1 = A_1, \quad T_2 = A_2, \quad T_n = A_n - A_{n-2}a^2 \quad$ for $\quad n \geqslant 3$. $\qquad (2·8, 4)$

Taking into account that, on the circle S, where $\zeta = a\,e^{i\theta}$,

$$\frac{dz}{d\zeta} = 1 - e^{-2i\theta} = 2i\,e^{-i\theta}\sin\theta,$$

and $\qquad\qquad \tau' = \frac{T_1}{a}e^{-i\theta} + \frac{T_2}{a^2}e^{-2i\theta} + \frac{T_3}{a^3}e^{-3i\theta} + \dots,$

we then obtain for $w' = \tau' \div \dfrac{dz}{d\zeta}$,

$$w' = -\frac{i}{2\sin\theta}\left(\frac{T_1}{a} + \frac{T_2}{a^2}e^{-i\theta} + \frac{T_3}{a^3}e^{-2i\theta} + \dots\right).$$

Writing $\qquad \dfrac{T_n}{a^n} = \tau_n + i\sigma_n \quad (n \geqslant 1), \qquad \tau_1 = \mathscr{R}\left(\frac{T_1}{a}\right) = \mathscr{R}\left(\frac{A_1}{a}\right) = 0,$

and separating real and imaginary parts, we then obtain for the total velocity $w = u - iv$

$$\left.\begin{aligned} u &= U\cos\alpha - \frac{1}{2\sin\theta}(-\sigma_1 + \tau_2\sin\theta - \sigma_2\cos\theta + \tau_3\sin 2\theta - \sigma_3\cos 2\theta + \dots), \\ v &= U\sin\alpha + \frac{1}{2\sin\theta}(\tau_2\cos\theta + \sigma_2\sin\theta + \tau_3\cos 2\theta + \sigma_3\sin 2\theta + \dots). \end{aligned}\right\}$$

$$(2·8, 5)$$

The coefficients τ_n, σ_n are determined by the boundary conditions. In applying these, we make the following simplifying assumptions. We take $u \doteqdot U\cos\alpha$, i.e. we assume that the longitudinal induced velocity is small compared with the free-stream velocity; and we refer the local slope of the aerofoil at a point R to the angle θ which corresponds to the vertical projection R' of R on to the chord of the aerofoil (Fig. 2·8, 1). Then

$$\frac{v}{U} \doteqdot \frac{v}{U\cos\alpha} = s(\theta) \quad \text{or} \quad v = Us(\theta)\cos\alpha. \qquad (2·8, 6)$$

Expressing this condition in terms of (2·8, 5),

$$U \sin \alpha + \frac{1}{2 \sin \theta} (\tau_2 \cos \theta + \sigma_2 \sin \theta + \tau_3 \cos 2\theta + \sigma_3 \sin 2\theta + \dots) = U s(\theta) \cos \alpha,$$

or

$$\tau_2 \cos \theta + \sigma_2 \sin \theta + \tau_3 \cos 2\theta + \sigma_3 \sin 2\theta + \dots = 2U \sin \theta (s(\theta) \cos \alpha - \sin \alpha).$$

Hence

$$\left. \begin{aligned} \tau_n &= \frac{2U}{\pi} \int_0^{2\pi} (s(\theta) \cos \alpha - \sin \alpha) \sin \theta \cos (n-1) \theta \, d\theta \\ \sigma_n &= \frac{2U}{\pi} \int_0^{2\pi} (s(\theta) \cos \alpha - \sin \alpha) \sin \theta \sin (n-1) \theta \, d\theta \end{aligned} \right\} \quad (n = 2, 3, \dots).$$

$$(2·8, 7)$$

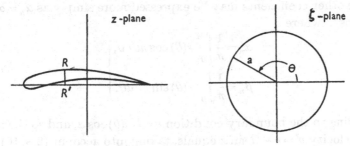

Fig. 2·8, 1

To determine σ_1 we apply the Joukowski condition according to which the velocity, and in particular u, must remain finite at the trailing edge, for $\theta = 0$. This shows that the sum in the brackets on the right-hand side of the first equation in (2·8, 5) vanishes for $\theta = 0$ or

$$\sigma_1 = -\sigma_2 - \sigma_3 - \dots. \qquad (2·8, 8)$$

Now

$$\sin \theta \cos (n-1) \theta = \tfrac{1}{2}[\sin n\theta - \sin (n-2) \theta],$$

$$\sin \theta \sin (n-1) \theta = -\tfrac{1}{2}[\cos n\theta - \cos (n-2) \theta],$$

and so

$$\left. \begin{aligned} \tau_n &= \frac{U}{\pi} \int_0^{2\pi} (s(\theta) \cos \alpha - \sin \alpha) (\sin n\theta - \sin (n-2) \theta) \, d\theta = U(\beta_n - \beta_{n-2}), \\ \sigma_n &= -\frac{U}{\pi} \int_0^{2\pi} (s(\theta) \cos \alpha - \sin \alpha) (\cos n\theta - \cos (n-2) \theta) \, d\theta = -U(\alpha_n - \alpha_{n-2}) \end{aligned} \right\}$$

$$(n = 2, 3, \dots), \quad (2·8, 9)$$

where

$$\left. \begin{aligned} \alpha_n &= \frac{1}{\pi} \int_0^{2\pi} (s(\theta) \cos \alpha - \sin \alpha) \cos n\theta \, d\theta \\ \beta_n &= \frac{1}{\pi} \int_0^{2\pi} (s(\theta) \cos \alpha - \sin \alpha) \sin n\theta \, d\theta \end{aligned} \right\} \quad (n = 0, 1, 2, \dots). \quad (2·8, 10)$$

By a general property of the coefficients of a Fourier series $\alpha_n \to 0$ as $n \to \infty$. Hence

$$\sigma_1 = -\sigma_2 - \sigma_3 - \sigma_4 - \ldots = U[(\alpha_2 - \dot\alpha_0) + (\alpha_3 - \alpha_1) + (\alpha_4 - \alpha_2) + \ldots]$$
$$= -U(\alpha_0 + \alpha_1). \tag{2·8, 11}$$

The coefficient α_0 can be further analysed into components depending on the shape of the aerofoil and on its incidence respectively,

$$\alpha_0 = \cos\alpha \frac{1}{\pi} \int_0^{2\pi} s(\theta)\,d\theta - \sin\alpha \frac{1}{\pi} \int_0^{2\pi} d\theta = \alpha_0' \cos\alpha - 2\sin\alpha, \quad (2·8, 12)$$

where

$$\alpha_0' = \frac{1}{\pi} \int_0^{2\pi} s(\theta)\,d\theta,$$

while the other coefficients may be expressed more simply as $\alpha_n = \alpha_n' \cos\alpha$, $\beta_n = \beta_n' \cos\alpha$, where

$$\left.\begin{array}{l} \alpha_n' = \dfrac{1}{\pi} \displaystyle\int_0^{2\pi} s(\theta)\cos n\theta\,d\theta, \\[2mm] \beta_n' = \dfrac{1}{\pi} \displaystyle\int_0^{2\pi} s(\theta)\sin n\theta\,d\theta. \end{array}\right\} \tag{2·8, 13}$$

According to the boundary condition $v = Us(\theta)\cos\alpha$, and so the induced normal velocity $v' = v - U\sin\alpha$ equals, taking into account (2·8, 10),

$$v' = U\left\{\tfrac{1}{2}\alpha_0 + \sum_{n=1}^{\infty} (\alpha_n \cos n\theta + \beta_n \sin n\theta)\right\}. \tag{2·8, 14}$$

The coefficients of this Fourier series are subject to the restriction that the corresponding aerofoil must close up. Now the equation of the aerofoil is given by

$$\frac{dy}{dx} = s(\theta) = \frac{v}{U\cos\alpha} = \frac{v' + U\sin\alpha}{U\cos\alpha}.$$

But at the aerofoil $\quad x = 2a\cos\theta, \quad \dfrac{dx}{d\theta} = -2a\sin\theta,$

and so $\quad \dfrac{dy}{d\theta} = -\dfrac{2a}{\cos\alpha}\left[\tfrac{1}{2}\alpha_0 + \sin\alpha + \sum_{n=1}^{\infty} (\alpha_n \cos n\theta + \beta_n \sin n\theta)\right]\sin\theta,$

$$y = -\frac{2a}{\cos\alpha}\int\left[\tfrac{1}{2}\alpha_0 + \sin\alpha + \sum_{n=1}^{\infty} (\alpha_n \cos n\theta + \beta_n \sin n\theta)\right]\sin\theta\,d\theta.$$

The condition that the aerofoil should close up then is

$$\int_0^{2\pi}\left[\tfrac{1}{2}\alpha_0 + \sin\alpha + \sum_{n=1}^{\infty} (\alpha_n \cos n\theta + \beta_n \sin n\theta)\right]\sin\theta\,d\theta = 0.$$

Integrating term by term we see that this condition reduces to

$$\beta_1 = 0.$$

On the other hand, the induced longitudinal velocity $u' = u - U \cos \alpha$, now becomes, by (2·8, 5) and (2·8, 9),

$$u' = -\frac{1}{2 \sin \theta}(-\sigma_1 + \tau_2 \sin \theta - \sigma_2 \cos \theta + \tau_3 \sin 2\theta - \sigma_3 \cos 2\theta + \ldots)$$

$$= -\frac{U}{2 \sin \theta}(\alpha_0 + \alpha_1 + \beta_2 \sin \theta + (\alpha_2 - \alpha_0) \cos \theta + (\beta_3 - \beta_1) \sin 2\theta$$
$$+ (\alpha_3 - \alpha_1) \cos 2\theta + \ldots)$$

$$= U\left(-\alpha_0 \frac{1 - \cos \theta}{2 \sin \theta} - \alpha_1 \frac{1 - \cos 2\theta}{2 \sin \theta} + \beta_1 \frac{\sin 2\theta}{2 \sin \theta} - \alpha_2 \frac{\cos \theta - \cos 3\theta}{2 \sin \theta}\right.$$
$$\left. - \beta_2 \frac{\sin \theta - \sin 3\theta}{2 \sin \theta} - \ldots\right)$$

$$= U(-\tfrac{1}{2}\alpha_0 \tan \tfrac{1}{2}\theta - \alpha_1 \sin \theta + \beta_1 \cos \theta - \alpha_2 \sin 2\theta + \beta_2 \cos 2\theta - \ldots),$$

or
$$u' = U\left(-\tfrac{1}{2}\alpha_0 \tan \tfrac{1}{2}\theta + \sum_{n=1}^{\infty} (\beta_n \cos n\theta - \alpha_n \sin n\theta)\right). \tag{2·8, 15}$$

It will be seen that if we neglect the terms involving α_0 then u' is conjugate to v', as defined in § 2·5, and, conversely, $-v'$ is conjugate to u'. It is not possible to derive (2·8, 14) and (2·8, 15) directly from the Laurent series of $w' = u' - iv'$ on the circle S, because the presence of the term $-\tfrac{1}{2}\alpha_0 \tan \tfrac{1}{2}\theta$ implies that, in general, u' does not possess a Fourier expansion on S.

If the shape of the aerofoil is known, α_n and β_n can be determined from (2·8, 13). Since v is given directly in the terms of the shape of the aerofoil, $v = Us(\theta) \cos \alpha$, it is necessary only to calculate u or u' as given by (2·8, 15). If it is known that $\alpha_0 = 0$, formula (2·5, 5) may be used for the calculation of u'. In particular, if the aerofoil is symmetrical and at zero incidence, $\alpha_0 = \alpha_1 = \ldots = 0$, so that (2·5, 10) may be used. To calculate the pressure distribution we use Bernoulli's equation. Substituting

$$q^2 = (U \cos \alpha + u')^2 + (U \sin \alpha + v')^2$$

in (2·3, 28), we obtain

$$p = p_0 - \rho[U(u' \cos \alpha + v' \sin \alpha) + \tfrac{1}{2}(u'^2 + v'^2)], \tag{2·8, 16}$$

where p_0 is the free-stream pressure. Now according to the present approximation u' and v' are small compared with U, so that their squares may be neglected in (2·8, 16). (This assumption may break down near the leading edge.) Then
$$p = p_0 - \rho U(u' \cos \alpha + v' \sin \alpha), \tag{2·8, 17}$$

where $u' \cos \alpha + v' \sin \alpha$ is the component of the induced velocity in the direction of the free stream. This again may be equated approximately to u', so that finally
$$p - p_0 = -\rho U u'. \tag{2·8, 18}$$

We might calculate the aerodynamic force acting on the wing by means of (2·8, 18). However, for the case of a flat plate, this would clearly be a resultant force normal to the plate, whereas, according to the general theory, the resultant force is always normal to the direction of flow. The discrepancy is due to the fact that the method does not take into account a finite 'suction force' which acts at the sharp leading edge. The magnitude of their force will be calculated shortly. In the present case, the effect of the suction force is taken into account implicitly if we use the general formulae (2·2, 10) and (2·2, 11).

We have

$$C_{-1} = 1, \quad C_0 = 0, \quad A_0 = U e^{-i\alpha},$$

$$A_1 = T_1 = ia\sigma_1 = -iUa(\alpha_0 + \alpha_1) = iU\alpha[2\sin\alpha - (\alpha_0' + \alpha_1')\cos\alpha],$$

$$A_2 = T_2 = a^2(\tau_2 + i\sigma_2) = Ua^2[\beta_2 + i(\alpha_0 - \alpha_2)]$$
$$= Ua^2[-2i\sin\alpha + (\beta_2' + i(\alpha_0' - \alpha_2'))\cos\alpha],$$

and so

$$X - iY = -2\pi\rho U^2 ai\, e^{-i\alpha}[2\sin\alpha - (\alpha_0' + \alpha_1')\cos\alpha]$$
$$= -\tfrac{1}{2}\pi\rho U^2 ci\, e^{-i\alpha}[2\sin\alpha - (\alpha_0' + \alpha_1')\cos\alpha].$$

Hence

$$\left.\begin{array}{l} L = \tfrac{1}{2}\pi\rho U^2 c[2\sin\alpha - (\alpha_0' + \alpha_1')\cos\alpha], \\[4pt] C_L = \pi[2\sin\alpha - (\alpha_0' + \alpha_1')\cos\alpha], \\[4pt] D = C_D = 0, \end{array}\right\} \tag{2·8, 19}$$

or, putting $\sin\alpha \doteqdot \alpha$, $\cos\alpha \doteqdot 1$, for small α,

$$C_L = \pi[2\alpha - (\alpha_0' + \alpha_1')] = 2\pi(\alpha + \epsilon_0), \tag{2·8, 20}$$

where

$$\epsilon_0 = -\tfrac{1}{2}(\alpha_0' + \alpha_1')$$

$$= -\frac{1}{2\pi}\int_0^{2\pi} s(\theta)(1 + \cos\theta)\, d\theta$$

$$= -\frac{1}{2\pi}\left[\int_0^\pi s_u(\theta)(1 + \cos\theta)\, d\theta - \int_0^{-\pi} s_l(\theta)(1 + \cos\theta)\, d\theta\right]$$

$$= -\frac{1}{2\pi}\int_0^\pi (s_u(\theta) + s_l(\theta))(1 + \cos\theta)\, d\theta$$

$$= -\frac{1}{\pi}\int_0^\pi s_m(\theta)(1 + \cos\theta)\, d\theta,$$

where, for given θ, $0 \leqslant \theta < \pi$, $s_u(\theta)$ is the slope of the upper surface and $s_l(\theta)$ is the slope of the corresponding point at the lower surface, i.e.

$$s_u(\theta) = s(\theta), \quad s_l(\theta) = s(-\theta).$$

s_m is

$$s_m(\theta) = \tfrac{1}{2}(s_u(\theta) + s_l(\theta)) = \frac{1}{2}\left(\frac{dy_u}{dx} + \frac{dy_l}{dx}\right) = \frac{d}{dx}(\tfrac{1}{2}(y_u + y_l)) = \frac{dy_m}{dx},$$

where $y = y_u(x)$ and $y = y_l(x)$ are the equations of the upper and lower surfaces of the aerofoil respectively, so that $y = y_m(x)$ is the equation of the mean camber line and s_m its slope.

Now

$$x = \tfrac{1}{2}c\cos\theta, \quad \frac{dx}{d\theta} = -\tfrac{1}{2}c\sin\theta,$$

and so
$$\epsilon_0 = -\frac{1}{\pi}\int_0^\pi \frac{dy_m}{dx}(1+\cos\theta)\,d\theta$$

$$= \frac{2}{\pi c}\int_0^\pi \frac{dy_m}{d\theta}\frac{1+\cos\theta}{\sin\theta}\,d\theta$$

$$= \frac{2}{\pi c}\left[y_m\cot\tfrac{1}{2}\theta\right]_0^\pi + \frac{1}{\pi c}\int_0^\pi y_m\frac{d\theta}{\sin^2\tfrac{1}{2}\theta},$$

or
$$\epsilon_0 = \frac{2}{\pi}\int_0^\pi \frac{y_m}{c}\frac{d\theta}{1-\cos\theta}, \tag{2·8, 21}$$

provided that as θ tends to 0, y_m tends to 0 more rapidly than $\tan\tfrac{1}{2}\theta$, i.e. provided $\lim_{\theta\to 0} y_m\cot\tfrac{1}{2}\theta = 0$. But

$$\cot\tfrac{1}{2}\theta = \sqrt{\frac{1+\cos\theta}{1-\cos\theta}} = \sqrt{\frac{c+2x}{c-2x}},$$

and so it is sufficient that $y_m(x)$ tends to 0 as x tends to $\tfrac{1}{2}c$ and that it has a finite slope (less than vertical) at that point. We may clearly assume that these assumptions are realized in practice. (2·8, 20) shows that $\alpha = -\epsilon_0$ is the angle of zero lift.

To find an expression for the moment, we apply (2·2, 11) and obtain

$$M = \pi\rho\mathscr{I}[2U^2ai\,e^{-i\alpha}(2\sin\alpha - (\alpha_0' + \alpha_1')\cos\alpha)\,r\,e^{i\psi}$$
$$- U^2a^2(2\sin\alpha - (\alpha_0' + \alpha_1')\cos\alpha)^2$$
$$+ 2U^2a^2\,e^{-i\alpha}(-2i\sin\alpha + [\beta_2' + i(\alpha_0' - \alpha_2')]\cos\alpha)],$$

where we have put $-z_0 = r\,e^{i\psi}$. Thus we have approximately

$$M = 2\pi\rho U^2a^2\mathscr{I}\left[(2\sin\alpha - (\alpha_0' + \alpha_1')\cos\alpha)\frac{r}{a}i\,e^{i(\psi-\alpha)} + \beta_2'\,e^{-i\alpha}\cos\alpha\right.$$
$$\left. + i\,e^{-i\alpha}(-2\sin\alpha + (\alpha_0' - \alpha_2')\cos\alpha)\right].$$

In particular, at the quarter chord $z_0 = -\tfrac{1}{4}c = -a$, so that $r = a$, $\psi = 0$ and the pitching moment becomes

$$M_0 = 2\pi\rho U^2a^2\mathscr{I}[e^{-i\alpha}(\beta_2' - i(\alpha_1' + \alpha_2'))\cos\alpha]$$
$$= -2\pi\rho U^2a^2\cos\alpha((\alpha_1' + \alpha_2')\cos\alpha + \beta_2'\sin\alpha)$$
$$= -\tfrac{1}{8}\pi\rho U^2c^2\cos\alpha((\alpha_1' + \alpha_2')\cos\alpha + \beta_2'\sin\alpha). \tag{2·8, 22}$$

Again, if we take the moment about a point on the x-axis at a distance $\tfrac{1}{8}\beta_2'c$ upstream of the quarter chord, $z_1 = -\tfrac{1}{4}c(1+\tfrac{1}{2}\beta_2^2)$, then we obtain by simple statics, taking into account that the lift direction is at an angle $(\tfrac{1}{2}\pi - \alpha)$ to the horizontal,

$$M_1 = M_0 + (z_0 - z_1)L\cos\alpha = M_0 + \tfrac{1}{8}c\beta_2'L\cos\alpha$$
$$= \tfrac{1}{8}\pi\rho U^2c^2\cos\alpha(-(\alpha_1' + \alpha_2')\cos\alpha - \beta_2'\sin\alpha + \beta_2'\sin\alpha - \tfrac{1}{2}\beta_2'(\alpha_0' + \alpha_1')\cos\alpha)$$
$$= -\tfrac{1}{8}\pi\rho U^2c^2\cos^2\alpha((\alpha_1' + \alpha_2') + \tfrac{1}{2}\beta_2'(\alpha_0' + \alpha_1')). \tag{2·8, 23}$$

Thus the moment about z_1 is independent of incidence except for terms of the second order of smallness in α, so that z_1 may be regarded as the aerodynamic centre according to the present approximate theory. For a very thin aerofoil $s(\theta)$ is an even function of θ and β_2' vanishes, and in that case z_1 coincides with the quarter chord, z_0, and M_1 becomes equal to M_0. In general, β_2' is due to thickness, and for the reasons stated in § 2·4, the effect of thickness on the moment and on the position of the aerodynamic centre is not described adequately by linearized theory. Thus it is customary to neglect β_2' in any case, and to take

$$M_0 = -\tfrac{1}{8}\pi\rho U^2 c^2(\alpha_1' + \alpha_2')\cos^2\alpha,$$

or
$$M_0 = -\tfrac{1}{8}\pi\rho U^2 c^2(\alpha_1' + \alpha_2'), \quad \text{approximately.} \tag{2·8, 24}$$

as the expression for the moment about the quarter chord. The corresponding moment coefficient is

$$C_{M_0} = -\tfrac{1}{4}\pi(\alpha_1' + \alpha_2') = -\tfrac{1}{4}\pi((\alpha_0' + \alpha_1') - (\alpha_0' - \alpha_2')),$$

or
$$C_{M_0} = \tfrac{1}{2}\pi\epsilon - 2\mu_0, \tag{2·8, 25}$$

where ϵ_0 is defined as in (2·8, 20) and

$$\mu_0 = -\tfrac{1}{8}\pi(\alpha_0' - \alpha_2') = -\frac{1}{8}\int_0^{2\pi} s(\theta)(1 - \cos 2\theta)\,d\theta$$

$$= \frac{1}{c}\int_0^{\pi} \frac{dy_m}{dx}\sin\theta\,d\theta$$

$$= \frac{1}{c}[y_m \sin\theta]_0^{\pi} - \frac{1}{c}\int_0^{\pi} y_m \cos\theta\,d\theta,$$

$$\mu_0 = -\int_0^{\pi} \frac{y_m}{c}\cos\theta\,d\theta. \tag{2·8, 26}$$

To obtain Glauert's expressions for ϵ_0 and μ_0,[‡] we assume $c = 1$, and introduce the new variable of integration $\xi = \tfrac{1}{2}(1 + \cos\theta)$, $\cos\theta = 2\xi - 1$. ξ varies from 0 at the leading edge to 1 at the trailing edge. Then

$$\sin\theta = 2\sqrt{\{\xi(1 - \xi)\}}, \quad 1 - \cos\theta = 2(1 - \xi),$$

$$d\theta = -\frac{d\xi}{\sqrt{\{\xi(1 - \xi)\}}},$$

and so
$$\left.\begin{aligned}
\epsilon_0 &= \int_0^1 y_m \frac{d\xi}{\pi(1 - \xi)\sqrt{\{\xi(1 - \xi)\}}}, \\
\mu_0 &= \int_0^1 y_m \frac{(1 - 2\xi)\,d\xi}{\sqrt{\{\xi(1 - \xi)\}}}.
\end{aligned}\right\} \tag{2·8, 27}$$

To find the lift and moment coefficients, (2·8, 20), (2·8, 22), for a given aerofoil we therefore only have to evaluate (2·8, 27).

‡ H. Glauert, *The Elements of Aerofoil and Airscrew Theory* (2nd ed., Cambridge University Press, 1947), p. 91.

Let us now consider the case of a flat plate in some more detail. For this case

$$s(\theta) = 0$$

identically, so that

$$\alpha_0 = -2\sin\alpha, \quad \alpha_n = 0 \quad \text{for} \quad n \geqslant 1,$$

and

$$\beta_n = 0 \quad \text{for} \quad n \geqslant 0.$$

According to (2·8, 9) and (2·8, 11) this corresponds to

$$\sigma_1 = 2U\sin\alpha, \quad \sigma_2 = -2U\sin\alpha, \quad \sigma_n = 0 \quad (n \geqslant 3);$$

$$\tau_n = 0 \quad (n \geqslant 2).$$

Then

$$T_1 = i\sigma_1 a = 2Uai\sin\alpha, \quad T_2 = -2Ua^2 i\sin\alpha,$$

and

$$T_n = 0 \quad \text{for} \quad n \geqslant 3.$$

Hence

$$\tau' = 2iU\sin\alpha\left(\frac{a}{\zeta} - \frac{a^2}{\zeta^2}\right)$$

by (2·8, 4), so that the induced complex velocity in the physical plane is given by

$$w' = \tau' \div \frac{dz}{d\zeta} = 2iU\sin\alpha\frac{a/\zeta - a^2/\zeta^2}{1 - a^2/\zeta^2} = 2iU\sin\alpha\frac{a}{a+\zeta}.$$

Now

$$z = \zeta + \frac{a^2}{\zeta},$$

so that

$$\zeta = \tfrac{1}{2}\{z + \sqrt{(z^2 - 4a^2)}\},$$

where the sign of the square root is chosen so as to make $\zeta = z$ at infinity. Hence

$$\frac{a}{a+\zeta} = \frac{2a}{z + 2a + \sqrt{(z^2 - 4a^2)}} = \frac{1}{2}\left(1 - \sqrt{\frac{z-2a}{z+2a}}\right),$$

and

$$w = Ue^{-i\alpha} + 2iU\sin\alpha\frac{a}{a+\zeta} = U\left(\cos\alpha - i\sin\alpha\sqrt{\frac{z-2a}{z+2a}}\right).$$

Writing $z' = z + 2a$ we obtain

$$w = U\left(\cos\alpha + \sin\alpha\sqrt{\frac{4a-z'}{z'}}\right),$$

so that w becomes infinite at the leading edge as $(z')^{-\frac{1}{2}}$. We may therefore put

$$w = f(z') + \frac{C}{\sqrt{z'}}, \qquad (2·8, 28)$$

where $f(z')$ is bounded at the leading edge ($z' = 0$) and C is a constant; more precisely in the present case

$$C = U\sqrt{c}\sin\alpha,$$

where $c = 4a$ is the chord. C can also be determined by means of the relation

$$C = \lim_{z'\to 0}\sqrt{z'}\, u, \qquad (2·8, 29)$$

where z' approaches 0 along the aerofoil, and we agree to take the square root with positive sign.

Now apply Blasius's formula (2·2, 3) to a small circle S which surrounds the leading edge,

$$X - iY = \tfrac{1}{2}i\rho \int w^2 dz = \tfrac{1}{2}i\rho \int \left(f(z') + \frac{C}{\sqrt{z'}} \right)^2 dz'.$$

In this formula, X and Y are the components of the aerodynamic force which is exerted by the air on the part of the aerofoil which is inside the circle S. As the radius of S approaches 0 the integrals $\int \{f(z')\}^2 dz'$ and $\int 2f(z')\frac{C}{\sqrt{z'}} dz'$ approach 0 since $f(z')$ is bounded. Hence in the limit

$$X - iY = \tfrac{1}{2}i\rho C^2 \int \frac{dz'}{z'} = -\pi\rho C^2, \tag{2·8, 30}$$

where $X - iY$ is an aerodynamic force which is concentrated at the leading edge. More particularly, since C is real in the case under consideration,

$$X = -\pi\rho C^2, \quad Y = 0.$$

The existence of a suction force at the leading edge of a flat plate may also be inferred indirectly from the fact that if we calculate the aerodynamic force which is due to the pressure distribution over the surface of the plate, we clearly obtain a resultant which is normal to the plate. On the other hand, it was shown earlier that the total aerodynamic force which acts on an aerofoil must be normal to the direction of flow. The discrepancy is accounted for by the existence of the suction force, which can also be calculated as the difference of the two resultant forces just mentioned. However, since the suction force is given by

$$X = -\pi\rho C^2 = -\pi\rho U^2 c \sin^2 \alpha, \quad Y = 0,$$

its component normal to the direction of flow is $X \sin\alpha = -\pi\rho U^2 c \sin^3 \alpha$, and this is of the second order of smallness compared with the lift, which is $L = \pi\rho U^2 c \sin\alpha$. On the other hand, the component of the suction force parallel to the direction of flow, precisely balances the component of the pressure integral in that direction.

M. J. Lighthill‡ has shown how the results of two-dimensional thin aerofoil theory can be modified so as to yield finite velocities and pressures at the leading edge. This eliminates the concentrated suction force at the leading edge.

We observe that even if the Fourier expansion of v', (2·8, 14), is arbitrary, the complex velocity will still be of the general form (2·8, 28), since the terms corresponding to α_n, β_n, $n \geqslant 1$, are all bounded near the leading edge. As a result, the suction force at the leading edge is still given by (2·8, 30), where C may be complex.

‡ M. J. Lighthill, 'A new approach to thin aerofoil theory', *Aero. Quart.* vol. 3 (1951), pp. 193–210.

Assume now that the pressure distribution is specified as a function of θ, for a given incidence α, and that it is desired to determine the shape of the corresponding aerofoil section. Adopting the approximate form of Bernoulli's equation (2·8, 18), we shall assume instead that the velocity components u or u' are specified, for given α. For example, for the case of a symmetrical aerofoil at zero incidence, equations (2·8, 14) and (2·8, 15) become

$$\left.\begin{aligned} v' &= U \sum_{n=2}^{\infty} \beta_n \sin n\theta, \\ u' &= U \sum_{n=2}^{\infty} \beta_n \cos n\theta, \end{aligned}\right\} \qquad (2\cdot8, 31)$$

so that $-v'$ is conjugate to u' and may be determined from it by the use of (2·5, 10). The shape of the aerofoil is then given by the relation

$$\frac{dy}{dx} = -\frac{1}{2a \sin \theta}\frac{dy}{d\theta} = s(\theta) = \frac{v'}{U}. \qquad (2\cdot8, 32)$$

However, the Fourier expansion of u' in (2·8, 31) does not include terms for $n = 0$ and $n = 1$ and so implies the conditions

$$\int_0^\pi u'\, d\theta = 0, \qquad \int_0^\pi u' \cos \theta\, d\theta = 0.$$

The second equation now corresponds to the condition that the aerofoil should close up.

To avoid this restriction we introduce the following modifications in the theory following (2·8, 5).

Let
$$u = UT(\theta)$$

be specified as a function of θ, for a given incidence α. According to the first equation of (2·8, 5) we then have

$$-\sigma_1 + \tau_2 \sin \theta - \sigma_2 \cos \theta + \tau_3 \sin 2\theta - \sigma_3 \cos 2\theta + \ldots = 2U(\cos \alpha - T(\theta)) \sin \theta.$$

Hence $\quad \tau_n = \dfrac{2U}{\pi} \displaystyle\int_0^{2\pi} (\cos \alpha - T(\theta)) \sin \theta \sin (n-1)\theta\, d\theta \quad (n = 2, 3, \ldots),$

$$\sigma_n = -\frac{2U}{\pi} \int_0^{2\pi} (\cos \alpha - T(\theta)) \sin \theta \cos (n-1)\theta\, d\theta \quad (n = 2, 3, \ldots)$$

and $\qquad\qquad \sigma_1 = -\dfrac{U}{\pi} \displaystyle\int_0^{2\pi} (\cos \alpha - T(\theta)) \sin \theta\, d\theta.$

Now $\qquad\qquad \sin \theta \sin (n-1)\theta = -\tfrac{1}{2}[\cos n\theta - \cos (n-2)\theta],$

and so $\qquad\quad \sin \theta \cos (n-1)\theta = \quad \tfrac{1}{2}[\sin n\theta - \sin (n-2)\theta],$

$$\left.\begin{aligned} \tau_n &= \frac{U}{\pi} \int_0^{2\pi} (T(\theta) - \cos \alpha)(\cos n\theta - \cos (n-2)\theta)\, d\theta = U(\gamma_n - \gamma_{n-2}), \\ \sigma_n &= \frac{U}{\pi} \int_0^{2\pi} (T(\theta) - \cos \alpha)(\sin n\theta - \sin (n-2)\theta)\, d\theta = U(\delta_n - \delta_{n-2}), \end{aligned}\right\}$$
$$(n = 2, 3, \ldots) \quad (2\cdot8, 33)$$

and
$$\sigma_1 = \frac{U}{\pi} \int_0^{2\pi} (T(\theta) - \cos\alpha) \sin\theta \, d\theta = U\delta_1,$$

where
$$\gamma_n = \frac{1}{\pi} \int_0^{2\pi} (T(\theta) - \cos\alpha) \cos n\theta \, d\theta,$$

$$\delta_n = \frac{1}{\pi} \int_0^{2\pi} (T(\theta) - \cos\alpha) \sin n\theta \, d\theta.$$

Thus γ_n, δ_n are the Fourier coefficients of

$$T(\theta) - \cos\alpha = \frac{1}{U}(UT(\theta) - U\cos\alpha) = \frac{u'}{U},$$

i.e. (provided that the series on the right-hand side converges)

$$u' = U\left(\tfrac{1}{2}\gamma_0 + \sum_{n=1}^{\infty} (\gamma_n \cos n\theta + \delta_n \sin n\theta)\right). \qquad (2\cdot 8, 34)$$

The corresponding expression for v' is obtained from the second equation of (2·8, 5),

$$v' = v - U\sin\alpha$$

$$= \frac{1}{2\sin\theta}(\tau_2\cos\theta + \sigma_2\sin\theta + \tau_3\cos 2\theta + \sigma_3\sin 2\theta + \ldots)$$

$$= \frac{U}{2\sin\theta}((\gamma_2 - \gamma_0)\cos\theta + \delta_2\sin\theta + (\gamma_3 - \gamma_1)\cos 2\theta + (\delta_3 - \delta_1)\sin 2\theta + \ldots)$$

$$= U\left(-\gamma_0\frac{\cos\theta}{2\sin\theta} - \gamma_1\frac{\cos 2\theta}{2\sin\theta} - \delta_1\frac{\sin 2\theta}{2\sin\theta} + \gamma_2\frac{\cos\theta - \cos 3\theta}{2\sin\theta}\right.$$

$$\left. + \delta_2\frac{\sin\theta - \sin 3\theta}{2\sin\theta} + \ldots\right)$$

$$= U\left\{-\gamma_0\frac{\cos\theta}{2\sin\theta} - \gamma_1\frac{\cos 2\theta}{2\sin\theta} - \delta_1\cos\theta + \sum_{n=2}^{\infty}(\gamma_n\sin n\theta - \delta_n\cos n\theta)\right\}. \quad (2\cdot 8, 35)$$

To determine the shape of the aerofoil, we have, by (2·8, 6),

$$\frac{dy}{d\theta} = -\frac{2a\sin\theta}{\cos\alpha}\left\{\sin\alpha - \gamma_0\frac{\cos\theta}{2\sin\theta} - \gamma_1\frac{\cos 2\theta}{2\sin\theta} - \delta_1\cos\theta\right.$$

$$\left. + \sum_{n=2}^{\infty}(\gamma_n\sin n\theta - \delta_n\cos n\theta)\right\},$$

and carrying out the integration term by term, we find that the corresponding aerofoil closes up.

Now

$$\gamma_0\frac{\cos\theta}{2\sin\theta} + \gamma_1\frac{\cos 2\theta}{2\sin\theta} + \gamma_1\sin\theta = (\gamma_0 - \gamma_1)\frac{\cos\theta - 1}{4\sin\theta} + (\gamma_0 + \gamma_1)\frac{\cos\theta + 1}{4\sin\theta}$$

$$= \frac{\gamma_1 - \gamma_0}{2}\tan\tfrac{1}{2}\theta + \frac{\gamma_0 + \gamma_1}{2}\cot\tfrac{1}{2}\theta.$$

It follows that v' may also be written in the form

$$v' = U\left(\frac{\gamma_0 - \gamma_1}{2}\tan\tfrac{1}{2}\theta - \frac{\gamma_0 + \gamma_1}{2}\cot\tfrac{1}{2}\theta + \sum_{n=1}^{\infty}(\gamma_n\sin n\theta - \delta_n\cos n\theta)\right). \quad (2\cdot 8, 36)$$

In this expression, the term $\frac{\gamma_0-\gamma_1}{2}\tan\frac{1}{2}\theta$ corresponds to infinite slope at

the leading edge and the term $\frac{\gamma_0+\gamma_1}{2}\cot\frac{1}{2}\theta$ corresponds to infinite slope at

the trailing edge. However, a blunt trailing edge produces the breakdown of potential flow and is incompatible with the Joukowski condition. Hence $\gamma_0+\gamma_1=0$, or, in terms of u',

$$\int_0^{2\pi} u'(1+\cos\theta)\,d\theta=0.$$

An infinite slope near the leading edge, on the other hand, simply implies that the nose is rounded, and this condition is usually satisfied in practice. Thus v' becomes

$$v'=U\left(-\gamma_1\tan\tfrac{1}{2}\theta+\sum_{n=1}^{\infty}(\gamma_n\sin n\theta-\delta_n\cos n\theta)\right) \qquad (2\cdot8,37)$$

for specified $\quad u'=U\left(\tfrac{1}{2}\gamma_0+\sum_{n=1}^{\infty}(\gamma_n\cos n\theta+\delta_n\sin n\theta)\right).$ $\qquad (2\cdot8,38)$

The shape of the aerofoil is then given by

$$y=\frac{2a}{\cos\alpha}\int\left(-\sin\alpha+\gamma_1\tan\tfrac{1}{2}\theta+\sum_{n=1}^{\infty}(\delta_n\cos n\theta-\gamma_n\sin n\theta)\right)\sin\theta\,d\theta. \quad (2\cdot8,39)$$

If v' becomes infinite at the leading edge ($\gamma_1\neq0$), then $(2\cdot8,30)$ can be used to calculate the magnitude of the concentrated force near the leading edge. In this case also the force acts along the x-axis, but in the positive direction of that axis.

Formulae $(2\cdot8,37)$ and $(2\cdot8,38)$ were based on the assumption that u' can be expanded into a Fourier series, while conversely $(2\cdot8,15)$ presupposes the existence of the Fourier expansion $(2\cdot8,14)$ for v'. Combining these four formulae we obtain the more general pair of equations

$$\left.\begin{aligned}u'&=U\left(-\alpha_0\tan\tfrac{1}{2}\theta-\tfrac{1}{2}\gamma_1+\sum_{n=1}^{\infty}(\gamma_n\cos n\theta+\delta_n\sin n\theta)\right),\\ v'&=U\left(\tfrac{1}{2}\alpha_0-\gamma_1\tan\tfrac{1}{2}\theta+\sum_{n=1}^{\infty}(\gamma_n\sin n\theta-\delta_n\cos n\theta)\right).\end{aligned}\right\} \quad (2\cdot8,40)$$

However, in this general case neither u' nor v' can be expanded into a Fourier series at the aerofoil.

It will be observed that owing to the linearization of the boundary conditions adopted in this aerofoil theory, the effect of thickness and camber can be calculated separately. Indeed, if we write $y_u(x)$ and $y_l(x)$ for the ordinates of the upper and lower surfaces of an aerofoil, as above, then

$$y_u(x)=y_m(x)+y_t(x), \quad y_l(x)=y_m(x)-y_t(x),$$

where $\qquad\qquad y_m(x)=\tfrac{1}{2}(y_u(x)+y_l(x))$

9

is the mean camber line and

$$y_t(x) = \tfrac{1}{2}(y_u(x) - y_l(x))$$

is one-half the local thickness.

We may then regard the equation $y = y_m(x)$ as defining a very thin cambered aerofoil, while $y = \pm y_t(x)$ defines the upper and lower surfaces of a symmetrical aerofoil. The pressure distributions around these aerofoils can then be studied individually and the results superimposed to obtain the pressure distribution about the original aerofoil.

2·9 Goldstein's approximate method

Both of the exact methods of solution described in § 2·6 and 2·7 involve extensive numerical work in the calculation of specific examples. Goldstein‡ has developed a series of systematic approximations to the exact theory which lead to sufficiently accurate results with little labour in computation. The first approximation corresponds to the thin-wing theory of § 2·8, and the higher-order approximations allow for greater thickness and camber of the aerofoil.

We use Theodorsen's approach as the basis for developing the theory (§ 2·6). Thus, for aerofoils of zero thickness and camber, $\delta = \omega - \theta$ and ψ as defined by $z' = d\,e^{\psi + i\omega}$ and $\zeta = d\,e^{\chi + i\theta}$ in the complex z'- and ζ-planes are zero, so that for thin aerofoils we may write equation (2·6, 15) for δ as

$$\delta(\omega_0) = \frac{1}{2\pi} \int_0^{2\pi} \psi(\omega) \cot \frac{\theta - \theta_0}{2} d\theta$$

$$= \frac{1}{2\pi} \int_0^{2\pi} \psi(\omega) \cot \frac{\omega - \delta - \omega_0}{2} (1 - \delta'(\omega)) \, d\omega$$

$$\doteqdot \frac{1}{2\pi} \int_0^{2\pi} \psi(\omega) \cot \frac{\omega - \omega_0}{2} \, d\omega, \tag{2·9, 1}$$

neglecting the product of ψ and $\delta'(\omega)$. Again, from (2·6, 14), we have approximately

$$\chi_0 = \frac{1}{2\pi} \int_0^{2\pi} \psi(\omega) \, d\theta$$

$$= \frac{1}{2\pi} \int_0^{2\pi} \left(\psi(\theta) + \delta\psi'(\theta) + \frac{\delta^2}{2!} \psi''(\theta) + \dots \right) d\theta$$

$$\doteqdot \frac{1}{2\pi} \int_0^{2\pi} \psi(\theta) \, d\theta. \tag{2·9, 2}$$

The transformation to the z'-plane (equation (2·6, 3)) simplifies to

$$\left. \begin{array}{l} x \doteqdot 2d \cos \omega, \\[2mm] y \doteqdot 2d\,\psi \sin \omega. \end{array} \right\} \tag{2·9, 3}$$

‡ S. Goldstein, 'A theory of aerofoils of small thickness', Parts 1–6, *Curr. Pap. Aero. Res. Coun., Lond.*, nos. 68–73 (1952).

The velocity q is given by equation (2·6, 26),

$$\frac{q}{U} = \frac{|e^{\chi_0}(1-\delta')|}{\sqrt{\{(1+\psi'^2)(\sinh^2\psi + \sin^2\omega)\}}}$$

$$\left| \sqrt{\left(1 - \frac{C_L^2}{a_0^2}\right)} \sin(\omega - \delta + \beta) - \frac{C_L}{a_0}\cos(\omega - \delta + \beta) + \frac{C_L}{2\pi} \right|. \quad (2·9, 4)$$

In the first-order approximation we assume that the lift is small, so that squares of C_L/a_0 and its product with δ may be neglected. The squares and products of δ and ψ and their derivatives are also neglected. Then (2·9, 4) is approximately

$$\frac{q}{U} = \frac{|1 + \chi_0 - \delta'|}{|\sin\omega|} \left| \sin\omega + (\beta - \delta)\cos\omega - \frac{C_L}{a_0}\cos\omega + \frac{C_L}{2\pi} \right| \quad (2·9, 5)$$

or

$$\frac{q}{U} = \left| 1 + \chi_0 - \delta'(\omega) + (\beta - \delta)\cot\omega - \frac{C_L}{a_0}\cot\omega + \frac{C_L}{2\pi}\operatorname{cosec}\omega \right|$$

$$= |1 + g|, \quad \text{say.} \quad (2·9, 6)$$

This is a poor approximation for small values of $\sin\omega$ (it will be seen that there is a singularity in (2·9, 6) at $\sin\omega = 0$, which does not occur in the exact equation (2·9, 4)), and a better approximation for small $\sin\omega$ is obtained by replacing the factor

$$|e^{\chi_0}| / \sqrt{\{(1+\psi'^2)(\sinh^2\psi + \sin^2\psi)\}}$$

in (2·9, 4) by

$$\frac{|1 + \chi_0 + \tfrac{1}{2}\chi_0^2|}{\sqrt{(\psi^2 + \sin^2\omega)}}.$$

This is approximately equivalent to multiplying equation (2·9, 6) by

$$\frac{(1 + \tfrac{1}{2}\chi_0^2)|\sin\omega|}{\sqrt{(\psi^2 + \sin^2\omega)}}, \quad (2·9, 7)$$

so that the next degree of approximation for q/U is

$$\frac{q}{U} = \frac{(1 + \tfrac{1}{2}\chi_0^2)|\sin\omega|}{\sqrt{(\psi^2 + \sin^2\omega)}} |1 + g|. \quad (2·9, 8)$$

For high values of lift neither (2·9, 6) nor (2·9, 8) is valid and we must return to the expression (2·9, 4) for the third-degree approximation. However, in this highest order approximation, as in the first and second orders given by equations (2·9, 6) and (2·9, 8), it is still assumed that δ and χ_0 are given by the approximate integrals (2·9, 1) and (2·9, 2).

The next step is to separate effects of thickness and camber, since these are often most conveniently calculated independently. The resulting velocity fields for the first degree of approximation may then be superimposed additively. In the notation of § 2·8 we write

$$\left. \begin{aligned} y_u &= y_m + y_t, \\ y_l &= y_m - y_t, \end{aligned} \right\} \quad (2·9, 9)$$

where y_u and y_l are the ordinates of the upper and lower surfaces of the aerofoil respectively, y_m is the mean ordinate of the upper and lower surfaces, and y_t is the local thickness of the aerofoil. We extend this notation to

$$\left.\begin{aligned}\psi_u &= \psi_m + \psi_t,\\ \psi_l &= \psi_m - \psi_t,\end{aligned}\right\} \tag{2·9, 10}$$

where ψ is given by equation (2·9, 3). Then ψ_t is an even function of ω and ψ_m is an odd function of ω. For the derivatives we obtain

$$\left.\begin{aligned}\psi_u' &= \psi_m' + \psi_t',\\ \psi_l' &= \psi_m' - \psi_t',\end{aligned}\right\} \tag{2·9, 11}$$

and here ψ_t' is an odd and ψ_m' an even function of ω. In precisely a similar manner we define

$$\left.\begin{aligned}\delta_u &= \delta_m + \delta_t,\\ \delta_l &= \delta_m - \delta_t,\end{aligned}\right\} \tag{2·9, 12}$$

where, since ψ_t and ψ_m are even and odd functions of ω respectively, from the results of § 2·5 we can write, using (2·9, 1),

$$\left.\begin{aligned}\delta_t(\omega_0) &= -\frac{\sin \omega_0}{\pi} \int_0^\pi \frac{\psi_t(\omega)}{\cos \omega - \cos \omega_0}\, d\omega,\\ \delta_m(\omega_0) &= -\frac{1}{\pi} \int_0^\pi \frac{\psi_m(\omega) \sin \omega}{\cos \omega - \cos \omega_0}\, d\omega.\end{aligned}\right\} \tag{2·9, 13}$$

Note that here we have employed the approximation (2·9, 1) to the exact equation for δ. From (2·9, 13), since δ is conjugate to ψ, we see that $\delta_t(\omega)$ is an odd and $\delta_m(\omega)$ an even function.

To obtain expressions for δ_t' and δ_m' we differentiate (2·9, 1). Changing the variable of integration to $\eta = \omega - \omega_0$, we find

$$\begin{aligned}\delta(\omega_0) &= \frac{1}{2\pi} \int_{-\omega_0}^{2\pi - \omega_0} \psi(\eta + \omega_0) \cot \tfrac{1}{2}\eta\, d\eta\\ &= \frac{1}{2\pi} \int_0^{2\pi} \psi(\eta + \omega_0) \cot \tfrac{1}{2}\eta\, d\eta,\end{aligned}$$

since ψ is periodic with period 2π. Hence

$$\begin{aligned}\delta'(\omega_0) &= \frac{1}{2\pi} \int_0^{2\pi} \psi'(\eta + \omega_0) \cot \tfrac{1}{2}\eta\, d\eta\\ &= \frac{1}{2\pi} \int_{\omega_0}^{2\pi + \omega_0} \psi'(\omega) \cot \frac{\omega - \omega_0}{2}\, d\omega\\ &= \frac{1}{2\pi} \int_0^{2\pi} \psi'(\omega) \cot \frac{\omega - \omega_0}{2}\, d\omega.\end{aligned}$$

When $\psi(\omega)$ is an even function of ω, so that $\psi'(\omega)$ is odd, it follows from (2·5, 11) that

$$\delta'(\omega_0) = -\frac{1}{\pi} \int_0^\pi \frac{\psi'(\omega) \sin \omega}{\cos \omega - \cos \omega_0}\, d\omega, \tag{2·9, 14}$$

and when $\psi(\omega)$ is odd, from (2·5, 10),

$$\delta'(\omega_0) = -\frac{\sin\omega}{\pi}\int_0^\pi \frac{\psi'(\omega)}{\cos\omega - \cos\omega_0}\,d\omega. \qquad (2\cdot9, 15)$$

Hence it follows that the derivatives defined by

$$\left.\begin{aligned}\delta'_u &= \delta'_m + \delta'_t, \\ \delta'_l &= \delta'_m - \delta'_t,\end{aligned}\right\} \qquad (2\cdot9, 16)$$

are related to ψ'_t and ψ'_m by

$$\left.\begin{aligned}\delta'_t(\omega_0) &= -\frac{1}{\pi}\int_0^\pi \frac{\psi'_t(\omega)\sin\omega}{\cos\omega - \cos\omega_0}\,d\omega, \\[2mm] \delta'_m(\omega_0) &= -\frac{\sin\omega_0}{\pi}\int_0^\pi \frac{\psi'_m(\omega)}{\cos\omega - \cos\omega_0}\,d\omega,\end{aligned}\right\} \qquad (2\cdot9, 17)$$

and so δ'_t is an even and δ'_m an odd function of ω.

From these results it follows that the quantity g, defined by equation (2·9, 6), can be separated into components depending on thickness alone and camber alone. We put

$$g(\omega) = g_t(\omega) + g_m(\omega) + g_L(\omega), \qquad (2\cdot9, 18)$$

where

$$\left.\begin{aligned}g_t(\omega) &= \chi_0 - \delta'_t(\omega) - \delta_t(\omega)\cot\omega, \\[2mm] g_m(\omega) &= \quad -\delta'_m(\omega) - (\delta_m(\omega) - \beta)\cot\omega, \\[2mm] g_L(\omega) &= \frac{C_L}{2\pi}\operatorname{cosec}\omega - \frac{C_L}{a_0}\cot\omega.\end{aligned}\right\} \qquad (2\cdot9, 19)$$

It will be seen that $g_t(\omega)$ is symmetric, and g_m and g_L antisymmetric about $\omega = 0$.

We consider first the thickness contributions to g. The first equation of (2·9, 19), together with equations (2·9, 2), (2·9, 13) and (2·9, 17), yields

$$g_t(\omega_0) = \frac{1}{2\pi}\int_0^{2\pi}\psi(\omega)\,d\omega + \frac{1}{\pi}\int_0^\pi \frac{\psi'_t(\omega)\sin\omega}{\cos\omega - \cos\omega_0}\,d\omega$$
$$+ \frac{\sin\omega_0}{\pi}\int_0^\pi \frac{\psi_t(\omega)}{\cos\omega - \cos\omega_0}\,d\omega\cot\omega_0,$$

or, since $\displaystyle\int_0^{2\pi}\psi_m\,d\omega = 0$,

$$g_t(\omega_0) = \frac{1}{\pi}\int_0^\pi \frac{d\omega}{\cos\omega - \cos\omega_0}\{\psi_t(\omega)\cos\omega + \psi'_t(\omega)\sin\omega\}$$

$$= \frac{1}{\pi}\int_0^\pi \frac{d\omega}{\cos\omega - \cos\omega_0}\frac{d}{d\omega}(\psi_t(\omega)\sin\omega). \qquad (2\cdot9, 20)$$

With the use of equation (2·9, 3), this leads to

$$g_t(\omega_0) = \frac{1}{2\pi d}\int_0^\pi \frac{y'_t(\omega)}{\cos\omega - \cos\omega_0}\,d\omega, \qquad (2\cdot9, 21)$$

expressing g_t as an integral involving the slope of the symmetric aerofoil.

Similarly, noting that β is the value of δ for $\omega = 0$, the contribution to g due to camber is found to be

$$g_m(\omega_0) = \frac{\sin \omega_0}{\pi} \int_0^\pi \frac{\psi_m'(\omega)}{\cos \omega - \cos \omega_0} d\omega + \frac{\cot \omega_0}{\pi} \int_0^\pi \frac{\psi_m(\omega) \sin \omega}{\cos \omega - \cos \omega_0} d\omega$$
$$+ \frac{\cot \omega_0}{2\pi} \int_0^{2\pi} \psi(\omega) \cot \tfrac{1}{2}\omega \, d\omega.$$

Since $\psi_l(\omega)$ is even, the last integral on the right-hand side can be written

$$\frac{\cot \omega_0}{2\pi} \int_0^{2\pi} \psi_m(\omega) \cot \tfrac{1}{2}\omega \, d\omega = \frac{\cot \omega_0}{\pi} \int_0^\pi \psi_m(\omega) \frac{1 + \cos \omega}{\sin \omega} d\omega,$$

and thus

$$g_m(\omega_0) = \frac{1}{\pi} \int_0^\pi \left(\frac{\psi_m'(\omega) \sin \omega_0 + \psi_m(\omega) \sin \omega \cot \omega_0}{\cos \omega - \cos \omega_0} + \psi_m(\omega) \frac{\cos \omega_0 (1 + \cos \omega)}{\sin \omega \sin \omega_0} \right) d\omega$$
$$= \frac{1}{\pi} \sin \omega_0 \int_0^\pi \frac{\psi_m'(\omega) + \psi_m(\omega) \cot \omega}{\cos \omega - \cos \omega_0} d\omega - \frac{\tan \tfrac{1}{2}\omega_0}{\pi} \int_0^\pi \frac{\psi_m(\omega)}{\sin \omega} d\omega. \quad (2\cdot9, 22)$$

The last step follows from the identity

$$\frac{\cos \omega_0 (1 + \cos \omega)}{\sin \omega \sin \omega_0} + \frac{\sin \omega \cot \omega_0}{\cos \omega - \cos \omega_0} = \frac{\sin \omega_0 \cot \omega}{\cos \omega - \cos \omega_0} - \frac{\tan \tfrac{1}{2}\omega_0}{\sin \omega}.$$

We can write this expression in terms of the slope of the mean camber line $s = dy_m/dx$. Thus from $(2\cdot9, 3)$,

$$s(\omega) = 2d \frac{d}{d\omega} (\psi \sin \omega) \frac{d\omega}{dx} = -\psi'(\omega) - \psi(\omega) \cot \omega,$$

whilst, integrating by parts,

$$\int_0^\pi \frac{\psi_m(\omega)}{\sin \omega} d\omega = \int_0^\pi \psi_m \left(\sin \omega + \frac{\cos^2 \omega}{\sin \omega} \right) d\omega$$
$$= \int_0^\pi (\psi_m' \cos \omega + \psi_m \cot \omega \cos \omega) \, d\omega$$
$$= -\int_0^\pi s(\omega) \cos \omega \, d\omega. \quad (2\cdot9, 23)$$

So that we obtain from $(2\cdot9, 22)$

$$g_m(\omega_0) = \frac{\tan \tfrac{1}{2}\omega_0}{\pi} \int_0^\pi s(\omega) \cos \omega \, d\omega - \frac{\sin \omega_0}{\pi} \int_0^\pi \frac{s(\omega) \, d\omega}{\cos \omega - \cos \omega_0}. \quad (2\cdot9, 24)$$

The no-lift angle β can be expressed in terms of the slope $s(\omega)$. Thus, since β is the value of $\delta(\omega)$ for $\omega = 0$, we obtain from $(2\cdot9, 1)$

$$\beta = \frac{1}{\pi} \int_0^\pi \psi_m(\omega) \frac{1 + \cos \omega}{\sin \omega} d\omega$$
$$= -\frac{1}{\pi} \int_0^\pi s(\omega) \cos \omega \, d\omega + \frac{1}{\pi} \int_0^\pi (\psi_m'(\omega) + \psi_m(\omega) \cot \omega) \, d\omega - \frac{1}{\pi} \int_0^\pi \psi_m'(\omega) \, d\omega,$$

or
$$\beta = -\frac{1}{\pi}\int_0^\pi (1+\cos\omega)\, s(\omega)\, d\omega, \qquad (2\cdot9, 25)$$

provided $\psi_m(0) = \psi_m(\pi) = 0$, i.e. provided $\dfrac{y_m}{\sqrt{\{(2d)^2 + x_m^2\}}} \to 0$ as $x \to \pm 2d$, a

condition which is satisfied for all usual camber lines. The value of β given by (2·9, 25) is the same as obtained in § 2·8 on thin aerofoil theory.

According to the first-order approximation (equation (2·9, 6)) there are infinite suction peaks at the leading and trailing edges, $\omega = \pi, 0$ (compare with the discussion on the leading edge suction force in § 2·8; the appearance of an infinite force at the trailing edge is due to omission of the Joukowski condition). Although these are not necessarily present in the second- and third-order approximations, there is still in practice the danger of a large suction peak near the leading edge (results at the trailing edge are largely modified by boundary-layer effects). For the elimination of early transition or separation (cf. § 1·13) it is usually desirable to avoid excessive negative velocity gradients following the large velocity maximum near the leading edge, and there is one specific value of the lift coefficient, designated as $C_{L\,\text{opt.}}$, the optimum lift coefficient, for which $1 + g$ remains finite at $\omega = \pi$. Now from equations (2·9, 20), (2·9, 24) and (2·9, 19), it will be seen that the only possible singularities of g at $\omega = \pi$, the leading edge, occur in g_m and g_L, hence g will remain finite at $\omega = \pi$ provided

$$g_m(\omega) + g_L(\omega)$$
$$= \frac{\tan\frac{1}{2}\omega}{\pi}\int_0^\pi s(\omega_0)\cos\omega_0\, d\omega_0 - \frac{\sin\omega}{\pi}\int_0^\pi \frac{s(\omega_0)\, d\omega_0}{\cos\omega_0 - \cos\omega} - \frac{C_L}{a_0}\cot\omega + \frac{C_L}{2\pi}\operatorname{cosec}\omega$$

remains finite at $\omega = \pi$, i.e. provided $C_L = C_{L\,\text{opt.}}$, where

$$C_{L\,\text{opt.}}\left(\frac{1}{a_0} + \frac{1}{2\pi}\right) = -\frac{2}{\pi}\int_0^\pi s(\omega)\cos\omega\, d\omega. \qquad (2\cdot9, 26)$$

The velocity can be expressed in terms of the optimum lift coefficient by writing
$$g_i(\omega) = -\frac{\sin\omega}{\pi}\int_0^\pi \frac{s(\omega_0)\, d\omega_0}{\cos\omega_0 - \cos\omega}, \qquad (2\cdot9, 27)$$

so that for the antisymmetric contribution to g we obtain

$$g_m(\omega) + g_L(\omega) = g_i(\omega) + C_L\left(\frac{1}{2\pi}\operatorname{cosec}\omega - \frac{1}{a_0}\cot\omega\right) - \tfrac{1}{2}C_{L\,\text{opt.}}\left(\frac{1}{2\pi} + \frac{1}{a_0}\right)\tan\frac{1}{2}\omega$$

$$= g_i(\omega) + \frac{1}{2}\left(\frac{1}{a_0} + \frac{1}{2\pi}\right)(C_L - C_{L\,\text{opt.}})\tan\frac{1}{2}\omega - \frac{1}{2}\left(\frac{1}{a_0} - \frac{1}{2\pi}\right)C_L\cot\frac{1}{2}\omega.$$
$$(2\cdot9, 28)$$

This shows that g_i is the value of $g_m + g_L$ when $C_L = C_{L\,\text{opt.}}$ and $\omega = \pi$. In designing an aerofoil for a specified pressure or velocity distribution, $g_m + g_L$ is found in the first-order approximation as the antisymmetric part of the velocity distribution, but if the first-order approximation is to be used to calculate the aerofoil shape, it is necessary to eliminate the singularity at $\omega = 0$. This can only be done by choosing a C_L value equal to $C_{L\,opt.}$ which will correspond to a specific incidence, called the design incidence. We shall next consider the design problem in a little more detail.

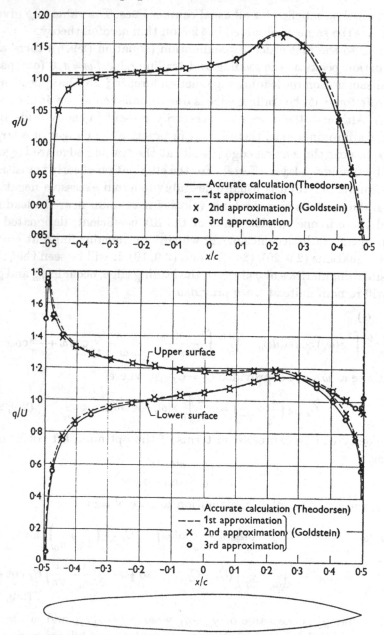

Fig. 2·9, 1. Velocity distributions about the symmetric low drag aerofoil EQH 1260 at zero incidence, and at a lift coefficient of $C_L = 0.4$, with $C_L = 4.4 \sin \alpha$. (*Courtesy H.M.S.O.*)

Having determined, from a given pressure distribution and the first-order equation (2·9, 6), the symmetric and antisymmetric contributions to g, i.e. the contributions due to thickness and camber respectively, the equations inverse to (2·9, 21) and (2·9, 27) are used to obtain the first-order approximation for the aerofoil shape. Thus, corresponding to equation (2·9, 21), we have

$$2y'_t(\omega) = -\frac{1}{\pi}\int_0^\pi \frac{g_t(\omega_0)\sin^2\omega_0}{\cos\omega_0 - \cos\omega}\,d\omega_0, \qquad (2·9, 29)$$

since $-g_t(\omega)\sin\omega$ is odd and is the conjugate of $2y'_t(\omega)$, and (referring back to § 2·5) there is no additive constant on the left-hand side since $g_t(\omega)\sin\omega$ is zero when $\omega = 0$. Similarly equation (2·9, 27) shows that $g_i(\omega)$ is odd and is the conjugate of $s(\omega) + s_0$, where s_0 is a constant, so that

$$s(\omega) + s_0 = \frac{1}{\pi}\int_0^\pi \frac{g_i(\omega_0)\sin\omega_0}{\cos\omega_0 - \cos\omega}\,d\omega_0, \qquad (2·9, 30)$$

and the constant s_0 is to be chosen so that $y_m = 0$ at $\omega = \pi$ and 0, the leading and trailing edges.

Integration of (2·9, 29) and (2·9, 30) will yield the thickness distribution and the shape of the mean camber line required according to the first-order approximations. With this solution the values of ψ and χ_0 can be calculated and used to find the values of

$$\frac{\sqrt{(\psi^2 + \sin^2\omega)}}{(1 + \tfrac{1}{2}\chi_0)^2\,|\sin\omega|},$$

which, when multiplied by the given values of q/U, will yield a modified velocity distribution to be used to get a better estimate of the aerofoil shape. This iterative process when repeated will finally give a result accurate according to the second approximation. If desired, the third-order approximation can be used to find the accurate velocity distribution which would exist with the shape of aerofoil calculated, and a final correction applied by using any theory which relates a small change in velocity with a small difference in contour.

Figures (2·9, 1) and (2·9, 2) show the result of calculations of the velocity distribution on the surfaces of two low drag aerofoils using the three degrees of approximation and also the exact formula, equation (2·9, 4).‡

2·10   The two-dimensional biplane

In the previous sections dealing with single aerofoils we have been able to derive all results by use of the complex representation and conformal transformations. Such methods can be extended to many-body problems, but the advantages of using the complex variable are then less manifest and we shall introduce in this section another approach which is also the

‡ These figures are taken from S. Goldstein, 'A theory of aerofoils of small thickness', Parts 1–6, *Curr. Pap. Aero. Res. Coun., Lond.*, nos. 68–73 (1952).

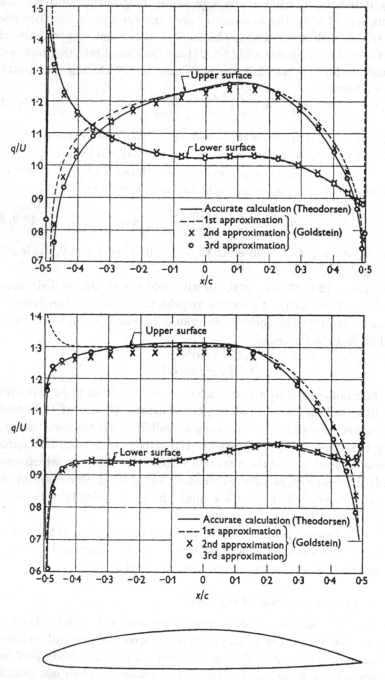

Fig. 2·9, 2.   Velocity distribution about the cambered low drag aerofoil EQH 1250/4050 at lift coefficients of $C_L = 0.2$ (upper figure) and $C_L = 0.6$, with $C_L = 4.8 \sin(\alpha + \beta)$. (*Courtesy H.M.S.O.*)

principal technique employed in Chapter 3 on three-dimensional aerofoils: the representation of the aerofoil or aerofoils by systems of singularities. However, before proceeding to this, we shall outline the method of solution of the biplane problem using transformations in the complex plane.‡

We shall consider one of the simplest configurations consisting of two straight parallel aerofoils of equal chord c and zero thickness. We suppose that the aerofoils are placed a distance h apart, symmetrically about the x-axis, as shown in Fig. 2·10, 1. As usual we suppose the incident velocity is U, making an angle α with the x-axis. A conformal transformation which will map this system into a 'tandem biplane' in the ζ-plane is given by

$$F(\zeta) = \frac{dz}{d\zeta} = i\,\frac{\zeta^2 - \lambda^2}{\sqrt{\{(\zeta^2 - 1)(\zeta^2 - k^2)\}}} \quad (1 > \lambda > k). \qquad (2\text{·}10, 1)$$

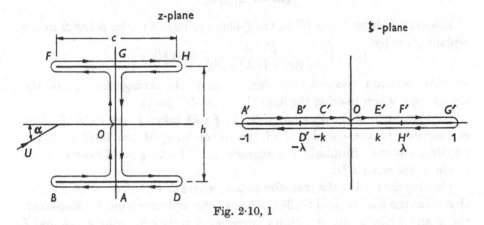

Fig. 2·10, 1

We choose $z = 0$ to correspond to $\zeta = 0$, and we take the sign of the square root such that for $1 > k > \xi > -k > -1$,

$$\sqrt{\{(\xi^2 - 1)(\xi^2 - k^2)\}} = -|\sqrt{\{(\xi^2 - 1)(\xi^2 - k^2)\}}|,$$

in order that the positive segment OE', $0 < \xi < k$, of the real axis of the ζ-plane should correspond to the upper positive segment $0 < z < i\dfrac{h}{2}$ of the imaginary axis in the z-plane. We then have

$$i\frac{h}{2} = \int_0^k F(\xi)\,d\xi = i\int_0^k \frac{\lambda^2 - \xi^2}{|\sqrt{\{(\xi^2 - 1)(\xi^2 - k^2)\}}|}\,d\xi, \qquad (2\text{·}10, 2)$$

Corresponding to the segment $E'F'$ of the ξ-axis we have the real quantity

$$\int_k^\lambda F(\xi)\,d\xi = -\frac{c}{2},$$

and moreover, since in going from the segment OE' to the segment $E'F'$

‡ The method is due to W. M. Kutta, 'Über ebene Zirkulationsströmungen nebst flugtechnischen Anwendungen', *S.B. bayer. Akad. Wiss.* (1911), p. 65.

in the ζ-plane we have passed through the singularity at E' ($\xi = k$) in a negative (clockwise) direction, it follows that in the latter segment

$$\sqrt{\{(\xi^2 - 1)(\xi^2 - k^2)\}} = -\frac{1}{i}\left|\sqrt{\{(\xi^2 - 1)(\xi^2 - k^2)\}}\right|,$$

(since $\sqrt{\{(\xi^2 - 1)(\xi^2 - k^2)\}} = -\left|\sqrt{\{(\xi^2 - 1)(\xi^2 - k^2)\}}\right|$ in the segment OE'). Hence

$$-\frac{c}{2} = \int_k^\lambda \frac{\xi^2 - \lambda^2}{\left|\sqrt{\{(\xi^2 - 1)(\xi^2 - k^2)\}}\right|}\, d\xi. \tag{2·10, 3}$$

Similarly we find that

$$\frac{c}{2} = \int_\lambda^1 F(\xi)\, d\xi = \int_\lambda^1 \frac{\lambda^2 - \xi^2}{\left|\sqrt{\{(\xi^2 - 1)(\xi^2 - k^2)\}}\right|}\, d\xi,$$

and adding the integrals corresponding to the segments OE', $E'F'$ and $F'G'$, we obtain

$$\tfrac{1}{2}ih = \int_0^1 F(\xi)\, d\xi.$$

Integration from G' to H' in the ζ-plane brings us to the point H in the z-plane given by

$$\tfrac{1}{2}ih + \int_1^\lambda \frac{\xi^2 - \lambda^2}{\left|\sqrt{\{(\xi^2 - 1)(\xi^2 - k^2)\}}\right|}\, d\xi = \tfrac{1}{2}ih + \tfrac{1}{2}c.$$

In this equation account has been taken of the change in sign of the square root when passing through the point G' ($\xi = 1$).

We thus see that the segment $E'G'$ ($1 > \xi > k$), taken twice, in the ζ-plane corresponds to the chord FH of the upper wing of the biplane in the physical z-plane. Similarly the segment $A'C'$ ($-k > \xi > -1$) corresponds to the lower wing BD.

The singularities of the transformation, where $dz/d\zeta = 0$, occur at $\xi = \pm\lambda$, that is, at the leading and trailing edges of the biplane aerofoils. Moreover, the x and y directions at infinity correspond to the directions $-\eta$ and ξ respectively. The factor λ is to be found from the condition that (2·10, 1) give a unique representation of z in terms of ζ, i.e. the integral of (2·10, 1) must be independent of the path of integration. This will be the case if the integral

$$\int \frac{\zeta^2 - \lambda^2}{\sqrt{\{(\zeta^2 - 1)(\zeta^2 - k^2)\}}}\, d\zeta, \tag{2·10, 4}$$

taken around each aerofoil, is zero. We can put this condition in more familiar form by the substitution, $\zeta = \sqrt{(1 - k'^2 t^2)}$, $k = \sqrt{(1 - k'^2)}$, so that it becomes

$$\int_0^1 \sqrt{\left(\frac{1 - k'^2 t^2}{1 - t^2}\right)}\, dt - \lambda^2 \int_0^1 \frac{dt}{\sqrt{\{(1 - t^2)(1 - k'^2 t^2)\}}} = 0,$$

or

$$\lambda^2 = \frac{E(k')}{K(k')}, \tag{2·10, 5}$$

where E and K are the complete elliptic integrals of the first and second kind,

$$E(k') = \int_0^1 \sqrt{\left(\frac{1 - k'^2 t^2}{1 - t^2}\right)}\, dt,$$

$$K(k') = \int_0^1 \frac{dt}{\sqrt{\{(1 - t^2)(1 - k'^2 t^2)\}}}.$$

To find the velocity distribution about the aerofoils we consider separately the effect of the horizontal and vertical components of the main-flow velocity U and the circulatory flow. The horizontal velocity $U \cos \alpha$ is not affected by the biplane system. The vertical component $U \sin \alpha$ is transformed into a horizontal velocity of equal magnitude in the ζ-plane, parallel to the tandem biplane system, so that in the ζ-plane this velocity component is unaffected by the aerofoils. The complex velocity in the z-plane without circulation is thus

$$U \cos \alpha + U \sin \alpha \frac{d\zeta}{dz} = U \cos \alpha - U \sin \alpha \frac{i \sqrt{\{(\zeta^2 - 1)(\zeta^2 - k^2)\}}}{\zeta^2 - \lambda^2}. \quad (2\cdot10, 6)$$

The circulatory flow can be considered as consisting of two components: equal circulations Γ_1 about each aerofoil of the biplane, and equal but opposite circulations Γ_2, which we call the counter-circulation. From considerations of symmetry and the behaviour at infinity we find that the velocities in the ζ-plane due to these circulatory flow, must be of the form

$$\left. \begin{array}{l} \tau_1 = \dfrac{\Gamma_1}{2\pi i \sqrt{\{(\zeta^2 - 1)(\zeta^2 - k^2)\}}} \zeta \\[3mm] \tau_2 = \dfrac{\Gamma_2}{2\pi i} \dfrac{A}{\sqrt{\{(\zeta^2 - 1)(\zeta^2 - k^2)\}}}, \end{array} \right\} \quad (2\cdot10, 7)$$

and

where A is a constant. The presence of the root factor ensures that the segments $-k > \xi > -1$, $1 > \xi > k$, are streamlines; the sign of the radical is determined in the fashion already introduced.

The values of Γ_1 and $\Gamma_2 A$ are obtained from the condition that the flow velocity w should remain finite at the trailing edge of the aerofoils. Here $dz/d\zeta = 0$, so that the velocity τ in the ζ-plane must be zero at the points $\zeta = \pm \lambda$ on the lower side of the tandem biplane. This means that

$$U \cos \alpha \frac{dz}{d\zeta} + U \sin \alpha + \frac{\Gamma_1 \zeta + \Gamma_2 A}{2\pi i} \frac{1}{\sqrt{\{(\zeta^2 - 1)(\zeta^2 - k^2)\}}} = 0$$

at $\zeta = \pm \lambda$ using (2·10, 6) and (2·10, 7). Hence

$$U \sin \alpha + \frac{1}{2\pi} \frac{1}{\sqrt{\{(1 - \lambda^2)(\lambda^2 - k^2)\}}} (\Gamma_1 \lambda + \Gamma_2 A) = 0$$

and

$$U \sin \alpha + \frac{1}{2\pi} \frac{1}{\sqrt{\{(1 - \lambda^2)(\lambda^2 - k^2)\}}} (\Gamma_1 \lambda - \Gamma_2 A) = 0.$$

Therefore $A = 0$ and the counter-circulation vanishes. Each aerofoil has an equal circulation

$$\Gamma_1 = \frac{-2\pi U \sin \alpha \sqrt{\{(1 - \lambda^2)(\lambda^2 - k^2)\}}}{\lambda}. \quad (2\cdot10, 8)$$

The factor $B = \dfrac{\sqrt{\{(1 - \lambda^2)(\lambda^2 - k^2)\}}}{\lambda}$ thus represents the reduction of the normal force acting on one wing of the biplane as compared with the force on the same aerofoil by itself at the same incidence. Numerical values B as a function of the gap to chord ratio h/c, calculated from equations (2·10, 2), (2·10, 3) and (2·10, 5), are given in Table 2·10, 1.

The forces acting on the individual aerofoils can be calculated by integration of the pressure distributions. The pressure is obtained from Bernoulli's equation in which the velocity is given by the sum of (2·10, 6) and (2·10, 7). When this is done it is found that the upper wing experiences a slightly greater lift than the lower, and also a smaller drag, the latter being balanced by an equal thrust on the lower aerofoil.

Table 2·10, 1

h/c	B
0·50	0·730
0·75	0·800
1·25	0·895
1·50	0·920

The exact method just described can be extended to cover the case of staggered aerofoils of unequal chord, but to calculate the effects of camber or thickness a series of conformal transformations has to be used. This has been done by Garrick[‡] using a method (analogous to Theodorsen's analysis for a single aerofoil (§ 2·6)) in which the region external to the biplane system is transformed into the annular region between two concentric circles. A simpler approach that leads to results that are problably of value equal to those obtained by conformal transformations, even though approximations are made, is one in which the aerofoils are replaced by a pair or a system of vortices, and their mutual interactions considered (see, for example, a report by Millikan[§]). We shall deal with the case of a pair of *thin* parallel aerofoils, taking them to be of equal chord and unstaggered, as shown in Fig. 2·10, 2. The method can readily be extended to cover the case when the aerofoils are staggered or have decalage (i.e. the aerofoils are not parallel). For very thin aerofoils, where thickness effects are neglected, the expansions (2·8, 14) and (2·8, 15) for the horizontal and vertical induced velocities on the aerofoil surface become

$$u' = -U\left(\tfrac{1}{2}\alpha_0 \tan \tfrac{1}{2}\theta + \sum_{n=1}^{\infty} \alpha_n \sin n\theta\right), \tag{2·10, 9}$$

$$v' = U\left(\tfrac{1}{2}\alpha_0 + \sum_{n=1}^{\infty} \alpha_n \cos n\theta\right), \tag{2·10, 10}$$

[‡] I. E. Garrick, 'Potential flow about arbitrary biplane wing sections', *N.A.C.A. Rep.* no. 542 (1936).

[§] C. B. Millikan, 'An extended theory of thin airfoils and its application to the biplane problem', *N.A.C.A. Rep.* no. 362 (1930).

since the coefficients β_n are zero. The variable θ is related to the chordwise coordinate x by

$$x = \tfrac{1}{2}c \cos \theta. \tag{2·10, 11}$$

Provided the local slope $s(\theta)$ of the aerofoil is small the induced velocity tangential to the aerofoil is also given by (2·10, 9), and the normal velocity is (2·10, 10). The tangential velocity therefore takes equal and opposite values on the upper and lower surfaces of the aerofoil. It follows from § 1·9 that the aerofoil can be replaced by a surface discontinuity in the form of a distribution of vorticity, the magnitude of which is given by

$$- [u'(\theta) - u'(-\theta)] = -2u'(\theta).$$

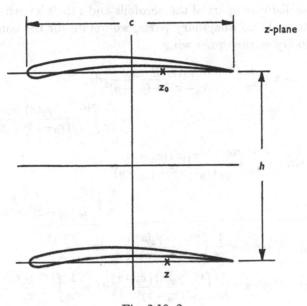

Fig. 2·10, 2

In the two-dimensional case the vorticity vector is always normal to the z-plane, and the complex velocity potential due to an isolated vortex of strength σ is given by the two-dimensional formula (1·10, 15)

$$\Pi(z) = \frac{i\sigma}{2\pi} \log (z - z_0).$$

The corresponding velocity is

$$w(z) = \frac{i\sigma}{2\pi} \frac{1}{z - z_0}. \tag{2·10, 12}$$

We shall assume that the same representation is possible for the two aerofoils forming the biplane, and, moreover, we shall assume that the expansion (2·10, 9) also holds for the self-induced velocity due to the vorticity distribution of each aerofoil. Thus, for the vorticity distribution on the upper aerofoil we put

$$\gamma_1 = -2u_1' = 2U\left(\tfrac{1}{2}\alpha_0 \tan\tfrac{1}{2}\theta_1 + \sum_{n=1}^{\infty} \alpha_n \sin n\theta_1\right) \quad (\pi \geqslant \theta_1 > 0), \quad (2\cdot10, 13)$$

suffix 1 denoting values on the upper aerofoil. A similar expression is taken for the lower aerofoil, using the suffix 2.

The induced velocity at a point z_0 of the upper aerofoil is found by integration of (2·10, 12) over the vorticity distributions on the two aerofoils:

$$w_1'(z_0) = \frac{1}{2\pi i}\int_{-\frac{1}{2}c+\frac{1}{2}ih}^{\frac{1}{2}c+\frac{1}{2}ih} \frac{\gamma_1(z)}{z-z_0}\,|\,dz\,| + \frac{1}{2\pi i}\int_{-\frac{1}{2}c-\frac{1}{2}ih}^{\frac{1}{2}c-\frac{1}{2}ih} \frac{\gamma_2(z)}{z-z_0}\,|\,dz\,|, \quad (2\cdot10, 14)$$

where h is the distance apart of the aerofoils and c their length. Separating (2·10, 14) into real and imaginary parts, we obtain for the components of the total velocity at the upper wing

$$u_1(z_0) = U\cos\alpha + \frac{1}{2\pi}\int_{-\frac{1}{2}c}^{\frac{1}{2}c} \frac{\gamma_1(x)\,(y-y_0)}{(x_0-x)^2+(y_0-y)^2}\,dx$$

$$+ \frac{1}{2\pi}\int_{-\frac{1}{2}c}^{\frac{1}{2}c} \frac{\gamma_2(x)\,(y-y_0)}{(x_0-x)^2+(y_0-y)^2}\,dx$$

and

$$v_1(z_0) = U\sin\alpha + \frac{1}{2\pi}\int_{-\frac{1}{2}c}^{\frac{1}{2}c} \frac{\gamma_1(x)\,(x_0-x)}{(x_0-x)^2+(y_0-y)^2}\,dx$$

$$+ \frac{1}{2\pi}\int_{-\frac{1}{2}c}^{\frac{1}{2}c} \frac{\gamma_2(x)\,(x_0-x)}{(x_0-x)^2+(y_0-y)^2}\,dx,$$

or

$$u_1(z_0) = U\cos\alpha - \frac{1}{2\pi}\int_{-\frac{1}{2}c}^{\frac{1}{2}c} \frac{\gamma_2(x)\,h}{(x_0-x)^2+h^2}\,dx - \frac{\gamma_1(z_0)}{2}$$

and

$$v_1(z_0) = U\sin\alpha + \frac{1}{2\pi}\int_{-\frac{1}{2}c}^{\frac{1}{2}c} \frac{\gamma_2(x)\,(x_0-x)}{(x_0-x)^2+h^2}\,dx + \frac{1}{2\pi}\int_{-\frac{1}{2}c}^{\frac{1}{2}c} \frac{\gamma_1(x)}{x_0-x}\,dx.$$

$$(2\cdot10, 15)$$

We can find the required vorticity distribution by applying the linearized boundary condition of thin aerofoil theory, as given in § 2·8,

$$\frac{v}{U\cos\alpha} = \frac{dy}{dx} = s(\theta).$$

Thus, the second equation of (2·10, 15) becomes

$$U\cos\alpha\,s(\theta_0) = U\sin\alpha + \frac{1}{2\pi}\int_0^{\pi} \frac{\gamma_2(\theta)\,(\cos\theta_0 - \cos\theta)}{(\cos\theta_0 - \cos\theta)^2 + (2h/c)^2}\sin\theta\,d\theta$$

$$+ \frac{1}{2\pi}\int_0^{\pi} \frac{\gamma_1(\theta)}{\cos\theta_0 - \cos\theta}\sin\theta\,d\theta. \quad (2\cdot10, 16)$$

Here we have used the fact that on the aerofoils $x = \tfrac{1}{2}c\cos\theta$.

Our next step would be to substitute for γ_1 and γ_2 series expansions such as (2·10, 13), then (2·10, 16) and the corresponding equation for the lower wing would enable us to find the coefficients α_n in the vorticity distributions. However, we shall first simplify the situation by assuming the gap between the aerofoils to be large compared with the chord length. Then (2·10, 16) becomes, neglecting powers of $(c/2h)^2$ greater than the first,

$$s(\theta_0)\cos\alpha \doteq \sin\alpha + \frac{1}{\pi}\left(\frac{c}{2h}\right)^2\int_0^\pi\left(\tfrac{1}{2}\alpha_0'\tan\tfrac{1}{2}\theta + \sum_{n=1}^\infty\alpha_n'\sin n\theta\right)(\cos\theta_0 - \cos\theta)\sin\theta\,d\theta$$

$$+\frac{1}{\pi}\int_0^\pi\frac{\tfrac{1}{2}\alpha_0\tan\tfrac{1}{2}\theta + \sum\limits_{n=1}^\infty\alpha_n\sin n\theta}{\cos\theta_0 - \cos\theta}\sin\theta\,d\theta,$$

where α and α' are the coefficients in the expansion (2·10, 13) for γ_1 and γ_2 respectively. Using the fact that

$$\int_0^\pi\frac{\cos n\theta}{\cos\theta_0 - \cos\theta}\,d\theta = -\pi\frac{\sin n\theta_0}{\sin\theta_0},$$

we can carry out the integrations indicated. We find that

$$s(\theta_0)\cos\alpha = \sin\alpha + \left(\frac{c}{2h}\right)^2\left(\tfrac{1}{2}\alpha_0'(\cos\theta_0 + \tfrac{1}{2}) + \tfrac{1}{2}\alpha_1'\cos\theta_0 - \tfrac{1}{4}\alpha_2'\right)$$

$$+ \tfrac{1}{2}\alpha_0 + \sum_{n=1}^\infty\alpha_n\cos n\theta_0. \qquad (2\cdot10, 17)$$

By symmetry it follows that for a biplane consisting of similar aerofoils $\alpha = \alpha'$, so that (2·10, 17) alone need be used to find the coefficients α and α'. In fact we obtain by integration

$$\cos\alpha\,\frac{1}{\pi}\int_0^{2\pi}s(\theta_0)\,d\theta_0 = 2\sin\alpha + \alpha_0\left(1 + \frac{1}{2}\left(\frac{c}{2h}\right)^2\right) - \alpha_2\frac{1}{2}\left(\frac{c}{2h}\right)^2,$$

$$\cos\alpha\,\frac{1}{\pi}\int_0^{2\pi}s(\theta_0)\cos\theta_0\,d\theta_0 = \alpha_1 + (\alpha_0 + \alpha_1)\frac{1}{2}\left(\frac{c}{2h}\right)^2,$$

$$\cos\alpha\,\frac{1}{\pi}\int_0^{2\pi}s(\theta_0)\cos n\theta_0\,d\theta_0 = \alpha_n \quad (n \geqslant 2).$$

$$(2\cdot10, 18)$$

Referring to § 2·8, equations (2·8, 12) and (2·8, 13), we find that

$$\alpha_0^* = \alpha_0\left(1 + \frac{1}{2}\left(\frac{c}{2h}\right)^2\right) - \alpha_2\frac{1}{2}\left(\frac{c}{2h}\right)^2,$$

$$\alpha_1^* = \alpha_1\left(1 + \frac{1}{2}\left(\frac{c}{2h}\right)^2\right) + \alpha_0\frac{1}{2}\left(\frac{c}{2h}\right)^2, \qquad (2\cdot10, 19)$$

$$\alpha_n^* = \alpha_n \quad (n \geqslant 2),$$

where α_k^* denotes the coefficient in the series expansion for the vorticity distribution of an isolated aerofoil, similar to the ones forming the biplane, situated in the same main-stream velocity at the same angle of attack. From (2·10, 19)

$$\alpha_0 + \alpha_1 = \frac{\alpha_0^* + \alpha_1^* + \frac{1}{2}\left(\frac{c}{2h}\right)^2 (\alpha_1^* + \alpha_2^*)}{\left(1 + \frac{1}{2}\left(\frac{c}{2h}\right)^2\right)^2},$$

or

$$-\pi(\alpha_0 + \alpha_1) = \frac{C_L^* + C_{M_0}^* 2\left(\frac{c}{2h}\right)^2}{\left(1 + \frac{1}{2}\left(\frac{c}{2h}\right)^2\right)^2}, \qquad (2·10, 20)$$

where C_L^* and $C_{M_0}^*$ are the lift coefficient and pitching moment coefficient about the quarter-chord for the isolated aerofoil, as given by equations (2·8, 20) and (2·8, 25). We can show that the left-hand side of equation (2·10, 20) is half the lift coefficient of the biplane. Thus, the total circulation is, with $\gamma_1 = \gamma_2 = \gamma$,

$$\Gamma = \int_{-\frac{1}{2}c}^{\frac{1}{2}c} (\gamma_1 + \gamma_2)\, dx = 2 \int_{-\frac{1}{2}c}^{\frac{1}{2}c} \gamma\, dx,$$

which becomes, on using the expression (2·10, 13) for γ,

$$\Gamma = 2Uc \int_0^{\pi} (\tfrac{1}{2}\alpha_0(1 - \cos\theta) + \alpha_1 \sin^2\theta)\, d\theta = Uc\pi(\alpha_0 + \alpha_1).$$

The law of Kutta–Joukowski as proved in §2·3 then yields for the lift coefficient

$$C_L = \frac{L}{\frac{1}{2}\rho U^2 c} = -\frac{2\Gamma}{cU} = -2\pi(\alpha_0 + \alpha_1)$$

$$= 2\frac{C_L^* + 2C_{M_0}^*\left(\frac{c}{2h}\right)^2}{\left(1 + \frac{1}{2}\left(\frac{c}{2h}\right)^2\right)^2}. \qquad (2·10, 21)$$

Retaining only powers of $(c/2h)^2$ less than the second, (2·10, 21) gives

$$C_L' = \tfrac{1}{2}C_L = C_L^* - \left(\frac{c}{2h}\right)^2 (C_L^* - 2C_{M_0}^*), \qquad (2·10, 22)$$

where C_L' is half the lift experienced by the biplane. C_L' approaches C_L^* as the gap h becomes larger. To compare the performance of the aerofoils when isolated and when in combination, we note that the lift coefficient C_L' for the biplane falls below the lift coefficient of the corresponding monoplane by an amount

$$\left(\frac{c}{2h}\right)^2 (C_L^* - 2C_{M_0}^*),$$

so that if the lift on the biplane is to equal the lift on the monoplane the incidence of the former must be increased by

$$\Delta\alpha = \frac{1}{2\pi}\left(\frac{c}{2h}\right)^2 (C_L^* - 2C_{M_0}^*). \qquad (2·10, 23)$$

The accuracy of this result may be judged by comparing it with the exact theory given at the start of this section for the case of straight aerofoils without decalage (in which case $C_{M_0}^* = 0$). Table 2·10, 2 gives the values of $\beta = \Delta\alpha/C_L^*$ for a few values of h/c according to (2·10, 23) and the exact theory.

Table 2·10, 2

h/c	β	
	Exact	Eqn. (2·10, 23)
0·50	0·059	0·159
0·75	0·039	0·071
1·00	0·027	0·040
1·25	0·019	0·026
1·75	0·014	0·018

2·11　Flow through a cascade of aerofoils

Consider the flow through a two-dimensional cascade consisting of an infinite set of similar aerofoils placed at the same incidence at equal intervals s on the y-axis (Fig. 2·11, 1). We suppose the flow at $x = \pm\infty$ to be uniform, having magnitudes U_1 and U_2 at $x = -\infty$ and $x = \infty$ respectively and making angles α_1 and α_2 with the x-axis. Then the flow pattern is repeated periodically in the direction of the cascade axis at intervals equal to the gap distance s as indicated by the set of similar streamlines drawn in Fig. 2·11, 1.

In order to find the lift force acting on the cascade we shall first have to establish the Kutta-Joukowski law for aerofoils in cascade. For this purpose we consider the equilibrium of a mass of fluid bounded by two similar streamlines AB and CD on either side of one aerofoil, and by two lines AC, BD parallel to the y-axis at large distances from the aerofoils. Then in the limit as AC and BD approach infinity the flow at AC and BD will be uniform. Doing this, the equation of continuity applied to the mass of fluid within the region shows at once that the horizontal flow velocities at $x = \pm\infty$ must be the same, and equal, say, to U. If the resultant force L on the aerofoil inside $ABDC$ acts at an angle α_m to the y-axis, we have, by equating the rate of change of momentum of the fluid within $ABDC$ to the force acting on the aerofoil

$$\left.\begin{array}{l} L = s\rho U(V + W - (V - W))\cos\alpha_m - (p_1 - p_2)\sin\alpha_m, \\ 0 = s\rho U(V + W - (V - W))\sin\alpha_m + (p_1 - p_2)\cos\alpha_m, \end{array}\right\} \quad (2·11, 1)$$

where p_1 and p_2 are the forces acting on AC and BD, and we have written the vertical components of U_1 and U_2 as $V + W$ and $V - W$ respectively, so that this component of velocity decreases by $2W$ in passing through the cascade. Bernoulli's equation (§ 1·6) applied to the region $ABDC$ gives

$$p_1 - p_2 = \frac{\rho s}{2}(U_2^2 - U_1^2). \quad (2·11, 2)$$

Substituting in the second equation of (2·11, 1) we get

$$\tan \alpha_m = \frac{U_1^2 - U_2^2}{4UW} = \frac{V}{U},$$ (2·11, 3)

since $U_1 = V + W$ and $U_2 = V - W$. Hence the lift L acts normal to the direction of the vector mean velocity $\mathbf{U}_m = \frac{1}{2}(\mathbf{U}_1 + \mathbf{U}_2) = \frac{1}{2}(\mathbf{U} + \mathbf{V})$. Using this result and equation (2·11, 2), the first equation of (2·11, 1) yields

$$L = \frac{2s\rho UW}{\sqrt{\{1 + (V/U)^2\}}} + \frac{2s\rho VW}{\sqrt{\{1 + (U/V)^2\}}}$$
$$= 2s\rho W \sqrt{(U^2 + V^2)},$$

or $$L = 2s\rho WU_m.$$ (2·11, 4)

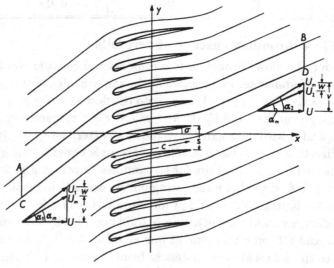

Fig. 2·11, 1

Integration round the boundary $ABDC$ shows that the circulation about one aerofoil is

$$\Gamma = -2sW,$$

and hence $$L = -\rho \Gamma U_m.$$ (2·11, 5)

Thus the law of Kutta–Joukowski, as derived in § 2·3 for an isolated aerofoil, is applicable to aerofoils in cascade if the vector mean velocity of the incoming and outgoing flow is used in place of the main-stream velocity.

Equation (2·11, 4) may be written in terms of the mean-flow direction α_m and the inflow and outflow angles α_1 and α_2 defined by

$$\tan \alpha_1 = \frac{V + W}{U}, \quad \tan \alpha_2 = \frac{V - W}{U},$$ (2·11, 6)

thus

$$L = 2s\rho WU_m = 2s\rho U_m^2 \frac{W/U}{\sqrt{\{1 + (V/U)^2\}}} = s\rho U_m^2 (\tan \alpha_1 - \tan \alpha_2) \cos \alpha_m.$$

Defining the lift coefficient for an aerofoil of the cascade as

$$C_L = L/\tfrac{1}{2}\rho U_m^2 c,$$

where c is the chord, we get

$$C_L = 2\frac{s}{c}(\tan\alpha_1 - \tan\alpha_2)\cos\alpha_m \Big\}$$

and $$\tan\alpha_m = \tfrac{1}{2}(\tan\alpha_1 + \tan\alpha_2),$$

$$(2·11, 7)$$

from which we see that the flow is always deflected in passing through the cascade, unless there is no force acting on it, and this can only occur if the circulation is zero.

Equation (2·11, 7) is the basic relationship for a cascade. From it the outlet flow angle can be calculated for a given inlet angle if the lift coefficient is known. To find the lift, the flow through the cascade must be investigated in greater detail. We shall deal first with a conformal transformation method for solving the problem.

The simplest case to consider is the lattice of straight aerofoils. Take the transformation from the z_1' to the ζ-plane to be given by

$$z_1' = \frac{s\cos\sigma}{2\pi}\left(\log\frac{b+\zeta}{b-\zeta} + \log\frac{\zeta+a^2/b}{\zeta-a^2/b}\right),\qquad (2·11, 8)$$

where a, $b > a$, are real. It can be expressed alternatively as

$$z_1' = \frac{s\cos\sigma}{2\pi}\log\left(\frac{\cosh\gamma + \cosh(\chi+i\theta)}{\cosh\gamma - \cosh(\chi+i\theta)}\right),\qquad (2·11, 9)$$

where $\zeta = a\,e^{\chi+i\theta}$ and $e^\gamma = b/a$, so that (2·11, 8) transforms the circle $\zeta = a\,e^{i\theta}$ into the straight lines given by the principal value of

$$z_1' = \frac{s\cos\sigma}{2\pi}\log\left(\frac{\cosh\gamma + \cos\theta}{\cosh\gamma - \cos\theta}\right),\qquad (2·11, 10)$$

and values differing from it by $\dfrac{s\cos\sigma}{2\pi}2k\pi i$, k taking integral values. Since the argument of the logarithm in equation (2·11, 10) is positive, the lines are parallel to the real axis as shown in Fig. 2·11, 2.

Correspondingly the transformation

$$z_2' = -\frac{is\sin\sigma}{2\pi}\left(\log\frac{b+\zeta}{b-\zeta} - \log\frac{\zeta+a^2/b}{\zeta-a^2/b}\right),\qquad (2·11, 11)$$

maps the circle $\zeta = a\,e^{i\theta}$ on to a series of straight lines lying on the real axis in the z_2'-plane with a period of $s\sin\sigma$ (see Fig. 2·11, 3).

Combining the transformations (2·11, 8) and (2·11, 11) we obtain

$$z' = z_1' + z_2' = \frac{s}{2\pi}\left(e^{-i\sigma}\log\frac{b+\zeta}{b-\zeta} + e^{i\sigma}\log\frac{\zeta+a^2/b}{\zeta-a^2/b}\right).\qquad (2·11, 12)$$

This maps the cascade of straight aerofoils of gap s and stagger σ in the z'-plane on to the circle $\zeta = a\,e^{i\theta}$ in the ζ-plane.

The end-points of the aerofoil chords are given by $dz'/d\zeta = 0$, i.e. by

$$e^{-i\sigma}\frac{2b}{b^2-\zeta^2} - e^{i\sigma}\frac{2a^2/b}{\zeta^2-(a^2/b)^2} = 0$$

with $\zeta = a\,e^{i\theta}$. This equation simplifies to

$$\tan\theta = \tanh\gamma\,\tan\sigma. \qquad (2·11, 13)$$

The two principal solutions of this equation θ_L and θ_T, differing by π, give the points in the ζ-plane corresponding to the end-points of the chord in the z'-plane. Hence we can find the chord length c by substituting $\zeta = a\,e^{i\theta_L}$

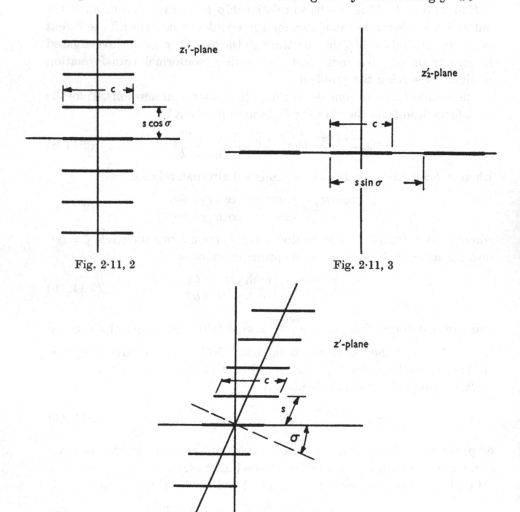

Fig. 2·11, 2　　　　　　　　　　　　　　　　　Fig. 2·11, 3

Fig. 2·11, 4

and $\zeta = a\,e^{i\theta}\tau$ in (2·11, 12) and subtracting the two values of z' thus obtained. After reduction we get

$$c = \frac{2s}{\pi}\left\{\cos\sigma\,\log\frac{\sqrt{(\cosh^2\gamma - \sin^2\sigma)} + \cos\sigma}{\sin\gamma} + \sin\sigma\,\tan^{-1}\left(\frac{\sin\sigma}{\sqrt{(\cosh^2\gamma - \sin^2\sigma)}}\right)\right\}.$$

$$(2·11, 14)$$

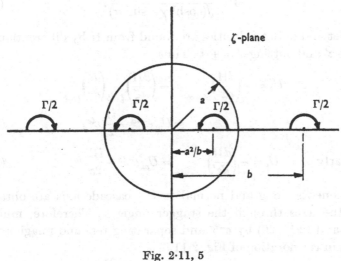

Fig. 2·11, 5

The singularities of the transformation (2·11, 12) are at $\zeta = \pm b$, $\zeta = \pm a^2/b$, located as shown in Fig. 2·11, 5, so that the potential function for the flow about the circle in the ζ-plane can have singularities only at these points. The velocity potential

$$\Pi' = -U_m z' = -\frac{U_m s}{2\pi}\left(e^{-i\sigma}\log\frac{b+\zeta}{b-\zeta} + e^{i\sigma}\frac{\zeta + a^2/b}{\zeta - a^2/b}\right) \qquad (2·11, 15)$$

satisfies these conditions, having sinks only at b, a^2/b, and sources at $-b$, $-a^2/b$, and the circle $\zeta = a\,e^{i\theta}$ as a streamline. In the z'-plane it represents uniform flow through the cascades. The modified potential

$$\Pi'' = -\frac{U_m s}{2\pi}\left(e^{-i(\sigma+\alpha)}\log\frac{b+\zeta}{b-\zeta} + e^{i(\sigma+\alpha)}\log\frac{\zeta + a^2/b}{\zeta - a^2/b}\right) \qquad (2·11, 16)$$

represents undeflected flow through the cascade, with incident and exit angle α, as we can readily verify by calculating the flow at $\pm\infty$ in the z'-plane where $\zeta = \pm b$. To introduce a circulation flow we add

$$\frac{i\Gamma}{4\pi}\log\frac{\zeta^2 - a^4/b^2}{\zeta^2 - b^2} \qquad (2·11, 17)$$

to (2·11, 16). This is the flow produced by two equal vortices at $\zeta = \pm a^2/b$, and opposite vortices at $\zeta = \pm b$, each being of strength $\frac{1}{2}\Gamma$, as shown in Fig. 2·11, 5. The circle $|\zeta| = a$ is a streamline and the circulation about it is Γ. To find the magnitude of Γ we apply the Joukowski condition that the velocity should be finite at the sharp trailing edges. Thus, if Π is the total velocity potential in the z'-plane (i.e. the sum of (2·11, 16) and (2·11, 17)), we must have $d\Pi/d\zeta = 0$ at the trailing edge, this being given by the solution of (2·11, 13). After some lengthy algebra we find that

$$\Gamma = \frac{-2sU\sin\alpha}{\sqrt{(\cosh^2\gamma - \sin^2\sigma)}}. \tag{2·11, 18}$$

The inlet and outlet velocities are found from Π by differentiating with respect to z' and putting $z' = \pm\infty$. Thus

$$U_2 = -\left(\frac{d\Pi}{dz'}\right)_{z'=\infty} = -\left(\frac{d\Pi}{d\zeta}\right)_{\zeta=b}\left(\frac{d\zeta}{dz}\right)_{\zeta=b}$$

$$= U_m e^{-i\alpha} - \frac{i\Gamma}{2s}e^{i\sigma}, \tag{2·11, 19}$$

and similarly
$$U_1 = -\left(\frac{d\Pi}{dz'}\right)_{z'=-\infty} = U_m e^{-i\alpha} + \frac{i\Gamma}{2s}e^{i\sigma}. \tag{2·11, 20}$$

The components along and normal to the cascade axis are obtained by rotating the axes through the stagger angle σ. Therefore, multiplying (2·11, 19) and (2·11, 20) by $e^{-i\sigma}$ and separating real and imaginary parts, we obtain in the notation of Fig. 2·11, 1

$$\left.\begin{aligned} V - W &= U_m \sin(\alpha + \sigma) + \frac{\Gamma}{2s}, \\ V + W &= U_m \sin(\alpha + \sigma) - \frac{\Gamma}{2s}, \\ U &= U_m \cos(\alpha + \sigma). \end{aligned}\right\} \tag{2·11, 21}$$

It follows that
$$\alpha = \alpha_m - \sigma,$$

and is the angle that the mean flow makes with the chord of an aerofoil of the cascade. It may be regarded as the effective angle of incidence of the aerofoils.

Equation (2·11, 18) is the relationship we require to determine the outlet angle from the general equation (2·11, 7). For from (2·11, 18) we can find the lift coefficient as
$$C_L = -\rho U_m \Gamma / \tfrac{1}{2}\rho U_m^2 c,$$

and substituting in (2·11, 7) we get

$$\frac{2\sin(\alpha_m - \sigma)}{\sqrt{(\cosh^2\gamma - \sin^2\sigma)}} = (\tan\alpha_1 - \tan\alpha_2)\cos\alpha_m,$$

or
$$2(\tan\alpha_m - \tan\sigma) = (\tan\alpha_1 - \tan\alpha_2)\cos\sigma\sqrt{(\cosh^2\gamma - \sin^2\sigma)}.$$

Writing
$$Q = \cos\sigma\sqrt{(\cosh^2\gamma - \sin^2\sigma)} - 1,$$

this becomes
$$\tan\alpha_2 = \frac{Q\tan\alpha_1 + 2\tan\sigma}{Q + 2}. \tag{2·11, 22}$$

(2·11, 22) gives the outlet flow angle in terms of the inlet angle and the geometry of the cascade. The quantity γ involved in Q has to be found by solution of equation (2·11, 14) (this must be done numerically, except for some special cases). The relationship is in fact quite general, as has been shown by Merchant,‡ and holds for any type of cascade with values of Q which are fixed by the cascade geometry alone.

The general case of an infinite straight cascade of similar aerofoils can also be solved in a variety of ways using conformal transformations. In particular, we refer to the method used by Garrick§ in which the transformation (2·11, 12) is used to map the aerofoils into an oval in the ζ-plane, which approximates a circle for thin aerofoils of small camber. Theodorsen's method (§ 2·6) is then followed to map this approximate circle into an exact circle in the ζ'-plane by the transformation

$$\frac{\zeta}{\zeta'} = \exp\left(\sum_{n=1}^{\infty} \frac{\alpha_n + i\beta_n}{\zeta'^n}\right).$$

The coefficients α_n, β_n are determined in a fashion analogous to that used by Theodorsen. Other methods employ an initial transformation which reduces the set of aerofoils to a single closed region; all these depend on the periodicity of the complex exponential function e^z. Thus, for example, Howell‖ uses the transformation

$$\zeta = \tanh z',$$

and follows this by three further transformations designed to transform the cascade finally into a circle.

In numerous cases the practical need is for cascades that turn the flow through large angles, so that the aerofoils of the cascade have to be highly cambered. In such cases as this the conformal mapping method of solution can become particularly lengthy and complicated. Another general method that is often simpler is one in which the aerofoils are replaced by vortex distributions, in a similar manner to that described in § 2·10 for the biplane. It is particularly suited to the inverse problem of designing aerofoils of a cascade for a specified pressure distribution.¶

As in § 2·10 on the biplane, we shall consider the case of thin aerofoils so that they can be represented by a distribution of vorticity $\gamma(z)$ along their

‡ S. M. Merchant, 'Flow of an ideal fluid past a cascade of blades', *Rep. Memor. Aero. Res. Comm., Lond.*, no. 1890 (1940).

§ I. E. Garrick, 'On the plane potential flow past a lattice of arbitrary airfoils', *N.A.C.A. Rep.* no. 778 (1944).

‖ A. R. Howell, 'Note on the theory of arbitrary aerofoils in cascade', *Roy. Aero. Est. Tech. Note*, no. E.3859 (1941).

¶ See A. W. Goldstein and M. Jerison, 'Isolated and cascade aerofoils with prescribed velocity distribution', *N.A.C.A. Rep.* no. 869 (1947). Also L. Diesendruck, 'Iterative interference methods in the design of thin cascade blades', *Tech. Notes Nat. Adv. Comm. Aero., Wash.*, no. 1254 (1947).

mean chord line. The complex velocity w' induced at a point z_0 by this vortex distribution is (see equation (1·10, 15))

$$w'(z_0) = \Sigma \frac{1}{2\pi i} \int \frac{\gamma(z)}{z_0 - z} |dz|, \qquad (2\cdot11, 23)$$

where the summation is over all the aerofoils, and the integrations are along the chords of the aerofoils. The axis of the cascade lies on the imaginary axis as shown in Fig. 2·11, 6. We can reduce (2·11, 23) to integration over a single aerofoil of the system, say the central one whose chord line passes through the origin, by putting

$$w'(z_0) = \sum_{n=-\infty}^{\infty} \frac{1}{2\pi i} \int \frac{\gamma(z)}{(z_0 - z) + ins} |dz|, \qquad (2\cdot11, 24)$$

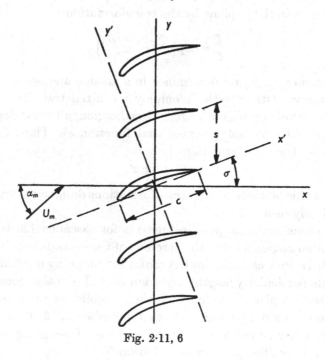

Fig. 2·11, 6

where z now lies on the central aerofoil and the integration is over its chord. We can write the equation as

$$w'(z_0) = \frac{1}{2\pi i} \int \frac{\gamma(z)}{z_0 - z} |dz| + \frac{1}{2\pi i} \sum_{n=1}^{\infty} \int \frac{2\gamma(z)(z_0 - z)}{(z_0 - z)^2 + (ns)^2} |dz|.$$

The sum of this series is given by Bromwich‡, and we obtain

$$w'(z_0) = \frac{1}{2\pi i} \int \frac{\pi}{s} \gamma(z) \coth\left(\pi \frac{z_0 - z}{s}\right) |dz|. \qquad (2\cdot11, 25)$$

‡ T. J. I'a. Bromwich, *An Introduction to the Theory of Infinite Series* (2nd ed., Macmillan and Co. 1942), p. 296.

The vorticity distributions γ on the aerofoils contribute the circulatory flow and the main-stream velocity component W (Fig. 2·11, 1). The remaining components U and V of the main flow can be represented in the complex plane as $U_m e^{-i\alpha_m}$. Thus the total velocity is

$$w(z_0) = U_m e^{-i\alpha_m} + \frac{1}{2\pi i} \int \gamma(z) \frac{\pi}{s} \coth\left(\pi \frac{z_0 - z}{s}\right) |dz|. \qquad (2·11, 26)$$

A rather more general form of equation (2·11, 26) has been given by Goldstein and Jerison,‡ in which the equation is rendered exact by using, not a distribution of vorticity along the mean chord lines as we have done, but a vorticity distribution along the surfaces of the aerofoils. The integral in (2·11, 26) along the chord is then replaced by a contour integral round the aerofoil boundary, and, moreover, the velocity at a point z_0 on the aerofoil surface is equal to the local vortex strength, $w(z_0) = \gamma(z_0)$ (for equation (2·11, 26), the corresponding equation would be $w_{1t} - w_{2t} = \gamma$, where w_{1t} and w_{2t} are the tangential velocities on either side of the aerofoil), so that the equation reduces to an integral equation for $w(z_0)$.

Our fundamental equation (2·11, 26) can be solved by expansion of γ in a suitable series, the coefficients of which are found by evaluating the integral and using the boundary condition that the flow should be tangential to the aerofoil boundary. However, instead of covering the general case, we shall outline the results obtained by use of thin aerofoil theory, in which case the boundary condition is taken to be (see § 2·8)

$$v' = U_m f'(\theta) \cos(\alpha_m - \sigma), \qquad (2·11, 27)$$

where θ is given by $x' = \frac{1}{2} c \cos \theta$, v' is the velocity normal to the chord and $f'(\theta) = dy'/dx'$ is the local slope of the aerofoil. The lines $y' = 0$, $x' = 0$, are here taken to lie along and normal to the chord of the central aerofoil, as shown in Fig. 2·11, 6. We shall assume the gap to chord ratio s/c to be large, in which case the analysis is quite simple. Now

$$w(z_0) = U_m e^{-i\alpha_m} + \frac{1}{2\pi i} \int \gamma(z) \frac{\pi}{s} \frac{e^{\pi(z_0-z)/s} + e^{-\pi(z_0-z)/s}}{e^{\pi(z_0-z)/s} - e^{-\pi(z_0-z)/s}} |dz|$$

$$= U_m e^{-i\alpha_m} + \frac{1}{2\pi i} \int \gamma(z) \frac{\pi}{s} \frac{1 + \frac{1}{2}(\pi/s)^2 (z_0-z)^2 + \dots}{(\pi/s)(z_0-z) + \frac{1}{3!}(\pi/s)^3 (z_0-z)^3 + \dots} |dz|,$$

and writing $z = \zeta c$, this becomes for large s/c,

$$w(\zeta_0) \doteqdot U_m e^{-i\alpha_m} + \frac{1}{2\pi i} \int \frac{\gamma(\zeta)}{\zeta_0 - \zeta} |d\zeta| + \frac{1}{6\pi i} \left(\frac{\pi c}{s}\right)^2 \int \gamma(z)(\zeta_0 - \zeta) |d\zeta|. \qquad (2·11, 28)$$

To apply the boundary conditions (2·11, 27), we need to find the components of velocity normal and tangential to the chord. This is most easily

‡ A. W. Goldstein and M. Jerison, 'Isolated and cascade aerofoils with prescribed velocity distribution', *N.A.C.A. Rep.* no. 869 (1947).

done by rotating the axes through the stagger angle σ, putting $z' = z\, e^{-i\sigma}$, $w' = w\, e^{i\sigma}$. Since in the z' frame of reference the chord line lies on the real x'-axis, the real and imaginary parts of the resulting equation for the velocity w' yield the velocity components on the aerofoil, parallel and normal to the chord,

$$u_0' = U_m \cos(\alpha_m - \sigma) - \frac{\gamma(\xi_0')}{2} + \frac{\sin 2\sigma}{6\pi}\left(\frac{\pi c}{s}\right)^2 \int_{-\frac12}^{\frac12} \gamma(\xi')\,(\xi_0' - \xi')\,d\xi',$$

$$v_0' = U_m \sin(\alpha_m - \sigma) + \frac{1}{2\pi}\int_{-\frac12}^{\frac12}\frac{\gamma(\xi')}{\xi_0' - \xi'}\,d\xi' + \frac{\cos 2\sigma}{6\pi}\left(\frac{\pi c}{s}\right)^2 \int_{-\frac12}^{\frac12} \gamma(\xi')\,(\xi_0' - \xi')\,d\xi',$$

$$(2\cdot11, 29)$$

where $\xi' = x'/c$.

As in § 2·10, we shall assume an expansion for the vorticity distribution of the form

$$\gamma(\xi') = 2U_m\left(\tfrac12\alpha_0 \tan\tfrac12\theta + \sum_{n=1}^{\infty}\alpha_n \sin n\theta\right), \qquad (2\cdot11, 30)$$

where $\cos\theta = 2\xi'$. Substituting in the second equation of $(2\cdot11, 29)$ and using the boundary condition $(2\cdot11, 27)$ we obtain

$$f'(\theta_0)\cos(\alpha_m - \sigma) = \sin(\alpha_m - \sigma)$$

$$+ \frac{1}{\pi}\int_0^{\pi}\left(\frac{\alpha_0}{2}\frac{1 - \cos\theta}{\cos\theta_0 - \cos\theta} + \sum_{n=1}^{\infty}\alpha_n \frac{\sin n\theta \sin\theta}{\cos\theta_0 - \cos\theta}\right)d\theta$$

$$+ \frac{\cos 2\sigma}{12\pi}\left(\frac{\pi c}{s}\right)^2\int_0^{\pi}\tfrac12\alpha_0(1 - \cos\theta)(\cos\theta_0 - \cos\theta)\,d\theta$$

$$+ \frac{\cos 2\sigma}{12\pi}\left(\frac{\pi c}{s}\right)^2\int_0^{\pi}\sum_{n=1}^{\infty}\alpha_n \sin n\theta \sin\theta(\cos\theta_0 - \cos\theta)\,d\theta$$

$$= \sin(\alpha_m - \sigma) + \tfrac12\alpha_0 + \sum_{n=1}^{\infty}\alpha_n \cos n\theta$$

$$+ \frac{\cos 2\sigma}{12}\left(\frac{\pi c}{s}\right)^2(\tfrac12\alpha_0(\cos\theta_0 + \tfrac12) + \tfrac12\alpha_1 \cos\theta_0 - \tfrac14\alpha_2).$$

$$(2\cdot11, 31)$$

From which we get, putting $\alpha_m - \sigma = \alpha$,

$$\cos\alpha\,\frac{1}{\pi}\int_0^{2\pi} f'(\theta_0)\,d\theta_0 = 2\sin\alpha + \alpha_0\left(1 + \frac{\cos 2\sigma}{24}\left(\frac{\pi c}{s}\right)^2\right) - \alpha_2\frac{\cos 2\sigma}{24}\left(\frac{\pi c}{s}\right)^2,$$

$$\cos\alpha\,\frac{1}{\pi}\int_0^{2\pi} f'(\theta_0)\cos\theta_0\,d\theta_0 = \alpha_1 + (\alpha_0 + \alpha_1)\frac{\cos 2\sigma}{24}\left(\frac{\pi c}{s}\right)^2,$$

$$\cos\alpha\,\frac{1}{\pi}\int_0^{2\pi} f'(\theta_0)\cos n\theta_0\,d\theta_0 = \alpha_n \quad (n \geqslant 2),$$

$$(2\cdot11, 32)$$

or, from § 2·8,

$$\alpha_0^* = \alpha_0\left(1 + \frac{\cos 2\sigma}{24}\left(\frac{\pi c}{s}\right)^2\right) - \alpha_2\frac{\cos 2\sigma}{24}\left(\frac{\pi c}{s}\right)^2,$$

$$\alpha_1^* = \alpha_0\frac{\cos 2\sigma}{24}\left(\frac{\pi c}{s}\right)^2 + \alpha_1\left(1 + \frac{\cos 2\sigma}{24}\left(\frac{\pi c}{s}\right)^2\right),$$

$$\alpha_n^* = \alpha_n \quad (n \geqslant 2),$$

$$(2\cdot11, 33)$$

where α_n^* are the coefficients for an isolated aerofoil at incidence to a main-stream flow. The set of equations (2·11, 33) correspond to equations (2·10, 19) for the biplane. That the same relationships between the lift and moment and the coefficients α_n hold for aerofoils in cascade as for the isolated aerofoil (equations (2·8, 20) and (2·8, 25)) can be demonstrated as follows. The force acting on an elementary vortex sheet lying on the x'-axis is

$$dL = -\rho q \gamma(x')\, dx' \tag{2·11, 34}$$

normal to the local velocity q. In thin-wing theory the induced velocities are assumed small in comparison with the main-stream velocity for deriving the boundary condition (2·11, 27). To the same order of approximation when integrating (2·11, 34) to obtain the total lift, we may replace q by the effective free-stream velocity U_m. The lift on an aerofoil of the cascade may then be written

$$L = -\rho U_m \int_{-\frac{1}{4}c}^{\frac{1}{4}c} \gamma(x')\, dx',$$

so that

$$C_L = -2 \int_0^\pi \left(\tfrac{1}{2}\alpha_0 (1 - \cos\theta) + \alpha_1 \sin^2\theta\right) d\theta$$

$$= -\pi(\alpha_0 + \alpha_1).$$

The moment about the quarter chord is

$$M_0 = -\rho U_m \int_{-\frac{1}{4}c}^{\frac{1}{4}c} \gamma(x')\,(x' + \tfrac{1}{4}c)\, dx',$$

and the corresponding moment coefficient

$$C_{M_0} = -\int_0^\pi \left(\alpha_2 \sin 2\theta \cos\theta \sin\theta + \tfrac{1}{2}\alpha_0 (1 - \cos\theta)\cos\theta \right.$$
$$\left. + \tfrac{1}{4}\alpha_0 (1 - \cos\theta) + \tfrac{1}{2}\alpha_1 \sin^2\theta\right) d\theta$$

$$= -\tfrac{1}{4}\pi(\alpha_1 + \alpha_2).$$

Using these results we find from (2·11, 33) that

$$C_L = -\pi \left(\frac{\alpha_0^* + \alpha_1^* + \alpha_2^* \chi}{1 + \chi} - \frac{\alpha_0^* \chi - \alpha_2^* \chi^2}{(1 + \chi)^2}\right)$$

$$= \frac{C_L^* + 4\chi C_{M_0}^*}{(1 + \chi)^2}, \tag{2·11, 35}$$

where we have written

$$\chi = \frac{\cos 2\sigma}{24}\left(\frac{\pi c}{s}\right)^2, \tag{2·11, 36}$$

and

$$C_{M_0} = -\frac{\pi}{4}\left(\frac{\alpha_1^*}{1 + \chi} - \frac{\alpha_0^* \chi + \alpha_2 \chi^2}{(1 + \chi)^2} + \alpha_2^*\right)$$

$$= \frac{C_{M_0}^* + \chi(2C_{M_0}^* - \tfrac{1}{4}C_L^*)}{(1 + \chi)^2}. \tag{2·11, 37}$$

These results give the corrections to be applied to the isolated lift and moment coefficients when the aerofoil is put in a cascade of low solidity (that is, a cascade with c/s small). The error involved in assuming the gap to chord ratio to be large is shown in Fig. 2·11, 7, in which equation (2·11, 35) is compared with the exact theory for a cascade of straight aerofoils, with $\sigma = 0$.

The velocity distribution can be found from equation (2·11, 27) and from the first equation of (2·11, 29) once the coefficients α_n have been calculated from (2·11, 32).

The particularly simple interference method just described for large gap cascades has been extended by Woolard ‡ to cover the case of thick aerofoils, following the analysis of Allen for isolated aerofoils.

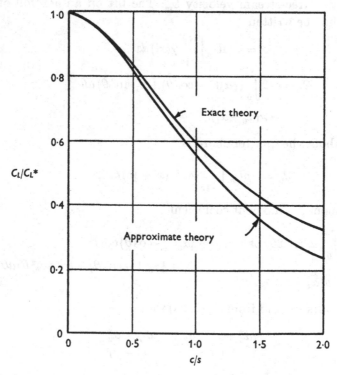

Fig. 2·11, 7. Lift on a cascade of straight aerofoils with zero stagger.

The solution of the potential-flow problem as discussed in this section must be modified in practical design by allowing for the losses due to boundary layers. The effect is especially important in large deflexion cascades where it is necessary to predict the stalling incidence. Generally the drag

‡ H. W. Woolard, 'The incompressible flow about a cascade of airfoils', *Cornell Aero. Lab. Inc. Rep.* no. AF-743-A-1 (1950).

is calculated semi-empirically or predicted from previous experimental results, and as such these methods lie outside the scope of this volume. Readers interested should consult Howell's paper on axial flow compressor design.‡

2·12   Profile drag

We have seen that, according to potential-flow theory, an aerofoil experiences no drag under two-dimensional conditions. This is contrary to experimental evidence. To explain and calculate the drag under such conditions we must have recourse to the theory of viscous flow. We showed in Chapter 1 how to calculate the drag of a flat plate at zero incidence on the assumptions of laminar boundary-layer theory. The determination of the drag of an aerofoil is rather more complicated, but an effective method which copes with this problem has been developed by Squire and Young.§

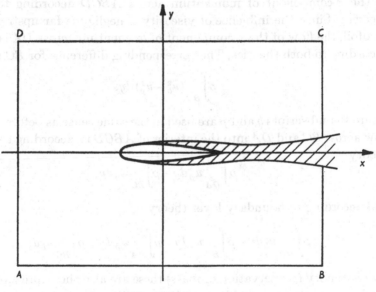

Fig. 2·12, 1

We consider the flow round an aerofoil in a uniform stream of fluid of velocity U which is directed along the x-axis (Fig. 2·12, 1). The field of flow round the aerofoil is described satisfactorily by potential theory, except in a thin layer of fluid V which passes, or has passed, close to the surface of the aerofoil. V constitutes the boundary layer of the aerofoil

‡ A. R. Howell, 'The present basis of axial flow compressor design', *Rep. Memor. Aero. Res. Comm., Lond.*, no. 2095 (1942).

§ H. B. Squire and A. D. Young, 'The calculation of the profile drag of aerofoils', *Rep. Memor. Aero. Res. Comm., Lond.*, no. 1838 (1937).

together with its viscous wake. Under usual conditions the general assumptions of boundary-layer theory are equally valid in the wake.

Let $ABCD$ be a large rectangle whose sides are parallel to the axes of coordinates, below, above, upstream and downstream of the aerofoil respectively, and at a great distance from it. Moreover, let the distances of AB and CD from the x-axis be equal. Since the conditions are supposed steady, the reaction of the aerofoil on the fluid equals the resultant stress of the volume of fluid within $ABCD$ on the surrounding fluid, plus the rate of flow of momentum across $ABCD$ in outward direction. Let T_p be the x-component of the reaction of the aerofoil on the fluid as calculated by potential-flow theory, and T_b the corresponding quantity as calculated by boundary-layer theory. According to the assumptions of that theory, the effect of viscosity on the normal pressure along BC and DA is negligible, as are the tangential stresses along AB and CD. It follows that the difference $T_b - T_p$ is equal to the difference between the expressions for the flow of the x-component of momentum across $ABCD$ according to the two theories. Since the influence of viscosity is negligible far upstream of the aerofoil, the flow of the x-component of momentum across DA is the same according to both theories. The corresponding difference for BC is

$$\rho \int_{BC} (u_b^2 - u_p^2)\, dy, \qquad (2·12, 1)$$

where the subscripts b and p are used in the same sense as before. The mass flow across BC and DA into the interior of $ABCD$ is, according to potential theory,

$$\rho \int_{DA} u_p\, dy - \rho \int_{BC} u_p\, dy, \qquad (2·12, 2)$$

and according to boundary-layer theory

$$\rho \int_{DA} u_b\, dy - \rho \int_{BC} u_b\, dy = \rho \int_{DA} u_p\, dy - \rho \int_{BC} u_b\, dy. \qquad (2·12, 3)$$

By continuity (conservation of mass) these are also the expressions for the outflow of mass across AB and CD according to the two theories. The x-component of the velocity along AB and CD is, to the first approximation, equal to V according to both theories, and so the expressions for the outflow of momentum across AB and CD are

$$\rho U \int_{DA} u_p\, dy - \rho U \int_{BC} u_p\, dy \qquad (2·12, 4)$$

and

$$\rho U \int_{DA} u_p\, dy - \rho U \int_{BC} u_b\, dy, \qquad (2·12, 5)$$

and their difference is $\qquad \rho U \int_{BC} (u_p - u_b)\, dy.$ $\qquad (2·12, 6)$

Adding (2·12, 1) and (2·12, 6), we obtain

$$T_b - T_p = \rho \int_{BC} u_b(u_b - U)\,dy - \rho \int_{BC} u_p(u_p - U)\,dy. \qquad (2·12, 7)$$

If there is no circulation, then the difference $u_p - U$ tends to zero inversely as the square of the distance from the wing and may therefore be neglected. If the circulation does not vanish then $u_p - U$ includes a component which tends to zero only inversely as the first power of the distance from the aerofoil. This component may be thought of as being due to a vortex in the region of the wing. It is therefore approximately equal to $-\gamma y/(x^2 + y^2)$ at large distances from the wing, where γ is the circulation. Now

$$\int_{BC} u_p(u_p - U)\,dy = \int_{BC} (u_p - U)^2\,dy + U \int_{BC} (u_p - U)\,dy,$$

and the first integral on the right-hand side of this equation vanishes in the limit as the boundary $ABCDA$ becomes infinite, because the integrand tends to zero inversely as the square of the distance from the aerofoil, while the second integral tends to zero because

$$\int_{BC} \frac{\gamma y}{x^2 + y^2}\,dy = 0$$

for any value of x. It follows that the second integral on the right-hand side of (2·12, 7) can be omitted,

$$T_b - T_p = \rho \int u_b(u_b - U)\,dy,$$

where the integral is taken across the wake. But, according to potential theory, $T_p = 0$, and so the drag on the aerofoil, $D = -T_b$, is given by the formula

$$D = \rho \int u(U - u)\,dy = \rho U^2 \vartheta_1, \qquad (2·12, 8)$$

where the integral is taken across the wake, and where we have omitted the subscripts. ϑ_1 is the momentum thickness as defined by (1·12, 11); (2·12, 8) is due to B. M. Jones and was first used for the experimental determination of the drag of a wing.‡

The formula involves only a region far downstream of the aerofoil, but in order to apply it theoretically we have to trace the development of the boundary layer beginning at the leading edge. To calculate the characteristics of the boundary layer in the laminar region, near the leading edge, we may use the so-called 'Pohlhausen method' which is based on von Kármán's momentum equation. We start from the momentum equation in the form (1·12, 18). However, it should be remembered that this equation

‡ Cambridge University Aeronautics Laboratory, 'Measurement of profile drag by the Pitot-traverse method', *Rep. Memor. Aero. Res. Comm., Lond.*, no. 1688 (1936).

is referred to a particular curvilinear system of coordinates (see Fig. 1·12, 2) along the top or bottom surface of the aerofoil as the case may be. We take the origin of the system at the forward stagnation point, i.e. at the point which would be the forward stagnation point according to potential-flow theory.

The function $f(\eta)$ is chosen so as to satisfy as many as possible of the boundary conditions (1·12, 22) and (1·12, 24). In Pohlhausen's method the function is assumed to be a polynomial of η, and in the present analysis we shall take it more particularly as a polynomial of fourth order which satisfies the first two conditions of (1·12, 22) and the first three conditions of (1·12, 24), so

$$\left.\begin{array}{ll} f(0)=0 & f''(0)=-\Lambda, \\ f(1)=1 & f'(1)=f''(1)=0, \end{array}\right\} \tag{2·12, 9}$$

where
$$\Lambda = \frac{du_1}{dx}\frac{\delta^2}{\nu}.$$

u_1, which is the velocity at the edge of the boundary layer, is equated to the value of the tangential velocity at the aerofoil as given by potential-flow theory.

The linearity of the boundary conditions shows that $f(\eta)$ can be represented in the form

$$f(\eta) = F(\eta) + \Lambda G(\eta), \tag{2·12, 10}$$

where $F(\eta)$ and $G(\eta)$ are fourth-order polynomials which satisfy the conditions

$$F(1)=1, \quad F(0)=F''(0)=F'(1)=F''(1)=0 \tag{2·12, 11}$$

and
$$G''(0)=-1, \quad G(0)=G(1)=G'(1)=G''(1)=0. \tag{2·12, 12}$$

(2·12, 11) yields five linear equations for the five coefficients of $F(\eta)$, and solving for these, we obtain

$$F(\eta) = 2\eta(1-\eta^2+2\eta^3). \tag{2·12, 13}$$

In a similar way, (2·12, 12) leads to

$$G(\eta) = \tfrac{1}{6}\eta(1-\eta)^3. \tag{2·12, 14}$$

The two functions are shown in Fig. 2·12, 2.

Using (1·12, 17), we find the following expressions for the displacement and momentum thicknesses, and for the shear stress at the surface of the aerofoil

$$\left.\begin{array}{l} \delta_1 = \tfrac{1}{120}(36-\Lambda)\,\delta, \\[2mm] \vartheta_1 = \tfrac{1}{315}(37-\tfrac{1}{3}\Lambda-\tfrac{5}{144}\Lambda^2), \\[2mm] \tau_0 = (2+\tfrac{1}{6}\Lambda)\,\mu\,\dfrac{u_1}{\delta}. \end{array}\right\} \tag{2·12, 15}$$

The momentum equation (1·12, 18) now becomes

$$(2+\tfrac{1}{6}\Lambda)\frac{\nu}{u_1\delta}=\frac{1}{315}\frac{d}{dx}\{(37-\tfrac{1}{3}\Lambda-\tfrac{5}{144}\Lambda^2)\,\delta\}$$

$$+\{\tfrac{2}{315}(37-\tfrac{1}{3}\Lambda-\tfrac{5}{144}\Lambda^2)+\tfrac{1}{120}(36-\Lambda)\}\frac{\delta}{u_1}\frac{du_1}{dx}. \quad (2·12, 16)$$

(2·12, 16) may be regarded as a differential equation for δ, but it should be remembered that Λ also involves δ (compare 2·12, 10). As a result, the equation is considerably more complicated than in the example considered in § 1·12, where there was no external pressure gradient. The equation can be brought into a more convenient form by the introduction of the new dependent variable

$$z=\frac{\delta^2}{\nu}=\Lambda\Big/\frac{du_1}{dx}. \quad (2·12, 17)$$

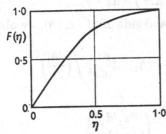
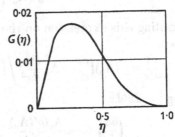

Fig. 2·12, 2

The result of the transformation is

$$\frac{dz}{dx}=\frac{1}{u_1}g(\Lambda)+\frac{d^2u_1}{dx^2}h(\Lambda)\,z^2, \quad (2·12, 18)$$

where
$$g(\Lambda)=\frac{15120-2784\Lambda+79\Lambda^2+\tfrac{5}{8}\Lambda^3}{(12-\Lambda)(37+\tfrac{25}{12}\Lambda)}$$

$$\quad (2·12, 19)$$

and
$$h(\Lambda)=\frac{8+\tfrac{5}{8}\Lambda}{(12-\Lambda)(37+\tfrac{25}{12}\Lambda)}.$$

We assume that the leading edge is rounded, so that $u_1=0$ at the forward stagnation point. Also, we may expect dz/dx to be finite at that point, and the right-hand side of (2·12, 18) shows that this condition implies, generally speaking, that

$$g(\Lambda)=0. \quad (2·12, 20)$$

The roots of this equation are found to be

$$\Lambda=-70,\ 7·052,\ 17·75. \quad (2·12, 21)$$

But
$$\Lambda=\frac{\delta^2}{\nu}\frac{du_1}{dx},$$

and du_1/dx must be positive just downstream of the stagnation point, so that Λ also must be positive there. This rules out the first root, which is negative. Moreover, du_1/dx must eventually become negative, and $du_1/dx = 0$ implies $\Lambda = 0$. If now we have $\Lambda = 17 \cdot 75$ at the forward stagnation point and $\Lambda = 0$ some way downstream, then for some point Λ would have to become equal to 12. But (2·12, 19) shows that both $g(\Lambda)$ and $h(\Lambda)$ become infinite for $\Lambda = 12$. Accordingly we reject also the third root in (2·12, 21) and conclude that

$$\Lambda = \Lambda_0 = 7 \cdot 052 \tag{2·12, 22}$$

is the appropriate value of Λ at the forward stagnation point. The value of $\frac{1}{u_1} g(\Lambda)$ at that point is found by means of de l'Hôpital's rule:

$$\lim_{x \to 0} \frac{1}{u_1} g(\Lambda) = \left(g'(\Lambda) \frac{d\Lambda}{dx} \middle/ \frac{du_1}{dx} \right)_{x=0}$$

$$= g'(\Lambda_0) \left(\frac{dz}{dx} + \Lambda_0 \frac{d^2 u_1}{dx^2} \middle/ \left(\frac{du_1}{dx} \right)^2 \right)_{x=0}.$$

Substituting this expression on the right-hand side of (2·12, 18), we obtain, for $x = 0$,

$$\frac{dz}{dx} = g'(\Lambda_0) \left(\frac{dz}{dx} + \Lambda_0 \frac{d^2 u_1}{dx^2} \middle/ \left(\frac{du_1}{dx} \right)^2 \right) + h(\Lambda_0) \Lambda_0^2 \frac{d^2 u_1}{dx^2} \middle/ \left(\frac{du_1}{dx} \right)^2.$$

Solving for dz/dx,

$$\left(\frac{dz}{dx} \right)_{x=0} = \frac{\Lambda_0 (g'(\Lambda_0) + h(\Lambda_0) \Lambda_0)}{1 - g'(\Lambda_0)} \left(\frac{d^2 u_1/dx^2}{(du_1/dx)^2} \right)_{x=0}, \tag{2·12, 23}$$

and substituting numerical values,

$$\left(\frac{dz}{dx} \right)_{x=0} = -5 \cdot 391 \left(\frac{d^2 u_1}{dx^2} \middle/ \left(\frac{du_1}{dx} \right)^2 \right)_{x=0}. \tag{2·12, 24}$$

(Compare *Modern Developments in Fluid Dynamics*,‡ where further references are given.) The value of z for $x = 0$ is simply

$$(z)_{x=0} = \Lambda_0 \middle/ \left(\frac{du_1}{dx} \right)_{x=0} = 7 \cdot 052 \middle/ \left(\frac{du_1}{dx} \right)_{x=0}. \tag{2·12, 25}$$

With these initial data (2·12, 18) can now be solved by a step-by-step method of integration, bearing in mind that $\Lambda = z(du_1/dx)$. Having determined z_1 and hence Λ and δ, we may then calculate δ_1, ϑ_1 and τ_0 by means of (2·12, 15).

The above procedure is based on the assumption that the flow in the boundary layer is laminar. However, we know from Chapter 1 that for the Reynolds numbers usually encountered in aeronautics, the flow in the neighbourhood of the aerofoil will eventually become turbulent. For the purpose of the present analysis it is assumed that the transition from laminar to turbulent flow occurs suddenly at the so-called 'transition

‡ *Modern Developments in Fluid Dynamics*, ed. S. Goldstein (Oxford University Press, 1938), pp. 156–63.

point'. This assumption, though not strictly true, does not introduce any appreciable errors. As a matter of experience, the Pohlhausen method actually ceases to be reliable a short distance upstream of the transition region, as soon as the pressure gradient becomes adverse (positive), but this again has little effect on the overall result.

The method used for the calculation of the flow in the laminar boundary layer is purely theoretical in the sense that no measurements are involved in the derivation of equations (2·12, 18), (2·12, 24) and (2·12, 25). Nevertheless, the approximations and simplifying assumptions involved are such that the method can be regarded as reliable only in view of the fact that it checks well with experimental evidence. For the turbulent region, a self-contained theory does not exist at all, and it will be sufficient here to sketch the semi-empirical procedure which has been adopted in its place.

The procedure is again based on von Kármán's momentum equation, whose meaning for a turbulent boundary layer was explained in § 1·13 (equation (1·13, 10)). The ratio $H = \delta_1/\vartheta_1$ which occurs in the equation is taken to be constant; more particularly it is assumed that $H = 1·4$, on the basis of measurements of Nikuradse and Buri of turbulent flow in converging and diverging channels.[‡] The remaining unknowns in (1·13, 10) are $\bar{\tau}_0$ and ϑ_1. To express ϑ_1 in terms of $\bar{\tau}_0$ we assume that the connexion between the two quantities is locally the same as in turbulent flow along a flat plate. For this case an analysis of theoretical and experimental results yields the relation[§]

$$\vartheta_1 = 0·2454 \frac{\nu}{u_1} \exp\left(0·3914 \frac{\rho u_1^2}{\bar{\tau}_0}\right). \tag{2·12, 26}$$

We substitute this expression for ϑ_1 in (1·13, 10) and introduce the new dependent variable

$$\zeta = \sqrt{(\rho u_1^2/\bar{\tau}_0)}.$$

The result is

$$\zeta^{-2} = \left(0·0960 \frac{\nu}{u_1} \frac{d\zeta}{dx} - 0·2454 \frac{\nu}{u_1^2} \frac{du_1}{dx} + 0·2454(H+2) \frac{\nu}{u_1^2} \frac{du_1}{dx}\right) \exp(0·3914\zeta^2). \tag{2·12, 27}$$

Substituting $H = 1·4$, we obtain after some modifications

$$\frac{d\zeta}{dx} + \frac{6·13}{u_1} \frac{du_1}{dx} = 10·41 \frac{u_1}{\nu} \zeta^{-2} \exp(-0·3914\zeta^2). \tag{2·12, 28}$$

Momentum considerations show that ϑ_1 must be continuous at the transition point; the contrary assumption would imply the existence of a finite

‡ See *Modern Developments in Fluid Dynamics*, ed. S. Goldstein (Oxford University Press, 1938), p. 375.

§ H. B. Squire and A. D. Young, 'The calculation of the profile drag of aerofoils', *Rep. Memor. Aero. Res. Comm., Lond.*, no. 1838 (1937); compare also Th. von Kármán, 'Turbulence and skin friction', *J. Aero. Sci.* vol. 1 (1934), pp. 1–20.

shear impulse at that point to balance the discontinuity of transport of momentum. Also, by (2·12, 26),

$$\zeta = 2 \cdot 557 \log \left(4 \cdot 075 \frac{u_1}{\nu} \vartheta_1 \right), \tag{2·12, 29}$$

and so ζ can be determined at the transition point in terms of ϑ_1, whose value at that point is known from the previous calculation. Step-by-step integration of (2·12, 28) then yields ζ, and hence ϑ_1, $\bar{\tau}_0$ and δ_1 along the entire boundary layer as far as the trailing edge. u_1 may again be taken from potential-flow theory, although according to Squire and Young some empirical modification of these data is desirable near the trailing edge, to take account of the presence of the boundary layer. The actual position of the transition point cannot be determined by theoretical means and, moreover, may vary with factors such as roughness and wind-tunnel turbulence, which we have not taken into consideration. Accordingly, it is appropriate to work out the numerical results for a complete range of likely positions of the transition point, say for points between the leading edge and the mid-chord.

The general approximations of boundary-layer theory may be applied equally well to the viscous wake. However, in the absence of a wing the shear term on the left-hand side of (1·13, 9) is to be replaced by zero. Thus the momentum equation becomes for the wake,

$$\frac{d\vartheta_1}{dx} + \frac{\vartheta_1}{u_1} \frac{du_1}{dx} (H+2) = 0, \tag{2·12, 30}$$

where ϑ_1 and δ_1, and hence H, now have to be taken across the entire wake. Thus, the momentum thickness of the wake at the trailing edge is obtained by adding the momentum thicknesses of the boundary layers on the upper and lower surfaces, as calculated by the method described above.

We rewrite (2·12, 30) in the form

$$\frac{d}{dx} \log \vartheta_1 = -(H+2) \frac{d}{dx} \log \frac{u_1}{U}, \tag{2·12, 31}$$

and integrate by parts on the right-hand side. This leads to

$$\log \vartheta_1 = -(H+2) \log \frac{u_1}{U} + \int \log \frac{u_1}{U} \frac{dH}{dx} dx. \tag{2·12, 32}$$

Evaluating this expression between the trailing edge and infinity, we obtain

$$\log \frac{\vartheta_1^\infty}{\vartheta_1^*} - (H^*+2) \log \frac{u_1^*}{U} = \int_{H^*}^{H^\infty} \log \frac{u_1}{U} dH. \tag{2·12, 33}$$

In this formula, the asterisks indicate values at the trailing edge, ϑ_1^∞ and H^∞ are the corresponding values at infinity, and we have made use of the fact that, at infinity, $u_1 = U$. Moreover, far behind the aerofoil u differs but little from U all across the wake, and so $u/u_1 \doteqdot 1$, $\delta_1 \doteqdot \vartheta_1$, and $H^\infty \doteqdot 1$. Thus we may substitute 1 for H^∞ as the upper limit of the integral in (2·12, 33). Also, $H^* = 1·4$, since this is the accepted value of H near the trailing edge on both surfaces of the aerofoil. To determine ϑ_1^∞ it therefore only remains to evaluate the integral in (2·12, 33). Experimental evidence tends to show that it is roughly a linear function of H,

$$\log\frac{u_1}{U} = aH + b.$$

The constants a and b are determined by conditions at the trailing edge, and at infinity. This yields

$$\log\frac{u_1}{U} = \frac{H-1}{H^*-1}\log\frac{u_1^*}{U}, \tag{2·12, 34}$$

and so

$$\int_{H^*}^{1} \log\frac{u_1}{U}dH = \left[\frac{(H-1)^2}{2(H^*-1)}\right]_{H^*}^{\infty}\log\frac{u_1^*}{U} = -\frac{H^*-1}{2}\log\frac{u_1^*}{U}.$$

Substituting on the right-hand side of (2·12, 33), we obtain, after some modification,

$$\log\frac{\vartheta_1^\infty}{\vartheta_1^*} = \frac{H^*+5}{2}\log\frac{u_1^*}{U}.$$

But $H^* = 1·4$, and so, finally

$$\vartheta_1^\infty = \vartheta_1^*\left(\frac{u_1^*}{U}\right)^{3·2}. \tag{2·12, 35}$$

Substituting in B. M. Jones's formula (2·12, 8) we obtain for the drag

$$D = \rho U^2\vartheta_1^*\left(\frac{u_1^*}{U}\right)^{3·2}, \tag{2·12, 36}$$

and for the drag coefficient

$$C_D = \frac{D}{\frac{1}{2}\rho U^2 c} = \frac{2\vartheta_1^*}{c}\left(\frac{u_1^*}{U}\right)^{3·2}, \tag{2·12, 37}$$

where c is the chord.

Fig. 2·12, 3 shows drag coefficients calculated by the above method for a number of wing thickness to chord ratios t/c, and Reynolds numbers Re. In general, the results are in good agreement with experiment. It is found that for moderate incidences the drag is sensibly independent of the lift.

The total drag can be analysed into two parts, the 'skin friction' which is due to the viscous shear stresses which act on the surface of the wing, and the form drag, which is due to the redistribution of normal pressures on the wing owing to the presence of the boundary layer. Since τ_0 (or $\bar\tau_0$) was determined incidentally in the above analysis, it is not difficult to calculate the skin friction separately; the form drag can then be obtained by subtraction. However, for practical purposes we require only the total drag— usually called profile drag, to distinguish it from its two components, skin friction and form drag—and this is the quantity which is obtained directly by the Squire–Young method.

The presence of the boundary-layer effects also the lift and moment of an aerofoil. Thus, the increase of lift coefficient with thickness predicted by potential-flow theory does not materialize, as stated already in § 2·8. For moderately thick aerofoils (thickness-chord ratio 0·12, say) and moderate incidences the measured lift coefficient may be smaller than the theoretical figure by as much as 10 % (compare Fig. 2·6, 2). Even more

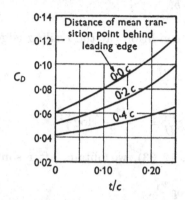
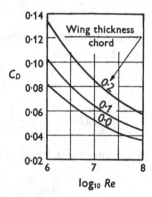

Fig. 2·12, 3. (*Courtesy H.M.S.O.*)‡

important is the effect of viscosity on the moment produced by the deflexion of a trailing edge control surface. In a theoretical approach to the problem,§ the presence of the boundary layer is taken into account as an effective change of the shape (camber, incidence) of the aerofoil. The circulation is determined by means of a trailing edge condition which replaces the Joukowski condition in potential theory.

‡ Taken from H. B. Squire and A. D. Young, 'The calculation of the profile drag of aerofoils', *Rep. Memor. Aero. Res. Coun., Lond.*, no. 1838 (1937).

§ See J. H. Preston, 'The calculation of lift taking account of the boundary layer', *Rep Memor. Aero. Res. Coun., Lond.*, no. 2725 (1953).

CHAPTER 3

AEROFOIL THEORY FOR STEADY FLOW IN THREE DIMENSIONS

3·1 Introduction

The general assumptions underlying the calculation of the flow around a finite wing were discussed in Chapter 1. We shall assume that the x-axis is parallel to the direction of flow, while the y-axis points to starboard and the z-axis upwards. The aerofoil will always be assumed to be situated approximately in the x, y-plane (Fig. 3·1, 1). As in thin aerofoil theory in two dimensions (§ 2·8), the boundary condition at a point Q at the surface of the aerofoil will be taken to be satisfied instead at the vertical projection of Q onto the x, y-plane, Q'. The vortex wake (if any) will be assumed to occupy a region of the x, y-plane downstream of the aerofoil and bounded by straight lines emanating from the tips, or more precisely from the vertical projections of the tips onto the x, y-plane.

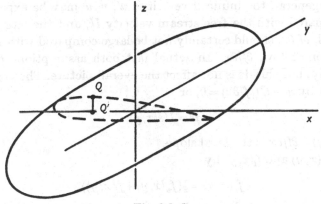

Fig. 3·1, 1

Let u, v, w be the velocity components, and Φ the velocity potential,

$$u = -\frac{\partial \Phi}{\partial x}, \quad v = -\frac{\partial \Phi}{\partial y}, \quad w = -\frac{\partial \Phi}{\partial z}.$$

Let U be the free-stream velocity and u' the induced longitudinal velocity, so that $u = U + u'$. Correspondingly, let Φ' be the induced velocity potential $\Phi = -Ux + \Phi'$, so that

$$u' = -\frac{\partial \Phi'}{\partial x}, \quad v = -\frac{\partial \Phi'}{\partial y}, \quad w = -\frac{\partial \Phi'}{\partial z}.$$

Both Φ and Φ' satisfy Laplace's equation

$$\frac{\partial^2 \Phi}{\partial x^2} + \frac{\partial^2 \Phi}{\partial y^2} + \frac{\partial^2 \Phi}{\partial z^2} = 0,$$

$$\frac{\partial^2 \Phi'}{\partial x^2} + \frac{\partial^2 \Phi'}{\partial y^2} + \frac{\partial^2 \Phi'}{\partial z^2} = 0.$$

The velocity is continuous everywhere in the fluid except possibly at the wake, and in accordance with § 1·15, it will be assumed that the vortex lines which constitute the wake are everywhere parallel to the x-axis. This is equivalent to the statement that v may be discontinuous across the wake while u and w are still continuous across it.

Let us assume that the aerofoil is given by an equation of the form $z = f(x, y)$, or more specifically by $z = f_u(x, y)$ for the upper surface and $z = f_l(x, y)$ for the lower. It should be noted that these equations are now supposed to include wing incidence. Then the boundary condition at any point on the aerofoil is that the flow should be tangential to the surface. This means analytically that the vectors $\mathbf{q} = (u, v, w)$ and $(-\partial f/\partial x, -\partial f/\partial y, 1)$ are normal to one another, or

$$w - u\frac{\partial f}{\partial x} - v\frac{\partial f}{\partial y} = 0. \tag{3·1, 1}$$

Now, in general, the induced velocities u', v, w may be expected to be small compared with the free-stream velocity U, and the lateral slope of the aerofoil, $\partial f/\partial y$, should certainly not be large compared with its slope in the direction of flow, $\partial f/\partial x$. In actual fact both assumptions may break down locally, but this does not affect the overall picture. Thus (3·1, 1) may be replaced by $w - U(\partial f/\partial x) = 0$, or

$$w = Us(x, y), \tag{3·1, 2}$$

where $s(x, y) = \partial f/\partial x$ is the local slope.

Define $f_c(x, y)$ and $f_t(x, y)$ by

$$f_c(x, y) = \tfrac{1}{2}(f_u(x, y) + f_l(x, y)),$$

and
$$f_t(x, y) = \tfrac{1}{2}(f_u(x, y) - f_l(x, y)).$$

$z = f_c(x, y)$ is the position of the 'mean camber surface', and $2f_t(x, y)$ is the local thickness. Then

$$\left.\begin{aligned} f_u(x, y) &= f_c(x, y) + f_t(x, y), \\ f_l(x, y) &= f_c(x, y) - f_t(x, y). \end{aligned}\right\} \tag{3·1, 3}$$

Since both the differential equation for Φ' and the boundary condition (3·1, 2) are linear, we may consider the effect of thickness and of the position of the mean camber line (which now includes the incidence) separately, by assuming $f_c(x, y) = 0$ and $f_t(x, y) = 0$ respectively, in (3·1, 3) and superimpose the results. The first case then represents the flow round a wing which is symmetrical and at zero incidence with respect to the x, y-plane. We conclude that in that case the induced velocity potential $\Phi'(x, y, z)$ also is symmetrical with respect to the x, y-plane

$$\Phi'(x, y, z) = \Phi'(x, y, -z), \tag{3·1, 4}$$

and

$$u(x, y, z) = U - \frac{\partial \Phi'}{\partial x} = u(x, y, -z), \quad v(x, y, z) = -\frac{\partial \Phi'}{\partial y} = v(x, y, -z),$$
$$w(x, y, z) = -\frac{\partial \Phi'}{\partial z} = -w(x, y, -z). \tag{3·1, 5}$$

In particular, Φ' and the velocity components are continuous everywhere in the fluid.

On the other hand, the second case $(f_t(x, y) = 0)$ represents the flow round a very thin wing, cambered or at incidence, or both, as the case may be. In that case we may expect the induced velocity potential $\Phi'(x, y, z)$ to be antisymmetrical with respect to the x, y-plane,

$$\Phi'(x, y, z) = -\Phi'(x, y, -z), \tag{3·1, 6}$$

and

$$u'(x, y, z) = -\frac{\partial \Phi'}{\partial x} = u'(x, y, -z), \quad v(x, y, z) = -\frac{\partial \Phi'}{\partial y} = -v(x, y, -z),$$
$$w(x, y, z) = -\frac{\partial \Phi'}{\partial z} = w(x, y, -z). \tag{3·1, 7}$$

v may now be discontinuous in the wake of the aerofoil, while u' must still be continuous across it, owing to the assumption that the vortex lines are parallel to the direction of flow. It follows that $u'(x, y, 0) = 0$, in other words $u(x, y, 0) = U$, except at the aerofoil.

3·2 Aerodynamic forces

Once the velocity potential, and hence the field of flow, have been determined for a given wing, the aerodynamic forces may be calculated in the following manner.

The pressure at any point of the wing is given by Bernoulli's equation (1·6, 7)

$$\tfrac{1}{2}q^2 + \frac{p}{\rho} = H, \tag{3·2, 1}$$

or

$$p + \tfrac{1}{2}\rho((U + u')^2 + v^2 + w^2) = p_0 + \tfrac{1}{2}\rho U^2, \tag{3·2, 2}$$

where p_0 is the free-stream pressure. Since the boundary conditions have been linearized, it is again in order, generally speaking, to neglect the second-order terms u'^2, v^2, w^2 in (3·2, 2). This yields the linearized Bernoulli equation

$$p - p_0 = -\rho U u', \tag{3·2, 3}$$

as in (2·8, 18). Also since the boundary condition is satisfied at the normal projection Q' of any point Q of the wing on the x, y-plane, it is appropriate to calculate the corresponding pressure distribution also at Q'. For the antisymmetrical case (thin cambered aerofoil at incidence), the value of the pressure difference between the top and bottom surfaces, $p_u - p_l$, is then independent of whether we use (3·2, 2) or (3·2, 3), since the quadratic terms cancel.

If we take the wing as the surface S in equation (1·4, 16), then we obtain

$$- \mathbf{T} = - \int_S p \, d\mathbf{S}, \qquad (3·2, 4)$$

where $- \mathbf{T}$ is the force exerted on the wing by the surrounding medium. Similarly, we obtain for the moment exerted on the wing

$$- \mathbf{N} = - \int_S p \mathbf{r} \wedge d\mathbf{S}. \qquad (3·2, 5)$$

However, these formulae are valid only on the assumption that the velocity remains finite (bounded) near the wing. In actual fact, as we saw in the preceding chapter, the field of flow calculated on the assumptions of thin aerofoil theory may give rise to infinite velocities at the leading edge, and in the preceding section these assumptions were adopted also for three-dimensional theory. It follows that the results furnished by (3·2, 4) and (3·2, 5) require a correction which takes into account the effect of the suction force. To calculate that force, we may adopt the results of § 2·8. We may assume that the slope of the leading edge of the wing plan-form is continuous, with the possible exception of a number of isolated points. Let $A(x_0, y_0, 0)$ be a point of the leading edge or, more precisely, the normal projection of a point of the leading edge on to the x, y-plane. We introduce a local system of coordinates $(\tilde{x}, \tilde{y}, \tilde{z})$ with origin at A, by choosing the $\tilde{x}$ and $\tilde{y}$ axes perpendicular and parallel to the leading edge in the $\tilde{x}, \tilde{y}$-plane respectively, while the $\tilde{z}$-axis is parallel to the z-axis such that the three axes form a right-handed system when taken in alphabetical order (Fig. 3·2, 1). In general, the $\tilde{y}$-component of the velocity will remain finite at A, so that

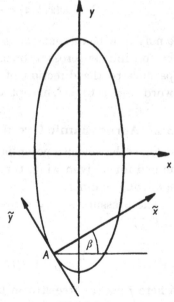

Fig. 3·2, 1

it does not affect the suction force. To calculate that force we may therefore confine ourselves to the velocity components in the $\tilde{x}, \tilde{z}$-plane. Introducing the complex variable

$$\tilde{\zeta} = \tilde{x} + i\tilde{z}$$

we assume that in the neighbourhood of A the velocity components $\tilde{u}, \tilde{w}$ in the $\tilde{x}, \tilde{z}$-plane are of the form

$$\tilde{u} - i\tilde{w} = \frac{\tilde{C}}{\sqrt{\tilde{\zeta}}} + \text{bounded terms,} \qquad (3\cdot2, 6)$$

where $\tilde{C}$ may be complex. Except for the notation, $(3\cdot2, 6)$ is equivalent to $(2\cdot8, 28)$. The assumption that the velocity variation in the neighbourhood of the leading edge is effectively two-dimensional must lead to the correct result for the suction force, which depends only on the local geometry. The formula for the suction force is now obtained from $(2\cdot8, 30)$ by an appropriate change of symbols

$$\tilde{X} - i\tilde{Z} = -\pi\rho\tilde{C}^2, \qquad (3\cdot2, 7)$$

where $\tilde{X}$ and $\tilde{Z}$ denote the force components in the direction of the $\tilde{x}$- and $\tilde{z}$-axes respectively.

In general $\tilde{C}$ will be complex,

$$\tilde{C} = \tilde{C}_1 - i\tilde{C}_2.$$

Then $(3\cdot2, 6)$ may be replaced by

$$\left.\begin{aligned}
\tilde{u} &= \frac{\tilde{C}_1}{\sqrt{\tilde{x}}} + \text{bounded terms,} \\
\tilde{w} &= \frac{\tilde{C}_2}{\sqrt{\tilde{x}}} + \text{bounded terms,}
\end{aligned}\right\} \qquad (3\cdot2, 8)$$

as A is approached along the positive $\tilde{x}$-axis on the upper surface of the wing. Also, in terms of $\tilde{C}_1$ and $\tilde{C}_2$, the force components are given by

$$\tilde{X} = -\pi\rho(\tilde{C}_1^2 - \tilde{C}_2^2), \quad \tilde{Z} = -2\pi\rho\tilde{C}_1\tilde{C}_2. \qquad (3\cdot2, 9)$$

We wish to refer these formulae to the original system of coordinates. Let β_0 be the angle between the x- and $\tilde{x}$-axes. Then from the second equation of $(3\cdot2, 8)$

$$w = \tilde{w} = \frac{\tilde{C}_2}{\sqrt{\tilde{x}}} + \text{bounded terms}$$

$$= \frac{\tilde{C}_2}{\sqrt{\{(x - x_0)\cos\beta_0\}}} + \text{bounded terms}$$

$$= \frac{C_2}{\sqrt{(x - x_0)}} + \text{bounded terms,}$$

where
$$C_2 = \tilde{C}_2/\sqrt{(\cos\beta_0)}.$$

To transform the first formula in $(3\cdot2, 8)$, we have to take into account that
$$u = \tilde{u}\cos\beta_0 + \text{bounded terms.}$$

Hence
$$u = \frac{\tilde{C}_1\cos\beta_0}{\sqrt{\tilde{x}}} + \text{bounded terms}$$

$$= \frac{C_1}{\sqrt{(x - x_0)}} + \text{bounded terms,}$$

where $\qquad\qquad\qquad C_1 = \bar{C}_1 \sqrt{(\cos\beta_0)}.$

Referring (3·2, 9) to the original system of coordinates, we then obtain for the x-, y- and z-components of the suction force, per unit length of the leading edge

$$\left.\begin{aligned} X &= -\pi\rho(C_1^2 - C_2^2\cos\beta_0),\\ Y &= -\pi\rho(C_1^2\tan\beta_0 - C_2^2\sin\beta_0\cos\beta_0),\\ Z &= -2\pi\rho C_1 C_2, \end{aligned}\right\} \qquad (3\cdot2, 10)$$

and

$$\left.\begin{aligned} u &= \frac{C_1}{\sqrt{(x-x_0)}} + \text{bounded terms},\\ w &= \frac{C_2}{\sqrt{(x-x_0)}} + \text{bounded terms}, \end{aligned}\right\} \qquad (3\cdot2, 11)$$

on the upper surface of the wing.

C_2 is different from zero only if the wing has a rounded leading edge. By (3·1, 2) the thickness t of the aerofoil near the leading edge is given by

$$\frac{\partial}{\partial x} t(x,y) = \frac{2C_2}{U\sqrt{(x-x_0)}},$$

with the initial condition $\qquad\qquad t(x_0, y) = 0.$

Hence $\qquad\qquad\qquad \tau = \tfrac{1}{2}t(x,y) = \frac{2C_2}{U}\sqrt{(x-x_0)}.$

This is the equation of a parabola, whose radius of curvature is the radius of curvature of the aerofoil at its leading edge. For given y, the equation of the parabola may be written as

$$x - x_0 = \frac{U^2\tau^2}{4C_2^2},$$

and the radius of curvature of the parabola for $x = x_0$ $(\tau = 0)$ is

$$r_0 = \left(\frac{(1 + (dx/d\tau)^2)^{\frac{3}{2}}}{d^2x/d\tau^2}\right)_{\tau=0} = \frac{2C_2^2}{U^2},$$

or $\qquad\qquad\qquad C_2^2 = \tfrac{1}{2}U^2 r_0.$ $\qquad\qquad\qquad\qquad (3\cdot2, 12)$

Thus we may replace (3·2, 10) by

$$\left.\begin{aligned} X &= -\pi\rho(C_1^2 - \tfrac{1}{2}U^2 r_0\cos^2\beta_0),\\ Y &= -\pi\rho(C_1^2\tan\beta_0 - \tfrac{1}{2}U^2 r_0\sin\beta_0\cos\beta_0),\\ Z &= -\sqrt{2}\,\pi\rho C_1 U\sqrt{r_0}. \end{aligned}\right\} \qquad (3\cdot2, 13)$$

C_1 is not related directly to the geometry of the leading edge, as appears already from the two-dimensional case (§ 2·8). To estimate the magnitude of the suction-force components, we observe that according to § 2·8,

$C_1 = U \sqrt{c} \sin \alpha$; we may conclude that more generally, the order of magnitude of C_1 is $U \sqrt{c} \, \alpha$ also in the three-dimensional case, where c is an average chord length.

To estimate the order of magnitude of C_2, we consider the elliptic wing section which is given by the equation

$$\frac{x^2}{a^2} + \frac{z^2}{b^2} = 1.$$

The magnitude of the curvature at the leading edge is

$$\frac{1}{r_0} = \left| a \left\{ \frac{d^2}{dz^2} \sqrt{\left(1 - \frac{z^2}{b^2}\right)} \right\}_{z=0} \right| = \frac{a}{b^2},$$

and so

$$r_0 = \frac{b^2}{a} = 2a \frac{b^2}{2a^2}.$$

We may identify $2a$ with the chord length c of the section and b/a with the thickness chord ratio t/c. Thus, we expect the radius of curvature to be of order $c(t/c)^2$, C_2^2 is of order $U^2 c(t/c)^2$, and the corresponding contribution to the drag is of order $\rho U^2 c(t/c)^2$. We may suppose that in general $\tan \beta_0$ is of order 1, thus neglecting the fact that $\tan \beta_0$ is unbounded near a rounded tip. Also, in many cases α and t/c are of the same order of magnitude. If this assumption holds, then the suction force components per unit length of the leading edge (3·2, 13) are all of order $\rho U^2 c \alpha^2$. To find the resultant suction force, we have to integrate along the leading edge. Thus the length of the path of integration is of the order of magnitude of the wing span and the resultant suction force is of the order of $\rho U^2 S \alpha^2$, where S is the wing area.

In order to determine the aerodynamic forces which act on the wing according to potential flow theory, we have to add the resultant of (3·2, 13) to (3·2, 4). Thus, we obtain for the z-component or lift

$$L = -\int p \, dx \, dy + \int Z \, dl, \tag{3·2, 14}$$

where the surface integral is taken over both sides of the wing, and the line integral is taken along the leading edge. The surface integral in (3·2, 14) may be replaced by

$$-\int (p_u - p_l) \, dx \, dy, \tag{3·2, 15}$$

where p_u and p_l denote the pressure at corresponding points on the upper and lower wing surface, and the integral is taken only once over the wing plan-form. Referring to the two-dimensional case, we see that the pressure difference $p_u - p_l$ is of order $\rho U^2 \alpha$, and this is confirmed by the three-dimensional cases which will be considered later. Thus, the surface integral in (3·2, 14) is of order $\rho U^2 S \alpha$, while the line integral is of order $\rho U^2 S \alpha^2$. Since α may be assumed small compared with unity, it follows that the line integral is small compared with the surface integral. In other words, the contribution of the suction force to the lift is small compared with the contribution of the surface pressure. Accordingly, we shall regard (3·2, 15) as the appropriate expression for the lift.

Using (3·2, 3), we may rewrite (3·2, 15) in the form

$$L = \rho U \int (u'_u - u'_l) \, dx \, dy, \tag{3·2, 16}$$

where u'_u and u'_l are the induced velocity components at the upper and lower surface of the wing. Then

$$L = -\rho U \left(\int \int \left(\frac{\partial \Phi'_u}{\partial x} - \frac{\partial \Phi'_l}{\partial x} \right) \right) dx \, dy$$

$$= -\rho U \int (\Phi'_u - \Phi'_l) \, dy, \tag{3·2, 17}$$

or
$$L = -\rho U \int \Delta \Phi \, dy, \tag{3·2, 18}$$

where
$$\Delta \Phi = \Phi'_u - \Phi'_l = \Phi_u - \Phi_l \tag{3·2, 19}$$

is the jump of both the velocity potential and the induced-velocity potential across the wake at a point of the trailing edge, and the integral is taken along the trailing edge. It will be seen that, by definition, $-\Delta\Phi$ is also the circulation round a closed curve which penetrates the wake just once at the trailing edge point in question. Moreover, since the x-component of the velocity is continuous across the wake (see § 1·15) it follows that the jump of the velocity potential across the wake is constant along lines parallel to the x-axis. Thus the integral in (3·2, 18) may be taken along any path which crosses the wake from port to starboard. Similarly, we may replace (3·2, 17) by

$$L = \rho U \int_C \Phi \, dy, \tag{3·2, 20}$$

where C is a closed curve drawn on the wake and surrounding it. The positive sign on the right-hand side of (3·2, 20) is due to the usual convention which implies that the integral is to be taken from starboard to port above the wing and from port to starboard below it.

If in (3·2, 16) we integrate only with respect to x, then we obtain the spanwise lift distribution function (lift per unit span) $l(y)$. Thus

$$l(y) = -\rho U \Delta \Phi, \tag{3·2, 21}$$

where $\Delta\Phi$ is defined as in (3·2, 19). (3·2, 21) is in agreement with the law of Kutta and Joukowski in two-dimensional flow (2·3, 5).

It will now be shown that (3·2, 10) and (3·2, 20) are also obtained from a more rigorous analysis of the aerodynamic forces which act on the wing according to potential theory. This implies that the errors introduced by the linearization of Bernoulli's equation and by the omission of the suction force exactly balance one another.

Let us apply (1·4, 16) to a large circular cylinder S which contains the wing in its interior such that the axis of the cylinder coincides with the x-axis, while its bases are parallel to the yz-plane far ahead of the wing and far downstream of it respectively. Then the lift on the wing is given by

$$L = -\int_S p\,dx\,dy - \int_S \rho w(u\,dy\,dz + v\,dz\,dx + w\,dx\,dy). \qquad (3\cdot2, 22)$$

Substituting the expression for the pressure from (3·2, 2) and omitting constants, for which the corresponding integrals over the closed surface S are known to vanish, we obtain

$$L = \rho\int_S (Uu' + \tfrac{1}{2}[u'^2 + v^2 + w^2])\,dx\,dy - \rho\int_S w([U + u']\,dy\,dz + v\,dz\,dx + w\,dx\,dy),$$

or

$$L = \rho U\int_S (u'\,dx\,dy - w\,dy\,dz)$$
$$- \rho\int_S (wu'\,dy\,dz + wv\,dz\,dx - \tfrac{1}{2}[u'^2 + v^2 - w^2]\,dx\,dy). \qquad (3\cdot2, 23)$$

To estimate the order of magnitude of the quantities which appear in (3·2, 23) we represent the field of flow in terms of distributions of doublets and vortices. Such representations can be realized in various ways and some give rise to concrete methods for the determination of the velocity field and will be considered later (§§ 3·6–3·8). However, for our present purposes it will be sufficient to sketch one convenient procedure.

We first represent the induced velocity vector $\mathbf{q}' = (u', v, w)$ as

$$\mathbf{q}' = \mathbf{q}_1 + \mathbf{q}_2,$$

where $\mathbf{q}_1$ includes the effect of the trailing vortex sheet W. The trailing vortices alone do not generate a physically possible field of flow, since vortex lines cannot end within the fluid (§ 1·5). We therefore supplement W by a system of vortex lines W' in the interior—or on the surface—of the wing in such a way that the vortex law in question is satisfied. The corresponding velocity field $\mathbf{q}_1$ can then be calculated by means of the formulae of Biot-Savart (1·9, 22). Then the flow $\int \mathbf{q}_1\,dS$ across the surface of the wing

is zero, and the same therefore applies to the flow $\int (\mathbf{q}' - \mathbf{q}_1)\,dS$. It follows (see § 1·8) that $\mathbf{q}_2 = \mathbf{q}' - \mathbf{q}_1$ can be represented by a distribution of doublets over the surface of the wing. Then the components of $\mathbf{q}_2$ tend to zero inversely as the third power of the distance from the wing. Also, writing $\mathbf{q}_1 = \mathbf{q}_3 + \mathbf{q}_4$, where $\mathbf{q}_3$ is due to W and $\mathbf{q}_4$ is due to W', we may determine the orders of magnitude of these vectors from the law of Biot-Savart. We find that the components of $\mathbf{q}_4$ tend to zero inversely as the square o the distance from the wing, while the components of $\mathbf{q}_3$ tend to zero inversely as the distance from the *wake*. As the distance of the cylinder from the wing, and hence its dimensions, increase indefinitely, and may therefore replace the second integral in (3·2, 23) by

$$\int_S (w_3 u_3\, dy\, dz + w_3 v_3\, dz\, dx - \tfrac{1}{2}(u_3^2 + v_3^2 - w_3^2)\, dx\, dy), \qquad (3\cdot 2, 24)$$

where $$\mathbf{q}_3 = (u_3, v_3, w_3).$$

Also, the x-component of $\mathbf{q}_3$, u_3 vanishes identically since the trailing vortices are parallel to the x-axis. It follows that the two bases of the cylinder S contribute nothing to the integral (3·2, 24) while the integral over the envelope of S reduces to

$$\int_S (w_3 v_3\, dz\, dx - \tfrac{1}{2}(v_3^2 - w_3^2)\, dx\, dy). \qquad (3\cdot 2, 25)$$

Finally, we shall have shown that this integral also vanishes in the limit if we can verify that v_3 and w_3 tend to zero on the envelope of S inversely as the *square* of the radius of S. To establish this fact without calculation, we replace each trailing vortex of W by a parallel vortex of equal strength along the x-axis. Since the distance of the trailing vortices from the x-axis are bounded, the difference between the velocities generated by the original vortices and the corresponding vortices along the x-axis tends to zero on the envelope inversely as the square of the radius of the cylinder. Thus it only remains to be shown that the same applies to the velocities generated by the resultant vortex along the x-axis. Now the strength of that vortex is, by its derivation, equal to the total strength of the trailing vortex sheet, i.e. the circulation round a curve C which surrounds W (Fig. 3·2, 2). But C bounds a surface S' which does not cross the wing or wake, and so the circulation round C is zero, by Stokes's theorem. It follows that the resultant vortex along the x-axis is of zero strength, i.e. the corresponding velocity vanishes identically. We conclude that v_3 and w_3 tend to zero inversely as the square of the radius of S, as asserted. Thus, for large S (3·2, 23) may be replaced by

$$L = \rho U \int_S (u\, dx\, dy - w\, dy\, dz). \qquad (3\cdot 2, 26)$$

Let R be the radius of the cylinder, supposed large. Then for given y_0, $|y_0| < R$, the plane $y = y_0$ intersects S in a rectangle, $C'(y_0)$. We may carry out the integration in (3·2, 26) first along the rectangle, and then with respect to y, so

$$L = \rho U \int_{-R}^{R} dy \int_{C'(y)} (u\, dx + w\, dz). \qquad (3\cdot 2, 27)$$

(Note the change of sign in the last term.) But $\displaystyle\int_{C'(y)} (u\, dx + w\, dz)$ is the circulation round C', and so this integral vanishes if $C'(y)$ does not intersect the wake, and equals $-\Delta\Phi$ if it does. Hence

$$L = -\rho U \int \Delta\Phi \, dy, \qquad (3·2, 28)$$

where the integral is taken across the wake. The formula has been established for the limiting case where the path of integration is far downstream of the wing, but since $\Delta\Phi$ is constant along lines parallel to the x-axis, (3·2, 28) must hold equally well for any other path across the wake. We have therefore established that (3·2, 18)—which is identical with (3·2, 28) —and (3·2, 20) can be derived on the assumptions of § 1·15 alone, without any further simplification. It should be noted that our argument does not extend to (3·2, 21).

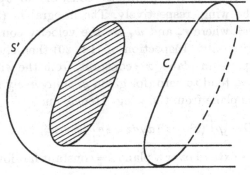

Fig. 3·2, 2

Next we consider the x-component, (or $drag$) D, of the aerodynamic force which acts on the wing according to potential theory. Since the pressures are of order $\rho U^2 \alpha$, where α is an average incidence, or slope, the corresponding components in the direction of the x-axis will be of order $\rho U^2 \alpha^2$ per unit area, i.e. of order $\rho U^2 S \alpha^2$ for the whole wing. Thus the contribution of the surface pressure is now only of the same order of magnitude as the contribution of the suction force, and it is no longer permissible to neglect the latter. Thus, (3·2, 13) may be used if it is desired to calculate the distribution of the drag forces on the wing. However, for most applications it is necessary to calculate only the overall drag, and this is again done more conveniently by means of the momentum theorem. Applying (1·4, 16) to the cylinder S defined above, we obtain for the drag,

$$D = -\int_S p \, dy \, dz - \int_S \rho u (u \, dy \, dz + v \, dz \, dx + w \, dx \, dy)$$

$$= \rho \int_S (Uu' + \tfrac{1}{2}[u'^2 + v^2 + w^2]) \, dy \, dz$$

$$\qquad\qquad - \rho \int_S (U + u')((U + u') \, dy \, dz + v \, dz \, dx + w \, dx \, dy),$$

or

$$D = -\rho U \int_S (u'\,dy\,dz + v\,dz\,dx + w\,dx\,dy)$$
$$+ \rho \int_S (\tfrac{1}{2}(-u'^2 + v^2 + w^2)\,dy\,dz - u'v\,dz\,dx - u'w\,dx\,dy). \quad (3\cdot2, 29)$$

Now the first integral in this expression is precisely the flow of the induced velocity $\mathbf{q}'$ across S, which is known to vanish. Considering orders of magnitude in the second integral, as before, we find that as the dimensions of the cylinder increase, $(3\cdot2, 29)$ reduces to

$$D = \tfrac{1}{2}\rho \int_S (v^2 + w^2)\,dy\,dz. \quad (3\cdot2, 30)$$

Since the envelope of the cylinder does not contribute to this integral, S will consist in the limit of two planes $x = $ constant far upstream and far downstream of the wing, respectively. The integral in $(3\cdot2, 30)$ may be replaced by $v_3^2 + w_3^2$, where v_3 and w_3 are the velocity components due to the trailing vortices only. Inspection of $(1\cdot9, 20)$ then shows that as the distance of the upstream plane, $x = $ constant, from the wing increases indefinitely, v_3 and w_3 tend to zero (for given y, z) inversely as the square of the distance of the plane from the wing. Hence, finally,

$$D = \tfrac{1}{2}\rho \int (v^2 + w^2)\,dy\,dz = \tfrac{1}{2}\rho \int (v_3^2 + w_3^2)\,dy\,dz, \quad (3\cdot2, 31)$$

where the integrals extend over a plane $x = $ constant far downstream of the wing.

The first integral in $(3\cdot2, 31)$ may be transformed further by the use of Green's formula for the plane. We have $v = -\partial\Phi'/\partial y$, $w = -\partial\Phi'/\partial z$, where Φ' is the induced velocity potential, and so

$$\int (v^2 + w^2)\,dy\,dz = \int \left(\left(\frac{\partial\Phi'}{\partial y}\right)^2 + \left(\frac{\partial\Phi'}{\partial z}\right)^2\right) dy\,dz.$$

Now by Green's formula

$$\int \left(\left(\frac{\partial\Phi'}{\partial y}\right)^2 + \left(\frac{\partial\Phi'}{\partial z}\right)^2 + \Phi'\left(\frac{\partial^2\Phi'}{\partial y^2} + \frac{\partial^2\Phi'}{\partial z^2}\right)\right) dy\,dz = \int \Phi'\frac{d\Phi'}{dn'}\,dl, \quad (3\cdot2, 32)$$

where $\partial/\partial n'$ refers to differentiation along the outward normal.

We apply this formula to a large area which is bounded internally by the line of intersection (or trace) of the wake, and externally by a large circle. As the radius of the circle increases indefinitely, the line integral along the circumference tends to zero. Also, the second term in the integral on the left-hand side of $(3\cdot2, 32)$ may be written as

$$\Phi'\left(\frac{\partial^2\Phi'}{\partial y^2} + \frac{\partial^2\Phi'}{\partial z^2}\right) = -\Phi'\frac{\partial^2\Phi'}{\partial x^2} = \Phi'\frac{\partial u'}{\partial x},$$

since Φ' is a solution of Laplace's equation, and as the distance of the plane from the wing increases indefinitely, $\Phi'(\partial u'/\partial x)$ tends to zero for given

y, z—at least inversely as the third power of the distance from the wing. Hence, in the limit,

$$\int \left(\left(\frac{\partial \Phi'}{\partial y} \right)^2 + \left(\frac{\partial \Phi'}{\partial z} \right)^2 \right) dy\,dz = \int_C \Phi' \frac{\partial \Phi'}{\partial n'}\,dl. \qquad (3·2, 33)$$

The path of integration on the right-hand side consists of two straight segments parallel to the y-axis and on either side of the wake.

The direction of normal differentiation $\partial/\partial n'$ is outward from the region bounded by the large circle and the trace of the wake, i.e. on the upper surface of the wake it is in the direction z decreasing and on lower surface it is in the direction of z increasing. Thus, $\partial \Phi'/\partial n' = -\partial \Phi'/\partial z = w$ above the wake, and $\partial \Phi/\partial n' = \partial \Phi'/\partial z = -w$ below it. Hence

$$D = -\tfrac{1}{2}\rho \int_C \Phi' w\,dy = \tfrac{1}{2}\rho \int \Delta\Phi w\,dy, \qquad (3·2, 34)$$

where the second integral is taken across the wake from port to starboard far behind the wing and $\Delta\Phi$ is the jump of the velocity potential across the wake as in (3·2, 18).

To determine D we may calculate one of the integrals in (3·2, 34) for planes which are at a finite distance aft of the wing and proceed to the limit afterwards, or we may first define

$$\Phi_\infty(y, z) = \lim_{x \to \infty} \Phi'(x, y, z),$$

$$w_\infty(y, z) = -\frac{\partial \Phi_\infty}{\partial z} = \lim_{x \to \infty} w(x, y, z),$$

and then calculate

$$D = -\tfrac{1}{2}\rho \int_C \Phi_\infty w_\infty\,dy = \tfrac{1}{2}\rho \int \Delta\Phi_\infty w_\infty\,dy. \qquad (3·2, 35)$$

In the second integral, $\qquad \Delta\Phi_\infty = \lim_{x \to \infty} \Delta\Phi$

is the discontinuity of Φ_∞ across the trace of the wake, but since $\Delta\Phi$ is constant along lines parallel to the x-axis, we might replace $\Delta\Phi_\infty$ by $\Delta\Phi$ in (3·2, 35). To calculate $\Phi_\infty(y, z)$ we only require a knowledge of the strength of the trailing vortices, since the contributions of the doublet distribution over the wing, and of the vortex distribution W', all tend to zero as the path of integration moves downstream. This remark will be found useful later on (§ 3·3).

By (3·2, 31), $\qquad DU = \tfrac{1}{2}\rho U \int (v_3^2 + w_3^2)\,dy\,dz. \qquad (3·2, 36)$

In this equation, the left-hand side represents the work done by the wing against the drag D in unit time. To interpret the right-hand side, we place ourselves in a system of reference which is at rest in the surrounding air, and consider the volume of fluid V bounded by two parallel planes which are normal to the direction of motion of the wing, and far behind it, and are a distance U units of length apart. Then the field of flow in V is due entirely to the trailing vortices, and its kinetic energy is expressed by the right-hand side of (3·2, 36). Now imagine that the wing started from rest at a point far behind the wing, and even far behind the volume V. Then the wake is extended by U units of length in every unit of time, and so its kinetic energy increases by the right-hand side of (3·2, 36). This shows that the work done by the wing against the drag D is converted entirely into the kinetic energy associated with the trailing vortices.

In order to obtain the total drag on a wing in a real fluid we have to add D to the profile drag which is due to the effect of viscosity (see § 2·12). By contradistinction, D is called the induced drag, since it is associated with, or induced by, the presence of a lifting force, and thus depends on the attitude of the wing.

It has been found that, in general, calculation and experiment are in good agreement as regards lift and induced drag, and what discrepancies there are can be explained satisfactorily. Unfortunately, this statement does not hold for the y-component of the aerodynamic force, or side force. For that force, the theoretical result is obtained by combining the contribution of the surface pressure (3·2, 4) and of the suction force (see (3·2, 13)), which are of the same order of magnitude, but the practical applicability of the result is in doubt.

We shall deal more briefly with the calculation of the aerodynamic moments. We shall refer to the moments about the x-, y- and z-axes briefly as rolling moment, pitching moment and yawing moment, respectively, although this agrees with the standard nomenclature only for a particular choice of the coordinate system. Then it is not difficult to see that the effect of the suction force on the rolling and pitching moments can be neglected. Moreover, the surface pressure acts approximately at points in the x, y-plane, and in the direction of the z-axis. Accordingly we may simplify the expressions for the rolling and pitching moments obtained from (3·2, 5) further to

$$L' = -\int (p_u - p_l)\, y\, dx\, dy, \qquad (3·2, 37)$$

and

$$M = \int (p_u - p_l)\, x\, dx\, dy, \qquad (3·2, 38)$$

respectively, where the integrals are taken over the wing plan-form. Using (3·2, 3) we may replace (3·2, 37) in turn by

$$L' = -\rho U \int \Delta\Phi y\, dy = \int l(y)\, y\, dy, \qquad (3·2, 39)$$

where the integrals are taken from port to starboard.

For the yawing moment, the contribution of the suction force is again of the same order of magnitude as that of the surface pressure, but, as in the case of the side force, the practical validity of the theoretical results is doubtful.

For a given wing S, let S_c be an infinitely thin wing which coincides with the mean camber surface of S, and S_t a wing which is symmetrical with respect to the x, y-plane and which has the same thickness distribution as S. It has been shown (§ 3·1) that the induced field of flow round S may be obtained by the superposition of the induced fields of flow round S_c and S_t. That is to say,

$$\Phi' = \Phi'_c + \Phi'_t, \qquad (3·2, 40)$$

where Φ', Φ'_c, Φ'_t are the induced velocity potentials of the fields of flow round S, S_c, S_t, the main-stream velocity U being directed along the x-axis in all cases. For S_t, the pressure differences at corresponding points above and below the wing, the discontinuity of the velocity potential across the wake, and the z-component of the velocity at the wake all vanish. It follows from (3·2, 18), (3·2, 34), (3·2, 37) and (3·2, 38) that the lift and induced drag, and the rolling and pitching moments of the wing S are equal to the corresponding quantities for S_c. In particular, these quantities all vanish if S is symmetrical with respect to the x, y-plane.

3·3   Lifting-line theory

Amongst the methods which have been devised for the calculation of the aerodynamic forces acting on an aerofoil of finite span, the most prominent is the lifting-line theory of L. Prandtl.‡ The theory has great inherent merit, being essentially simple and at the same time very effective. Beyond that, it is of considerable historical importance, for it was the first three-dimensional wing theory to be put forward at all, and it was in this connexion that the existence of a system of trailing vortices was first realized. It must be mentioned that Prandtl's ideas were anticipated to some extent by Lanchester,§ who did not, however, develop a quantitative analysis.

Lifting-line theory can be justified rigorously by a passage to the limit from lifting-plane theory (§ 3·6 below). Such a justification, comprising even the unsteady case, is implicit in the work of Küssner‖ (see § 5·6). However, as appears from the above remarks, the theory was introduced

---

‡ L. Prandtl, 'Tragflügeltheorie', 1st part, *Nachr. Ges. Wiss. Göttingen*, Math. Phys. Kl. (1918), pp. 451–77; 2nd part, pp. 107–37. An alternative approach is given by W. R. Sears, 'A new treatment of the lifting line wing theory with applications to rigid and elastic wings', *Quart. Appl. Math.* vol. 16 (1948), pp. 239–56.

§ F. W. Lanchester, *Aerodynamics* (Constable, London, 1907), ch. 4.

‖ H. G. Küssner, 'Allgemeine Tragflächentheorie', *Luftfahrtforsch.* vol. 17 (1940), pp. 370–8. Translated as *Tech. Memor. Nat. Adv. Comm. Aero., Wash.*, no. 979.

originally in a less formal manner. It will be appropriate to use such an approach also in the present exposition.

We know from Chapter 2 (compare (2·3, 18), (2·3, 19) and (2·4, 6)) that the circulation round an aerofoil in two dimensions may be written in the form

$$\Gamma = -\tfrac{1}{2}a_0\,Uc\sin\alpha, \tag{3·3, 1}$$

where U is the free-stream velocity and c is the chord of the aerofoil. α is the angle of attack or incidence, measured from the zero-lift angle, and a_0 is a constant which is equal to 2π for a flat plate. So far as the present theory is concerned, we may use the best value available for a_0, either from two-dimensional theory or from experiment. (3·3, 1) holds in the system of coordinates used in Chapter 2. In our present system of coordinates—the x-axis in the direction of flow, the y-axis pointing to starboard, and the z-axis upward—the sign on the right-hand side of (3·3, 1) has to be reversed. This is due to the fact that the circulation is now positive in the opposite sense. The incidence is regarded as positive when the aerofoil is in a 'nose-up' attitude, and this is in agreement with Chapter 2. Also, replacing $\sin\alpha$ by α for small angles of attack, we obtain instead of (3·3, 1)

$$\Gamma = \tfrac{1}{2}a_0\,Uc\alpha. \tag{3·3, 2}$$

The law of Kutta–Joukowski is, in the present notation,

$$l = \rho U\Gamma, \tag{3·3, 3}$$

and so

$$l = \tfrac{1}{2}a_0\rho U^2 c\alpha, \tag{3·3, 4}$$

where we now use l to denote the lift per unit span. Formulae (3·3, 2) and (3·3, 4) are strictly true only under two-dimensional conditions, but if the span of the wing is large compared with the chord, and if its sectional characteristics vary only slowly with the span, then we may use two-dimensional theory as a crude approximation to calculate the pressure distribution, and hence the lift on the wing for any given span-station (spanwise coordinate). This procedure is known as 'simple strip-theory'. In order to improve upon this procedure we try to isolate the most significant features which distinguish three-dimensional from two-dimensional conditions.

We know that under two-dimensional conditions the flow behaviour at some distance from the wing is as if it were due to a vortex of strength Γ which extends along the wing in spanwise direction and which we may locate, for example, along the y-axis. If the strength of the vortex is known, then the field of flow at some distance from the wing may therefore be calculated by means of the law of Biot-Savart ((1·9, 20) or (1·9, 21)). Now if the characteristics of the wing, and hence the circulation, vary along the span, then we may still calculate the velocity field due to a vortex of varying strength $\Gamma(y)$ along the y-axis according to the formula (1·9, 20). We note that according to that formula, a vortex element along the y-axis at A induces zero velocity (i.e. does not induce any flow at all) at all points B which are situated on the y-axis at a finite distance from B (Fig. 3·3, 1). It follows that the velocity field at B is due entirely to the vortex elements in the neighbourhood of B, so that to this extent the fact that the strength of the vortex $\Gamma(y)$ is variable, does not affect conditions at B. However, this picture is incomplete in itself, since an isolated vortex must be of constant strength, according to the vortex laws (§ 1·5). Thus, we have to superimpose on the velocity field due to the vortex which replaces the wing (or 'bound vortex') the velocities due to the trailing vortex sheet (compare §§ 1·15 and 3·2). Since the trailing vortices are parallel to the x-axis, the velocities induced by them anywhere in the x, y-plane are, by the law of Biot-Savart, parallel to the z-axis. This has the effect of deflecting the main stream in the neighbourhood of the wing in an upward or downward direction.

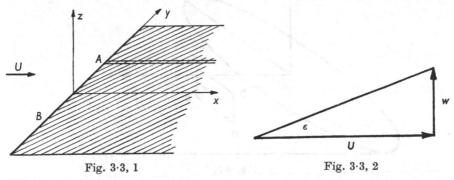

Fig. 3·3, 1   Fig. 3·3, 2

Let the velocity induced by the trailing vortices at a particular span station be w. Then the main stream is deflected by an angle

$$\epsilon = \tan^{-1}\frac{w}{U} \doteqdot \frac{w}{U} \qquad (3\cdot3, 5)$$

(see Fig. 3·3, 2). Since the induced velocity is small compared with the free-stream velocity, the resultant velocity is still approximately equal to U. However, the angle of attack of the aerofoil relative to the deflected main stream is

$$\alpha = \alpha_0 + \epsilon \doteqdot \alpha_0 + \frac{w}{U}, \qquad (3\cdot3, 6)$$

where α_0 is the 'geometrical angle of attack', or angle of attack relative to the free stream. In general, w is negative for positive α_0, and its numerical value is called the 'downwash velocity'. The numerical value of ϵ is then called the 'downwash angle'. The angle α which, in general, is numerically smaller than α_0, is called the 'effective angle of attack'. Introducing the effective angle of attack in (3·3, 2), we obtain

$$\Gamma = \tfrac{1}{2}a_0 Uc\left(\alpha_0 + \frac{w}{U}\right) = \tfrac{1}{2}a_0 c(U\alpha_0 + w). \qquad (3\cdot3, 7)$$

By the definition, of the velocity potential, Γ equals (except for sign) the discontinuity of the velocity potential at the trailing edge

$$\Gamma = -\Delta\Phi, \tag{3·3, 8}$$

where $\Delta\Phi$ is defined as in (3·2, 19). To express the strength of the trailing vortices in terms of the velocity potential, we consider a small rectangular circuit $ABCD$, whose sides AB and CD are parallel to the y-axis and close to the wake on either side of it, and the sides BC and DA are parallel to the z-axis (Fig. 3·3. 3). The circulation round this circuit is given by

$$-\{(\Delta\Phi)_D - (\Delta\Phi)_C\} \doteq \left(\frac{d}{dy}\Delta\Phi\right)\delta y, \tag{3·3, 9}$$

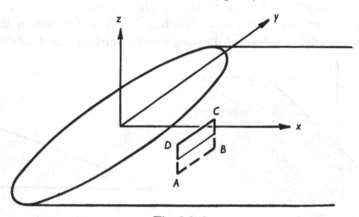

Fig. 3·3, 3

where δy is the length of CD, supposed small. Let $\gamma(y)$ be the strength of the trailing vortex sheet per unit span, then the circulation round $ABCD$ must also be equal to the strength of the vortices enclosed by the circuit, $\gamma(y)\,\delta y$. Hence

$$\gamma(y) = \frac{d}{dy}\Delta\Phi, \tag{3·3, 10}$$

or, in terms of the circulation round the wing,

$$\gamma(y) = -\frac{d\Gamma}{dy}. \tag{3·3, 11}$$

A similar argument shows that $\Gamma(y)$ must drop to zero at the tips of the wing, otherwise we should have to accept the existence of trailing vortices of finite strength behind the tips, and these in turn would give rise to infinite downwash velocities at the wing.

To calculate the downwash distribution along the span of the wing, we use the law of Biot-Savart. Since the wing has been replaced by a bound vortex along the y-axis, we have to assume that the trailing vortices extend from the y-axis downstream. We may also assume, without any essential limitation, that the wing extends from $y = -s_0$ to $y = s_0$. Then the z-component of the velocity induced by the trailing vortices at a point $(0, y, z)$ is (see (1·9, 20))

$$w = -\frac{1}{4\pi} \int_{-s_0}^{s_0} \frac{\gamma(y_1)(y_1 - y)}{z^2 + (y_1 - y)^2} dy_1. \tag{3·3, 12}$$

This integral diverges for $z = 0$, $-s_0 < y < s_0$, and the correct value of w is then given by

$$w = \lim_{z \to 0} \left\{ -\frac{1}{4\pi} \int_{-s_0}^{s_0} \frac{\gamma(y_1)(y_1 - y)}{z^2 + (y_1 - y)^2} dy_1 \right\}. \tag{3·3, 13}$$

For given y, $-s_0 < y < s_0$, we write $\gamma(y_1)$ in the form

$$\gamma(y_1) = \gamma(y) + \gamma'(y_2)(y_1 - y), \tag{3·3, 14}$$

where y_2 is an intermediate value between y and y_1. Such representation will be possible in all concrete cases, except perhaps at a number of isolated points, and we may assume that $\gamma'(y_2)$ is continuous when regarded as a function of y_1. Then

$$\lim_{z \to 0} \int_{-s_0}^{s_0} \frac{\gamma(y_1)(y_1 - y)}{z^2 + (y_1 - y)^2} dy_1$$

$$= \lim_{z \to 0} \left\{ \gamma(y) \int_{-s_0}^{s_0} \frac{y_1 - y}{z^2 + (y_1 - y)^2} dy_1 + \int_{-s_0}^{s_0} \gamma'(y_2) \frac{(y_1 - y)^2}{z^2 + (y_1 - y)^2} dy_1 \right\}$$

$$= \lim_{z \to 0} \left\{ \gamma(y) \tfrac{1}{2} [\log (z^2 + (y_1 - y)^2)]_{-s_0}^{s_0} + \int_{-s_0}^{s_0} \gamma'(y_2) \frac{(y_1 - y)^2}{z^2 + (y_1 - y)^2} dy_1 \right\},$$

and so

$$w = -\frac{1}{4\pi} \left\{ \gamma(y) \log \frac{s_0 - y}{s_0 + y} + \int_{-s_0}^{s_0} \gamma'(y_2) dy_1 \right\}. \tag{3·3, 15}$$

On the other hand, consider the principal value of the integral

$$\int_{-s_0}^{s_0} \frac{\gamma(y_1)}{y_1 - y} dy_1.$$

This is equal to

$$\lim_{\epsilon \to 0} \left\{ \left(\int_{-s_0}^{y-\epsilon} + \int_{y+\epsilon}^{s_0} \right) \frac{\gamma(y_1)}{y_1 - y} dy_1 \right\} = \lim_{\epsilon \to 0} \left\{ \gamma(y) \left(\int_{-s_0}^{y-\epsilon} + \int_{y+\epsilon}^{s_0} \right) \frac{dy_1}{y_1 - y} \right.$$

$$\left. + \left(\int_{-s_0}^{y-\epsilon} + \int_{y+\epsilon}^{s_0} \right) \gamma'(y_2) dy_1 \right\}$$

$$= \gamma(y) \log \frac{s_0 - y}{s_0 + y} + \int_{-s_0}^{s_0} \gamma'(y_2) dy_1. \tag{3·3, 16}$$

Comparing this result with (3·3, 15), we see that w can also be written in the form

$$w = -\frac{1}{4\pi} \int_{-s_0}^{s_0} \frac{\gamma(y_1)}{y_1 - y} dy_1, \tag{3·3, 17}$$

where the principal value of the integral is to be taken on the right-hand side. Replacing $\gamma(y_1)$ by $-\Gamma'(y_1)$, in accordance with (3·3, 11), and substituting the result in (3·3, 7), we obtain

$$\Gamma(y) = \tfrac{1}{2} a_0 c \left(U\alpha_0 + \frac{1}{4\pi} \int_{-s_0}^{s_0} \frac{\Gamma'(y_1)}{y_1 - y} dy_1 \right). \qquad (3\cdot3, 18)$$

This is the fundamental lifting-line equation. It is a singular linear integro-differential equation, which determines $\Gamma(y)$ subject to the boundary conditions

$$\Gamma(-s_0) = \Gamma(s_0) = 0. \qquad (3\cdot3, 19)$$

Formal integration by parts on the right-hand side of (3·3, 18) transforms it into an integral equation,‡ but the result is really meaningless since the integral which appears in the transformed equation is properly divergent.

To determine $\Gamma(y)$ we adopt a procedure due to Trefftz,§ which actually does not use (3·3, 18) directly.

The function $\Phi_\infty(y, z)$ was defined in § 3·2 by

$$\Phi_\infty(y, z) = \lim_{x \to \infty} \Phi'(x, y, z). \qquad (3\cdot3, 20)$$

Φ_∞ may be said to be the induced velocity potential in a plane which is normal to the x-axis far behind the wing (the 'Trefftz plane'). According to a remark in § 3·2, Φ_∞ is due entirely to the trailing vortex sheet. Hence, it is an odd function of z, and

$$\Gamma(y) = -\Delta\Phi = -(\Phi_\infty(y, +0) - \Phi_\infty(y, -0)) = -2\Phi_\infty(y, +0). \qquad (3\cdot3, 21)$$

To express the downwash at the wing in terms of Φ_∞, we observe that by (1·9, 20), or more simply, by considerations of symmetry, the velocity induced by the trailing vortices in the y, z-plane is just one-half the velocity induced in the Trefftz plane. Thus

$$w(0, y, z) = \tfrac{1}{2} \lim_{x \to \infty} w(x, y, z) = -\frac{1}{2} \frac{\partial}{\partial z} \Phi_\infty(y, z). \qquad (3\cdot3, 22)$$

Substituting in (3·3, 7) from (3·3, 21) and (3·3, 22), we obtain

$$4\Phi_\infty(y, +0) = a_0 c \left(\frac{1}{2} \frac{\partial}{\partial z} \Phi_\infty(y, +0) - U\alpha_0 \right) \quad (-s_0 < y < s_0), \qquad (3\cdot3, 23)$$

and similarly

$$4\Phi_\infty(y, -0) = -a_0 c \left(\frac{1}{2} \frac{\partial}{\partial z} \Phi_\infty(y, -0) - U\alpha_0 \right) \quad (-s_0 < y < s_0). \qquad (3\cdot3, 24)$$

Φ_∞ is a solution of Laplace's equation in the cut y, z-plane whose boundary is made up by the two 'banks' of the segment $\langle -s_0, s_0 \rangle$ on the y-axis. Denoting normal differentiation across the boundary in a direction out-

‡ For example see H. Glauert, *The Elements of Aerofoil and Airscrew Theory* (2nd ed., Cambridge University Press, 1947), p. 135.

§ E. Trefftz, 'Prandtlsche Tragflächen und Propellertheorie', *Z. angew. Math. Mech.* vol. 1 (1921), p. 206.

ward from the region exterior to the wake, as in (3·2, 33), by $\partial/\partial n'$, we may rewrite (3·3, 23) and (3·3, 24) as

$$
\left.
\begin{aligned}
4\Phi_\infty + \tfrac{1}{2} a_0 c \frac{\partial \Phi_\infty}{\partial n'} &= -a_0 U c \alpha_0 \quad \text{on the upper bank,} \\
4\Phi_\infty + \tfrac{1}{2} a_0 c \frac{\partial \Phi_\infty}{\partial n'} &= a_0 U c \alpha_0 \quad \text{on the lower bank.}
\end{aligned}
\right\}
\tag{3·3, 25}
$$

In this form the problem belongs to potential theory. (3·3, 25) is said to be a mixed boundary condition or, a boundary condition of the third kind.

We introduce the complex variable

$$
\chi = y + iz. \tag{3·3, 26}
$$

Since Φ_∞ is a solution of Laplace's equation, it can be written as the real part of a complex potential Ω

$$
\Phi_\infty(y, z) = \mathscr{R}(\Omega(\chi)). \tag{3·3, 27}
$$

It was pointed out earlier that the circulation round a closed curve which embraces the wake far behind the wing, vanishes, and this applies in particular to curves in the Trefftz plane. It is also easy to see that the flow across such a curve vanishes, since the entire field in the Trefftz plane is due to a distribution of vortices. It follows that Ω is a one-valued function, and we may assume that it vanishes at infinity.

The transformation

$$
\chi = \frac{s_0}{2}\left(\zeta + \frac{1}{\zeta}\right) \tag{3·3, 28}
$$

maps the cut χ-plane on the region outside the unit circle in the ζ-plane in such a way that the point at infinity is transformed into the point at infinity. Then Ω can be represented in the region outside the ζ-plane by the Laurent series

$$
\Omega = -i\left(\frac{a_1}{\zeta} + \frac{a_2}{\zeta^2} + \dots\right). \tag{3·3, 29}
$$

Now Φ_∞ is an odd function of $z = \mathscr{I}(\chi)$ in the χ-plane, i.e. it must vanish along the real axis in the χ-plane outside the segment $\langle -s_0, s_0 \rangle$. It follows that Φ_∞ vanishes also along the real axis of the ζ-plane, outside the unit circle, and hence that the coefficients $a_1, a_2, \dots$ in (3·3, 29) are all real. At the unit circle where $\zeta = e^{i\theta}$, Φ_∞ is given by

$$
\Phi_\infty = -\mathscr{R}\{i(a_1 e^{-i\theta} + a_2 e^{-2i\theta} + \dots)\} = -(a_1 \sin\theta + a_2 \sin 2\theta + \dots). \tag{3·3, 30}
$$

To calculate $\partial \Phi_\infty / \partial z$ at the trailing vortex sheet, we make use of the relation

$$
\frac{\partial \Phi_\infty}{\partial z} = -\mathscr{I}\left(\frac{d\Omega}{d\chi}\right) = -\mathscr{I}\left(\frac{d\Omega}{d\zeta} \Big/ \frac{d\chi}{d\zeta}\right).
$$

At the unit circle, which corresponds to the trailing vortex sheet,

$$\frac{d\Omega}{d\zeta} = i\left(\frac{a_1}{\zeta^2} + \frac{2a_2}{\zeta^3} + \dots\right)$$

$$= i(a_1 e^{-2i\theta} + 2a_2 e^{-3i\theta} + \dots),$$

and

$$\frac{d\chi}{d\zeta} = \frac{s_0}{2}\left(1 - \frac{1}{\zeta^2}\right) = \frac{s_0}{2}(1 - e^{-2i\theta}) = i e^{-i\theta} s_0 \sin\theta,$$

and so

$$\frac{d\Phi_\infty}{dz} = -\mathscr{I}\left(\frac{a_1 e^{-i\theta} + 2a_2 e^{-2i\theta} + \dots}{s_0 \sin\theta}\right)$$

$$= \frac{a_1 \sin\theta + 2a_2 \sin\theta + \dots}{s_0 \sin\theta}. \tag{3·3, 31}$$

Substituting in (3·3, 23) from (3·3, 30) and (3·3, 31), we obtain

$$4\sum_{n=1}^{\infty} a_n \sin n\theta = a_0 c\left(U\alpha_0 - \frac{1}{2s_0 \sin\theta}\sum_{n=1}^{\infty} n a_n \sin n\theta\right). \tag{3·3, 32}$$

Substituting in (3·3, 24) leads to an equivalent result. Putting

$$A_n = \frac{a_n}{2s_0 U} \quad (n = 1, 2, \dots) \quad \text{and} \quad \mu = \frac{a_0 c}{8s_0}, \tag{3·3, 33}$$

we obtain, after rearrangement,

$$\sum_{n=1}^{\infty} A_n \sin n\theta(n\mu + \sin\theta) = \mu\alpha_0 \sin\theta \quad (0 < \theta < \pi), \tag{3·3, 34}$$

where μ may be regarded as a function of θ. (3·3, 34) is known as the Prandtl–Glauert equation, because of the numerous applications to which it was put by Glauert at an early stage.‡ The determination of the A_n from (3·3, 34) will be discussed presently. In terms of these coefficients, we have

$$\Phi_\infty = -2s_0 U(A_1 \sin\theta + A_2 \sin 2\theta + \dots), \tag{3·3, 35}$$

and so (compare (3·3, 21))

$$\Delta\Phi = -4s_0 U(A_1 \sin\theta + A_2 \sin 2\theta + \dots) \quad (0 < \theta < \pi), \tag{3·3, 36}$$

and

$$\frac{\partial\Phi_\infty}{\partial z} = \frac{2U}{\sin\theta}(A_1 \sin\theta + 2A_2 \sin 2\theta + \dots). \tag{3·3, 37}$$

Also, at the vortex sheet,

$$y = \tfrac{1}{2}s_0(e^{i\theta} + e^{-i\theta}) = s_0 \cos\theta. \tag{3·3, 38}$$

Substituting in (3·2, 18), we obtain for the total lift on the wing

$$L = 4\rho U^2 s_0^2 \int_0^\pi (A_1 \sin\theta + A_2 \sin 2\theta + \dots) \sin\theta \, d\theta,$$

or

$$L = 2\pi\rho U^2 s_0^2 A_1. \tag{3·3, 39}$$

‡ Consult H. Glauert, *The Elements of Aerofoil and Airscrew Theory* (2nd ed., Cambridge University Press, 1947), ch. 9.

The spanwise lift distribution is, by (3·2, 21),

$$l = 4\rho U^2 s_0 (A_1 \sin \theta + A_2 \sin 2\theta + \ldots), \tag{3·3, 40}$$

from which (3·3, 39) may also be obtained directly by integration.

We calculate the induced drag D_i by means of (3·2, 34) and observe that in that formula, w denotes the z-component of the velocity far behind the wing, i.e. in the Trefftz plane. Thus

$$D_i = -\tfrac{1}{2}\rho \int \Delta\Phi \frac{\partial \Phi_\infty}{\partial z} dy, \tag{3·3, 41}$$

i.e. $D_i = 4\rho U^2 s_0^2 \displaystyle\int_0^\pi (A_1 \sin \theta + A_2 \sin 2\theta + \ldots)(A_1 \sin \theta + 2A_2 \sin 2\theta + \ldots) d\theta,$

or

$$D_i = 2\pi\rho U^2 s_0^2 \sum_{n=1}^\infty n A_n^2. \tag{3·3, 42}$$

Equation (3·3, 41) does not depend on the approximations of the present section. However, lifting-line theory provides an interesting interpretation of the formula. Since the flow round any particular aerofoil section is supposed to be two-dimensional, the resultant force on the section should be normal to the (corrected) main-stream direction. It is therefore inclined by an angle w/U relative to the z-axis, where w is the z-component of the velocity induced by the trailing vortices at the wing. It follows that the force has a component

$$d_i(y) = -l(y)\frac{w}{U} = -\tfrac{1}{2}\rho\Delta\Phi\frac{\partial \Phi_\infty}{\partial z} \tag{3·3, 43}$$

in the direction of the x-axis. The total induced drag can then be obtained from (3·3, 43) by integration across the span, and the result agrees with (3·3, 41). Moreover, (3·3, 43) provides an expression for the spanwise distribution of the induced drag.

Next, we may find a simple expression for the rolling moment about the x-axis. Using the expression given in (3·2, 39), we obtain

$$L' = \int l(y) y \, dy = 4\rho U^2 s_0^3 \int_0^\pi (A_1 \sin \theta + A_2 \sin 2\theta + \ldots) \cos \theta \sin \theta \, d\theta.$$

Hence

$$L' = \pi\rho U^2 s_0^3 A_2 \tag{3·3, 44}$$

is the required expression. There is no simple formula for the pitching moment except in certain special cases.

If we specify the spanwise lift distribution as a function of y or θ, for a given free-stream velocity, then this determines the downwash. Theoretically speaking we may, in addition, specify $a_0 c$ arbitrarily, and then determine α_0 from (3·3, 7) or (3·3, 32), or, conversely, we may specify α_0 and then determine $a_0 c$ from these equations. (Physically speaking, the choice of $a_0 c$ or α_0 is of course subject to the conditions that α_0 must be reasonably small, and $a_0 c$ must be positive.) However, in general, the choice of the wing slope is influenced by structural and other considerations, so that in practice the important problem is how to determine the constants A_n and hence the aerodynamic forces for a given wing. A variety of methods have been devised to cope with this problem and three of these are described below. The reader may consult papers by Lotz,‡ Pugsley§ and the book of Fuchs–Hopf–Seewald.‖ The relative efficiency of the different methods depends to some extent on the characteristics of the wing and on the data required (lift and drag distribution or resultant forces).

Glauert's method¶ consists in neglecting all coefficients A_n of order greater than some specific n ($n = 8$, say). The remaining A_n are then determined by the condition that (3·3, 34) be satisfied for n distinct values of θ, corresponding to n points along the span of the wing. In practice the problem can be simplified by the separate consideration of symmetrical and antisymmetrical conditions (with respect to the x, z-plane). Let us consider the symmetrical case in more detail. The assumption of symmetry implies that $\Phi_\infty(-y, 0) = \Phi_\infty(y, 0)$, and this in turn entails that the coefficients of even order vanish, $A_2 = A_4 = A_6 = \dots = 0$. To determine the coefficients of odd order, $A_1, A_3, \dots, A_{2m-1}$, we attempt to satisfy (3·3, 34) for

$$\theta_k = \frac{k}{2m}\pi \quad (k = 1, 2, \dots, m).$$

(3·3, 34) is then satisfied automatically also for

$$\theta_k = \frac{k}{2m}\pi \quad (k = m+1, \dots, 2m-1).$$

Thus for the case $n = 8$ ($m = 4$), we obtain the following system of four linear equations for the four unknowns A_1, A_3, A_5, A_7,

$$\sum_{j=1}^{4} A_{2j-1} \sin(2j-1)\theta_k((2j-1)\mu + \sin\theta_k) = \mu\alpha\sin\theta_k \quad (k = 1, 2, 3, 4). \quad (3·3, 45)$$

Note that, in general, μ and α also depend on k.

In the procedure suggested by Gates,‡‡ one tries to satisfy (3·3, 34) approximately in the least mean-square sense. Retaining only the coefficients $A_1, \dots, A_n$, as before, one now minimizes the expression

$$\int_0^\pi \left(\sum_{j=1}^{n} A_j \sin j\theta (j\mu + \sin\theta) - \mu\alpha\sin\theta \right)^2 d\theta = F(A_1, \dots, A_n). \quad (3·3, 46)$$

The conditions for a minimum are

$$\frac{\partial F}{\partial A_j} = 0 \quad (j = 1, 2, \dots, n). \quad (3·3, 47)$$

‡ I. Lotz, 'Berechnung der Auftriebsverteilung beliebig geformter Flügel', *Z. Flugtech.* vol. 22 (1931), pp. 189–95.

§ A. G. Pugsley, 'Approximate method of determining aerodynamic loading on wings of monoplane', *Rep. Memor. Aero. Res. Comm., Lond.*, no. 1643 (1934).

‖ R. Fuchs, L. Hopf and Fr. Seewald, *Aerodynamik* (Julius Springer, Berlin, 1934), vol. 2.

¶ H. Glauert, *The Elements of Aerofoil and Airscrew Theory* (2nd ed., Cambridge University Press, 1947), ch. 9.

‡‡ S. B. Gates, 'An analysis of a rectangular monoplane with hinged tips', *Rep. Memor. Aero. Res. Comm., Lond.*, no. 1175 (1928).

Now

$$F(A_1, A_2, ..., A_n) = \sum_{j=1}^{n} A_j^2 \int_0^{\pi} \sin^2 j\theta (j\mu + \sin\theta)^2 \, d\theta$$

$$+ 2\sum_{j=1}^{n} \sum_{k=1}^{n} A_j A_k \int_0^{\pi} \sin j\theta \sin k\theta (j\mu + \sin\theta)(k\mu + \sin\theta) \, d\theta$$

$$- 2\sum_{j=1}^{n} A_j \int_0^{\pi} \sin j\theta (j\mu + \sin\theta) \mu\alpha \sin\theta \, d\theta$$

$$+ \int_0^{\pi} \mu^2 \alpha^2 \sin^2\theta \, d\theta.$$

Then

$$\frac{\partial F}{\partial A_j} = 2\sum_{k=1}^{n} \alpha_{jk} A_k - 2\beta_0 \quad (j = 1, 2, ..., n), \qquad (3\cdot3, 48)$$

where

$$\left. \begin{array}{l} \alpha_{jk} = \displaystyle\int_0^{\pi} \sin j\theta \sin k\theta (j\mu + \sin\theta)(k\mu + \sin\theta) \, d\theta \\[2mm] \beta_0 = \displaystyle\int_0^{\pi} \sin j\theta (j\mu + \sin\theta) \mu\alpha \sin\theta \, d\theta \end{array} \right\} \quad (j, k = 1, 2, ..., n). \quad (3\cdot3, 49)$$

Hence from (3·3, 47)
$$\sum_{k=1}^{n} \alpha_{jk} A_k = \beta_j \quad (j = 1, 2, ..., n), \qquad (3\cdot3, 50)$$

which is a set of linear equations for the coefficients $A_1, ..., A_n$. For a given wing, the coefficients α_{jk} can be calculated once and for all, while the β_j depend also on the incidence (which may be variable). The advantage of the method is that it does not depend only on the characteristics of the wing at a small number of isolated points, but leads to an approximate solution of (3·3, 34) along the entire span. It can therefore be applied most conveniently to problems which involve discontinuities, as in the case for a wing with deflected flaps or ailerons.

Gates's procedure is based on a variational principle, viz. the obvious fact that the solution of (3·3, 34) corresponds to a minimum of the integral of the square of the differences between the two sides of that equation. Another variational principle which may be applied to the same problem will be derived later (§ 3·5).

Multhopp's method‡ is based on the following trigonometrical identities which may be derived by mathematical induction:

$$s_n(\alpha, \beta) = \sum_{j=1}^{n} \sin j\alpha \sin j\beta$$

$$= \begin{cases} \dfrac{1}{2} \dfrac{\sin(n+1)\alpha \sin n\beta - \sin(n+1)\beta \sin n\alpha}{\cos\alpha - \cos\beta}, & \text{when} \quad \alpha \neq \beta, \\[4mm] \dfrac{1}{2}\left\{(n+1) - \dfrac{\sin(n+1)\alpha \cos n\alpha}{\sin\alpha}\right\} & \text{when} \quad \alpha = \beta, \qquad (3\cdot3, 51) \end{cases}$$

where
$$0 < \alpha < \pi, \quad 0 < \beta < \pi, \quad n = 1, 2, ...$$

‡ H. Multhopp, 'Die Berechnung der Auftriebsverteilung von Tragflügeln', *Luftfahrtforsch.* vol. 15 (1938), pp. 153–69. Translated as *R.T.P. Transl.* no. 2392.

For given n, define θ_k by

$$\theta_k = \frac{k\pi}{n+1} \quad (k = 1, \ldots, n),$$

then

$$s_n(\theta_j, \theta_k) = \begin{cases} 0 & j \neq k \\ \frac{1}{2}(n+1) & j = k \end{cases} \quad (j, k = 1, 2, \ldots, n). \tag{3·3, 52}$$

Let $\Gamma_1, \Gamma_2, \ldots, \Gamma_n$ be a given set of real numbers, and consider the trigonometrical polynomial

$$f(\theta) = \frac{2}{n+1} \sum_{k=1}^{n} s_n(\theta_k, \theta) \Gamma_k$$

$$= \frac{2}{n+1} \sum_{k=1}^{n} \left(\sum_{j=1}^{n} \sin j\theta_k \sin j\theta \right) \Gamma_k. \tag{3·3, 53}$$

Using the relations (3·3, 52), we find that

$$f(\theta_k) = \Gamma_k \quad (k = 1, 2, \ldots, n). \tag{3·3, 54}$$

Assume now that the circulation Γ in (3·3, 7), regarded as a function of θ, is equal to $2s_0 U f(\theta)$, where $f(\theta)$ is defined by (3·3, 53). Then the corresponding Φ_∞ (see (3·3, 21)) is equal to $-s_0 U f(\theta)$. To obtain $\partial\Phi_\infty/\partial z$, we observe that the jth term in (3·3, 37) is equal to the jth term in (3·3, 35) multiplied by $-j/s_0 \sin\theta$ $(j = 1, 2, \ldots)$. Hence

$$\frac{\partial\Phi_\infty}{\partial z} = \frac{2U}{(n+1)\sin\theta} \sum_{k=1}^{n} \left(\sum_{j=1}^{n} j \sin j\theta_k \sin\theta \right) \Gamma_k, \tag{3·3, 55}$$

and further, by (3·3, 22),

$$w = -\frac{U}{(n+1)\sin\theta} \sum_{k=1}^{n} \left(\sum_{j=1}^{n} j \sin j\theta_k \sin j\theta \right) \Gamma_k, \tag{3·3, 56}$$

or

$$w = -\frac{U}{(n+1)} \sum_{j=1}^{n} t_n(\theta_j, \theta) \Gamma_j, \tag{3·3, 57}$$

where

$$t_n(\alpha, \beta) = \frac{1}{\sin\beta} \sum_{j=1}^{n} j \sin j\alpha \sin j\beta. \tag{3·3, 58}$$

It is again not difficult to evaluate (3·3, 58), by mathematical induction or otherwise. The result is

$$t_n(\alpha, \beta) = \begin{cases} \dfrac{1}{2(\cos\alpha - \cos\beta)} \{(n+1)\sin(n+1)\alpha \sin n\beta - n \sin n\alpha \sin(n+1)\beta\} \\[2mm] \quad - \dfrac{\sin\alpha}{2(\cos\alpha - \cos\beta)^2 \sin\beta} \{\cos(n+1)\alpha \sin\beta - \cos n\alpha \sin(n+1)\beta \\[1mm] \quad\quad\quad + \sin\beta\} \quad \text{when} \quad \alpha \neq \beta \\[3mm] \dfrac{1}{4\sin\alpha} \left\{ n(n+1) + \dfrac{1}{2\sin^2\alpha}(1 - \cos 2n\alpha - 2n \sin(2n+1)\alpha \sin\alpha) \right\} \\[2mm] \quad\quad\quad\quad\quad\quad \text{when} \quad \alpha = \beta, \tag{3·3, 59} \end{cases}$$

where $0 < \alpha < \pi$, $0 < \beta < \pi$, $n = 1, 2, \ldots$. Hence

$$t_n(\theta_j, \theta_k) = \begin{cases} \dfrac{-\sin\theta_j}{(\cos\theta_j - \cos\theta_k)^2} & (|j-k| = 1, 3, 5, \ldots) \\ 0 & (|j-k| = 2, 4, 6, \ldots) \\ \dfrac{(n+1)^2}{4\sin\theta_k} & (j = k) \end{cases} \quad (j, k = 1, 2, \ldots, n).$$

$$(3\cdot3, 60)$$

Substituting $(3\cdot3, 57)$ in $(3\cdot3, 7)$, for $\theta = \theta_k$, we obtain

$$2s_0 U\Gamma_k = \tfrac12 a_0 Uc\left(\alpha_0 - \frac{1}{n+1}\sum_{j=1}^{n} t_n(\theta_j, \theta_k)\,\Gamma_j\right) \quad (k = 1, \ldots, n). \quad (3\cdot3, 61)$$

This may also be written as

$$\Gamma_k = 2\mu_k\left\{\alpha_k - \left(b_{kk}\Gamma_k - \sum_{j=1}^{n}{}' b_{kj}\Gamma_j\right)\right\},$$

where the dash affixed to the sign of summation indicates that the term $j = k$ is to be omitted, and where, for given n,

$$\left.\begin{aligned} b_{kk} &= \frac{1}{n+1}\,t_n(\theta_k, \theta_k), \\ b_{kj} &= -\frac{1}{n+1}\,t_n(\theta_j, \theta_k) \quad (j \neq k), \end{aligned}\right\} \quad (3\cdot3, 62)$$

and μ_k, α_k are the values of μ, α_0 for $\theta = \theta_k$. Finally, putting

$$b_k = \frac{1}{2\mu_k} + b_{kk} \quad (k = 1, \ldots, n), \quad (3\cdot3, 63)$$

we obtain

$$b_k\Gamma_k = \alpha_k + \sum_{j=1}^{n}{}' b_{kj}\Gamma_j \quad (k = 1, \ldots, n). \quad (3\cdot3, 64)$$

This is Multhopp's fundamental system of linear equations.

Note that the coefficients b_{kj} are independent of the particular wing under consideration. Tables of these coefficients are given by Multhopp in the above-mentioned paper.

For sufficiently small n, the system can be solved in the usual way by successive elimination. Since b_{kj} vanishes for even $k-j$, $k \neq j$, we may first express the Γ_k with odd subscripts in terms of the Γ_k with even subscripts, thus reducing the number of unknowns by half. Further simplifications are obtained if symmetrical and antisymmetrical cases are considered separately, as before.

An alternative procedure, suitable for large n, consists in solving $(3\cdot3, 64)$ by iteration. We first put

$$b_k\Gamma_k^{(0)} = \alpha_k, \quad (3\cdot3, 65)$$

and we then calculate the quantities $\Gamma_k^{(m)}$ successively from

$$\Gamma_k^{(m+1)} = \alpha_k + \sum_{j=1}^{n}{}' b_{kj}\Gamma_j^{(m)} \quad (m = 0, 1, \ldots). \quad (3\cdot3, 66)$$

We note that for a wing of very large aspect ratio and constant characteristics, $\Gamma = 2s_0 U \Gamma_k^{(0)}$ is already the correct value of the circulation. Indeed, in that case,

$$\Gamma = 2s_0 U \Gamma_k^{(0)} = 2s_0 U \frac{\alpha_k}{b_k} \doteq 4s_0 U \mu \alpha_0 = \tfrac{1}{2} a_0 U c \alpha_0$$

in agreement with two-dimensional theory.

For given n, Γ is represented by a sine polynomial of order n which satisfies (3·3, 7)—and hence (3·3, 34)—exactly for $\theta_1, \ldots, \theta_n$. The solution to this problem is unique, and it follows that (rounding-off errors apart) the same result is obtained if instead of Multhopp's procedure we use Glauert's method for the same points $\theta_1, \ldots, \theta_k$. However, the former method leads directly to the calculation of the circulation and hence of the spanwise lift distribution, while the latter yields in the first instance the coefficients A_k, which are involved in the formulae for the resultant aerodynamic forces (3·3, 39), (3·3, 42) and (3·3, 44). Thus, preference for one method or the other may depend on what data are required. Apart from this consideration, Multhopp's method is regarded as superior from the computational point of view. To obtain the coefficients A_k by this method, we write

$$\Phi_\infty = - s_0 U f(\theta) = - \frac{2s_0 U}{n+1} \sum_{k=1}^{n} \left(\sum_{j=1}^{n} \sin j\theta_k \sin j\theta \right) \Gamma_k$$

$$= - 2s_0 U \sum_{j=1}^{n} \left(\frac{1}{n+1} \sum_{k=1}^{n} \Gamma_k \sin j\theta_k \right) \sin j\theta.$$

Comparing this formula with (3·3, 35), we see that

$$A_j = \frac{1}{n+1} \sum_{k=1}^{n} \Gamma_k \sin j\theta_k \quad (j = 1, \ldots, n). \tag{3·3, 67}$$

Table 3·3, 1

	Glauert	Gates	Multhopp
A_1/α_0	0·2320	0·2321	0·2320
A_3/α_0	0·0287	0·0289	0·0286
A_5/α_0	0·0058	0·0062	0·0058
A_7/α_0	0·0010	0·0017	0·0010

Table 3·3, 1 shows the results of calculations made for a wing at constant incidence for which $\mu = \tfrac{1}{4}$.‡ For constant a_0, $a_0 = 2\pi$ say, this is a rectangular wing of aspect ratio

$$\mathcal{R} = \frac{\text{span}^2}{\text{area}} = \frac{4s_0^2}{2s_0 c} = a_0 = 6 \cdot 28.$$

‡ Taken from S. Kirkby and A. Robinson, *Note on Computational Methods*, no. 7, *Numerical Solution of Integral Equations* (College of Aeronautics, Cranfield, 1949), pp. 48–57.

It will be seen that the results provided by the three methods are in good agreement with each other, and that the coefficients decrease rapidly after the second. In general, unless it is desired to take account of the effect of sudden local variations of the wing characteristics, the retention of four coefficients (A_1, A_3, A_5, A_7 for the symmetrical case, A_2, A_4, A_6, A_8 for the antisymmetrical case) is sufficient to provide reliable results.

We shall now consider some special cases. First suppose that

$$A_2 = A_3 = \ldots = 0,$$

so that only the first term appears in the expression for Φ_∞, and hence for the circulation and lift distribution, so

$$\left.\begin{aligned}
\Phi_\infty &= -2s_0 U A_1 \sin \theta, \\
w &= -\frac{1}{2}\frac{\partial \Phi_\infty}{\partial z} = -U A_1, \\
l &= 4\rho U^2 s_0 A_1 \sin \theta, \\
L &= 2\pi \rho U^2 s_0^2 A_1, \\
D_i &= 2\pi \rho U^2 s_0^2 A_1^2.
\end{aligned}\right\} \tag{3·3, 68}$$

The second formula in (3·3, 68) shows that in this case the downwash is constant. It is not difficult to see that if the downwash is constant then $A_2 = A_3 = \ldots = 0$. Again, (3·3, 39) shows that for a given span, A_1 is proportional to the lift in any case. Also, for given A_1, the induced drag as given by (3·3, 42) is clearly a minimum when $A_2 = A_3 = \ldots = 0$. Thus, the case under consideration is characterized also by the fact that it is 'most favourable' in the sense that it furnishes a given lift for minimum drag. The corresponding condition imposed on the slope of the wing is found by means of (3·3, 34), which now reads

$$A_1 \sin \theta (\mu + \sin \theta) = \mu \alpha_0 \sin \theta. \tag{3·3, 69}$$

This requires that the ratio

$$\frac{\mu + \sin \theta}{\mu \alpha_0} = \text{constant} \tag{3·3, 70}$$

along the span of the wing. (3·3, 70) can be satisfied in various ways, but if α_0 is constant, then it implies that μ is proportional to $\sin \theta$, $\mu = \mu_0 \sin \theta$, where $\mu = \mu_0$ at the centre line. If α_0 is constant, then this relation in turn entails $c = c_0 \sin \theta$, where c_0 is the length of the chord at the centre line. Now $\sin \theta = \sqrt{\{1 - (y/s_0)^2\}}$, and so

$$c = \frac{c_0}{s_0} \sqrt{(s_0^2 - y^2)}. \tag{3·3, 71}$$

This law of variation for the chord is realized by a wing of elliptic planform whose equation is

$$\frac{4x^2}{c_0^2} + \frac{y^2}{s_0^2} = 1. \tag{3·3, 72}$$

However, so far as lifting-line theory is concerned, the same results are obtained if the wing sections are displaced parallel to the direction of the x-axis. The lift distribution is given by

$$l = 4\rho U^2 A_1 \sqrt{(s_0^2 - y^2)}, \tag{3·3, 73}$$

and the graph of this curve against the y-coordinate is again a (semi-) ellipse.

The area of the wing is, by (3·3, 71) or (3·3, 72), $S = \frac{1}{2}\pi c_0 s_0$, and so its aspect ratio is, in terms of s_0 and c_0,

$$Æ = \frac{4s_0^2}{S} = \frac{8}{\pi}\frac{s_0}{c_0}. \tag{3·3, 74}$$

It follows that μ may be written in the form

$$\mu = \frac{a_0 c}{8s_0} = \frac{a_0 c_0}{8s_0}\sin\theta = \frac{a_0}{\pi Æ}\sin\theta. \tag{3·3, 75}$$

Substituting in (3·3, 69), we obtain for A_1,

$$A_1 = \frac{\alpha_0}{1 + \dfrac{\pi}{a_0}Æ}. \tag{3·3, 76}$$

Hence

$$C_L = \frac{L}{\frac{1}{2}\rho U^2 S} = \pi Æ A_1 = \frac{\pi Æ \alpha_0}{1 + \dfrac{\pi}{a_0}Æ}. \tag{3·3, 77}$$

Table 3·3, 2

$Æ$	∞	10	8	6	4
a (per radian)	6·28	5·24	5·02	4·70	4·18

As the aspect ratio approaches infinity, C_L tends to its two-dimensional value, $a_0\alpha_0$, as might be expected. We write a for the lift-curve slope $dC_L/d\alpha_0$, then

$$\frac{a}{\pi} = \frac{Æ}{1 + \dfrac{\pi}{a_0}Æ}. \tag{3·3, 78}$$

In particular, if a_0 attains its theoretical value for the flat plate, $a_0 = 2\pi$, then

$$a = 2\pi\frac{Æ}{2 + Æ} \tag{3·3, 79}$$

(see Table 3·3, 2). (3·3, 79) shows that for a given lift coefficient, the incidence is

$$\alpha_0 = \frac{C_L}{a} = \frac{1}{2\pi}\left(1 + \frac{2}{Æ}\right)C_L. \tag{3·3, 80}$$

If α_0' is the incidence for another aspect ratio $\mathcal{R}'$ and for the same lift coefficient, then

$$\alpha_0' - \alpha_0 = \frac{1}{\pi}\left(\frac{1}{\mathcal{R}'} - \frac{1}{\mathcal{R}}\right)C_L. \tag{3·3, 81}$$

Although this relation has been derived only for elliptic wings, it may be used as an approximate formula also for other wings which are obtained from each other by the scaling up of the spanwise and chordwise dimensions in different proportions. If α_0 is known from experiment, (3·3, 81) may then be used to calculate α_0'.

Substituting (3·3, 76) in (3·3, 68), we obtain for the induced drag coefficient C_{D_i},

$$C_{D_i} = \frac{D_i}{\frac{1}{2}\rho U^2 S} = \pi \mathcal{R} A_1^2 = \frac{\pi \mathcal{R}}{\left(1 + \dfrac{\pi}{a_0}\mathcal{R}\right)^2}\alpha_0^2, \tag{3·3, 82}$$

or

$$C_{D_i} = \frac{1}{\pi \mathcal{R}}C_L^2. \tag{3·3, 83}$$

The total drag is $D = D_0 + D_i$, where D_0 is the profile drag, and the corresponding drag coefficient is

$$C_D = C_{D_0} + C_{D_i} = C_{D_0} + \frac{1}{\pi \mathcal{R}}C_L^2, \tag{3·3, 84}$$

where

$$C_D = D/\tfrac{1}{2}\rho U^2 S, \quad C_{D_0} = D_0/\tfrac{1}{2}\rho U^2 S.$$

More generally, for a wing of arbitrary plan-form,

$$A_1 = \frac{L}{2\pi\rho U^2 s_0^2} = \frac{1}{\pi \mathcal{R}}C_L,$$

and

$$C_{D_i} = \frac{D_i}{\frac{1}{2}\rho U^2 S} = \pi \mathcal{R}\sum_{n=1}^{\infty} nA_n^2 = \pi \mathcal{R}A_1^2\left(1 + \sum_{n=2}^{\infty} nA_n^2/A_1^2\right)$$

$$= \frac{1}{\pi \mathcal{R}}C_L^2\left(1 + \sum_{n=2}^{\infty} nA_n^2/A_1^2\right),$$

or

$$C_{D_i} = \frac{1}{\pi \mathcal{R}}(1 + \delta)C_L^2, \tag{3·3, 85}$$

where

$$\delta = \sum_{n=2}^{\infty} nA_n^2/A_1^2.$$

δ is independent of the incidence α_0 provided the latter is constant across the span. Figs. 3·3, 4a and b show δ plotted against aspect ratio, for rectangular wings, and against taper ratio λ for wings of a given aspect ratio for which the chord length varies linearly from the root centre line to the tips. For such wings the taper ratio is defined as (tip chord) $\div$ (root chord). It will be seen that δ is small compared with unity in all these cases.

The total drag coefficient is now given by

$$C_{D_i} = C_{D_0} + \frac{1}{\pi \mathcal{R}}(1 + \delta)C_L^2. \tag{3·3, 86}$$

WING THEORY

This is one of the fundamental formulae of the theory of aircraft performance. In level flight, the lift must be equal to the weight W of the aeroplane, thus

$$W = L = C_L \tfrac{1}{2}\rho U^2 S,$$

and this relation determines the lift coefficient for given air density (or altitude), forward speed and wing area. It then appears from (3·3, 86) that

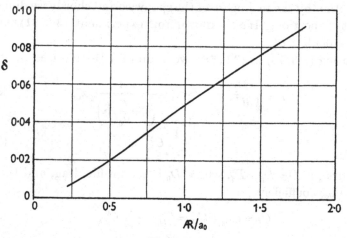

Fig. 3·3, 4 (a)

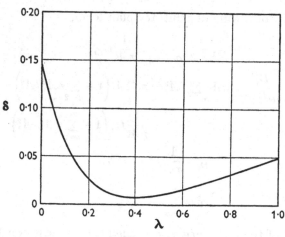

Fig. 3·3, 4 (b)

in order to reduce the drag, it is advisable to make the aspect ratio of the wing as large as possible. However, other considerations such as the question of structure weight, set a limit to the application of this idea.

As the aspect ratio of the wing decreases, lifting-line theory becomes progressively less reliable. No useful analytical bound is known for the errors involved, but experience has shown that the theory remains applicable for aspect ratios as low as $R = 4$.

The general expression for the rolling moment is given by (3·2, 39) as

$$L' = \int l(y)\,y\,dy,\qquad (3\cdot3, 87)$$

which in lifting-line theory becomes equal to (3·3, 44). However, for plane wings of symmetric plan-form both (3·3, 44) and (3·3, 87) are zero.

If we consider a wing with small dihedral (Fig. 3·3, 5) the same formula (3·3, 87) is applicable (it is a special case of a 'compound lifting unit', to be covered in detail in the next section). Moreover, if the wing is yawed relative to the free-stream velocity U, L' has a non-zero value which may be evaluated as follows.

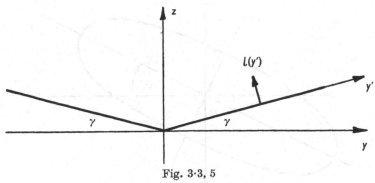

Fig. 3·3, 5

We shall calculate L' about an axis Ox', fixed relative to the wing, that is, perpendicular to the span Oy' (see Figs. 3·3, 5 and 3·3, 6), and we shall assume the angle of yaw is ψ, measured from the line $x = 0$. In this case

$$L' = \int_{-s_0}^{s_0} l(y')\,y'\,dy' \qquad (3\cdot3, 88)$$

and $l(y')$ is the lift force acting normal to the wing.

The main-stream velocity can be resolved into components lying in planes normal to Oy'. They are

$$U \cos\alpha \cos\psi, \qquad (3\cdot3, 89)$$

and

$$U \sin\alpha \cos\psi \cos\gamma \pm U \sin\psi \sin\gamma \qquad (3\cdot3, 90)$$

parallel and normal to the wing respectively (Fig. 3·3, 7). The positive sign in the second term of (3·3, 90) applies for the starboard wing, and the negative sign for the port wing. The 'effective' angle of incidence is therefore

$$\alpha' = \tan^{-1}(\tan\alpha \cos\gamma \pm \sec\alpha \tan\psi \sin\gamma). \qquad (3\cdot3, 91)$$

For small incidence and small values of γ and ψ, this reduces to

$$\alpha'_0 - \alpha_0 \doteqdot \pm\psi\gamma. \qquad (3\cdot3, 92)$$

To find the rolling moment we can now use (3·3, 88), with $l(y')$ calculated for an incidence angle given by (3·3, 91) and a free-stream velocity that is the resultant of (3·3, 89) and (3·3, 90). Let us consider the case of a wing of symmetric plan-form with both the dihedral γ and the yaw ψ small. Since $l(y')$ is proportional to the incidence, we see from (3·3, 92) that the change in rolling moment due to dihedral and yaw is approximately

$$\Delta L' = \int_0^{s_0} \psi\gamma \frac{\partial l}{\partial \alpha_0} y'\,dy' - \int_{-s_0}^0 \psi\gamma \frac{\partial l}{\partial \alpha_0} y'\,dy'.$$

Fig. 3·3, 6

Fig. 3·3, 7

For a symmetric plan-form L' is zero if there is no dihedral, hence we may write the total rolling moment as

$$L' = 2\psi\gamma \int_0^{s_0} \frac{\partial l}{\partial \alpha_0} y'\,dy'. \tag{3·3, 93}$$

(3·3, 93) may be evaluated for any chosen lift loading $l(y')$. For example, with an elliptic lift distribution we find, using (3·3, 68) and (3·3, 76),

$$L' = -\psi\gamma \frac{8\rho U^2 s_0^3}{1 + \pi R/a_0} \int_{\frac{1}{2}\pi}^{0} \sin^2\theta \cos\theta \, d\theta$$

$$= \psi\gamma \frac{8\rho U^2 s_0^3}{3(1 + \pi R/a_0)}.$$

Defining the rolling moment coefficient by

$$C_{L'} = \frac{L'}{\rho U^2 S s_0},$$

we find

$$C_{L'} = \frac{2\psi\gamma R}{3(1 + R/a_0)}, \qquad (3·3, 94)$$

since the aspect ratio $R = 4s_0^2/S$.

3·4   Compound lifting units

By a 'compound lifting unit', we mean any system of aerofoils; for example, a tail unit with one or more fins, a wing with large dihedral, or the two wings of a biplane.

The unit may be either connected, as in the first and second example, or disconnected, as in the last case. If the lateral dimensions of the unit (its dimensions in directions normal to the x-axis) are reasonably large compared with its chordwise dimensions (parallel to the x-axis) then the general ideas of lifting-line theory are still applicable. Thus, with an appropriate choice of signs, we now have, in place of (3·3, 7),

$$\Gamma = \tfrac{1}{2}a_0 U c\left(\alpha_0 + \frac{q_n}{U}\right) = \tfrac{1}{2}a_0 c(U\alpha_0 + q_n), \qquad (3·4, 1)$$

where q_n is the normal velocity at any point of the lifting unit and the other symbols are defined as before. The induced velocity potential in the Trefftz-plane, $\Phi_\infty(y, z)$, is again due solely to the trailing vortices which are supposed to extend downstream in straight lines parallel to the x-axis. Then

$$\Gamma = -\Delta\Phi = -\Delta\Phi_\infty, \qquad (3·4, 2)$$

where $\Delta\Phi = \Delta\Phi_\infty$ is the jump of the velocity potential across the vortex wake at the trailing edge or in the Trefftz plane.

It is assumed that the components of the lifting unit all occupy approximately the same region in the fore-and-aft direction. It follows that the bound vortices which replace these components do not affect the induced velocity at the unit. Hence, as in (3·3, 22),

$$q_n = -\frac{1}{2}\frac{\partial\Phi_\infty}{\partial n}, \qquad (3·4, 3)$$

where the normal direction n is taken from the lower to the upper surface. And so (3·4, 1) may be replaced by

$$\Delta\Phi_\infty = -\tfrac{1}{2}a_0 c\left(U\alpha_0 - \frac{1}{2}\frac{\partial\Phi_\infty}{\partial n}\right). \qquad (3·4, 4)$$

However, it is no longer possible to express Γ in terms of Φ_∞ on one side of the wing only, as in (3·3, 21).

The force per unit span in a direction normal to the wing in the y, z-plane is, at any point of the unit,

$$l = \rho U\Gamma = -\rho U\Delta\Phi = -\rho U\Delta\Phi_\infty, \qquad (3·4, 5)$$

where the degree of accuracy is the same as in (3·2, 21). Thus the y- and z-components on an element (dy, dz) of the wing (see Fig. 3·4, 1) are

$$-l\,dz = \rho U\Delta\Phi_\infty dz$$

and

$$l\,dy = -\rho U\Delta\Phi_\infty dy.$$

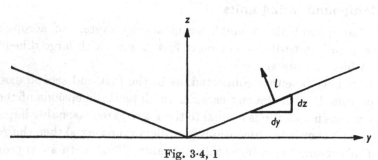

Fig. 3·4, 1

It follows that the y- and z-components of the resultant force on the lifting unit are given by

$$(Y, Z) = \int l(-dz, dy) = \rho U\int\Gamma(-dz, dy) = -\rho U\int\Delta\Phi_\infty(-dz, dy), \qquad (3·4, 6)$$

where the integrals are taken over the entire unit. (3·4, 6) may also be written as

$$(Y, Z) = \rho U\int\Phi(-dz, dy) = \rho U\int\Phi_\infty(-dz, dy), \qquad (3·4, 7)$$

where the integral is taken round the wake of the unit if the latter is connected, or else, round its several connected components, in a mathematically positive direction. (3·4, 7) includes (3·2, 20) as a special case.

An expression for the induced drag can be derived after the same fashion as in § 3·2. The result is

$$D_i = \tfrac{1}{2}\rho\int\Phi_\infty\frac{\partial\Phi_\infty}{\partial n'}\,dl, \qquad (3·4, 8)$$

where the integral is again taken round the wake of the unit, or round its several connected components, as the case may be. This is in agreement with (3·2, 34). D_i may also be expressed in terms of the circulation Γ by

$$D_i = -\rho \int \Gamma q_n \, dl, \qquad (3·4, 9)$$

where the integral is now taken only once along each wing of the unit, in the appropriate direction.

To solve the problem for a given lifting unit one may represent the circulation Γ or the potential Φ_∞ as linear combinations of suitable particular functions and then determine the coefficients by satisfying (3·4, 1) or (3·4, 4) at a finite number of points, or by a least-square method, as in § 3·3. Robinson‡ prescribed particular functions for the circulation round a tail unit with end-fins in a more or less arbitrary manner, and then calculated the normal velocity q_n by means of Biot-Savart's law. However, this procedure is very cumbersome, as is also Falkner's method.§

We shall now describe a more convenient manner‖ which provides suitable families of functions for Φ_∞, subject to the condition that the unit is simply connected (e.g. a normal tail unit).

Let $\chi = y + iz$ be the complex variable in the Trefftz plane. Map the region outside the trace T of the wake in the Trefftz plane (Fig. 3·4, 2) on the region outside the unit circle in an auxiliary plane, whose complex variable will be denoted by ζ, in such a way that the points at infinity correspond. Then the mapping function is of the form

$$\chi = f(\zeta) = b_{-1}\zeta + b_0 + \frac{b_1}{\zeta} + \frac{b_2}{\zeta^2} + \dots . \qquad (3·4, 10)$$

We have to determine potentials $\Phi_\infty(y, z)$ which are real parts of complex functions $\Omega(\chi)$, such that the Ω are one-valued and regular in the χ-plane, except at the trace T of the wake, and such that Ω vanishes, or tends to a finite value at infinity. Moreover, the normal derivative $\partial\Phi_\infty/\partial n$ must be continuous across T.

Let

$$\chi^n = (f(\zeta))^n = b^{(n)}_{-n}\zeta^n + b^{(n)}_{-n+1}\zeta^{n-1} + \dots + b^{(n)}_0 + \frac{b^{(n)}_1}{\zeta} + \dots \quad (n = 1, 2, \dots). \quad (3·4, 11)$$

Then the required conditions are satisfied by the functions

$$\Phi_\infty = \Phi_r^{(n)} = \mathscr{R}(\Omega_r^{(n)}) \quad (n = 1, 2, \dots),$$

where

$$\Omega_r^{(n)} = \chi^n - \sum_{k=1}^n \left(b^{(n)}_{-k}\zeta^k + \frac{\bar{b}^{(n)}_{-k}}{\zeta^k} \right). \qquad (3·4, 12)$$

In this formula $\bar{b}^{(n)}_{-k}$ is the conjugate complex number to $b^{(n)}_{-k}$.

‡ A. Robinson, 'The aerodynamic loading of wings with endplates', Rep. Memor. Aero. Res. Comm., Lond., no. 2342 (1945).

§ V. M. Falkner, 'The design of minimum drag tip fins', Rep. Memor. Aero. Res. Comm., Lond., no. 2279 (1945).

‖ A. Robinson, 'Flow round compound lifting units', Proc. Symposium on High Speed Aerodynamics, Ottawa (1953), pp. 26–9.

Indeed, if we express χ^n by the right-hand side of (3·4, 11), then it is evident that the Laurent series of $\Omega_r^{(n)}$ in the ζ-plane does not contain any positive powers of ζ. This implies that $\Omega_r^{(n)}$ is regular at infinity in the ζ-plane, and hence also in the χ-plane. Also, at the unit circle in the ζ-plane,

$$\sum_{k=1}^{n}\left(b_{-k}^{(n)}\zeta^k + \frac{\bar{b}_{-k}^{(n)}}{\zeta^k}\right) = \sum_{k=1}^{n}\left(b_{-k}^{(n)}e^{ik\theta} + \bar{b}_{-k}^{(n)}e^{-ik\theta}\right)$$

$$= \sum_{k=1}^{n}\left(b_{-k}^{(n)}e^{ik\theta} + \overline{b_{-k}^{(n)}e^{ik\theta}}\right). \tag{3·4, 13}$$

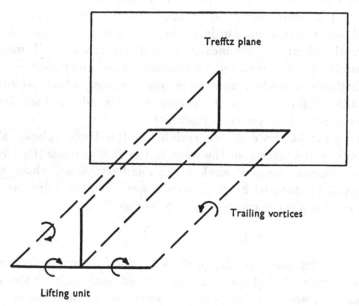

Fig. 3·4, 2

This shows that the expression is real at the unit circle in the ζ-plane, and hence at the trace T of the wake in the χ-plane. It follows that if we regard (3·4, 13) as the complex potential of a field of flow in the χ-plane, then T is a streamline of that flow, i.e. the velocity component normal to T vanishes. Thus, in order to calculate the normal velocity component corresponding to $\Omega_r^{(n)}$ at T, we may disregard the finite sum on the right-hand side of (3·4, 12). In other words, the velocity component normal to T, $-\partial\Phi_\infty/\partial n$, is the component of the vector

$$(-\mathscr{R}(n\chi^{n-1}),\ \mathscr{I}(n\chi^{n-1})) \tag{3·4, 14}$$

in that direction, and this is clearly continuous across T. On the other hand, the tangential component of the velocity at a point of T cannot be obtained from (3·4, 11) alone, and is, in general, discontinuous across T.

To calculate the jump $\Delta\Phi_\infty$ across T, we may write, for the points of T,

$$\Phi_\infty = \Phi_r^{(n)} = \mathscr{R}(\Omega_r^{(n)}) = \mathscr{R}(\chi^n) - \mathscr{R}\sum_{k=1}^{n}(b_{-k}^{(n)}e^{ik\theta} + \bar{b}_{-k}^{(n)}e^{-ik\theta}). \qquad (3\cdot4, 15)$$

Now it is clear that $\mathscr{R}(\chi^n)$ is continuous across T, and so

$$\Delta\Phi_\infty = \Delta\Phi_r^{(n)} = -\Delta\mathscr{R}\sum_{k=1}^{n}(b_{-k}^{(n)}e^{ik\theta} + \bar{b}_{-k}^{(n)}e^{-ik\theta})$$

$$= -2\Delta\left(\sum_{k=1}^{n}B_{-k}^{(n)}\cos k\theta - C_{-k}^{(n)}\sin k\theta\right), \qquad (3\cdot4, 16)$$

where $b_{-k}^{(n)} = B_{-k}^{(n)} + iC_{-k}^{(n)}$, and the symbol Δ indicates the jump across T in all cases.

Similarly, if we define a second family of velocity potentials in the Trefftz plane by

$$\Phi_\infty = \Phi_i^{(n)} = \mathscr{I}(\Omega_i^{(n)}) = \mathscr{R}(-i\Omega_i^{(n)}) \qquad n = 1, 2, \ldots,$$

where

$$\Omega_i^{(n)} = \chi^n - \sum_{k=1}^{n}\left(b_{-k}^{(n)}\zeta^k - \frac{\bar{b}_{-k}^{(n)}}{\zeta^k}\right), \qquad (3\cdot4, 17)$$

then it is not difficult to see that the required conditions are again satisfied. The velocity component normal to T, $-\partial\Phi_\infty/\partial n$, now is the component of the vector

$$(-\mathscr{R}(-in\chi^{n-1}), \mathscr{I}(-in\chi^{n-1})) = (-\mathscr{I}(n\chi^{n-1}), -\mathscr{R}(n\chi^{n-1})) \qquad (3\cdot4, 18)$$

in that direction, and the jump of the velocity potential is given by

$$\Delta\Phi_\infty = \Delta\Phi_i^{(n)} = -2\Delta\sum_{k=1}^{n}(C_{-k}^{(n)}\cos k\theta + B_{-k}^{(n)}\sin k\theta). \qquad (3\cdot4, 19)$$

We see that, provided the mapping function of $f(\zeta)$ has been found—exactly or approximately—we have here a convenient method for the derivation of suitable particular solutions. Having found a number of these, we may then complete the calculation, as indicated above. It will be instructive to take as a first example the case of a simple wing which was considered in the preceding section. In this case T is the segment $\langle -s_0, s_0\rangle$ on the y-axis in the χ-plane, and the mapping function is

$$\chi = f(\zeta) = \frac{s_0}{2}\left(\zeta + \frac{1}{\zeta}\right), \qquad (3\cdot4, 20)$$

Then

$$\chi^n = \frac{s_0^n}{(2\zeta)^n}(\zeta^2 + 1)^n = \frac{s_0^n}{2^n}\left(\zeta^n + \binom{n}{1}\zeta^{n-2} + \binom{n}{2}\zeta^{n-4} + \ldots + \binom{n}{m}\zeta^{n-2m} + \ldots\right),$$

and so

$$b_{-k}^{(n)} = \begin{cases} \binom{n}{\frac{1}{2}(n-k)}\dfrac{s_0^n}{2^n} & \text{for} \quad n-k \text{ even} \\ 0 & \text{for} \quad n-k \text{ odd} \end{cases} \quad (1 \leqslant k \leqslant n), \qquad (3\cdot4, 21)$$

where $\binom{n}{m}$ is the binomial coefficient $\dfrac{n!}{m!\,(n-m)!}$. Hence

$$
B^{(n)}_{-k}=
\begin{cases}
\left(\dfrac{n}{\frac{1}{2}(n-k)}\right)\dfrac{s_0^n}{2^n} & \text{for } n-k \text{ even}\\[2mm]
0 & \text{for } n-k \text{ odd}
\end{cases}
\quad (1\leqslant k\leqslant n). \qquad (3{\cdot}4,22)
$$

$$
C^{(n)}_{-k}=0 \qquad \text{in all cases}
$$

Every point $(y,0)$ on T corresponds to two points $e^{i\theta}$, $e^{i\theta'}$ in the ζ-plane. These points are given by the equation

$$
y=s_0\cos\theta=s_0\cos\theta', \qquad (3{\cdot}4,23)
$$

so that $\theta=-\theta'$, $\cos n\theta=\cos n\theta'$, and, for the functions $\Phi_r^{(n)}$,

$$
\Delta\Phi_\infty=\Delta\Phi_r^{(n)}=-2\left(\sum_{k=1}^{n} B^{(n)}_{-k}\cos k\theta-\sum_{k=1}^{n} B^{(n)}_{-k}\cos k\theta'\right)=0.
$$

Also, $(3{\cdot}4,14)$ becomes $(-ny^{n-1},0)$, and the component of this vector normal to the wing vanishes identically. Thus the functions $\Phi_r^{(n)}$ are useless and may be discarded. It is not difficult to verify that they all reduce to constants.

On the other hand, for the functions $\Phi_i^{(n)}$,

$$
\Delta\Phi=\Delta\Phi_i^{(n)}=-2\left(\sum_{k=1}^{n} B^{(n)}_{-k}\sin k\theta-\sum_{k=1}^{n} B^{(n)}_{-k}\sin k\theta'\right)
$$

$$
=-4\sum_{k=1}^{n} B^{(n)}_{-k}\sin k\theta
$$

$$
=-\frac{s_0^n}{2^{n-2}}\sum_{m=0}^{[\frac{1}{2}n]}\binom{n}{m}\sin(n-2m)\,\theta, \qquad (3{\cdot}4,24)
$$

where $[\frac{1}{2}n]$ is the greatest integer included in $\frac{1}{2}n$, i.e. $[\frac{1}{2}n]=\frac{1}{2}n$ if n is even and $[\frac{1}{2}n]=\frac{1}{2}(n-1)$ if n is odd.

The vector $(3{\cdot}4,18)$ is, for the case in hand, $(0,-ny^{n-1})$, and so the normal velocity at T is equal to

$$
-\frac{\partial\Phi_\infty}{\partial n}=-ny^{n-1}. \qquad (3{\cdot}4,25)
$$

It will be seen from $(3{\cdot}4,24)$ that $\Delta\Phi_i^{(n)}$ is a sine polynomial of order n. Thus it is a linear combination of the individual terms of $(3{\cdot}3,36)$, and conversely the individual terms of $(3{\cdot}3,36)$ can be represented as linear combinations of the functions $(3{\cdot}4,24)$. It follows that if we use the first n functions in $(3{\cdot}4,24)$ and then satisfy $(3{\cdot}4,4)$ at n points along the span, the result is exactly the same as is obtained by applying Glauert's method with respect to the same points.

Even if we confine ourselves to simply connected lifting units, the number of configurations which may occur in practice is so great that a comprehensive discussion is impossible. Instead, we shall consider only the derivation of the functions $\Phi_r^{(n)}$ and $\Phi_i^{(n)}$ for a tail unit with central fin, as an example.

We assume that the fin is attached to the tailplane from above, so that the unit has the shape of an inverted T when viewed in the direction of flow.

The conformal mapping of the region outside a closed polygonal line on the region outside the unit circle can be carried out by a modified version of the Schwartz–Christoffel method.‡ However, if the polygonal line consists of a number of straight segments emanating from a single point, then it is not difficult to write down the mapping function $f(\zeta)$ directly. Thus, for the present case,

$$\chi = f(\zeta) = ia(1-\zeta)\left(1-\frac{e^{i\psi}}{\zeta}\right)^{\frac{1}{2}}\left(1-\frac{e^{-i\psi}}{\zeta}\right)^{\frac{1}{2}}$$

$$= ia(1-\zeta)\left(1-\frac{2\cos\psi}{\zeta}+\frac{1}{\zeta^2}\right)^{\frac{1}{2}}, \tag{3·4, 26}$$

where a and ψ are positive constants which depend on the shape of the unit, and where the square roots are supposed to approach $+1$ for large ζ.

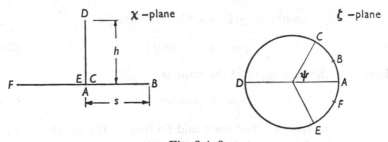

Fig. 3·4, 3

At the unit circle,

$$\chi = ia(1-e^{i\theta})(1+e^{-2i\theta}-2\cos\psi\, e^{-i\theta})^{\frac{1}{2}}$$

$$= ia(e^{-\frac{1}{2}i\theta}-e^{\frac{1}{2}i\theta})(e^{i\theta}+e^{-i\theta}-2\cos\psi)^{\frac{1}{2}},$$

or

$$\chi = 2a\sin\tfrac{1}{2}\theta\,\sqrt{\{2(\cos\theta-\cos\psi)\}}. \tag{3·4, 27}$$

As θ varies from 0 to ψ, χ varies from 0 along the positive real axis and back again to 0. As θ varies from ψ to $2\pi-\psi$, χ varies from 0 along the positive imaginary axis, and back again to 0. Finally, as θ varies from $2\pi-\psi$ to 2π, χ varies from 0 along the negative real axis, and back again to 0. Thus, the starboard half of the tailplane corresponds to $0 \leqslant \theta \leqslant \psi$, the port half to $2\pi-\psi \leqslant \theta \leqslant 2\pi$, and the fin to $\psi \leqslant \theta \leqslant 2\pi-\psi$. Excepting the joint and tips, just two values, θ, θ', correspond to each point of the tail unit. The relation between θ and θ' can be written out directly. By (3·4, 27),

$$\sin\tfrac{1}{2}\theta\,\sqrt{(\cos\theta-\cos\psi)} = \sin\tfrac{1}{2}\theta'\,\sqrt{(\cos\theta'-\cos\psi)},$$

‡ See P. Frank and R. von Mises, *Die Differential- und Integralgleichungen der Mechanik und Physik* (F. Vieweg, 1930), p. 717.

which yields $(\cos\theta - \cos\theta')(1 + \cos\psi - (\cos\theta + \cos\theta')) = 0.$ (3·4, 28)

This condition is satisfied at the fin by $\cos\theta = \cos\theta'$, $\theta + \theta' = 2\pi$, as is otherwise obvious, for reasons of symmetry. The tip of the fin is given by $\theta = \pi$

$$\chi = 2a\sqrt{(2\cos\psi)}\,i, \tag{3·4, 29}$$

so that the height of the fin is

$$h = \sqrt{(8\cos\psi)}\,a. \tag{3·4, 30}$$

At the tailplane, the first factor of (3·4, 28) cannot vanish, except at the joint and tips, and so the condition becomes

$$\cos\theta + \cos\theta' = 1 + \cos\psi. \tag{3·4, 31}$$

For the tips, set $\theta = \theta'$ in (3·4, 31). Then

$$\cos\theta = \tfrac{1}{2}(1 + \cos\psi),$$

$$\sin^2\tfrac{1}{2}\theta = \tfrac{1}{2}(1 - \cos\theta) = \tfrac{1}{4}(1 - \cos\psi),$$

and so $\chi = \pm a(1 - \cos\psi).$ (3·4, 32)

It follows that the semi-span of the wing is

$$s_0 = (1 - \cos\psi)\,a. \tag{3·4, 33}$$

Conversely, for specified tail span and fin height, the constants a and ψ may be determined from (3·4, 30) and (3·4, 33).

Write $\mu = \cos\psi$, then by (3·4, 26),

$$\chi = -ia\zeta\left(1 - \frac{1}{\zeta}\right)\left(1 - \frac{2\mu}{\zeta} + \frac{1}{\zeta^2}\right)^{\frac{1}{2}},$$

and so, for even n, $n = 2m$, $m = 1, 2, \ldots$,

$$\chi^{2m} = (-1)^m a^{2m}\zeta^{2m}\left(1 - \frac{1}{\zeta}\right)^{2m}\left(1 - \frac{2\mu}{\zeta} + \frac{1}{\zeta^2}\right)^m, \tag{3·4, 34}$$

and for odd n, $n = 2m - 1$, $m = 1, 2, \ldots$,

$$\chi^{2m-1} = i(-1)^m a^{2m-1}\zeta^{2m-1}\left(1 - \frac{1}{\zeta}\right)^{2m-1}\left(1 - \frac{2\mu}{\zeta} + \frac{1}{\zeta^2}\right)^m$$

$$\times\left(1 + \frac{P_1(\mu)}{\zeta} + \frac{P_2(\mu)}{\zeta^2} + \frac{P_3(\mu)}{\zeta^3} + \ldots\right). \tag{3·4, 35}$$

The coefficients $P_k(\mu)$ in the last bracket are the Legendre polynomials,

$$P_1(\mu) = \mu, \quad P_2(\mu) = \tfrac{1}{2}(3\mu^2 - 1), \quad P_3(\mu) = \tfrac{1}{2}(5\mu^3 - 3\mu), \quad \text{etc.}$$

and the expansion converges for sufficiently large $|\zeta|$.

To write down $\Omega_r^{(n)}$ and $\Omega_i^{(n)}$, we require the coefficients of the positive powers of χ^n at infinity. For example, for $n = 1, 2, 3, 4$,

$$\left. \begin{aligned}
\chi &= -ia\zeta + \dots, \\
\chi^2 &= -a^2(\zeta^2 - 2(1+\mu)\,\zeta + \dots), \\
\chi^3 &= ia^3(\zeta^3 - 3(1+\mu)\,\zeta^2 + \tfrac{1}{2}(3\mu^2 + 18\mu + 9)\,\zeta + \dots), \\
\chi^4 &= a^4(\zeta^4 - 4(1+\mu)\,\zeta^3 + (4\mu^2 + 16\mu + 8)\,\zeta^2 \\
&\qquad - (16\mu^2 + 28\mu + 12)\,\zeta + \dots).
\end{aligned} \right\} \qquad (3\cdot4, 36)$$

It appears from $(3\cdot4, 14)$ that for the $\Phi_r^{(n)}$ functions, the normal velocity at the tailplane is always zero, and the normal velocity at the fin is also zero for even n. It follows that for such n the $\Phi_r^{(n)}$'are useless—in fact they reduce to constants. For odd n, $n = 2m - 1$, the corresponding normal velocity at the fin is equal to

$$(-1)^m (2m - 1)\, z^{2m-2}.$$

For the $\Phi_i^{(n)}$ functions, $(3\cdot4, 18)$ shows that the normal velocity at the tailplane is ny^{n-1}, and the normal velocity at the fin is zero for odd n, while it is

$$(-1)^m\, 2m z^{2m-1}$$

for $n = 2m$. Thus we are left with the following non-trivial solutions for $n = 1, 2, 3, 4$:

$$\left. \begin{aligned}
\Omega_r^{(1)} &= \chi + ia\left(\zeta - \frac{1}{\zeta}\right), \\[4pt]
\Omega_i^{(1)} &= \chi + ia\left(\zeta + \frac{1}{\zeta}\right), \\[4pt]
\Omega_i^{(2)} &= \chi + a^2\left\{\left(\zeta^2 - \frac{1}{\zeta^2}\right) - 2(1+\mu)\left(\zeta - \frac{1}{\zeta}\right)\right\}, \\[4pt]
\Omega_r^{(3)} &= \chi - ia^3\left\{\left(\zeta^3 - \frac{1}{\zeta^3}\right) - 3(1+\mu)\left(\zeta^2 - \frac{1}{\zeta^2}\right)\right. \\
&\qquad\qquad \left. + \tfrac{1}{2}(3\mu^2 + 18\mu + 9)\left(\zeta - \frac{1}{\zeta}\right)\right\}, \\[4pt]
\Omega_i^{(3)} &= \chi - ia^3\left\{\left(\zeta^3 + \frac{1}{\zeta^3}\right) - 3(1+\mu)\left(\zeta^2 + \frac{1}{\zeta^2}\right)\right. \\
&\qquad\qquad \left. + \tfrac{1}{2}(3\mu^2 + 18\mu + 9)\left(\zeta + \frac{1}{\zeta}\right)\right\}, \\[4pt]
\Omega_i^{(4)} &= \chi - a^4\left\{\left(\zeta^4 - \frac{1}{\zeta^4}\right) - 4(1+\mu)\left(\zeta^3 - \frac{1}{\zeta^3}\right)\right. \\
&\quad \left. + (4\mu^2 + 16\mu + 8)\left(\zeta^2 - \frac{1}{\zeta^2}\right) - (16\mu^2 + 28\mu + 12)\left(\zeta - \frac{1}{\zeta}\right)\right\}.
\end{aligned} \right\} \qquad (3\cdot4, 37)$$

Hence, taking into account (3·4, 16) and (3·4, 19),

$$\Delta\Phi_r^{(1)} = \Delta(-2a\sin\theta),$$

$$\left.\begin{array}{ll} \text{normal velocity at tailplane} & = 0, \\ \text{normal velocity at fin} & = -1; \end{array}\right\} \quad (3\cdot4, 38)$$

$$\Delta\Phi_i^{(1)} = \Delta(2a\cos\theta),$$

$$\left.\begin{array}{ll} \text{normal velocity at tailplane} & = 1, \\ \text{normal velocity at fin} & = 0; \end{array}\right\} \quad (3\cdot4, 39)$$

$$\Delta\Phi_i^{(2)} = \Delta(2a^2[\sin 2\theta - 2(1+\mu)\sin\theta]),$$

$$\left.\begin{array}{ll} \text{normal velocity at tailplane} & = 2y, \\ \text{normal velocity at fin} & = -2z; \end{array}\right\} \quad (3\cdot4, 40)$$

$$\Delta\Phi_r^{(3)} = \Delta(2a^3[\sin 3\theta - 3(1+\mu)\sin 2\theta + \tfrac{1}{2}(3\mu^2 + 18\mu + 9)\sin\theta]),$$

$$\left.\begin{array}{ll} \text{normal velocity at tailplane} & = 0, \\ \text{normal velocity at fin} & = 3z^2; \end{array}\right\} \quad (3\cdot4, 41)$$

$$\Delta\Phi_i^{(3)} = \Delta(-2a^3[\cos 3\theta - 3(1+\mu)\cos 2\theta + \tfrac{1}{2}(3\mu^2 + 18\mu + 9)\cos\theta]),$$

$$\left.\begin{array}{ll} \text{normal velocity at tailplane} & = 3y^2, \\ \text{normal velocity at fin} & = 0; \end{array}\right\} \quad (3\cdot4, 42)$$

$$\Delta\Phi_i^{(4)} = \Delta(-2a^4[\sin 4\theta - 4(1+\mu)\sin 3\theta$$
$$+ (4\mu^2 + 16\mu + 8)\sin 2\theta - (16\mu^2 + 28\mu + 12)\sin\theta]),$$

$$\left.\begin{array}{ll} \text{normal velocity at tailplane} & = 4y^3, \\ \text{normal velocity at fin} & = 4z^3. \end{array}\right\} \quad (3\cdot4, 43)$$

It will be seen from (3·4, 1) and (3·4, 2) that at the tailplane the correct interpretation of $\Delta\Phi_r^{(n)}$ is

$$\Delta\Phi_r^{(n)} = \Phi_r^{(n)}(y, +0) - \Phi_r^{(n)}(y, -0),$$

while at the fin $$\Delta\Phi_r^{(n)} = \Phi_r^{(n)}(+0, z) - \Phi_r^{(n)}(-0, z).$$

Similar definitions apply wherever the symbol Δ appears in (3·4, 38)–(3·4, 43).

In the absence of a comprehensive series of numerical calculations, it is difficult to say how many potentials $\Phi_r^{(n)}$, $\Phi_i^{(n)}$ would be required to provide a satisfactory answer for a given aerofoil. However, the study of the individual solutions (3·4, 38)–(3·4, 43) is of considerable interest by itself. If we substitute any of these solutions in (3·4, 4) then we obtain a condition from which $a_0 c$ may be determined for given α_0, or α_0 for given $a_0 c$. So far, systematic series of calculations, for the case of a tail unit with central fin and for other compound lifting units, have been carried out only for the case $n = 1$.‡ For this case, the normal velocity $-\partial\Phi_\infty/\partial n$ at any point of the

‡ W. Mangler, 'Die Auftriebsverteilung am Tragflügel mit Endscheiben', *Luftfahrtforsch.* vol. 14 (1937), pp. 564–9. Earlier references are given in *Aerodynamic Theory* (ed. Durand, Julius Springer, Berlin, 1934), vol. 2, p. 359. See also J. Rotta, 'Aerodynamic forces on a wing fitted with a lateral plate.' *R.T.P. Translation* 2039 (1942).

unit is the projection of a constant vector in the direction normal to the
wing at that point. Thus the problem is mathematically equivalent to the
flow of a rigidly connected system of flat plates in uniform rectilinear motion.
By the use of this idea, the required velocity potentials Φ_∞ were determined without reference to a more general theory such as is outlined above.

A qualititative idea of the conditions which exist near a joint in a compound lifting unit can be obtained by a simple physical argument. Consider
again the case of a tailplane with central fin (Fig. 3·4, 3), and assume that
the fin is deflected relative to the main stream. This leads in the first instance
to a side force on the fin (a force which acts in the direction of the y-axis).
Thus, there will be an excess of pressure on one side of the fin, over the port
half of the tailplane, say, and a deficiency of pressure (or 'suction') over the
starboard half. As a result, there will be an upward force on the starboard
half of the tailplane and a downward force on the port half, and the two forces
together produce a moment round the joint. Exactly the same effect occurs
near the tip of a wing which is fitted with an end-plate. The presence of end-
plates usually prevents the lift at the tip from dropping to zero, unless the
end-plates happen to be twisted in a suitable fashion, for a given incidence.

The method of particular solutions may also be used for compound
lifting units in which parts of the unit are separate, as in a biplane. In this
case the transformation to the ζ-plane becomes difficult. Thus, for a biplane,
the wake divides the Trefftz plane into a doubly connected region and the
transformation is to a ring-shaped area in the ζ-plane. However, if the lift
loading on the biplane wings is assumed or is known, simple consideration
of the effect of the vortex wake is sufficient to find the influence of aspect
ratio on the overall forces acting on a biplane. Thus we start by considering
the induced drag of the biplane, which is given by (3·4, 9) as

$$D_i = -\tfrac{1}{2}\rho \int \Gamma v_n \, dl \qquad (3·4, 44)$$

where $v_n = 2q_n$ is the velocity normal to the wake in the Trefftz plane, and
the integral is taken along each wing of the biplane. Taking the span
direction of the biplane to be parallel to the y-axis, the velocity v_{n1} induced at the trace of the upper wing in the Trefftz plane is from (3·3, 12)
and (3·3, 17)

$$v_{n1} = -\frac{1}{2\pi} \int_{-s_1}^{s_1} \frac{\gamma_1}{\eta_1 - y_1} \, d\eta_1 - \frac{1}{2\pi} \int_{-s_2}^{s_2} \frac{\gamma_2(\eta_2 - y_1)}{h^2 + (\eta_2 - y_1)^2} \, d\eta_2, \qquad (3·4, 45)$$

where the integrals on the right-hand side extend over the upper and lower
wing span respectively, and we have taken the suffix 1 to refer to the upper
wing and the suffix 2 to the lower wing. h is the normal distance between the
wings of the biplane (it being assumed that the chord lines of the wings are
parallel), and η is the variable of integration used in place of y. We may
write (3·4, 45) as

$$v_{n1} = v_{n11} + v_{n12}, \qquad (3·4, 46)$$

the velocities v_{n11} and v_{n12} standing for the self-induced velocity at the trace of the upper wing and the velocity induced by the lower at the upper wing respectively. Thus v_{n11} and v_{n12} are the first and second integrals on the right-hand side of (3·4, 45). For the lower wing we obtain correspondingly

$$v_{n2} = v_{n22} + v_{n21}$$

$$= -\frac{1}{2\pi} \int_{-s_2}^{s_2} \frac{\gamma_2}{\eta_2 - y_2} d\eta_2 - \frac{1}{2\pi} \int_{-s_1}^{s_1} \frac{\gamma_1(\eta_1 - y_2)}{h^2 + (\eta_1 - y_2)^2} d\eta_1. \qquad (3·4, 47)$$

We find it convenient to put

$$D_i = D_{11} + D_{12} + D_{22} + D_{21}, \qquad (3·4, 48)$$

where

$$D_{jk} = -\tfrac{1}{2}\rho \int \Gamma_j v_{njk} d\eta_j \quad (j, k = 1, 2).$$

Now it may be shown readily by partial integration that

$$v_{n12} = -\frac{1}{2\pi} \int_{-s_2}^{s_2} \Gamma_2 \frac{h^2 - (\eta_2 - y_1)^2}{(h^2 - (\eta_2 - y_1)^2)^2} d\eta_2,$$

Γ_2 being given by $\gamma_2 = -d\Gamma_2/d\eta_2$ (see equation (3·3, 11)), so that

$$D_{12} = \frac{1}{4\pi}\rho \int_{-s_1}^{s_1} \int_{-s_2}^{s_2} \Gamma_1 \Gamma_2 \frac{h^2 - (\eta_2 - \eta_1)^2}{(h^2 - (\eta_2 - \eta_1)^2)^2} d\eta_1 d\eta_2, \qquad (3·4, 49)$$

and this is symmetric with respect to the suffices 1 and 2. It follows therefore that

$$D_{12} = D_{21}, \qquad (3·4, 50)$$

a result which will be proved for a general system in § 3·5. Γ_1 and Γ_2 are related to the lift l per unit span by

$$l_j = \rho U \Gamma_j \quad (j = 1, 2),$$

(equation (3·3, 3) above), and (3·4, 49) can be written in the form

$$D_{12} = \sigma \frac{L_1 L_2}{2\pi\rho U^2 s_1 s_2}, \qquad (3·4, 51)$$

where L_1 and L_2 are integrated values of the lift loading l, and σ is a non-dimensional constant that is a function of the wing-span loading distribution, the span ratio s_1/s_2, and the average gap to span ratio $h/(s_1 + s_2)$. For any given lift distribution σ can be found from (3·4, 49). Thus for an elliptic distribution,

$$l = \frac{2L^2}{\pi s_0^2} \sqrt{(s_0^2 - y^2)}$$

(equations (3·3, 73) and (3·3, 68)), and the corresponding values of σ are as given in Table 3·4, 1.

The self-induced drags D_{11} and D_{22} are calculated in the manner of § 3·3, so we may write, as in equation (3·3, 85),

$$D_{11} = (1+\delta_1)\frac{L_1^2}{2\pi\rho U^2 s_1^2},$$
$$D_{22} = (1+\delta_2)\frac{L_2^2}{2\pi\rho U^2 s_2^2}, \qquad (3\cdot4, 52)$$

and hence the total drag is

$$D_i = \frac{1}{2\pi\rho U^2}\left\{(1+\delta_1)\frac{L_1^2}{s_1^2} + 2\sigma\frac{L_1 L_2}{s_1 s_2} + (1+\delta_2)\frac{L_2^2}{s_2^2}\right\}. \qquad (3\cdot4, 53)$$

Table 3·4, 1. *Values of σ for elliptic loading*

$h/(s_1+s_2)$	0	0·05	0·10	0·15	0·20	0·30	0·40
$s_1/s_2 = 1\cdot0$	1·000	0·780	0·655	0·561	0·485	0·370	0·290
0·8	0·800	0·690	0·600	0·523	0·459	0·355	0·282
0·6	0·600	0·540	0·485	0·437	0·394	0·315	0·255

For a biplane with equal span wings, $s_1 = s_2 = s_0$, the induced drag as given by (3·4, 53) is a minimum when the lifts on the two wings are the same, $L_1 = L_2$. In this case

$$D_i = D_{i\,\text{min.}} = \frac{L^2}{\pi\dfrac{8s_0^2}{S}\tfrac{1}{2}\rho U^2 S}(1+\delta+\sigma), \qquad (3\cdot4, 54)$$

where $L = L_1 + L_2$ is the total lift on the biplane, and S is the area of each wing. If we define the aspect ratio of the biplane as $\mathcal{R} = 8s_0^2/S$, we find for the minimum induced drag coefficient

$$C_{Di\,\text{min.}} = \frac{C_L^2}{\pi\mathcal{R}}(1+\delta+\sigma). \qquad (3\cdot4, 55)$$

This relationship holds also with fair accuracy when the lift is distributed unequally between the two wings.

Let us suppose (3·4, 55) is applicable to a fictitious monoplane aerofoil at a lift coefficient C_L. The relationship derived for a single wing from lifting-line theory is

$$C_D = \frac{C_L^2}{\pi\mathcal{R}}(1+\delta) \qquad$$

(§ 3·3). Hence we may say that the aspect ratio $\mathcal{R}'$ of the equivalent monoplane is smaller than the value of the apect ratio $\mathcal{R}$ of the biplane, and

$$\mathcal{R}' = \mathcal{R}\frac{1+\delta}{1+\delta+\sigma} \doteqdot \mathcal{R}(1-\sigma), \qquad (3\cdot4, 56)$$

δ and σ being regarded as small correction factors.

The relationship between the lift coefficient and the incidence of the biplane can be written, using equation (3·3, 80),

$$\alpha_0 = \frac{C_L^*}{a_0^*} = \alpha_0^* + \frac{1}{\pi R'} C_L^*,$$

or, from (3·4, 56), $\alpha_0 \doteq \alpha_0^* + \dfrac{1}{\pi R} (1 + \sigma) C_L^*,$ (3·4, 57)

where we have used the asterisks to denote two-dimensional values for the equivalent isolated aerofoil. Since (3·4, 56) was derived only for elliptic lift distributions, (3·4, 57) holds strictly only for such cases.

In § 2·10 it was shown that there is also a direct interference between the wings of the biplane in two-dimensional flow which alters their lift characteristics. The effect is due to the non-zero chord lengths of the wings, and is not taken into account by lifting-line theory. For the same lift coefficient C_L^* the total change in the incidence of the biplane from the incidence of the aerofoil sections in two-dimensional flow is thus

$$\Delta \alpha = \frac{1}{\pi R} (1 + \sigma) C_L^* + \frac{1}{2\pi} \left(\frac{c}{2h}\right)^2 (C_L^* - 2 C_{M_0}^*),$$ (3·4, 58)

where we have used (2·10, 23) to obtain the second term, which is an approximation valid for small values of $c/2h$.

We have discussed only the case of an unstaggered biplane; if the wings of the biplane are staggered, so that they do not lie in the same fore-and-aft-plane, the technique developed for compound lifting units does not apply exactly. However, by a theorem due to Munk, which will be proved in the next section, the induced drag is unchanged when the stagger is altered, provided the lift distribution (and hence the circulation) is maintained the same. It follows that the calculations made above for the unstaggered biplane hold also in the more general case.

3·5 Variational principles

It was shown in § 3·3 that, for given total lift on a simple wing, the induced drag is smallest when the downwash (induced normal velocity) is constant across the span. The corresponding problem was considered for general systems of wings, and a complete solution provided by M. Munk in his thesis.‡

We consider an arbitrary system of wings in a uniform stream of velocity U in the direction of the x-axis. The system may be staggered, that is to say, there is no restriction on the relative position of the components

‡ M. Munk, 'Isoperimetrische Aufgaben aus der Theorie des Fluges', Inaug.-Dissertation, Göttingen (1919). English version in *N.A.C.A. Rep.* no. 121 (1921). Compare also L. Prandtl, 'Tragflügeltheorie', 2nd part, *Nachr. Ges. Wiss. Göttingen* (1918), p. 274.

parallel to the direction of flow. Then the trailing vortices mark a system of curves T in the Trefftz plane, and it is possible that a single curve in the Trefftz plane represents several wings belonging to the system. This will be the case, for example, for a pair of equal wings 'in tandem' (one behind the other). The argument used in § 3·2 for the derivation of the induced drag may again be applied, and the result is

$$D_i = \tfrac{1}{2}\rho \int \Phi_\infty \frac{\partial \Phi_\infty}{\partial n'} dl, \qquad (3\cdot5, 1)$$

where the integral is taken round the components of T as in (3·4, 8). This is equivalent to

$$D_i = -\tfrac{1}{2}\rho \int_T \Gamma v_n dl, \qquad (3\cdot5, 2)$$

where the integral is now taken only once along each component wing. In this formula, v_n is the velocity component normal to the wing at any point P of T, and $\Gamma = -\Delta\Phi_\infty$ is the total circulation round the span-station (or span-stations) of the system ahead of P. Alternatively, if there are several wings corresponding to a single curve in the Trefftz plane, then we may also take Γ as the circulation round the individual wings, and integrate along the curve repeatedly, once for each wing. However, v_n must still be taken as the total normal velocity at the corresponding curve of T.

If the total circulation Γ is specified at all points of T, then the strength of the trailing vortex sheet at any point Q at T is

$$\gamma = -\frac{d\Gamma}{dl}, \qquad (3\cdot5, 3)$$

as in (3·3, 11), where the right-hand side indicates differentiation along the curve of T which passes through Q. The velocity field in the Trefftz plane, and hence the potential Φ_∞, can then be calculated by means of the law of Biot-Savart. Finally, the corresponding induced drag may be determined from (3·5, 1) or (3·5, 2).

If we displace any of the components of the system in the direction of the x-axis while adjusting the incidence in such a way that the distribution of the circulation remains the same at all points of the individual wings, then none of the quantities which are involved in (3·5, 1) or (3·5, 2) are affected, and so the induced drag remains itself unchanged throughout the operation. This conclusion is known as 'Munk's stagger theorem'.

Let Γ_1 and Γ_2 be the circulation functions for two cases, for the same system of wings (i.e. for the same T), and let the corresponding potentials in the Trefftz plane be $\Phi_\infty = \Phi_1$ and $\Phi_\infty = \Phi_2$. Then the induced drag for the first case is

$$D_{11} = \tfrac{1}{2}\rho \int \Phi_1 \frac{\partial \Phi_1}{\partial n'} dl = \tfrac{1}{2}\rho \int \left\{ \left(\frac{\partial \Phi_1}{\partial y}\right)^2 + \left(\frac{\partial \Phi_1}{\partial z}\right)^2 \right\} dy\,dz, \qquad (3\cdot5, 4)$$

where the integral in the last member is taken over the entire plane. Similarly, the induced drag for the second case is

$$D_{22} = \tfrac{1}{2}\rho \int \Phi_2 \frac{\partial \Phi_2}{\partial n'} dl = \tfrac{1}{2}\rho \iint \left\{ \left(\frac{\partial \Phi_2}{\partial y} \right)^2 + \left(\frac{\partial \Phi_2}{\partial z} \right)^2 \right\} dy\,dz. \qquad (3·5, 5)$$

These expressions show that $D_{11} > 0$, $D_{22} > 0$, unless $\Phi_1 \equiv 0$ or $\Phi_2 \equiv 0$, respectively, at least if we consider only continuous Φ_1, Φ_2.

We now define the 'cross-drags' D_{12} and D_{21} by

$$D_{jk} = \tfrac{1}{2}\rho \int \Phi_j \frac{\partial \Phi_k}{\partial n'} dl = -\tfrac{1}{2}\rho \int_T \Gamma_j v_k\, dl$$

for $j = 1$, $k = 2$ or $j = 2$, $k = 1$, where at any point of T, v_1 and v_2 are the normal velocity components which correspond to Γ_1 and Γ_2, and where the integrals are taken round T or along T respectively, as in (3·5, 1) and (3·5, 2).

Now

$$\int \Phi_1 \frac{\partial \Phi_2}{\partial n'} dl = \iint \left(\frac{\partial \Phi_1}{\partial y} \frac{\partial \Phi_2}{\partial y} + \frac{\partial \Phi_1}{\partial z} \frac{\partial \Phi_2}{\partial z} \right) dx\,dy = \int \Phi_2 \frac{\partial \Phi_1}{\partial n'} dl, \qquad (3·5, 6)$$

and so

$$D_{12} = D_{21}. \qquad (3·5, 7)$$

This relation is fundamental in the further development of the theory.

Let us resolve the given system of wings into a number of segments by subdividing the individual wings which constitute the system along the chord, at an arbitrary number of span-stations. Let the individual segments which are obtained by this procedure be numbered consecutively $\Delta_1, \Delta_2, \ldots, \Delta_m$, where we shall use the symbol Δ_k also to denote the length of the segment in question, measured along the span.

For a given distribution of circulation Γ over the entire system, we define

$$\Gamma_k = \begin{Bmatrix} \Gamma & \text{on } \Delta_k \\ 0 & \text{elsewhere} \end{Bmatrix} \quad (k = 1, 2, \ldots, m) \qquad (3·5, 8)$$

(see Fig. 3·5, 1). In any physically possible case, Γ must be continuous, and so the definition (3·5, 8) must be regarded as a limiting case. For example, in Fig. 3·5, 1 Γ_k rises rapidly at A, from 0 to its ultimate value, and then drops rapidly to 0 at B.

Let v_j be the normal velocity component corresponding to Γ_j ($j = 1, \ldots, m$). Then the total induced drag is given by

$$D_i = \sum_{j,\,k=1}^{m} D_{jk}, \qquad (3·5, 9)$$

where D_{jk} is defined as in (3·5, 5). If Δ_j and Δ_k are approximately in the same transverse plane (the same plane normal to the x-axis), then the velocity induced by Γ_j at the segment Δ_k is approximately equal to $q_n = \tfrac{1}{2}v_n$. It follows that in this case, the drag on Δ_k produced by Γ_j is equal to D_{kj}, to the degree of approximation of (3·4, 3). However, D_{kj} does not always represent the induced drag on Δ_k due to the circulation round Δ_j, since we calculate the induced velocity in the Trefftz plane and not at the wings. For example, if the incidence of both the wings of a tandem biplane is positive, then the bound vortex at the front wing induces a downward velocity at the rear wing. This tilts the resultant force on the rear wing backward and increases the drag. On the other hand, the rear wing induces an upward velocity at the front wing, this contributing negative drag (thrust). Munk's stagger theorem shows that the two terms balance each other in the calculation of the total induced drag.

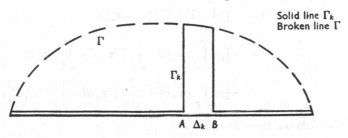

Fig. 3·5, 1

Let β be the angle between the normal $\mathbf{n}$ at a point of T and a fixed direction $\mathbf{m}$ in the y, z-plane. Then the aerodynamic force on the system in the direction of $\mathbf{m}$ is

$$F = \rho U \int_T \Gamma \cos \beta \, dl \qquad (3\cdot5, 10)$$

(compare (3·4, 5), (3·4, 6)).

Assume now that the force F in the direction $\mathbf{m}$ is specified in advance, i.e. for a given velocity U and density ρ, the integral $\int \Gamma \cos \beta \, dl$ is fixed. We inquire under what conditions the corresponding induced drag is a minimum.

Let Γ be the circulation round T for the case for which the drag is a minimum, and let v_n be the corresponding normal velocity at T. Also, let P_1 and P_2 be two arbitrary points of T, and let Δ_1, Δ_2 be two segments of T which include the points P_1 and P_2 respectively. We define a new circulation function Γ' on T by $\Gamma' = 0$ at all points not belonging to Δ_1 or Δ_2, $\Gamma' = \Gamma_j$ at $\Delta_j, j = 1, 2$, where Γ_1, Γ_2 are constants. Let v_n' be the corresponding normal velocity at T. Then by (3·5, 7)

$$\int_T \Gamma v_n' \, dl = \int_T \Gamma' v_n \, dl. \qquad (3\cdot5, 11)$$

But Γ' vanishes everywhere except at the two small intervals Δ_1 and Δ_2. It follows that the right-hand side of (3·5, 11) is equal to

$$\int_{\Delta_1} \Gamma_1 v_n \, dl + \int_{\Delta_2} \Gamma_2 v_n \, dl \doteq \Delta_1 \Gamma_1 v_1 + \Delta_2 \Gamma_2 v_2,$$

where v_1 and v_2 are the values of v_n at P_1 and P_2 respectively. Hence

$$\int_T \Gamma' v_n \, dl = \Delta_1 \Gamma_1 v_1 + \Delta_2 \Gamma_2 v_2. \tag{3·5, 12}$$

Now put $\qquad\qquad \Gamma'' = \Gamma + \Gamma', \quad v_n'' = v_n + v_n',$

so that v_n'' is the normal velocity due to Γ''. The induced drag which corresponds to Γ'' is

$$D_i'' = -\tfrac{1}{2}\rho \int_T \Gamma'' v_n \, dl = -\tfrac{1}{2}\rho \int_T (\Gamma + \Gamma')(v_n + v_n') \, dl$$

$$= -\tfrac{1}{2}\rho \left\{ \int_T \Gamma v_n \, dl + \int_T \Gamma' v_n \, dl + \int_T \Gamma v_n' \, dl + \int_T \Gamma' v_n' \, dl \right\}$$

$$= -\tfrac{1}{2}\rho \left\{ \int_T \Gamma v_n \, dl + 2\int_T \Gamma' v_n \, dl + \int_T \Gamma' v_n' \, dl \right\}.$$

Using (3·5, 12) we then obtain

$$D_i'' - D_i = -\rho(\Delta_1 \Gamma_1 v_1 + \Delta_2 \Gamma_2 v_2) + D_i', \tag{3·5, 13}$$

where D_i is the induced drag due to Γ and D_i' is the induced drag due to Γ',

$$D_i' = -\tfrac{1}{2}\rho \int_T \Gamma' v_n' \, dl \doteq -\tfrac{1}{2}\rho(\Delta_1 \Gamma_1' v_1' + \Delta_2 \Gamma_2' v_2').$$

Now v_n' is proportional to Γ', and so D_i' is of the second order of smallness for small Γ', and may be neglected compared with the remaining terms on the right-hand side of (3·5, 13), which are of the first order of smallness. Hence, for small Γ_1, Γ_2,

$$D_i'' - D_i = -\rho(\Delta_1 \Gamma_1 v_1 + \Delta_2 \Gamma_2 v_2). \tag{3·5, 14}$$

We now require that the aerodynamic force due to Γ'' in the direction of **m** be the same as for Γ. Thus

$$\rho U \int_T (\Gamma + \Gamma') \cos\beta \, dl = \rho U \int_T \Gamma \cos\beta \, dl,$$

which implies

$$\int_T \Gamma' \cos\beta \, dl = \int_{\Delta_1} \Gamma_1 \cos\beta \, dl + \int_{\Delta_2} \Gamma_2 \cos\beta \, dl = 0. \tag{3·5, 15}$$

Denoting by β_1, β_2 the value of β at P_1, P_2, we obtain, to the first approximation,

$$\Delta_1 \Gamma_1 \cos\beta_1 + \Delta_2 \Gamma_2 \cos\beta_2 = 0. \tag{3·5, 16}$$

Subject to this condition, the left-hand side of (3·5, 14) must be positive, or at least non-negative. But if $D_i'' - D_i$ is positive for some admissible values of Γ_1, Γ_2, then by reversing the sign of Γ_1, Γ_2, we may make this expression negative. We therefore obtain as a necessary condition for a minimum

$$\Delta_1 \Gamma_1 v_1 + \Delta_2 \Gamma_2 v_2 = 0. \tag{3·5, 17}$$

(3·5, 16) together with (3·5, 17) then implies

$$\begin{vmatrix} \cos\beta_1 & \cos\beta_2 \\ v_1 & v_2 \end{vmatrix} = 0, \tag{3·5, 18}$$

and this relation holds for all pairs of points P_1, P_2 on T. We first rule out the case that T is everywhere parallel to $\mathbf{m}$, and so we may assume $\cos\beta \neq 0$ for $\beta = \beta_1$. Putting

$$v_0 = \frac{v_1}{\cos\beta_1},$$

we then have at any other point $P = P_2$ of T,

$$v_n = v_0 \cos\beta. \tag{3·5, 19}$$

One concludes from the positive character of the expression for the drag that (3·5, 17) does in fact correspond to a minimum. In the exceptional case that $\cos\beta = 0$ throughout, the only admissible value for F is $F = 0$. In that case the minimum drag is obtained when $\Gamma = 0$ identically. It then follows that $v'_n = 0$ identically, so that (3·5, 19) is satisfied for arbitrary v_0.

If we imagine that T is a rigid two-dimensional system of plates (flat or curved) in the y, z-plane, then (3·5, 19) is precisely the boundary condition which is determined when the system moves with velocity v_0 in the direction $\mathbf{m}$. For an unstaggered simply-connected system of wings, the solution of this problem is given by a special case of the general theory of § 3·4. If staggered lifting units are admitted then it follows immediately from Munk's stagger theorem that by staggering the lifting unit while keeping the total circulation at any point of T the same, one can obtain an unlimited number of solutions to the same problem.

To reduce a given system by one of its components is mathematically equivalent to the condition that the circulation round this component is identically 0. Given that the aerodynamic force F in a direction $\mathbf{m}$ has a specified value, the function(s) Γ for which the system possesses minimum drag does not in general include the possibility that $\Gamma = 0$ at the component in question. It follows that after the abstraction of the component the minimum drag, for given F, will in general be greater, and certainly not smaller, than before. Thus we may state briefly that, for a given force F, the addition of lifting surfaces to a given system will usually reduce the minimum drag.

Let Γ, Γ' be two functions for the circulation on a given T, and let v, v' be the corresponding normal velocities at T, and D_i, D'_i the corresponding values of the induced drag,

$$D_i = -\tfrac{1}{2}\rho \int_T \Gamma v_n \, dl,$$

$$D'_i = -\tfrac{1}{2}\rho \int_T \Gamma' v'_n \, dl.$$

Then
$$D_i' - D_i = \tfrac{1}{2}\rho \int_T (\Gamma v_n - \Gamma' v_n')\, dl$$

$$= -\tfrac{1}{4}\rho\left(\int_T (\Gamma' - \Gamma)(v_n' + v_n)\, dl + \int_T (\Gamma' + \Gamma)(v_n' - v_n)\, dl \right).$$
$$\text{(3·5, 20)}$$

Applying (3·5, 7) for $\Gamma_1 = \Gamma' - \Gamma$, $\Gamma_2 = \Gamma' + \Gamma$, we obtain

$$\int_T (\Gamma' - \Gamma)(v_n' + v_n)\, dl = \int_T (\Gamma' + \Gamma)(v_n' - v_n)\, dl. \qquad \text{(3·5, 21)}$$

Hence, by (3·5, 20),

$$D_i' - D_i = -\tfrac{1}{2}\rho \int_T (\Gamma' + \Gamma)(v_n' - v_n)\, dl, \qquad \text{(3·5, 22)}$$

and similarly
$$D_i' - D_i = -\tfrac{1}{2}\rho \int_T (\Gamma' - \Gamma)(v_n' + v_n)\, dl. \qquad \text{(3·5, 23)}$$

(3·5, 23) may also be written as

$$D_i' - D_i = -\tfrac{1}{2}\rho\left\{ \int_T (\Gamma' - \Gamma)(v_n' - v_n)\, dl + 2\int_T (\Gamma' - \Gamma) v_n\, dl \right\}. \quad \text{(3·5, 24)}$$

Now assume that the circulation is specified for a part T_1 of T, so that $\Gamma' = \Gamma$ for all points of T_1. Then D_i is a minimum if $v_n = 0$ at all the remaining points of T.

Indeed, in that case, the integrand in the last integral of (3·5, 24) vanishes throughout, and so

$$D_i' - D_i = -\tfrac{1}{2}\rho \int_T (\Gamma' - \Gamma)(v_n' - v)\, dl. \qquad \text{(3·5, 25)}$$

But the right-hand side of (3·5, 25) is the induced drag due to the circulation $\Gamma' - \Gamma$. It is therefore positive, unless $\Gamma' = \Gamma$ identically. We conclude that if the circulation is specified over a part T_1 of T, then the induced drag is a minimum when the induced normal velocity v_n vanishes at all the remaining points of T.

Now assume that the aerodynamic force F on a part T_1 of T in a fixed direction $\mathbf{m}$ is specified in advance. Thus, F is now given by

$$F = \rho U \int_{T_1} \Gamma \cos\beta\, dl. \qquad \text{(3·5, 26)}$$

The argument used for the derivation of (3·5, 19) shows that in the present case there exists a constant v_0 such that

$$v_n = v_0 \cos\beta \qquad \text{(3·5, 27)}$$

for all points of T_1 (provided that T_1 is not parallel to $\mathbf{m}$ everywhere).

Let Γ, Γ' be the circulation for two cases such that both Γ and Γ' satisfy (3·5, 26). Then

$$\int_{T_1} (\Gamma' - \Gamma)\cos\beta\, dl = 0. \qquad \text{(3·5, 28)}$$

Now if Γ is the circulation for the case in which the induced drag is a minimum, then we may modify (3·5, 24) as follows:

$$D_i' - D_i = -\tfrac{1}{2}\rho\Big\{ \int_T (\Gamma' - \Gamma)(v_n' - v_n)\,dl$$
$$+ 2\int_{T_1} (\Gamma' - \Gamma)\cos\beta\, v_0\, dl + 2\int_{T-T_1} (\Gamma' - \Gamma)v_n\, dl\Big\},$$

or $$D_i' - D_i = -\tfrac{1}{2}\rho\Big\{ \int_T (\Gamma' - \Gamma)(v_n' - v_n)\,dl + 2\int_{T-T_1} (\Gamma' - \Gamma)v_n\, dl\Big\}. \tag{3·5, 29}$$

This formula shows, as before, that if the normal velocity v_n satisfies, in addition to (3·5, 27), also

$$v_n = 0 \tag{3·5, 30}$$

for all points of $T - T_1$, then $D_i' - D_i > 0$,

unless $\Gamma' = \Gamma$ identically. We conclude that, for given F, the case for which the induced drag is a minimum satisfies (3·5, 30) as well as (3·5, 27).

More generally, if it is specified that the aerodynamic forces on certain disjoint overlapping parts T_1, T_2, ... of T in the directions $\mathbf{m}_1$, $\mathbf{m}_2$, ... be F_1, F_2, ..., then the induced drag becomes a minimum when conditions of the type (3·5, 27) are satisfied separately for T_1, T_2, ..., while at the same time (3·5, 30) holds for all the remaining points of T. If some T_k is everywhere parallel to $\mathbf{m}_k$, then the only admissible value for F_k is $F_k = 0$. The conditions for a minimum remain the same as for the general case.

To conclude this section, we shall establish a variational principle which is equivalent to the lifting-line equation. This principle is due to Ziller[‡] who derived it for the case of a single wing. In the present context, we shall use a different method[§] by means of which the principle can be established for all compound lifting units such as are considered in §3·4. Thus, it will be assumed that (3·4, 1) applies at all points of the unit, and that

$$q_n = \tfrac{1}{2}v_n,$$

where v_n is the normal velocity at T, as before. Equation (3·4, 1) then becomes

$$\Gamma = \tfrac{1}{2}a_0 c(U\alpha_0 + \tfrac{1}{2}v_n), \tag{3·5, 31}$$

and the induced drag is given by

$$D_i = D_i(v_n) = -\tfrac{1}{2}\rho\int_T \Gamma v_n\, dl. \tag{3·5, 32}$$

In this formula, we have used the notation $D_i = D_i(v_n)$ to indicate that the drag may be regarded as a functional of v_n. D_i is positive unless Γ, or (which amounts to the same) v_n, vanishes identically.

‡ F. Ziller, 'Beitrag zur Theorie des Tragflügels von endlicher Spannweite', *Ingen.-Arch.* vol. 11 (1940), pp. 239–59.

§ A Robinson, 'A minimum energy theorem in aerodynamics', *Royal Aircraft Establishment Tech. Note*, no. S.M.E. 298 (1945).

We define three other quantites E, P and W by

$$E(v_n) = \tfrac{1}{8}\rho \int_T a_0 c v_n^2 \, dl, \tag{3·5, 33}$$

$$P(v_n) = \tfrac{1}{2}\rho U \int_T a_0 c \alpha_0 v_n \, dl, \tag{3·5, 34}$$

$$W(v_n) = D_i + E + P. \tag{3·5, 35}$$

Then E also is positive unless v_n vanishes identically.

Let Γ, Γ' be two circulation functions defined on T, and v_n, v_n' the corresponding expressions for the normal velocity. Then

$$D_i(v_n') = D_i((v_n' - v_n) + v_n) = D_i(v_n' - v_n) + D_i(v_n)$$
$$- \tfrac{1}{2}\rho \int_T (v_n' - v_n)\, \Gamma \, dl - \tfrac{1}{2}\rho \int_T v_n(\Gamma' - \Gamma)\, dl.$$

Or, using (3·5, 7),

$$D_i(v_n') = D_i(v_n' - v_n) + D_i(v_n) - \rho \int_T (v_n' - v_n)\, \Gamma \, dl. \tag{3·5, 36}$$

Also $\quad E(v_n') = E(v_n' - v_n) + E(v_n) + \tfrac{1}{4}\rho \int_T (v_n' - v_n)\, a_0 c v_n \, dl,$

and $\quad P(v_n') = P(v_n) + \tfrac{1}{2}\rho \int_T (v_n' - v_n)\, a_0 c U \alpha_0 \, dl.$

Hence

$$W(v_n') - W(v_n) = D_i(v_n' - v_n) + E(v_n' - v_n)$$
$$+ \rho \int_T (v_n' - v_n)\{-\Gamma + \tfrac{1}{2}a_0 c(U\alpha_0 + \tfrac{1}{2}v_n)\}\, dl. \tag{3·5, 37}$$

Now assume that Γ is the solution of (3·5, 31). Then the expression in the curly brackets on the right-hand side of (3·5, 37) vanishes, and so

$$W(v_n') - W(v_n) = D_i(v_n' - v_n) + E(v_n' - v_n) \geqslant 0.$$

Or $\qquad\qquad\qquad W(v_n') \geqslant W(v_n), \tag{3·5, 38}$

where the sign of equality applies only if v_n' is identically equal to v_n. We conclude that the expression

$$W = \rho \int_T \{-\tfrac{1}{2}\Gamma v_n + \tfrac{1}{8}a_0 c v_n^2 + \tfrac{1}{2}a_0 c U\alpha_0 v_n\}\, dl \tag{3·5, 39}$$

possesses an absolute minimum for the solution of (3·5, 31).

For the application of this principle, one may employ the method of Rayleigh–Ritz. That is to say, one assumes that Γ and the corresponding v_n can be written as linear combinations of suitable particular functions, with undetermined coefficients $A_1, \ldots, A_m$. Then W is a quadratic form of $A_1, \ldots, A_m$, and the condition for a minimum is

$$\frac{\partial W}{\partial A_k} = 0 \quad (k = 1, \ldots, m). \tag{3·5, 40}$$

(3·4, 40) constitutes a system of m linear equations for the determination of the unknowns $A_1, ..., A_m$. For the case of a simply-connected lifting unit we may use the particular solutions developed in § 3·4. However, for the case of a single wing the particular solutions used in § 3·3,

$$\Gamma = 4s_0 U \sin k\theta, \quad v_n = -U \frac{\sin k\theta}{\sin \theta},$$

are more convenient. Detailed calculations will be found in Ziller's paper, where the last-mentioned particular solutions are employed. Ziller considers also the case of an elastically deformable wing and derives a similar principle.

The method just described possesses advantages similar to Gates's procedure (§ 3·3), in that it takes into account the characteristics of the wing along the entire span and not only at a relatively small number of points. It is evident that a variational principle such as that used by Gates can be derived for any other functional equation, whereas Ziller's principle depends more intimately on the physical circumstances of the case.

3·6   Source and doublet representations

The foregoing sections have dealt with a particular method of representing the aerofoil as a line of bound vorticity, with a set of trailing vortices forming the wake. We now consider a more detailed approach in which the entire wing as well as the wake are replaced by continuous surface distributions of singularities. In § 3·1 it was shown how, by linearization for small induced velocities, the velocity potential may be calculated by the superposition of the flow fields about a symmetric wing at zero incidence and a thin cambered wing. We shall take first the case of a symmetric wing at zero incidence.

Let S be the plan-form of the wing, and let

$$z = f_u(x, y) = f_l(x, y) \tag{3·6, 1}$$

be the equation of the upper surface, and

$$z = f_l(x, y) = -f_l(x, y) \tag{3·6, 2}$$

be the equation of the lower surface. The local slope $s(x, y)$ is then given by

$$s(x, y) = \pm \frac{\partial f_l}{\partial x}.$$

The corresponding induced velocity potential Φ' is symmetrical with respect to the x, y-plane (see 3·1, 4).

This suggests that Φ' can be replaced by a distribution of sources and sinks over S,

$$\Phi'(x, y, z) = \frac{1}{4\pi} \int_S \frac{\sigma(x_0, y_0) \, dx_0 \, dy_0}{\sqrt{\{(x - x_0)^2 + (y - y_0)^2 + z^2\}}}, \tag{3·6, 3}$$

where $\sigma(x_0, y_0)$ is the local surface density of the distribution (equation 1·7, 5). According to (1·7, 8), $\sigma(x_0, y_0)$ is linked with the jump of the normal derivative of Φ' across S by the relation

$$\left(\frac{\partial \Phi'}{\partial z}\right)_{z=+0} - \left(\frac{\partial \Phi'}{\partial z}\right)_{z=-0} = -\sigma(x_0, y_0). \tag{3·6, 4}$$

Now, by (3·1, 5),

$$w(x_0, y_0, +0) = -\left(\frac{\partial \Phi'}{\partial z}\right)_{z=+0} = \left(\frac{\partial \Phi'}{\partial z}\right)_{z=-0},$$

and so, by (3·1, 2),

$$\sigma(x_0, y_0) = 2w(x_0, y_0, +0) = 2U s_u(x_0, y_0), \tag{3·6, 5}$$

where $s_u(x, y)$ is the local slope of the upper surface of the aerofoil

$$s_u(x, y) = \frac{\partial f_t}{\partial x}.$$

Substituting in (3·6, 3), we obtain the formula

$$\Phi'(x, y, z) = \frac{U}{2\pi} \int_S \frac{s_u(x_0, y_0)\, dx_0\, dy_0}{\sqrt{\{(x-x_0)^2 + (y-y_0)^2 + z^2\}}}. \tag{3·6, 6}$$

In particular, the induced longitudinal velocity at the upper surface of the aerofoil is given by

$$u' = -\left(\frac{\partial \Phi'}{\partial x}\right)_{z=+0} = -\frac{U}{2\pi} \frac{\partial}{\partial x} \int_S \frac{s_u(x_0, y_0)\, dx_0\, dy_0}{\sqrt{\{(x-x_0)^2 + (y-y_0)^2\}}}. \tag{3·6, 7}$$

(3·6, 7) in conjunction with (3·2, 3) then determines the pressure distribution round the aerofoil. In actual fact, the evaluation of the integral in (3·3, 7) and its subsequent differentiation may still present considerable practical difficulties. A number of examples have been worked out by Neumark and Collingbourne.‡

It is not difficult to show that the total strength of the distribution for a given span station equals zero. In fact, by (3·6, 5),

$$\int_a^b \sigma(x, y)\, dx = 2U \int_a^b s_u(x, y)\, dx = \int_a^b \frac{\partial f_t}{\partial x} dx = [f_t]_a^b = 0, \tag{3·6, 8}$$

since the aerofoil is closed, where $x = a$ and $x = b$ corresponds to the leading and trailing edge respectively.

Suppose now that the aerofoil S is of zero thickness, but cambered, and possibly at a finite incidence to the main stream. Let the equation of the surface of the aerofoil be

$$z = f_u(x, y) = f_l(x, y) = f_m(x, y), \tag{3·6, 9}$$

so that the slope of the aerofoil at corresponding points of both upper and lower surfaces is

$$s(x, y) = \frac{\partial f_m}{\partial x}. \tag{3·6, 10}$$

‡ S. Neumark and J. Collingbourne, 'Velocity distribution on untapered sheared and swept-back wings of small thickness and finite aspect ratio at zero incidence', *Rep. Memor. Aero. Res. Comm.*, Lond., no. 2717 (1949).

By (3·1, 6), Φ' is antisymmetrical with respect to the x, y-plane, which suggests that Φ' can be represented by a distribution of doublets of density τ in the x, y-plane. However, since the wake is now a surface of discontinuity, we must assume that the distribution extends over the wake W as well as over S. Thus (see (1·7, 12))

$$\Phi'(x, y, z) = -\frac{1}{4\pi} \int_{S+W} \tau(x_0, y_0) \frac{\partial}{\partial z}\left(\frac{1}{\sqrt{\{(x-x_0)^2 + (y-y_0)^2 + z^2\}}}\right) dx_0 dy_0$$

$$= -\frac{1}{4\pi} \frac{\partial}{\partial z} \int_{S+W} \frac{\tau(x_0, y_0)\, dx_0 dy_0}{\sqrt{\{(x-x_0)^2 + (y-y_0)^2 + z^2\}}}, \tag{3·6, 11}$$

where the suffix $S + W$ indicates that the integral is taken over aerofoil and wake. It is now more difficult to find Φ' for a given aerofoil because $\tau(x_0, y_0)$ is no longer related directly to the local slope. Indeed, it has been shown in Chapter 1 (equation (1·7, 14)) that

$$\Phi'(x_0, y_0, +0) - \Phi'(x_0, y_0, -0) = \tau(x_0, y_0). \tag{3·6, 12}$$

But (3·1, 6) shows that Φ' is an odd function of z and so

$$\tau(x_0, y_0) = 2\Phi'(x_0, y_0, +0). \tag{3·6, 13}$$

By (3·1, 7) u' is an odd function of z, and since it is continuous across the wake, it must vanish at the wake,

$$u'(x, y, 0) = \left(\frac{\partial \Phi'}{\partial x}\right)_{z=0} = 0. \tag{3·6, 14}$$

It follows that Φ' is constant along lines parallel to the direction of flow in the wake, although it will normally be discontinuous across the wake. Thus

$$\Phi'(x, y, +0) = \Phi'(b, y, +0), \tag{3·6, 15}$$

for $-s_0 \leqslant y \leqslant s_0$, $x > b$, where $x = b(y)$ is the equation of the trailing edge, and where the aerofoil is supposed to extend from $-s_0$ to s_0 in lateral direction. Hence, by (3·6, 13),
$$\tau(x_0, y_0) = \tau(b, y_0) \tag{3·6, 16}$$

for
$$-s_0 \leqslant y_0 \leqslant s_0, \quad x_0 > b = b(y_0).$$

Consider the contribution of the wake to the integral in (3·6, 11),

$$\int_W \tau(x_0, y_0) \frac{\partial}{\partial z}\left(\frac{1}{\sqrt{\{(x-x_0)^2 + (y-y_0)^2 + z^2\}}}\right) dx_0 dy_0$$

$$= -\int_{-s_0}^{s_0} \tau(b, y_0)\, dy_0 \int_b^\infty \frac{z}{\{(x-x_0)^2 + (y-y_0)^2 + z^2\}^{\frac{3}{2}}}\, dx_0$$

$$= \int_{-s_0}^{s_0} \tau(b, y_0) \frac{z}{(y-y_0)^2 + z^2}\left[\frac{x - x_0}{\sqrt{\{(x-x_0)^2 + (y-y_0)^2 + z^2\}}}\right]_{x_0=b}^{x_0=\infty} dy_0$$

$$= z\int_{-s_0}^{s_0} \frac{\tau(b, y_0)}{(y-y_0)^2 + z^2}\left(1 + \frac{x - b}{\sqrt{\{(x-b)^2 + (y-y_0)^2 + z^2\}}}\right) dy_0.$$

Thus, (3·6, 11) may also be written as

$$\Phi'(x,y,z) = -\frac{1}{4\pi}\frac{\partial}{\partial z}\int_S \frac{\tau(x_0,y_0)\,dx_0\,dy_0}{\sqrt{\{(x-x_0)^2+(y-y_0)^2+z^2\}}}$$
$$-\frac{z}{4\pi}\int_{-s_0}^{s_0}\frac{\tau(b,y_0)}{(y-y_0)^2+z^2}\left\{1+\frac{x-b}{\sqrt{\{(x-b)^2+(y-y_0)^2+z^2\}}}\right\}dy_0.$$

$$(3·6, 17)$$

The corresponding component w, the 'downwash' velocity, is

$$w(x,y,z) = \frac{1}{4\pi}\frac{\partial^2}{\partial z^2}\int_S \frac{\tau(x_0,y_0)\,dx_0\,dy_0}{\sqrt{\{(x-x_0)^2+(y-y_0)^2+z^2\}}}$$
$$+\frac{\partial}{\partial z}\frac{z}{4\pi}\int_{-s_0}^{s_0}\frac{\tau(b,y_0)}{(y-y_0)^2+z^2}\left\{1+\frac{x-b}{\sqrt{\{(x-b)^2+(y-y_0)^2+z^2\}}}\right\}dy_0.$$

$$(3·6, 18)$$

For a given aerofoil, w is determined by (3·1, 2), and so we may regard (3·6, 18) as an integral equation for $\tau(x,y)$. However, the closed analytical solution of (3·6, 18) is beyond our resources even for simple wing plan-forms. In order to solve the problem approximately, but to any required degree of accuracy, it is therefore convenient to assume a suitable expansion, and to determine its coefficients. In expounding such a method, we follow a paper by W. P. Jones.‡

Let $x = a(y)$ be the equation of the leading edge, then

$$x_m = \tfrac{1}{2}(a(y)+b(y)) \quad (-s_0 \leqslant y \leqslant s_0)$$

is the longitudinal coordinate of the mid-chord of the aerofoil for given y, and let $c = c(y)$ be its chord, $c(y) = b(y) - a(y)$. We put

$$x = x_m - \tfrac{1}{2}c(y)\cos\theta \quad (0 \leqslant \theta \leqslant \pi) \tag{3·6, 19}$$

for any point of the aerofoil, so that $\theta = 0$ corresponds to the leading edge, and $\theta = \pi$ to the trailing edge, and (θ, y) may be regarded as coordinates of the points of aerofoil within the given limits of variation. It should be remarked here that this method of defining θ is different from the one used in Chapter 2, where the leading edge corresponded to $\theta = \pi$ and the trailing edge to $\theta = 0$. Equation (3·6, 19) agrees with the previous definition if we replace θ by $\pi - \theta$.

We now assume that the induced longitudinal velocity at the aerofoil

$$u'(x,y,+0) = \tfrac{1}{2}\gamma(\theta,y), \quad \text{say}, \tag{3·6, 20}$$

can be expressed in the form of an expansion

$$\gamma(\theta,y) = U\sum_{n=0}^{\infty}C_n(y)\,\gamma_n(\theta), \tag{3·6, 21}$$

where

$$\gamma_0 = 2\cot\tfrac{1}{2}\theta = \gamma_0' + \gamma_0'',$$
$$\gamma_0' = 2\operatorname{cosec}\theta,$$
$$\gamma_0'' = 2\cot\theta,$$

‡ W. P. Jones, 'Theoretical determination of the pressure distribution on a finite wing in steady motion', *Rep. Memor. Aero. Res. Comm., Lond.*, no. 2145 (1943).

(since $\operatorname{cosec} \theta + \cot \theta = \cot \frac{1}{2}\theta$)

$$\gamma_1 = -2\sin\theta + \cot \tfrac{1}{2}\theta$$

$$\gamma_n = -2\sin n\theta \quad (n \geqslant 2).$$

We shall see presently why $\gamma_1(\theta)$ has been chosen in that particular fashion. The choice of the functions $C_n(y)$ depends on the wing plan-form, and will be discussed later.

The induced velocity potential $\Phi'(x, y, z)$ is continuous upstream of the aerofoil and is antisymmetrical with respect to the x, y-plane, i.e. it is an odd function of z (see (3·1, 6)). Hence

$$\Phi'(x, y, 0) = 0 \quad (-s_0 \leqslant y \leqslant s_0; \; x \leqslant a(y)), \tag{3·6, 22}$$

and so

$$\Phi'(x, y, +0) = -\int_a^x u'(\xi, y, +0)\, d\xi \quad (-s_0 \leqslant y \leqslant s_0; \; a(y) \leqslant x \leqslant b(y)). \tag{3·6, 23}$$

Using (3·6, 13) and (3·6, 20), we then obtain for $\tau(x, y)$,

$$\tau(x, y) = -\tfrac{1}{2}c\int_0^\theta \gamma(\psi, y)\sin\psi\, d\psi \quad (-s_0 \leqslant y \leqslant s_0; \; a(y) \leqslant x \leqslant b(y)), \tag{3·6, 24}$$

where

$$x = x_m - \tfrac{1}{2}c(y)\cos\theta, \quad \xi = x_m - \tfrac{1}{2}c(y)\cos\psi,$$

or

$$\tau(x, y) = -Uc\sum_{n=0}^{\infty} C_n(y)\tau_n(\theta), \tag{3·6, 25}$$

where

$$\tau_n(\theta) = \frac{1}{2}\int_0^\theta \gamma_n(\psi)\sin\psi\, d\psi \quad (n = 0, 1, 2, \ldots).$$

In particular,

$$\tau_0(\theta) = \tau_0'(\theta) + \tau_0''(\theta),$$

where

$$\tau_0'(\theta) = \frac{1}{2}\int_0^\theta \frac{2}{\sin\psi}\sin\psi\, d\psi = \theta,$$

$$\tau_0''(\theta) = \frac{1}{2}\int_0^\theta \frac{2\cos\psi}{\sin\psi}\sin\psi\, d\psi = \sin\theta,$$

so that

$$\tau_0(\theta) = \theta + \sin\theta.$$

Also

$$\frac{1}{2}\int_0^\theta -2\sin\psi\sin\psi\, d\psi = -\tfrac{1}{2}(\theta - \tfrac{1}{2}\sin 2\theta),$$

and so

$$\tau_1(\theta) = \frac{1}{2}\int_0^\theta \left(-2\sin\psi + \cot\frac{\psi}{2}\right)\sin\psi\, d\psi$$

$$= -\tfrac{1}{2}(\theta - \tfrac{1}{2}\sin 2\theta) + \tfrac{1}{2}(\theta + \sin\theta)$$

$$= \tfrac{1}{2}(\sin\theta + \tfrac{1}{2}\sin 2\theta).$$

For $n \geqslant 2$,

$$\tau_n(\theta) = \frac{1}{2}\int_0^\theta -2\sin n\psi\sin\psi\, d\psi = \frac{1}{2}\int_0^\theta (\cos(n+1)\psi - \cos(n-1)\psi)\, d\psi$$

$$= \frac{1}{2}\left\{\frac{\sin(n+1)\theta}{n+1} - \frac{\sin(n-1)\theta}{n-1}\right\}.$$

At the trailing edge, $\theta = \pi$, and so

$$\tau_0(\pi) = \pi, \quad \tau_0'(\pi) = \pi, \quad \tau_0''(\pi) = 0, \quad \tau_n(\pi) = 0 \quad (n = 1, 2, \ldots). \quad (3\cdot6, 26)$$

Thus, by $(3\cdot6, 24)$, $\qquad \tau(b, y) = -\pi U c(y) C_0(y)$.

Thus, the contribution of the wake to the induced velocity potential and to the normal velocity (see $(3\cdot6, 17)$ and $(3\cdot6, 18)$) depends only on $C_0(y)$, and is independent of $C_1(y)$, $C_2(y)$, ...

We now represent the functions $C_n(y)$ again by expansions

$$C_n(y) = \sum_{m=1}^{\infty} C_{nm} A_m(y) \quad (n = 0, 1, 2, \ldots), \qquad (3\cdot6, 27)$$

where the C_{nm} are constants and the $A_m(y)$ depend on the wing plan-form and will be discussed later. Then, by $(3\cdot6, 25)$,

$$\tau(x, y) = -U c(y) \sum_{n=0}^{\infty} \sum_{m=1}^{\infty} C_{nm} A_m(y) \tau_n(\theta). \qquad (3\cdot6, 28)$$

Substituting this expression for $\tau(x, y)$ in $(3\cdot6, 18)$, and inverting the order of summation and integration, we obtain

$$w(x, y, z) = U \sum_{n=0}^{\infty} \sum_{m=1}^{\infty} C_{nm} w_{nm}, \qquad (3\cdot6, 29)$$

where $\quad w_{0m} = -\dfrac{1}{4\pi} \dfrac{\partial^2}{\partial z^2} \displaystyle\int_S \dfrac{c(y_0) A_m(y_0) \tau_0(\theta_0)\, dx_0\, dy_0}{\sqrt{\{(x-x_0)^2 + (y-y_0)^2 + z^2\}}}$

$$-\frac{\partial}{\partial z} \frac{z}{4} \int_{-s_0}^{s_0} \frac{c(y_0) A_m(y_0)}{(y-y_0)^2 + z^2} \left\{ 1 + \frac{x-b}{\sqrt{\{(x-b)^2 + (y-y_0)^2 + z^2\}}} \right\} dy_0,$$
$$(3\cdot6, 30)$$

for $m = 1, 2, \ldots$, and

$$w_{nm} = -\frac{1}{4\pi} \frac{\partial^2}{\partial z^2} \int_S \frac{c(y_0) A_m(y_0) \tau_n(\theta_0)}{\sqrt{\{(x-x_0)^2 + (y-y_0)^2 + z^2\}}}\, dx_0\, dy_0, \qquad (3\cdot6, 31)$$

for $n = 1, 2, \ldots$; $m = 1, 2, \ldots$. In these integrals θ_0 is linked with x_0, y_0 by the relation

$$x_0 = x_m(y_0) - \tfrac{1}{2} c(y_0) \cos \theta_0.$$

Thus the w_{nm} are constants which are independent of the slope function $s(x, y)$, although they have to be calculated separately for each plan-form. Having calculated these constants for a number of points on the aerofoil, say for NM points, we may try to satisfy the boundary condition $(3\cdot1, 2)$ at these points, by retaining just NM terms in the expansion of $\tau(x, y)$ $(0 \leqslant n \leqslant N-1; 1 \leqslant m \leqslant M)$. Combining $(3\cdot1, 2)$ and $(3\cdot6, 29)$, we then obtain a system of NM linear equations of the form

$$\sum_{n=0}^{N-1} \sum_{m=1}^{M} C_{nm} w_{nm} = s(x, y) \qquad (3\cdot6, 32)$$

for the constants C_{nm} $(0 \leqslant n \leqslant N-1; 1 \leqslant m \leqslant M)$. Having determined the

C_{nm}, we obtain the longitudinal induced velocity from (3·6, 20), (3·6, 21) and (3·6, 27),

$$u'(x, y, +0) = \frac{U}{2} \sum_{n=0}^{N-1} \sum_{m=1}^{M} C_{nm} A_m(y) \gamma_n(\theta). \tag{3·6, 33}$$

The corresponding pressure increment is, using (3·2, 3),

$$p - p_0 = -\tfrac{1}{2}\rho U^2 \sum_{n=0}^{N-1} \sum_{m=1}^{M} C_{nm} A_m(y) \gamma_n(\theta). \tag{3·6, 34}$$

The pressure difference between top and bottom surfaces, $p_u - p_l$, is twice the right-hand side of (3·6, 34), whether we employ (3·2, 2) or (3·2, 3). The lift per unit span, $l(y)$ is therefore given by

$$l(y) = -\int_a^b (p_u - p_l)\,dx = \tfrac{1}{2}\rho U^2 c(y) \int_0^\pi \sum_{n=0}^{N-1} \sum_{m=1}^{M} C_{nm} A_m(y) \gamma_n(\theta) \sin\theta\,d\theta,$$

or, by (3·6, 26), $$l(y) = \pi\rho U^2 c(y) \sum_{m=1}^{M} C_{0m} A_m(y). \tag{3·6, 35}$$

Finally, the total lift is obtained by integrating (3·6, 35) across the span,

$$L = \int_{-s_0}^{s_0} l(y)\,dy.$$

It will be seen that the choice of the functions $\gamma_n(\theta)$ permits the representation of an arbitrary, but sufficiently regular (e.g. differentiable, except for a finite number of ordinary discontinuities) $u'(x, y, +0)$, in the form of a sine series. In addition, $\gamma_0(\theta)$ caters for the unbounded distribution corresponding to a flat plate at incidence, which cannot be expanded into a Fourier series. The particular form chosen for $\gamma_1(\theta)$ does not modify the generality of the expansion but it ensures that only functions with suffix $n = 0$ occur in the expression for the span-wise lift distribution, and for the doublet density at the trailing edge. It will be seen that the selection of the $\gamma_n(\theta)$ is suggested by thin aerofoil theory. The choice of the functions $A_m(y)$ is still arbitrary to a large extent. It is desirable to ensure that every one of these functions individually represents a possible solution of a real problem (i.e. that all the coefficients $C_{nm'}$ may vanish for $m' \neq m$, for a suitable incidence distribution). This implies, owing to the continuity of the pressure, that the corresponding lift distribution must tend to zero at the tips. Hence, by (3·6, 27), we must have $c(y) A_m(y) = 0$ for $y = \pm s_0$. If the tips are rounded it is therefore sufficient to assume that the functions $A_m(y)$ are bounded, while $A_m(-s_0) = A_m(s_0) = 0$ for bluff (e.g. rectangular) tips. For this case, suitable functions are

$$A_m(y) = \sin n\psi, \tag{3·6, 36}$$

where $$y = -s_0 \cos\psi \quad (0 \leqslant \psi \leqslant \pi).$$

The corresponding lift distributions $l(y)$ possess infinite slope at the tips, but although under the influence of the results of lifting-line theory (see

§ 3·3) it was considered that this condition is essential, the known exact solutions (§ 3·9) do not bear out this assumption. Thus, simpler expressions like

$$A_m(y) = \eta^{m+2} - \eta^m, \quad \eta = \frac{y}{s_0} \quad (-1 \leqslant \eta \leqslant 1)$$

might be used with equal justification. Yet another series which is closely akin to (3·6, 36) is provided by

$$A_m(y) = \eta^m \sqrt{(1 - \eta^2)}, \quad \eta = \frac{y}{s_0} \quad (-1 \leqslant \eta \leqslant 1). \tag{3·6, 37}$$

3·7 The rectangular wing

For the case of a rectangular aerofoil, the constants w_{nm} in equation (3·6, 29) for the velocity w may be determined in the following way.

By (3·6, 11), the expression for the normal velocity may also be written as

$$w(x, y, z) = \frac{1}{4\pi} \frac{\partial^2}{\partial z^2} \int_{S+W} \frac{\tau(x_0, y_0) \, dx_0 \, dy_0}{\sqrt{\{(x - x_0)^2 + (y - y_0)^2 + z^2\}}}. \tag{3·7, 1}$$

$w(x, y, z)$ is a solution of Laplace's equation, and so, differentiating under the integral sign on the right-hand side of (3·7, 1) and writing

$$r = \sqrt{\{(x - x_0)^2 + (y - y_0)^2 + z^2\}},$$

we may replace (3·7, 1) by

$$w = -\frac{1}{4\pi} \int_{S+W} \tau(x_0, y_0) \left(\frac{\partial^2}{\partial x_0^2} \left(\frac{1}{r} \right) + \frac{\partial^2}{\partial y_0^2} \left(\frac{1}{r} \right) \right) dx_0 \, dy_0. \tag{3·7, 2}$$

Now for a rectangular wing, the expansion (3·6, 28) shows that we may assume not only $\tau(x, y) = 0$ along the finite boundaries of $S + W$, but also $\partial \tau / \partial y = 0$ along the leading edge, and $\partial \tau / \partial x = 0$ along the lines $y_0 = \pm s_0$. Hence, integrating by parts with respect to x_0, we obtain

$$\int_{S+W} \tau \frac{\partial^2}{\partial x_0^2} \left(\frac{1}{r} \right) dx_0 \, dy_0 = \int_{-s_0}^{s_0} dy_0 \int_a^\infty \tau \frac{\partial^2}{\partial x_0^2} \left(\frac{1}{r} \right) dx_0$$

$$= \int_{-s_0}^{s_0} dy_0 \left\{ \left[\tau \frac{\partial}{\partial x_0} \left(\frac{1}{r} \right) \right]_{x_0=a}^{x_0=\infty} - \int_a^\infty \frac{\partial \tau}{\partial x_0} \frac{\partial}{\partial x_0} \left(\frac{1}{r} \right) dx_0 \right\}$$

$$= \int_{-s_0}^{s_0} \tau(b, y_0) \left[\frac{\partial}{\partial x_0} \left(\frac{1}{r} \right) \right]_{x_0=\infty} dy_0 - \int_S \frac{\partial \tau}{\partial x_0} \frac{\partial}{\partial x_0} \left(\frac{1}{r} \right) dx_0 \, dy_0.$$

In this expression, the first integral is taken across the wake in a region far behind the aerofoil and therefore vanishes in the limit. The second integral extends only over the wing S since $\partial \sigma / \partial x = 0$ in the wake. Now

$$\int \frac{\partial}{\partial x_0} \left(\frac{1}{r} \right) dy_0 = \int \frac{x - x_0}{((x - x_0)^2 + (y - y_0)^2 + z^2)^{\frac{3}{2}}} \, dy_0$$

$$= -\frac{(x - x_0)(y - y_0)}{((x - x_0)^2 + z^2) \sqrt{\{(x - x_0)^2 + (y - y_0)^2 + z^2\}}} + \text{constant},$$

and so another integration by parts yields

$$\int_S \frac{\partial \tau}{\partial x_0} \frac{\partial}{\partial x_0}\left(\frac{1}{r}\right) dx_0\, dy_0 = \int_a^b dx_0 \int_{-s_0}^{s_0} \frac{\partial \tau}{\partial x_0} \frac{\partial}{\partial x_0}\left(\frac{1}{r}\right) dy_0$$

$$= \int_a^b dx_0 \left\{ -\left[\frac{\partial \tau}{\partial x_0} \frac{(x-x_0)(y-y_0)}{((x-x_0)^2+z^2)\, r}\right]_{y=-s_0}^{y_0=s_0} \right.$$

$$\left. + \int_{-s_0}^{s_0} \frac{\partial^2 \tau}{\partial x_0 \partial y_0} \frac{(x-x_0)(y-y_0)}{((x-x_0)^2+z^2)\, r}\, dy_0 \right\}$$

$$= \int_S \frac{\partial^2 \tau}{\partial x_0 \partial y_0} \frac{(x-x_0)(y-y_0)}{((x-x_0)^2+z^2)\, r}\, dx_0\, dy_0.$$

Thus
$$\int_{S+W} \tau \frac{\partial^2}{\partial x_0^2}\left(\frac{1}{r}\right) dx_0\, dy_0 = -\int_S \frac{\partial^2 \tau}{\partial x_0 \partial y_0} \frac{(x-x_0)(y-y_0)}{((x-x_0)^2+z^2)\, r}\, dx_0\, dy_0. \quad (3\cdot 7,3)$$

Again

$$\int_{S+W} \tau \frac{\partial^2}{\partial y_0^2}\left(\frac{1}{r}\right) dx_0\, dy_0 = \int_a^\infty dx_0 \int_{-s_0}^{s_0} \tau \frac{\partial^2}{\partial y_0^2}\left(\frac{1}{r}\right) dy_0$$

$$= \int_a^\infty dx_0 \left\{ \left[\sigma \frac{\partial}{\partial y_0}\left(\frac{1}{r}\right)\right]_{-s_0}^{s_0} - \int_{-s_0}^{s_0} \frac{\partial \tau}{\partial y_0} \frac{\partial}{\partial y_0}\left(\frac{1}{r}\right) dy_0 \right\}$$

$$= -\int_{S+W} \frac{\partial \tau}{\partial y_0} \frac{\partial}{\partial y_0}\left(\frac{1}{r}\right) dx_0$$

$$= -\int_{-s_0}^{s_0} dy_0 \left\{ -\left[\frac{\partial \tau}{\partial y_0} \frac{(x-x_0)(y-y_0)}{((y-y_0)^2+z^2)\, r}\right]_{x_0=a}^{x_0=\infty} \right.$$

$$\left. + \int_a^\infty \frac{\partial^2 \tau}{\partial x_0 \partial y_0} \frac{(x-x_0)(y-y_0)}{((y-y_0)^2+z^2)\, r}\, dx_0 \right\}.$$

Now
$$\lim_{x_0 \to \infty} \frac{(x-x_0)(y-y_0)}{((y-y_0)^2+z^2)\, r} = -\frac{(y-y_0)}{(y-y_0)^2+z^2}.$$

Also $\partial^2 \tau/\partial x_0\, \partial y_0 = 0$ in the wake, and so

$$\int_{S+W} \tau \frac{\partial^2}{\partial y_0^2}\left(\frac{1}{r}\right) dx_0\, dy_0 = -\int_{-s_0}^{s_0} \frac{\partial \tau(b, y_0)}{\partial y_0} \frac{y-y_0}{(y-y_0)^2+z^2}\, dy_0$$

$$- \int_S \frac{\partial^2 \tau}{\partial x_0 \partial y_0} \frac{(x-x_0)(y-y_0)}{((x-x_0)^2+z^2)\, r}\, dx_0\, dy_0. \quad (3\cdot 7,4)$$

Hence, from $(3\cdot 7,2)$, $(3\cdot 7,3)$ and $(3\cdot 7,4)$,

$$w(x,y,z) = \frac{1}{4\pi} \int_S \frac{\partial^2 \tau}{\partial x_0 \partial y_0} \left(\frac{1}{(x-x_0)^2+z^2} + \frac{1}{(y-y_0)^2+z^2}\right)$$

$$\times \frac{(x-x_0)(y-y_0)}{\sqrt{\{(x-x_0)^2+(y-y_0)^2+z^2\}}}\, dx_0\, dy_0$$

$$+ \frac{1}{4\pi} \int_{-s_0}^{s_0} \frac{\partial \tau(b, y_0)}{\partial y_0} \frac{(y-y_0)}{(y-y_0)^2+z^2}\, dy_0$$

$$= \frac{1}{4\pi} \int_{-s_0}^{s_0} \left\{ \int_a^b \frac{\partial^2 \tau}{\partial x_0 \partial y_0} \left(\frac{1}{(x-x_0)^2+z^2} + \frac{1}{(y-y_0)^2+z^2}\right) \right.$$

$$\times \frac{(x-x_0)(y-y_0)}{\sqrt{\{(x-x_0)^2+(y-y_0)^2+z^2\}}}\, dx_0 + \left. \frac{\partial \tau(b, y_0)}{\partial y_0} \frac{(y-y_0)}{(y-y_0)^2+z^2} \right\} dy_0. \quad (3\cdot 7,5)$$

The integrals in (3·7, 5) are singular for $z = 0$. It can be shown that in this case the correct limiting value is obtained by taking the principal values of the integrals. Thus

$$w(x, y, 0) = \frac{1}{4\pi} \int_{-s_0}^{s_0} \left\{ \int_a^b \frac{\partial^2 \tau}{\partial x_0 \partial y_0} \frac{\sqrt{\{(x - x_0)^2 + (y - y_0)^2\}}}{(x - x_0)(y - y_0)} dx_0 + \frac{\partial \tau(b, y_0)}{\partial y_0} \frac{1}{y - y_0} \right\} dy_0,$$

$$(3·7, 6)$$

where $\int_a^b$ and $\int_{-s_0}^{s_0}$ are to be interpreted as

$$\int_a^b = \lim_{\epsilon \to 0} \left(\int_a^{x - \epsilon} + \int_{x + \epsilon}^b \right), \qquad \int_{-s_0}^{s_0} = \lim_{\epsilon \to 0} \left(\int_{-s_0}^{y - \epsilon} + \int_{y + \epsilon}^{s_0} \right).$$

To show that (3·7, 6) represents the correct limit of (3·7, 5), consider first the integral

$$J(z) = \int_{-s_0}^{s_0} f(y_0) \frac{y - y_0}{(y - y_0)^2 + z^2} dy_0 \quad (-s_0 < y < s_0), \qquad (3·7, 7)$$

where $f(y_0)$ is arbitrary for the time being. We assume that $f(y_0)$ possesses a continuous derivative in the (closed) range of integration, so that

$$f(y_0) = f(y) + (y_0 - y) f'(y_1),$$

where y_1 is some intermediate value between y_0 and y, which may be regarded as a function of y_0 for given y. Hence

$$J(z) = \int_{-s_0}^{s_0} f(y) \frac{y - y_0}{(y - y_0)^2 + z^2} dy_0 - \int_{-s_0}^{s_0} f'(y_1) \frac{(y - y_0)^2}{(y - y_0)^2 + z^2} dy_0$$

$$= J_1(z) - J_2(z), \quad \text{say},$$

where

$$J_1(z) = f(y) \int_{-s_0}^{s_0} \frac{y - y_0}{(y - y_0)^2 + z^2} dy_0 = \tfrac{1}{2} f(y) \log \frac{(y + s_0)^2 + z^2}{(y - s_0)^2 + z^2}, \qquad (3·7, 8)$$

while

$$J_2(z) = \int_{-s_0}^{s_0} f'(y_1) \frac{(y - y_0)^2}{(y - y_0)^2 + z^2} dy_0. \qquad (3·7, 9)$$

Hence

$$\lim_{z \to 0} J_1(z) = f(y) \log \frac{s_0 + y}{s_0 - y}, \qquad (3·7, 10)$$

and

$$\lim_{z \to 0} J_2(z) = J_2(0) = \int_{-s_0}^{s_0} f'(y_1) dy_0. \qquad (3·7, 11)$$

The last relation, which affirms that $J_2(z)$ is a continuous function of z for $z = 0$, follows from the fact that we can make the modulus of

$$J_2(z) - J_2(0) = \int_{-s_0}^{s_0} f'(y) \left(\frac{(y - y_0)^2}{(y - y_0)^2 + z^2} - 1 \right) dy_0$$

as small as we please by choosing $|z|$ sufficiently small. Indeed let M be an upper bound for $f'(y_1)$, $|f'(y_1)| \leqslant M$ in the range of integration. Then

$$\left| \int_{-s_0}^{s_0} f'(y_1) \left(\frac{(y - y_0)^2}{(y - y_0)^2 + z^2} - 1 \right) dy_0 \right| \leqslant M \int_{-s_0}^{s_0} \frac{z^2}{(y - y_0)^2 + z^2} dy_0$$

$$= M \int_{-s_0}^{s_0} \frac{|z|}{1 + \left(\frac{y_0 - y}{z} \right)^2} d\left(\frac{y_0 - y}{|z|} \right) = M |z| \int_{-(s_0 + y)/z}^{(s_0 - y)/z} \frac{d\eta}{1 + \eta^2} \leqslant \pi M |z|.$$

Combining (3·7, 10) and (3·7, 11), we obtain

$$\lim_{z \to 0} J(z) = f(y) \log \frac{s_0 + y}{s_0 - y} - \int_{-s_0}^{s_0} f'(y_1) \, dy_0. \qquad (3\cdot7, 12)$$

On the other hand, define $K(\epsilon)$ by

$$K(\epsilon) = \left(\int_{-s_0}^{y-\epsilon} + \int_{y+\epsilon}^{s_0} \right) f(y_0) \frac{dy_0}{y - y_0}. \qquad (3\cdot7, 13)$$

Then

$$K(\epsilon) = f(y) \left(\int_{-s_0}^{y-\epsilon} + \int_{y+\epsilon}^{s_0} \right) \frac{dy_0}{y - y_0} - \left(\int_{-s_0}^{y-\epsilon} + \int_{y+\epsilon}^{s_0} \right) f'(y_1) \, dy_0$$

$$= f(y) \left(\log \frac{y + s_0}{\epsilon} - \log \frac{s_0 - y}{\epsilon} \right) - \left(\int_{-s_0}^{y-\epsilon} + \int_{y+\epsilon}^{s_0} \right) f'(y_1) \, dy_0$$

$$= f(y) \log \frac{s_0 + y}{s_0 - y} - \left(\int_{-s_0}^{y-\epsilon} + \int_{y+\epsilon}^{s_0} \right) f'(y_1) \, dy_0,$$

and so

$$\lim_{z \to 0} K(\epsilon) = f(y) \log \frac{s_0 + y}{s_0 - y} - \int_{-s_0}^{s_0} f'(y_1) \, dy_0. \qquad (3\cdot7, 14)$$

Comparing (3·7, 14) with (3·7, 12) we see that

$$\lim_{\epsilon \to 0} K(\epsilon) = \lim_{z \to 0} J(z),$$

i.e. in detail

$$\lim_{z \to 0} \int_{-s_0}^{s_0} f(y_0) \frac{y - y_0}{(y - y_0)^2 + z^2} \, dy_0 = \lim_{\epsilon \to 0} \left(\int_{-s_0}^{y-\epsilon} + \int_{y+\epsilon}^{s_0} \right) \frac{f(y_0)}{y - y_0} \, dy_0. \qquad (3\cdot7, 15)$$

Now we may write (3·7, 5) in the form

$$w(x, y, z) = \frac{1}{4\pi} \int_{-s_0}^{s_0} \frac{y - y_0}{(y - y_0)^2 + z^2}$$

$$\times \left(\int_a^b \frac{\sqrt{\{(x - x_0)^2 + (y - y_0)^2 + z^2\}}\,(x - x_0)}{(x - x_0)^2 + z^2} \frac{\partial^2 \tau}{\partial x_0 \, \partial y_0} \, dx_0 \right) dy_0$$

$$+ \frac{1}{4\pi} \int_{-s_0}^{s_0} \frac{y - y_0}{(y - y_0)^2 + z^2}$$

$$\times \left(\int_a^b \frac{z^2}{\sqrt{\{(x - x_0)^2 + (y - y_0)^2 + z^2\}}} \frac{x - x_0}{(x - x_0)^2 + z^2} \frac{\partial^2 \tau}{\partial x_0 \, \partial y_0} \, dx_0 \right) dy_0$$

$$\cdot \frac{1}{4\pi} \int_{-s_0}^{s_0} \frac{\partial \tau(b, y_0)}{\partial y_0} \frac{y - y_0}{(y - y_0)^2 + z^2} \, dy_0. \qquad (3\cdot7, 16)$$

Putting $f(y_0) = \dfrac{\partial \tau(b, y_0)}{\partial y_0}$ in (3·7, 15), we see that the last integral in (3·7, 16)

tends to the principal value of $\displaystyle\int_{-s_0}^{s_0} \frac{\partial \tau(b, y_0)}{\partial y_0} \frac{dy_0}{y - y_0}$, in accordance with (3·7, 6).

Furthermore, it is not difficult to show that the second double integral in (3·7, 16) tends to 0 as z tends to 0. To find the limit of the first double integral in (3·7, 16) we can show by an argument similar to that used for the derivation of (3·7, 15), that the limit of the integral

$$\int_a^b \frac{\sqrt{\{(x-x_0)^2+(y-y_0)^2+z^2\}}(x-x_0)}{(x-x_0)^2+z^2} \frac{\partial^2 \tau}{\partial x_0 \partial y_0} dx_0,$$

as z tends to 0 is the principal value of

$$\int_a^b \frac{\sqrt{\{(x-x_0)^2+(y-y_0)^2\}}}{x-x_0} \frac{\partial^2 \tau}{\partial x_0 \partial y_0} dx_0.$$

This is a function of y_0 to which we may apply (3·7, 15), and so

$$\lim_{z\to 0} \int_{-s_0}^{s_0} \frac{y-y_0}{(y-y_0)^2+z^2} \left(\int_a^b \frac{\sqrt{\{(x-x_0)^2+(y-y_0)^2+z^2\}}(x-x_0)}{(x-x_0)^2+z^2} \frac{\partial^2 \tau}{\partial x_0 \partial y_0} dx_0 \right) dy_0$$

$$= \int_{-s_0}^{s_0} \int_a^b \frac{\sqrt{\{(x-x_0)^2+(y-y_0)^2\}}}{(x-x_0)(y-y_0)} \frac{\partial^2 \tau}{\partial x_0 \partial y_0} dx_0 dy_0,$$

where principal values are to be taken on the right-hand side. Equation (3·7, 6) is thus confirmed.

Using (3·1, 2) and the formula $\dfrac{\partial \tau(b, y_0)}{\partial y_0} = \int_a^b \dfrac{\partial^2 \tau}{\partial x_0 \partial y_0} dx_0$, we may write (3·7, 6) as an integral equation of the first kind for $\partial^2 \tau / \partial x_0 \partial y_0$,

$$Us(x, y) = \frac{1}{4\pi} \int_{-s_0}^{s_0} \int_a^b \frac{\partial^2 \tau}{\partial x_0 \partial y_0} \left(\frac{\sqrt{\{(x-x_0)^2+(y-y_0)^2\}}}{(x-x_0)(y-y_0)} + \frac{1}{y-y_0} \right) dx_0 dy_0. \quad (3·7, 17)$$

However, even in this particular case (of a rectangular wing) a closed analytical solution of (3·7, 17) appears to be beyond our mathematical resources.

We may assume that the mid-chord coordinate $x_m = 0$, so that the leading and trailing edges of the rectangular wing correspond to $x = -\frac{1}{2}c$ and $x = \frac{1}{2}c$, respectively. Referring to (3·6, 28) and (3·6, 29), we see that (3·7, 6) applies not only to the total downwash and doublet distribution, but also to the individual w_{nm} and $-UcA_m(y)\tau_n(\theta)$, respectively. Hence

$$w_{nm} = -\frac{c}{4\pi} \int_{-s_0}^{s_0} \int_{-\frac{1}{2}c}^{\frac{1}{2}c} \frac{dA_m}{dy_0} \frac{d\tau_n}{dx_0}$$

$$\times \frac{\sqrt{\{(x-x_0)^2+(y-y_0)^2\}}}{(x-x_0)(y-y_0)} dx_0 dy_0 \quad (n=1, 2, \ldots;\ m=1, 2, \ldots) \quad (3·7, 18)$$

since the $\tau_n(\theta)$ are now independent of y and $\tau_n(\pi) = 0$ $(n = 1, 2, \ldots)$.

Substituting θ_0 for x_0 in the integral on the right-hand side of (3·7, 18),

$$x_0 = x_m - \tfrac{1}{2}c \cos \theta_0, \quad dx_0 = \tfrac{1}{2}c \sin \theta_0 d\theta_0,$$

we obtain

$$w_{nm} = \frac{c}{2\pi} \int_{-s_0}^{s_0} \int_0^\pi \frac{dA_m}{dy_0} \frac{d\tau_n}{d\theta_0} \frac{\sqrt{\{c^2(\cos \theta - \cos \theta_0) + 4(y-y_0)^2\}}}{c(\cos \theta - \cos \theta_0)(y-y_0)} d\theta_0 dy_0.$$

Let the integrals I_n be defined by

$$I_n = -\int_0^\pi \frac{d\tau_n}{d\theta_0} \frac{\sqrt{\{c^2(\cos\theta - \cos\theta_0)^2 + 4(y - y_0)^2\}}}{(\cos\theta - \cos\theta_0)(y - y_0)} d\theta_0 \quad (n = 1, 2, \ldots). \quad (3\cdot7, 19)$$

Then $\qquad w_{nm} = -\frac{1}{2\pi} \int_{-s_0}^{s_0} \frac{dA_m}{dy_0} I_n dy_0 \quad (n = 1, 2, \ldots; \; m = 1, 2, \ldots). \quad (3\cdot7, 20)$

It is shown by W. P. Jones in the report mentioned above that the functions I_n can be reduced to elliptic integrals.

To calculate the functions w_{0m}, it is convenient to divide them into w'_{0m} and w''_{0m} corresponding to τ'_0 and τ''_0 respectively, $w_{0m} = w'_{0m} + w''_{0m}$. Since $\tau''_0(\pi) = 0$, we have

$$w''_{0m} = -\frac{1}{2\pi} \int_{-s_0}^{s_0} \frac{dA_m}{dy_0} I''_0 dy_0 \quad (m = 1, 2, \ldots), \qquad (3\cdot7, 21)$$

where I''_0 is obtained by replacing τ_n by τ''_0 in $(3\cdot7, 19)$. On the other hand, $\tau'_0(\pi) = \pi$, and so

$$w'_{0m} = -\frac{c}{4\pi} \int_{-s_0}^{s_0} \int_{-\frac{1}{2}c}^{\frac{1}{2}c} \frac{dA_m}{dy_0} \frac{d\tau_0}{dx_0} \frac{\sqrt{\{(x - x_0)^2 + (y - y_0)^2\}}}{(x - x_0)(y - y_0)} dx_0 dy_0$$
$$-\frac{c}{4} \int_{-s_0}^{s_0} \frac{dA_m}{dy_0} \frac{dy_0}{y - y_0}.$$

This may be written as

$$w'_{0m} = -\frac{1}{2\pi} \int_{-s_0}^{s_0} \frac{dA_m}{dy_0} I'_0 dy_0 - \frac{c}{4} \int_{-s_0}^{s_0} \frac{dA_m}{dy_0} \frac{dy_0}{y - y_0}, \qquad (3\cdot7, 22)$$

where I'_0 is obtained by replacing τ_n in $(3\cdot7, 19)$ by τ'_0.

Having calculated a number of the w_{nm} we determine the coefficients C_{nm} so as to satisfy the specified boundary condition $(3\cdot1, 2)$ at a number of points, as indicated in the preceding section.

3·8   Vortex theory

An alternative approach to the problem of the thin (possibly cambered and twisted) wing at incidence is based on the fact that the flow round the wing may be considered to be due to a vortex distribution over the wing as well as in the wake. We retain the approximations of §§ 3·6 and 3·7, and assume in particular that the vortices which correspond to wing and wake are located in the x, y-plane. In these circumstances, as will be shown below, vortex theory leads to exactly the same results as the theory based on doublets which was described in § 3·6. However, vortex theory also leads to a number of important approximate methods, which do not present themselves readily when the problem is approached from a different point of view. Moreover, the theory is physically more intuitive than the doublet method and, indeed, preceded all others dealing with three-dimensional problems, in the historical order of development.

Let U be the free-stream velocity and Φ and Φ' the total and induced velocity potential, as before. We are dealing with the case of a thin wing at incidence so that (3·1, 6) and (3·1, 7) apply. In particular, at the wing and in the wake,

$$\left.\begin{aligned}
u'(x, y, +0) &= -u'(x, y, -0), \\
v(x, y, +0) &= -v(x, y, -0), \\
w(x, y, +0) &= w(x, y, -0),
\end{aligned}\right\} \qquad (3·8, 1)$$

where $(x, y, +0)$ and $(x, y, -0)$ denote points just above and just below the wing or the wake, respectively. Moreover, for points of the wake (3·6, 14) holds,

$$u'(x, y, +0) = u'(x, y, -0) = 0. \qquad (3·8, 2)$$

(3·8, 1) and (3·8, 2) shows that we may regard the combined area of wing and wake as a vortex sheet with a surface distribution of vorticity $\gamma(x, y)$ which is given by

$$\gamma = (-(v_+ - v_-), (u_+ - u_-), 0) = (-2v_+, 2u'_+, 0), \qquad (3·8, 3)$$

where $u_+ = u(x, y, +0)$, $u_- = u(x, y, -0)$, $u'_+ = u'(x, y, +0)$, etc.

In particular, in the wake $\gamma = (-2v_+, 0, 0)$, $\qquad (3·8, 4)$

in agreement with the assumption made in § 1·15, that the vortex lines in the wake are parallel to the direction of flow.

Also, since the flow is irrotational, we have

$$\frac{\partial v_+}{\partial x} - \frac{\partial u_+}{\partial y} = 0, \quad \frac{\partial v_-}{\partial x} - \frac{\partial u_-}{\partial y} = 0.$$

Hence

$$\frac{\partial (v_+ - v_-)}{\partial x} - \frac{\partial (u_+ - u_-)}{\partial y} = 0. \qquad (3·8, 5)$$

This equation is equivalent to div $\gamma = 0$ as required for a vorticity vector. In particular in the wake

$$\frac{d(v_+ - v_-)}{\partial x} = 2\frac{dv_+}{dx} = 0. \qquad (3·8, 6)$$

The velocity $\mathbf{q}' = (u', v, w)$ due to a three-dimensional distribution of vorticity $\boldsymbol{\omega} = (\xi, \eta, \zeta)$ is given by (cf. § 1·9)

$$\mathbf{q}' = \operatorname{curl} \boldsymbol{\Psi}, \qquad (3·8, 7)$$

where

$$\boldsymbol{\Psi}(x, y, z) = \frac{1}{4\pi} \int_{S+W} \frac{\boldsymbol{\omega}(x_0, y_0, z_0)\, dx_0\, dy_0\, dz_0}{\sqrt{\{(x - x_0)^2 + (y - y_0)^2 + (z - z_0)^2\}}}. \qquad (3·8, 8)$$

In the case under consideration, the vortex layer becomes infinitely thin and occupies a portion of the x, y-plane. Thus (3·8, 8) applies where $\boldsymbol{\omega}(x_0, y_0, z_0)\, dz_0$ remains finite and equals $\gamma(x_0, y_0)$ in the limit. $\boldsymbol{\Psi}$ is now given by

$$\begin{aligned}
\boldsymbol{\Psi}(x, y, z) &= \frac{1}{4\pi} \int_{S+W} \frac{\gamma(x_0, y_0)}{\sqrt{\{(x - x_0)^2 + (y - y_0)^2 + z^2\}}}\, dx_0\, dy_0 \\
&= \frac{1}{2\pi} \int_{S+W} \frac{(-v_+, u'_+, 0)}{\sqrt{\{(x - x_0)^2 + (y - y_0)^2 + z^2\}}}\, dx_0\, dy_0. \qquad (3·8, 9)
\end{aligned}$$

Now, with the notation of (3·6, 13),

$$u'_+ = -\frac{\partial\Phi'(x_0, y_0, +0)}{\partial x_0} = -\frac{1}{2}\frac{\partial\tau(x_0, y_0)}{\partial x_0},$$
$$v_+ = -\frac{\partial\Phi'(x_0, y_0, +0)}{\partial y_0} = -\frac{1}{2}\frac{\partial\tau(x_0, y_0)}{\partial y_0},$$

(3·8, 10)

and so

$$\Psi(x, y, z) = \frac{1}{4\pi}\int_{S+W}\left(\frac{\partial\tau(x_0, y_0)}{\partial y_0}, -\frac{\partial\tau(x_0, y_0)}{\partial x_0}, 0\right)\frac{dx_0\,dy_0}{\sqrt{\{(x-x_0)^2 + (y-y_0)^2 + z^2\}}}.$$

Hence

(3·8, 11)

$$w = \frac{\partial\Psi_y}{\partial x} - \frac{\partial\Psi_x}{\partial y} = -\frac{1}{4\pi}\int_{S+W}\left\{\frac{\partial\tau}{\partial x_0}\frac{\partial}{\partial x}\left(\frac{1}{\sqrt{\{(x-x_0)^2 + (y-y_0)^2 + z^2\}}}\right)\right.$$
$$\left. + \frac{\partial\tau}{\partial y_0}\frac{\partial}{\partial y}\left(\frac{1}{\sqrt{\{(x-x_0)^2 + (y-y_0)^2 + z^2\}}}\right)\right\}dx_0\,dy_0$$

or

$$w = \frac{1}{4\pi}\int_{S+W}\left\{\frac{\partial\tau}{\partial x_0}\frac{\partial}{\partial x_0}\left(\frac{1}{r}\right) + \frac{\partial\tau}{\partial y_0}\frac{\partial}{\partial y_0}\left(\frac{1}{r}\right)\right\}dx_0\,dy_0.$$

(3·8, 12)

Integrating (3·8, 12) by parts, and taking into account that $\tau = 0$ on the finite boundaries of $S + W$, while $\frac{\partial}{\partial x_0}\left(\frac{1}{r}\right)$ and $\frac{\partial}{\partial y_0}\left(\frac{1}{r}\right)$ tend to 0 far behind the wing, we obtain

$$w = -\frac{1}{4\pi}\int_{S+W}\tau(x_0, y_0)\left(\frac{\partial^2}{\partial x_0^2}\left(\frac{1}{r}\right) + \frac{\partial^2}{\partial y_0^2}\left(\frac{1}{r}\right)\right)dx_0\,dy_0$$
$$= -\frac{1}{4\pi}\int_{S+W}\tau(x_0, y_0)\left(\frac{\partial^2}{\partial x^2}\left(\frac{1}{r}\right) + \frac{\partial^2}{\partial y^2}\left(\frac{1}{r}\right)\right)dx_0\,dy_0.$$

Since $1/r$ is a solution of Laplace's equation, the last expression may also be written as

$$w = \frac{1}{4\pi}\int_{S+W}\tau(x_0, y_0)\frac{\partial^2}{\partial z^2}\left(\frac{1}{r}\right)dx_0\,dy_0$$

(3·8, 13)

or

$$w = \frac{1}{4\pi}\frac{\partial^2}{\partial z^2}\int_{S+W}\tau(x_0, y_0)\frac{dx_0\,dy_0}{r}.$$

(3·8, 14)

The same expression for w is obtained by differentiating $-\Phi'$ as given by (3·6, 11) with respect to z. This confirms that the present theory leads to precisely the same results as the doublet approach.

Thus, for a given aerofoil at incidence, we still have to find $\tau(x, y)$ as a solution of the equation

$$\lim_{z\to 0}\frac{1}{4\pi}\frac{\partial^2}{\partial z^2}\int_{S+W}\tau(x_0, y_0)\frac{dx_0\,dy_0}{r} = Us(x, y),$$

(3·8, 15)

which is obtained by combining (3·8, 14) and (3·1, 2). By (3·2, 2) or (3·2, 3), the pressure difference between the top and bottom surfaces of the wing is then given by

$$p_u - p_l = -2\rho U u'_+ = 2\rho U\frac{\partial\Phi'(x, y, +0)}{\partial x} = \rho U\frac{\partial\tau(x, y)}{\partial x}.$$

(3·8, 16)

$\tau(x, y)$ was defined in §3·6 as the local strength of a doublet. However, by (3·8, 3) and (3·8, 10) it is also related directly to the vorticity vector $\gamma(x, y)$, viz.

$$\gamma(x, y) = \left(\frac{\partial \tau}{\partial y}, -\frac{\partial \tau}{\partial x}, 0\right). \tag{3·8, 17}$$

Let $Q(x, y, 0)$ be a point on the wing or in the wake, and let C be a closed circuit of the type shown in Fig. 3·8, 1, or more generally any circuit which penetrates the aerofoil at Q and is simply interlaced with the wing.

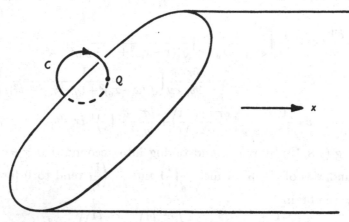

Fig. 3·8, 1

Integrating round C from the point Q regarded as a point of the upper surface to Q regarded as a point of the lower surface, we find that the circulation round C is

$$\int_C \mathbf{q} \, dl = -\int_C \left(\frac{\partial \Phi}{\partial x} dx + \frac{\partial \Phi}{\partial y} dy + \frac{\partial \Phi}{\partial z} dz\right)$$

$$= -\int_C \left(\frac{\partial \Phi'}{\partial x} dx + \frac{\partial \Phi'}{\partial y} dy + \frac{\partial \Phi'}{\partial z} dz\right)$$

$$= -\{\Phi'(x, y, -0) - \Phi'(x, y, +0)\}$$

$$= \tau(x, y),$$

where we have taken into account (3·6, 12). Thus $\tau(x, y)$ equals the circulation round C (and therefore is also equal to the integral of the vorticity over a surface bounded by C, by Stokes's theorem).

We may solve (3·8, 14) by means of a suitably chosen expansion for $\tau(x, y)$, and this is essentially the method followed by Blenk,‡ who approached the problem from the point of view of the present section. However, since vortex theory has been shown to be completely equivalent to doublet theory,

‡ H. Blenk, 'Der Eindecker als tragende Wirbelfläche', *Z. angew. Math. Mech.* vol. 5 (1925), pp. 36–47.

this of course leads precisely to the developments of §§ 3·6 and 3·7. On the other hand, we may replace the continuous vorticity distribution $\gamma(x, y)$ approximately by a distribution of line vortices which is discrete in one or in both directions. The latter method has been developed by Falkner‡ and has been used in calculations for numerous types of wings.

The expansion chosen by Falkner for the vorticity distribution per unit chord, γ, is

$$\frac{\gamma}{U}\frac{c}{8s_0} = \sqrt{(1-\eta^2)}\{\cot \tfrac{1}{2}\theta(a_0 + b_0\eta + c_0\eta^2 + \ldots) + \sin\theta(a_1 + b_1\eta + c_1\eta^2 + \ldots)$$
$$+ \sin 2\theta(a_2 + b_2\eta + c_2\eta^2 + \ldots) + \ldots\}, \quad (3\cdot8, 18)$$

where θ is defined as in (3·6, 19) by

$$x = x_m - \tfrac{1}{2}c(y)\cos\theta \quad (0 \leqslant \theta \leqslant \pi), \quad (3\cdot8, 19)$$

x_m being the mid-chord coordinate. η is the non-dimensional spanwise coordinate given by (3·6, 37), i.e.

$$\eta = y/s_0, \quad (3\cdot8, 20)$$

where s_0 is the semi-span of the wing. (3·8, 18) is analogous to the expansions (3·6, 33) with (3·6, 37), used in the doublet representation method. It is a combination of the two-dimensional thin aerofoil series expansion (2·8, 15) (the first term of which, $\cot \tfrac{1}{2}\theta$ corresponds to the flat plate solution—notice that in order to make the definition (3·8, 19) agree with Chapter 2, θ in (2·8, 15) should be replaced by $\pi - \theta$), with the spanwise terms in η, where the first term $\sqrt{(1-\eta^2)}$ is the elliptic loading solution of lifting-line theory (cf. equation (3·3, 73)).

For purposes of calculation (3·8, 18) is split into two parts, the distribution at zero lift, which retains the form of equation (3·8, 18), and the distribution additional to this at an incidence α relative to the angle of zero lift. The latter we write as

$$\gamma_\alpha \frac{c}{8s_0 U \tan\alpha_0} = \sqrt{(1-\eta^2)}\{\cot \tfrac{1}{2}\theta(a_0 + b_0\eta + c_0\eta^2 + \ldots)$$
$$+ \sin\theta(a_1 + b_1\eta + c_1\eta^2 + \ldots) + \ldots\}. \quad (3\cdot8, 21)$$

The continuous vorticity distribution represented by (3·8, 18) and (3·8, 21) is now supposed to be replaced by a set of discrete vortex lines, of strengths $\Gamma_1, \Gamma_2, \ldots$. The particular system used by Falkner consists of four chordwise vortex lines, which are taken to be of constant strength within a number of spanwise sections of the wing. Falkner divided the semi-span into ten such sections, each of width $0\cdot1s_0$, and placed the spanwise vortex lines at 0·125, 0·375, 0·625 and 0·875 of the local chord. From the edges of each spanwise section emanate trailing vortex lines, so that the wing is divided up into a network of straight vortex lines in the manner shown in

‡ V. M. Falkner, 'The calculation of aerodynamic loading on surfaces of any shape', *Rep. Memor. Aero. Res. Comm., Lond.*, no. 1910 (1943).

Fig. 3·8, 2. The system may be regarded as made up of a number of horse-shoe vortices, each formed by one of the chordwise vortex line segments and the trailing vortices emanating from its ends. The velocity field for a horseshoe vortex has been derived in Chapter 1 (see equation (1·9, 32)).

Our object is to find the values of the coefficients $a_0, b_0, \ldots$ in the continuous representation, (3·8, 18) and (3·8, 21), from the aerofoil shape. This can be done by relating them to the strength of the individual vortex lines $\Gamma_1, \Gamma_2, \ldots$, and then solving the relatively simple problem of determining the Γ_i by applying the boundary conditions at a discrete number of chosen points of the aerofoil.

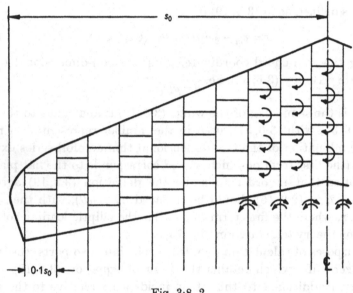

Fig. 3·8, 2

To establish the relationship between the coefficients $a_0, b_0, \ldots$ and $\Gamma_1, \Gamma_2, \ldots$, we first of all use the condition that the total circulation at a given span station for the continuous distribution is equal to the total strength of the four discrete vortex lines at that station. Now, the circulation is the integral of the γ over the chord, and is therefore equal to

$$4\pi s_0 U \tan \alpha_0 \sqrt{(1 - \eta^2)}\, (a_0 + b_0 \eta + \ldots).$$

Equating this to the sum of the strength of the vortex lines we obtain

$$\Gamma_1(\eta) + \Gamma_2(\eta) + \Gamma_3(\eta) + \Gamma_4(\eta) = 4\pi s_0 U \tan \alpha_0 \sqrt{(1 - \eta^2)}\, (a_0 + b_0 \eta + c_0 \eta^2 + \ldots)$$

$$= \tfrac{1}{2}\pi U c A_0(\eta), \tag{3·8, 22}$$

where
$$A_0(\eta) = \frac{8 s_0}{c} \tan \alpha_0 \sqrt{(1 - \eta^2)}\, (a_0 + b_0 \eta + c_0 \eta^2 + \ldots). \tag{3·8, 23}$$

We find three further relations by equating the downwash velocity w at three points on the wing due to the discrete and continuous distributions.

The three points chosen are at the mid-points between the vortex lines, i.e. at 0·25, 0·5 and 0·75 of the chord. The contributions of the series terms in θ of equations (3·8, 18) and (3·8, 21) are then calculated most conveniently separately. Thus for the first term, $\cot \frac{1}{2}\theta$, the downwash at the quarter chord point, for example, is

$$\frac{4s_0 U}{c} \tan \alpha_0 \sqrt{(1-\eta^2)} (a_0 + b_0 \eta + \ldots) = \frac{1}{2} U A_0(\eta),$$

which we equate to the effect of the four spanwise vortex lines at the particular span section considered, regarded as being of infinite length. By this means we obtain four simultaneous equations for the values of Γ_1, Γ_2, Γ_3, Γ_4, corresponding to the term $\cot \frac{1}{2}\theta$, and the solutions of these equations are in fact

$$\left.\begin{aligned}
\Gamma_1(\eta) &= 0\cdot2734\pi c U A_0, \\
\Gamma_2(\eta) &= 0\cdot1172\pi c U A_0, \\
\Gamma_3(\eta) &= 0\cdot0703\pi c U A_0, \\
\Gamma_4(\eta) &= 0\cdot0391\pi c U A_0.
\end{aligned}\right\} \tag{3·8, 24}$$

The total strengths of the vortex lines are found by repeating this procedure for as many terms of the series in θ as is necessary, and adding the individual vortex strengths thus obtained.

Finally, the downwash velocities w due to these vortex lines are calculated at a number of 'pivotal' points, where the linearized boundary conditions given by (3·1, 2),

$$\frac{w}{U} = s(\theta, \eta), \tag{3·8, 25}$$

are applied. In this instance the complete set of horseshoe vortices of the discrete system are used to find the velocities, employing the expressions calculated in equation (1·9, 32). The pivotal points selected are at the mid-points of the spanwise segments of the vortex lines, so that, with the lattice system of vortices illustrated in Fig. 3·8, 2, we have in all seventy-six linear equations for the coefficients a_0, b_0, c_0, ..., involving the slope s of the aerofoil. Thus a maximum of seventy-six of these coefficients can be calculated and, if less than this are desired (so that there are more equations than unknowns) a least-squares method can be used to find them (cf. Gates's method in §3·3).

This theory has proved to be rather more accurate in practice than might have been expected from the approximations made. For, in particular, the choice of pivotal points is, on the one hand, arbitrary, and on the other hand, all important. Another criticism can be made in the choice of the conditions used to relate the discrete vortex distribution to the continuous. Thus it would appear just as logical and important to apply equality of pitching moment at a given span section, as equality of the lift (equation (3·8, 22)). A detailed criticism of Falkner's method has been

made by Schlichting and Thomas,‡ and they suggest a modified approach
in which the wing is replaced by a set of four spanwise vortex lines, with a
continuous spanwise distribution of vorticity in the form of equation
(3·8, 22)

$$\frac{\Gamma_n}{2s_0 U} = \sqrt{(1-\eta^2)} \, (a_n + b_n \eta + c_n \eta^2 + \ldots) \quad (n = 0, 1, 2, 3). \quad (3·8, 26)$$

The arrangement of the vortices for a swept-back wing is illustrated in
Fig. 3·8, 3.

In this case the continuous spanwise vorticity distribution is retained, and an infinite set of trailing
vortices leave the spanwise vortex
lines. The situation of the vortex
lines is determined only by the
condition that the position of the
aerodynamic centre is the same
for the discrete and continuous
(two-dimensional) distributions.
They are thus each placed at the
quarter chord point of the four
strips formed by dividing the wing
into four chordwise sections.

The pivotal points used are
chosen from the following considerations. For a two-dimensional

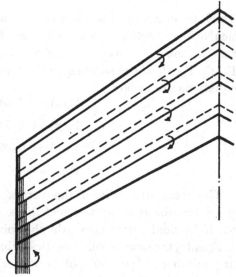

Fig. 3·8, 3

flat plate aerofoil represented by a vortex line of strength Γ at its quarter
chord point, the downwash velocity due to the vortex line equals the true
velocity, $U\alpha_0$, at a distance d from the quarter chord of the aerofoil, where

$$U\alpha_0 = \frac{\Gamma}{2\pi d}. \quad (3·8, 27)$$

From the Kutta–Joukowski theorem Γ is equal in magnitude to

$$L/\rho U = \pi U c\alpha_0,$$

and hence from (3·8, 27) $$U\alpha_0 = \frac{\pi U c\alpha_0}{2\pi d},$$

or $$d = \tfrac{1}{2}c, \quad (3·8, 28)$$

so that the downwash velocities according to the accurate theory and single-
line vortex representations are equal at the three-quarter chord point of the
aerofoil. For this reason the pivotal points for calculating the downwash

‡ H. Schlichting and H. H. B. M. Thomas, 'Note on the calculation of lift distribution of swept wings', *Royal Aircraft Establishment Rep.* no. Aero. 2236 (1947).

velocities due to the vortex lines and applying the boundary condition (3·8, 25) are selected at the three-quarter chord points of the four chordwise sections.

A further degree of simplification of the method of representing the aerofoil by a set of discrete vortex lines has been made by Weissinger.‡ In his approach a single spanwise vortex line only is retained, and in this respect it is similar to lifting-line theory (§ 3·3).

On the same basis as Schlichting and Thomas's modification of Falkner's method, Weissinger places the vortex line at the aerofoil quarter chord point and applies the boundary condition at the three-quarter chord point. The downwash field of the vortex line and its trailing vortices is found by applying the Biot-Savart law (equation (1·9, 21)). Thus, considering a trapezoidal wing with a straight vortex line of strength Γ, swept back at an angle ϕ, as illustrated in Fig. 3·8, 4, the vertical velocity at a point (x, y) due to the vortex line on the starboard wing is

$$w_{A_1}(x, y) = -\frac{1}{4\pi} \int_0^{s_0} \Gamma(\eta) \frac{\sin \beta'}{r^2} \frac{d\eta}{\cos \phi}, \qquad (3·8, 29)$$

where r is the distance of the point (x, y) to the point $(\eta \tan \phi, \eta)$ on the vortex, and β' is the angle between the line joining (x, y) and $(\eta \tan \phi, \eta)$ and the vortex line. s_0 is the semi-span measured along the y-axis. Since

$$\frac{\sin \beta'}{\cos \phi} = \frac{x - y \tan \phi}{r},$$

we can write (3·8, 29) as

$$w_{A_1}(x, y) = -\frac{1}{4\pi} \int_0^{s_0} \Gamma(\eta) \frac{x - y \tan \phi}{[(x - \eta \tan \phi)^2 + (y - \eta)^2]^{\frac{3}{2}}} d\eta. \qquad (2·8, 30)$$

Assuming the vortex line on the port wing has the same sweep-back angle ϕ, we can find its contribution to the downwash velocity by replacing ϕ by $-\phi$ in (3·8, 30), thus

$$w_{A_2}(x, y) = -\frac{1}{4\pi} \int_{-s_0}^0 \Gamma(\eta) \frac{x + y \tan \phi}{[(x + \eta \tan \phi)^2 + (y - \eta)^2]^{\frac{3}{2}}} d\eta. \qquad (3·8, 31)$$

The effect of the trailing vortices is found from (1·9, 30) as the velocity due to a sheet of semi-infinite vortex lines. The strength of the vortex shed from an elementary segment of the bound vortex is $-d\Gamma$, and therefore the vertical velocity at a point $(x, y, 0)$ induced by the trailing vortices emanating from the starboard wing is

$$w_{B_1} = \frac{1}{4\pi} \int \frac{1}{\eta - y} \left(1 + \frac{x - \eta \tan \phi}{\sqrt{\{(x - \eta \tan \phi)^2 + (y - \eta)^2\}}} \right) d\Gamma(\eta), \qquad (3·8, 32)$$

‡ J. Weissinger, 'Über die Auftriebsverteilung von Pfeilflügeln' (Zentrale für wissenschaftliches Berichtswesen der Luftfahrtforschung des Generalluftzeugmeisters Berlin-Aldershof), *Forschungsbericht*, no. 1553 (1942). Translation: *Tech. Memor. Nat. Adv. Comm. Aero., Wash.*, no. 1120.

where the integration extends over the starboard vortex line, and it is understood that the principal value of the integral is to be taken. We may write (3·8, 32) as

$$w_{B_1} = \frac{1}{4\pi} \int_0^{s_0} \frac{\Gamma'(\eta)}{\eta - y} \left(1 + \frac{x - \eta \tan\phi}{\sqrt{\{(x - \eta \tan\phi)^2 + (y - \eta)^2\}}} \right) d\eta. \quad (3·8, 33)$$

Similarly, the contribution to the velocity due to the trailing vortex sheet from the port wing is

$$w_{B_2} = \frac{1}{4\pi} \int_{-s_0}^0 \frac{\Gamma'(\eta)}{\eta - y} \left(1 + \frac{x + \eta \tan\phi}{\sqrt{\{(x + \eta \tan\phi)^2 + (y - \eta)^2\}}} \right) d\eta. \quad (3·8, 34)$$

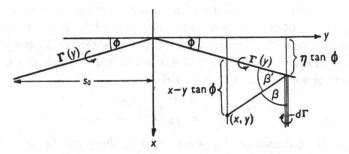

Fig. 3·8, 4

If we integrate (3·8, 30) by parts, we find

$$w_{A_1} = -\frac{1}{4\pi} \left[\Gamma(\eta) \frac{(1 + \tan^2\phi)\eta - (y + x\tan\phi)}{(x - y\tan\phi)\sqrt{\{(x - \eta\tan\phi)^2 + (y - \eta)^2\}}} \right]_0^{s_0}$$
$$+ \frac{1}{4\pi} \int_0^{s_0} \Gamma'(\eta) \frac{(1 + \tan^2\phi)\eta - (y + x\tan\phi)}{(x - y\tan\phi)\sqrt{\{(x - \eta\tan\phi)^2 + (y - \eta)^2\}}} d\eta \quad (3.8, 35)$$

with the similar equation for (3·8, 31),

$$w_{A_2} = -\frac{1}{4\pi} \left[\Gamma(\eta) \frac{(1 + \tan^2\phi)\eta - (y - x\tan\phi)}{(x + y\tan\phi)\sqrt{\{(x + \eta\tan\phi)^2 + (y - \eta)^2\}}} \right]_{-s_0}^0$$
$$+ \frac{1}{4\pi} \int_{-s_0}^0 \Gamma'(\eta) \frac{(1 + \tan^2\phi)\eta - (y - x\tan\phi)}{(x + y\tan\phi)\sqrt{\{(x + \eta\tan\phi)^2 + (y - \eta)^2\}}} d\eta. \quad (3·8, 36)$$

Adding (3·8, 33), (3·8, 34), (3·8, 35) and (3·8, 36), we find for the total normal induced velocity

$$w(x, y, 0) = \frac{1}{4\pi} \int_0^{s_0} \frac{\Gamma'(\eta)}{\eta - y} \left(1 + \frac{\sqrt{\{(x - \eta\tan\phi)^2 + (y - \eta)^2\}}}{x - y\tan\phi} \right) d\eta$$
$$+ \frac{1}{4\pi} \int_{-s_0}^0 \frac{\Gamma'(\eta)}{\eta - y} \left(1 + \frac{\sqrt{\{(x + \eta\tan\phi)^2 + (y - \eta)^2\}}}{x + y\tan\phi} \right) d\eta$$
$$- \frac{1}{4\pi} \Gamma(0) \left\{ \frac{x\tan\phi + y}{(x - y\tan\phi)\sqrt{(x^2 + y^2)}} + \frac{x\tan\phi - y}{(x + y\tan\phi)\sqrt{(x^2 + y^2)}} \right\}.$$
$$(3·8, 37)$$

The last term of (3·8, 37) may be incorporated into either of the two other integral terms. If we chose the second, we obtain

$$
w(x, y, 0) = \frac{1}{4\pi} \int_{-s_0}^{0} \Gamma'(\eta) \left\{ \frac{1}{\eta - y} \left(1 + \frac{\sqrt{\{(x + \eta \tan \phi)^2 + (y - \eta)^2\}}}{x + y \tan \phi} \right) \right.
$$

$$
\left. - 2 \tan \phi \, \frac{\sqrt{(x^2 + y^2)}}{x^2 - y^2 \tan^2 \phi} \right\} d\eta
$$

$$
+ \frac{1}{4\pi} \int_{0}^{s_0} \frac{\Gamma'(\eta)}{\eta - y} \left\{ 1 + \frac{\sqrt{\{(x - \eta \tan \phi)^2 + (y - \eta)^2\}}}{x - y \tan \phi} \right\} d\eta. \qquad (3·8, 38)
$$

When the boundary condition (3·8, 25) is applied, (3·8, 38) gives an integral equation for $\Gamma'(y)$, which in Weissinger's method is satisfied just at the three-quarter chord points, i.e. in (3·8, 38) we put

$$
x = |y| \tan \phi + \tfrac{1}{2} c(y), \qquad (3·8, 39)
$$

where $c(y)$ is the local chord of the aerofoil. Then (3·8, 38) may be written (assuming $\Gamma'(\eta) = \Gamma'(-\eta)$)

$$
w(y) = -\frac{1}{2\pi} \int_{-s_0}^{s_0} \frac{\Gamma'(\eta)}{y - \eta} d\eta - \frac{1}{2\pi} \int_{-s_0}^{s_0} \frac{L(y, \eta)}{c(y)} \Gamma'(\eta) d\eta, \qquad (3·8, 40)
$$

where, for $y > 0$,

$$
L(y, \eta) = \frac{1}{2} \frac{c}{y - \eta} \left\{ \frac{\sqrt{\{(\tfrac{1}{2}c + (y + \eta) \tan \phi)^2 + (y - \eta)^2\}}}{2y \tan \phi + \tfrac{1}{2} c} - 1 \right\}
$$

$$
+ 2 \tan \phi \, \frac{\sqrt{\{(\tfrac{1}{2}c + y \tan \phi)^2 + y^2\}}}{2y \tan \phi + \tfrac{1}{2} c} \qquad (\eta < 0),
$$

and $L(y, \eta) = \dfrac{1}{y - \eta} \{ \sqrt{[(\tfrac{1}{2}c + (y - \eta) \tan \phi)^2 + (y - \eta)^2]} - \tfrac{1}{2}c \} \quad (\eta > 0),$

with similar expressions for $y < 0$.

Equation (3·8, 38) should be compared with the lifting-line equation (3·3, 17). Multhopp's method for solving the latter equation can be extended to solve (3·8, 38). Thus, in accordance with the analysis given in § 3·3, the first integral in (3·8, 40) is

$$
2U \left(b_{kk} \Gamma_k - \sum_{j=1}^{n} {}' b_{kj} \Gamma_j \right), \qquad (3·8, 41)
$$

where b_{kk} and b_{kj} are defined by (3·3, 62). The second integral in (3·8, 40) can be written in terms of $\theta = \cos^{-1} y/s_0$ and $\omega = \cos^{-1} \eta/s_0$ as

$$
\frac{1}{2\pi c(\theta)} \int_{0}^{\pi} L(\theta, \omega) \Gamma'(\omega) d\omega. \qquad (3·8, 42)
$$

From the expansion (3·3, 53), which we can rewrite as

$$
\Gamma(\omega) = \frac{4s_0 U}{n + 1} \sum_{j=1}^{n} \left(\sum_{k=1}^{n} \sin k\theta_j \sin k\omega \right) \Gamma_j,
$$

we find
$$\Gamma'(\omega) = \frac{4s_0 U}{n+1} \sum_{j=1}^{n} \left(\sum_{k=1}^{n} k \sin k\theta_j \cos k\omega \right) \Gamma_j$$

$$= 2s_0 U \sum_{k=1}^{n} \Gamma_j f_j(\omega), \tag{3·8, 43}$$

where
$$f_j(\omega) = \frac{2}{n+1} \sum_{j=1}^{n} k \cos k\theta_j \cos k\omega.$$

(3·8, 42) now becomes

$$\sum_{j=1}^{n} \frac{s_0 U}{\pi c(\theta)} \int_0^{\pi} L(\theta, \omega)\, \Gamma_j f_j(\omega)\, d\omega, \tag{3·8, 44}$$

which can be written approximately as

$$\frac{U s_0}{c(\theta)} \sum_{j=1}^{n} \Gamma_j \frac{1}{N+1} \left\{ \frac{L(\theta,0)f_j(0) + L(\theta,\omega_{N+1})f_j(\omega_{N+1})}{2} + \sum_{n=1}^{N} L(\theta,\omega_n)f_j(\omega_n) \right\}, \tag{3·8, 45}$$

the integral having now been replaced by a summation over $N+1$ intervals between $\omega = 0$ and $\omega = \pi$ determined by the values

$$\omega_n = \frac{n\pi}{N+1} \quad (n = 0, 1, ..., N+1).$$

Taking the argument θ in (3·8, 45) to refer specifically to the point $\theta_k = k\pi/(n+1)$, and putting

$$g_{kj} = \frac{1}{2(N+1)} \left\{ \frac{L(\theta_k,0)f_j(0) + L(\theta_k,\omega_{N+1})f_j(\omega_{N+1})}{2} + \sum_{n=1}^{N} L(\theta_k,\omega_n)f_j(\omega_n) \right\},$$

$$\tag{3·8, 46}$$

we obtain finally for (3·8, 40)

$$\frac{w(\theta_k)}{U} = 2 \left(b_{kk}\Gamma_k - \sum_{j=1}^{n}{}' b_{kj}\Gamma_j \right) + \lambda_k \sum_{j=1}^{n} g_{kj}\Gamma_j,$$

where
$$\lambda_k = \frac{2s_0}{c(\theta_k)}.$$

The equation can be written alternatively as

$$\frac{w(\theta_k)}{U} = b_k^*\Gamma_k - \sum_{j=1}^{n}{}' b_{kj}^*\Gamma_j, \tag{3·8, 47}$$

where
$$\begin{aligned} b_k^* &= 2b_{kk} + \lambda_k g_{kk} \\ b_{kj}^* &= 2b_{kj} - \lambda_k g_{kj} \end{aligned} \bigg\} \quad (k \neq j). \tag{3·8, 48}$$

The left-hand side of (3·8, 47) is known from the boundary conditions at the wing; viz. $w(\theta_k)/U$ is taken to be equal to the slope of the mean camber line of the wing at the three-quarter chord point. Hence (3·8, 47) is a set of n simultaneous equations for the spanwise circulation functions Γ_k.

Comparison of Weissinger's method with Falkner's when applied to specific cases shows that Weissinger's method involves much less computational work, and at the same time appears to yield satisfactory results. Thus, as indicated by the curves drawn in Fig. 3·8, 5, both theories are generally in good agreement with experiment.‡

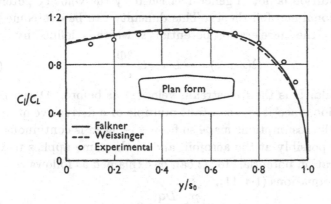

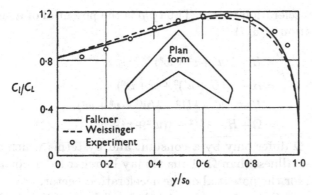

Fig. 3·8, 5. Spanwise distribution of the local lift coefficient C_l.

3·9   The acceleration potential, circular wings

The methods of source, doublet or vortex distributions described in the preceding sections involve a number of integrals whose evaluation may prove awkward, even where the choice of the distributions is to some extent arbitrary. This position compares unfavourably with two-dimensional theory (Chapter 2). The success of that theory depended to a large extent on the availability of particular solutions involving trigonometric functions, whose properties are thoroughly familiar. Now the methods of potential theory provide similar solutions of Laplace's equation in three dimensions, for surfaces which can be fitted into particular systems of orthogonal curvi-

‡ These results are taken from N. H. Van Dorn and J. DeYoung, 'A comparison of three theoretical methods of calculating span load distribution on swept wings', *Tech. Notes, Nat. Adv. Comm. Aero. Wash.*, no. 1476 (1947).

linear coordinates. However, the standard solutions of Laplace's equation in potential theory refer to functions which are continuous everywhere except at the surface (in our case the wing) in question, and, as we have seen, this condition is not in general satisfied by the velocity potential in three-dimensional aerofoil theory. This difficulty can be overcome by the introduction of the 'acceleration potential' Ω which is defined by

$$\Omega(x, y, z) = - Uu = U \frac{\partial \Phi}{\partial x}. \qquad (3·9, 1)$$

In this formula, U is the free-stream velocity, as before. Ω satisfies Laplace's equation since it is a constant multiple of a derivative of Φ. Also, according to the assumptions made so far $u = - \partial \Phi / \partial x$ is continuous everywhere except possibly at the aerofoil, and so the same applies to Ω. The name 'acceleration potential' for Ω can be explained as follows.

By Euler's equations (1·4, 11),

$$- \operatorname{grad} \frac{p}{\rho} = \frac{D\mathbf{q}}{Dt} = \mathbf{a},$$

where $\mathbf{a}$ is the acceleration vector. Thus p/ρ is the potential of $\mathbf{a}$. Now, by Bernoulli's equation (1·6, 7)

$$\begin{aligned}
\frac{p}{\rho} &= H - \tfrac{1}{2}(u^2 + v^2 + w^2) \\
&= H - \tfrac{1}{2}((U + u')^2 + v^2 + w^2) \\
&= H - Uu' - \tfrac{1}{2}U^2 - \tfrac{1}{2}(u'^2 + v^2 + w^2) \\
&= \Omega + H - \tfrac{1}{2}U^2 - \tfrac{1}{2}(u'^2 + v^2 + w^2),
\end{aligned}$$

so that p/ρ and Ω differ only by a constant and by terms which are of a higher order of smallness than Ω. Thus Ω may be regarded as an approximate expression for the potential of the acceleration vector.

Prandtl, who was the first to point out the usefulness of the acceleration potential,[‡] actually took $\Omega = p/\rho$, and then attempted to show that Ω satisfies Laplace's equation approximately. With this definition the condition that Ω be continuous throughout the fluid is satisfied exactly. Also, Prandtl's approach points the way to the corresponding definition for unsteady conditions.

We may also define the induced acceleration potential $\Omega'(x, y, z)$ by

$$\Omega = - U^2 + \Omega', \qquad (3·9, 2)$$

so that $\Omega' = U(\partial \Phi'/\partial x)$. Given the induced acceleration potential, we may regain the induced velocity potential by means of the formula

$$\Phi'(x, y, z) = \frac{1}{U} \int_{-\infty}^{x} \Omega'(\xi, y, z) \, d\xi. \qquad (3·9, 3)$$

‡ L. Prandtl, 'Beitrag zur Theorie der tragenden Fläche', *Z. angew. Math. Mech.* vol. 16 (1936), pp. 360–61; also, 'Theorie des Flugzeugtragflügels im zusammendrückbaren Medium', *Luftfahrtforsch.* vol. 13 (1936), pp. 313–19.

In this integral, the lower limit is chosen so as to make Φ' equal to zero in regions far upstream of the wing. Similarly, the vertical component of the velocity is given by

$$w = -\frac{\partial \Phi'}{\partial z} = -\frac{1}{U}\int_{-\infty}^{x} \frac{\partial \Omega'}{\partial z}\, d\xi. \qquad (3·9, 4)$$

This integral may diverge if the path of integration crosses the leading edge, since u', and hence Ω' and $\partial \Omega'/\partial z$, may become infinite there. Accordingly, if the value of w is required for points on the surface of the aerofoil ($z = 0$), (3·9, 4) should be evaluated first for $z \neq 0$. The desired value of w for $z = 0$ may then be obtained by a passage to the limit.

The concept of the acceleration potential has been applied by Krienes[‡] to the calculation of the flow round a thin elliptic wing at incidence. The particular case of a circular wing had been considered earlier by Kinner.[§] Subsequently, Kochin dealt with the same problem by means of a Green's function method.[||] Although circular aerofoils are only rarely met with in practice, their consideration enables us to elucidate various points (such as the conditions near a rounded tip) which are intractable by other methods. The analysis given in the present section may serve as an introduction to the methods of Kinner and Kochin, although it is developed along different lines. We shall require certain particular solutions of Laplace's equation which are known as spheroidal harmonics, and which involve Legendre functions of the first and second kind. The standard work on these functions is Hobson;[¶] there are relevant chapters in many other books on analysis or on the differential equations of Mathematical Physics.[‡‡]

Assume that the circular wing W of radius R is situated (approximately) in the x, y-plane with centre at the origin, while the free-stream velocity is parallel to the x-axis.

We introduce spheroidal coordinates by

$$x = R \sin \chi \cosh \eta \cos \zeta = R \sqrt{(1 - \mu^2)} \sqrt{(1 + \nu^2)} \cos \zeta,$$

$$y = R \sin \chi \cosh \eta \sin \zeta = R \sqrt{(1 - \mu^2)} \sqrt{(1 + \nu^2)} \sin \zeta,$$

$$z = R \cos \chi \sinh \eta \qquad = R \mu \nu,$$

where
$$\mu = \cos \chi, \quad \nu = \sinh \eta,$$

‡ K. Krienes, 'Die elliptische Tragfläche auf potentialtheoretischer Grundlage', *Z. angew. Math. Mech.* vol. 20 (1940), pp. 65–88. Translated in *Tech. Memor. Nat. Adv. Comm. Aero., Wash.*, no. 971.

§ W. Kinner, 'Die Kreisförmige Tragfläche auf potentialtheoretischer Grundlage', *Ingen.-Arch.* vol. 8 (1937), p. 47.

|| N. E. Kochin, 'Theory of wing of circular plan form', *Tech. Memor. Nat. Adv. Comm. Aero., Wash.*, no. 1324. Translated from *Prikladnaya Matematika i Mekhanika*, vol. 4 (1940), pp. 3–32.

¶ E. W. Hobson, *Theory of Spherical and Ellipsoidal Harmonics* (Cambridge University Press, 1931).

‡‡ For example, E. T. Whittaker and G. N. Watson, *A Course of Modern Analysis* (4th ed., Cambridge University Press, 1927).

with the limits of variation

$$0 \leqslant \chi \leqslant \pi, \quad 0 \leqslant \eta < \infty, \quad 0 \leqslant \zeta < 2\pi, \quad 1 \geqslant \mu \geqslant -1, \quad 0 \leqslant \nu < \infty.$$

(χ, η, ζ) and (μ, ν, ζ) are orthogonal coordinates for the (x, y, z) space. The surfaces $\eta = $ constant (or $\nu = $ constant) are ellipsoids of revolution about the z-axis which degenerate into the two-sided circular disk

$$x^2 + y^2 \leqslant R^2,$$

$z = 0$ (i.e. the wing W) for $\eta = 0$. The surfaces $\chi = $ constant (or $\mu = $ constant) are one-sheeted hyperboloids of revolution about the z-axis which degenerate into the area $x^2 + y^2 \geqslant 0$, $z = 0$ for $\chi = \pi$. As χ, η and ζ vary within the specified intervals, all the points of x, y, z-space are obtained just once with the exception of the points which belong to the x, y-plane, or to the z-axis.

The direct transformation of Laplace's equation into arbitrary orthogonal curvilinear coordinates is laborious, but standard methods (given in the above mentioned books) show that for any scalar ϕ,

$$\frac{\partial^2 \phi}{\partial x^2} + \frac{\partial^2 \phi}{\partial y^2} + \frac{\partial^2 \phi}{\partial z^2} = \Delta \phi$$

$$= \frac{1}{h_1 h_2 h_3} \left[\frac{\partial}{\partial \mu} \left(\frac{h_2 h_3}{h_1} \frac{\partial \phi}{\partial \mu} \right) + \frac{\partial}{\partial \nu} \left(\frac{h_3 h_1}{h_2} \frac{\partial \phi}{\partial \nu} \right) + \frac{\partial}{\partial \zeta} \left(\frac{h_1 h_2}{h_3} \frac{\partial \phi}{\partial \zeta} \right) \right],$$

where

$$h_1 = \sqrt{\left\{ \left(\frac{\partial x}{\partial \mu} \right)^2 + \left(\frac{\partial y}{\partial \mu} \right)^2 + \left(\frac{\partial z}{\partial \mu} \right)^2 \right\}}, \quad h_2 = \sqrt{\left\{ \left(\frac{\partial x}{\partial \nu} \right)^2 + \left(\frac{\partial y}{\partial \nu} \right)^2 + \left(\frac{\partial z}{\partial \nu} \right)^2 \right\}},$$

$$h_3 = \sqrt{\left\{ \left(\frac{\partial x}{\partial \zeta} \right)^2 + \left(\frac{\partial y}{\partial \zeta} \right)^2 + \left(\frac{\partial z}{\partial \zeta} \right)^2 \right\}},$$

the square roots being taken with positive sign. Straightforward calculation shows that in the present case more particularly

$$h_1^2 = R^2 \left(\mu^2 \frac{1+\nu^2}{1-\mu^2} \cos^2 \zeta + \mu^2 \frac{1+\nu^2}{1-\mu^2} \sin^2 \zeta + \nu^2 \right)$$

$$= \frac{R^2}{1-\mu^2} \left(\mu^2 (1+\nu^2) + \nu^2 (1-\mu^2) \right)$$

$$= R^2 \frac{\mu^2 + \nu^2}{1-\mu^2},$$

$$h_2^2 = R^2 \left(\nu^2 \frac{1-\mu^2}{1+\nu^2} \cos^2 \zeta + \nu^2 \frac{1-\mu^2}{1+\nu^2} \sin^2 \zeta + \mu^2 \right)$$

$$= R^2 \frac{\mu^2 + \nu^2}{1+\nu^2},$$

$$h_3^2 = R^2 ((1-\mu^2)(1+\nu^2) \sin^2 \zeta + (1-\mu^2)(1+\nu^2) \cos^2 \zeta)$$

$$= R^2 (1-\mu^2)(1+\nu^2).$$

Hence

$$\Delta\phi = \frac{1}{R^2(\mu^2+\nu^2)}\left[\frac{\partial}{\partial\mu}\left((1-\mu^2)\frac{\partial\phi}{\partial\mu}\right) + \frac{\partial}{\partial\nu}\left((1+\nu^2)\frac{\partial\phi}{\partial\nu}\right)\right.$$
$$\left. + \frac{\partial}{\partial\zeta}\left(\frac{\mu^2+\nu^2}{(1-\mu^2)(1+\nu^2)}\frac{\partial\phi}{\partial\zeta}\right)\right],$$

and Laplace's equation becomes

$$\frac{\partial}{\partial\mu}\left((1-\mu^2)\frac{\partial\phi}{\partial\mu}\right) + \frac{\partial}{\partial\nu}\left((1+\nu^2)\frac{\partial\phi}{\partial\nu}\right) + \left(\frac{1}{1-\mu^2}-\frac{1}{1+\nu^2}\right)\frac{\partial^2\phi}{\partial\zeta^2} = 0. \quad (3\cdot9, 5)$$

Normal solutions of $(3\cdot9, 5)$—solutions of the form $\phi = F(\mu)G(\nu)H(\zeta)$—which are bounded at infinity are found by the method of separation of coordinates. This leads to the equations

$$\frac{d}{d\mu}\left((1-\mu^2)\frac{dF}{d\mu}\right) + \left(n(n+1)-\frac{m^2}{1-\mu^2}\right)F = 0, \quad (3\cdot9, 6)$$

$$\frac{d}{d\nu}\left((1+\nu^2)\frac{dG}{d\nu}\right) - \left(n(n+1)-\frac{m^2}{1+\nu^2}\right)G = 0, \quad (3\cdot9, 7)$$

$$\frac{d^2H}{d\zeta^2} + m^2H = 0. \quad (3\cdot9, 8)$$

If $(3\cdot9, 6)$ is Legendre's equation for $m = 0$, or one of the associated equations, for $m = 1, 2, \ldots$. We obtain the same set of equations if we make the transformation $\nu' = i\nu$ in $(3\cdot9, 7)$. Finally, independent solutions of $(3\cdot9, 8)$ are $\cos m\zeta$ and $\sin m\zeta$. Thus, the following are normal solutions of $(3\cdot9, 5)$ which are bounded at infinity:

$$\phi = i^{n-m+1}P_n^m(\mu)Q_n^m(i\nu)\begin{cases}\cos m\zeta \\ \sin m\zeta\end{cases} \quad (n = 0, 1, \ldots; \; m = 0, 1, \ldots, n), \quad (3\cdot9, 9)$$

where the coefficient i^{n-m+1} is introduced in order to make ϕ real.

For
$$m = 0, \quad P_n^0(\mu) = P_n(\mu) \quad \text{and} \quad Q_n^0(\mu) = Q_n(\mu)$$

are the Legendre functions of the first and second kind respectively. The corresponding functions for $m \geqslant 1$ are the associated Legendre functions.

To calculate the gradient of ϕ, we require the partial derivatives of μ, ν, ζ with respect to the space coordinates x, y, z. They are

$$\frac{\partial\mu}{\partial x} = -\frac{\mu\sqrt{(1-\mu^2)}\sqrt{(1+\nu^2)}}{R(\mu^2+\nu^2)}\cos\zeta, \quad \frac{\partial\mu}{\partial y} = -\frac{\mu\sqrt{(1-\mu^2)}\sqrt{(1+\nu^2)}}{R(\mu^2+\nu^2)}\sin\zeta,$$

$$\frac{\partial\mu}{\partial z} = \frac{\nu(1-\mu^2)}{R(\mu^2+\nu^2)},$$

$$\frac{\partial\nu}{\partial x} = \frac{\nu\sqrt{(1-\mu^2)}\sqrt{(1+\nu^2)}}{R(\mu^2+\nu^2)}\cos\zeta, \quad \frac{\partial\nu}{\partial y} = \frac{\nu\sqrt{(1-\mu^2)}\sqrt{(1+\nu^2)}}{R(\mu^2+\nu^2)}\sin\zeta,$$

$$\frac{\partial\nu}{\partial z} = \frac{\mu(1+\nu^2)}{R(\mu^2+\nu^2)},$$

$$\frac{\partial\zeta}{\partial x} = -\frac{\sin\zeta}{R\sqrt{(1-\mu^2)}\sqrt{(1+\nu^2)}}, \quad \frac{\partial\zeta}{\partial y} = \frac{\cos\zeta}{R\sqrt{(1-\mu^2)}\sqrt{(1+\nu^2)}}, \quad \frac{\partial\zeta}{\partial z} = 0.$$

$$(3\cdot9, 10)$$

Hence, at the wing W,

$$\frac{\partial \phi}{\partial z} = \frac{\partial \phi}{\partial \mu} \frac{\partial \mu}{\partial z} + \frac{\partial \phi}{\partial \nu} \frac{\partial \nu}{\partial z} + \frac{\partial \phi}{\partial \zeta} \frac{\partial \zeta}{\partial z}$$

$$= \frac{1}{R\mu} \left(\frac{\partial \phi}{\partial \nu} \right)_{\nu = 0}$$

$$= -\frac{i^{n-m}}{R\mu} P_n^m(\mu) Q_n'^m(0) \begin{cases} \cos m\zeta, \\ \sin m\zeta, \end{cases} \qquad (3\cdot9, 11)$$

where $Q_n'^m$ is the derivative of Q_n^m with respect to its argument $\nu' = i\nu$.

The upper surface of the wing corresponds to $\mu \geqslant 0$, $0 \leqslant \chi \leqslant \frac{1}{2}\pi$ and the lower surface to $\mu \leqslant 0$, $\frac{1}{2}\pi \leqslant \chi \leqslant \pi$. Points of the wing which are given by μ_1, μ_2 or χ_1, χ_2 coincide if $\mu_1 = -\mu_2$, i.e. $\theta_1 + \theta_2 = \pi$. Thus, for any point of the wing, $\partial \phi / \partial z$ is the same on either side if $P_n^m(\mu)$ is odd, and is numerically the same, but of different sign if $P_n^m(\mu)$ is even. These two cases correspond to $n - m$ odd, and $n - m$ even, respectively.

If ϕ stands for an induced velocity potential ($\phi = \Phi'$) then the boundary condition at the wing is

$$w = -\frac{\partial \phi}{\partial z} = Us(x, y), \qquad (3\cdot9, 12)$$

where $s(x, y)$ is the local slope. Thus, if $\phi = \Phi'$ is given by (3·9, 9) or by a linear combination of such functions, we may investigate the circular wings whose slopes are, except for a constant of proportionality, given by (3·9, 11). In view of the fact that μ appears in the denominator of (3·9, 11), the slope may become infinite at the edge of the wing. Thus, if we consider functions for which $n - m$ is odd, or linear combinations of such functions, then we shall find that the corresponding wings are (infinitely) thin and warped in both directions. However, the specified velocity potentials do not represent the physically correct solution for the flow round these wings, since the Joukowski condition is not, in general, satisfied at the trailing edge. We may verify this by calculating $\partial \phi / \partial x$ at the aerofoil, but it is also indicated by the fact that the velocity potential is continuous across the wake, so that the spanwise lift distribution $l(y)$—which is proportional to the discontinuity of the velocity potential across the wake—must vanish identically across the span.

Let us now assume that the induced *acceleration potential* Ω' is given by a linear combination of functions (3·9, 9), $n - m$ odd. The corresponding wing is again infinitely thin and its local slope function is given by (3·9, 4) and (3·9, 12). The pressure difference between the upper and lower surface of the wing is proportional to $\partial \Phi' / \partial x$, and hence to the induced acceleration potential. But at the edge of the aerofoil $\mu = 0$, and $P_n^m(0) = 0$ for $n - m$ odd. It follows that for these solutions the pressure difference vanishes not only at the trailing edge, as we may expect in accordance with the Joukowski condition, but also at the leading edge. On the other hand, we know from two-dimensional theory, as well as from three-dimensional theory as developed so far, that in the general case we must admit the possibility that the pressure difference becomes infinite at the leading edge. Accordingly, we are impelled to introduce further particular solutions which satisfy this condition. They correspond roughly to the $\tan \frac{1}{2}\theta$ term of two-dimensional thin aerofoil theory. Kinner‡ derives such solutions by a limiting process from the normal solutions, but once these functions are obtained it is not difficult to verify directly that they do indeed satisfy Laplace's equation. They are

$$\phi = \frac{\mu}{\mu^2 + \nu^2}\left(\frac{1-\mu^2}{1+\nu^2}\right)^{\frac{1}{2}n}\begin{Bmatrix}\cos n\zeta \\ \sin n\zeta\end{Bmatrix}\quad (n = 0, 1, 2, \ldots). \qquad (3\cdot9, 13)$$

These functions become infinite along the edge of the wing, where $\mu = \nu = 0$.
Let

$$r = \sqrt{\frac{1-\mu^2}{1+\nu^2}}, \quad (r \geqslant 0),$$

and

$$\tau = r(\cos \zeta + i \sin \zeta) = r\, e^{i\zeta}. \qquad (3\cdot9, 14)$$

Then r varies in the range $\langle 0, 1\rangle$ for the specified ranges of μ and ν, and τ varies in the unit circle of the complex plane, $|\tau| \leqslant 1$, as μ, ν, ζ take all admissible values. To each point in the unit circle there correspond an infinite number of points (μ, ν, ζ) or (x, y, z). At the wing

$$\nu = 0, \quad x = R\sqrt{(1-\mu^2)}\cos \zeta, \quad y = R\sqrt{(1-\mu^2)}\sin \zeta, \qquad (3\cdot9, 15)$$

and so $\quad \tau = r(\cos \zeta + i \sin \zeta) = \sqrt{(1-\mu^2)}\,(\cos \zeta + i \sin \zeta) = \dfrac{1}{R}(x + iy).$

Also, for such points, $\quad \mu = \dfrac{1}{R}\sqrt{(R^2 - x^2 - y^2)}.$

The solutions $(3\cdot9, 13)$ may now be written in the form

$$\phi = \frac{\mu}{\mu^2 + \nu^2}\mathscr{R}(\tau^n) \quad \text{or} \quad \phi = \frac{\mu}{\mu^2 + \nu^2}\mathscr{I}(\tau^n), \qquad (3\cdot9, 16)$$

where $\mathscr{R}$ and $\mathscr{I}$ indicated the real and imaginary part of a complex number, as usual, $\tau = \mathscr{R}(\tau) + i\mathscr{I}(\tau)$. It follows that for every complex function $f(\tau)$ which is analytic in the interior of the unit circle, and which therefore can be written as a power series $\Sigma c_n \tau^n$, the functions

$$\phi = \frac{\mu}{\mu^2 + \nu^2}\mathscr{R}(f(\tau)) \quad \text{and} \quad \phi = \frac{\mu}{\mu^2 + \nu^2}\mathscr{I}(f(\tau)) \qquad (3\cdot9, 17)$$

are solutions of Laplace's equation.

For given y, $-R < y < R$, the spanwise lift distribution is given by

$$l(y) = -2\rho U \int_{-\sqrt{(R^2-y^2)}}^{\sqrt{(R^2-y^2)}} \frac{\partial \Phi'}{\partial x}\,dx = -2\rho \int_{-\sqrt{(R^2-y^2)}}^{\sqrt{(R^2-y^2)}} \Omega'\,dx,$$

‡ W. Kinner, 'Die kreisförmige Tragfläche auf potentialtheoretischer Grundlage,' *Ingen.-Arch.* vol. 8 (1937), p. 47.

where the integral is then over the upper surface of the wing. Thus, if Ω' is given by the first formula in (3·9, 17), the corresponding lift distribution is

$$l(y) = -2\rho \int_{-\sqrt{(R^2-y^2)}}^{\sqrt{(R^2-y^2)}} \frac{1}{\mu} \mathscr{R}(f(\tau))\, dx \quad (\mu \geqslant 0). \tag{3·9, 18}$$

Now $x = R\sqrt{(1-\mu^2)}\cos\zeta = R\tau - iy$ at the wing so that $dx = R\,d\tau$. Also, the limits of integration, $x = -\sqrt{(R^2-y^2)}$ and $x = \sqrt{(R^2-y^2)}$, correspond to $\tau = \dfrac{1}{R}\{iy - \sqrt{(R^2-y^2)}\}$ and $\tau = \dfrac{1}{R}\{iy + \sqrt{(R^2-y^2)}\}$, and, for all points of the wing

$$x^2 + y^2 = R^2(1-\mu^2),$$

i.e.

$$(R\tau - iy)^2 + y^2 = R^2(1-\mu^2),$$

$$\mu = \sqrt{(1 + 2iY\tau - \tau^2)},$$

where $Y = y/R$. Hence

$$l(y) = -2\rho R\mathscr{R}\int_{iY-\sqrt{(1-Y^2)}}^{iY+\sqrt{(1-Y^2)}} \frac{f(\tau)}{\sqrt{(1 + 2iY\tau - \tau^2)}}\, d\tau,$$

or

$$l(y) = -2\rho R\mathscr{I}\int_{Y-i\sqrt{(1-Y^2)}}^{Y+i\sqrt{(1-Y^2)}} \frac{f(i\omega)}{\sqrt{(1 - 2Y\omega + \omega^2)}}\, d\omega, \tag{3·9, 19}$$

where $\tau = i\omega$.

For the particular case $f(\tau) = \tau^n$, these integrals can be expressed in terms of Legendre polynomials in view of the standard formula

$$P_n(Y) = \frac{1}{\pi i}\int_{Y-i\sqrt{(1-Y^2)}}^{Y+i\sqrt{(1-Y^2)}} \frac{\omega^n\, d\omega}{\sqrt{(1 - 2\omega Y + \omega^2)}}. \tag{3·9, 20}$$

Thus, if

$$\Omega' = \frac{\mu}{\mu^2 + \nu^2}\mathscr{R}(\tau^{2k}) = \frac{\mu}{\mu^2 + \nu^2}\left(\frac{1-\mu^2}{1+\nu^2}\right)^k \cos 2k\zeta \quad (k = 0, 1, 2, \ldots), \tag{3·9, 21}$$

then

$$l(y) = 2\rho R\mathscr{I}((-1)^{k+1} i\pi P_{2k}(Y))$$

$$= 2\pi\rho R(-1)^{k+1} P_{2k}(Y), \tag{3·9, 22}$$

while for

$$\Omega' = \frac{\mu}{\mu^2 + \nu^2}\mathscr{R}(\tau^{2k+1}) = \frac{\mu}{\mu^2 + \nu^2}\left(\frac{1-\mu^2}{1+\nu^2}\right)^{k+\frac{1}{2}} \cos(2k+1)\zeta \quad (k = 0, 1, 2, \ldots), \tag{3·9, 23}$$

the corresponding lift distribution vanishes identically,

$$l(y) = 0. \tag{3·9, 24}$$

Using (3·9, 20), we find that (3·9, 24) also holds if

$$\Omega' = \frac{\mu}{\mu^2 + \nu^2}\mathscr{R}(-i\tau^{2k}) = \frac{\mu}{\mu^2 + \nu^2}\left(\frac{1-\mu^2}{1+\nu^2}\right)^k \sin 2k\zeta \quad (k = 0, 1, 2, \ldots), \tag{3·9, 25}$$

while for

$$\Omega' = \frac{\mu}{\mu^2 + \nu^2}\mathscr{R}(-i\tau^{2k+1}) = \frac{\mu}{\mu^2 + \nu^2}\left(\frac{1-\mu^2}{1+\nu^2}\right)^{k+\frac{1}{2}} \sin(2k+1)\zeta \quad (k = 0, 1, 2, \ldots), \tag{3·9, 26}$$

the spanwise lift distribution is given by

$$l(y) = 2\pi\rho R(-1)^{k+1} P_{2k+1}(Y). \tag{3·9, 27}$$

Reverting to the more general case

$$\Omega' = \frac{\mu}{\mu^2 + \nu^2} \mathscr{R}(f(\tau)), \tag{3·9, 28}$$

we have at the wing

$$\Omega' = \frac{1}{\mu} \mathscr{R}(f(\tau)) = \frac{R\mathscr{R}(f(\tau))}{\sqrt{(R^2 - x^2 - y^2)}}. \tag{3·9, 29}$$

In order that Ω' and therefore $\partial\Phi'/\partial x$ remain finite along the trailing edge, in accordance with the Joukowski condition, we therefore require that

$$\mathscr{R}(f(\tau)) = 0 \quad \text{for} \quad |\tau| = 1, \quad -\tfrac{1}{2}\pi < \arg\tau < \tfrac{1}{2}\pi. \tag{3·9, 30}$$

Let $F(s)$ be any continuous real-valued function which is defined in the closed interval $\langle -1, 1 \rangle$. We consider the induced acceleration potential (3·9, 29) for which $f(\tau)$ is defined by

$$f(\tau) = \frac{1}{\pi} \int_{-1}^{1} F(s) \left\{ \frac{\tau^{-1}}{1 + \tau^{-2}s^2} - \frac{\tau}{1 + \tau^2 s^2} \right\} ds \tag{3·9, 31}$$

for τ in the positive half of the unit circle, $\mathscr{R}(\tau) > 0$. For given τ, the singularities of the kernel of this integral,

$$\frac{\tau^{-1}}{1 + \tau^{-2}s^2} - \frac{\tau}{1 + \tau^2 s^2},$$

are given by $\tau^2 + s^2 = 0$, and $1 + \tau^2 s^2 = 0$, i.e. by

$$s = \pm\tau, \quad s = \pm\tau^{-1}. \tag{3·9, 32}$$

Thus the integral exists for $\mathscr{R}(\tau) > 0$. Now for $\tau = e^{i\zeta} (-\tfrac{1}{2}\pi < \zeta < \tfrac{1}{2}\pi)$, the kernel equals

$$\frac{e^{-i\zeta}}{1 + s^2 e^{-2i\zeta}} - \frac{e^{i\zeta}}{1 + s^2 e^{2i\zeta}} = 2i \frac{(s^2 - 1)\sin\zeta}{1 + 2s^2 \cos\zeta + s^4}.$$

That is to say, the kernel, and hence $f(\tau)$, are pure imaginary for the points of the trailing edge. At the same time the denominator of (3·9, 29) becomes infinite at the edge of the wing as

$$\frac{1}{\sqrt{(R^2 - x^2 - y^2)}} = \frac{1}{R\sqrt{(1 - |\tau|^2)}}.$$

But $f(\tau)$ is regular at the trailing edge and so we may conclude that Ω' also vanishes at the trailing edge, in accordance with the Joukowski condition. The argument does not apply to the leading edge since in general $f(\tau)$ is not represented by (3·9, 31) for $\mathscr{R}(\tau) < 0$.

Assume in particular that

$$F(s) = (-s^2)^k \quad (k = 0, 1, 2, \ldots). \tag{3·9, 33}$$

The following expansion is valid for $-1 \leqslant s \leqslant 1$, $|\tau| < 1$:

$$\frac{\tau}{1+s^2\tau^2} = \tau(1 - (s\tau)^2 + (s\tau)^4 - (s\tau)^6 + \ldots).$$

Hence

$$\int_{-1}^{1} s^{2k} \frac{\tau\, ds}{1+s^2\tau^2} = \sum_{\lambda=0}^{\infty} \int_{-1}^{1} (-1)^\lambda \tau^{2\lambda+1} s^{2(k+\lambda)}\, ds$$

$$= 2 \sum_{\lambda=0}^{\infty} \frac{(-1)^\lambda}{2\lambda+2k+1} \tau^{2\lambda+1}. \qquad (3\cdot9, 34)$$

Next, we shall require the value of the integral

$$I = \int_{-1}^{1} s^{2k} \frac{\tau}{\tau^2+s^2}\, ds \quad \text{for} \quad \mathscr{R}(\tau) > 0, \quad |\tau| < 1,$$

where the path of integration is the real interval $\langle -1, 1 \rangle$. We deform this interval into a path C which consists of segments of the real axis and of a large semicircle in the lower half of the complex s-plane (Fig. 3·9, 1).

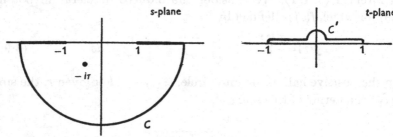

Fig. 3·9, 1

Now the integral of I has a pole at the point $s = -i\tau$ which is in the interior of the region bounded by the two paths. The residue of the integral at that point is

$$r_0 = -\frac{(-1)^k}{2i} \tau^{2k}.$$

Hence

$$I = \int_C s^{2k} \frac{\tau}{\tau^2+s^2}\, ds - 2\pi i r_0$$

$$= \int_C s^{2k} \frac{\tau}{\tau^2+s^2}\, ds + \pi(-1)^k \tau^{2k}$$

$$= -\int_{C'} t^{-2k} \frac{\tau}{1+t^2\tau^2}\, dt + \pi(-1)^k \tau^{2k},$$

where we have put

$$s = \frac{1}{t}, \quad ds = -\frac{1}{t^2}\, dt,$$

and the path of integration C' now consists of segments of the interval $\langle -1, 1 \rangle$ and of a small semicircle round the origin in the upper half of the complex t-plane (Fig. 3·9, 1). Expanding the integrand as before, we obtain

$$I = -2 \sum_{\lambda=0}^{\infty} \frac{(-1)^\lambda}{2\lambda-2k+1} \tau^{2\lambda+1} + \pi(-1)^k \tau^{2k}. \qquad (3\cdot9, 35)$$

Combining (3·9, 34) and (3·9, 35), we see that (3·9, 31) becomes, for the particular choice of $F(s)$ as given by (3·9, 33),

$$f(\tau) = \tau^{2k} - \frac{2}{\pi}(-1)^k \sum_{\lambda=0}^{\infty} (-1)^\lambda \left(\frac{1}{2\lambda + 2k + 1} + \frac{1}{2\lambda - 2k + 1} \right) \tau^{2\lambda+1}.$$

$$(3·9, 36)$$

The corresponding spanwise lift distribution function is, by (3·9, 22) and (3·9, 24),
$$l(y) = 2\pi\rho R(-1)^{k+1} P_{2k}(Y). \qquad (3·9, 37)$$

We wish to express $f(\tau)$ directly in terms of the corresponding $l(y)$. For this purpose we first represent s^n as an integral transform of $P_n(Y)$. Introducing the generating function for the Legendre polynomials,

$$\frac{1}{\sqrt{(1 - 2Ys + s^2)}} = \sum_{n=0}^{\infty} s^n P_n(Y),$$

we obtain

$$\frac{1-s^2}{2(1-2Ys+s^2)^{\frac{3}{2}}} = \frac{d}{ds} \frac{s}{\sqrt{(1-2Ys+s^2)}} - \frac{1}{2\sqrt{(1-2Ys+s^2)}} = \sum_{n=0}^{\infty} \frac{2n+1}{2} s^n P_n(Y).$$

$$(3·9, 38)$$

We integrate with respect to Y and use the relations of orthogonality for Legendre polynomials

$$\int_{-1}^{1} P_n(Y) P_m(Y) dY = \begin{cases} 0 & (m \neq n), \\ \dfrac{2}{2n+1} & (m = n). \end{cases} \qquad (3·9, 39)$$

Then
$$\int_{-1}^{1} P_n(Y) \frac{(1-s^2) dY}{2(1-2Ys+s^2)^{\frac{3}{2}}} = s^n \quad (n = 0, 1, 2, \ldots). \qquad (3·9, 40)$$

Hence $F(s)$, as given by (3·9, 33), is represented in terms of $l(y)$, as given by (3·9, 37), by the formula
$$F(s) = -\frac{1}{2\pi\rho R} \int_{-1}^{1} l(RY) \frac{(1-s^2) dY}{2(1-2Ys+s^2)^{\frac{3}{2}}}. \qquad (3·9, 41)$$

More generally, if $l(y) = l(RY)$ is a sufficiently regular (e.g. sectionally differentiable) even function of Y, then it can be expanded into a Legendre series involving the even Legendre polynomials, and so the corresponding $F(s)$ is still given by (3·9, 41). Thus the formulae (3·9, 41) and (3·9, 31) together enable us to express $f(\tau)$ and hence the induced acceleration potential directly in terms of the spanwise lift distribution, provided we know that Ω' can indeed be represented in the form (3·9, 28). It is evident that (3·9, 31) and (3·9, 41) can be combined into a single formula.

Next we shall determine the z-component of the velocity which corresponds to an Ω' as given by (3·9, 28). For this purpose, we first calculate $d\Omega'/dz$. Now

$$\frac{\partial}{\partial z} = \frac{\partial \mu}{\partial z} \frac{\partial}{\partial \mu} + \frac{\partial \nu}{\partial z} \frac{\partial}{\partial \nu} = \frac{1}{R(\mu^2 + \nu^2)} \left(\nu(1-\mu^2) \frac{\partial}{\partial \mu} + \mu(1+\nu^2) \frac{\partial}{\partial \nu} \right), \qquad (3·9, 42)$$

by (3·9, 10), and

$$\frac{\partial}{\partial \mu}\left(\frac{\mu}{\mu^2+\nu^2}\right)=\frac{\nu^2-\mu^2}{(\mu^2+\nu^2)^2}, \quad \frac{\partial}{\partial \nu}\left(\frac{\mu}{\mu^2+\nu^2}\right)=-\frac{2\mu\nu}{(\mu^2+\nu^2)^2},$$

$$\frac{\partial \tau}{\partial \mu}=-\frac{\mu}{1-\mu^2}\tau, \quad \frac{\partial \tau}{\partial \nu}=-\frac{\nu}{1+\nu^2}\tau.$$

Hence

$$\frac{d\Omega'}{\partial z}=\frac{\nu}{R(\mu^2+\nu^2)^2}\mathscr{R}\left\{\frac{(1-\mu^2)(\nu^2-\mu^2)-2\mu^2(1+\nu^2)}{(\mu^2+\nu^2)}f(\tau)-2\mu^2\tau f'(\tau)\right\}.$$

In particular, at the wing $\nu=0$, and so
$$\tag{3·9, 43}$$

$$\frac{d\Omega'}{dz}=0. \tag{3·9, 44}$$

The vertical velocity at any point is then given by (3·9, 4), and, more generally, the difference between the vertical velocities at two points (x_1, y, z) and (x_2, y, z) is

$$w(x_2, y, z)-w(x_1, y, z)=-\frac{1}{U}\int_{x_1}^{x_2}\frac{d\Omega'}{dz}dx. \tag{3·9, 45}$$

It follows that at the wing, where (3·9, 44) holds, the vertical velocity, and hence the slope, are constant in the direction of flow (i.e. along the chords). Thus the wings which are given by (3·9, 38) are uncambered but in general twisted, $s(x, y)=S(y)$, say. For given $f(\tau)$, $S(y)$ can be found either by determining w first for $z \neq 0$ and then proceeding to the limit, or by a complex variable method of the type used above for the calculation of the lift. The result is‡

$$w=\frac{(-1)^k}{U}\frac{dQ_{2k}}{dY} \quad \text{for} \quad f(\tau)=\tau^{2k}, \quad k=0, 1, 2, \ldots,$$

and

$$w=\frac{(-1)^{k+1}}{U}\frac{\pi}{2}\frac{dP_{2k+1}}{dY} \quad \text{for} \quad f(\tau)=\tau^{2k+1}, \quad k=0, 1, 2, \ldots,$$

where $Y=y/R$ as before. The corresponding slope functions $S(y)$ are

$$S(y)=\begin{cases}\dfrac{(-1)^k}{U^2}\dfrac{dQ_{2k}}{dY} & \text{for} \quad f(\tau)=\tau^{2k} \\[2mm] \dfrac{(-1)^{k+1}}{U^2}\dfrac{\pi}{2}\dfrac{dP_{2k+1}}{dY} & \text{for} \quad f(\tau)=\tau^{2k+1}\end{cases} \quad (k=0, 1, 2, \ldots). \tag{3·9, 46}$$

It follows that if $f(\tau)$ is given by (3·9, 36), then the corresponding slope function is

$$S(y)=\frac{(-1)^k}{U^2}\frac{d}{dY}\left[Q_{2k}(Y)+\sum_{\lambda=0}^{\infty}\left(\frac{1}{2\lambda+2k+1}+\frac{1}{2\lambda-2k+1}\right)P_{2\lambda+1}(Y)\right]. \tag{3·9, 47}$$

‡ See W. Kinner, 'Die Kreisförmige Tragfläche auf potentialtheoretischer Grundlage', *Ingen.-Arch.* vol. 8 (1937), p. 62.

We shall express $S(y)$ also as an integral transform of $F(s)$. For this purpose, we multiply the generating function

$$\frac{1}{\sqrt{(1-2sY+s^2)}} = \sum_{n=0}^{\infty} s^n P_n(Y)$$

by s^{2k-1} ($k=1, 2, ...$) and integrate with respect to s between the limits -1 and 1. Formal term-by-term integration yields

$$\int_{-1}^{1} \frac{s^{2k-1}}{\sqrt{(1-2sY+s^2)}}\, ds = 2 \sum_{\lambda=0}^{\infty} \frac{P_{2\lambda+1}(Y)}{2\lambda+2k+1}, \tag{3·9, 48}$$

and this result can also be justified more rigorously. Similarly, for $k=1, 2, 3, ...$,

$$\int_{-1}^{1} \frac{t^{-2k-1}}{\sqrt{(1-2tY+t^2)}}\, dt = 2 \sum_{\lambda=0}^{\infty} \frac{P_{2\lambda+1}(Y)}{2\lambda-2k+1} - i\pi P_{2k}(Y), \tag{3·9, 49}$$

where the integral on the left-hand side is taken along the real axis except for a small semicircle round the origin in the upper half of the complex t-plane (see Fig. 3·9, 1). The last term on the right-hand side of (3·9, 49) is, except for a factor of $-i\pi$, the residue of the integrand at the origin.

Substitute $s=1/t$. Then the integral on the left-hand side of (3·9, 49) becomes $-\int_{-1}^{1} \frac{s^{2k}\, ds}{\sqrt{(1-2sY+s^2)}}$, where the path of integration C is as in Fig. 3·9, 1, above. Some reflexion shows that the sign of the root must be positive near $s=1$. The integrand has a branch point at

$$s = s_0 = Y - i\sqrt{(1-Y^2)},$$

and this is in the interior of the region bounded by the path of integration C together with the real interval $\langle -1, 1 \rangle$.

We now deform C into a path C' which consists of the following three parts (Fig. 3·9, 2): (i) the real interval $\langle -1, 0 \rangle$; (ii) a loop which starts at the origin and, while varying in clockwise direction, encircles the singular point s_0 and then returns to the origin; (iii) the real interval $\langle 0, 1 \rangle$. Denote the

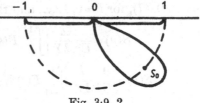

Fig. 3·9, 2

three corresponding integrals by I_1, I_2, I_3, so that

$$I_1 + I_2 + I_3 = \int_C \frac{s^{2k}\, ds}{\sqrt{(1-2sY+s^2)}}.$$

Then $\quad I_1 = -\int_{-1}^{0} \frac{s^{2k}\, ds}{\sqrt{(1-2sY+s^2)}}, \quad I_3 = \int_{0}^{1} \frac{s^{2k}\, ds}{\sqrt{(1-2sY+s^2)}}. \tag{3·9, 50}$

The minus sign in the formula for I_1 is explained by the fact that, beginning at $s=1$ where the square root is positive, we encircle s_0 before reaching the path of integration of I_1. Again, by a standard formula for Legendre functions of the second kind,‡

$$I_2 = -2(Q_{2k}(Y) - \tfrac{1}{2}i\pi P_{2k}(Y)). \tag{3·9, 51}$$

Substituting $-(I_1 + I_2 + I_3)$ for the left-hand side of (3·9, 49) and taking real parts on both sides, we then obtain

$$\left(\int_{-1}^{0} - \int_{0}^{1}\right) \frac{s^{2k}\,ds}{\sqrt{(1 - 2sY - s^2)}} + 2Q_{2k}(Y) = 2\sum_{\lambda=0}^{\infty} \frac{P_{2\lambda+1}(Y)}{2\lambda - 2k + 1}. \tag{3·9, 52}$$

Hence, taking into account (3·9, 48),

$$\int_{-1}^{1} s^{2k} K(s, Y)\,ds + 2Q_{2k}(Y)$$

$$= Q_{2k}(Y) + \sum_{\lambda=0}^{\infty} \left(\frac{1}{2\lambda + 2k + 1} + \frac{1}{2\lambda - 2k + 1}\right) P_{2\lambda+1}(Y),$$

$$\tag{3·9, 53}$$

where
$$K(s, Y) = \begin{cases} \dfrac{s^{-1} + 1}{2\sqrt{(1 - 2sY + s^2)}} & (s < 0), \\[2ex] \dfrac{s^{-1} - 1}{2\sqrt{(1 - 2sY + s^2)}} & (s > 0). \end{cases}$$

(3·9, 53) has been derived for positive k only ($k = 1, 2, \ldots$). However, the same formula can be established for $k = 0$, provided it is understood that the principal value of the integral at the origin is to be taken on the left-hand side,

$$\int_{-1}^{1} = \lim_{\epsilon \to 0} \left(\int_{-1}^{\epsilon} + \int_{\epsilon}^{1}\right).$$

In this sense,

$$S(y) = \frac{(-1)^k}{U^2} \frac{d}{dY} \left\{ \int_{-1}^{1} s^{2k} K(s, Y)\,ds + 2Q_{2k}(Y) \right\}$$

by (3·9, 47), for $k = 0, 1, \ldots$, and this may also be written as

$$S(y) = \frac{1}{U^2} \frac{d}{dY} \left\{ \int_{-1}^{1} F(s) K(s, Y)\,ds + 2(-1)^k Q_{2k}(Y) \right\},$$

or
$$S(y) = \frac{1}{U^2} \frac{d}{dY} \left\{ \int_{-1}^{1} F(s) K(s, Y)\,ds + (-1)^k \int_{-1}^{1} \frac{P_{2k}(Y')}{Y - Y'}\,dY' \right\},$$

$$\tag{3·9, 54}$$

where the principal value is to be taken for the second integral at $Y' = Y$, by Neumann's formula for the Legendre functions of the second kind. Referring to (3·9, 37) we see that (3·9, 54) may again be replaced by

$$S(y) = \frac{1}{U^2} \frac{d}{dY} \left\{ \int_{-1}^{1} F(s) K(s, Y)\,ds - \frac{1}{2\pi\rho R} \int_{-1}^{1} \frac{l(RY')}{Y - Y'}\,dY' \right\}. \tag{3·9, 55}$$

This formula represents the slope function $S(y) = S(RY)$ for $k = 0, 1, \ldots$. But since the representation is actually independent of k, it follows that

‡ E. W. Hobson, *Theory of Spherical and Ellipsoidal Harmonics* (Cambridge University Press, 1931), pp. 52 and 66.

(3·9, 55) is valid more generally if $l(y) = l(RY)$ is any sufficiently regular even function of y or Y which can be expanded into a series of Legendre polynomials, as in (3·9, 41). Using that formula, we may, moreover, replace the first integral in (3·9, 55) by a direct integral transform of $l(RY)$, so

$$\int_{-1}^{1} F(s) K(s, Y) ds = -\frac{1}{2\pi\rho R} \int_{-1}^{1} \left(\int_{-1}^{1} l(RY') \frac{(1-s^2) dY'}{2(1-2Y's+s^2)^{\frac{3}{2}}} \right) K(s, Y) ds$$

$$= -\frac{1}{2\pi\rho R} \int_{-1}^{1} l(RY') K^*(Y, Y') dY',$$

where
$$K^*(Y, Y') = \int_{-1}^{1} \frac{(1-s^2) K(s, Y) ds}{2(1-2sY'+s^2)^{\frac{3}{2}}}. \qquad (3·9, 56)$$

The integral on the right-hand side is again singular, and the principal value has to be taken at the origin. Then

$$S(RY) = -\frac{1}{2\pi\rho RU^2} \frac{d}{dY} \int_{-1}^{1} l(RY') \left(K^*(Y, Y') + \frac{1}{Y-Y'} \right) dY',$$
$$(3·9, 57)$$

or, integrating with respect to Y,

$$\int_{-1}^{1} l(RY') \left(K^*(Y, Y') + \frac{1}{Y-Y'} \right) dY' = -2\pi\rho RU^2 \int_{0}^{Y} S(RY) dY.$$
$$(3·9, 58)$$

The lower limit on the right-hand side is zero, for the integral on the left-hand side vanishes for $Y = 0$, because of the fact that $l(RY)$ is an even function of Y.

For a given uncambered but possibly twisted thin circular wing which is symmetrical with respect to the x, z-plane (e.g. for a flat wing) we might try to determine $l(RY)$ from (3·9, 58), and then calculate $F(s)$ and Ω' directly by means of (3·9, 41), (3·9, 31) and (3·9, 28). However, no exact method of solution for (3·9, 58), which is a singular integral equation of the first kind, is known at present. Instead, we proceed by the following alternative method. We choose a spanwise lift distribution $l(RY)$ which is known to correspond to a solution of the given problem according to an approximate method, and we then calculate the corresponding wing shape and pressure distribution as given by (3·9, 57) and (3·9, 28). This method is of the type called 'semi-inverse' in various branches of applied mathematics.

Let us apply this procedure to a flat circular wing at incidence. This is a special case of an elliptic wing at constant incidence, and the approximate lifting-line theory (§ 3·3) predicts that the corresponding spanwise lift distribution is elliptic. Thus we assume

$$l(y) = l(RY) = 2\pi\rho RU^2 C \sqrt{(1 - Y^2)}, \qquad (3·9, 59)$$

where C is an arbitrary constant.

We first calculate the corresponding $F(s)$ as given by (3·9, 41)

$$F(s) = -U^2 C \int_{-1}^{1} \frac{(1-s^2)\sqrt{(1-Y^2)}\,dY}{2(1-2Ys+s^2)^{\frac{3}{2}}}. \qquad (3·9, 60)$$

Assume $s \geqslant 0$ and put $\eta = \sqrt{\dfrac{1+Y}{2}}$, so that $Y = 2\eta^2 - 1$, $dY = 4\eta\,d\eta$ and

$$1 - 2sY + s^2 = (1+s)^2 - 4s\eta^2 = (1+s)^2(1-r^2\eta^2), \text{ where } r^2 = \frac{4s}{(1+s)^2} \quad (0 \leqslant r \leqslant 1).$$

Then

$$F(s) = -4U^2 C \frac{1-s^2}{(1+s)^3} \int_0^1 \frac{\sqrt{(1-\eta^2)}\,\eta^2\,d\eta}{(1-r^2\eta^2)^{\frac{3}{2}}}$$

$$= -4U^2 C \frac{1-s^2}{(1+s)^3} \frac{1}{r} \frac{d}{dr} \int_0^1 \frac{\sqrt{(1-\eta^2)}\,d\eta}{\sqrt{(1-r^2\eta^2)}}$$

$$= -4U^2 C \frac{1-s^2}{(1+s)^3} \frac{1}{r} \frac{d}{dr} \left[\int_0^1 \frac{d\eta}{\sqrt{\{(1-\eta^2)(1-r^2\eta^2)\}}} - \int_0^1 \frac{\eta^2\,d\eta}{\sqrt{\{(1-\eta^2)(1-r^2\eta^2)\}}} \right]$$

$$= -4U^2 C \frac{1-s^2}{(1+s)^3} \frac{1}{r} \frac{d}{dr} \left(K(r) + \frac{1}{r} \frac{dE}{dr} \right),$$

where K and E are the complete elliptic integrals of modulus r, of the first and second kind respectively,

$$K = \int_0^1 \frac{d\eta}{\sqrt{\{(1-\eta^2)(1-r^2\eta^2)\}}}, \quad E = \int_0^1 \sqrt{\left(\frac{1-r^2\eta^2}{1-\eta^2}\right)}\,d\eta.$$

Carrying out the differentiation we obtain

$$F(s) = -4U^2 C \frac{1-s^2}{(1+|s|)^3} \frac{1}{r^4} \{2(K(r) - E(r)) - r^2 K(r)\}, \qquad (3·9, 61)$$

where $r^2 = \dfrac{4|s|}{(1+|s|)^2}$. This formula is valid also for negative s, since $F(s)$ must be an even function of s. r^2 may also be written in the form

$$r^2 = 1 - \frac{(1-|s|)^2}{(1+|s|)^2}.$$

Hence, by Landen's transformation of elliptic integrals‡

$$F(s) = -U^2 C \frac{1-s^2}{s^2} (K(|s|) - E(|s|)). \qquad (3·9, 62)$$

K and E are even functions of their arguments which are analytic in the unit circle, and

$$\lim_{s \to 0} \frac{K(s) - E(s)}{s^2} = \frac{\pi}{4}.$$

---

‡ E. T. Whittaker and G. N. Watson, *A Course of Modern Analysis* (4th ed., Cambridge University Press, 1927) p. 507.

Hence $F(s)$ is an analytic function of s for $|s| < 1$. To calculate $S(y)$ it is convenient to use $(3·9, 55)$. The first integral in that formula may be modified as follows:

$$\int_{-1}^{1} F(s)\, K(s, Y)\, ds$$

$$= \lim_{\epsilon \to 0} \left(\int_{-1}^{-\epsilon} + \int_{\epsilon}^{1} \right) F(s)\, K(s, Y)\, ds$$

$$= \lim_{\epsilon \to 0} \int_{\epsilon}^{1} (F(s)\, K(s, Y) + F(-s)\, K(-s, Y))\, ds$$

$$= \lim_{\epsilon \to 0} \frac{1}{2} \int_{\epsilon}^{1} F(s)\, (s^{-1} - 1) \left(\frac{1}{\sqrt{(1 - 2sY + s^2)}} - \frac{1}{\sqrt{(1 + 2sY + s^2)}} \right) ds,$$

$$(3·9, 63)$$

where we have taken into account that F is an even function of s. Now

$$\lim_{s \to 0} \left(\frac{1}{\sqrt{(1 - 2sY + s^2)}} - \frac{1}{\sqrt{(1 + 2sY + s^2)}} \right) = 0,$$

and so the product

$$(s^{-1} - 1) \left(\frac{1}{\sqrt{(1 - 2sY + s^2)}} - \frac{1}{\sqrt{(1 + 2sY + s^2)}} \right)$$

remains finite for $s = 0$. We may therefore replace the limit in the last member of $(3·9, 63)$ by the ordinary integral from 0 to 1. Hence $(3·9, 55)$ becomes

$$S(y) = \frac{1}{U^2} \frac{d}{dY} \left\{ \frac{1}{2} \int_{0}^{1} F(s)\, (s^{-1} - 1) \left(\frac{1}{\sqrt{(1 - 2sY + s^2)}} - \frac{1}{\sqrt{(1 + 2sY + s^2)}} \right) ds \right.$$

$$\left. - \frac{1}{2\pi\rho R} \int_{-1}^{1} \frac{l(RY')}{Y - Y'}\, dY' \right\},$$

$$(3·9, 64)$$

or, in our particular case,

$$S(y) = - C \left\{ \frac{1}{2} \frac{d}{dY} \int_{0}^{1} \frac{(1 - s)^2 (1 + s)}{s^3}\, (K(s) - E(s)) \right.$$

$$\left. \times \left(\frac{1}{\sqrt{(1 - 2sY + s^2)}} - \frac{1}{\sqrt{(1 + 2sY + s^2)}} \right) ds + \frac{d}{dY} \int_{-1}^{1} \frac{\sqrt{(1 - Y'^2)}}{Y - Y'}\, dY' \right\}.$$

$$(3·9, 65)$$

The second integral on the right-hand side is singular and we have to take its principal value. To evaluate it, we substitute $Y' = \cos \theta$. Then

$$\frac{d}{dY} \int_{-1}^{1} \frac{\sqrt{(1 - Y'^2)}}{Y - Y'}\, dY' = \frac{d}{dY} \int_{0}^{\pi} \frac{\sin^2 \theta\, d\theta}{Y - \cos \theta} = \frac{d}{dY} (\pi Y) = \pi, \quad (3·9, 66)$$

where the details of the calculation have been omitted. Carrying out the differentiation under the integral sign in the first term on the right-hand side of (3·9, 65), we then obtain

$$S(y) = -C\left\{ \int_0^1 \frac{(1-s)^2(1+s)}{2s^2} \, (K(s) - E(s)) \right.$$
$$\left. \times \left(\frac{1}{(1-2sY+s^2)^{\frac{3}{2}}} + \frac{1}{(1+2sY+s)^{\frac{3}{2}}} \right) ds + \pi \right\}.$$
$$(3·9, 67)$$

The integrand on the right-hand side is bounded except at $Y = 1$ and the integral can therefore be evaluated numerically without difficulty except for values of Y very near to unity. The resulting $S(y)$ is plotted in Fig. 3·9, 3. It will be seen that $S(y)$ is approximately constant except near the tips. Thus, the corresponding wing is almost flat, but its tips are tilted upward. A more detailed investigation shows that $S(y)$ becomes infinite at the tips, but that the infinity is weak. The mean slope of the wing is given by

$$\int_0^1 S(RY) \, dY = -C\left\{ \frac{1}{2} \int_0^1 \frac{(1-s)^2(1+s)}{s^3} \, (K(s) - E(s)) \left(\frac{1}{1-s} - \frac{1}{1+s} \right) ds + \pi \right\}$$
$$= -C\left\{ \int_0^1 \frac{1-s}{s^2} \, (K(s) - E(s)) \, ds + \pi \right\}. \qquad (3·9, 68)$$

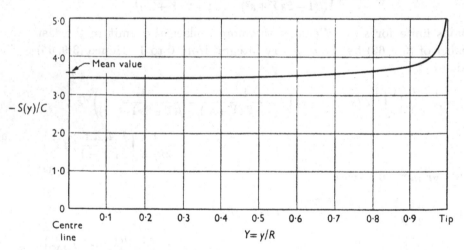

Fig. 3·9, 3

Now
$$\int_0^1 \frac{K(s) - E(s)}{s^2} \, ds = \left[sK + \frac{E-K}{s} \right]_0^1 = 1,$$

and
$$\int_0^1 \frac{K(s) - E(s)}{s} \, ds = [-E]_0^1 = \tfrac{1}{2}\pi - 1.$$

Hence
$$\int_0^1 S(RY) \, dY = -C(2 + \tfrac{1}{2}\pi). \qquad (3·9, 69)$$

It will be seen that the mean incidence is only about 4 % higher numerically than the smallest incidence, which occurs at the centre line of the wing. Thus, we may regard the mean incidence, approximately, as the constant incidence of the wing.

The total lift is, by (3·9, 59),

$$\int_{-1}^{1} l(y)\, dy = 4\pi\rho R^2 U^2 C \int_{0}^{1} \sqrt{(1 - Y^2)}\, dY = \pi^2 \rho R^2 U^2 C,$$

so that the lift coefficient becomes·

$$C_L = 2\pi C. \tag{3·9, 70}$$

This corresponds to an incidence $\alpha = C(2 + \tfrac{1}{2}\pi)$, and so

$$\frac{dC_L}{d\alpha} = \frac{4\pi}{4 + \pi} = 1\cdot77. \tag{3·9, 71}$$

A somewhat different method for the solution of the same problem was adopted by Kinner.‡ Kinner used a linear combination of the first four cosine functions (3·9, 21) and the first four sine functions (3·9, 25), in order to obtain the correct incidence at a number of points along the span of the wing. The Joukowski condition is satisfied only for a finite number of points at the trailing edge. The resulting lift distribution is roughly elliptical, and the lift curve slope is found to be 1·82, which is in quite good agreement with (3·9, 71). The corresponding figure obtained by Kolchin§ is 1·79, which is in even better agreement with (3·9, 71). The value furnished by lifting-line theory is 2·45 (see (3·3, 79)).

The pressure distribution can be calculated by means of (3·9, 41). The result may be expected to yield the pressure distribution on a flat wing at incidence adequately (within the limits of potential flow theory) except very near the tips. However, one may be interested particularly in the flow round the tips of the wing. For this purpose it is necessary to study an acceleration potential such that the lift vanishes along the trailing edge, while at the same time the incidence of the wing remains finite at the tips. It can be shown that these conditions are satisfied, for example, by the field of flow given by

$$F(s) = -1 + \tfrac{18}{13}s^2 - \tfrac{5}{13}s^4 \tag{3·9, 72}$$

(compare (3·9, 31)). The corresponding spanwise lift distribution is

$$l(y) = l(RY) = 2\pi\rho R(P_0(Y) - \tfrac{18}{13}P_2(Y) + \tfrac{5}{13}P_4(Y)). \tag{3·9, 73}$$

We note that the slope of the lift distribution curve, $dl(y)/dy$ is finite at the tip

$$\left(\frac{dl(y)}{dy}\right)_{y=R} = 2\pi\rho(\tfrac{175}{26}Y^2 - \tfrac{183}{26}Y)_{Y=1} = -\frac{8\pi\rho}{13}. \tag{3·9, 74}$$

‡ W. Kinner, 'Die Kreisförmige Tragfläche auf potentialtheoretischer Grundlage', *Ingen.-Arch.* vol. 8 (1937), pp. 47–80.

§ N. E. Kolchin, 'Theory of wing of circular plan form', *Tech. Memor. Nat. Adv. Comm. Aero.*, Wash., no. 1324. Translated from *Prikladnaya Matematika i Mekhanika*, vol. 4 (1940), pp. 3–32.

This result is of some importance, because the particular solutions which arise in lifting-line theory all possess an infinite slope at the tips. It has therefore been suggested that only solutions with that property are acceptable in three-dimensional wing theory. Equation (3·9, 74) shows that such a restriction is unwarranted.

The acceleration potential, and hence the pressure distribution corresponding to (3·9, 72), can also be expressed in finite terms. In fact the appropriate function of $f(\tau)$ is a linear combination with constant coefficients of the functions (3·9, 36) for $k = 0, 1, 2$. Now

$$\sum_{\lambda=0}^{\infty} (-1)^{\lambda} \left(\frac{1}{2\lambda + 2k + 1} + \frac{1}{2\lambda - 2k + 1} \right) \tau^{2\lambda+1}$$

$$= \sum_{\nu=k}^{\infty} (-1)^{\nu-k} \frac{\tau^{2(\nu-k)+1}}{2\nu+1} + \sum_{\nu=-k}^{\infty} (-1)^{\nu+k} \frac{\tau^{2(\nu+k)+1}}{2\nu+1}$$

$$= (-1)^{k} \left\{ (\tau^{-2k} + \tau^{2k}) \sum_{\nu=0}^{\infty} (-1)^{\nu} \frac{\tau^{2\nu+1}}{2\nu+1} - \tau^{-2k} \sum_{\nu=0}^{k-1} (-1)^{\nu} \frac{\tau^{2\nu+1}}{2\nu+1} \right.$$
$$\left. + \tau^{2k} \sum_{\nu=-k}^{-1} (-1)^{\nu} \frac{\tau^{2\nu+1}}{2\nu+1} \right\}$$

$$= (-1)^{k} \left\{ (\tau^{-2k} + \tau^{2k}) \tan^{-1} \tau - \tau^{-2k} \sum_{\nu=0}^{k-1} (-1)^{\nu} \frac{\tau^{2\nu+1}}{2\nu+1} + \tau^{2k} \sum_{\nu=-k}^{-1} (-1)^{\nu} \frac{\tau^{2\nu+1}}{2\nu+1} \right\}.$$

Hence (3·9, 36) may also be written in the form

$$f(\tau) = \tau^{2k} - \frac{2}{\pi} \left\{ (\tau^{-2k} + \tau^{2k}) \tan^{-1} \tau - \tau^{-2k} \sum_{\nu=0}^{k-1} (-1)^{\nu} \frac{\tau^{2\nu+1}}{2\nu+1} \right.$$
$$\left. + \tau^{2k} \sum_{\nu=-k}^{-1} (-1)^{\nu} \frac{\tau^{2\nu+1}}{2\nu+1} \right\}.$$

$$(3·9, 75)$$

Taking real parts, and multiplying by $\mu/(\mu^2 + \nu^2)$, we obtain a finite expression for the corresponding induced acceleration potential.

In developing the analysis of this section we have confined ourselves largely to wings which are symmetrical with respect to the x, z-plane. Similar formulae can be derived for the antisymmetrical case.

Kinner's method was extended by Krienes‡ to the case of a general elliptic wing. The analysis requires the separation of Laplace's equation in terms of confocal coordinates and leads to solutions in terms of Lamé functions. Fig. 3·9, 4, which is taken from Krienes's paper, shows that the lift curve slope calculated on this basis, together with the curves predicted by lifting-line theory.

‡ K. Krienes, 'Die elliptische Tragfläche auf potentialtheoretischer Grundlage', *Z. angew. Math. Mech.* vol. 20 (1940), pp. 65–88. Translated in *Tech. Memor. Nat. Adv. Comm. Aero., Wash.*, no. 971.

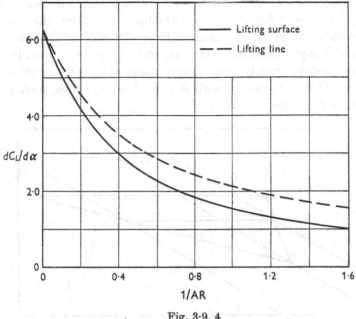

——— Lifting surface

— — — Lifting line

Fig. 3·9, 4

3·10   Theory of aerofoils of small aspect ratio

For wings of very small aspect ratio, there exists a theory which is comparable in scope to the lifting-line theory for wings of large aspect ratio. The fundamental idea goes back to Munk,‡ who used it implicitly in the theory of airship hulls. Its adoption to aerofoil problems is due to R. T. Jones§ (see also a later paper by Lawrence improving on Jones's theory‖). The theory will be developed for thin, possibly cambered, aerofoils. The effect of thickness may be calculated separately (see § 3·6) and the total result obtained by superposition. As a rough rule, the theory provides acceptable results for wings whose aspect ratio does not exceed 1·5. However, for reasons which will appear presently, the quality of the results depends also on the shape of the wings. The theory is most suitable for wings with sweptback leading edges ending in straight trailing edges which are normal to the direction of flow. An important example of this kind is provided by a wing of isosceles triangular plan-form (Delta wing), but wings with curved leading edges are also included. For this case, the theory ceases to give valid results only in the immediate neighbourhood of the trailing edge. On the

‡ M. M. Munk, 'The aerodynamic forces on airship hulls', *N.A.C.A. Rep.* no. 184 (1924).

§ R. T. Jones, 'Properties of low-aspect-ratio pointed wings at speeds below and above the speed of sound', *N.A.C.A. Rep.* no. 835 (1946).

‖ H. R. Lawrence, 'The lift distribution on low aspect ratio wings at subsonic speeds', *J. Aero. Sci.* vol. 18, no. 10 (1951), pp. 683–704.

other hand, for wings whose width first increases and then decreases gradually in the direction of flow, e.g. wings of elliptic planform, the results are unsatisfactory so far as conditions aft of the position of maximum span are concerned, although the overall data are still acceptable. Finally, as we shall see, the theory can only provide indications of the general trend of the pressure distribution for wings with straight leading edges, such as wings of rectangular plan-form.

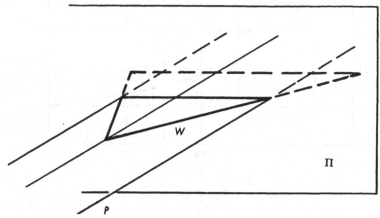

Fig. 3·10, 1

To introduce the fundamental idea, we consider a wing W of the first type, e.g. a Delta wing which advances with a uniform velocity U and at a small incidence α in the negative direction of the x-axis. We place ourselves in a frame of reference which is at rest relative to the surrounding air, and we consider the field of flow in a plane Π which is normal to the direction of motion of the wing. As the wing advances across the plane Π, it communicates a downward component of velocity, $-w = U \tan \alpha \doteq U\alpha$ to the fluid. Thus, the field of flow in the plane Π is, to the first approximation, the same as that produced by a flat plate which moves downward with velocity $U\alpha$, and whose trace in Π coincides with the intersection of the wing with Π at the moment under consideration (Fig. 3·10, 1). This implies that, so far as the calculation of the field of flow is concerned, we neglect the fact that the width of the wing is variable. It will be seen that the error introduced by this procedure is small if the slope dy/dx of the leading edge is small numerically.

In analytical terms, the replacement of the wing W by the flat plate P, so far as the calculation of the flow in the (arbitrary) transverse plane Π is concerned, is tantamount to the assumption that the field of flow in Π is due to a velocity potential which satisfies the two-dimensional Laplace equation

$$\frac{\partial^2 \Phi}{\partial y^2} + \frac{\partial^2 \Phi}{\partial z^2} = 0. \tag{3·10, 1}$$

It follows that the same equation holds, both for the induced velocity potential and for the total velocity potential, in the usual frame of reference which is at rest relative to the wing.

The argument leading to (3·10, 1) still applies if the slope of the wing is variable in both the spanwise and chordwise direction, provided the variation of the slope in the direction of the chord is gradual. If the equation of surface of the wing is

$$z = f(x, y) \tag{3·10, 2}$$

in the system of coordinates which is at rest relative to the wing, then the boundary condition is, with the usual approximations (see § 3·1),

$$w = U \frac{\partial f}{\partial x} = U s(x, y) = - U \alpha(x, y), \tag{3·10, 3}$$

where $s(x, y) = \partial f / \partial x$ is the local slope of the wing in the direction of the x-axis, and $\alpha(x, y)$ is its local incidence, $s(x, y) = - \tan \alpha(x, y)$. Since the wing is supposed very thin, f, s and α are respectively the same at corresponding points of the upper and lower surfaces.

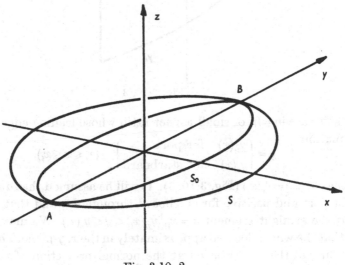

Fig. 3·10, 2

One can justify the use of (3·10, 1) more formally in the following way. Let S_0 be an area in the x, y-plane which penetrates the y, z-plane along a segment AB of the y-axis, and let $\tau_0(x, y)$ define the strength of a distribution of doublets over S_0 (Fig. 3·10, 2). For any $\mu > 0$, let S be the area which is obtained from S_0 by displacing the points (x, y) on the circumference of S_0 to $(x/\mu, y)$. Also, let the doublet distribution $\tau(x, y)$ be defined by

$$\tau(x, y) = \tau_0(\mu x, y).$$

As μ tends to 0, the dimensions of S in the direction of the x-axis increase indefinitely, while its dimensions in the direction of the y-axis remain constant. It is not then difficult to show analytically that at the same time the field of flow in the y, z-plane approaches the field of flow due to a two-dimensional distribution of doublets along the segment AB, such that the strength of the distribution is given by $\tau(y) = \tau_0(0, y)$. Thus, $(3\cdot10, 1)$ is justified in the limit if we regard the y, z-plane as the plane Π. The argument does not provide a bound for the errors introduced by using $(3\cdot10, 1)$ everywhere, for a wing of given small aspect ratio. In fact, here as elsewhere in aerofoil theory, one tends to rely, for an estimate of the quality of the theory, on comparison with the results provided by more exact theories and by experiment.

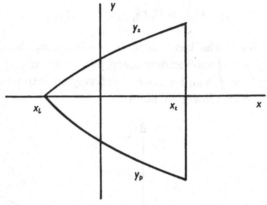

Fig. 3·10, 3

Consider a wing W of small aspect ratio whose leading edge is given by the equations

$$y = \begin{cases} y_p(x) & \text{for port} \\ y_s(x) & \text{for starboard} \end{cases} \quad (x_l \leqslant x \leqslant x_t),$$

such that $y_p(x_l) = y_s(x_l)$ (Fig. 3·10, 3). It will be assumed that dy/dx is negative for port and positive for starboard throughout, and that the trailing edge is the straight segment $x = x_t$, $y_p(x_t) \leqslant y \leqslant y_s(x_t)$. We may suppose as usual that the wing is located approximately in the x, y-plane. The boundary condition will then be satisfied at the normal projection of the points of the wing on to that plane. The intersection of the wing with any plane $x = $ constant is represented by the segment $z = 0$, $y_p(x) \leqslant y \leqslant y_s(x)$, which will be denoted by T_x $(x_l \leqslant x \leqslant x_t)$. For the same range of x, define

$$y_m(x) = \tfrac{1}{2}(y_s(x) + y_p(x)), \quad a(x) = \tfrac{1}{2}(y_s(x) - y_p(x)), \tag{3·10, 4}$$

so that $y_m(x)$ is the mid-point of T_x and $a(x)$ is its length.

Put $\xi = y + iz$. For given x $(x_l \leqslant x \leqslant x_t)$, map the region outside T_x on the region outside the unit circle in an auxiliary complex plane by means of the transformation

$$\xi = y_m + \frac{a}{2}\left(\zeta + \frac{1}{\zeta}\right), \tag{3·10, 5}$$

where $y_m = y_m(x)$ and $a = a(x)$ are defined by $(3\cdot10, 4)$.

Let $\Phi(x, y, z)$ be the induced velocity potential. Φ is a solution of (3·10, 1). It follows that there exists a function $\Omega(x, \xi)$, analytic with respect to ξ, such that $\Phi = \mathscr{R}(\Omega)$. Then the velocity components v, w are given by

$$v - iw = -\left(\frac{\partial \Phi}{\partial y} + i \frac{\partial \Phi}{\partial z}\right) = -\frac{d\Omega}{d\xi}, \qquad (3 \cdot 10, 6)$$

and $d\Omega/d\xi$ is a one-valued function which is regular and vanishes at infinity. The same applies to $\dfrac{d\Omega}{d\zeta} = \dfrac{d\Omega}{d\xi} \dfrac{d\xi}{d\zeta}$, which may therefore be expanded into a Laurent series:

$$\frac{d\Omega}{d\zeta} = i\left(\frac{A_1}{\zeta} + \frac{A_2}{\zeta^2} + \frac{A_3}{\zeta^3} + \dots\right). \qquad (3 \cdot 10, 7)$$

In this formula, the coefficients $A_1, A_2, A_3, \dots$ depend on x. They are all real because Φ is an odd function of z.

Let C be a closed curve which surrounds T_x in a given plane $x = \text{constant}$. An argument used in §3·2 (see Fig. §3·2, 2) then shows that the circulation round C vanishes,

$$\int_C (v\, dy + w\, dz) = 0. \qquad (3 \cdot 10, 8)$$

Now

$$\int_C (v\, dy + w\, dz) = \mathscr{R} \int_C (v - iw)(dy + i\, dz) = -\mathscr{R} \int_C \frac{d\Omega}{d\xi} d\xi = -\mathscr{R} \int_C \frac{d\Omega}{d\zeta} d\zeta$$

$$= -\mathscr{I} \int_C \left(\frac{A_1}{\zeta} + \frac{A_2}{\zeta^2} + \dots\right) d\zeta = -\mathscr{I}(2\pi i A_1) = -2\pi A_1,$$

and so $A_1 = 0.$ $\qquad (3 \cdot 10, 9)$

Put $\zeta = e^{i\theta}$ at the unit circle. For such values of ζ,

$$\frac{d\xi}{d\zeta} = \frac{a}{2}\left(1 - \frac{1}{\zeta^2}\right) = \frac{a}{2}(1 - e^{-2i\theta}) = ia\, e^{-i\theta} \sin\theta. \qquad (3 \cdot 10, 10)$$

Hence $\dfrac{d\Omega}{d\xi} = \dfrac{d\Omega}{d\zeta} \Big/ \dfrac{d\xi}{d\zeta} = \displaystyle\sum_{n=2}^{\infty} iA_n\, e^{-i(n-1)\theta} / ia \sin\theta$

$$= \frac{1}{a \sin\theta} \sum_{n=2}^{\infty} A_n(\cos(n-1)\theta - i \sin(n-1)\theta)$$

$$= \frac{1}{a \sin\theta} \sum_{n=2}^{\infty} A_n \cos(n-1)\theta - \frac{i}{a \sin\theta} \sum_{n=2}^{\infty} A_n \sin(n-1)\theta. \qquad (3 \cdot 10, 11)$$

The imaginary part of this expression is

$$w = -\frac{1}{a \sin\theta} \sum_{n=2}^{\infty} A_n \sin(n-1)\theta. \qquad (3 \cdot 10, 12)$$

Now at T_x, $y = y_m + a \cos\theta$, and so by (3·10, 3)

$$\sum_{n=2}^{\infty} A_n \sin(n-1)\theta = -Ua \sin\theta\, s(x, y_m + a \cos\theta). \qquad (3 \cdot 10, 13)$$

Using the Euler–Fourier formulae, we obtain from (3·10, 13),

$$A_n = -\frac{2U}{\pi}\int_0^\pi as(x, y_m + a\cos\theta)\sin(n-1)\theta\sin\theta\,d\theta \quad (n = 2, 3, \ldots).$$

(3·10, 14)

Integrating (3·10, 7) with respect to ζ, we have

$$\Omega = a_0 - i\left(\frac{A_2}{\zeta} + \frac{A_3}{2\zeta^2} + \frac{A_4}{3\zeta^3} + \ldots\right),$$

(3·10, 15)

where a_0 is a constant of integration. Writing $A_0 = \mathscr{R}(a_0)$, $\zeta = r\,e^{i\theta}$, and taking real parts in (3·10, 15), we then obtain

$$\Phi = A_0 - \sum_{n=1}^{\infty}\frac{r^{-n}}{n}A_{n+1}\sin n\theta.$$

Now we may assume that the induced velocity potential vanishes upstream of $x = x_l$. The condition that the pressure increment induced by the pressure of the wing vanish far away from the wing in the transverse planes $x = \mathrm{constant}$ is then satisfied by making $A_0 = 0$. Thus, Φ is now given by

$$\Phi = -\sum_{n=1}^{\infty}\frac{r^{-n}}{n}A_{n+1}\sin n\theta.$$

(3·10, 16)

To determine the pressure at the wing, we use the linearized Bernoulli equation (3·2, 3)

$$p = p_0 - \rho U u' = p_0 + \rho U\frac{\partial\Phi}{\partial x},$$

(3·10, 17)

where p_0 is the free-stream pressure. The pressure difference between the top and bottom surfaces at any point of the wing is

$$\Delta p = 2\rho U\left(\frac{\partial\Phi}{\partial x}\right)_{z=+0}.$$

(3·10, 18)

This relation holds even if we use the exact Bernoulli equation. To evaluate (3·10, 18) we require $\partial\Phi/\partial x$ for $r = 1$,

$$\frac{\partial\Phi}{\partial x} = -\sum_{n=1}^{\infty}\left(\frac{dA_{n+1}}{dx}\sin n\theta + \frac{\partial\theta}{\partial x}A_{n+1}\cos n\theta\right).$$

(3·10, 19)

In this formula, $\partial\theta/\partial x$ denotes the partial derivative of θ with respect to x, when y is kept constant. To calculate $\partial\theta/\partial x$, we differentiate the relation

$$y = y_m + a\cos\theta$$

with respect to x. Then

$$\frac{\partial\theta}{\partial x} = \frac{1}{a\sin\theta}\left(\frac{dy_m}{dx} + \frac{da}{dx}\cos\theta\right).$$

(3·10, 20)

The derivatives dA_n/dx are obtained from (3·10, 14),

$$\frac{dA_n}{dx} = -\frac{2U}{\pi}\int_0^\pi\frac{d}{dx}(as(x, y_m + a\cos\theta))\sin(n-1)\theta\,d\theta \quad (n = 2, 3, \ldots).$$

(3·10, 21)

The spanwise lift distribution is given by (3·2, 21), where

$$\Delta\Phi = -2(A_2\sin\theta + \tfrac{1}{2}A_3\sin 2\theta + \tfrac{1}{3}A_4\sin 3\theta + \ldots)$$

(3·10, 22)

from $(3·10, 16)$. In this formula $A_2, A_3, \ldots$ denote the values of these coefficients at the trailing edge, for $x = x_l$. Then

$$l = 2\rho U(A_2 \sin \theta + \tfrac{1}{2}A_3 \sin 2\theta + \tfrac{1}{3}A_4 \sin 3\theta + \ldots). \qquad (3·10, 23)$$

The total lift on the wing is given by $(3·2, 18)$

$$L = 2\rho Ua \int_0^\pi (A_2 \sin \theta + \tfrac{1}{2}A_3 \sin 2\theta + \ldots) \sin \theta \, d\theta,$$

where $a = a(x_l)$. Or, using $(3·10, 14)$,

$$L = \pi \rho Ua A_2 = -2\rho U^2 a^2 \int_0^\pi s(x, y_m + a \cos \theta) \sin^2 \theta \, d\theta, \quad (3·10, 24)$$

or, in terms of the incidence,

$$L = 2\rho U^2 a^2 \int_0^\pi \alpha(x, y_m + a \cos \theta) \sin^2 \theta \, d\theta. \qquad (3·10, 25)$$

Inspection of $(3·10, 23)$–$(3·10, 25)$ shows the remarkable fact that, according to the present theory, the spanwise distribution and the total lift depend only on the characteristics of the wing, y_m, a and $s(x, y)$ or $\alpha(x, y)$ at the trailing edge. In particular, the integral in $(3·10, 25)$ may be regarded as a weighted mean, or average, of the values of the incidence along the trailing edge, since

$$\int_0^\pi \frac{2}{\pi} \sin^2 \theta \, d\theta = 1.$$

Thus, if α is constant along the trailing edge, then

$$L = \pi \rho U^2 a^2 \alpha. \qquad (3·10, 26)$$

Note that in order that α be constant along the span, it is necessary and sufficient that $A_n = 0$ for $n \geqslant 3$. The semi-span is equal to a at the trailing edge, and so the aspect ratio of the wing is given by

$$\mathcal{R} = 4a^2/S,$$

where S is the wing area. Hence

$$\left. \begin{array}{l} L = \tfrac{1}{4}\pi \rho U^2 S \mathcal{R} \alpha, \\ C_L = L/\tfrac{1}{2}\rho U^2 S = \tfrac{1}{2}\pi \mathcal{R} \alpha. \end{array} \right\} \qquad (3·10, 27)$$

In particular, the formulae hold if α is constant all over the wing. Then

$$\frac{dC_L}{d\alpha} = \frac{\pi}{2}\mathcal{R}. \qquad (3·10, 28)$$

$(3·10, 28)$ holds, for example, for a flat Delta wing which moves in a direction normal to its base. For that case we may assume

$$y_m = 0, \quad a = a_0 x, \quad x_l = 0, \quad x_l = c,$$

so that $a_0 = \tan \gamma$, where γ is the semi-apex angle of the wing. Then,

$$A_2 = Ua_0 x \alpha, \quad A_3 = A_4 = \ldots = 0, \quad \frac{\partial \theta}{\partial x} = \frac{1}{x}\cot \theta$$

and
$$\frac{\partial \Phi}{\partial x} = - U a_0 x (\sin \theta + \cot \theta \cos \theta) = - U a_0 \alpha \operatorname{cosec} \theta. \qquad (3\cdot10, 29)$$

The pressure difference between the top and bottom surfaces now becomes

$$\Delta p = 2 \rho U \frac{\partial \Phi}{\partial x} = - 2 \rho U^2 a_0 \alpha \operatorname{cosec} \theta, \qquad (3\cdot10, 30)$$

or
$$\Delta p = - 2 \rho U^2 \alpha \frac{a_0^2 x}{\sqrt{(a_0^2 x^2 - y^2)}}. \qquad (3\cdot10, 31)$$

Thus, the pressure difference becomes infinite at the leading edges. Integrating $(3\cdot10, 30)$ across the local span for a given x, we obtain for the chordwise lift distribution $\lambda(x)$ $(0 \leqslant x \leqslant x_l)$,

$$\lambda(x) = - \int_{-a_0 x}^{a_0 x} \Delta p \, dy = 2 \rho U^2 a_0^2 x \alpha \int_0^{\pi} \operatorname{cosec} \theta \sin \theta \, d\theta,$$

or
$$\lambda(x) = 2 \pi \rho U^2 a_0^2 x \alpha. \qquad (3\cdot10, 32)$$

$(3\cdot10, 32)$ shows that $\lambda(x)$ is proportional to the local width of the wing. It follows that for any line $x = \text{constant}$ the lift on the wing ahead of $x = \text{constant}$ is proportional to the wing area ahead of that line. We conclude that the centre of pressure of the lift coincides with the centroid of the wing, i.e. it is at a distance $d = \frac{2}{3} c$ aft of the apex. Thus the pitching moment M about the y-axis can be written down immediately

$$M = - L d = - \tfrac{2}{3} \pi \rho U^2 a_0^2 c^3 \alpha. \qquad (3\cdot10, 33)$$

Finally, the spanwise lift distribution now is

$$l(y) = 2 \rho U^2 a_0 c \alpha \sin \theta = 2 \rho U^2 \alpha \sqrt{(a_0^2 c^2 - y^2)}. \qquad (3\cdot10, 34)$$

In this formula, $a_0 c = a(c)$ is the semi-span and so the distribution is elliptic.

We still have to calculate the induced drag for the general case.

The vortex wake is bounded by the two straight lines $y = y_p(x_l)$ and $y = y_s(x_l)$, which are parallel to the x-axis in the x, y-plane. Thus, the intersection T_x of the wake with any plane $x = \text{constant} > x_l$ has the same dimensions as the trailing edge. Moreover, Φ is constant at the wake along straight lines parallel to the x-axis. It follows that if we map the region outside T_x on the region outside the unit circle in the ζ-plane by $(3\cdot10, 5)$, then we obtain precisely the same velocity potential $(3\cdot10, 16)$ as for the transverse plane at the trailing edge. To obtain the induced drag, we may therefore evaluate $(3\cdot2, 34)$ at the trailing edge, so

$$D_i = \tfrac{1}{2} \rho \int_0^{\pi} 2 (A_2 \sin \theta + \tfrac{1}{2} A_3 \sin 2\theta + \dots)$$

$$\times \frac{1}{a \sin \theta} (A_2 \sin \theta + A_3 \sin 2\theta + \dots) a \sin \theta \, d\theta. \qquad (3\cdot10, 35)$$

Integrating, we obtain

$$D_i = \tfrac{1}{2} \pi \rho (A_2^2 + \tfrac{1}{2} A_3^2 + \tfrac{1}{3} A_4^2 + \dots). \qquad (3\cdot10, 36)$$

In particular, if α is constant along the trailing edge,

$$A_2 = -U a \alpha, \quad A_3 = A_4 = \ldots = 0,$$

and so

$$D_i = \tfrac{1}{2} \pi \rho U^2 a^2 \alpha^2 = \tfrac{1}{8} \pi \rho U^2 S \mathcal{R} \alpha^2 = \tfrac{1}{2} L \alpha. \tag{3·10, 37}$$

The last formula shows that if W is a flat wing at incidence α, then the resultant force points in a direction midway between the direction of the z-axis and the direction normal to the wing. The induced drag coefficient is given by

$$C_{D_i} = D_i / \tfrac{1}{2} \rho U^2 S = \tfrac{1}{4} \pi \mathcal{R} \alpha^2 = \frac{1}{\pi \mathcal{R}} C_L^2. \tag{3·10, 38}$$

We shall now consider some wings which are not of the type investigated above. Let W be a wing whose leading edges are defined as before for $0 \leqslant x \leqslant x_t$, and whose trailing edges are given by

$$y = y_p^*(x), \quad y = y_s^*(x), \quad x_t \leqslant x \leqslant x_{tt},$$

for port and starboard respectively, such that dy/dx is positive on port and negative on starboard and

$$y_p(x_t) = y_p^*(x_t), \quad y_s(x_t) = y_s^*(x_t)$$

(see Fig. 3·10, 4).

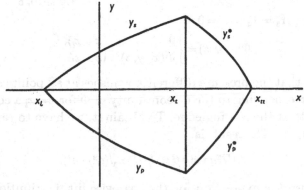

Fig. 3·10, 4

These conditions are satisfied, for instance, by a wing of elliptic planform. We confine ourselves to the assumption that the wing is flat so that α is constant all over the wing. The analysis given above remains applicable in the region $x_l \leqslant x \leqslant x_t$. Suppose that for $x > x_t$, Φ is given by the same expression as at the trailing edge,

$$\Phi(x, y, z) = \Phi(x_t, y, z) \quad (x > x_t). \tag{3·10, 39}$$

Then the z-component of the induced velocity is $w = -U \alpha$ for all points $x > x_t$, $y_p(x_t) \leqslant y \leqslant y_s(x_t)$, in the x, y-plane. This implies in particular that the required boundary condition is satisfied at all points of the wing also aft of x_t, and so we conclude that (3·10, 39) is the appropriate expression. We deduce that $\partial \Phi / \partial x = 0$ for $x > x_t$, so that the pressure increment and the pressure difference between the upper and lower surfaces of W vanish for all points in that region. Thus, no lift is carried by the part of the wing aft of $x = x_t$. It follows that the lift and induced drag and the lift and induced drag coefficients are still given by (3·10, 26)–(3·10, 28), (3·10, 37), (3·10, 38). The solution is physically unsatisfactory for $x = x_t$, where it implies the occurrence of a discontinuity in the pressure.

Next, let W be a flat rectangular wing at incidence under symmetrical conditions, i.e. such that two of the sides are normal to the direction of flow (Fig. 3·10, 5). Let the leading edge be given by $x = x_l$. If we put $y_m = 0$ and define a as the semi-span of the wing, then the velocity potential is again given by (3·10, 16) in all transverse planes $x = $ constant aft of the leading edge, with

Fig. 3·10, 5

$$A_2 = Ua\alpha, \quad A_3 = A_4 = \ldots = 0.$$

This implies

$$\Phi(x, y, z) = \begin{cases} 0 & (x < x_l), \\ \Phi(x_l, y, z) & (x > x_l). \end{cases} \tag{3·10, 40}$$

We conclude that the pressure difference vanishes at all points of the wing. However, corresponding to the discontinuity of Φ for $x = x_t$ a concentrated force now acts at the leading edge. To obtain it, we have to replace $\partial\Phi/\partial x$ by Φ in (3·10, 18). The result is

$$2\rho U^2 a\alpha \sin \theta = 2\rho U^2 \alpha \sqrt{(a^2 - y^2)}, \tag{3·10, 41}$$

and this is also the expression for the spanwise lift distribution since the entire lift is now carried by the leading edge. The formulae for the lift and induced drag, and for the corresponding coefficients are the same as before. All that this analysis tells us with regard to the chordwise pressure distribution on a real rectangular wing is that the smaller the aspect ratio, the more will the lift tend to be concentrated near the leading edge. This conclusion has been confirmed by other methods.‡

The flow round a Delta wing under symmetrical conditions was considered earlier (see equations (3·10, 29)–(3·10, 34)). Assume now that the direction of flow is yawed relative to the plane of symmetry of the wing (Fig. 3·10, 6). So long as the angle of yaw, β, is smaller than the semi-apex angle of the

‡ K. Wieghardt, 'Über die Auftriebsverteilung des einfachen Rechteckflügels über die Tiefe', *Z. angew. Math. Mech.* vol. 19 (1939), pp. 257–70. Translated in *Tech. Memor. Nat. Adv. Comm. Aero., Wash.,* no. 963.

wing, γ, the general analysis given earlier in this section still applies upstream of the foremost point of the trailing edge, B. And since the aspect ratio of the wing is supposed small, it will be sufficient to use a rough estimate for the aerodynamic forces which act on the area downstream of B (the shaded area $BA'A$ in Fig. 3·10, 6). This can be done, for example, by replacing the trailing edge AB by a trailing edge $A''B''$ which is parallel to the y-axis, such that the area $OA''B''$ equals the area OAB. Some simple trigonometry shows that the aspect ratio of the new wing is

$$\mathcal{R}' = \frac{(\text{span})^2}{\text{area}} = \frac{4\tan\gamma}{(\cos\beta + \sin\beta\,\tan\gamma)^2}. \tag{3·10, 42}$$

With this expression for the aspect ratio, we may still apply equations (3·10, 27), (3·10, 28) and (3·10, 38) for the lift and induced drag.

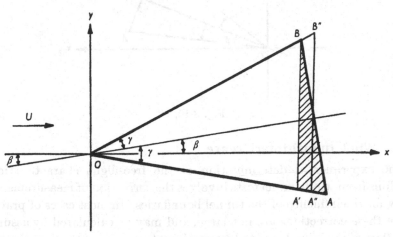

Fig. 3·10, 6

Suppose next, that the angle of yaw β exceeds the semi-apex angle γ. Assuming that the wing is deflected to starboard (Fig. 3·10, 7), it then follows that the line OA which under symmetrical conditions constitutes part of the leading edge, now becomes part of the trailing edge. Accordingly, the physically correct solution to this problem will satisfy the Joukowski condition along OA. Such a solution will appear as the limiting case of a problem considered in Chapter 4 (see §4·9). We note that the Joukowski condition was not satisfied for the aerofoils considered at the beginning of the section, but this was an inevitable consequence of the fact that, according to our present theory, the pressure at any point of the aerofoil is independent of conditions downstream of that point.

The small aspect-ratio theory may also be applied to compound lifting units. For simply-connected units, the particular functions $\Phi_r^{(n)}$, $\Phi_i^{(n)}$ of

§ 3·4 may be combined to yield a solution. The theory has also been employed to calculate the aerodynamic forces on wing-body combinations. Also some of the results are quoted only for supersonic flow, but they apply equally well at other speeds (§ 4·2). Problems of wing-body interference are of considerable practical importance, and a great amount of theoretical work has been done on the subject.‡ However, the phenomenon depends to a large extent on viscosity effects which are difficult to assess, and as a result no satisfactory theory on the subject exists at present.

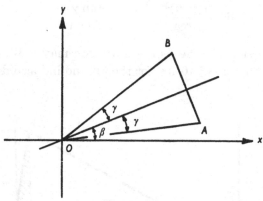

Fig. 3·10, 7

3·11   Wind-tunnel interference

The experimental determination of the free-flight characteristics of aerofoils from wind-tunnel tests involves the correction of measurements to allow for the influence of the tunnel boundaries. In most cases of practical value these corrections are not large, and may be calculated by assuming the effect of the boundaries to be a small perturbation of the flow about the aerofoil when in an infinite unbounded stream. We shall see that when the small perturbation approximation is valid, solutions can be developed readily for aerofoils in all wind tunnels of conventional shape.

For the purposes of analysis we define the wind tunnel as a region of parallel uniform flow of velocity U in the x-direction, contained within an infinite cylindrical surface whose generators are parallel to the x-axis. The boundary of the cylinder may be solid, so that the usual boundary condition (1·2, 10) holds, or it may be 'free', in the sense that outside the cylinder the air is at rest, but there is no rigid boundary between the interior moving air and exterior stationary air. We also admit types of tunnel in which the boundary is part solid and part free, but in this case we

‡ We may mention two papers: G. N. Ward, 'Supersonic flow past slender pointed bodies', *Quart. J. Mech. Appl. Math.* vol. 2 (1949), pp. 75–97; J. R. Spreiter, 'The aerodynamic forces on slender plane- and cruciform-wing and body combinations', *N.A.C.A. Rep.* no. 962 (1950).

require that the boundary of every cross-section of the tunnel be geometric-ally the same, so that the edges of the solid parts of the tunnel surface are all parallel to the x-axis and of infinite extent. With this formulation it is possible to derive the free-surface boundary condition as follows.

Let us apply Bernoulli's equation (1·6, 6) to a stream tube within the tunnel adjacent to the free surface of the boundary. When there are no body forces acting, this is

$$p + \tfrac{1}{2}\rho((U+u')^2 + v'^2 + w'^2) = p_0 + \tfrac{1}{2}\rho U^2, \qquad (3·11, 1)$$

where p_0 is the pressure far upstream and (u', v', w') is the velocity induced by the presence of the aerofoil in the tunnel. Neglecting squares of the induced velocity, we obtain

$$p + \rho U u' = p_0. \qquad (3·11, 2)$$

Outside the tunnel there will also be an induced flow, however, in accordance with the approximations already made in deriving (3·11, 2), this will not affect the outside pressure, which is therefore still constant everywhere. The pressure across the free surface must be continuous; it follows that $p = \text{constant} = p_0$ adjacent to the free boundary. Hence the boundary condition becomes simply

$$u' = 0. \qquad (3·11, 3)$$

Alternatively, this condition may be expressed as $\partial \Phi'/\partial x = 0$ at the boundary, where Φ' is the induced velocity potential, or

$$\Phi' = 0 \qquad (3·11, 4)$$

on the surface, since Φ' is zero everywhere far upstream of the aerofoil. It should be remarked that equation (3·11, 4), in contrast to the boundary condition for a solid wall, $\partial \Phi'/\partial n = 0$, which is exact, involves approxima-tions whose validity, in the last resort, can only be verified by experiment. Such verification has been obtained,‡ and the use of (3·11, 4) for the free jet boundary condition is now well established.

In deriving the free-surface boundary condition we have already assumed that the induced velocities at the tunnel wall are small compared with the main-stream velocity. The complementary assumption, that the velocities induced at the aerofoil by the tunnel boundary are small, has also to be made if excessive complications in analysis are to be avoided. Glauert§ has considered the effects of this approximation for the case of an aerofoil at the centre of a circular closed tunnel, when the aerofoil is represented

‡ See M. Knight and T. A. Harris, 'Experimental determination of jet boundary corrections for airfoil tests in four open wind tunnel jets of different shape', *N.A.C.A. Rep.* no. 361 (1930), and also J. E. Adamson, 'An experimental investigation of wind tunnel interference in the R.A.E. 5 ft. open jet circular tunnel', *Rep. Memor. Aero. Res. Comm., Lond.*, no. 1897 (1941).

§ H. Glauert, 'The interference on the characteristics of an aerofoil in a wind tunnel of circular section', *Rep. Memor. Aero. Res. Comm., Lond.*, no. 1453 (1931).

by a lifting line in the manner of § 3·3. This case illustrates also the use of the method of images in aerodynamics, and we shall consider it as an example.

We shall need to use again the notion of the Trefftz plane, introduced in § 3·3, as a plane far downstream of the aerofoil at which the flow is effectively two-dimensional. Similar results hold for wind-tunnel flow as can be shown by replacing the tunnel boundary by a surface distribution of singularities and considering the symmetry of the configuration. It can be seen, however, that, unless the tunnel boundaries are symmetric about the plane containing the wake, a velocity may be induced at the bound vortex in a direction normal to the transverse yz-plane. In any event, the velocities induced

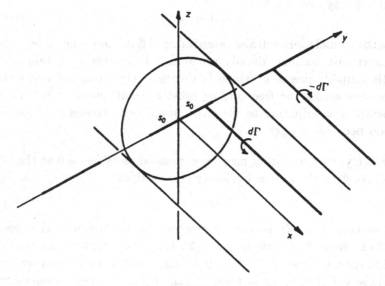

Fig. 3·11, 1

by the bound and trailing vortices and the corresponding surface distributions on the boundary will be at right angles, and for calculating the flow in the transverse plane only the two-dimensional Trefftz-plane problem need be considered. Thus, in particular, the downwash velocity at the wing is one-half of the downwash velocity at the wake in the Trefftz plane.

Consider the lifting line, of length $2s_0$, placed on the y-axis at the centre of the circular tunnel, as illustrated in Fig. 3·11, 1. Suppose that, corresponding to the trailing vortex of strength $d\Gamma$ emanating from the spanwise position A, we place an image vortex line of strength $-d\Gamma$ at the point A' that is inverse to A with respect to the circle in the Trefftz plane (Fig. 3·11, 2). The stream function for these two vortices is (see equation (1·10, 16))

$$\Psi = \frac{d\Gamma}{2\pi} \log \frac{|AP|}{|A'P|}$$

at a point P. If P lies on the tunnel boundary, we have, since triangles OAP and OPA' are similar,

$$\frac{AP}{A'P} = \frac{OA}{OP}, \quad \text{a constant,}$$

where O is the centre of the tunnel. Hence it follows that the stream function Ψ is constant over the tunnel wall, and the boundary condition for a solid wall is satisfied.

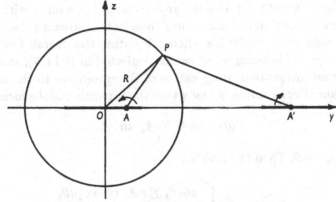

Fig. 3·11, 2

To find the flow inside the tunnel we have to calculate the flow field for the complete distribution Γ over the wing and the image system obtained as the inverse of Γ. If the vortex element $d\Gamma$ is situated at $y = \eta$, and its inverse at $y = \eta'$,

$$\eta\eta' = R^2,$$

where R is the radius of the circle. Therefore the velocity induced by this image at a point (O, y) at the aerofoil is

$$dw' = -\frac{d\Gamma}{4\pi}\left(\frac{R^2}{\eta} - y\right)^{-1}, \tag{3·11, 5}$$

where, instead of the factor $1/2\pi$ of equation (1·10, 15) the factor $1/4\pi$ has been used, since the velocity at the aerofoil is one-half of the velocity in the Trefftz plane.

The total perturbation velocity w' is obtained from (3·11, 5) by integration over the wing span. Thus

$$w'(y) = -\frac{1}{4\pi}\int_{-s_0}^{s_0}\frac{d\Gamma}{d\eta}\frac{1}{(R^2/\eta - y)}\,d\eta. \tag{3·11, 6}$$

The exact solution of this equation must take into account that the distribution Γ is affected by the tunnel boundary. Thus we must have, using (3·3, 7)

$$\Gamma = \tfrac{1}{2}a_0 c(U\alpha_0 + w + w'). \tag{3·11, 7}$$

Here c is the local chord, a_0 is the two-dimensional lift coefficient, and $w+w'$ is the total downwash velocity (strictly speaking in our coordinate system $w+w'$ is an 'upwash' velocity, however, since this term is used little, we shall retain the word 'downwash' as a useful expression for the vertical velocity at the wing in the z-direction).

(3·11, 7) substituted in (3·11, 6) yields an integral equation for w', which may be solved for specified aerofoils by expansion of Γ in a suitable trigonometric series, such as that given by (3·3, 36). This is the method followed by Glauert‡ for elliptic and rectangular wings, with the important conclusion that the change in Γ from its free-stream value due to the tunnel boundaries is negligible when calculating the overall forces acting on the aerofoil. This being so, w' can be neglected in (3·11, 7), and (3·11, 6) reduces to an integration using values of Γ appropriate to the aerofoil in a free stream. For example, we may take the expansion of the form (3·3, 36)

$$\Gamma(\theta) = 4s_0 U \sum_{n=1}^{\infty} A_n \sin n\theta, \tag{3·11, 8}$$

where $y = s_0 \cos \theta$. Then (3·11, 6) becomes

$$\frac{w'(\theta)}{U} = \frac{s_0^2}{\pi R^2} \int_0^{\pi} \frac{\cos\theta_1 \sum_{n=1}^{\infty} nA_n \cos n\theta_1 d\theta_1}{\left(1 - \frac{s_0^2}{R^2}\cos\theta\cos\theta_1\right)}$$

$$= \frac{2\pi s_0^2}{C} \sum_{n=1}^{\infty} nA_n B_n, \tag{3·11, 9}$$

where

$$B_n = \frac{1}{2} \int_0^{\pi} \frac{\cos n\theta_1 \cos\theta_1 d\theta_1}{\left(1 - \frac{s_0^2}{R^2}\cos\theta\cos\theta_1\right)}, \tag{3·11, 10}$$

and $C = \pi R^2$ is the tunnel cross-sectional area.

The lift coefficient obtained from (3·11, 8) is

$$C_L = \frac{4\pi s_0^2}{S} A_1 \tag{3·11, 11}$$

(see equation (3·3, 39)), where S is the wing area. Defining the interference factor δ by means of the equation

$$\frac{w'}{U} = \delta C_L \frac{S}{C}, \tag{3·11, 12}$$

we find, using (3·11, 9) and (3·11, 11),

$$\delta = \frac{1}{2A_1} \sum_{n=1}^{\infty} nA_n B_n. \tag{3·11, 13}$$

‡ H. Glauert, 'The interference on the characteristics of an aerofoil in a wind tunnel of circular section', *Rep. Memor. Aero. Res. Comm., Lond.*, no. 1453 (1931).

Values of δ given by (3·11, 13) have been calculated by Glauert, and Table 3·11, 1 gives values of the average interference factor $\bar{\delta}$ defined as

$$\bar{\delta} = \frac{1}{L} \int_{\text{span}} \delta \, dL \qquad (3·11, 14)$$

(L being the lift) for uniform and elliptic loading Γ. It will be seen that the difference between the two sets of figures (which we may consider as the two extreme cases of possible lift distributions) is quite small up to $s_0/R \doteqdot 0·7$, and in most practical applications the form of the lift distribution over the span is not important in evaluating wind-tunnel interferences effects. A more detailed analysis given by Glauert in the same paper shows that if no allowance is to be made in the calculations for the change of lift distribution due to the tunnel walls, it is most accurate to assume an elliptic lift distribution in all cases; the values of $\bar{\delta}$ in Table 3·11, 1 corresponding to elliptic loading should therefore be used.

Table 3·11, 1. *Values of the interference factor $\bar{\delta}$ for a wing at the centre of a circular closed tunnel*

s_0/R	0	0·1	0·2	0·3	0·4	0·5	0·6	0·7	0·8
Elliptic loading	0·125	0·125	0·125	0·125	0·126	0·127	0·128	0·131	0·137
Uniform loading	0·125	0·125	0·125	0·126	0·126	0·127	0·130	0·136	0·148

From values of $\bar{\delta}$ or δ correction of measured quantities to allow for the effect of the tunnel boundary can be made. Thus if C_L is the (measured) lift coefficient, the effective incidence is changed by an amount

$$\Delta\alpha = \bar{\delta} C_L \frac{S}{C} \qquad (3·11, 15)$$

(cf. (3·11, 12)), and from the results of lifting-line theory (§ 3·3, equation (3·3, 43)), there is a correction also to the induced drag

$$\Delta C_{D_i} = -\bar{\delta} C_L^2 \frac{S}{C}. \qquad (3·11, 16)$$

Formulae (3·11, 15) and (3·11, 16), with different values for the constant $\bar{\delta}$, apply to all types of tunnel. In particular, for a wing at the centre of a rectangular tunnel the method of images is again applicable, and leads to the same equations (3·11, 15) and (3·11, 16). Values of $\bar{\delta}$ for rectangular tunnels as calculated by Glauert‡ are shown in Table 3·11, 2. In this instance the image system is doubly infinite, as illustrated in Fig. 3·11, 3 for the limiting case of a wing of zero span. The wing is obtained by reducing

‡ H. Glauert, 'The interference on the characteristics of an aerofoil in a wind tunnel of rectangular section', *Rep. Memor. Aero. Res. Comm., Lond.*, no. 1459 (1932).

the distance $2s_0$ between the arms of a horseshoe vortex of strength γ' to zero while maintaining the product $\tau = 2s_0\gamma'$ constant. It can be seen by comparing (1·9, 37) with (1·10, 13) that in the Trefftz plane this singularity is equivalent to a two-dimensional doublet of strength τ, which explains the significance of the arrow method of representation used in Fig. 3·11, 3. The direction of the arrow is from sink to source.

Table 3·11, 2. *Values of $\bar{\delta}$ for a wing at the centre of closed rectangular tunnels*

Ratio of wing span to tunnel width	0	0·2	0·4	0·5	0·6	0·7	0·8	0·9
Square tunnel:								
Elliptic loading	0·137	0·138	0·140	0·143	0·148	0·154	0·164	0·180
Uniform loading	0·137	0·138	0·142	0·146	0·152	0·163	0·181	0·218
Breadth = 2 × height:								
Elliptic loading	0·137	0·129	0·112	0·104	0·097	0·092	0·092	0·094
Uniform loading	0·137	0·127	0·107	0·098	0·092	0·090	0·094	0·110

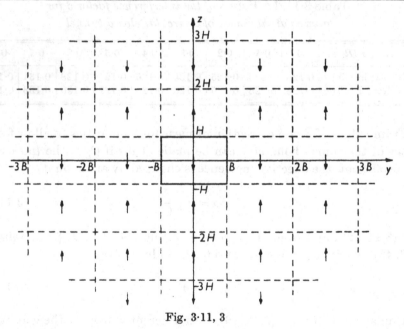

Fig. 3·11, 3

From the expression (1·10, 13) for the complex potential due to a doublet at the origin of the complex χ-plane,

$$\Pi(z) = \frac{\tau\, e^{i\alpha}}{2\pi\chi},\qquad (3\cdot11, 17)$$

it is readily seen that the system of images in Fig. 3·11, 3, together with the doublet representing the wing itself, produce a potential, the imaginary part of which is zero on the lines $z = \pm H$, $y = \pm B$. These lines are therefore streamlines of the flow and may be taken as the solid boundaries of a rectangular tunnel. With appropriate changes in signs of the images the real part Φ of Π can be made to vanish along one or more of the lines $z = \pm H$, $y = \pm B$, which then could correspond to free-surface boundaries of a tunnel. In this manner the interference can be found for any type of rectangular tunnel, in which each wall is either free or solid. Fig. 3·11, 4 illustrates the

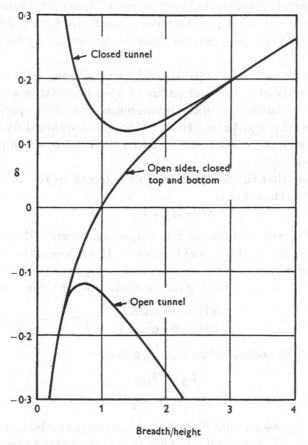

Fig. 3·11, 4. Interference factor for rectangular tunnels. (*Courtesy H.M.S.O.*)

results in the form of variation of the interference factor δ for small wings with tunnel geometry. It is taken from a report by Glauert‡ which contains also a useful and fairly detailed survey of wind-tunnel interference in general.

The representation of a wing by a doublet is of special interest, for it leads to solution of problems that are difficult or cumbersome to deal with when the aerofoil is represented by a lifting line. Moreover, it will be seen

‡ H. Glauert, 'Wind tunnel interference on wings, bodies and airscrews', *Rep. Memor. Aero. Res. Comm., Lond.*, no. 1566 (1933).

from Tables 3·11, 1 and 3·11, 2 that in most cases the values of $\bar{\delta}$ for zero span are also good approximations for wings of span appreciable compared with the tunnel dimensions. If a solution has been obtained for a doublet singularity at a general position in the tunnel, the result for a wing of finite span can be found by superposition.

If we represent the aerofoil by a doublet it is possible to prove a theorem (again due to Glauert‡) which enables us to calculate the interference factor for many types of open (free jet) tunnel from the factor for a closed tunnel, and vice versa. In its most general form the theorem may be stated as follows:

The interference velocities experienced by a small wing W in a tunnel C that consists partly of closed and partly of open boundaries are of equal magnitude but opposite sign to the interference velocities experienced by W when rotated through $\frac{1}{2}\pi$ in a tunnel C' which is obtained by replacing the solid boundaries of C by free boundaries, and the free boundaries by solid boundaries.

Let us suppose that the complex velocity potential in the Trefftz plane $(\chi = y + iz)$ for the tunnel C is

$$\Pi_1 = \Pi_{1r} + \Pi_1', \qquad (3·11, 18)$$

where Π_{1r} is the potential due to the wing W alone and Π_1' is the interference potential due to the tunnel boundary. If we denote the sections of the boundary which, in the first case (Fig. 3·11, 5 (a)), are solid by A, and the sections that are free by B, the boundary conditions for C may be taken as

$$\left.\begin{array}{ll} \mathscr{R}\Pi_1 = 0 & \text{on } B, \\ \mathscr{I}\Pi_1 = 0 & \text{on } A. \end{array}\right\} \qquad (3·11, 19)$$

Consider now the potential function Π_2 given by

$$\Pi_2 = i\Pi_1, \qquad (3·11, 20)$$

and suppose also that $\qquad \Pi_2 = \Pi_{2r} + \Pi_2'. \qquad (3·11, 21)$

Taking $\Pi_{2r} = i\Pi_{1r}$, we see that if Π_{1r} is the velocity potential (3·11, 17) due to a doublet, then Π_{2r} is the potential due to a doublet of the same strength rotated through $\frac{1}{2}\pi$ (Fig. 3·11, 5 (b)). Furthermore, by (3·11, 20) we then require

$$\Pi_2' = i\Pi_1'. \qquad (3·11, 22)$$

The boundary conditions (3·11, 19) give for the new potential Π_2,

$$\left.\begin{array}{ll} \mathscr{R}(i\Pi_2) = 0 & \text{on } B, \\ \mathscr{I}(i\Pi_2) = 0 & \text{on } A, \end{array}\right\}$$

or

$$\left.\begin{array}{ll} \mathscr{I}(\Pi_2) = 0 & \text{on } B, \\ \mathscr{R}(\Pi_2) = 0 & \text{on } A, \end{array}\right\} \qquad (3·11, 23)$$

‡ H. Glauert, 'Some general theorems concerning wind tunnel interference on aerofoils', *Rep. Memor. Aero. Res. Comm., Lond.*, no. 1470 (1932).

and these are satisfied if we replace the solid sections B of C by solid walls, and the sections A by free walls, as shown in Fig. 3·11, 5 (b). The interference velocity components (v', w') in the second configuration can be found immediately from (3·11, 22). In fact

$$\left.\begin{aligned} v' &= w, \\ w' &= -v. \end{aligned}\right\} \tag{3·11, 24}$$

Relative to the wing W' position in C' the interference velocities are therefore equal and opposite to their values (v, w) for W in C.

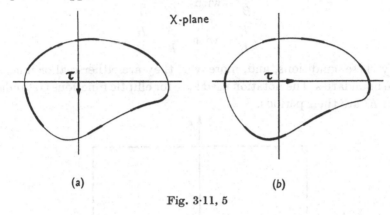

X-plane

(a)   (b)

Fig. 3·11, 5

The general theorem shows at once that the interference velocities for closed and open tunnels are of opposite sign. This suggests that by a suitable combination of closed and free walls the interference might be altogether eliminated. Indeed, Fig. 3·11, 4 already shows that for a rectangular tunnel with one pair of walls fixed and the other pair free, the interference is zero if the tunnel breadth equals its height. More complicated situations may be studied by using the complex representation introduced above, and as an example we shall treat the case of a rectangular tunnel with walls that are slotted in the tunnel direction,‡ so that a cross-section (in the Trefftz plane) will appear as shown in Fig. 3·11, 6. We shall make use of the boundary conditions that on a solid wall the normal velocity is zero and that on a free boundary the induced velocity potential $\Phi' = 0$ (equation (3·11, 4)). The latter condition is equivalent to stating that the tangential velocity component is zero. For a rectangular tunnel a necessary condition is therefore that, on a vertical wall, the complex velocity $v - iw$ be alternately real and imaginary on the free and fixed boundaries respectively, and vice versa for a horizontal wall. In addition, we require that the velocity potential Φ have the same constant value on each free surface. The problem is thus to find a complex velocity function w, analytic within the boundaries $-B < y < B$, $-H < z < H$, whose only singularities occur at the junctions

‡ J. A. Laurmann, 'Interference on lifting surfaces in slotted rectangular tunnels' *Nat. Aero. Est. Ottawa, Rep.* no. LR–85 (1953).

of the free and fixed boundaries and at the wing, and which takes either pure imaginary or real values on the boundaries. These requirements lead us to suppose that it is possible to represent the solution in terms of doubly-periodic functions, possessing a single pole within the rectangle $-B \leqslant y \leqslant B$, $-H \leqslant z \leqslant H$. The Jacobian elliptic functions of modulus k

$$\text{ns}\,(P\chi, k), \quad \text{cs}\,(P\chi, k), \quad \text{ds}\,(P\chi, k), \tag{3·11, 25}$$

where P takes one of the values

$$\frac{K}{B}, \quad \text{with} \quad \frac{K}{K'} = \frac{B}{H}$$

or

$$i\frac{K}{H}, \quad \text{with} \quad \frac{K}{K'} = \frac{H}{B}$$

satisfy these conditions, and, moreover, they are either real or imaginary on the boundaries. The notation used here for elliptic functions is standard.‡ K and K' are their periods.

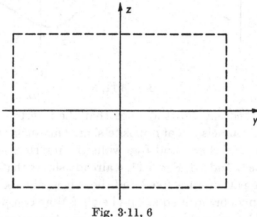

Fig. 3·11, 6

The alternating form of the boundary condition leads to the introduction of terms of the type $\sqrt{\{c^2 - \text{ds}^2\,(P\chi)\}}$,

where c is a real constant, although this implies that the geometric configuration is symmetric about $y = 0$, $z = 0$.

The only poles of (3·11, 25) within the rectangular region are at $\chi = 0$, and the order of the singularity there must be made to correspond to the doublet representing the wing. However, zeros of (3·11, 25) occur at $P\chi = \pm iK'$ and therefore terms of order $\text{ns}^n\,(P\chi)$, $\text{cs}^n\,(P\chi)$, $\text{ds}^n\,(P\chi)$ at $P\chi = \pm iK'$ with $n < 0$ must be avoided in the expression for the complex velocity. Such considerations show that the velocity must be of the form

$$w_1\left(\frac{K_1}{B}\chi\right) w_2\left(\frac{K_2}{H}\chi\right), \quad \frac{K_1}{K_1'} = \frac{B}{H}, \quad \frac{K_2}{K_2'} = \frac{H}{B}, \tag{3·11, 26}$$

‡ See Chapter 22 of E. T. Whittaker and G. N. Watson, *A Course of Modern Analysis* (4th ed., Cambridge University Press, 1927).

$$\text{where} \quad w_1 = \frac{\sum_{n=0}^{n_1+1}\left\{A_1 + \left[A_{1n}\,\text{ns}\left(\frac{K_1}{B}\chi\right) + B_{1n}\,\text{cs}\left(\frac{K_1}{B}\chi\right)\right]\text{ns}^n\left(\frac{K_1}{B}\chi\right)\right\}}{\prod_{n=1}^{n_1}\sqrt{\left\{c_{1n}^2 - \text{ds}^2\left(\frac{K_1}{B}\chi\right)\right\}}},$$

$$\text{or} \quad w_1 = \frac{\sum_{n=0}^{n_1}\text{ds}\left(\frac{K_1}{B}\chi\right)\left\{A_1 + \left[A_{1n}\,\text{ns}\left(\frac{K_1}{B}\chi\right) + B_{1n}\,\text{cs}\left(\frac{K_1}{B}\chi\right)\right]\text{ns}^n\left(\frac{K_1}{B}\chi\right)\right\}}{\prod_{n=1}^{n_1}\sqrt{\left\{c_{1n}^2 - \text{ds}^2\left(\frac{K_1}{B}\chi\right)\right\}}},$$

$$(3\cdot11, 27)$$

$$\text{and} \quad w_2 = \frac{\sum_{n=0}^{n_2+1}\left\{A_2 + \left[A_{2n}\,\text{ns}\left(\frac{K_2}{H}i\chi\right) + B_{2n}\,\text{cs}\left(\frac{K_2}{H}i\chi\right)\right]\text{ns}^n\left(\frac{K_2}{H}i\chi\right)\right\}}{\prod_{n=1}^{n_2}\sqrt{\left\{c_{2n}^2 - \text{ds}^2\left(\frac{K_2}{H}i\chi\right)\right\}}},$$

$$\text{or} \quad w_2 = \frac{\sum_{n=0}^{n_2}\text{ds}\left(\frac{K_2}{H}i\chi\right)\left\{A_2 + \left[A_{2n}\,\text{ns}\left(\frac{K_2}{H}i\chi\right) + B_{2n}\,\text{cs}\left(\frac{K_2}{H}i\chi\right)\right]\text{ns}^n\left(\frac{K_2}{H}i\chi\right)\right\}}{\prod_{n=1}^{n_2}\sqrt{\left\{c_{2n}^2 - \text{ds}^2\left(\frac{K_2}{H}i\chi\right)\right\}}},$$

$$(3\cdot11, 28)$$

where n_1 is the number of slots (free portions) in a vertical wall, and n_2 the number in a horizontal wall.

Equation $(3\cdot11, 26)$ has no singularities in the rectangle other than at the slot edges and at the centre. At the centre there are poles of order one and two, of which the first (representing a source or a sink) must be eliminated by suitable choice of the constants A_n and B_n, and the residue of the second must be made to agree with that for the potential of a doublet, representing the wing. The remaining constants occurring in $(3\cdot11, 27)$ and $(3\cdot11, 28)$ are found from the set of simultaneous equations obtained by using the condition that the velocity potential on each free portion of the boundary is the same.

As a particularly simple case of $(3\cdot11, 26)$ we may quote the results for a tunnel with solid horizontal walls and a single slot of width $2a$ at the centre of each vertical wall. The complex velocity is then found to be

$$\frac{\tau}{2\pi}\frac{\text{ds}\left(\frac{K}{B}\chi\right)\text{cs}\left(\frac{K}{B}\chi\right)\text{ns}\left(\frac{K}{B}\chi\right)}{\sqrt{\left\{c^2 - \text{ds}^2\left(\frac{K}{B}\chi\right)\right\}}}, \quad \frac{K}{K'} = \frac{B}{H}, \qquad (3\cdot11, 29)$$

where

$$c = k'\,\text{cn}\left(\frac{aK}{B}, k'\right)$$

and we have supposed the wing to be represented by a vertical doublet of strength τ at the origin. The interference factor in this case is

$$\delta = \frac{KK'}{2\pi}\frac{1}{3!}\left(3k'^2\,\text{cn}^2\left(\frac{aK}{B}, k'\right) + 2k^2 - 1\right). \qquad (3\cdot11, 30)$$

A similar analysis can be given for a circular slotted tunnel; the appropriate functions to consider in this case are the singly periodic trigonometric functions.‡

Thus far we have dealt exclusively with the two-dimensional problem as represented in the Trefftz plane. The three-dimensional problems, such as the determination of interference velocities ahead or behind a wing, and the interference due to thickness effects, are naturally more difficult. An axial doublet may be considered as the basic type of singularity for a thick non-lifting wing, corresponding to the vertical two-dimensional doublet for the lifting wing of zero thickness. Thus for a symmetric aerofoil the induced velocity potential is given by (3·6, 3) as

$$\Phi'(x, y, z) = \frac{1}{4\pi} \int_S \frac{\sigma(x_0, y_0)\, dx_0\, dy_0}{\sqrt{\{(x-x_0)^2 + (y-y_0)^2 + z^2\}}}, \qquad (3\cdot11, 31)$$

where σ is the local source-sink density distribution on the aerofoil, and the integral extends over the wing surface. For points (x, y, z) far away from the wing compared with its dimensions (e.g. at the tunnel walls), we may replace (3·11, 31) approximately by

$$\Phi'(x, y, z) \doteq \frac{1}{4\pi} \int_S \sigma(x_0, y_0) \frac{1}{r}\left(1 + \frac{xx_0 + yy_0}{r^2}\right) dx_0\, dy_0, \qquad (3\cdot11, 32)$$

where $\qquad\qquad r = \sqrt{(x^2 + y^2 + z^2)}.$

There can be no inflow or outflow across the wing surface, and hence

$$\int_S \sigma(x_0, y_0)\, dx_0\, dy_0 = 0.$$

So that, assuming the wing to be symmetric about the xz-plane, we find

$$\Phi'(x, y, z) = \frac{x}{4\pi r^3} \int_S \sigma(x_0, y_0)\, x_0\, dx_0\, dy_0, \qquad (3\cdot11, 33)$$

approximately. But (3·11, 33) is the potential due to a doublet at the origin with its axis on $x = 0$ and of strength

$$\tau = \int_S \sigma(x_0, y_0)\, x_0\, dx_0\, dy_0. \qquad (3\cdot11, 34)$$

This explains the use of a doublet to represent a non-lifting wing. Note that the derivation of (3·11, 34) assumes that the distribution is unaffected by the tunnel boundaries.

It is customary to write $\qquad \tau = \frac{2}{\sqrt{\pi}} KVU, \qquad (3\cdot11, 35)$

‡ In fact the first work on slotted tunnels was done with reference to circular tunnels; see E. Pistolesi, 'Sull' Interferenza di una Galleria Aerodinamica a Contorno Misto', *Comment. pont. Acad. Sci.* no. 9, vol. 4 (1940); also translation no. 66 of the Cornell Aeronautical Laboratory Inc.

where V is the volume of the wing and K is a constant that is independent of both the tunnel geometry and the main-stream velocity U. Some values of K are given in Table 3·11, 3.

Table 3·11, 3. *Values of K‡*

Thickness	Wing section		
Chord	Elliptic	Joukowski	Conventional N.A.C.A.– 00XX
0·06	0·938	—	0·941
0·09	0·965	0·991	0·972
0·12	0·993	1·016	1·005
0·15	1·019	1·045	1·035
0·18	1·046	1·068	1·063
0·21	1·072	1·083	1·090
0·25	1·108	—	1·128
0·30	1·152	—	—
0·35	1·196	—	—
0·50	1·329	—	—
1·00	1·772	—	—

‡ Both this table and Table 3·11, 4 are taken from J. G. Herriot, 'Blockage corrections for three-dimensional-flow closed throat wind tunnels, with consideration of the effect of compressibility', *N.A.C.A. Res. Memor.* no. A 7 B 28 (1947).

An interference factor λ can be introduced to correspond to the factor δ for the lifting wing. The defining equation is

$$\frac{u'}{U} = \frac{V}{C^{\frac{3}{2}}} K\lambda, \tag{3·11, 36}$$

where u' is the axial interference velocity at the wing and C is the cross-sectional area of the tunnel.

For rectangular tunnels in which each wall is either completely open or closed, λ can be found by the method of images, as in the case of a lifting wing in a rectangular tunnel. Values of λ for several shapes of tunnel are given in Table 3·11, 4.

The method of images is not applicable to the three-dimensional case in a circular tunnel, and we must consider the problem of solving Laplace's equation in cylindrical coordinates

$$\frac{\partial^2 \Phi_1}{\partial r^2} + \frac{1}{r}\frac{\partial \Phi_1}{\partial r} + \frac{1}{r^2}\frac{\partial^2 \Phi_1}{\partial \theta^2} + \frac{\partial^2 \Phi_1}{\partial x^2} = 0, \tag{3·11, 37}$$

for the interference potential Φ_1 due to the tunnel walls. In a closed tunnel the boundary condition to be satisfied is

$$\frac{\partial}{\partial r}(\Phi_0 + \Phi_1) = 0 \tag{3·11, 38}$$

at the walls. Here Φ_0 is the potential due to the wing at the origin in an unbounded stream

$$\Phi_0 = \frac{\tau}{4\pi} \frac{x}{(x^2 + r^2)^{\frac{3}{2}}}.$$

Separating variables by putting $\Phi_1 = P(r) X(x)$, and assuming axial symmetry, we obtain for (3·11, 37)

$$P'' + \frac{1}{r} P' + P\frac{X''}{X} = 0, \qquad (3·11, 39)$$

Table 3·11, 4. *Values of λ for a wing represented by an axial doublet at the centre of several types of tunnel*

Tunnel shape	Wing span : tunnel width				
	0‡	0·25	0·50	0·75	1·00
Circular closed	0·797	0·812	0·828	0·859	—
Circular open	−0·206	—	—	—	—
Square closed	0·812	0·818	0·836	0·874	0·951
Square open	−0·238	—	—	—	—
Rectangular closed:					
Breadth/height = 10/7	0·863	0·864	0·866	0·884	0·916
7/4	0·946	0·941	0·930	0·923	0·937
2	1·028	1·017	0·990	0·967	0·962
7/2	1·729	1·630	1·436	1·204	1·160
2/7	1·729	1·783	1·896	2·196	2·665
1/2	1·028	—	—	—	1·180
7/10	0·863	—	—	—	1·110

‡ This corresponds to a body of revolution.

and hence X satisfies the equation $X''/X = $ constant. In virtue of the boundary condition (3·11, 38) X must be an odd function of x, and therefore we choose as a fundamental solution

$$X = \sin\frac{k\pi x}{l}, \qquad (3·11, 40)$$

where k is an integral and l is a length which for the present shall be kept arbitrary. However, in order to satisfy the boundary condition (3·11, 38) for all x, the limiting case as l approaches infinity must be taken, and this will be done at a later stage.

The equation for $P(r)$ now becomes

$$P'' + \frac{1}{r} P' - \frac{k^2\pi^2}{l^2} P = 0. \qquad (3·11, 41)$$

The solution is

$$P = I_0\!\left(\frac{k\pi r}{l}\right), \qquad (3·11, 42)$$

a modified Bessel function of the first kind and zero order; the corresponding function of the second kind does not enter since Φ_1 must be regular. The general solution of (3·11, 39) is thus

$$\Phi_1 = \sum_{k=1}^{\infty} c_k I_0\left(\frac{k\pi\rho}{\lambda}\right) \sin\frac{n\pi\xi}{\lambda}, \qquad (3·11, 43)$$

where the c_k are constants, and we have put

$$\xi = \frac{x}{R}, \quad \rho = \frac{r}{R}, \quad \lambda = \frac{l}{R},$$

R being the radius of the tunnel, centre at the origin.

The coefficients c_k are found by applying the boundary conditions (3·11, 38) using (3·11, 42); thus, with $\rho = 1$,

$$\sum_{k=1}^{\infty} c_k \frac{k\pi}{\lambda} I_0'\left(\frac{k\pi}{\lambda}\right) \sin\frac{k\pi\xi}{\lambda} = \frac{3\tau}{4\pi R^2} \frac{\xi}{(\xi^2+1)^{\frac{3}{2}}}. \qquad (3·11, 44)$$

Expanding the right-hand side in a Fourier series and equating coefficients, we find

$$c_k = \frac{\tau}{2\pi^2 R^2} \frac{\dfrac{\pi}{\lambda}\displaystyle\int_0^{\lambda} \dfrac{\cos(n\pi\alpha/\lambda)}{(\alpha^2+1)^{\frac{3}{2}}}\,d\alpha}{I_0'\left(\dfrac{k\pi}{\lambda}\right)}:$$

Substituting in (3·11, 43) and taking the limit $\lambda \to \infty$, we obtain after some calculations

$$\Phi_1 = \frac{\tau}{2\pi^2 R^2} \int_0^{\infty} \frac{\sin(q\xi)\, I_0(q\rho)\, q K_1(q)}{I_0'(q)}\,dq, \qquad (3·11, 45)$$

where K_1 is the modified Bessel function of the second kind and first order. From (3·11, 45) we get for the perturbation velocity in the axial direction

$$u' = \frac{\tau}{2\pi^2 R^3} \int_0^{\infty} \cos(q\xi)\, I_0(q\rho) \frac{q^2 K_1(q)}{I_1(q)}\,dq. \qquad (3·11, 46)$$

The above analysis is due to Baranoff‡ (though the solution was given first by Lamb)§ and shown in Fig. 3·11, 7 are the results of his calculations for u' using (3·11, 46). He covered also the linearized compressible flow case, but, as will be shown in § 4·2, according to linearized theory the effect of compressibility can be allowed for by a simple transformation of coordinates.

Mention might also be made in this section of two further wind-tunnel corrections which are sometimes necessary. The effect of the viscous wake of the aerofoil is to alter the downwash at the aerofoil from its value in an unbounded stream, and this effect can become large for thick wings. We

‡ A. v. Baranoff, 'Zur Frage der Kanalkorrektur bei kompressibler Unterschall-strömung', *Dtsch. Luftfahrtforsch.* Forschungsbericht, no. 1272 (1947).

§ H. Lamb, 'On the effect of the walls of an experimental tank on the resistance of a model', *Rep. Memor. Aero. Res. Comm., Lond.*, no. 1010 (1926).

shall not enter into this aspect further than to state that the wake is equivalent to an effective source of strength

$$\sigma = D/U,$$

where D is the drag of the aerofoil. The resulting interference velocity at the wing is

$$w' = U\frac{C_D S}{4C},$$

where C is the tunnel cross-sectional area, C_D is the drag coefficient of the wing, and S is the area on which this coefficient is based.

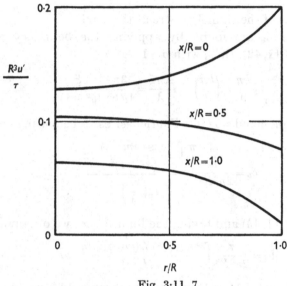

Fig. 3·11, 7

Another possibility is that the wind-tunnel flow, before the aerofoil is introduced, is itself non-uniform. The most common type of non-uniformity is an axial flow acceleration or deceleration, normally specified by the pressure gradient dp/dx. This case, as well as more general types of non-uniformity, is covered in an article by G. I. Taylor,[‡] and results in corrections to the drag and pressure on the aerofoil proportional to the pressure gradient. The two-dimensional problem has been studied more recently in more detail by Goldstein[§] and by E. E. Jones,[||] whilst the effect of a span-wise non-uniformity has been calculated by von Kármán and Tsien.[¶]

‡ G. I. Taylor, 'The force acting on a body placed in a curved and converging stream of fluid', *Proc. Roy. Soc.* A, vol. 120 (1928), pp. 260–83.

§ S. Goldstein, 'Steady two-dimensional flow past a solid cylinder in a non-uniform stream', *Rep. Memor. Aero. Res. Comm., Lond.*, no. 1902 (1942).

|| E. E. Jones, 'The effect of the non-uniformity of the stream on the aerodynamic characteristics of a moving aerofoil', *Quart. J. Mech. Appl. Math.* vol. 4 (1951), pp. 64–77.

¶ Th. von Kármán and H. S. Tsien, 'Lifting-line theory for a wing in non-uniform flow', *Quart. Appl. Math.* vol. 3 (1945), pp. 1–11.

We have been concerned only with approximate solutions to the interference problem, and it is found experimentally that in every case of practical importance (at least for incompressible flow), this approach is adequate. The only case in which complete solutions of the flow about a general aerofoil have been found is for two-dimensional tunnels,‡ and the only exact explicit solution in this restricted field is for a flat plate.§

‡ See S. Goldstein, 'Two-dimensional wind tunnel interference', *A.R.C. Rep. Ae.* 2075 (1942), and H. E. Moses, 'Velocity distributions on arbitrary airfoils in closed tunnels by conformal mapping', *Tech. Notes Nat. Adv. Comm. Aero., Wash.*, no. 1899 (1949). Also a series of papers by S. Tomotika in *Rep. Aero. Res. Inst., Tokyo*, nos. 97, 100, 120, 182 (1933–9).

§ L. Rosenhead, 'The lift on a flat plate between parallel walls', *Proc. Roy. Soc. A*, vol. 132 (1931), p. 127.

CHAPTER 4

AEROFOILS IN COMPRESSIBLE FLOW

4·1 Introduction

All real fluids are compressible to a greater or lesser extent; in particular, the variation of the density of a gas, such as air, with the pressure is a familiar phenomenon. It was stated in Chapter 1 that under normal conditions the effects of compressibility are unimportant at low forward speeds, and consequently the density of the air was assumed constant in Chapters 2 and 3. We shall now drop this restriction and assume instead that the density ρ is a function of the pressure p,

$$\rho = f(p). \tag{4·1, 1}$$

In particular, it is shown in thermodynamics that if the entropy is constant throughout a perfect gas, then (4·1, 1) can be replaced by the more specific

$$\frac{p}{p_0} = \left(\frac{\rho}{\rho_0}\right)^\gamma. \tag{4·1, 2}$$

In this equation p_0 and ρ_0 are the pressure and density at some reference point, e.g. in a region where the fluid is at rest. γ is the ratio of specific heats which depends on the nature of the gas. For air, the constant γ may be taken as 1·40, although other values between 1·40 and 1·41 are also quoted.

The theory of aerofoils in compressible fluids, like the theory developed in the preceding chapters, is based on the assumption that the fluid is inviscid, and the effects of viscosity are treated separately. On this assumption, the addition of (4·1, 1) or (4·1, 2) to the fundamental equations of perfect fluid flow yields an adequate basis for the theory. However, it is found that at forward speeds of about 550 m.p.h. (depending on the shape and incidence of the wing) sharp discontinuities or shock waves appear in the field of flow. The occurrence of shock waves will be discussed later (§ 4·12). It implies that (4·1, 1) and (4·1, 2) are no longer strictly valid, although they may still be accepted as a first approximation. The analysis given below depends on (4·1, 1) only, but since in practice the dependence of the density on the pressure always takes the form (4·1, 2), we shall assume more specifically that (4·1, 2) is satisfied.

Since ρ is a function of the pressure, Kelvin's theorem on the constancy of the circulation round a circuit which moves with the fluid is valid, and with it Lagrange's theorem and the vortex theorems of § 1·5. It follows that the field of flow may again be assumed irrotational except possibly in the wake of the aerofoil where the particles have passed through the boundary layer. Then Bernoulli's equation (1·6, 3) applies, where

$$\int \frac{dp}{\rho} - \int \frac{dp}{d\rho}\frac{d\rho}{\rho} = \frac{p_0}{\rho_0^\gamma}\gamma \int \rho^{\gamma-2}\,d\rho = \frac{\gamma}{\gamma-1}\frac{p_0}{\rho_0^\gamma}\rho^{\gamma-1}.$$

Substituting in (1·6, 3) we obtain

$$\rho^{\gamma-1} = \frac{\gamma-1}{\gamma}\frac{\rho_0^\gamma}{p_0}\left\{h(t) + \frac{\partial\Phi}{\partial t} - \tfrac{1}{2}q^2 - W\right\}. \tag{4·1, 3}$$

Since Φ is indeterminate to the extent of a constant in space we may set

$$h(t) = \text{constant} = H.$$

Then $\qquad \dfrac{dp}{d\rho} = \gamma \dfrac{p_0}{\rho_0^\gamma}\rho^{\gamma-1} = (\gamma-1)\left(H + \dfrac{\partial\Phi}{\partial t} - \tfrac{1}{2}q^2 - W\right) = a^2, \qquad$ (4·1, 4)

where a is a function of x, y, z, t which may be assumed positive. We note that a has the dimensions of a velocity. From (4·1, 4),

$$\frac{1}{\rho}\frac{\partial\rho}{\partial x} = \frac{1}{(\gamma-1)a^2}\frac{\partial}{\partial x}(a^2) = \frac{1}{a^2}(\Phi_{xt} - \Phi_x\Phi_{xx} - \Phi_y\Phi_{xy} - \Phi_z\Phi_{xz} - W_x),$$

where the subscripts indicate partial differentiation. Similarly,

$$\frac{1}{\rho}\frac{\partial\rho}{\partial y} = \frac{1}{(\gamma-1)a^2}\frac{\partial}{\partial y}(a^2) = \frac{1}{a^2}(\Phi_{yt} - \Phi_x\Phi_{xy} - \Phi_y\Phi_{yy} - \Phi_z\Phi_{zy} - W_y),$$

$$\frac{1}{\rho}\frac{\partial\rho}{\partial z} = \frac{1}{(\gamma-1)a^2}\frac{\partial}{\partial z}(a^2) = \frac{1}{a^2}(\Phi_{zt} - \Phi_x\Phi_{xz} - \Phi_y\Phi_{yz} - \Phi_z\Phi_{zz} - W_z),$$

$$\frac{1}{\rho}\frac{\partial\rho}{\partial t} = \frac{1}{(\gamma-1)a^2}\frac{\partial}{\partial t}(a^2) = \frac{1}{a^2}(\Phi_{tt} - \Phi_x\Phi_{xt} - \Phi_y\Phi_{yt} - \Phi_z\Phi_{zt} - W_t).$$

To obtain a differential equation for the velocity potential Φ, we write the equation of continuity (1·3, 4) in the form

$$-\frac{1}{\rho}\frac{\partial\rho}{\partial x}u - \frac{1}{\rho}\frac{\partial\rho}{\partial y}v - \frac{1}{\rho}\frac{\partial\rho}{\partial z}w - \frac{\partial u}{\partial x} - \frac{\partial v}{\partial y} - \frac{\partial w}{\partial z} = \frac{1}{\rho}\frac{\partial\rho}{\partial t}.$$

Substituting the expressions first obtained for $\dfrac{1}{\rho}\dfrac{\partial\rho}{\partial x}$, etc., and obtain, bearing in mind that $\mathbf{q} = -\operatorname{grad}\Phi$,

$$\left(1 - \frac{\Phi_x^2}{a^2}\right)\Phi_{xx} + \left(1 - \frac{\Phi_y^2}{a^2}\right)\Phi_{yy} + \left(1 - \frac{\Phi_z^2}{a^2}\right)\Phi_{zz} - 2\frac{\Phi_x\Phi_y}{a^2}\Phi_{xy}$$

$$- 2\frac{\Phi_z\Phi_x}{a^2}\Phi_{zx} - 2\frac{\Phi_y\Phi_z}{a^2}\Phi_{yz}$$

$$= \frac{1}{a^2}(\Phi_{tt} - 2\Phi_x\Phi_{xt} - 2\Phi_y\Phi_{yt} - 2\Phi_z\Phi_{zt} + W_x\Phi_x + W_y\Phi_y + W_z\Phi_z - W_t). \tag{4·1, 5}$$

By (4·1, 4), a^2 is a quantity which involves the first derivatives of Φ and a known function of x, y, z, t. It follows that (4·1, 5) is a quasi-linear partial differential equation of the second order. That is to say, the highest partial derivatives in (4·1, 5) are of the second order and they appear linearly with coefficients which depend on the derivatives of lower order only. It is shown in the general theory of these equations‡ that the general character of the solution depends only on the coefficients of the second-order derivatives.

In practice the external body force is the force of gravity, so that

$$W_x\Phi_x + W_y\Phi_y + W_z\Phi_z - W_t$$

on the right-hand side of (4·1, 5) reduces to $g\Phi_z$. It is usual to neglect the effect of gravity, so that (4·1, 5) becomes

$$\left(1-\frac{\Phi_x^2}{a^2}\right)\Phi_{xx} + \left(1-\frac{\Phi_y^2}{a^2}\right)\Phi_{yy} + \left(1-\frac{\Phi_z^2}{a^2}\right)\Phi_{zz} - 2\frac{\Phi_x\Phi_y}{a^2}\Phi_{xy}$$

$$-2\frac{\Phi_z\Phi_x}{a^2}\Phi_{zx} - 2\frac{\Phi_y\Phi_z}{a^2}\Phi_{yz}$$

$$= \frac{1}{a^2}(\Phi_{tt} - 2\Phi_x\Phi_{xt} - 2\Phi_y\Phi_{yt} - 2\Phi_z\Phi_{zt}). \quad (4·1, 6)$$

Now assume that a solid body which was initially at rest follows a prescribed motion in the fluid, which is itself at rest at infinity. Let p_0, ρ_0 be the initial pressure and density. Denoting the initial value of a by a_0, we then have, by (4·1, 4),

$$a_0^2 = (\gamma - 1)H = \gamma\frac{p_0}{\rho_0}. \quad (4·1, 7)$$

Under standard conditions a_0 is equal to about 763 m.p.h. Then at any time

$$a^2 = a_0^2 + (\gamma - 1)\left(\frac{\partial\Phi}{\partial t} - \tfrac{1}{2}(\Phi_x^2 + \Phi_y^2 + \Phi_z^2)\right). \quad (4·1, 8)$$

Substituting this expression for a^2 in (4·1, 6) we see that the only constants which occur in the equation for Φ are γ and a_0. Also, Φ satisfies a boundary condition such as that given by (1·2, 8)

$$\frac{\partial\Phi}{\partial x}\frac{\partial F}{\partial x} + \frac{\partial\Phi}{\partial y}\frac{\partial F}{\partial y} + \frac{\partial\Phi}{\partial z}\frac{\partial F}{\partial z} - \frac{\partial F}{\partial t} = 0, \quad (4·1, 9)$$

where $F(x, y, z, t) = 0$ is the equation of the surface of the body.

We now introduce the following substitutions: $\Phi = \lambda\Phi'$, $x = \mu x'$, $y = \mu y'$, $z = \mu z'$, $t = \nu t'$, $a_0 = \sigma a_0'$. Then equations (4·1, 6), (4·1, 8) and (4·1, 9) are still satisfied for the primed variables provided

$$\nu = \mu^2/\lambda = \mu/\sigma, \quad \sigma = \lambda/\mu.$$

Now $\dfrac{\partial\Phi}{\partial x} = \dfrac{\lambda}{\mu}\dfrac{\partial\Phi'}{\partial x'}$, so that the velocity component $\Phi_{x'}'$ is obtained from Φ_x by scaling the latter down in the ratio $1 : \mu/\lambda = \sigma : 1$. The same applies to

‡ See, for example, R. Courant and D. Hilbert, *Methoden der Mathematischen Physik* (Julius Springer, Berlin, 1937), ch. 5.

all the other velocity components of the fluid flow and also to the velocity components of the surface of the body at corresponding points. It follows that the ratio of any velocity component and of a_0 remains constant on passing to the primed system of variables, and the same applies to the ratio of any velocity component and of a, by (4·1, 8). The ratios just considered are called Mach numbers.

The primed variables determine the motion of a body through a compressible inviscid fluid with the same constant γ as before such that the position of the body in the second case is obtained from the position of the body in the first case by applying the scale factors $1/\mu$ and $1/\nu = \sigma/\mu$ to the space coordinates and to the time respectively. We conclude that in that case, all the velocity components are related in the ratio $\sigma = a_0/a_0'$ at corresponding points, at corresponding instants of time, and hence that the corresponding Mach numbers are identical.

Assume now that the body is rigid. Then we may reformulate our result in a more effective manner, as follows:

Given only that γ is the same in both cases, that the two bodies are geometrically similar and similarly disposed initially, and that the Mach numbers of the linear velocity components of the rigid bodies with respect to a_0, a_0' are the same at corresponding moments of time, while the angular velocity components are related in the ratio $\omega'/\omega = a_0'l/a_0l'$—we conclude that the velocities in the second field of flow are related to the corresponding velocities in the first field of flow by the ratio a_0'/a_0 throughout.

In this connexion 'corresponding moments of time' are related to each other by $t'/t = a_0l'/a_0'l$, where l and l' are typical linear dimensions of the two bodies. The same results may be obtained by abstract dimensional analysis. Thus, for steady rectilinear motion of a rigid body, the flow depends only on γ and on the Mach number U/a_0, where U is the forward velocity of the body. a_0 may be taken as the value of $\sqrt{\left(\gamma \dfrac{p}{\rho}\right)}$ in an undisturbed region of the fluid, e.g. far upstream of the body.

We shall now assume that the velocities induced by the presence of the solid body, or of any other disturbance, are small. That is to say, these quantities are small compared with a_0 which is essentially the only velocity which is given by the initial data. The assumption implies in the first instance that products of derivatives of Φ can be omitted compared with terms which are linear in these derivatives. Thus (4·1, 6) and (4·1, 8) become

$$\Phi_{xx} + \Phi_{yy} + \Phi_{zz} = \frac{1}{a^2}\Phi_{tt}, \qquad (4·1, 10)$$

$$a^2 = a_0^2 + (\gamma - 1)\frac{\partial \Phi}{\partial t}. \qquad (4·1, 11)$$

It is not strictly a consequence of the assumptions made so far that $\partial \Phi / \partial t$ also must be made small compared with a_0^2. We assume that this condition also is satisfied, so that a^2 may be replaced by a_0^2 in (4·1, 10). Then

$$\Phi_{xx} + \Phi_{yy} + \Phi_{zz} = \frac{1}{a_0^2} \Phi_{tt}. \qquad (4\cdot1, 12)$$

This is the wave equation in three dimensions. We shall give a second derivation of the equation which emphasizes some other aspects of the approximations involved.

We now begin with the assumption that the density ρ never varies greatly from its value under undisturbed conditions ρ_0. This assumption was in fact implicit in the replacement of a^2 by a_0^2 in (4·1, 12). Thus we write

$$\rho = \rho_0(1+s), \qquad (4\cdot1, 13)$$

where the 'condensation' $s = s(x, y, z, t)$ is small compared with unity. Then

$$p = p_0 \left(\frac{\rho}{\rho_0}\right)^\gamma = p_0(1+s)^\gamma \doteqdot p_0(1+\gamma s), \qquad (4\cdot1, 14)$$

and $\qquad a^2 = \dfrac{dp}{d\rho} = \gamma \dfrac{p_0}{\rho_0^\gamma} \rho^{\gamma-1} \doteqdot \gamma \dfrac{p_0}{\rho_0}(1+(\gamma-1)s) \doteqdot \gamma \dfrac{p_0}{\rho_0} = a_0^2.$

Assuming that the velocity components are small and neglecting products, as before, we obtain Euler's equations (1·4, 9) in the simplified form

$$\left.\begin{aligned}
\frac{\partial u}{\partial t} &= -\frac{1}{\rho}\frac{\partial p}{\partial x} = -\frac{a^2}{\rho}\frac{\partial \rho}{\partial x}, \\[2mm]
\frac{\partial v}{\partial t} &= -\frac{1}{\rho}\frac{\partial p}{\partial y} = -\frac{a^2}{\rho}\frac{\partial \rho}{\partial y}, \\[2mm]
\frac{\partial w}{\partial t} &= -\frac{1}{\rho}\frac{\partial p}{\partial z} = -\frac{a^2}{\rho}\frac{\partial \rho}{\partial z}.
\end{aligned}\right\} \qquad (4\cdot1, 15)$$

Hence $\dfrac{1}{\rho}\dfrac{\partial \rho}{\partial x} = -\dfrac{1}{a^2}\dfrac{\partial u}{\partial t}$, etc. Substituting these expressions in the equation of continuity (1·3, 4) we obtain

$$\frac{\partial \rho}{\partial t} - \frac{\rho}{a^2}\left(u\frac{\partial u}{\partial t} + v\frac{\partial v}{\partial t} + w\frac{\partial w}{\partial t}\right) + \rho\left(\frac{\partial u}{\partial x} + \frac{\partial v}{\partial y} + \frac{\partial w}{\partial z}\right) = 0.$$

The second term now contains products of the velocities and of their derivatives which are small compared with a^2, and may therefore be omitted. Hence

$$\frac{\partial \rho}{\partial t} = -\rho\left(\frac{\partial u}{\partial x} + \frac{\partial v}{\partial y} + \frac{\partial w}{\partial z}\right).$$

But $\dfrac{\partial \rho}{\partial t} = \rho_0 \dfrac{\partial s}{\partial t}$, $\rho = \rho_0(1+s)$, $u = -\Phi_x$, etc., and so

$$\frac{\partial s}{\partial t} = \Phi_{xx} + \Phi_{yy} + \Phi_{zz}. \qquad (4\cdot1, 16)$$

Again, we may write the equations of motion (4·1, 15) in the form

$$\frac{\partial}{\partial x}\frac{\partial \Phi}{\partial t} = \frac{\partial}{\partial x}\int\frac{dp}{\rho},$$

$$\frac{\partial}{\partial y}\frac{\partial \Phi}{\partial t} = \frac{\partial}{\partial y}\int\frac{dp}{\rho},$$

$$\frac{\partial}{\partial z}\frac{\partial \Phi}{\partial t} = \frac{\partial}{\partial z}\int\frac{dp}{\rho}.$$

Hence

$$\frac{\partial \Phi}{\partial t} - \int_{p_0}^{p}\frac{dp}{\rho} = h(t). \qquad (4·1, 17)$$

This equation can also be obtained directly from Bernoulli's equation by neglecting squares of the velocity components. Now

$$\int_{p_0}^{p}\frac{dp}{\rho} = \int_{\rho_0}^{\rho}\frac{1}{\rho}\frac{dp}{d\rho}\,d\rho \doteqdot a_0^2\int_{\rho_0}^{\rho}\frac{1}{\rho}\,d\rho = a_0^2\log\frac{\rho}{\rho_0} \doteqdot a_0^2 s,$$

and so

$$\frac{\partial \Phi}{\partial t} = a_0^2 s, \qquad (4·1, 18)$$

where we have assumed that the function $h(t)$ is absorbed by $\partial\Phi/\partial t$. Combining (4·1, 16) with (4·1, 18) we then obtain (4·1, 12).

The wave equation (4·1, 12) is applicable in compressible aerofoil theory, at least as a first approximation, because we may assume that the disturbance velocities set up by a wing which is thin and at small incidence relative to the direction of motion, are small compared with a_0. Except in the region of the leading edge, this is quite compatible with the assumption that the forward velocity of the wing is of the order of a_0. Assume now that a rigid wing advances with constant velocity U in the negative direction of the x-axis. We refer (4·1, 12) to a system of coordinates fixed in the wing which is given by

$$x' = x + Ut,$$

$$y' = y,$$

$$z' = z,$$

$$t' = t.$$

Then

$$\frac{\partial^2 \Phi}{\partial x^2} = \frac{\partial^2 \Phi}{\partial x'^2}, \quad \frac{\partial^2 \Phi}{\partial y^2} = \frac{\partial^2 \Phi}{\partial y'^2}, \quad \frac{\partial^2 \Phi}{\partial z^2} = \frac{\partial^2 \Phi}{\partial z'^2},$$

$$\frac{\partial^2 \Phi}{\partial t^2} = \left(U\frac{\partial}{\partial x'} + \frac{\partial}{\partial t'}\right)^2 \Phi = U^2\frac{\partial^2 \Phi}{\partial x'^2} + 2U\frac{\partial^2 \Phi}{\partial x'\,\partial t'} + \frac{\partial^2 \Phi}{\partial t'^2}.$$

Substituting in (4·1, 12) and omitting the primes, we obtain

$$\left(1 - \frac{U^2}{a_0^2}\right)\frac{\partial^2 \Phi}{\partial x^2} + \frac{\partial^2 \Phi}{\partial y^2} + \frac{\partial^2 \Phi}{\partial z^2} = \frac{1}{a_0^2}\frac{\partial^2 \Phi}{\partial t^2} - 2\frac{U}{a_0^2}\frac{\partial^2 \Phi}{\partial x\,\partial t}. \qquad (4·1, 19)$$

In particular, if the flow is steady relative to the wing, (4·1, 19) becomes

$$\left(1 - \frac{U^2}{a_0^2}\right)\frac{\partial^2\Phi}{\partial x^2} + \frac{\partial^2\Phi}{\partial y^2} + \frac{\partial^2\Phi}{\partial z^2} = 0. \tag{4·1, 20}$$

This is the linearized equation of steady compressible flow. U is now the velocity of the fluid at infinity (free-stream velocity). The exact equation of steady compressible flow is obtained by setting the right-hand side equal to zero in (4·1, 6), so

$$\left(1 - \frac{\Phi_x^2}{a^2}\right)\Phi_{xx} + \left(1 - \frac{\Phi_y^2}{a^2}\right)\Phi_{yy} + \left(1 - \frac{\Phi_z^2}{a^2}\right)\Phi_{zz} - 2\frac{\Phi_x\Phi_y}{a^2}\Phi_{xy}$$
$$- 2\frac{\Phi_z\Phi_x}{a^2}\Phi_{zx} - 2\frac{\Phi_y\Phi_z}{a^2}\Phi_{yz} = 0. \tag{4·1, 21}$$

(4·1, 20) can then be obtained from (4·1, 21) by setting $a = a_0$, $\Phi_x = 0$, $\Phi_y = \Phi_z = 0$ in that formula. Finally, (4·1, 20) can also be derived directly from the equations of motion and the equation of continuity, by assuming that the velocities induced by the presence of the wing u', v, w are small compared with the free-stream velocity U, where $u = U + u'$, and where U is supposed to be of the order of magnitude of a. Then a^2 can again be replaced by a constant, a_0^2. Omitting products of small quantities, we then obtain, in place of Euler's equations of motion,

$$\left.\begin{aligned}U\frac{\partial u'}{\partial x} &= -\frac{1}{\rho}\frac{\partial p}{\partial x} = -\frac{a_0^2}{\rho}\frac{\partial p}{\partial x}, \\[4pt] U\frac{\partial v}{\partial x} &= -\frac{1}{\rho}\frac{\partial p}{\partial y} = -\frac{a_0^2}{\rho}\frac{\partial p}{\partial y}, \\[4pt] U\frac{\partial w}{\partial x} &= -\frac{1}{\rho}\frac{\partial p}{\partial z} = -\frac{a_0^2}{\rho}\frac{\partial p}{\partial z}.\end{aligned}\right\} \tag{4·1, 22}$$

Substituting the resulting expressions for $\frac{1}{\rho}\frac{\partial\rho}{\partial x}$, etc., in the equation of continuity for steady compressible flow, we obtain

$$-\frac{U}{a_0^2}\left((U + u')\frac{\partial u'}{\partial x} + v\frac{\partial v}{\partial x} + w\frac{\partial w}{\partial x}\right) + \frac{\partial u'}{\partial x} + \frac{\partial v}{\partial y} + \frac{\partial w}{\partial z} = 0,$$

or, omitting products of small quantities,

$$\left(1 - \frac{U^2}{a_0^2}\right)\frac{\partial u'}{\partial x} + \frac{\partial v}{\partial y} + \frac{\partial w}{\partial z} = 0. \tag{4·1, 23}$$

In this equation, we may replace $\partial u'/\partial x$ by $\partial u/\partial x$, since u differs from u' only by a constant. (4·1, 20) is then obtained by expressing in terms of Φ, $u = -\partial\Phi/\partial x$, etc.

To derive a corresponding formula for the pressure, we use (4·1, 2) and (4·1, 3), with $h(t) = \text{constant} = H$, $\partial\Phi/\partial t = 0$, $W = 0$,

$$p = \frac{p_0}{\rho_0^\gamma}\rho^\gamma = \left(\frac{\gamma - 1}{\gamma}\right)^{\gamma/(\gamma-1)}\left(\frac{\rho_0^\gamma}{p_0}\right)^{1/(\gamma-1)}(H - \tfrac{1}{2}(u^2 + v^2 + w^2))^{\gamma/(\gamma-1)}. \tag{4·1, 24}$$

Now p_0 and ρ_0 may be taken as the pressure and density in the free stream far ahead of the wing, where $u = U$, $v = w = 0$. Hence H is given by the condition

$$p_0 = \left(\frac{\gamma - 1}{\gamma}\right)^{\gamma/(\gamma-1)} \left(\frac{\rho_0^\gamma}{p_0}\right)^{1/(\gamma-1)} (H - \tfrac{1}{2}U^2)^{\gamma/(\gamma-1)}$$

$$\frac{\gamma - 1}{\gamma}(H - \tfrac{1}{2}U^2) = \frac{p_0}{\rho_0}. \tag{4·1, 25}$$

Writing $u = U + u'$ in (4·1, 24) and making use of (4·1, 25), we then obtain

$$p = p_0\left[1 - \frac{\gamma - 1}{2\gamma}\frac{\rho_0}{p_0}(2Uu' + u'^2 + v^2 + w^2)\right]^{\gamma/(\gamma-1)}. \tag{4·1, 26}$$

Now it is assumed that p never differs greatly from p_0, and so the second term in the square brackets is small compared with unity. Hence

$$p \doteqdot p_0\left[1 - \frac{\rho_0}{p_0}(Uu' + \tfrac{1}{2}(u'^2 + v^2 + w^2))\right] \doteqdot p_0\left(1 - \frac{\rho_0}{p_0}Uu'\right),$$

or $$p = p_0 - \rho_0 Uu' = p_0 - \rho_0 U(u - U). \tag{4·1, 27}$$

(4·1, 27) is the linearized form of Bernoulli's equation for steady compressible flow. A quicker, though less instructive, derivation of (4·1, 27) is based on the integration of (4·1, 22). If we wish to retain also the terms of second order in the expression for p, we have to continue the binomial expansion of the right-hand side of (4·1, 26). Thus, we now write

$$p \doteqdot p_0\left[1 - \frac{\rho_0}{p_0}(Uu' + \tfrac{1}{2}(u'^2 + v^2 + w^2))\right.$$
$$\left. + \frac{\rho_0^2}{p_0^2}\frac{1}{2}\frac{\gamma}{\gamma - 1}\left(\frac{\gamma}{\gamma - 1} - 1\right)\left(\frac{\gamma - 1}{2\gamma}\right)^2 (2Uu' + u'^2 + v^2 + w^2)^2\right].$$

Omitting terms of third or higher order of smallness and taking into account that

$$\frac{1}{2\gamma}\frac{\rho_0}{p_0}U^2 u'^2 = \frac{U^2}{2a_0^2}u'^2,$$

we obtain $$p = p_0 - \rho_0(Uu' + \tfrac{1}{2}((1 - M^2)\,u'^2 + v^2 + w^2)). \tag{4·1, 28}$$

In this formula we have put $M = U/a_0$. M is a Mach number in the sense defined above. We call it the Mach number of the flow. For a given Mach number, geometrically similar wings, which are similarly disposed towards the free stream, yield velocity fields which are obtained from each other by changing the velocity components everywhere in the ratio of the free-stream velocities.

If the velocity potential is determined by means of (4·1, 20), it appears reasonable to use (4·1, 27) for the determination of the pressure at the wing, since both formulae depend on approximations of the same order. On the other hand, the momentum principle involves essentially squares of the velocities, and accordingly (4·1, 28) will be used in a consistent formulation of that principle, even when the velocity potential is determined by means of (4·1, 20).

The momentum principle for steady compressible flow (1·4, 16) states that the force exerted by the fluid on a solid body immersed in it is

$$-\mathbf{T} = -\left(\int_S p \, d\mathbf{S} + \int_S \rho \mathbf{q}(\mathbf{q} \, d\mathbf{S}) \right), \qquad (4\cdot1, 29)$$

where S is a closed surface surrounding the body, $\mathbf{q}$ is the velocity vector, $\mathbf{q} = (U + u', v, w)$, and p is the pressure as given by (4·1, 28). Now

$$\int_S \rho \mathbf{q}(\mathbf{q} \, d\mathbf{S}) = U \int_S \rho(\mathbf{q} \, d\mathbf{S}) + \int_S \rho(u', v, w)(\mathbf{q} \, d\mathbf{S})$$

$$= \int_S \rho(u', v, w) [(U + u') \, dy \, dz + v \, dz \, dx + w \, dx \, dy],$$

since the first integral in the second member indicates the amount of fluid which crosses S in unit time and so vanishes identically. Also, by (4·1, 13) and (4·1, 18),

$$\rho = \rho_0(1 + s) \doteq \rho_0 \left(1 - \frac{U}{a_0^2} u' \right) = \rho_0 \left(1 - M^2 \frac{u'}{U} \right),$$

where $\partial \Phi / \partial t$ in (4·1, 18) has been replaced by $U(\partial \Phi / \partial x) = -Uu'$ on passing to a coordinate system which moves with the body. Hence

$$\int_S \rho \mathbf{q}(\mathbf{q} \, d\mathbf{S}) \doteq \rho_0 \int_S \left(1 - M^2 \frac{u'}{U} \right) (u', v, w) [(U + u') \, dy \, dz + v \, dz \, dx + w \, dx \, dy]$$

$$\doteq \rho_0 \int_S (U - M^2 u')(u', v, w) \, dy \, dz$$

$$+ \rho_0 \int_S (u', v, w)(u' \, dy \, dz + v \, dz \, dx + w \, dx \, dy),$$

where we have omitted terms of third or higher order of smallness. Substituting the value of p from (4·1, 28) in (4·1, 29), we obtain

$$-\mathbf{T} = \rho_0 \left\{ \int_S (Uu' + \tfrac{1}{2}[(1 - M^2) u'^2 + v^2 + w^2]) (dy \, dz, dz \, dx, dx \, dy) \right.$$

$$- \int_S (U - M^2 u')(u', v, w) \, dy \, dz$$

$$\left. - \int_S (u', v, w)(u' \, dy \, dz + v \, dz \, dx + w \, dx \, dy) \right\}. \qquad (4\cdot1, 30)$$

This is the momentum theorem to the order of accuracy required in linearized theory.

Although the equations (4·1, 20) and (4·1, 21) can be investigated directly without reference to (4·1, 12) or (4·1, 6), we shall find it easier to understand the physical nature of compressible flow if we consider first some simple cases of unsteady motion.

Let $P(x_0, y_0, z_0)$ be an arbitrary fixed point and let

$$r = \sqrt{\{(x - x_0)^2 + (y - y_0)^2 + (z - z_0)^2\}},$$

where the point (x, y, z) is regarded as variable. Then it is easy to verify by direct differentiation that the function

$$\Phi(x, y, z, t) = \frac{f(r - a_0 t)}{r} \qquad (4·1, 31)$$

is a solution of the linearized equation $(4·1, 12)$. In this formula, $f(\tau)$ is an arbitrary function of one variable τ. Similarly

$$\Phi(x, y, z, t) = \frac{g(r + a_0 t)}{r} \qquad (4·1, 32)$$

is a solution $(4·1, 12)$ where g is again arbitrary.

To interpret the meaning of $(4·1, 31)$, we compare conditions at times t and $t + \Delta t$, where Δt is some finite positive increment. Then the argument in the numerator of $(4·1, 31)$ in the second case will be the same as, for a given r, in the first case, if we add the increment $\Delta r = a_0 \Delta t$ to r. At the same time the denominator of course increases in the ratio $(r + \Delta r)/r$. That is to say, $(4·1, 31)$ represents a disturbance which travels outward from P with velocity a_0, while being attenuated all the time in the ratio of the distances from P. Φ is therefore said to be the potential of a diverging wave, or, more briefly, a retarded potential. The fact that a_0 is the velocity of propagation of the disturbance can be seen more clearly if we consider the particular case given by

$$f(\tau) = 1 \quad \text{for} \quad \tau \leqslant 0, \\ f(\tau) = 0 \quad \text{for} \quad \tau > 0. \qquad (4·1, 33)$$

Then in the neighbourhood of P, i.e. for small r, the argument of f in $(4·1, 31)$ becomes negative for small positive t and for all subsequent t. Hence $\Phi = 1/r$ for such values of t. On the other hand, at a point P_0, which is at a finite distance r_0 from P, the argument of f is positive for $t < r_0/a_0$ and negative for $t > r_0/a_0$. Hence $\Phi = 0$ for $t < r_0/a_0$ and $\Phi = 1/r$ for $t \geqslant r_0/a_0$. In other words, the fluid in the neighbourhood of P_0 is at rest up to the time $t = r_0/a_0$, which is therefore the time required by the disturbance to travel from P to P_0, a distance r_0. It follows that a_0 is indeed the velocity of propagation of the disturbance. Small disturbances in air are familiar as sound waves, so that the constant a_0 is the velocity of sound.

A similar argument shows that $(4·1, 32)$ represents a wave which converges towards P with the velocity a_0. In that case Φ is also said to be an advanced potential.

Let S be a closed surface, which bounds a volume V. The rate of fluid flow across S in outward direction is given by

$$\int_S \rho \mathbf{q} \, d\mathbf{S}. \qquad (4·1, 34)$$

According to our present assumptions, $\rho = \rho_0(1+s)$, so that ρ differs only by a small quantity from ρ_0. We may therefore replace (4·1, 34) by

$$\rho_0 \int_S \mathbf{q}\, d\mathbf{S} = -\rho_0 \int_S \operatorname{grad} \Phi\, d\mathbf{S}. \tag{4·1, 35}$$

If Φ is finite (and possesses continuous second-order derivatives) within S, then we may replace the right-hand side of (4·1, 35) by

$$-\rho_0 \int_V \operatorname{div} \operatorname{grad} \Phi\, dV = -\frac{\rho_0}{a_0^2} \int_V \frac{\partial^2 \Phi}{\partial t^2}\, dV = -\rho_0 \int_V \frac{\partial s}{\partial t}\, dV = -\int_V \frac{\partial \rho}{\partial t}\, dV.$$

Thus

$$\rho_0 \int_S \mathbf{q}\, d\mathbf{S} = -\int_V \frac{\partial \rho}{\partial t}\, dV, \tag{4·1, 36}$$

which is merely a verification of the fact that the rate of outward flow of fluid across S is balanced by the rate of change of fluid mass within S.

Now assume that Φ is given by (4·1, 31), and that S is a sphere of radius R whose centre is at P. Then (4·1, 36) does not apply since Φ becomes infinite within S. In order to calculate the flow directly, we observe that

$$\frac{\partial \Phi}{\partial r} = \frac{\partial}{\partial r} \frac{f(r-a_0 t)}{r} = \frac{1}{r} f'(r-a_0 t) - \frac{1}{r^2} f(r-a_0 t).$$

Hence

$$\rho_0 \int_S \mathbf{q}\, d\mathbf{S} = -\rho_0 \int_S \operatorname{grad} \Phi\, d\mathbf{S} = -\rho_0 \int_S \frac{\partial \Phi}{\partial r}\, dS = 4\pi\rho_0(f(r-a_0 t) - rf'(r-a_0 t)).$$

Letting r tend to zero, we obtain on the right-hand side $4\pi\rho_0 f(-a_0 t)$. It follows that this must be the rate at which fluid is created at P. On the other hand, (4·1, 36) shows that if we take S as a small sphere which does not contain P in its interior, then $\rho_0 \int_S \mathbf{q}\, d\mathbf{S}$ vanishes in the limit. For these reasons (4·1, 31) is also said to represent a source (sometimes, an acoustic source) at P whose strength at time t is $4\pi f(-a_0 t)$. Thus, in quoting the strength of a source, we shall omit the constant factor ρ_0, and similarly, we shall refer to $\int_S \mathbf{q}\, d\mathbf{S}$ rather than $\rho_0 \int_S \mathbf{q}\, d\mathbf{S}$ as the flux across S.

Consider a distribution of sources along a straight line parallel to the x-axis, as given by $y = y_0$, $z = z_0$, say. Let the strength of the source at a point x_0 at a time t be given by $F(x_0, t)$ per unit length of the line. The total effect of the sources at a point (x, y, z) at a time t then is

$$\Phi(x, y, z, t) = \int_{-\infty}^{\infty} \frac{f(x_0, r-a_0 t)}{r}\, dx_0, \tag{4·1, 37}$$

where

$$r = \sqrt{\{(x-x_0)^2 + (y-y_0)^2 + (z-z_0)^2\}},$$

and

$$f(x_0, \tau) = \frac{1}{4\pi} F\left(x_0, -\frac{\tau}{a_0}\right),$$

taking account of the above definition of the strength of a source. Hence

$$\Phi(x, y, z, t) = \frac{1}{4\pi} \int_{-\infty}^{\infty} \frac{F(x_0, t - r/a_0)}{r} dx_0. \qquad (4·1, 38)$$

This formula shows that the contribution to Φ at a point (x, y, z) at a time t, which is due to the source at $x = x_0$ depends only on the strength of the source at time $t - r/a_0$.

Assume in particular that $F(x_0, t)$ is of the form

$$F(x_0, t) = G(x_0 + Ut), \qquad (4·1, 39)$$

where U is a positive constant. $(4·1, 39)$ may be said to represent a system of sources which travel with velocity U in the negative direction of the x-axis. The corresponding velocity potential is

$$\Phi(x, y, z, t) = \frac{1}{4\pi} \int_{-\infty}^{\infty} \frac{G(x_0 + Ut - Ur/a_0)}{r} dx_0. \qquad (4·1, 40)$$

We now refer the velocity potential to a system of coordinates which moves with the sources. If the two systems coincide at time $t = 0$, then the co-ordinate transformation is given by

$$x' = x + Ut, \quad y' = y, \quad z' = z, \quad t' = t. \qquad (4·1, 41)$$

Then
$$\Phi(x', y', z', t') = \frac{1}{4\pi} \int_{-\infty}^{\infty} \frac{G(x_0' - Ur'/a_0)}{r'} dx_0', \qquad (4·1, 42)$$

where $x_0' = x_0 + Ut$ and $r' = \sqrt{\{(x' - x_0')^2 + (y' - y_0)^2 + (z' - z_0)^2\}}$,

so that Φ is independent of time, as might be expected.

Consider the special case

$$G(\tau) = \begin{cases} 0 & \text{for} \quad \tau < \tau_0, \\ \text{constant} = G_0 & \text{for} \quad \tau_0 \leqslant \tau \leqslant \tau_0 + \delta, \\ 0 & \text{for} \quad \tau > \tau_0 + \delta, \end{cases}$$

where τ_0 is an arbitrary constant, and δ is a small positive quantity. Then

$$\Phi(x', y', z') = \frac{G_0}{4\pi} \int \frac{dx_0'}{r}, \qquad (4·1, 43)$$

where the integral on the right-hand side ranges over the values of x_0' for which

$$\tau_0 \leqslant x_0' - \frac{Ur'}{a_0} \leqslant \tau_0 + \delta. \qquad (4·1, 44)$$

To interpret $(4·1, 44)$, we determine the roots of the equation

$$x_0' - \frac{Ur'}{a_0} = c, \quad c \text{ arbitrary.} \qquad (4·1, 45)$$

We may write this equation in the form

$$x_0' - c = Mr', \qquad (4·1, 46)$$

where we have introduced the Mach number $M = U/a_0$.

Since r' is always positive (or zero), we see that the roots of (4·1, 46) are precisely those roots of

$$(x_0' - c)^2 = M^2 r'^2 = M^2 \{(x_0' - x')^2 + (y_0 - y')^2 + (z_0 - z')^2\} \quad (4·1, 47)$$

for which $x_0' - c \geqslant 0$. The roots of (4·1, 47) are

$$x_0' = c + \frac{-M^2(x' - c) \pm \sqrt{\{M^4(x' - c)^2 + (1 - M^2) M^2((x' - c)^2 + r_0^2)\}}}{1 - M^2}, \quad (4·1, 48)$$

where

$$r_0^2 = (y_0 - y')^2 + (z_0 - z')^2.$$

We consider first the case $M < 1$. Then x_0' possesses just one root for which $x_0' - c \geqslant 0$, and this corresponds to the positive square root in (4·1, 48). It is a single root except when $x' = c$, $y' = y_0$, $z' = z_0$, and this condition will be excluded from now on. (4·1, 48) may also be written in the form

$$x_0' = \frac{c - M^2 x' \pm M \sqrt{\{(x' - c)^2 + (1 - M^2) r_0^2\}}}{1 - M^2}, \quad (4·1, 49)$$

and the derivative of the right-hand side with respect to c is

$$\frac{1}{1 - M^2} \left(1 \pm M \frac{c - x'}{\sqrt{\{(x' - c)^2 + (1 - M^2) r_0^2\}}}\right). \quad (4·1, 50)$$

Note that (4·1, 50) is positive so that the root x_0' increases with increasing c. Also, the left-hand side of (4·1, 45) is clearly numerically large and negative for large negative x_0'. It follows that the lower limit of the range of x_0' which corresponds to (4·1, 44) is obtained by putting $c = \tau_0$ in (4·1, 49), while the upper limit is given by $c = \tau_0 + \delta$, so

$$\frac{\tau_0 - M^2 x' \pm M \sqrt{\{(x' - \tau_0)^2 + (1 - M^2) r_0^2\}}}{1 - M^2} \leqslant x_0'$$

$$\leqslant \frac{\tau_0 + \delta - M^2 x' \pm M \sqrt{\{(x' - \tau_0 - \delta)^2 + (1 - M^2) r_0^2\}}}{1 - M^2}. \quad (4·1, 51)$$

For small δ the length d of this interval is obtained by putting $c = \tau_0$ in the derivative (4·1, 50) and multiplying the result by δ. Also, since this is the interval of integration for (4·1, 43), the expression for r' in the integral is given approximately by (4·1, 46) and (4·1, 49), where $c = \tau_0$. Hence

$$r' \doteqdot \frac{1}{M}(x_0' - \tau_0) = \frac{M(\tau_0 - x') \pm \sqrt{\{(x' - \tau_0)^2 + (1 - M^2) r_0^2\}}}{1 - M^2},$$

and

$$\Phi(x', y', z') \doteqdot \frac{G_0}{4\pi} \frac{d}{r'} = \frac{G_0 \delta}{4\pi} \frac{1}{\sqrt{\{(x_0' - \tau_0)^2 + (1 - M^2) r_0^2\}}}. \quad (4·1, 52)$$

We now let δ tend to zero while keeping the product $G_0 \delta$ constant, $G_0 \delta = \sigma$. Then, in the limit,

$$\Phi(x', y', z') = \frac{\sigma}{4\pi \sqrt{\{(x' - \tau_0)^2 + (1 - M^2) r_0^2\}}}. \quad (4·1, 53)$$

Omitting the primes and replacing τ_0 by x_0 we obtain finally

$$\Phi(x, y, z) = \frac{\sigma}{4\pi \sqrt{\{(x - x_0)^2 + (1 - M^2)[(y - y_0)^2 + (z - z_0)^2]\}}}. \quad (4·1, 54)$$

Assume now that $M > 1$. Then the roots of (4·1, 47) as given by (4·1, 48) and (4·1, 49) are conjugate complex if $(x' - c)^2 - (M^2 - 1) r_0^2 < 0$, real coinciding if $(x' - c)^2 - (M^2 - 1) r_0^2 = 0$, and real different if $(x' - c)^2 - (M^2 - 1) r_0^2 > 0$. Since the left-hand side of (4·1, 45) is negative and numerically as large as we please for large negative x_0', it follows that if $(x' - c)^2 - (M^2 - 1) r_0^2 < 0$, then the condition (4·1, 44) is not satisfied by any x_0'. Accordingly, $\Phi(x', y', z')$ as given by (4·1, 43) vanishes in that case. If $(x' - c)^2 - (M^2 - 1) r_0^2 > 0$ and at the same time $x' - c < 0$, then $x_0' - c$ as given by (4·1, 48) is negative, and so neither root of (4·1, 48) is a root of (4·1, 46). Thus $\Phi(x', y', z')$ vanishes in this case also.

Finally, if $(x' - c)^2 - (M^2 - 1) r_0^2 > 0$, and at the same time $x' - c > 0$, then (4·1, 47) has two distinct real roots which are also roots of (4·1, 46). Then the left-hand side of (4·1, 45), $x_0' - Mr'$, is greater than c in the finite interval determined by these roots and smaller than c outside the interval. It follows that (4·1, 44) now corresponds to the following two intervals of x_0':

$$\frac{M^2 x' - \tau_0 - M \sqrt{\{(x' - \tau_0)^2 - (M^2 - 1) r_0^2\}}}{M^2 - 1} \leqslant x_0'$$

$$\leqslant \frac{M^2 x' - \tau_0 - \delta - M \sqrt{\{(x' - \tau_0 - \delta)^2 - (M^2 - 1) r_0^2\}}}{M^2 - 1} \quad (4·1, 55)$$

and $$\frac{M^2 x' - \tau_0 - \delta + M \sqrt{\{(x' - \tau_0 - \delta)^2 - (M^2 - 1) r_0^2\}}}{M^2 - 1} \leqslant x_0'$$

$$\leqslant \frac{M^2 x' - \tau_0 + M \sqrt{\{(x' - \tau_0)^2 - (M^2 - 1) r_0^2\}}}{M^2 - 1}. \quad (4·1, 56)$$

For small δ, the approximate lengths of these intervals are obtained by multiplying δ by the modulus of (4·1, 50) for $c = \tau_0$. Thus the approximate length of the interval (4·1, 55) is

$$d_1 = \frac{\delta}{M^2 - 1} \left(-1 + M \frac{x' - \tau_0}{\sqrt{\{(x' - \tau_0)^2 - (M^2 - 1) r_0^2\}}} \right),$$

and the length of the interval (4·1, 56) is

$$d_2 = \frac{\delta}{M^2 - 1} \left(1 + M \frac{x' - \tau_0}{\sqrt{\{(x' - \tau_0)^2 - (M^2 - 1) r_0^2\}}} \right).$$

The approximate value of r' corresponding to the first interval is

$$r_1' = \frac{M(x' - \tau_0) - \sqrt{\{(x' - \tau_0)^2 - (M^2 - 1) r_0^2\}}}{M^2 - 1},$$

and, corresponding to the second interval,

$$r_2' = \frac{M(x' - \tau_0) + \sqrt{\{(x' - \tau_0)^2 - (M^2 - 1) r_0^2\}}}{M^2 - 1}.$$

Accordingly $\Phi(x', y', z')$ is given by

$$\Phi(x', y', z') \doteqdot \frac{G_0}{4\pi} \left(\frac{d_1}{r_1'} + \frac{d_2}{r_2'} \right) = \frac{G_0 \delta}{2\pi} \frac{1}{\sqrt{\{(x' - \tau_0)^2 - (M^2 - 1) r_0^2\}}}.$$

If δ tends to zero while the product $G_0 \delta$ is kept constant, $G_0 \delta = \sigma$, then, in the limit

$$\Phi(x', y', z') = \frac{\sigma}{2\pi \sqrt{\{(x' - \tau_0)^2 - (M^2 - 1) r_0^2\}}}.$$

Again, omitting the primes and replacing τ_0 by x_0, we obtain the final formula

$$\Phi(x, y, z) = \begin{cases} 0 \quad \text{for} \quad (x - x_0)^2 - (M^2 - 1)\{(y - y_0)^2 + (z - z_0)^2\} < 0, \\ \quad \text{and for} \\ \quad\quad (x - x_0)^2 - (M^2 - 1)\{(y - y_0)^2 + (z - z_0)^2\} > 0, \quad x - x_0 < 0, \\ \dfrac{\sigma}{2\pi \sqrt{\{(x - x_0)^2 - (M^2 - 1)\{(y - y_0)^2 + (z - z_0)^2\}\}}} \\ \quad \text{for} \quad (x - x_0)^2 - (M^2 - 1)\{(y - y_0)^2 + (z - z_0)^2\} > 0, \quad x - x_0 > 0. \end{cases}$$

$$(4\cdot1, 57)$$

The third region, in which $\Phi(x, y, z)$ is different from zero, constitutes the interior of one-half of the cone

$$(x - x_0)^2 - (M^2 - 1)\{(y - y_0)^2 - (z - z_0)^2\} = 0. \qquad (4\cdot1, 58)$$

This is called the Mach cone which emanates from (x_0, y_0, z_0). The two halves of this cone are distinguished as the fore-cone $(x < x_0)$ and the after-cone $(x > x_0)$.

$(4\cdot1, 54)$ and $(4\cdot1, 57)$ were obtained by the superposition of solutions of the wave equation and they are expressed in terms of a system of coordinates in which the wave equation is given by $(4\cdot1, 19)$. Also, $(4\cdot1, 54)$ and $(4\cdot1, 57)$ are independent of time and so, more particularly, these functions must be solutions of $(4\cdot1, 20)$. This can be seen also more directly by writing

$$\sqrt{(1 - M^2)}\, y, \quad \sqrt{(1 - M^2)}\, z, \quad \sqrt{(1 - M^2)}\, y_0, \quad \sqrt{(1 - M^2)}\, z_0$$

for y, z, y_0, z_0 in the potential of a source $(1\cdot7, 2)$. This transforms $(1\cdot7, 2)$ into $(4\cdot1, 54)$ or, except for a factor of $\frac{1}{2}$, into the last formula on the right-hand side of $(4\cdot1, 57)$. At the same time, Laplace's equation $(1\cdot7, 1)$ is transformed into $(4\cdot1, 20)$. Since $(1\cdot7, 2)$ is a solution of $(1\cdot7, 1)$, $(4\cdot1, 54)$ and $(4\cdot1, 57)$ must therefore be solutions of $(4\cdot1, 20)$. However, this purely formal procedure does not explain why $\Phi(x, y, z)$ is different from zero only within the after-cone emanating from (x_0, y_0, z_0), nor does it account for the difference between the numerical factors in $(4\cdot1, 54)$ and $(4\cdot1, 57)$.

The field of flow due to a moving acoustic source is of fundamental importance in compressible aerofoil theory, because the motion of an aerofoil through a compressible fluid sets up disturbances which are propagated into the fluid. If the incidence of the aerofoil is small, then we may assume that the disturbances are governed approximately by the wave equation, and a simple disturbance of this type is that due to a moving acoustic source. Moreover, within the approximations adopted here, an aerofoil can actually be replaced by a system of sources, or of solutions derived from such sources by differentiation. This was shown earlier for the limting case $M = 0$ when (4·1, 20) reduces to Laplace's equation (§ 1·7). We shall see presently that the same is still true for $M > 0$. For $M < 1$ this can in fact be shown without difficulty by means of the formal transformation mentioned above (write $\sqrt{(1 - M^2)}\,y$ for y, $\sqrt{(1 - M^2)}\,z$ for z; compare § 4·2). However, for $M > 1$ this transformation involves the imaginary quantity $\sqrt{(M^2 - 1)}\,i$, so that in this case it does not possess a direct physical or geometrical interpretation. However, in this case also, the representation of the field of flow by means of distributions of sources or of solutions derived from these, proves possible (§ 4·5).

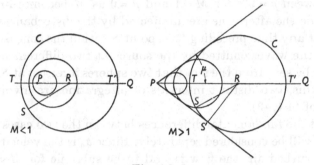

Fig. 4·1, 1

In order to understand more fully the physical difference between the cases $M < 1$ and $M > 1$, we consider a source travelling along QP which reaches P at a time t_0 (Fig. 4·1, 1). Let Δt be some finite positive interval of time. Then the source was at a point R at time $t_0 - \Delta t$ where $PR = U\,\Delta t$. The wave which was set up by the source when at R travels with velocity a_0, and so at time t_0 it forms a sphere C of radius $a_0\,\Delta t$ about R. C meets the line PQ in two points T and T' which are given by

$$(U \pm a_0)\,\Delta t = a_0(M \pm 1)\,\Delta t,$$

in terms of their distance from P, measured against the direction of motion of the source. Then for $M < 1$ the points T and T' are situated on either side of P and C contains P in its interior. Thus, the effect of the disturbance is felt at any point in the path of the source before the source actually reaches that point. Moreover, if we compare two spheres C_1 and C_2 corresponding to two different intervals of time Δt_1 and Δt_2, $\Delta t_1 < \Delta t_2$, then C_2 contains C_1 in its interior. It follows that, for $M < 1$, an arbitrary point S is at time t_0, affected only by the wave emitted by the source at a single moment of time preceding t_0. This accounts for the single interval of integration which appeared in the evaluation of (4·1, 43). At the same time, it is not difficult to see that for varying positive Δt the spheres C cover the entire space, so that no region of space remains unaffected by the disturbance.

Conditions are quite different if $M > 1$. In that case, the points T and T' are both on the same side of P as R. It follows that no disturbance is felt at P until the source actually reaches that point. Also, it is easy to show geometrically that for varying positive Δt, the spheres C will fill the interior of a circular cone with vertex at P, axis PQ and semi-apex angle

$$\mu = \sin^{-1}\frac{S'R}{PR} = \sin^{-1}\frac{a_0\Delta t}{U\Delta t} = \sin^{-1}\frac{1}{M}. \tag{4·1, 59}$$

This is the after-cone emanating from P as defined earlier. The angle μ defined by (4·1, 59) is called the Mach angle. It decreases with increasing M, varying between $\mu = 90°$ for $M = 1$ and $\mu = 0$ as M becomes infinite. The points outside the after-cone are unaffected by the disturbances set up by the source at any time preceding t_0. A point S within the cone is, at time t_0, affected by the waves emitted by the source at two different moments of time, as can be seen from the fact that two spheres intersect at S. In agreement with this, two distinct intervals of integration are involved in the evaluation of (4·1, 43).

In view of the fundamental differences between the two cases $M < 1$ and $M > 1$, they will be considered separately. Since a_0 is the velocity of sound in the undisturbed air, the flow is said to be subsonic for $M < 1$, $U < a_0$, and supersonic for $M > 1$, $U > a_0$. According to this classification, there would appear to occur a sudden change of type as the Mach number increases beyond $M = 1$. However, this conclusion is based on the approximations which lead to the wave equation, and in actual fact the transition from subsonic to supersonic flow is more gradual. For low free-stream Mach numbers, the velocity is everywhere smaller than the velocity of sound. However, for sufficiently great free-stream velocities, the flow will become supersonic (i.e. the local Mach number q/a will be greater than 1) somewhere in the fluid, even though the free-stream Mach number M is still smaller than 1. For a given wing with a given attitude relative to the main stream, the smallest free-stream Mach number for which the local Mach number reaches unity somewhere in the fluid, is called the critical Mach number $M_{\text{crit.}}$. For a normal wing at small incidence $M_{\text{crit.}}$ may be expected to be in the region 0·80–0·90. The critical Mach number is the lower limit of the so-called transonic flow régime. As the free-stream Mach number increases further, there are rapid changes in the character of the flow until it 'settles down' to the supersonic régime as the free-stream Mach number reaches the values 1·10–1·15. The lowest free-stream Mach number for which the flow is supersonic throughout the medium has been suggested as the exact dividing point between transonic and supersonic flow, but it appears that this definition is too artificial for most practical purposes.

A detailed discussion of the influence of compressibility on boundary-layer theory and profile drag is beyond the scope of the present book. It turns out that the thermodynamic conditions, which could be left out of account during the calculation of the drag for the incompressible case, play an important part in determining the profile drag on a body in compressible flow. For Mach numbers beyond the critical, the situation is complicated further by the interaction of the boundary layer with the shock waves which arise under transonic and supersonic conditions (see §§ 4·12 and 4·13). The reader is referred to *Modern Developments in Fluid Dynamics*‡ for the determination of the drag due to viscosity in compressible flow. This has to be added to the drag calculated according to the perfect fluid theory of the present chapter. However, it is important to note that the general considerations which led to the derivation of the Joukowski condition (§ 1·14) apply also to compressible flow.

To conclude this section we shall show that the law of Kutta and Joukowski (§ 2·3) still holds for subsonic compressible flow. The momentum principle (4·1, 30) becomes, for two-dimensional conditions,

$$(X, Y) = -\mathbf{T} = \rho_0 \bigg\{ \int_S (U u' + \tfrac{1}{2}[(1 - M^2) u'^2 + v^2]) \, (dy, -dx)$$
$$- \int_S (U - M^2 u') (u', v) \, dy - \int_S (u', v) (u' \, dy - v \, dx) \bigg\}, \quad (4·1, 60)$$

where S is a simple closed curve which surrounds the aerofoil. In particular, if S is a large circle of radius R whose centre is in the neighbourhood of the aerofoil, then u', v vary as R^{-1} with increasing R, while the length of the path of integration varies as R. It follows that for large R, the right-hand side of (4·1, 60) may be replaced by

$$\rho_0 U \bigg(\int_S u' (dy, -dx) - \int_S (u', v) \, dy \bigg) = -\rho_0 U \bigg(0, \int_S (u' \, dx + v \, dy) \bigg)$$
$$= -\rho_0 U \bigg(0, \int_S (u \, dx + v \, dy) \bigg). \quad (4·1, 61)$$

The integral on the right-hand side of (4·1, 61) is the circulation round the wing, Γ, which is independent of R. Hence

$$D = X = 0, \quad L = Y = -\rho_0 U \Gamma, \quad (4·1, 62)$$

which is the law of Kutta–Joukowski.

The derivation of (4·1, 62) was based on the second-order approximation to the momentum principle (4·1, 30), and the terms of which were neglected in the derivation of that principle cannot affect the result. Thus, (4·1, 62)

‡ *Modern Developments in Fluid Dynamics. High Speed Flow*, ed. Howarth (Oxford University Press, 1953).

depends only on the assumption that the velocity components are regular at infinity and that the flow is irrotational. Due to the appearance of shock waves (see § 4·12), these assumptions do not hold for supersonic flow, and the Kutta–Joukowski law is no longer applicable.

4·2   Linearized theory of subsonic flow

The governing equation of the linearized theory of steady subsonic flow is (4·1, 20). This may also be written as

$$(1 - M^2)\frac{\partial^2 \Phi}{\partial x^2} + \frac{\partial^2 \Phi}{\partial y^2} + \frac{\partial^2 \Phi}{\partial z^2} = 0, \tag{4·2, 1}$$

where $M = U/a_0 < 1$. A solution of (4·2, 1) is (4·1, 54), and may be said to be the velocity potential of a steady subsonic source. The theory of steady subsonic flow can be based on (4·1, 54), just as the general theory of incompressible flow, which is governed by Laplace's equation (1·7, 1), was based to a large extent on the solution (1·7, 2). However, there are simple transformations which reduce (4·2, 1) to (1·7, 1) and vice versa, and we may use these in order to derive the forces which act on an aerofoil in subsonic compressible flow directly from the corresponding results for incompressible flow. This procedure can be applied even if the data for the incompressible flow case are known only from experiment. The first derivations of one of these transformations (for the two-dimensional case) were given by Prandtl[‡] and by Glauert[§] and the three-dimensional transformations were given by Göthert[||] and by Goldstein and Young.[¶] The general transformation can be obtained most readily by the following argument.

Let us introduce the affine coordinate transformation

$$\left.\begin{aligned} x^* &= ax, \\ y^* &= by, \\ z^* &= cz, \end{aligned}\right\} \tag{4·2, 2}$$

where a, b, c are constant, and let us associate with the (x, y, z) and (x^*, y^*, z^*) coordinate systems the velocity potentials

$$\Phi = -Ux + \phi \quad \text{and} \quad \Phi^* = -Ux^* + \phi^*$$

---

‡ L. Prandtl, 'Über Strömungen deren Geschwindigkeiten mit der Schallgeschwindigkeit vergleichbar sind', *J. Aero. Res. Inst., Tokyo*, no. 65 (1930), p. 14 (translated as *Tech. Memor. Nat. Adv. Comm. Aero., Wash.*, no. 805).

§ H. Glauert, 'The effect of compressibility on the lift of an aerofoil', *Proc. Roy. Soc.* A, vol. 118 (1928), p. 113.

|| B. Göthert, 'Einige Bemerkungen zur Prandtl'schen Regel in Bezug auf ebene und räumliche Strömung (ohne Auftrieb)', F.B.1165, Institut für Aerodynamik, Berlin-Adlershof (30 Dec. 1939).

¶ S. Goldstein and A. D. Young, 'The linear perturbation theory of compressible flow with applications to wind tunnel interference', *A.R.C. Rep. Memor. Aero. Res. Comm., Lond.*, no. 1909 (1943).

respectively, where ϕ and ϕ^* are the induced velocity potentials. Here Φ and ϕ are assumed to satisfy the linearized compressible flow equation (4·2, 1), and we take

$$\phi^* = d\phi, \qquad (4\cdot2, 3)$$

where d is a constant. We inquire what values of the constant a, b, c and d made Φ^* (and ϕ^*) satisfy the incompressible flow equation (1·7, 1)

$$\frac{\partial^2\Phi^*}{\partial x^{*2}} + \frac{\partial^2\Phi^*}{\partial y^{*2}} + \frac{\partial^2\Phi^*}{\partial z^{*2}} = 0. \qquad (4\cdot2, 4)$$

Substituting in (4·2, 1) for (x, y, z) and ϕ from (4·2, 2) and (4·2, 3), we find that

$$a^2(1 - M^2)\frac{\partial^2\Phi^*}{\partial x^{*2}} + b^2\frac{\partial^2\Phi^*}{\partial y^{*2}} + c^2\frac{\partial^2\Phi^*}{\partial z^{*2}} = 0,$$

which reduces to (4·2, 4) if $\qquad a\beta = b = c, \qquad (4\cdot2, 5)$

where we have put $\sqrt{(1 - M^2)} = \beta < 1$.

Further conditions to be satisfied by the constants a, b, c and d are found by considering the boundary conditions at the aerofoil. Writing (4·2, 3) as

$$d\phi(x, y, z) = \phi^*(x^*, y^*, z^*) = \phi^*(ax, by, cz),$$

and using (4·2, 5), we see that the induced velocities are related by

$$\left. \begin{aligned} u'(x, y, z) &= \frac{a}{d}u^{*\prime}(x^*, y^*, z^*), \\ v'(x, y, z) &= \frac{b}{d}v^{*\prime}(x^*, y^*, z^*) = \beta\frac{a}{d}v^{*\prime}(x^*, y^*, z^*), \\ w'(x, y, z) &= \frac{c}{d}w^{*\prime}(x^*, y^*, z^*) = \beta\frac{a}{d}w^{*\prime}(x^*, y^*, z^*). \end{aligned} \right\} \qquad (4\cdot2, 6)$$

More briefly (4·2, 6) can be written as

$$\left. \begin{aligned} u' &= \frac{a}{d}u^{*\prime}, \\ v &= \beta\frac{a}{d}v^*, \\ w &= \beta\frac{a}{d}w^*, \end{aligned} \right\} \qquad (4\cdot2, 7)$$

since $v' = v$, $v^{*\prime} = v^*$, $w' = w$, $w^{*\prime} = w^*$, in the linear approximation.

In accordance with the usual assumptions of linearized theory (§ 3·1), the boundary conditions at the surfaces of aerofoil in the x, y, z and x^*, y^*, z^*-spaces are respectively

$$\frac{w}{U + u'} \doteqdot \frac{w}{U} = \frac{\partial f}{\partial z}, \qquad \frac{w^*}{U + u^{*\prime}} \doteqdot \frac{w^{*\prime}}{U} = \frac{\partial f^*}{\partial z^*}, \qquad (4\cdot2, 8)$$

in which we have written the equations of the aerofoil surfaces in the two coordinate systems as $\qquad z = f(x, y), \qquad z^* = f^*(x^*, y^*).$

If these equations are to represent the same aerofoil, we must have, from (4·2, 2) and (4·2, 5),

$$\frac{\partial f^*}{\partial z^*} = \frac{a}{c}\frac{\partial f}{\partial z} = \beta\frac{\partial f}{\partial z}. \tag{4·2, 9}$$

Substituting in (4·2, 8) we get $\dfrac{w}{U} = \dfrac{1}{\beta}\dfrac{w^*}{U}.$

Comparing with the last equation of (4·2, 6) we find

$$\frac{d}{a} = \beta^2,$$

so that

$$\left.\begin{aligned}
\phi &= \frac{1}{a\beta^2}\phi^*, \\[4pt]
u' &= \frac{1}{\beta^2}u^{*\prime}, \\[4pt]
v &= \frac{1}{\beta}v^*, \\[4pt]
w &= \frac{1}{\beta}w^*.
\end{aligned}\right\} \tag{4·2, 10}$$

These are the equations relating the velocity potential and the velocities at corresponding points of the incompressible flow about an aerofoil in the x^*, y^*, z^*-space and a corresponding aerofoil in compressible flow in the x, y, z-space. The correspondence between the x, y, z and x^*, y^*, z^*-spaces is given by equations (4·2, 2) combined with (4·2, 5), i.e. by

$$x^* = ax, \quad y^* = a\beta y, \quad z^* = a\beta z. \tag{4·2, 11}$$

The constant a affects only the relative scales of the flow and may be equated to any constant value without loss of generality. The information contained in (4·2, 10) and (4·2, 11) may be combined into the single set of equations

$$\left.\begin{aligned}
\phi(x, y, z) &= \frac{1}{a\beta^2}\phi^*(ax, a\beta y, a\beta z), \\[4pt]
u'(x, y, z) &= \frac{1}{\beta^2}u^{*\prime}(ax, a\beta y, a\beta z), \\[4pt]
v(x, y, z) &= \frac{1}{\beta}v^*(ax, a\beta y, a\beta z), \\[4pt]
w(x, y, z) &= \frac{1}{\beta}w^*(ax, a\beta y, a\beta z).
\end{aligned}\right\} \tag{4·2, 12}$$

From the first equation of (4·2, 12) and the linearized Bernoulli equation (4·1, 27), we find for the pressure on the aerofoil

$$p(x, y, z) = \frac{1}{\beta^2}p^*(ax, a\beta y, a\beta z). \tag{4·2, 13}$$

With the aid of (4·2, 12) and (4·2, 13) the aerodynamic forces acting on an aerofoil in subsonic flow can be found from a corresponding incompressible flow, provided the limitations of linearized theory are not exceeded. Thus, the lift, drag, pitching and rolling moments can be found by using equations (3·2, 16), (3·2, 30), (3·2, 37) and (3·2, 38), for these equations hold equally well in subsonic compressible flow about thin wings at small incidence, where the linearized Bernoulli equation (4·1, 27) is still valid. In fact

$$L = \frac{1}{a^2\beta^3} L^*, \quad D = \frac{1}{a^2\beta^4} D^*, \quad M = \frac{1}{a^3\beta^3} M^*, \quad L' = \frac{1}{a^3\beta^4} L'^*,$$

where the shapes of the two wings are related by the transformation (4·2, 11). It should also be observed that, since the downwash is in general determined by the strength of the trailing vortex sheet, and since by (4·1, 62) the circulation Γ (and hence the strength $\gamma = -d\Gamma/dy$ of the trailing vortices) is still given uniquely in terms of the lift by the Kutta-Joukowski law, it follows that, for a *given* lift distribution on the wing, the downwash velocity is independent of the Mach number and has its incompressible value.

Equations (4·2, 12) and (4·2, 13) are the only transformations in linearized theory that can be applied to wings of arbitrary shape. (They are also valid for bodies other than aerofoils such as slender bodies of revolution.) In special cases other more useful transformations are possible. Let us take, in the first instance, a two-dimensional thin wing section in the x, y-plane. Putting $a = 1$ in (4·2, 12) and (4·2, 13), we now obtain

$$\left.\begin{aligned}
\phi(x, y) &= \frac{1}{\beta^2} \phi^*(x, \beta y), \\[2mm]
u'(x, y) &= \frac{1}{\beta^2} u^{*\prime}(x, \beta y), \\[2mm]
v(x, y) &= \frac{1}{\beta} v^*(x, \beta y), \\[2mm]
p(x, y) &= \frac{1}{\beta^2} p^*(x, \beta y),
\end{aligned}\right\} \tag{4·2, 14}$$

which states that the velocities induced by an aerofoil in two-dimensional compressible flow along and normal to the main-flow direction are increased by factors $1/\beta^2$ and $1/\beta$ above their values at corresponding points in the incompressible flow about an aerofoil which is obtained from the original by reducing the normal coordinates in the ratio $\beta : 1$.

We may use (4·2, 14) to study the suction force at the leading edge of a thin aerofoil in compressible flow, using the development given in § 2·8 for incompressible flow. Thus, the suction force in two-dimensional incompressible flow according to (2·8, 30) is

$$X^* - iY^* = -\pi\rho C^{*2}, \tag{4·2, 15}$$

where
$$u^* - iv^* = \frac{C^*}{\sqrt{(x^* + iy^*)}} + \text{bounded terms,} \qquad (4\cdot2, 16)$$

near the leading edge. We have introduced the asterisk to refer to values in incompressible flow; plain symbols will be used for the compressible flow, in agreement with the notation used in the rest of this section. In (4·2, 16) it is assumed that the leading edge coincides with the origin of coordinates, and that the x-axis is tangential to the (mean camber line of the) wing at the leading edge. C is also given by the relation

$$u^{*\prime} - iv^* = \frac{C^*}{\sqrt{(x^* + iy^*)}} + \text{bounded terms,} \qquad (4\cdot2, 17)$$

since the main-stream velocity may be included amongst the bounded terms on the right-hand side. Let

$$C^* = C_1^* - iC_2^*. \qquad (4\cdot2, 18)$$

Then
$$\left. \begin{array}{l} u^{*\prime} = \dfrac{C_1^*}{\sqrt{x^*}} + \text{bounded terms,} \\[2mm] v^* = \dfrac{C_2^*}{\sqrt{x^*}} + \text{bounded terms,} \end{array} \right\} \qquad (4\cdot2, 19)$$

on approaching the leading edge along the wing. Similarly we may write, for compressible flow,

$$\left. \begin{array}{l} u' = \dfrac{C_1}{\sqrt{x}} + \text{bounded terms,} \\[2mm] v = \dfrac{C_2}{\sqrt{x}} + \text{bounded terms.} \end{array} \right\} \qquad (4\cdot2, 20)$$

Substituting in (4·2, 20) from (4·2, 11) and (4·2, 14), with $a = 1$, we obtain

$$\frac{1}{\beta^2} u^{*\prime} = \frac{C_1}{\sqrt{x^*}} + \text{bounded terms,}$$

$$\frac{1}{\beta} v^* = \frac{C_2}{\sqrt{x^*}} + \text{bounded terms,} \qquad (4\cdot2, 21)$$

and comparing (4·2, 19) with (4·2, 21) we then see that

$$C_1^* = \beta^2 C_1, \quad C_2^* = \beta C_2. \qquad (4\cdot2, 22)$$

Separating real and imaginary parts in (4·2, 15) we obtain

$$X^* = -\pi\rho(C_1^{*2} - C_2^{*2}), \quad Y^* = -2\pi\rho C_1^* C_2^*. \qquad (4\cdot2, 23)$$

To find the corresponding forces for the compressible case, we may avoid an explicit calculation by merely scaling formulae (4·2, 23) in the following way.

We apply the momentum principle (4·1, 30) in two dimensions to a small circle of radius r round the leading edge of the wing in the x, y-plane. We may omit the terms which involve U, since the length of the path of integration tends to zero as r, while u' and v tend to infinity as $r^{-\frac{1}{2}}$ only. Then the components of the suction force in the x, y-plane are given by

$$X = \rho_0 \left[\frac{1}{2} \int \{(1 - M^2)\, u'^2 + v^2\}\, dy - \int \{(1 - M^2)\, u'^2\, dy - u'v\, dx\} \right]$$

$$= -\tfrac{1}{2}\rho_0 \int \{(\beta^2 u'^2 - v^2)\, dy - 2u'v\, dx\}$$

$$= -\frac{\rho_0}{2\beta^3} \int \{(u^{*\prime 2} - v^{*2})\, dy^* - 2u^{*\prime} v^*\, dx^*\}$$

$$= \frac{X^*}{\beta^3},$$

and

$$Y = \rho_0 \left[-\frac{1}{2} \int \{(1 - M^2)\, u'^2 + v^2\}\, dx - \int \{(1 - M^2)\, u'v\, dy - v^2\, dx\} \right]$$

$$= -\tfrac{1}{2}\rho_0 \int \{(\beta^2 u'^2 - v^2)\, dx + 2\beta^2 u'v\, dy\}$$

$$= -\frac{\rho_0}{2\beta^2} \int \{(u^{*\prime 2} - v^{*2})\, dx^* + u^{*\prime} v^*\, dy^*\}$$

$$= \frac{Y^*}{\beta^2},$$

where it is understood that $\rho = \rho_0$ for the incompressible case. Hence, by (4·2, 22) and (4·2, 23),

$$X = -\pi\rho\left(\beta C_1^2 - \frac{1}{\beta} C_2^2\right), \quad Y = -2\pi\rho_0 \beta C_1 C_2. \qquad (4\cdot2, 24)$$

It remains to generalize these expressions for three-dimensional conditions (Fig. 4·2, 1). We may use a procedure completely analogous to that given in § 3·2. Since the suction force at any point $A\ (x_0, y_0, 0)$ of the leading edge depends only on the local conditions, we can replace the given leading edge, L, by a straight line $\tilde{L}$, which is tangential to L at A. $\tilde{L}$ may be regarded as the leading edge of a wing of infinite span which is yawed at any angle β_0 relative to the direction of the free stream. We introduce a system of coordinates $(\tilde{x}, \tilde{y}, \tilde{z})$ with origin at A, such that the $\tilde{x}$ and $\tilde{y}$ axes are normal to, and coinciding with, $\tilde{L}$ in the x, y-plane, while the $\tilde{z}$-axis is parallel to the z-axis. Then the components

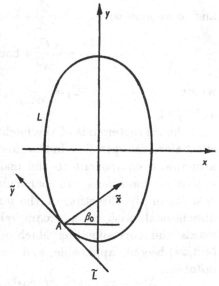

Fig. 4·2 ,1

of the free-stream velocity in the $(\tilde{x}, \tilde{y}, \tilde{z})$ system are $(U \cos\beta_0, -U \sin\beta_0, 0)$.

We assume that
$$u = \frac{C_1}{\sqrt{(x - x_0)}} + \text{bounded terms,}$$
$$w = \frac{C_2}{\sqrt{(x - x_0)}} + \text{bounded terms,}$$
$$(4 \cdot 2, 25)$$

as A is approached along the surface of the wing (compare equation $(3 \cdot 2, 11)$). In general, the velocity component parallel to the leading edge will be bounded, and so we have, in the neighbourhood of A,

$$u = \tilde{u} \cos \beta_0 - \tilde{v} \sin \beta_0$$
$$= \tilde{u} \cos \beta_0 + \text{bounded terms,}$$

and
$$w = \tilde{w},$$

where $\tilde{u}, \tilde{v}, \tilde{w}$ are the velocity components in the $(\tilde{x}, \tilde{y}, \tilde{z})$ system. Taking into account the fact that the angle between the x-axis and the $\tilde{x}$-axis is β_0 we obtain from $(4 \cdot 2, 25)$

$$\tilde{u} = \frac{C_1}{\cos \beta_0 \sqrt{(x - x_0)}} + \text{bounded terms,}$$
$$\tilde{w} = \frac{C_2}{\sqrt{(x - x_0)}} + \text{bounded terms.}$$
$$(4 \cdot 2, 26)$$

This implies
$$\tilde{u} = \frac{C_1}{\cos \beta_0} \sqrt{\frac{\cos \beta_0}{\tilde{x}}} + \text{bounded terms,}$$

$$\tilde{w} = C_2 \sqrt{\frac{\cos \beta_0}{\tilde{x}}} + \text{bounded terms,}$$

and so we may write
$$\tilde{u} = \frac{\tilde{C}_1}{\sqrt{\tilde{x}}} + \text{bounded terms,}$$
$$\tilde{w} = \frac{\tilde{C}_2}{\sqrt{\tilde{x}}} + \text{bounded terms,}$$
$$(4 \cdot 2, 27)$$

where
$$\tilde{C}_1 = \frac{C_1}{\sqrt{\cos \beta_0}}, \quad \tilde{C}_2 = C_2 \sqrt{\cos \beta_0}, \tag{4 \cdot 2, 28}$$

as in § 3·2.

If the characteristics of the modified wing with leading edge $\tilde{L}$ are constant along the span (i.e., in the direction of the y-axis), then the addition of a spanwise component to the main-stream velocity will not affect the boundary conditions. In particular, if we superimpose a component $U \sin \beta_0$ in the direction of the $\tilde{y}$-axis, then the resultant force is two-dimensional, with main-stream velocity $U \cos \beta_0$ in the direction of the x-axis and corresponding Mach number $M \cos \beta_0$. Thus, the formulae $(4 \cdot 2, 24)$ become applicable, and we have, with the appropriate change in notation,

$$\tilde{X} = -\pi \rho_0 \left(\sqrt{(1 - M^2 \cos^2 \beta_0)} \, \tilde{C}_1^2 - \frac{1}{\sqrt{(1 - M^2 \cos^2 \beta_0)}} \, \tilde{C}_2^2 \right),$$
$$\tilde{Y} = 0,$$
$$\tilde{Z} = -2\pi \rho_0 \sqrt{(1 - M^2 \cos^2 \beta_0)} \, \tilde{C}_1 \tilde{C}_2.$$
$$(4 \cdot 2, 29)$$

Using (4·2, 28), we then obtain for the components of the suction force in the (x, y, z)-system of coordinates

$$
\left.
\begin{aligned}
X = \tilde{X} \cos \beta_0 &= -\pi \rho_0 \left(\sqrt{(1 - M^2 \cos^2 \beta_0)}\, C_1^2 - \frac{\cos^2 \beta_0}{\sqrt{(1 - M^2 \cos^2 \beta_0)}}\, C_2^2 \right), \\
Y = \tilde{X} \sin \beta_0 &= -\pi \rho_0 \left(\sqrt{(1 - M^2 \cos^2 \beta_0)}\, C_1^2 \tan^2 \beta_0 \right. \\
&\qquad \left. - \frac{1}{\sqrt{(1 - M^2 \cos^2 \beta_0)}}\, C_2^2 \sin \beta_0 \cos \beta_0 \right), \\
Z &= -2\pi \rho \sqrt{(1 - M^2 \cos^2 \beta_0)}\, C_1 C_2.
\end{aligned}
\right\}
$$

$$(4·2, 30)$$

We may modify these formulae by introducing the radius of curvature of the leading edge, r_0, by means of (3·2, 12). Then

$$
\left.
\begin{aligned}
X &= -\pi \rho_0 \left(\sqrt{(1 - M^2 \cos^2 \beta_0)}\, C_1^2 - \frac{U^2 r_0 \cos^2 \beta_0}{2 \sqrt{(1 - M^2 \cos^2 \beta_0)}} \right), \\
Y &= -\pi \rho_0 \left(\sqrt{(1 - M^2 \cos^2 \beta_0)}\, C_1^2 \tan \beta_0 - \frac{U^2 r_0 \sin \beta_0 \cos \beta_0}{2 \sqrt{(1 - M^2 \cos^2 \beta_0)}} \right), \\
Z &= -\sqrt{2}\, \pi \rho_0 \sqrt{(1 - M^2 \cos^2 \beta_0)}\, C_1 U \sqrt{r_0},
\end{aligned}
\right\}
$$

$$(4·2, 31)$$

are the suction-force components per unit length of the leading edge. We note that for vanishing Mach number (4·2, 31) reduces to (3·2, 13).

It is possible, using the results of incompressible thin aerofoil theory (§ 2·8), to modify the two-dimensional transformation (4·2, 13) so as to hold for aerofoils of similar section in both the compressible and incompressible flows. Thus the effect of the coordinate transformation (4·2, 11) from the (x, y, z) system to the (x^*, y^*, z^*) system in the two-dimensional case is to reduce the local aerofoil slopes in the ratio $\beta : 1$, viz.

$$
s^* = \frac{df^*}{dx^*} = \beta \frac{df}{dx} = \beta s, \tag{4·2, 32}
$$

and according to § 2·8 the induced velocities $u^{*\prime}$ and v^* on the aerofoil are proportional to s^*. Hence if we now introduce the inverse coordinate transformation from the (x^*, y^*, z^*) system to the (x, y, z) system, for the incompressible flow velocities, we find

$$
\left.
\begin{aligned}
\frac{u^{\dagger\prime}}{u^{*\prime}} &= \frac{s}{s^*} = \frac{1}{\beta}, \\
\frac{v^{\dagger}}{v^*} &= \frac{s}{s^*} = \frac{1}{\beta},
\end{aligned}
\right\}
\tag{4·2, 33}
$$

where the superscript † refers now to incompressible flow on the *same* aerofoil as in compressible flow in the (x, y, z)-space. From (4·2, 14), for aerofoils of the same section, we therefore find

$$u'(x, y) = \frac{1}{\beta} u^{+\prime}(x, y), \Bigg\}$$
$$v(x, y) = v^{+}(x, y), \Bigg\} \tag{4·2, 34}$$

so that only the induced velocity component u' is affected by compressibility, increasing as $\dfrac{1}{\beta} = \dfrac{1}{\sqrt{(1 - M^2)}}$ with Mach number.

(4·2, 34) can be used to find the effect of compressibility according to linearized theory on the aerodynamic force coefficients. Thus, the linearized Bernoulli equation (4·1, 27) is

$$p = p_0 - \rho_0 U u',$$

and the pressure coefficient is therefore

$$C_p = \frac{p - p_0}{\frac{1}{2}\rho_0 U^2} = -2\frac{u'}{U}. \tag{4·2, 35}$$

So that from (4·2, 34) $C_p = \dfrac{1}{\beta} C_p^{+},$ (4·2, 36)

where we have again used the dagger to signify incompressible values for the same aerofoil. From (4·2, 36) it follows at once that

$$C_L = \frac{1}{\beta} C_L^{+} \tag{4·2, 37}$$

and $C_M = \dfrac{1}{\beta} C_M^{+},$ (4·2, 38)

C_L and C_M being the lift and pitching moment coefficients. Similarly, for the lift curve slope $a_0 = \partial C_L / \partial \alpha_0$,

$$a_0 = \frac{1}{\beta} a_0^{+}. \tag{4·2, 39}$$

(4·2, 37) is known as the Prandtl–Glauert law. Values of the compressibility factor $\dfrac{1}{\beta} = \dfrac{1}{\sqrt{(1 - M^2)}}$ are shown in Table 4·2, 1. Equations (4·2, 36)–(4·2, 39) are found experimentally to give the correct type of variation with Mach number for values not too close to the critical Mach number and for low values of C_p. However, in general, experimental values lies above the values predicted by the theory, at least for Mach numbers below the critical. An example is given in Fig. 4·2, 2, which shows the theoretical and experimental values of the pressure coefficient at two points on the surface of an N.A.C.A. 4412 aerofoil. It will be seen that the theory gives a value for the free-stream Mach number M at which the local speed at the aerofoil first becomes sonic, as indicated by the intersection of the 'sonic line' with the curves. It is therefore possible to predict the critical Mach number of a given aerofoil using the Prandtl–Glauert law if the low-speed pressure distribution is known.

It is clear that for high enough subsonic Mach numbers the theory must break down, since it predicts infinite values for the force coefficients at $M = 1$. Thus, even if the assumption of linearized theory that the induced velocities are small still holds at main-stream Mach numbers close to one, purely subsonic linearized theory (as dealt with here) cannot be applied for Mach numbers greater than the critical, when the flow over the aerofoil becomes partly supersonic.

Table 4·2, 1

M	0	0·1	0·2	0·3	0·4	0·5	0·6	0·7	0·8	0·9	1·0
$1/\beta$	1	1·0050	1·0206	1·0483	1·0911	1·1547	1·2500	1·4003	1·6667	2·2942	∞

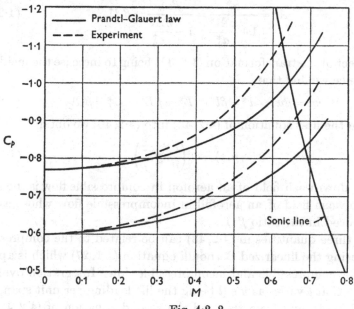

Fig. 4·2, 2

Corresponding results for the three-dimensional aerofoil may be derived by using the results of lifting-line theory (§ 3·3). In this case simple relations can be obtained between properties of the wing in compressible flow and a wing of reduced aspect ratio in incompressible flow. We write (4·2, 12) as

$$u'(x, y, z) = \frac{1}{\beta^2} u^{*\prime}\left(\frac{x}{\beta}, y, z\right),$$

$$v(x, y, z) = \frac{1}{\beta} v^*\left(\frac{x}{\beta}, y, z\right), \qquad (4·2, 40)$$

$$w(x, y, z) = \frac{1}{\beta} w^*\left(\frac{x}{\beta}, y, z\right),$$

by putting $a = 1/\beta$, and consider the effect in the incompressible space of increasing the z^* coordinate so that

$$x^\dagger = x^*, \quad y^\dagger = y^*, \quad z^\dagger = \frac{1}{\beta} z^*. \tag{4·2, 41}$$

By this transformation the local aerofoil sections in the compressible (x, y, z)-space and the $(x^\dagger, y^\dagger, z^\dagger)$-space are made similar, although the aspect ratio in the $(x^\dagger, y^\dagger, z^\dagger)$-space is still β times the aspect ratio in the compressible space. We now study the effect of (4·2, 41) on the quantities w, l, L, and C_L according to lifting-line theory. By equation (3·3, 68) each of these quantities is proportional to the coefficient A_1, for given semi-span s_0. Writing the asterisk to refer to values in the (x^*, y^*, z^*)-space and the dagger to refer to values in the $(x^\dagger, y^\dagger, z^\dagger)$-space, we have by (3·3, 76) for elliptic wings,

$$A_1^* = \frac{\alpha_0^*}{1 + \dfrac{\pi \bar{R}^*}{a_0^*}} = \frac{\beta \alpha_0^\dagger}{1 - \dfrac{\pi \bar{R}^\dagger}{a_0^\dagger}} = \beta A_1^\dagger, \tag{4·2, 42}$$

the only effect of the transformation (4·2, 41) being to increase the incidence by $1/\beta$. Hence we find that

$$w^* = \beta w^\dagger, \quad l^* = \beta l^\dagger, \quad L^* = \beta L^\dagger, \quad C_L^* = \beta C_L^\dagger. \tag{4·2, 43}$$

Introducing the first equation of (4·2, 43) into (4·2, 40) we obtain

$$w(x, y, z) = w^\dagger\left(\frac{x}{\beta}, y, \frac{z}{\beta}\right), \tag{4·2, 44}$$

so that the downwash field of an aerofoil in compressible flow is the same as the downwash field of an aerofoil in incompressible flow whose aspect ratio is reduced in the ratio $\beta : 1$.

The last three quantities in (4·2, 43) can be related to the compressible values by using the linearized Bernoulli equation (4·1, 27) which is applicable both for compressible and incompressible flow. Integration over the chord shows that l varies as $u'c$ (l being the lift loading per unit span, and c the chord load), and from this and the second equation of (4·2, 43) we deduce immediately that

$$u^{*\prime} = \beta u^{\dagger\prime},$$

and applying (4·2, 40) we get

$$u'(x, y, z) = \frac{1}{\beta} u^{\dagger\prime}\left(\frac{x}{\beta}, y, \frac{z}{\beta}\right). \tag{4·2, 45}$$

This result may be used in turn to find the compressibility effect on lift and lift loading, thus, keeping in mind that

$$l \propto u'c, \quad L \propto u's_0 c, \quad C_L \propto u' \frac{s_0 c}{S}, \tag{4·2, 46}$$

where s_0 is the semi-span and S the wing area, we obtain

$$l \propto c u'(x, y, z) = c \frac{1}{\beta} u^{\dagger\prime}\left(\frac{x}{\beta}, y, \frac{z}{\beta}\right), \quad l^\dagger \propto c^\dagger u^{\dagger\prime}\left(\frac{x}{\beta}, y, \frac{z}{\beta}\right),$$

and therefore
$$l = \frac{1}{\beta} \frac{c}{c^\dagger} l^\dagger = l^\dagger. \qquad (4\cdot2, 47)$$

Similarly
$$L = L^\dagger, \quad C_L = \frac{1}{\beta} C_L^\dagger. \qquad (4\cdot2, 48)$$

It should be emphasized that equations $(4\cdot2, 45)$, $(4\cdot2, 47)$ and $(4\cdot2, 48)$ do not refer to the same aerofoil for compressible and incompressible flow, the aspect ratios being different in the ratio $1:\beta$. If we wish to obtain similar aerofoils in the two flows, an allowance must be made in using equation $(3\cdot3, 76)$ for this difference in aspect ratio. The procedure is similar to that first carried out for wings of different aspect ratios, but using this time the transformation of coordinates

$$x^\dagger = \beta x^*, \quad y^\dagger = y^*, \quad z^\dagger = z^*.$$

Then
$$A_1^* = \frac{\beta a_0^\dagger}{1 + \dfrac{\pi \beta \mathcal{R}^\dagger}{a_0^\dagger}},$$

and
$$l^* = \beta \frac{1 + \dfrac{\pi \mathcal{R}^\dagger}{a_0^\dagger}}{1 + \beta \dfrac{\pi \mathcal{R}^\dagger}{a_0^\dagger}} l^\dagger, \quad C_L^* = \beta^2 \frac{1 + \dfrac{\pi \mathcal{R}^\dagger}{a_0^\dagger}}{1 + \beta \dfrac{\pi \mathcal{R}^\dagger}{a_0^\dagger}} C_L^\dagger.$$

Using the fact that l varies as $u'c$, since

$$u'(x, y, z) = \frac{1}{\beta^2} u^{*\prime} \left(\frac{x}{\beta}, y, z \right),$$

we find
$$l = \frac{1}{\beta^2} \frac{c}{c^*} l^* = \frac{1}{\beta} l^*, \qquad (4\cdot2, 49)$$

and
$$C_L = \frac{1}{\beta} \frac{S^*}{S} \frac{s_0}{s_0^*} C_L^* = \frac{1}{\beta^2} C_L^*. \qquad (4\cdot2, 50)$$

Hence, since now $\mathcal{R}^\dagger = \mathcal{R}$,

$$l = \frac{1 + \dfrac{\pi \mathcal{R}}{a_0^\dagger}}{1 + \beta \dfrac{\pi \mathcal{R}}{a_0^\dagger}} l^\dagger, \quad C_L = \frac{1 + \dfrac{\pi \mathcal{R}}{a_0^\dagger}}{1 + \beta \dfrac{\pi \mathcal{R}}{a_0^\dagger}} C_L^\dagger. \qquad (4\cdot2, 51)$$

The lift curve slope, $a = dC_L/d\alpha$, is obtained from $(4\cdot2, 51)$ as

$$a = \frac{1 + \dfrac{\pi \mathcal{R}}{a_0^\dagger}}{1 + \beta \dfrac{\pi \mathcal{R}}{a_0^\dagger}} a^\dagger, \qquad (4\cdot2, 52)$$

and this relationship is shown graphically in Fig. $4\cdot2, 3$ for a value of $a_0^\dagger = 6\cdot0$.

For a given aspect ratio $\mathcal{R} = \mathcal{R}^\dagger$ it is to be expected that $(4\cdot2, 51)$ and $(4\cdot2, 52)$ become increasingly inaccurate as M approaches 1, since the aspect ratio $Æ^*$ in the (x^*, y^*, z^*)-space, which of necessity must be assumed large for lifting-line theory to apply, tends to zero as M tends to 1.

For large aspect ratios (4·2, 51) and (4·2, 52) yield

$$C_L = \frac{1}{\beta} C_L^\dagger, \quad a = \frac{1}{\beta} a^\dagger,$$

in agreement with the two-dimensional result. For small aspect ratios however we get

$$C_L = C_L^\dagger, \quad a = a^\dagger.$$

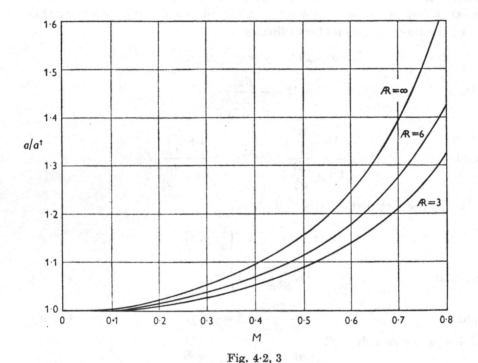

Fig. 4·2, 3

The lift coefficient and lift curve slope are therefore independent of Mach number for small aspect ratios. Although derived using lifting-line theory (in which it is assumed that the aspect ratio is large), this result is correct, though the actual magnitudes of C_L and $dC_L/d\alpha$ are wrong. We may verify it by using the small aspect ratio theory developed in § 3·10. Thus equation (3·10, 27) is

$$C_L = \frac{\pi}{2} Æ\alpha,$$

so that by the transformation $x^\dagger = \beta x^*$, $y^\dagger = y^*$, $z^\dagger = z^*$,

$$C_L^* = \frac{\pi}{2} \beta Æ^\dagger \beta a^\dagger = \beta^2 C_L^\dagger,$$

where the affix † again refers to incompressible values for the same aerofoil as for compressible flow. From (4·2, 50) it follows that

$$C_L = C_L^\dagger, \qquad (4·2, 53)$$

and it is readily seen from this that u' and the lift curve slope a are the same in both compressible and incompressible flow.

This result can also be obtained by extension to compressible flows of the general incompressible theory of § 3·10 on small aspect-ratio wings. It is found that the governing equation is still

$$\frac{\partial^2 \Phi}{\partial y^2} + \frac{\partial^2 \Phi}{\partial x^2} = 0,$$

and since this equation is independent of Mach number, it follows that for wings of small aspect ratio the pressure distribution and the aerodynamic forces may be expected to be (approximately) independent of M. This has been confirmed by experiment.‡ Furthermore, as M approaches 1 the error introduced by the particular assumptions of small aspect-ratio theory is reduced. At the same time linearized theory itself becomes less reliable. However, since the result for a given wing is independent of Mach number, the theory provides at least a valuable guide even for transonic speeds.

As another application of linearized subsonic theory, we shall consider the problem of wind-tunnel interference for compressible flow. (3·11, 36) and (3·11, 12) are the equations used to determine the solid blockage and lifting surface interference in incompressible flow. With the aid of (4·2, 12) we can transform them to compressible flow. Thus, in the equation for solid-body interference

$$\frac{u'}{U} = \frac{V}{C^{\frac{3}{2}}} K\lambda, \qquad (4·2, 54)$$

where u' is the perturbation velocity due to the presence of the tunnel walls. In the transformation (4·2, 12) with $a = 1/\beta$ to incompressible flow, the volume V of the body is increased by $1/\beta$, whilst the cross-sectional area C of the tunnel remains the same. λ depends only on the relative shape of the tunnel section and the cross-section of the body, and this remains the same under the transformation. K, which is a function of the base profile, varies so little (see Table 3·11, 3) that it may be regarded as constant. Hence, using the first equation of (4·2, 40), we find that

$$\frac{u'}{U} = \frac{1}{\beta^2} \frac{V^*}{C^{*\frac{3}{2}}} K^* \lambda^* = \frac{1}{\beta^2} \frac{(V/\beta)}{C^{\frac{3}{2}}} K\lambda = \frac{1}{\beta^3} \frac{V}{C^{\frac{3}{2}}} K\lambda. \qquad (4·2, 55)$$

Equation (3·11, 12) $\qquad \dfrac{w'}{U} = \delta \dfrac{S}{C} C_L, \qquad (4·2, 56)$

‡ See R. T. Jones, 'Properties of low-aspect-ratio pointed wings at speeds below and above the speed of sound', *N.A.C.A. Rep.* no. 835 (1946).

for lifting surface interference, on the other hand, remains the same in compressible flow. Thus, from (4·2, 40), it follows that

$$\frac{w'}{U} = \frac{1}{\beta}\frac{w'^*}{U} = \frac{1}{\beta}\delta^*\frac{S^*}{C^*}C_L^* \doteq \frac{1}{\beta}\delta\frac{S}{C}C_L^*,$$

since S and C remain the same under the transformation, and δ is approximately independent of aspect ratio, provided the span of the wing is not too large (see Tables 3·11, 1 and 3·11, 2). By both lifting-line theory and small aspect-ratio theory $C_L^* = \beta C_L$, for in the former case (4·2, 48) holds, and in the latter there is no effect of compressibility on C_L, but because C_L varies as the aspect ratio, the effect of the transformation $x^* = x/\beta$ is to reduce the lift coefficient by a factor β, since it varies inversely as the wing area. Hence in both cases

$$\frac{w'}{U} = \delta\frac{C}{S}C_L, \tag{4·2, 57}$$

and this holds for all wings to a fair degree of approximation.

(4·2, 55) indicates a fundamental difficulty in wind-tunnel testing at high Mach numbers, for the axial interference velocity increases rapidly as an incident Mach number of 1 is approached. The same effect is predicted in the complete non-linearized formulation.‡ A possible solution of this problem is indicated by the results given in § 3·11, where it was shown that by the use of mixed open and closed tunnel boundaries the interference velocities could be eliminated altogether.§

4·3　The hodograph method for two-dimensional subsonic flow

In two-dimensional irrotational flow the full non-linearized equation (4·1, 21) for steady motion can be dealt with most effectively in the hodograph plane with the velocity components taken as the independent variables. The method is of particular value for those cases in which the linearized approach of § 4·2 is inadequate, for example, at very high subsonic speeds.

In two dimensions, the equation of continuity is

$$\frac{\partial}{\partial x}(\rho u) + \frac{\partial}{\partial y}(\rho v) = 0, \tag{4·3, 1}$$

and we may introduce a stream function Ψ defined by

$$u\frac{\rho}{\rho_0} = -\frac{\partial\Psi}{\partial y}, \quad v\frac{\rho}{\rho_0} = \frac{\partial\Psi}{\partial x}, \tag{4·3, 2}$$

‡ H. W. Emmons, 'Flow of a compressible fluid past a symmetrical airfoil in a wind tunnel and in free air', *Tech. Notes, Nat. Adv. Comm. Aero., Wash.*, no. 1746 (1948).

§ In fact a number of wind tunnels are operating transonically with the use of slotted walls. See, for example, R. H. Wright and V. G. Ward, 'N.A.C.A. transonic wind tunnel test sections', *Res. Memor. Nat. Adv. Comm. Aero., Wash.*, no. L 8 J 06 (1948).

so that Ψ is related to the velocity potential Φ through

$$\frac{\rho}{\rho_0}\frac{\partial\Phi}{\partial x} = \frac{\partial\Psi}{\partial y}, \quad \frac{\rho}{\rho_0}\frac{\partial\Phi}{\partial y} = -\frac{\partial\Psi}{\partial x}. \tag{4·3, 3}$$

We wish to find the equations satisfied by Ψ and Φ with the velocity components as independent variables. This is most conveniently done by introducing the complex representation

$$w = u - iv = q\,e^{-i\theta}. \tag{4·3, 4}$$

We then have, using (4·3, 2) and (4·3, 4), in differential notation,

$$
\begin{aligned}
d\Phi + i\frac{\rho_0}{\rho}d\Psi &= -(u\,dx + v\,dy) + i(v\,dx - u\,dy)\\
&= -(u - iv)(dx + i\,dy)\\
&= -q\,e^{-i\theta}\,dz,
\end{aligned}
\tag{4·3, 5}
$$

and hence, with q and θ as independent variables, and ρ a function of q only,

$$
\left.
\begin{aligned}
\left(\frac{\partial z}{\partial\theta}\right)_q &= -\frac{1}{q}e^{i\theta}\left(\frac{\partial\Phi}{\partial\theta} + i\frac{\rho_0}{\rho}\frac{\partial\Psi}{\partial\theta}\right),\\
\left(\frac{\partial z}{\partial q}\right)_\theta &= -\frac{1}{q}e^{i\theta}\left(\frac{\partial\Phi}{\partial q} + i\frac{\rho_0}{\rho}\frac{\partial\Psi}{\partial q}\right).
\end{aligned}
\right\}
\tag{4·3, 6}
$$

Differentiating the first equation of (4·3, 6) with respect to q and the second with respect to θ and equating, we get

$$\frac{i}{q}e^{i\theta}\left(\frac{\partial\Phi}{\partial q} + i\frac{\rho_0}{\rho}\frac{\partial\Psi}{\partial q}\right) = e^{i\theta}\left(-\frac{1}{q^2}\frac{\partial\Phi}{\partial\theta} + i\frac{d}{dq}\left(\frac{\rho_0}{\rho q}\right)\frac{\partial\Psi}{\partial\theta}\right),$$

or

$$\frac{\partial\Phi}{\partial\theta} = \left(\frac{q\rho_0}{\rho}\right)\frac{\partial\Psi}{\partial q}, \quad \frac{\partial\Phi}{\partial q} = q\frac{d}{dq}\left(\frac{\rho_0}{\rho q}\right)\frac{\partial\Psi}{\partial\theta}. \tag{4·3, 7}$$

The assumption used in deriving (4·3, 6) that ρ is a function of q only results from Bernoulli's equation (1·6, 4) and the relation (4·1, 2). Differentiation of the former equation yields

$$\frac{d}{dq}\left(\frac{\rho_0}{\rho q}\right) = -\frac{\rho_0}{\rho q^2}(1 - M^2).$$

We then get in place of (4·3, 7)

$$\frac{\rho}{\rho_0}\frac{\partial\Phi}{\partial\theta} = q\frac{\partial\Psi}{\partial q}, \quad \frac{\rho}{\rho_0}q\frac{\partial\Phi}{\partial q} = -(1 - M^2)\frac{\partial\Psi}{\partial\theta}. \tag{4·3, 8}$$

These are the hodograph equations for Φ and Ψ. It will be seen that they are linear. The equations satisfied by Φ and Ψ individually are

$$
\left.
\begin{aligned}
(1 - M^2)\frac{\partial^2\Psi}{\partial\theta^2} + q(1 + M^2)\frac{\partial\Psi}{\partial q} + q^2\frac{\partial^2\Psi}{\partial q^2} &= 0,\\
q^2\frac{\partial^2\Phi}{\partial q^2} + q\frac{(1 + \gamma M^4)}{1 - M^2}\frac{\partial\Phi}{\partial q} + (1 - M^2)\frac{\partial^2\Phi}{\partial\theta^2} &= 0.
\end{aligned}
\right\}
\tag{4·3, 9}
$$

Since the equations are linear there is usually no particular difficulty in finding solutions of (4·3, 9), by numerical methods if necessary, to any required degree of accuracy. However, the boundary conditions have to be specified in the hodograph plane, i.e. Φ and Ψ' (or their derivatives with respect to the velocity) have to be specified along boundaries in the u, v-plane. For a given problem in the physical x, y-plane we are therefore immediately confronted with the task of transcribing the boundary conditions into the hodograph plane, and this can lead to situations of considerable complexity. As an example, we may consider the mapping into the hodograph of a flow which has equal velocities at a number of different points in the physical plane. These will all correspond to the same point in the hodograph plane, and the hodograph plane must be considered as a Riemann surface of many (possibly infinitely many) sheets. The solution in the hodograph plane on each of these sheets will correspond to different areas of the physical plane, and these have to be matched by analytic continuation.

Because of such difficulties, the main application thus far of the hodograph equations has been to study special solutions—in particular, those involving mixed subsonic and supersonic flows—and to investigate by approximate methods cases in which the perturbations from uniform flow are small. The former provide an insight into the nature of solutions of the exact flow equations when sonic speeds occur, but are of little direct interest in aerofoil theory. In the remainder of this chapter, we shall be concerned almost exclusively with approximate methods.

The inverse transformation from the hodograph to the physical plane is given by (4·3, 5), which we can write as

$$\left.\begin{aligned} dx &= -\frac{\cos\theta}{q}\,d\Phi + \frac{\rho_0}{\rho}\frac{\sin\theta}{q}\,d\Psi', \\ dy &= -\frac{\sin\theta}{q}\,d\Phi - \frac{\rho_0}{\rho}\frac{\cos\theta}{q}\,d\Psi'. \end{aligned}\right\} \tag{4·3, 10}$$

The Jacobian of the transformation is

$$J = \frac{\partial(x, y)}{\partial(q, \theta)} = \frac{\partial(\Phi, \Psi')}{\partial(q, \theta)}\left(\frac{\partial(\Phi, \Psi')}{\partial(x, y)}\right)^{-1}. \tag{4·3, 11}$$

Using (4·3, 10) and (4·3, 7) we find that

$$J = \frac{\rho_0^2}{\rho q}\left\{\left(\frac{\partial\Psi'}{\partial q}\right)^2 - \frac{1}{q^2}(M^2 - 1)\left(\frac{\partial\Psi'}{\partial\theta}\right)^2\right\}. \tag{4·3, 12}$$

The mapping from the hodograph to the physical plane is non-singular provided $J \neq 0$. Except in trivial cases $J = 0$ only when $M \geqslant 1$, i.e. when the flow is partly or wholly sonic or supersonic. The locus of points for which $J = 0$ is known as a limiting line, and forms a boundary between regions of subsonic and supersonic flow. A discussion of the significance of limit lines as well as a survey of the more advanced use of hodograph methods can be found in *Modern Developments in Fluid Dynamics.*‡ It will be assumed in this section that the flow is entirely subsonic, so that $J \neq 0$.

It is convenient to introduce the new variable

$$\omega = \int \sqrt{(1 - M^2)} \frac{dq}{q} \tag{4·3, 13}$$

in place of the velocity q. Thus the hodograph equations (4·3, 8) become

$$\left. \begin{aligned} \frac{\partial \Phi}{\partial \omega} &= -\frac{\rho_0}{\rho} \sqrt{(1 - M^2)} \frac{\partial \Psi}{\partial \theta}, \\ \frac{\partial \Phi}{\partial \theta} &= \frac{\rho_0}{\rho} \sqrt{(1 - M^2)} \frac{\partial \Psi}{\partial \omega}. \end{aligned} \right\} \tag{4·3, 14}$$

If the flow is incompressible, these reduce to the Cauchy–Riemann equations

$$\left. \begin{aligned} \frac{\partial \Phi}{\partial \omega} &= -\frac{\partial \Psi}{\partial \theta}, \\ \frac{\partial \Phi}{\partial \theta} &= \frac{\partial \Psi}{\partial \omega}, \end{aligned} \right\} \tag{4·3, 15}$$

and Φ and Ψ satisfy Laplace's equation in the θ, ω-plane. However, we see that if

$$\frac{\rho_0}{\rho} \sqrt{(1 - M^2)} = c, \quad \text{a constant}, \tag{4·3, 16}$$

Φ and Ψ again satisfy Laplace's equation. From the isentropic flow relationship (cf. equation (4·1, 4))

$$\frac{\rho_0}{\rho} = \left(1 + \frac{\gamma - 1}{2} M^2 \right)^{1/(\gamma - 1)},$$

we see that

$$\frac{\rho_0}{\rho} \sqrt{(1 - M^2)} = 1 + 0(M^4),$$

so that $\dfrac{\rho_0}{\rho} \sqrt{(1 - M^2)} = 1$ is a good approximation for low speeds. Also (4·3, 16) is approximately true if the changes in density and Mach number are small, and hence should be applicable to flow over thin aerofoils. In this case we put

$$\frac{\rho_0}{\rho} \sqrt{(1 - M^2)} = \frac{\rho_0}{\rho_1} \sqrt{(1 - M_1^2)} = c, \tag{4·3, 17}$$

where the subscript 1 refers to values in the undisturbed stream. Then

$$\sqrt{(1 - M^2)} = c \frac{\rho}{\rho_0} = c \left(1 + \frac{\gamma - 1}{2} M^2 \right)^{-1/(\gamma - 1)} \doteqdot c \left(1 - \frac{1}{2} \frac{q^2}{a_0^2} \right), \tag{4·3, 18}$$

neglecting terms of order greater than q^2/a_0^2.

Let us define

$$\Omega = \int \frac{dQ}{Q}, \tag{4·3, 19}$$

‡ *Modern Developments in Fluid Dynamics. High Speed Flow*, ed. Howarth (Oxford University Press, 1953), ch. 7, by M. J. Lighthill.

corresponding to (4·3, 13), where Q is the velocity in incompressible flow. Then, accepting (4·3, 16), both the compressible and incompressible flows are governed by the Cauchy–Riemann equations with ω, θ and Q, θ respectively as independent variables. From a given incompressible flow, it is therefore possible to find a compressible flow velocity q which is given by

$$\frac{dQ}{Q} = \sqrt{(1 - M^2)}\frac{dq}{q}. \tag{4·3, 20}$$

Using (4·3, 18) we find that

$$\int \frac{dQ}{Q} = c \int \left(1 - \frac{1}{2}\frac{q^2}{a_0^2}\right)\frac{dq}{q},$$

which on integration yields

$$\left(\frac{Q}{Q_1}\right)^{1/c} = \frac{q}{q_1}\exp\left\{-\frac{q_1^2}{4a_0^2}\left(\frac{q^2}{q_1^2} - 1\right)\right\}, \tag{4·3, 21}$$

where q_1 and Q_1 are the free-stream velocities.

It should be observed that the boundaries in the compressible and incompressible flows are not the same; their correspondence can be found from (4·3, 11) using (4·3, 21), and this will be discussed later.

The relationship (4·3, 21) was established by von Kármán,‡ an earlier and better known equation is due to Kármán and Tsien§ and this is based on the pioneering work of Chaplygin.|| We suppose that the isentropic relationship

$$\frac{p}{\rho^\gamma} = \text{constant} \tag{4·3, 22}$$

is replaced by the equation of the tangent to the curve of p versus $1/\rho$ at a point $(p_a, 1/\rho_a)$, and this is clearly valid if the variations in p and $1/\rho$ about $(p_a, 1/\rho_a)$ are small. Chaplygin took $(p_a, 1/\rho_a)$ to correspond to stagnation values, but a better approximation for flow about a body in a uniform stream is $(p_1, 1/\rho_1)$, the main-stream values. This was the choice made by Kármán and Tsien. In place of (4·3, 22) we thus have a linear relation between the pressure and volume, which we shall write in terms of the pressure and density as

$$p - p_1 = \left(\frac{dp}{d\left(\frac{1}{\rho}\right)}\right)_1 \left(\frac{1}{\rho} - \frac{1}{\rho_1}\right)$$

$$= -a_1^2\rho_1^2\left(\frac{1}{\rho} - \frac{1}{\rho_1}\right),$$

‡ Th. von Kármán, 'Compressibility effects in aerodynanics', *J. Aero. Sci.*, vol. 8 (1941), pp. 337–56.

§ H. S. Tsien, 'Two-dimensional subsonic flow of compressible fluids', *J. Aero. Sci.* vol. 6 (1939), pp. 399–407.

|| S. Chaplygin, 'Gas jets', *Tech. Memor. Nat. Adv. Comm. Aero., Wash.*, no. 1063. Translated from *Sci. Memor. Math. Phys. Soc.* no. 21, Moscow University (1902), pp. 1–121.

where we have put $a_1^2 = (dp/d\rho)_1$. Applying Bernoulli's equation

$$\int \frac{dp}{\rho} + \tfrac{1}{2}q^2 = \text{constant},$$

and using $a^2 = dp/d\rho = \gamma p_1 \rho_1 / \rho^2$, we find that

$$q^2 - a^2 = -a_0^2, \tag{4·3, 23}$$

so that

$$\frac{\rho_0}{\rho}\sqrt{(1 - M^2)} = 1, \tag{4·3, 24}$$

which is the condition that the hodograph relations reduce to the Cauchy–Riemann equations. It is of interest to note that from the isentropic flow relationship

$$\frac{\rho_0}{\rho} = \left(1 + \frac{\gamma - 1}{2} M^2\right)^{1/(\gamma - 1)},$$

condition (4·3, 24) is equivalent to taking $\gamma = -1$, and this is the original assumption made by Chaplygin. We see from (4·2, 23) that $q \leqslant a$, so that the flow must always be subsonic, the governing differential equations are therefore elliptic in type, and for this reason, as will be seen presently, a complex representation of the flow functions is possible.

As in the von Kármán method, we may again relate the incompressible and compressible flows by equation (4·3, 20). But now,

$$\frac{dQ}{Q} = \sqrt{(1 - M^2)}\frac{dq}{q} = \frac{\rho}{\rho_0}\frac{dq}{q} \tag{4·3, 25}$$

from (4·3, 24). Since, with $\gamma = -1$,

$$\frac{\rho}{\rho_0} = \left(1 + \frac{\gamma - 1}{2} M^2\right)^{-1/(\gamma - 1)} = \left(1 - \frac{q^2}{a^2}\right)^{\frac{1}{2}} = \left(1 + \frac{q^2}{a_0^2}\right)^{-\frac{1}{2}}, \tag{4·3, 26}$$

we can integrate (4·3, 25) to give

$$Q = \text{constant}\frac{q}{a_0 + \sqrt{(a_0^2 + q^2)}}.$$

Since $q \to Q$ as $a_0 \to \infty$, we obtain finally

$$Q = \frac{2a_0 q}{a_0 + \sqrt{(a_0^2 + q^2)}}. \tag{4·3, 27}$$

The density in terms of the incompressible velocity can be found using (4·3, 26), thus

$$\frac{\rho_0}{\rho} = \frac{4a_0^2 + Q^2}{4a_0^2 - Q^2}. \tag{4·3, 28}$$

Defining the pressure coefficient as $C_p = (p - p_1)/\tfrac{1}{2}\rho_1 q_1^2$, and using (4·3, 22), we find, omitting some algebra, that

$$\frac{C_{pc}}{C_{pi}} = 1 \bigg/ \left\{\sqrt{(1 - M_1^2)} + \frac{M_1^2}{\sqrt{(1 - M_1^2)} + 1}\frac{C_{pi}}{2}\right\}, \tag{4·3, 29}$$

where the suffices c and i refer to corresponding compressible and incompressible flow. This equation should be compared with the Prandtl–Glauert equation (4·2, 36) obtained from linearized theory. (4·3, 29) reduces to (4·2, 36) for small values of C_{p_i} (i.e. sufficiently small perturbations of the flow relative to the main incident flow). Fig. 4·3, 1 compares the Kármán–Tsien formula (4·3, 29) with Prandtl–Glauert law. For high Mach numbers, the hodograph methods give results better in accord with experiment than those of linearized theory, which underestimates the effect of compressibility.

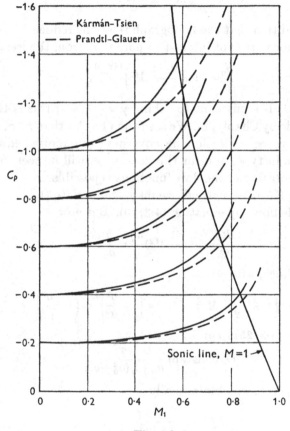

Fig. 4·3, 1

As in von Kármán's method, when applying the Kármán–Tsien results a correction should be applied for the difference in contour between the compressible and incompressible cases. In practice, however, this is rarely done; the calculation is not easy and in fact it appears that for the two approximations involved in using (4·3, 29) (approximating the isentrope by a straight line and neglecting the change in boundaries) have a tendency to cancel each other, so that the result obtained in using (4·3, 29) without correcting for the change in profile are often remarkably close to more exact solutions.

Using (4·3, 27), the transformation relations (4·3, 10) become

$$
\left.
\begin{aligned}
-dx &= \frac{\cos\theta}{4a_0^2 Q}(4a_0^2 - Q^2)\,d\Phi - \frac{\sin\theta}{4a_0^2 Q}(4a_0^2 + Q^2)\,d\Psi, \\
-dy &= \frac{\sin\theta}{4a_0^2 Q}(4a_0^2 - Q^2)\,d\Phi + \frac{\cos\theta}{4a_0^2 Q}(4a_0^2 + Q^2)\,d\Psi,
\end{aligned}
\right\}
\tag{4·3, 30}
$$

and putting

$$Q^* = Q\,e^{-i\theta},$$

$$\Phi^* = \Phi + i\Psi,$$

these can be written

$$-dz = \frac{d\Phi^*}{Q^*} - \frac{\overline{Q^*\,d\Phi^*}}{4a_0^2}, \tag{4·3, 31}$$

where the bar over $Q^*\,d\Phi^*$ denotes the conjugate complex. This equation relates corresponding elements in the Φ, Ψ-plane and the physical x, y-plane in terms of the incompressible solution $Q^*(\Phi^*)$ in the hodograph plane. If we take $\zeta = \xi + i\eta$ as the coordinates of the incompressible flow in the physical plane, we have

$$\frac{d\Phi^*}{d\zeta} = -Q^*,$$

so that

$$dz = d\zeta - \frac{\overline{Q^{*2}\,d\zeta}}{4a_0^2},$$

and writing

$$G = \frac{1}{Q_1^*}\Phi^*, \quad \text{so that} \quad \frac{dG}{d\zeta} = -\frac{1}{Q_1^*}Q^*,$$

we obtain finally

$$z = \zeta - \frac{Q_1^{*2}}{4a_0^2}\int \overline{\left(\frac{dG}{d\zeta}\right)^2\,d\zeta}. \tag{4·3, 32}$$

Equations (4·3, 31) and (4·3, 32) are signifiant in the first place because they relate the complex functions Φ^* and Q^* in the complex z- and ζ-planes. The powerful methods of the complex variable are therefore available to deal with the problem. The magnitude of the parameter

$$\lambda = \frac{Q_1^{*2}}{4a_0^2} \tag{4·3, 33}$$

is a measure of the distortion of the profile from the compressible to the incompressible plane. From (4·3, 27) we can write λ as

$$\lambda = \frac{M_1^2}{(1 + \sqrt{(1 - M_1^2)})^2}, \tag{4·3, 34}$$

where M_1 is the undisturbed flow Mach number in the compressible plane; the relationship (4·3, 34) is plotted in Fig. 4·3, 2. Except for Mach numbers close to unity the term $\lambda\int\overline{\left(\frac{dG}{d\zeta}\right)^2\,d\zeta}$ in (4·3, 32) is sufficiently small to be regarded as a correction factor when applied to normal aerofoil contours.

In general, the expansion of G in the incompressible plane will contain a term of order $1/\zeta$, corresponding to circulatory flow about the body, and on integration this leads to a logarithmic term in ζ which is not a single-valued function of the complex variable ζ. The solution of (4·3, 32) therefore is not unique if there is circulation and lift on the body. We may derive this result from another viewpoint. Thus, on the profile, $d\Psi = 0$, and $d\Phi/d\zeta = -Q^*$ so that

$$\overline{Q^*} d\Phi^* = \overline{Q^*} d\Phi$$
$$= -Q^2(\cos\theta + i\sin\theta)(\cos\theta - i\sin\theta)\,d\zeta$$
$$= -Q^2 d\zeta,$$

and so
$$dz = d\zeta - \frac{1}{4a_0^2}Q^2 d\zeta.$$

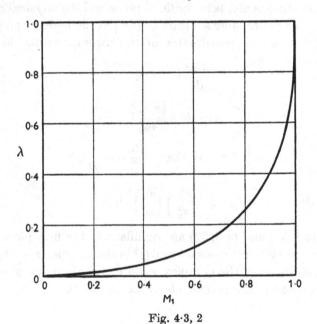

Fig. 4·3, 2

From Bernoulli's equation for incompressible flow, $P + \frac{1}{2}\rho_0 Q^2 = \text{constant}$, where P is the pressure in incompressible flow, we get therefore

$$z = \zeta + \frac{1}{2a_0^2\rho_0}\int P\,d\zeta.$$

If there is circulation $\int_C P\,d\zeta$ round the profile C is finite, and hence z and ζ are not uniquely related. Thus if, for example, the profile is closed in the incompressible plane, and we integrate from the leading edge, the integral over the top and bottom surfaces will have different values at the trailing edge, and the profile in the compressible plane will not close. This defect in the method may be overcome by introducing an additional transformation.‡ We put

$$d\zeta = k(Z)\, dZ, \tag{4·3, 35}$$

so that (4·3, 32) becomes

$$dz = k(Z)\, dZ - \lambda \frac{(dG/dZ)^2 \, dZ}{k(Z)}. \tag{4·3, 36}$$

In order that the profile in the Z-plane be closed we require that

$$\int_C k(Z)\, dZ - \lambda \int_C \frac{(G'(Z))^2 \, dZ}{k(Z)} = 0, \tag{4·3, 37}$$

where the integration is taken round a contour enclosing the profile. The transformation from the z- to the Z-plane will be regular provided

$$J = \frac{\partial(x, y)}{\partial(X, Y)} \neq 0 \quad \text{or} \quad \infty, \tag{4·3, 38}$$

where $Z = X + iY$. But

$$J = (x_X y_Y - x_Y y_X) = \mathscr{I}(\bar{z}_X z_Y)$$

$$= |k(Z)|^2 \left(1 - \lambda^2 \left|\frac{G'(Z)}{k(Z)}\right|^4\right) \tag{4·3, 39}$$

from (4·3, 37). Hence (4·3, 38) is satisfied if

$$|k(Z)| \neq 0,$$

and

$$\left|\frac{Q_1^*}{2a_0} G'(Z)\right| < |k(Z)| < \infty. \tag{4·3, 40}$$

Within the limitations of (4·3, 37) and (4·3, 40) the choice of $k(Z)$ is arbitrary, and a number of functions have been used. If the profiles in the compressible and incompressible z- and ζ-planes do not differ greatly, $k(Z) \doteq 1$, and this will be true as long as the thickness to chord ratio is not too large, or the Mach number not very close to 1. As an example we choose the case considered by K. Jaeckel,§ in which k is expanded as

$$k(Z) = 1 + \frac{ik_1}{Z} + \dots, \tag{4·3, 41}$$

and the expansion for $G'(Z)$ is taken as

$$\frac{dG}{dZ} = 1 + \frac{ia_1}{Z} + \frac{b_2 + ia_2}{Z^2}.$$

These give for z, from (4·3, 36)

$$z = Z + ik_1 \log Z - \lambda(\bar{Z} - i(2a_1 - k_1)\log \bar{Z}) + O\!\left(\frac{1}{Z}\right),$$

‡ The idea is due to C. C. Lin, 'On an extension of the von Kármán–Tsien method to two-dimensional subsonic flows with circulation around closed profiles', *Quart. Appl. Math.* vol. 4 (1946), pp. 291–7.

§ See J. W. Craggs, 'The application of the hodograph method to problems of subsonic compressible flow in two dimensions', *Rep. Memor. Aero. Res. Comm., Lond.*, no. 2273 (1946).

which is single-valued if
$$k_1 = \frac{2a_1\lambda}{1+\lambda},$$

though the conditions of regularity (4·3, 40) are not assured.

An alternative function used by Bers‡ is

$$k = Q_1^2[G'(Z)]^{n-1},$$

with
$$n = \frac{1-\lambda}{1+\lambda}.$$

A method somewhat similar to Jaeckel's due to Gelbart,§ takes an expansion analogous to (4·3, 41) for $k(Z)$, but the coefficients are chosen so as to approximate the incompressible profile in the Z-plane to the compressible profile, as well as to make the term in $\log Z$ zero.

These methods using transformations between compressible and incompressible flows, with a linearized pressure-volume relationship, have been applied to the calculation of subsonic flows about circular and elliptic cylinders,‖ and to a few Joukowski aerofoils.¶ Up to the present, however, they have not been of wide application in aerofoil theory, apart from the use of the Kármán–Tsien equation (4·3, 29) for the pressure coefficient, and here, as stated previously, the profile distortion is customarily neglected.

We shall next consider briefly the general solution of the hodograph equations (4·3, 9). This was given first by Chaplygin.‡‡ He introduced the new variable

$$\tau = \frac{\gamma-1}{2}\frac{q^2}{a_0^2}, \tag{4·3, 42}$$

in terms of which the hodograph equations (4·3, 8) become

$$\left.\begin{aligned}
\frac{\partial\Phi}{\partial\theta} &= 2\tau(1-\tau)^{-\beta}\frac{\partial\Psi}{\partial\tau}, \\
\frac{\partial\Phi}{\partial\tau} &= \frac{\tau(2\beta+1)-1}{2\tau(1-\tau)^{\beta+1}}\frac{\partial\Psi}{\partial\theta},
\end{aligned}\right\} \tag{4·3, 43}$$

where
$$\beta = 1/(\gamma-1). \tag{4·3, 44}$$

‡ L. Bers, 'On a method of constructing two-dimensional subsonic compressible flows around closed profiles', *Tech. Notes Nat. Adv. Comm. Aero., Wash.*, no. 969 (1945).

§ A. Gelbart, 'On a function-theory method of obtaining potential flow patterns of a compressible fluid', *N.A.C.A. Wartime Rep.* L460 (1943).

‖ S. Bartnoff and A. Gelbart, 'On subsonic compressible flows by a method of correspondence. II. Application of methods to studies of flow with circulation about a circular cylinder', *Tech. Notes Nat. Adv. Comm. Aero., Wash.*, no. 1171 (1947); L. Bers, 'On the circulatory flow of a compressible fluid past a circular cylinder', *Tech. Notes Nat. Adv. Comm. Aero., Wash.*, no. 970 (1945).

¶ A. Gelbart and D. Resch, 'A method of computing subsonic flows around given airfoils', *Tech. Notes Nat. Adv. Comm. Aero., Wash.*, no. 2057 (1950).

‡‡ S. Chaplygin, 'Gas jets', *Tech. Memor. Nat. Adv. Comm. Aero., Wash.*, no. 1063. Translated from *Sci. Mem. Math. Phys. Soc.* no. 21, Moscow University (1902), pp. 1–121.

The equation satisfied by Ψ' is then

$$\frac{\partial}{\partial \tau}\left(2\tau(1-\tau)^{-\beta}\frac{\partial \Psi'}{\partial \tau}\right) + \frac{1-(2\beta+1)\tau}{2\tau(1-\tau)^{\beta+1}}\frac{\partial^2 \Psi'}{\partial \theta^2} = 0. \tag{4·3, 45}$$

Separation of variables by putting

$$\Psi' = \begin{cases} \Psi'_\nu(\tau)\sin\nu\theta, \\ \Psi'_\nu(\tau)\cos\nu\theta, \end{cases} \tag{4·3, 46}$$

yields

$$\frac{d}{d\tau}\left(2\tau(1-\tau)^{-\beta}\frac{d\Psi'_\nu}{d\tau}\right) - \nu^2\frac{1-(2\beta+1)\tau}{2\tau(1-\tau)^{\beta+1}}\Psi'_\nu = 0, \tag{4·3, 47}$$

and if we introduce $\qquad F_\nu(\tau) = \tau^{-\frac{1}{2}\nu}\Psi'_\nu(\tau),$ $\qquad$ (4·3, 48)

F_ν satisfies the hypergeometric equation

$$\tau(1-\tau)F''_\nu + [(\nu+1)-(\nu+1-\beta)\tau]F'_\nu + \tfrac{1}{2}\beta\nu(\nu+1)F_\nu = 0. \tag{4·3, 49}$$

The solution of (4·3, 49) is the hypergeometric function

$$F_\nu(a_\nu, b_\nu; \nu+1; \tau) = 1 + \frac{a_\nu b_\nu}{1!\,(\nu+1)}\tau + \frac{a_\nu b_\nu(a_\nu+1)(b_\nu+1)}{2!(\nu+1)(\nu+2)}\tau^2 + \dots, \tag{4·3, 50}$$

where $\qquad a_\nu + b_\nu = \nu - \beta,$

$$a_\nu b_\nu = -\tfrac{1}{2}\beta\nu(\nu+1),$$

or

$$\left.\begin{aligned} a_\nu &= \tfrac{1}{2}\{(\nu-\beta)+\sqrt{\{(2\beta+1)\nu^2+\beta^2\}}\}, \\ b_\nu &= \tfrac{1}{2}\{(\nu-\beta)-\sqrt{\{(2\beta+1)\nu^2+\beta^2\}}\}. \end{aligned}\right\} \tag{4·3, 51}$$

If the flow is incompressible $\beta = 0$, and $F_\nu = 1$ for all ν. Hence in this event

$$\Psi'_\nu = \tau^{\frac{1}{2}\nu}, \tag{4·3, 52}$$

and the basic solutions (4·3, 46) become

$$\Psi' = \begin{cases} \tau^{\frac{1}{2}\nu}\sin\nu\theta, \\ \tau^{\frac{1}{2}\nu}\cos\nu\theta. \end{cases} \tag{4·3, 53}$$

As in the special case considered above with a linear pressure-volume relationship we can establish a correspondence between the compressible flow and an incompressible flow about a related profile. We write

$$\Phi^* = \Phi^*\left(\tfrac{1}{2}\log\left(\frac{\tau_1}{\tau}\right)+i\theta\right) = \Phi^*\left(\log\left(\frac{q}{q_1}e^{-i\theta}\right)\right),$$

where τ_1 is the main-stream value, and we assume an expansion for the incompressible case

$$\Phi^* = B\left(\tfrac{1}{2}\log\left(\frac{\tau_1}{\tau}\right)+i\theta\right) + \sum_{n=1}^{\infty} C_n\left(\frac{\tau}{\tau_1}\right)^{\frac{1}{2}n}e^{in\theta},$$

with $\qquad C_n = B_n \exp(i\alpha_n), \quad B_n$ real.

Then $\qquad \Psi = \mathscr{I}\Phi^* = B\theta + \sum_{n=1}^{\infty} B_n\sin(n\theta+\alpha_n)\left(\frac{\tau}{\tau_1}\right)^{\frac{1}{2}n}.$ $\qquad$ (4·3, 54)

Comparison of (4·3, 48) with (4·3, 52) shows that from the incompressible solution (4·3, 54) we can obtain a solution of a compressible flow by replacing $\left(\tau/\tau_1\right)^{\frac{1}{2}n}$ in (4·3, 54) with $\left(\dfrac{\tau}{\tau_1}\right)^{\frac{1}{2}n} \dfrac{F_n(\tau)}{F_n(\tau_1)}$. Thus the compressible solution is

$$\Psi = B\theta + \sum_{n=1}^{\infty}\left[B_n\left(\frac{\tau}{\tau_1}\right)^{\frac{1}{2}n} \frac{F_n(\tau)}{F_n(\tau_1)} \sin\left(n\theta + \alpha_n\right)\right]. \qquad (4\cdot3, 55)$$

Chaplygin has shown that this series is convergent for $\tau < \tau_1$ when the corresponding incompressible series (4·3, 54) is convergent.

We now consider the approximation to these solutions made by Temple and Yarwood for subsonic flows.‡ Since in incompressible flow

$$\Psi_\nu = (\tau)^{\frac{1}{2}\nu}, \qquad (4\cdot3, 56)$$

Ψ_ν in compressible flow is approximated by

$$\Psi_\nu = [\eta(\tau)]^\nu, \qquad (4\cdot3, 57)$$

and there is a similar approximation for the velocity potential Φ. (4·3, 57) is to be chosen so that it tends to $(\tau)^{\frac{1}{2}}$ as M tends to zero. At corresponding points in the compressible and incompressible flows the stream functions are equal, so we write

$$\Psi\left(\frac{\eta(\tau)}{\eta(\tau_1)}, \theta\right) = \Psi\left(\left(\frac{\tau}{\tau_1}\right)^{\frac{1}{2}}, \theta\right),$$

and at the same points θ on the compressible and incompressible profiles

$$\frac{\eta(\tau)}{\eta(\tau_1)} = \left(\frac{\tau}{\tau_1}\right)^{\frac{1}{2}} = \frac{Q}{Q_1}, \qquad (4\cdot3, 58)$$

with the suffix referring to main-stream values.

Temple and Yarwood were able to approximate $\eta(\tau)$ by finding upper and lower bounds to the functions $\Psi(\tau)$ and $\Phi(\tau)$. They found as a reasonable average value

$$\frac{\Psi_\nu(\tau)}{\Psi_\nu(\tau_1)} = \left(\frac{\tau^{\frac{1}{2}}(1 - 1\cdot2\tau)}{\tau_1^{\frac{1}{2}}(1 - 1\cdot2\tau_1)}\right)^\nu. \qquad (4\cdot3, 59)$$

Using (4·3, 57) and (4·3, 58) this leads to

$$\frac{Q}{Q_1} = \left(\frac{\tau}{\tau_1}\right)^{\frac{1}{2}} \frac{1 - 1\cdot2\tau}{1 - 1\cdot2\tau_1} = \frac{q}{q_1} \frac{1 - \mu(q/q_1)^2}{1 - \mu}, \qquad (4\cdot3, 60)$$

where

$$\mu = 1\cdot2\tau_1 = \tfrac{1}{4}M_1^2 / (1 + \tfrac{1}{5}M_1^2) \qquad (4\cdot3, 61)$$

(taking $\gamma = 1\cdot40$). Finally, inversion of (4·3, 60) gives

$$\frac{q}{q_1} = \frac{2}{\sqrt{(3\mu)}} \cos\left(\tfrac{1}{3}(\pi + \delta)\right), \qquad (4\cdot3, 62)$$

where

$$\cos\delta = \frac{Q}{Q_1} \frac{3\sqrt{3\mu}\,(1 - \mu)}{2} \qquad (0 \leqslant \delta \leqslant \tfrac{1}{2}\pi).$$

‡ G. Temple and J. Yarwood, 'The approximate solution of the hodograph equations for compressible flow', *Royal Aircraft Establishment Rep.* no. S.M.E. 3201 (1942).

In general (4·3, 62) predicts a slightly greater compressibility effect than the Kármán–Tsien theory, but is not believed to be in better accord with experiment.

The hodograph methods described in this section are the principal ones used in applied subsonic aerofoil theory. Several authors have taken a more general approach than those used in the approximate methods given here, and this is necessary in considering mixed subsonic and supersonic flows, but the mathematical complexity of such methods has so far restricted their use in pure subsonic problems where the approximate methods are usually adequate. For readers interested in the general theory it may be mentioned that Lighthill's method appears to be one of the more promising approaches for flow about closed profiles.‡

4·4 Higher order theories in subsonic flow

Section 4·2 considered the first-order approximation to the complete compressible flow equation (4·1, 21) for small perturbations of the flow from the main-stream value. The hodograph method (§ 4·3) gave theoretical methods of solving the two-dimensional problem, but here again the more practical procedures assumed small perturbations of the main flow or low Mach numbers. By considering an expansion for the potential

$$\Phi = - Ux + \epsilon\Phi_1 + \epsilon^2\Phi_2 + \ldots \tag{4·4, 1}$$

in powers of the thickness parameter ϵ (such as the thickness to chord ratio of an aerofoil), we can extend these methods to higher orders of perturbation in three-dimensional flow. However, the evaluation of particular solutions is difficult and lengthy, so that few applications have yet been made. We shall therefore restrict ourselves to outlining the principles involved in the method.

If we substitute (4·4, 1) in the full non-linear equation (4·1, 21), use (4·1, 8) to obtain a^2 in terms of Φ, and equate coefficient of successive powers of ϵ to zero, we find that

$$\left.\begin{aligned}
(1 - M_1^2)\frac{\partial^2\Phi_1}{\partial x^2} + \frac{\partial^2\Phi_1}{\partial y^2} + \frac{\partial^2\Phi_1}{\partial z^2} &= 0, \\[2mm]
(1 - M_1^2)\frac{\partial^2\Phi_2}{\partial x^2} + \frac{\partial^2\Phi_2}{\partial y^2} + \frac{\partial^2\Phi_2}{\partial z^2} &= F_2(\Phi_1), \\[2mm]
(1 - M_1^2)\frac{\partial^2\Phi_3}{\partial x^2} + \frac{\partial^2\Phi_3}{\partial y^2} + \frac{\partial^2\Phi_3}{\partial z^2} &= F_3(\Phi_1, \Phi_2), \\[2mm]
\vdots \quad \vdots \quad \vdots \quad \vdots \quad &
\end{aligned}\right\} \tag{4·4, 2}$$

‡ *Modern Developments in Fluid Dynamics. High Speed Flow*, ed. Howarth (Oxford University Press, 1953), ch. 7, by M. J. Lighthill.

where M_1 is the free-stream Mach number and the right-hand sides are functions only of the solutions of the previous equations. For example, the first such function is

$$F_2(\Phi_1) = M_1^2 \left[(\gamma+1) \frac{\partial \Phi_1}{\partial x} \frac{\partial^2 \Phi_1}{\partial x^2} + (\gamma-1) \frac{\partial \Phi_1}{\partial x} \left\{ \frac{\partial^2 \Phi_1}{\partial y^2} + \frac{\partial^2 \Phi_1}{\partial z^2} \right\} \right. $$
$$\left. + 2 \frac{\partial \Phi_1}{\partial y} \frac{\partial^2 \Phi_1}{\partial x \partial y} + 2 \frac{\partial \Phi_1}{\partial z} \frac{\partial^2 \Phi_1}{\partial x \partial z} \right].$$

The first equation of (4·4, 2) is the linearized equation (4·2, 1). Each successively higher order equation can be solved from a knowledge of the solutions of the previous ones.

Equations (4·4, 2) have been solved for a number of problems in two-dimensional flow using relaxation methods‡ and complex variable representation.§ With the latter method Kaplan has derived particular integrals of the second and third iteration equations of (4·4, 2), from which a solution can be obtained in particular cases by adding solutions of the homogeneous equations, having regard to the boundary conditions of the problem. If the new variables

$$\left. \begin{array}{l} X = x, \\ Y = y/\sqrt{(1-M_1^2)}, \end{array} \right\} \tag{4·4, 3}$$

are introduced into (4·4, 2), the first equation reduces to Laplace's equation, and the second becomes

$$\Delta \Phi_2 = 2 M_1^2 \left[(1+\sigma) \frac{\partial \Phi_1}{\partial X} \frac{\partial^2 \Phi_1}{\partial X^2} + \frac{\partial \Phi_1}{\partial Y} \frac{\partial^2 \Phi_1}{\partial X \partial Y} \right], \tag{4·4, 4}$$

where $\quad \sigma = \dfrac{\gamma+1}{2} \dfrac{M_1^2}{1-M_1^2} \quad$ and $\quad \Delta \Phi_2 = \dfrac{\partial^2 \Phi_2}{\partial X^2} + \dfrac{\partial^2 \Phi_2}{\partial Y^2},$

which is Poisson's equation with the right-hand side a function determined from the solution of the previous equation. Similar equations are found for the higher orders, the left-hand side always being the Laplacian $\Delta \Phi_n$ and the right-hand side a function of lower order solutions. The complex variables

$$\left. \begin{array}{l} Z = X + iY, \\ \bar{Z} = X - iY, \end{array} \right\} \tag{4·4, 5}$$

are now introduced as independent variables. Then the equation for Φ_1 becomes

$$4 \frac{\partial^2 \Phi_1}{\partial Z \partial \bar{Z}} = 0,$$

‡ H. W. Emmons, 'The numerical solution of compressible fluid flow problems', *Tech. Notes Nat. Adv. Comm. Aero., Wash.*, no. 932 (1944).

§ There are several papers by C. Kaplan using the complex variable, for example: 'The flow of a compressible fluid past a curved surface', *N.A.C.A. Rep.* no. 768 (1943); 'Effect of compressibility at high subsonic velocities on the lifting force acting on an elliptic cylinder', *N.A.C.A. Rep.* no. 834 (1946); 'On the particular integrals of the Prandtl–Busemann iteration equations for the flow of a compressible fluid', *Tech. Notes Nat. Adv. Comm. Aero., Wash.*, no. 2159 (1950).

of which the general solution is

$$\Phi_1 = \tfrac{1}{2}[w_1(Z) + \overline{w}_1(\overline{Z})], \tag{4·4, 6}$$

where w_1 is an arbitrary analytic function of Z, and $\overline{w}_1$ is the complex conjugate of w_1. In terms of this solution it is not difficult to show that the equation (4·4, 4) for Φ_2 becomes

$$(\Phi_2)_{Z\overline{Z}} = \tfrac{1}{4}M_1^2 \mathscr{R}[\tfrac{1}{2}\sigma\,(w_{1Z}^2)_Z + (2+\sigma)\,\overline{w}_{1\overline{Z}}\overline{w}_{1ZZ}], \tag{4·4, 7}$$

so that if we take Φ_2 to be the real part of a (non-analytic) function $w(Z, \overline{Z})$ we get

$$w_{2Z\overline{Z}} = \tfrac{1}{4}M_1^2[\tfrac{1}{2}\sigma\,(w_{1Z}^2)_Z + (2+\sigma)\,(w_{1Z}\overline{w}_{1\overline{Z}})_Z]. \tag{4·4, 8}$$

This equation may be solved immediately to give a particular integral

$$w_2 = \tfrac{1}{4}M_1^2[\tfrac{1}{2}\sigma\overline{Z}w_{1Z}^2 + (2+\sigma)\,\overline{w}_1 w_{1Z}]. \tag{4·4, 9}$$

By taking the real part of (4·4, 9) we find as a particular integral of (4·4, 3)

$$\Phi_2 = M_1^2[(1 + \tfrac{1}{2}\sigma)\,\Phi_1 - \tfrac{1}{2}\sigma Y\Phi_{1Y}]\,\Phi_{1X}. \tag{4·4, 10}$$

The homogeneous equation corresponding to (4·4, 4) is simply

$$\Delta\Phi_2 = 0, \tag{4·4, 11}$$

for which there are standard methods of solution for particular boundary conditions. The formal solution of the second-order equation is thus complete; from a solution of the linearized first-order equation of (4·4, 2) we find the particular solution for Φ_2 given by (4·4, 10), and combine it with the general solution of (4·4, 11) to satisfy the boundary conditions of the problem in the X, Y-plane, obtained through the transformation (4·4, 3).

In a similar manner a particular integral can be obtained for $\Phi_3(X, Y)$ in terms of Φ_1 and Φ_2. However, the algebra is lengthy and we will not discuss that case here.

An approach complementary to expansion in powers of a thickness parameter is the Rayleigh–Janzen method of expansion in powers of the main-stream Mach number.‡ Whereas the former is apposite in calculations of flow about thin profiles at high subsonic speeds, the latter is suitable for calculating relatively slow flows about thick bodies. Thus, while in the perturbation method the first-order approximation is the linearized compressible flow equation and gives accuracy at moderate or high Mach numbers for thin profiles, the first-order solution in the Rayleigh–Janzen expansion is Laplace's equation for incompressible flow, and is therefore valid for small Mach numbers only, although the profile may in this case be thick. As far as conventional aerofoils are concerned it follows that the extended small perturbation method is of greater use.

‡ The original papers are: Lord Rayleigh, 'On the flow of a fluid past an obstacle', *Phil. Mag.* vol. 6, no. 32 (1916), pp. 1–6; O. Jenzen, 'Beitrag zu einer Theorie der stationären Strömung kompressibler Flüssigkeiten', *Phys. Z.* vol. 14 (1913), pp. 639–43.

It can be seen from the basic equations (4·1, 21) and (4·1, 8) that the Mach number can only enter as a squared term, so the Rayleigh–Janzen expansion in powers of Mach number is taken as

$$\Phi = \Phi_0 + M_1^2 \Phi_1 + M_1^4 \Phi_2 + \dots, \qquad (4\cdot4, 12)$$

where M_1 is the Mach number in the main flow. The velocity q may be expressed conveniently as

$$q^2 = q_0^2 + M_1^2 q_1^2 + M_1^4 q_2^2 + \dots, \qquad (4\cdot4, 13)$$

in which case it follows readily from (4·4, 12), and using $q = -\operatorname{grad} \Phi$, that

$$\left.\begin{aligned}
q_0^2 &= (\Phi_{0x})^2 + (\Phi_{0y})^2 + (\Phi_{0z})^2, \\
q_1^2 &= 2(\Phi_{0x}\Phi_{1x} + \Phi_{0y}\Phi_{1y} + \Phi_{0z}\Phi_{1z}), \\
q_2^2 &= 2(\Phi_{1x}\Phi_{0x} + \Phi_{1y}\Phi_{0y} + \Phi_{1z}\Phi_{0z} + (\Phi_{1x})^2 + (\Phi_{1y})^2 + (\Phi_{1z})^2),
\end{aligned}\right\} \quad (4\cdot4, 14)$$

Proceeding as in the previous method, viz. substituting into (4·1, 21), using (4·1, 8), and equating coefficients of successive powers of M_1^2 to zero, we obtain the set of equations

$$\left.\begin{aligned}
\Delta\Phi_0 &= 0, \\
\Delta\Phi_1 &= \frac{1}{2}\left(\Phi_{0x}\frac{q_{0x}^2}{U^2} + \Phi_{0y}\frac{q_{0y}^2}{U^2} + \Phi_{0z}\frac{q_{0z}^2}{U^2}\right), \\
\Delta\Phi_2 &= \frac{\gamma-1}{2}\left(1 - \frac{q_0^2}{U^2}\right)\Delta\Phi_1 + \frac{1}{2}\left(\Phi_{0x}\frac{q_{1x}^2}{U^2} + \Phi_{0y}\frac{q_{1y}^2}{U^2} + \Phi_{0z}\frac{q_{1z}^2}{U^2}\right) \\
&\qquad\qquad + \Phi_{1x}\frac{q_{0x}^2}{U^2} + \Phi_{1y}\frac{q_{0y}^2}{U^2} + \Phi_{1z}\frac{q_{0z}^2}{U^2},
\end{aligned}\right\} \quad (4\cdot4, 15)$$

where U is the main-stream velocity.

The right-hand side of each equation of (4·4, 15) is a function only of the solutions of the previous equations. Starting with the solution of the incompressible flow it is possible therefore in principle to solve successive equations as Poisson equations (compare (1·7, 10)) for Φ_1, Φ_2, …. However, the problem is quite difficult in practice, and as with the perturbation method, application has been chiefly to two-dimensional flow. We shall outline an approach to the two-dimensional case that is due to Poggi.‡

We may regard each equation of (4·4, 15) as applying for an incompressible flow in which there is a continuous source density distribution given by -1 times the right-hand side, which is known from the solutions of the previous equations (see § 1·7). The solution is then to be found by integrating over the distribution according to (1·7, 6). Suppose we have mapped the flow field on to the region outside a circle of radius R, centre the origin, in the z-plane (Fig. 4·4, 1). Then the complex potential function for the field due

‡ L. Poggi, 'Campo di Velocità in una Correnta Piana di Fluido Compressibile', *Aerotecnica, Roma*, vol. 12 (1932), pp. 1579–93.

to a source of unit strength at the point z_1 outside the circle is (cf. equation (1.10, 11))

$$\Pi'(z) = -\frac{1}{2\pi} \log (z - z_1) - \frac{1}{2\pi} \log (z - z_2) + \frac{1}{2\pi} \log (z), \qquad (4·4, 16)$$

where z_2 is the point inverse to z_1 with respect to the circle. (4·4, 16) represents the effect of a unit source at z_1, of a unit source at z_2 and a source of strength -1 at the origin. We may verify that the circle is a streamline by introducing $z = r e^{i\theta}$, $z_1 = r_1 e^{i\theta_1}$ and $z_2 = \dfrac{R^2}{r_1} e^{i\theta_1}$. After reduction we find for the complex potential

$$\Pi'(z) = \Phi' + i\Psi'' = -\frac{1}{4\pi} \log (r_1^2 - 2rr_1 \cos (\theta_1 - \theta) + r^2)$$

$$-\frac{1}{4\pi} \log \left(1 - \frac{2R^2}{rr_1} \cos (\theta_1 - \theta) + \left(\frac{R^2}{rr_1} \right)^2 \right)$$

$$+ \frac{i}{2\pi} \left\{ \tan^{-1} \left(\sin (\theta_1 - \theta) \frac{r(R^2 + r_1^2) - 2r_1 R^2 \cos (\theta_1 - \theta)}{\begin{array}{c} r_1 r^2 + r_1 R^2 (\cos^2 (\theta_1 - \theta) - \sin^2 (\theta_1 - \theta)) \\ - r(R^2 + r_1^2) \cos (\theta_1 - \theta) \end{array}} \right) - \theta \right\}.$$

$$(4·4, 17)$$

When $r = R$ the stream function $\Psi'' = -\theta_1$, a constant, so that the circle $|z| = R$ is indeed a streamline of the flow.

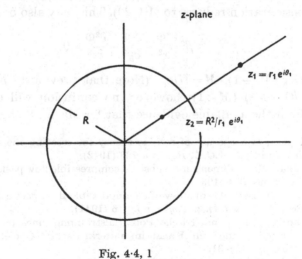

Fig. 4·4, 1

For the source distribution given by the right-hand side of (4·4, 15) we thus obtain the potential, taking, for example, the second equation of (4·4, 15),

$$\Phi_1 = -\frac{1}{2} \int \Phi' \left(\Phi_{0x} \frac{q_{0x}^2}{U^2} + \Phi_{0y} \frac{q_{0y}^2}{U^2} \right) dx\, dy, \qquad (4·4, 18)$$

where Φ' is given by (4·4, 17), and the integration extends over the region exterior to the circle.

(4·4, 18) may be calculated by double integration in polar coordinates, or it is possible to introduce complex representation for the potential and the velocity q, in which case the equation can be reduced to a line integral in the complex plane, and then may be evaluated using the theory of residues.‡ Most of the standard types of profile have been considered, elliptic and circular cylinders,§ Joukowski aerofoils,|| and circular cylinders in channels.¶ In three dimensions results for spheres and ellipsoids are available.‡‡

Variational methods have been used by Wang§§ and by Thom|||| to solve the exact equations by numerical procedure. Such techniques are of particular interest because the exact solutions can be approximated to any degree of accuracy by a routine computational procedure without additional algebraic analysis, such as is introduced when higher order terms in the expansions (4·4, 1) or (4·4, 12) are used.

4·5 Linearized theory of supersonic flow

For Mach numbers greater than unity, the flow is again tractable by detailed theoretical methods. Most of the extensive literature which has been published in recent years on the subject of supersonic aerofoil theory is based on the wave equation (4·1, 10) or (4·1, 19), which for the case now under consideration reduces to (4·1, 20). This may also be written as

$$\beta^2 \frac{\partial^2 \Phi}{\partial x^2} - \frac{\partial^2 \Phi}{\partial y^2} - \frac{\partial^2 \Phi}{\partial z^2} = 0, \qquad (4·5, 1)$$

where $\beta = \sqrt{(M^2 - 1)}$, $M = U/a_0$. (Note that previously β was defined to denote $\sqrt{(1 - M^2)}$ $(M < 1)$; however, no confusion will arise.) It would appear from the available evidence that for Mach numbers which are not

‡ C. Kaplan, 'On the use of residue theory for treating the subsonic flow of a compressible fluid', *N.A.C.A. Rep.* no. 728 (1942).

§ C. Kaplan, 'Two-dimensional subsonic compressible flow past elliptic cylinders', *N.A.C.A. Rep.* no. 624 (1938).

|| I. Imai, 'On the subsonic flow of a compressible fluid past a general Joukowski profile', *Rep. Aero. Res. Inst. Toyko*, no. 216 (1941).

¶ E. Lamla, 'Die symmetrische Potentialströmung eines kompressibeln Gases um einen Kreiszylinder im Kanal im unterkritischen Gebiet', *Luftfahrtforsch.* vol. 17 (1940), pp. 329–31.

‡‡ E. Lamla, 'Die symmetrische Potentialströmung eines kompressibeln Gases um Kreiszylinder und Kugel im unterkritischen Gebiet', *Dtsch. Luftfahrtforsch.* FB no. 1014 (1939).

‡‡ C. Schmeiden, 'Die kompressible Strömung um ein Rotationsellipsoid nach der Methode von Janzen–Rayleigh', *Jb. 1942 Dtsch. Luftfahrtforsch.* pp. I72–9.

§§ C. T. Wang, 'Two-dimensional subsonic compressible flows past arbitrary bodies by the variational method', *Tech. Notes Nat. Adv. Comm. Aero., Wash.*, no. 2326 (1951).

|||| A. Thom, 'The method of influence factors in arithmetical solutions of certain field problems', *Rep. Memor. Aero. Res. Comm., Lond.*, no. 2440 (1946).

too close to unity ($M > 1·15$, say) the results obtained on the basis of (4·5, 1) are either in reasonable agreement with experiment, or, where this is not the case, the reason may be found in a local breakdown of the flow which is due to the effects of viscosity rather than in the simplifications which lead from the general equations of inviscid flow to (4·5, 1).

The particular solution of (4·5, 1) which was obtained earlier (4·1, 57) may be written as

$$\Phi(x,y,z) = \begin{cases} \dfrac{\sigma}{2\pi \sqrt{[(x-x_0)^2 - \beta^2\{(y-y_0)^2 + (z-z_0)^2\}]}} \\ \qquad \text{for} \quad (x-x_0)^2 - \beta^2\{(y-y_0)^2 + (z-z_0)^2\} > 0, \quad x-x_0 > 0, \\ 0 \quad \text{elsewhere.} \end{cases} \tag{4·5, 2}$$

This may be called the velocity potential of a steady supersonic source of strength σ, located (or with origin) at (x_0, y_0, z_0). On the other hand, the velocity potential Φ which is given by

$$\Phi(x,y,z) = \begin{cases} \dfrac{\sigma}{2\pi \sqrt{[(x-x_0)^2 - \beta^2\{(y-y_0)^2 + (z-z_0)^2\}]}} \\ \qquad \text{for} \quad (x-x_0)^2 - \beta^2\{(y-y_0)^2 + (z-z_0)^2\} > 0, \quad x-x_0 < 0, \\ 0 \quad \text{elsewhere,} \end{cases} \tag{4·5, 3}$$

will be said to be the velocity potential of a counter-source.

We recall that (4·5, 2) was obtained originally as the velocity potential of a travelling acoustic source. The further investigation of the field of flow due to (4·5, 2) might be based on this derivation, i.e. we might base the present consideration of the steady case on a more general unsteady theory (compare § 5·7). On the other hand, it is of practical importance to develop direct methods for dealing with (4·5, 1), which may be regarded as a companion to, or counterpart of, Laplace's equation (1·7, 1). In the present theory, as in the theory of Laplace's equation, an important part is played by the concept of flux across a closed surface.

Let S be a closed surface. Then the mass of fluid which crosses S in the outward direction, in unit time, is given by

$$\int_S \rho \mathbf{q}\, d\mathbf{S} = \int_S (\rho u, \rho v, \rho w)\, d\mathbf{S}. \tag{4·5, 4}$$

According to our present assumptions, ρ never differs greatly from the free-stream density ρ_0, but on the other hand v, w and u' (where $u' = u - U$) are all small compared with the free-stream velocity U. Accordingly, the variation of ρ must be taken into account in the calculation of ρu in (4·5, 4). Now

$$\frac{\rho - \rho_0}{p - p_0} = \frac{d\rho}{dp} = \frac{1}{a_0^2} \tag{4·5, 5}$$

for small variations of ρ and so, by (4·1, 27),

$$\rho - \rho_0 = \frac{1}{a_0^2}(p - p_0) = -\rho_0 \frac{Uu'}{a_0^2},$$

or
$$\rho = \rho_0\left(1 - \frac{Uu'}{a_0^2}\right). \qquad (4\cdot5, 6)$$

Hence ρu becomes in the second approximation

$$\rho u = \rho_0\left(1 - \frac{Uu'}{a_0^2}\right)(U + u') \doteqdot \rho_0\left(U + \left(1 - \frac{U^2}{a_0^2}\right)u'\right) = \rho_0(U - \beta^2 u').$$
$$(4\cdot5, 7)$$

Now the integral of the constant $\rho_0 U$ over the closed surface S vanishes, and so the expression for the rate of fluid flow across S becomes

$$\rho_0\int_S(-\beta^2 u', v, w)\,d\mathbf{S} = \rho_0\int_S \mathbf{q}^*\,d\mathbf{S}, \qquad (4\cdot5, 8)$$

where the vector $\mathbf{q}^* = (-\beta^2 u', v, w)$ may be called the flow vector. The integral $\int_S \mathbf{q}^*\,d\mathbf{S}$, where the factor ρ_0 is now omitted, will be called the flux across S. We note that the linearized equation of continuity states that

$$\operatorname{div}\mathbf{q}^* = 0. \qquad (4\cdot5, 9)$$

Accordingly, if we take the flux over a closed surface S within which $\mathbf{q}$ and hence $\mathbf{q}^*$ are bounded and continuous, we have

$$\int_S \mathbf{q}^*\,d\mathbf{S} = \int_V \operatorname{div}\mathbf{q}^*\,dV = 0 \qquad (4\cdot5, 10)$$

by the divergence theorem and $(4\cdot5, 9)$, where V is the volume enclosed by S.

Now consider the field of flow which is due to the velocity potential $(4\cdot5, 2)$. It will be seen that $(4\cdot5, 2)$ describes only the field of flow induced by the presence of the source. In order to obtain the complete field of flow we have to add to $(4\cdot5, 2)$ the term

$$\Phi_0 = -Ux,$$

which is the velocity potential due to the free-stream velocity. Hence

$$u = -\frac{\partial}{\partial x}(\Phi_0 + \Phi) = U - \frac{\partial\Phi}{\partial x},$$

so that
$$u' = -\frac{\partial\Phi}{\partial x}, \quad v = -\frac{\partial\Phi}{\partial y}, \quad w = -\frac{\partial\Phi}{\partial z},$$

where Φ is given by $(4\cdot5, 2)$. It follows that the flux across a closed surface is

$$\int_S \mathbf{q}^*\,d\mathbf{S} = \int_S\left(\beta^2\frac{\partial\Phi}{\partial x}, -\frac{\partial\Phi}{\partial y}, -\frac{\partial\Phi}{\partial z}\right)d\mathbf{S}. \qquad (4\cdot5, 11)$$

If the surface S does not meet the after-cone of (x_0, y_0, z_0), then this integral exists and, moreover, its value is zero, by $(4\cdot5, 10)$. However, if S intersects the after-cone of (x_0, y_0, z_0), then the integral $(4\cdot5, 11)$ does not exist at all, since Φ becomes infinite at the cone as the square root of the inverse distance from the cone and so its derivatives become infinite as $\epsilon^{-\frac{3}{2}}$, where ϵ is the distance from the cone. Thus (4·5, 11) does not even exist as an improper integral. This situation is both undesirable and paradoxical, since (4·5, 2) is merely the velocity potential due to a travelling acoustic source, and as we saw in § 4·1 such a source emits fluid at a finite rate. It has been known for some time that difficulties of this sort may arise in the treatment of hyperbolic partial differential equations such as (4·5, 1), and different methods have been put forward for their resolution. One is the method of 'finite parts' of Hadamard.‡ A second method was put forward by Riesz§ and has been developed by him in more detail recently. Finally, a method which disposes of the infinities which occur in the present theory, as well as many other singularities which occur elsewhere in pure and applied mathematics, has been developed in recent years by Schwartz.‖ It is basically simpler but at the same time more abstract than Riesz's method, and depends on the ideas of modern functional analysis. Our treatment will be based on Hadamard's method which is understood and applied most easily and which is at the same time quite as effective as the other two methods, at least so far as the problems of the present section are concerned.

Let $D(x, y, z)$ be a polynomial of three variables so that the equation $D(x, y, z) = 0$ determines a surface Σ in the three-dimensional space. Σ divides the space (excluding the points of Σ) into disconnected components V_j in which $D(x, y, z)$ is of constant sign; moreover, $D(x, y, z)$ will be supposed to change sign across any ordinary point of Σ.

Let $f(x, y, z)$ be a real function defined in a region of space R such that

$$f(x, y, z) = g(x, y, z) + g_0(x, y, z) D^{-\frac{1}{2}n} + g_1(x, y, z) D^{-\frac{1}{2}n+1} + \dots$$
$$+ g_k(x, y, z) D^{-\frac{1}{2}} + g_{k+1}(x, y, z) D^{\frac{1}{2}} + \dots + g_{k+m}(x, y, z) D^{m-\frac{1}{2}},$$
$$(4·5, 12)$$

where n is a positive odd integer, and the functions $g(x, y, z)$, $g_0(x, y, z)$, ... $g_{k+m}(x, y, z)$, are either all analytic everywhere except on Σ, or at least have derivatives of sufficiently high order which are bounded in the neighbourhood of Σ. However, we admit the possibility that the functions $g(x, y, z)$, ..., $g_{k+m}(x, y, z)$ may be given by different analytic functions in the different regions V_j. A function $f(x, y, z)$ which satisfies these conditions will be called an admissible function.

Given a small positive quantity ϵ, we denote by $N(\epsilon)$ the set of all points P for which there exists a point P_0 on Σ such that the distance PP_0 does not exceed ϵ. The boundary (bounding surface) of $N(\epsilon)$ will be called $B(\epsilon)$, while $R(\epsilon)$ will be defined as the region which is obtained by excluding from

‡ J. Hadamard, *Lectures on Cauchy's Problem* (Yale University Press, 1923).

§ M. Riesz, 'Intégral de Riemann–Liouville et solution invariative du problème de Cauchy pour l'équation des ondes', *C.R. Congr., Int. Math. Oslo*, vol. 2 (1936), pp. 44–5.

‖ L. Schwartz, 'Théorie des distributions', *Actualités sci. industr.* vols. 1 and 2 (Hermann, Paris, 1950, 1951).

R all the points of $N(\epsilon)$. Furthermore, given a curve C, a surface S on a volume V in R, we shall denote by $C(\epsilon)$, $S(\epsilon)$ and $V(\epsilon)$ the subsets of C, S and V respectively which are obtained by the exclusion of the points of $N(\epsilon)$.

The concept of the finite part *J of a (finite or infinite) integral J—where J is any line, surface or volume integral of $f(x, y, z)$ on a bounded curve, surface, or volume, e.g.

$$\int_C f\,dx, \quad \int_S f\,dx\,dy, \quad \int_V f\,dx\,dy\,dz,$$

will now be defined as follows.

Given the formal expression for J, we denote by $J(\epsilon)$ the integral

$$\int_{C(\epsilon)} f\,dx, \quad \int_{S(\epsilon)} f\,dx\,dy \quad \text{or} \quad \int_{V(\epsilon)} f\,dx\,dy\,dz.$$

However, if C or S have (line or surface) segments in common with Σ, then we replace the integral by

$$\int_{C(\epsilon)} (f-g)\,dx + \int_C g\,dx \quad \text{or} \quad \int_{S(\epsilon)} (f-g)\,dx\,dy + \int_S g\,dx\,dy.$$

Subject to the specified conditions of regularity, $J(\epsilon)$ will be finite and of the form

$$J(\epsilon) = a_0 \epsilon^{-\frac{1}{2}n+1} + \ldots + a_{k-1} \epsilon^{-\frac{1}{2}} + H(\epsilon),$$

where $H(\epsilon)$ denotes a function which remains finite as ϵ tends to zero. We then define *J by

$$*J = \lim_{\epsilon \to 0} (J(\epsilon) - a_0 \epsilon^{-\frac{1}{2}n+1} - \ldots - a_{k-1} \epsilon^{-\frac{1}{2}}) = \lim_{\epsilon \to 0} H(\epsilon). \qquad (4\cdot5, 13)$$

It will be seen that if the integral J is finite then *$J = J$. There will be no occasion for confusion if in future we refer to the finite part of a (finite or infinite) integral simply as 'a finite part'. The concept of the finite part is invariant with respect to a transformation of coordinates provided the Jacobian of the transformation does not vanish on Σ. In particular, if we are dealing with the finite part of integrals involving vector quantities, the result is independent of a rotation of coordinates.

The finite parts of m-fold integrals in n-dimensional space, $n > 3$, $m \leqslant n$, can be defined in a strictly analogous manner.

The rules of calculation with finite parts, such as the rules of addition, are the same as for ordinary integrals. Also, if f depends on a parameter λ, but D is fixed, then, provided the functions g, g_0, ... are sufficiently regular (e.g. if they are analytic in the various V_j), it is not difficult to show that we may differentiate under the sign of the integral. For example,

$$\frac{d}{d\lambda} *\!\int_C f\,dx = *\!\int_C \frac{\partial f}{\partial \lambda}\,dx. \qquad (4\cdot5, 14)$$

Under similar conditions, the finite part of a multiple integral may be obtained by successive integration (including the operation of taking the finite part) with respect to the independent variables involved, taken in any arbitrary order. Thus, we have, for instance,

$$\overset{*}{\int_C} f\,dx\,dy\,dz = \overset{*}{\int}\left(\overset{*}{\int}\left(\overset{*}{\int} f\,dx\right) dy\right) dz, \qquad (4\cdot5,\,15)$$

where the integrals on the right-hand side are taken with appropriate limits.

More generally, we shall encounter cases where D and Σ depend algebraically on one or more parameters. We shall show (i) that even in that case we may 'differentiate under the integral sign', and (ii) that if a given integral, or finite part, involves integration with respect to such parameters as well as with respect to one or more of the space coordinates, then we may exchange the order of integration without affecting the value of the integral.

To see this, we increase the dimensions of the given space (x, y, z if the space is three-dimensional) by the parameter or parameters involved. Then, in the augmented space, the surface $D = 0$ is again fixed. To prove our assertions it is sufficient to show (i)′ that in order to find the derivative of a finite part in n-dimensional space with respect to any variable which is not involved in the integration we may differentiate under the sign of integration, and (ii)′ that for any multiple integral in n-dimensional space ($1 < m \leqslant n$), we have

$$\overset{*}{\int}\left(\overset{*}{\int} f\,dx_1\right) dx_2 \ldots dx_m = \overset{*}{\int} f\,dx_1\,dx_2 \ldots dx_m$$

taken over the appropriate regions. It is clear that (ii)′ will prove (ii) by induction.

We may reduce (i)′ to (ii)′. In fact, (i)′ states explicitly that

$$\frac{\partial}{\partial x_m} \overset{*}{\int} f\,dx_1 \ldots dx_{m-1} = \overset{*}{\int} \frac{\partial f}{\partial x_m}\,dx_1 \ldots dx_{m-1},$$

and this will be proved if it can be shown that

$$\overset{*}{\int} f\,dx_1 \ldots dx_{m-1} = \overset{*}{\int}\left(\overset{*}{\int} \frac{\partial f}{\partial x_m}\,dx_1 \ldots dx_{m-1}\right) dx_m + C,$$

where the lower limit of the integral with respect to x_m is arbitrary, and C is independent of x_m. Putting $F = \partial f/\partial x_m$ we have

$$f = f_0 + \overset{*}{\int} F\,dx_m,$$

where f_0 is the value of f for an arbitrary but definite value of x_m, for given $x_1, \ldots, x_{m-1}$, and the integral is taken with that particular value of x_m as lower limit. Now, assuming that (ii)′ has been proved, we have

$$\overset{*}{\int}\left(\overset{*}{\int} F\,dx_m\right) dx_1 \ldots dx_{m-1} = \overset{*}{\int}\left(\overset{*}{\int} F\,dx_1 \ldots dx_{m-1}\right) dx_m,$$

and so $\qquad \overset{*}{\int}(f-f_0)\,dx_1\ldots dx_{m-1} = \overset{*}{\int}\left(\overset{*}{\int}\int\frac{\partial f}{dx_m}\,dx_1\ldots dx_{m-1}\right)dx_m,$

i.e. $\quad \overset{*}{\int}f\,dx_1\ldots dx_{m-1} = \overset{*}{\int}\left(\overset{*}{\int}\int\frac{\partial f}{\partial x_m}\,dx_1\ldots dx_{m-1}\right)dx_m + \overset{*}{\int}f_0\,dx_1\ldots dx_{m-1},$

$$(4\cdot5,16)$$

and the last term on the right-hand side of (4·5, 16) is independent of x_m, as required.

To establish (ii)′ we have to prove that

$$\overset{*}{\int}\left(\overset{*}{\int}f\,dx_m\right)dx_1\ldots dx_{m-1} = \overset{*}{\int}f\,dx_1\ldots dx_m. \qquad (4\cdot5,17)$$

Putting $\overset{*}{\int}f\,dx_m = F$, we see that (4·5, 17) becomes

$$\overset{*}{\int}_S F\,dx_1\ldots dx_{m-1} = \overset{*}{\int}_V \frac{\partial F}{\partial x_m}\,dx_1\ldots dx_m, \qquad (4\cdot5,18)$$

taken over a certain region on the right-hand side and over its boundary S on the left-hand side, respectively. This is a special case of the divergence theorem for finite parts. We shall now prove the theorem for $m=3$; the proof for general m is quite similar.

A vector function $\mathbf{f}$ will be called admissible if its components are admissible, according to the conditions laid down above. This definition is independent of the particular system of coordinates to which $\mathbf{f}$ is referred. If all the first derivatives of the components of $\mathbf{f}$ are admissible, then div $\mathbf{f}$ also is admissible. We are going to show that under these conditions we have, for any volume V bounded by a surface S such as is considered in the ordinary divergence theorem,

$$\overset{*}{\int}_S \mathbf{f}\,d\mathbf{S} = \overset{*}{\int}_V \operatorname{div}\mathbf{f}\,dV. \qquad (4\cdot5,19)$$

By equation (4·5, 12), the vector $\mathbf{f}$ can be divided into two parts,

$$\mathbf{f}=\mathbf{F}+\mathbf{G}, \quad \mathbf{F}=(F_1,F_2,F_3), \quad \mathbf{G}=(G_1,G_2,G_3),$$

such that the F_i are of the form

$$g(x,y,z)+g_{k+1}(x,y,z)\,D^{\frac12}+\ldots+g_{k+m}(x,y,z)\,D^{m-\frac12},$$

while the G_i are given by expressions such as

$$g_0(x,y,z)\,D^{-\frac12 n}+g_1(x,y,z)\,D^{-\frac12 n+1}+\ldots+g_k(x,y,z)\,D^{-\frac12}. \qquad (4\cdot5,20)$$

Thus, the F_i are finite and continuous on Σ while the G_i become infinite there. Moreover, div $\mathbf{F}$ is either finite and continuous everywhere in its domain of definition or it becomes infinite of order $\frac12$ on Σ. Even then $\int_V \operatorname{div}\mathbf{F}\,dV$ exists as an ordinary improper integral and

$$\int_S \mathbf{F}\,d\mathbf{S} = \int_V \operatorname{div}\mathbf{F}\,dV.$$

Assume first that S has no finite area in common with Σ. Since

$$\operatorname{div} \mathbf{f} = \operatorname{div} \mathbf{F} + \operatorname{div} \mathbf{G},$$

it is therefore sufficient to show that

$$*\!\!\int_S \mathbf{G}\,d\mathbf{S} = *\!\!\int_V \operatorname{div} \mathbf{G}\,dV. \tag{4·5, 21}$$

Let $S'(\epsilon)$ be the intersection of (the set of points common to) V and $B(\epsilon)$. Then $V(\epsilon)$ is bounded by $S(\epsilon) + S'(\epsilon)$, the union of $S(\epsilon)$ and $S'(\epsilon)$. Hence, applying the divergence theorem for ordinary integrals to the volume $V(\epsilon)$, we obtain

$$\int_{S(\epsilon)} \mathbf{G}\,d\mathbf{S} + \int_{S'(\epsilon)} \mathbf{G}\,d\mathbf{S} = \int_{V(\epsilon)} \operatorname{div} \mathbf{G}\,dV \tag{4·5, 22}$$

for sufficient small positive ϵ.

Now $\displaystyle\int_{S(\epsilon)} \mathbf{G}\,d\mathbf{S}$ is of the form

$$\int_{S(\epsilon)} \mathbf{G}\,d\mathbf{S} = a_0 \epsilon^{-\frac{1}{2}n+1} + \ldots + a_{k-1}\epsilon^{-\frac{1}{2}} + *\!\!\int_S \mathbf{G}\,d\mathbf{S} + H(\epsilon),$$

where $H(\epsilon)$ is a function which tends to zero as ϵ tends to zero, and similarly $\displaystyle\int_{V(\epsilon)} \operatorname{div} \mathbf{G}\,dV$ will be seen to be of the form

$$\int_{V(\epsilon)} \operatorname{div} \mathbf{G}\,dV = b_0 \epsilon^{-\frac{1}{2}n} + \ldots + b_k \epsilon^{-\frac{1}{2}} + *\!\!\int_V \operatorname{div} \mathbf{G}\,dV + K(\epsilon),$$

where $K(\epsilon)$ also is a function which tends to zero as ϵ tends to zero. In other words, $\displaystyle\int_{S(\epsilon)} \mathbf{G}\,d\mathbf{S}$ and $\displaystyle\int_{V(\epsilon)} \operatorname{div} \mathbf{G}\,dV$ differ from $*\!\!\displaystyle\int_S \mathbf{G}\,d\mathbf{S}$ and $*\!\!\displaystyle\int_V \operatorname{div} \mathbf{G}\,dV$ respectively only by functions of ϵ which vanish in the limit and by fractional infinities of ϵ. Hence, in order to prove (4·5, 21), it is, according to (4·5, 22), sufficient to show that

$$\int_{S'(\epsilon)} \mathbf{G}\,d\mathbf{S} = c_0 \epsilon^{-\frac{1}{2}n} + \ldots + c_k \epsilon^{-\frac{1}{2}} + L(\epsilon), \tag{4·5, 23}$$

where $L(\epsilon)$ is a function which tends to zero as ϵ tends to zero. And (4·5, 23) can readily be deduced from the fact that the components of $\mathbf{G}$ are of the type indicated by (4·5, 20). In fact (4·5, 20) implies that on $S'(\epsilon)$, G is of the type

$$G_1 = C_0 \epsilon^{-\frac{1}{2}n} + \ldots + C_k \epsilon^{-\frac{1}{2}},$$

where the coefficients C_j depend on the parameters of $S'(\epsilon)$, and similar expressions hold for the other components of $\mathbf{G}$. It follows that $\displaystyle\int_{S'(\epsilon)} \mathbf{G}\,d\mathbf{S}$ is of the form indicated by the right-hand side of (4·5, 23). This proves the divergence theorem for finite parts provided S has no finite area in common with Σ. If S has a finite area in common with Σ, then we first displace S by small distance δ, obtaining another surface S_δ for which this is no longer true. S_δ bounds a volume V_δ for which we may then prove the theorem. Finally, we let δ tend to zero. Then (4·5, 19) still holds in the limit.

Similar arguments lead to the proof of the divergence theorem in spaces of higher dimensions, and in the plane, so that (i) and (ii) are established.

Next, we shall show that Stokes's curl theorem also is valid for finite parts. Let f be a vector function of the same description as before, and let S be an open surface bounded by a curve C such as is considered in the ordinary curl theorem. Under these conditions, we are going to show that

$$*\!\!\int_C f\,dl = *\!\!\int_S \operatorname{curl} f\,dS. \tag{4·5, 24}$$

For the proof, we shall assume that C does not have a curve of finite length in common with Σ, and at the same time that S has not got a finite area in common with Σ. For these exceptional conditions, the theorem will then follow from considerations of continuity, as in the proof of the divergence theorem.

Splitting f into two parts F and G as before, we shall establish first that

$$*\!\!\int_C G\,dl = *\!\!\int_S \operatorname{curl} G\,dS. \tag{4·5, 25}$$

Let $C'(\epsilon)$ be the intersection of S and $B(\epsilon)$. Then $S(\epsilon)$ is bounded by $C(\epsilon) + C'(\epsilon)$, and so applying Stokes's theorem for ordinary integrals we obtain, for sufficiently small ϵ,

$$\int_{C(\epsilon)} G\,dl + \int_{C'(\epsilon)} G\,dl = \int_{S(\epsilon)} \operatorname{curl} G\,dS. \tag{4·5, 26}$$

In order to deduce (4·5, 25) from (4·5, 26) we show similarly, as in the proof of the divergence theorem, that

$$\int_{C'(\epsilon)} G\,dl = c_0 \epsilon^{-\frac{1}{2}n} + \ldots + c_k \epsilon^{-\frac{1}{2}} + L(\epsilon),$$

where $L(\epsilon)$ tends to zero with ϵ. (4·5, 25) then follows as before. It only remains to show that

$$*\!\!\int_C F\,dl = *\!\!\int_S \operatorname{curl} F\,dS, \tag{4·5, 27}$$

and this is true by Stokes's theorem for ordinary integrals both if $\operatorname{curl} F$ remains finite everywhere and if $\operatorname{curl} F$ becomes infinite on Σ, in which case the right-hand side of (4·5, 27) may be an ordinary improper integral.

The above sketch of the theory of the finite part will be sufficient for our purposes. It should be clearly understood in the applications which are to follow that our procedure does not attempt to attribute a finite meaning to expressions which are naturally meaningless, but rather that the theory of the finite part enables us to regain the physical meaning of certain quantities, for which the ordinary analytical expressions have become meaningless owing to various indiscriminate passages to the limit (concentration of a source at a point, differentiation under the integral sign).

Before we proceed further it will be convenient to introduce some formal definitions. These are suggested by the comparison of (4·5, 1) with Laplace's equation, and more particularly by the flow vector $\mathbf{q}^* = (-\beta^2 u, v, w)$, which was introduced earlier in this section (see (4·5, 8)).

The operator ∇h_β (read 'hyperbolic nabla of index β') is introduced by

$$\nabla h_\beta = \left(-\beta^2 \frac{\partial}{\partial x}, \frac{\partial}{\partial y}, \frac{\partial}{\partial z}\right), \tag{4·5, 28}$$

and gradh_β and divh_β (read 'hyperbolic gradient of index β', 'hyperbolic divergence of index β') are then defined as the two modes of ∇h_β which apply to scalars, and to vectors in scalar multiplication respectively. Finally, the operator Δh_β 'hyperbolic Laplacian of index β') is defined by

$$\Delta h_\beta = \mathrm{div}\,\mathrm{gradh}_\beta = \mathrm{divh}_\beta\,\mathrm{grad}.$$

Equation (4·5, 1) may now be written in the form

$$\Delta h_\beta \Phi = \mathrm{div}\,\mathrm{gradh}_\beta\,\Phi = \mathrm{divh}_\beta\,\mathrm{grad}\,\Phi = 0, \tag{4·5, 29}$$

and the flow vector $\mathbf{q}^*$ is given by $\mathbf{q}^* = -\mathrm{gradh}_\beta\,\Phi$.

The index β is usually fixed throughout the discussion and will therefore be omitted in the sequel. In the language of differential geometry, the operators just defined are associated with the metric imposed by the differential equation (4·5, 1). In terms of this metric, the 'hyperbolic distance' r_h between two points (x, y, z) and (x_0, y_0, z_0) is defined by

$$r_h^2 = (x - x_0)^2 - \beta^2((y - y_0)^2 + (z - z_0)^2). \tag{4·5, 30}$$

(4·5, 2) may now be written simply as

$$\Phi(x, y, z) = \begin{cases} \dfrac{\sigma}{2\pi r_h} & \text{for } r_h^2 > 0, \quad x - x_0 > 0, \\ 0 & \text{elsewhere,} \end{cases} \tag{4·5, 31}$$

where the positive value of r_h is taken on the right-hand side.

By the divergence theorem we have, for any function which is sufficiently regular on and inside a closed surface S bounding a volume V,

$$\int_S \mathrm{gradh}\,\Phi\,d\mathbf{S} = \int_V \Delta h\,\Phi\,dV. \tag{4·5, 32}$$

If, in addition, Φ is a solution of (4·5, 29), then

$$\int_S \mathrm{gradh}\,\Phi\,d\mathbf{S} = 0. \tag{4·5, 33}$$

If we replace ordinary integrals by finite parts, then equation (4·5, 32) holds even when infinities are involved, provided $\operatorname{gradh} \Phi$ and $\Delta h\, \Phi$ are admissible with respect to a given $D(x,y,z)$ or Σ. Thus, in that case

$$^*\!\!\int_S \operatorname{gradh} \Phi\, d\mathbf{S} = {}^*\!\!\int_V \Delta h\, \Phi\, dV, \qquad (4·5, 34)$$

and (4·5, 33) becomes

$$^*\!\!\int_S \operatorname{gradh} \Phi\, d\mathbf{S} = 0. \qquad (4·5, 35)$$

We notice that if a function Ψ is admissible with respect to a surface Σ, and another function Φ is analytic, or has derivatives of sufficiently high order on Σ, then the product $\Psi\Phi$ is admissible. (The product of two admissible functions is not, in general, admissible.) We also notice that if a finite number of functions are involved, admissible with respect to different surfaces $\Sigma^{(n)}$, then they will all be admissible with respect to the surface Σ which is the union of the different $\Sigma^{(n)}$ and is given by the product D of the polynomials $D^{(n)}$ which define the $\Sigma^{(n)}$.

Let Φ be a function which has derivatives of sufficiently high order on and inside a volume V bounded by a surface S, and let Ψ be a function which, together with its first and second derivatives, is admissible in V, with respect to some algebraic surface Σ. Applying the divergence theorem for finite parts to $\mathbf{f} = \Psi\operatorname{gradh} \Phi$, we obtain

$$^*\!\!\int_S \Psi\operatorname{gradh} \Phi\, d\mathbf{S} = {}^*\!\!\int_V \operatorname{grad} \Psi\operatorname{gradh} \Phi\, dV + {}^*\!\!\int_V \Psi\, \Delta h\, \Phi\, dV, \quad (4·5, 36)$$

and similarly

$$^*\!\!\int_S \Phi\operatorname{gradh} \Psi\, d\mathbf{S} = {}^*\!\!\int_V \operatorname{grad} \Phi\operatorname{gradh} \Psi\, dV + {}^*\!\!\int_V \Phi\, \Delta h\, \Psi\, dV. \quad (4·5, 37)$$

Now

$$\operatorname{grad} \Psi\operatorname{gradh} \Phi = -\beta^2 \frac{\partial \Psi}{\partial x}\frac{\partial \Phi}{\partial x} + \frac{\partial \Psi}{\partial y}\frac{\partial \Phi}{\partial y} + \frac{\partial \Psi}{\partial z}\frac{\partial \Phi}{\partial z} = \operatorname{gradh} \Psi\operatorname{grad} \Phi.$$

Hence, subtracting (4·5, 37) from (4·5, 36), we obtain

$$^*\!\!\int_S (\Psi\operatorname{gradh} \Phi - \Phi\operatorname{gradh} \Psi)\, d\mathbf{S} = {}^*\!\!\int_V (\Psi\, \Delta h\, \Phi - \Phi\, \Delta h\, \Psi)\, dV. \quad (4·5, 38)$$

This is Green's formula for the differential equation (4·5, 1) or (4·5, 29) extended to include finite parts.

We shall now evaluate a concrete example of a finite part. Let $D(x,y,z)$ be the polynomial $D = x^2 - \beta^2(y^2 + z^2)$ and let the function $f(x,y,z)$ be defined by

$$f(x,y,z) = \begin{cases} \dfrac{\sigma x}{2\pi(x^2 - \beta^2(y^2 + z^2))^{\frac{3}{2}}} & \text{for} \quad x^2 - \beta^2(y^2 + z^2) > 0, \quad x > 0, \\[2mm] 0 & \text{elsewhere,} \end{cases}$$

where σ is an arbitrary constant. The function f is admissible in any region which does not include the origin. The surface Σ is in the present case the Mach cone which emanates from the origin.

Let the surface S be given by $x = \alpha$, $y^2 + z^2 \leqslant b^2$, where $\alpha > 0$ and $b > \alpha/\beta$. S is a circular area which includes the circle $x = \alpha$, $y^2 + z^2 = \alpha^2/\beta^2$ on which f becomes infinite of order $\frac{3}{2}$. We are going to evaluate $*J = \overset{*}{\int_S} f \, dy \, dz$.

Given $\epsilon > 0$, let $S_1(\epsilon)$ be the set of points of $S(\epsilon)$ for which $\alpha^2 > \beta^2(y^2 + z^2)$ and $S_2(\epsilon)$ the set of points of $S(\epsilon)$ which do not belong to $S_1(\epsilon)$. Then f vanishes on $S_2(\epsilon)$ and so

$$J(\epsilon) = \int_{S(\epsilon)} f \, dy \, dz = \int_{S_1(\epsilon)} f \, dy \, dz = \frac{\sigma}{2\pi} \int_{S_1(\epsilon)} \frac{\alpha \, dy \, dz}{(\alpha^2 - \beta^2(y^2 + z^2))^{\frac{3}{2}}}$$
$$= \frac{\sigma}{2\pi\beta^2} \int_0^{2\pi} d\theta \int_0^{\eta\beta/\alpha} \frac{\tau \, d\tau}{(1 - \tau^2)^{\frac{3}{2}}},$$

where $y = (\alpha/\beta)\tau \cos\theta$, $z = (\alpha/\beta)\tau \sin\theta$ and η is the radius of the circle which bounds $S_1(\epsilon)$. It is easy to deduce from the definition of $S(\epsilon)$ that (Fig. 4·5, 1)

$$\eta = \frac{1}{\beta}(\alpha - \epsilon\sqrt{(1 + \beta^2)}).$$

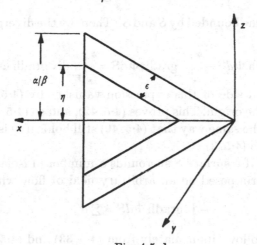

Fig. 4·5, 1

Then

$$\frac{\sigma}{\beta^2}\left[\frac{1}{(1 - \tau^2)^{\frac{1}{2}}}\right]_0^{\eta\beta/\alpha} = \frac{\sigma}{\beta^2}\left\{\frac{1}{\left(\left(2 - \epsilon\frac{\sqrt{(1 + \beta^2)}}{\alpha}\right)\epsilon\frac{\sqrt{(1 + \beta^2)}}{\alpha}\right)^{\frac{1}{2}}} - 1\right\}$$
$$= \frac{\sigma}{\beta^2}\{\alpha^{\frac{1}{2}}(2\sqrt{(1 + \beta^2)})^{-\frac{1}{2}}\epsilon^{-\frac{1}{2}} - 1 + O(\epsilon^{\frac{1}{2}})\},$$

where $O(\epsilon^{\frac{1}{2}})$ is a quantity which tends to zero with ϵ as $\epsilon^{\frac{1}{2}}$. Hence

$$*J = \frac{\sigma}{2\pi}\overset{*}{\int_S} \frac{x \, dy \, dz}{(x^2 - \beta^2(y^2 + z^2))^{\frac{3}{2}}} = -\frac{\sigma}{\beta^2}. \tag{4·5, 39}$$

This result is of fundamental importance for the development of the theory.

Let Φ be the potential of a source as defined by (4·5, 2) and let S be a closed surface which surrounds (x_0, y_0, z_0). Then the flux across S is given by

$$F = -{}^*\!\!\int_S \operatorname{gradh} \Phi \, d\mathbf{S} = \sigma. \tag{4·5, 40}$$

To prove (4·5, 40), we may assume, without loss of generality, that $x_0 = y_0 = z_0 = 0$. Assume first that S is a small cylindrical surface which is given by the two planes $x = \pm \alpha$ and by the cylinder $y^2 + z^2 = \tau^2$, where $\alpha > 0$, $\tau > \alpha/\beta$. Then the integrand of (4·5, 40) vanishes everywhere on S except at the base of the cylinder which belongs to the plane $x = \alpha$. Hence F reduces to

$$F = -\frac{\sigma}{2\pi} {}^*\!\!\int_{y^2 + z^2 \leqslant \tau^2} \frac{\alpha \beta^2 \, dy \, dz}{(\alpha^2 - \beta^2(y^2 + z^2))^{\frac{3}{2}}} = \sigma \tag{4·5, 41}$$

by (4·5, 39).

Next let S be an arbitrary surface surrounding the source. We may then find a small cylindrical surface S' of the above description inside S, and we only have to show that

$${}^*\!\!\int_S \operatorname{gradh} \Phi \, d\mathbf{S} = {}^*\!\!\int_{S'} \operatorname{gradh} \Phi \, d\mathbf{S}. \tag{4·5, 42}$$

Let V' be the volume bounded by S and S'. Then, by the divergence theorem for finite parts,

$${}^*\!\!\int_S \operatorname{gradh} \Phi \, d\mathbf{S} - {}^*\!\!\int_{S'} \operatorname{gradh} \Phi \, d\mathbf{S} = {}^*\!\!\int_{V'} \operatorname{div} \operatorname{gradh} \Phi \, dV,$$

and the right-hand side of this expression vanishes by (4·5, 29), since V' does not include the origin. This proves (4·5, 42), and so (4·5, 40). It can be shown in exactly the same way that (4·5, 40) still holds if Φ is the potential of a counter-source (4·5, 3).

More generally, if a surface S surrounds a number of isolated sources of strengths σ_n superimposed on an arbitrary field of flow which is regular inside S, then

$$-\int_S \operatorname{gradh} \Phi \, d\mathbf{S} = \sum_n \sigma_n. \tag{4·5, 43}$$

In fact, (4·5, 43) follows immediately from (4·5, 33) and (4·5, 40). There is a similar theorem for counter-sources.

Other solutions of (4·5, 1) can be obtained from (4·5, 2) or (4·5, 3) by differentiation or integration.

By analogy with the definition for incompressible flow, we obtain a doublet by differentiating the potential of a source with respect to length in any given direction, the differentiation being performed with respect to the origin of the source, (x_0, y_0, z_0). Thus the velocity potential of a doublet whose axis is parallel to the z-axis, is given by

$$\Phi = \begin{cases} \dfrac{-\sigma \beta^2 (z - z_0)}{\{(x - x_0)^2 - \beta^2((y - y_0)^2 + (z - z_0)^2)\}^{\frac{3}{2}}} \\ \qquad\qquad \text{for} \quad (x - x_0)^2 - \beta^2((y - y_0)^2 + (z - z_0)^2) > 0, \quad x - x_0 > 0, \\ 0 \qquad\qquad \text{elsewhere.} \end{cases}$$

Λ counter-doublet is obtained by applying a similar procedure to the potential of a counter-source. The velocity potentials of sources, counter-sources, doublets, etc., and all their derivatives, are admissible functions in all regions excluding their respective origins, the polynomial D being given by

$$D(x, y, z) = (x - x_0)^2 - \beta^2((y - y_0)^2 + (z - z_0)^2).$$

The corresponding surface Σ is the Mach cone through (x_0, y_0, z_0). The potentials of sources and counter-sources at a point (x, y, z) tend to zero as the inverse distance from (x_0, y_0, z_0) provided (x, y, z) tends to infinity in any direction not asymptotic to Σ, while the potentials of doublets and counter-doublets tend to zero as the square of the inverse distance from (x_0, y_0, z_0) under these conditions.

We denote by Φ_P^* the velocity potential of a source of unit strength with origin at the points $P(x_0, y_0, z_0)$. Then we obtain the velocity potentials due to specified line, surface or volume distributions of sources by evaluating integrals of the type

$$\int \sigma \Phi_P^* dl, \quad \int \sigma \Phi_P^* dS, \quad \int \sigma \Phi_P^* dV, \qquad (4·5, 44)$$

where σ denotes the (variable) line, surface or volume density. For sufficiently regular distribution functions σ (e.g. if σ has bounded first derivatives) these integrals exist as ordinary improper integrals at points which do not belong to the region of integration. For example, if the surface S is defined by the equations

$$x_0 = x_0(\lambda, \mu), \quad y_0 = y_0(\lambda, \mu), \quad z_0 = z_0(\lambda, \mu),$$

where λ, μ are parameters, then σ also must be specified as a function of λ and μ. The corresponding velocity potential is

$$\Phi(x, y, z) = \frac{1}{2\pi} \int \frac{\sigma \, dS}{\sqrt{\{(x - x_0)^2 - \beta^2((y - y_0)^2 + (z - z_0)^2)\}}}, \qquad (4·5, 45)$$

where the integral is taken over those parts of the surface S for which

$$(x - x_0)^2 - \beta^2((y - y_0)^2 + (z - z_0)^2) > 0 \quad (x > x_0).$$

We may write down the integrals for the velocity potentials due to line, surface or volume distributions of doublets in a similar way, but these integrals are, in general, infinite. Thus, for a surface distribution of doublets whose axes are all parallel to the z-axis, we obtain the integral

$$-\int \frac{\sigma \beta^2 (z - z_0) \, dS}{\{(x - x_0)^2 - \beta^2((y - y_0)^2 + (z - z_0)^2)\}^{\frac{3}{2}}}, \qquad (4·5, 46)$$

which is, in general, infinite. However, provided σ is sufficiently regular, the finite part of the integral still exists, and we then regard this finite part as the velocity potential due to the given distribution of doublets (with similar definitions for the other doublet distributions). An alternative method which avoids the use of the finite part and which has been employed by Schlichting‡ and others, is to consider first the corresponding integral for sources (4·5, 45) and then to differentiate with respect to $-z$. From a physical point of view this means that we calculate the potential due to two infinitely near source distributions of opposite strength. The final result is the same since, according to the rules given in the preceding section, finite parts can always be differentiated under the sign of the integral. It is precisely the possibility of carrying out all the necessary operations directly, without fear of obtaining meaningless symbols which makes the finite part such a convenient concept. It will be seen that the alternative method is applicable only when all the doublets have parallel axes.

In order to define the potential due to a volume distribution of sources (or of counter-sources) at a point within the distribution, we must have recourse to a limit process, since the integrand $\sigma\Phi_P^*$ tends to infinity as the inverse distance on approaching the point for which the potential is calculated. We surround the point with a small sphere of radius ϵ, evaluate the integral excluding the interior of the sphere, and then let ϵ tend to zero. For finite ϵ, the integrals in question exist as ordinary improper integrals, and the limit of these integrals as ϵ tends to zero will be seen to exist, since the volume of the sphere tends to zero as ϵ^3.

The total strength of a number of isolated sources or of a continuous distribution of sources is defined as in § 1·7. (4·5, 43) states that the flux across a surface which surrounds a number of isolated sources equals the total strength of these sources. Similarly, using the rule (ii) for the finite part, it can be shown, as in § 1·7, that

$$-{}^*\!\!\int_S \operatorname{gradh} \Phi\, d\mathbf{S} = \int \sigma. \qquad (4·5, 47)$$

In this formula $\int \sigma$ is the total strength of the sources enclosed by S, which may be given in the form of a line, surface or volume distribution.

We may also deduce, taking into account rule (i), that the flux across a surface which surrounds an isolated doublet vanishes, and, more generally, that the flux across a closed surface is not affected by the superposition of doublets either inside the surface or outside it.

Applying (4·5, 47) to a small surface surrounding a point P inside a volume distribution of sources, we obtain

$$-{}^*\!\!\int_S \operatorname{gradh} \Phi\, d\mathbf{S} = \int_V \sigma\, dV, \qquad (4·5, 48)$$

where V is the volume enclosed by S. The left-hand side of (4·5, 48) may be transformed by means of the divergence theorem, so

$$-{}^*\!\!\int_V \operatorname{div} \operatorname{gradh} \Phi\, dV = \int_V \sigma\, dV.$$

‡ H. Schlichting, 'Tragflügeltheorie bei Überschallgeschwindigkeit', *Luftfahrt-forsch.* vol. 13 (1936), pp. 320–35.

Since this is true for any small volume containing P we must have, at the arbitrary point P inside the distribution,

$$\Delta h \, \Phi = \operatorname{div} \operatorname{grad} h \, \Phi = -\sigma. \tag{4·5, 49}$$

Conversely, given the partial differential equation (4·5, 49) over a certain volume R, a particular solution of it is

$$\Phi = \int_R \sigma \Phi_P^* dV. \tag{4·5, 50}$$

The general solution then is

$$\Phi = \int_R \sigma \Phi_P^* dV + \Phi_1, \tag{4·5, 51}$$

where Φ_1 is an arbitrary solution of (4·5, 29).

Given a surface distribiton of sources, it can be shown that the components of the gradient, and therefore of the hyperbolic gradient of the velocity potential, remain finite and continuous on either side of the surface S. Also, Φ, and therefore its tangential derivatives, are continuous across S.

In order to find the discontinuity of the normal derivative across S, we apply (4·5, 47) to a small cylinder whose bases are parallel to the surface on either side of it, and whose height is small compared with its lateral dimensions. Letting first the height of the cylinder tend to zero, we find that

$$\int_{S_+'} \operatorname{grad} h \, \Phi \, d\mathbf{S} - \int_{S_-'} \operatorname{grad} h \, \Phi \, d\mathbf{S} = - \int_{S'} \sigma \, dS,$$

where S' denotes the part of S inside the cylinder and S_+' and S_-' denote the two bases of the cylinder respectively, S_+' being the base whose outside normal coincides with the chosen direction of normal differentiation across S. Denoting the direction cosines of that direction by λ, μ, ν, and letting the area of S_+', S_-', S' tend to zero, we then obtain, at any point of S,

$$-\lambda \beta^2 \left(\frac{\partial \Phi}{\partial x_+} - \frac{\partial \Phi}{\partial x_-} \right) + \mu \left(\frac{\partial \Phi}{\partial y_+} - \frac{\partial \Phi}{\partial y_-} \right) + \nu \left(\frac{\partial \Phi}{\partial z_+} - \frac{\partial \Phi}{\partial z_-} \right) = -\sigma. \tag{4·5, 52}$$

In this equation the subscripts $+$ and $-$ distinguish the derivatives on either side of S, as before. Since the tangential components of the gradient of Φ are continuous across S, it follows that the vector

$$\mathbf{q}_+ - \mathbf{q}_- = -(\operatorname{grad} \Phi_+ - \operatorname{grad} \Phi_-)$$

is parallel to the normal unit vector to S, $\mathbf{n} = (\lambda, \mu, \nu)$

$$\mathbf{q}_+ - \mathbf{q}_- = -\left(\frac{\partial \Phi}{\partial n_+} - \frac{\partial \Phi}{\partial n_-} \right) \mathbf{n} = \Delta q \mathbf{n}. \tag{4·5, 53}$$

Then $\quad \dfrac{\partial \Phi}{\partial x_+} - \dfrac{\partial \Phi}{\partial x_-} = -\lambda \Delta q, \quad \dfrac{\partial \Phi}{\partial y_+} - \dfrac{\partial \Phi}{\partial y_-} = -\mu \Delta q, \quad \dfrac{\partial \Phi}{\partial z_+} - \dfrac{\partial \Phi}{\partial z_-} = -\nu \Delta q.$

Substituting these expressions in (4·5, 52) we obtain

$$\Delta q = \sigma / (-\beta^2 \lambda^2 + \mu^2 + \nu^2). \tag{4·5, 54}$$

The above developments as far as (4·5, 52) are taken, with some alterations, from a paper by one of the authors.‡ However, the formula given in that paper in place of (4·5, 54) depended on an arithmetical mistake, as was pointed out by Ward.§

The use to which Hademard puts his concept of the finite part is related to the above applications but is the outcome of a rather different approach. Hadamard's purpose is the solution of Cauchy's problem (the initial value problem) for a very general class of hyperbolic partial differential equations which includes (4·5, 29) as a special case. Hadamard's result for this case will now be obtained.

It is assumed that the unknown function Φ is a solution of (4·5, 29), and that Φ and its first derivatives take specified values on a closed surface S. Since the derivatives of Φ in directions tangential to S are determined by S, it is in general only necessary to know the derivative of Φ in a direction which is not tangential to S. More particularly, we shall assume that the derivative of Φ is specified in a direction parallel to the vector $(-\lambda\beta^2, \mu, \nu)$. If this direction is tangential to S, then the derivative is already determined by Φ.

To find the value of Φ at a point $P(x_0, y_0, z_0)$ inside S, let Ψ_P^* be the velocity potential of a counter-source of unit strength located at P. The function $\Phi\Psi_P^*$ is admissible in the volume V' which is bounded by S and by a small sphere S' surrounding P, provided Φ is sufficiently regular inside S. Φ and Ψ_P^* are both solutions of (4·5,29), so that $\Delta h\, \Phi = \Delta h\, \Psi_P^* = 0$. Applying (4·5, 38) to the volume V', we then obtain

$$\overset{*}{\int}_{S+S'} (\Psi_P^* \operatorname{gradh} \Phi - \Phi \operatorname{gradh} \Psi_P^*)\, d\mathbf{S} = 0, \qquad (4·5, 55)$$

$$\text{or } \overset{*}{\int}_{S'} (\Psi_P^* \operatorname{gradh} \Phi - \Phi \operatorname{gradh} \Psi_P^*)\, d\mathbf{S} = \overset{*}{\int}_{S} (\Psi_P^* \operatorname{gradh} \Phi - \Phi \operatorname{gradh} \Psi_P^*)\, d\mathbf{S},$$

$$(4·5, 56)$$

where the directed surface element $d\mathbf{S}$ is now supposed to point away from P', towards the outside of the surfaces S and S'. In particular, if we let S' be a small cylindrical surface as in the derivation of (4·5, 40), we obtain on the left-hand side of (4·5, 56), approximately

$$\overset{*}{\int}_{S'} \Psi_P^* \operatorname{gradh} \Phi\, d\mathbf{S} - \overset{*}{\int}_{S'} \Phi \operatorname{gradh} \Psi_P^*\, d\mathbf{S} \doteqdot \Phi(x_0, y_0, z_0)\left(-\overset{*}{\int}_{S'} \operatorname{gradh} \Psi_P^*\, d\mathbf{S}\right)$$

$$= \Phi(x_0, y_0, z_0). \qquad \text{by } (4·5, 40)$$

Hence, in the limit,

$$\Phi(x_0, y_0, z_0) = \overset{*}{\int}_{S} (\Psi_P^* \operatorname{gradh} \Phi - \Phi \operatorname{gradh} \Psi_P^*)\, d\mathbf{S}. \qquad (4·5, 57)$$

‡ A. Robinson, 'On source and vortex distribution in the linearised theory of steady supersonic flow', *Rep. Coll. Aero Cranfield*, no. 9 (1947); also in *Quart. J. Mech. Appl. Math.* vol. 1 (1948), pp. 408–32.

§ G. N. Ward, 'On the integration of some vector differential equations', *Quart. J. Mech. Appl. Math.* vol. 5 (1952), p. 446.

This may also be written as

$$\Phi(x_0, y_0, z_0) = {}^*\!\!\int_S \left(\Psi_P^* \left(-\lambda\beta^2 \frac{\partial\Phi}{\partial x} + \mu \frac{\partial\Phi}{\partial y} + \nu \frac{\partial\Phi}{\partial z} \right) \right.$$
$$\left. - \Phi \left(-\lambda\beta^2 \frac{\partial\Psi_P^*}{\partial x} + \mu \frac{\partial\Psi_P^*}{\partial y} + \nu \frac{\partial\Psi_P^*}{\partial z} \right) \right) dS.$$

$$(4\cdot5, 58)$$

It appears from the formula that the value of Φ at the point P depends only on the specified data on S within the force-cone of P. Thus, the formula may be applied also in certain regions if S is open.

The methods developed above can also be applied to the calculation of the field of flow due to a distribution of vorticity in the fluid. As usual, we denote by ξ, η, ζ the components of the vorticity vector curl $\mathbf{q}$

$$\xi = \frac{\partial w}{\partial y} - \frac{\partial v}{\partial z}, \quad \eta = \frac{\partial u}{\partial z} - \frac{\partial w}{\partial x}, \quad \zeta = \frac{\partial v}{\partial x} - \frac{\partial u}{\partial y}.$$

The differential equation of the system of vortex lines is

$$\frac{dx}{\xi} = \frac{dy}{\eta} = \frac{dz}{\zeta},$$

as before, and the strength of a slender vortex tube equals the product of the cross-section A into the resultant vorticity $\omega = (\xi^2 + \eta^2 + \zeta^2)^{\frac{1}{2}}$ and is the same at all points of the vortex. All these results and definitions are in fact equally valid for compressible and incompressible fluids, except that in the present case it may be necessary to consider the finite parts of integrals of the type $\int_C \mathbf{q}\, d\mathbf{l}$ and $\int_S \operatorname{curl} \mathbf{q}\, d\mathbf{S}$ in cases where the ordinary integrals do not exist.

Applying the vector ∇h in cross-multiplication to $\mathbf{q}$, we obtain a vector which will be called curlh $\mathbf{q}$ (read 'hyperbolic curl of $\mathbf{q}$'). Thus

$$\operatorname{curlh} \mathbf{q} = \left(\frac{\partial w}{\partial y} - \frac{\partial v}{\partial z}, \frac{\partial u}{\partial z} + \beta^2 \frac{\partial w}{\partial x}, -\beta^2 \frac{\partial v}{\partial x} - \frac{\partial u}{\partial y} \right). \qquad (4\cdot5, 59)$$

Direct calculations show that

$$\left. \begin{array}{l} \operatorname{divh} \operatorname{curlh} \mathbf{q} = 0, \\ \operatorname{curl} \operatorname{curlh} \mathbf{q} = \operatorname{gradh} \operatorname{div} \mathbf{q} - \operatorname{div} \operatorname{gradh} \mathbf{q}. \end{array} \right\} \qquad (4\cdot5, 60)$$

Let the vector $\mathbf{q}$ be defined in a region R and admissible in that region. We propose to represent $\mathbf{q}$ as the sum of three vectors $\mathbf{q}_1$, $\mathbf{q}_2$, $\mathbf{q}_3$ such that

$$\left. \begin{array}{ll} \operatorname{divh} \mathbf{q}_1 = \operatorname{divh} \mathbf{q}, & \operatorname{curl} \mathbf{q}_1 = 0 \\ \operatorname{curl} \mathbf{q}_2 = \operatorname{curl} \mathbf{q}, & \operatorname{divh} \mathbf{q}_2 = 0 \\ \operatorname{divh} \mathbf{q}_3 = 0, & \operatorname{curl} \mathbf{q}_3 = 0 \end{array} \right\} \text{ in } R. \qquad (4\cdot5, 61)$$

Assuming that vectors $\mathbf{q}_1$ and $\mathbf{q}_2$ which satisfy these conditions have already been found, we put

$$\mathbf{q}_3 = \mathbf{q} - \mathbf{q}_1 - \mathbf{q}_2.$$

Then $\qquad\qquad \operatorname{divh} \mathbf{q}_3 = \operatorname{divh} \mathbf{q} - \operatorname{divh} \mathbf{q}_1 - \operatorname{divh} \mathbf{q}_2 = 0$

and $\qquad\qquad \operatorname{curl} \mathbf{q}_3 = \operatorname{curl} \mathbf{q} - \operatorname{curl} \mathbf{q}_1 - \operatorname{curl} \mathbf{q}_2 = 0,$

so that $\mathbf{q}_3$ also satisfies the required conditions.

Define the scalar function Φ by

$$\Phi = {}^*\!\!\int_R \sigma \Phi_P^* dV, \qquad\qquad (4\cdot5, 62)$$

where $\sigma = \operatorname{divh} \mathbf{q}$. Then $\Delta \mathrm{h}\, \Phi = -\sigma$ according to (4·5, 49), i.e.

$$\Delta \mathrm{h}\, \Phi = -\operatorname{divh} \mathbf{q}.$$

It follows that the vector $\mathbf{q}_1 = -\operatorname{grad} \Phi$ satisfies the conditions of (4·5, 61).

To find $\mathbf{q}_2$ we assume that this vector is given as the hyperbolic curl of a vector $\mathbf{\Psi} = (F, G, H)$, $\mathbf{q}_2 = \operatorname{curlh} \mathbf{\Psi}$. Then, by (4·5, 60),

$$\operatorname{curl} \mathbf{q}_2 = \operatorname{curl} \operatorname{curlh} \mathbf{\Psi} = \operatorname{gradh} \operatorname{div} \mathbf{\Psi} - \operatorname{div} \operatorname{gradh} \mathbf{\Psi}. \qquad (4\cdot5, 63)$$

We now restrict $\mathbf{\Psi}$ by the condition $\operatorname{div} \mathbf{\Psi} = 0$. Then we must have

$$\operatorname{div} \operatorname{gradh} \mathbf{\Psi} = -\operatorname{curl} \mathbf{q}_2 = -\operatorname{curl} \mathbf{q},$$

or, in scalar notation,

$$\Delta \mathrm{h}\, F = -\xi, \qquad \Delta \mathrm{h}\, G = -\eta, \qquad \Delta \mathrm{h}\, H = -\zeta. \qquad (4\cdot5, 64)$$

According to (4·5, 50), (4·5, 64) is satisfied by

$$F = {}^*\!\!\int_R \xi \Phi_P^* dV, \quad G = {}^*\!\!\int_R \eta \Phi_P^* dV, \quad H = {}^*\!\!\int_R \zeta \Phi_P^* dV, \quad (4\cdot5, 65)$$

or

$$\mathbf{\Psi} = {}^*\!\!\int_R \operatorname{curl} \mathbf{q}\, \Phi_P^* dV. \qquad\qquad (4\cdot5, 66)$$

Then $\mathbf{q}_2 = \operatorname{curlh} \mathbf{\Psi}$ satisfies (4·5, 61), provided it can be shown that indeed $\operatorname{div} \mathbf{\Psi} = 0$. And this can be established exactly as in §1·9, provided the vorticity distribution is continuous and extends over a finite region only, and also if $\operatorname{curl} \mathbf{q}$ is discontinuous at the surface of a vortex filament. Moreover, in the present case the convergence of the integrals is assured if $\operatorname{curl} \mathbf{q}$ vanishes for sufficiently large negative x, since Φ_P^* vanishes for sufficiently large positive x.

It is clear from the above construction that $\mathbf{q}_1$ respresents the flow due to a given source distribution in R, while $\mathbf{q}_2$ represents the flow due to the vorticity distribution in R. The components of $\mathbf{q}_2 = \operatorname{curlh} \mathbf{\Psi}$, where $\mathbf{\Psi}$ is given by (4·5, 66), are in more detail

$$
\left.\begin{aligned}
u &= \frac{\beta^2}{2\pi} {}^*\!\!\int_{R'} ((y - y_0)\,\zeta - (z - z_0)\,\eta) \frac{dx_0\,dy_0\,dz_0}{\{(x - x_0)^2 - \beta^2((y - y_0)^2 + (z - z_0)^2)\}^{\frac{3}{2}}}, \\
v &= \frac{\beta^2}{2\pi} {}^*\!\!\int_{R'} ((z - z_0)\,\xi - (x - x_0)\,\zeta) \frac{dx_0\,dy_0\,dz_0}{\{(x - x_0)^2 - \beta^2((y - y_0)^2 + (z - z_0)^2)\}^{\frac{3}{2}}}, \\
w &= \frac{\beta^2}{2\pi} {}^*\!\!\int_{R'} ((x - x_0)\,\eta - (y - y_0)\,\xi) \frac{dx_0\,dy_0\,dz_0}{\{(x - x_0)^2 - \beta^2((y - y_0)^2 + (z - z_0)^2)\}^{\frac{3}{2}}}
\end{aligned}\right\}
$$

$$(4\cdot5, 67)$$

This may also be written as

$$q_2 = \frac{\beta^2}{2\pi} {}^*\!\!\int_{R'} (\mathbf{r} \wedge \operatorname{curl} \mathbf{q}) \frac{dV}{r_h^3}, \tag{4·5, 68}$$

where $\mathbf{r} = (x - x_0, y - y_0, z - z_0)$, while r_h is the hyperbolic distance as defined earlier in this section,

$$r_h = \sqrt{\{(x - x_0)^2 - \beta^2((y - y_0)^2 + (z - z_0)^2)\}}.$$

The region of integration R' in both (4·5, 67) and (4·5, 68) is defined as the subdomain of R whose points satisfy the condition

$$(x - x_0)^2 - \beta^2((y - y_0)^2 + (z - z_0)^2) > 0 \quad (x > x_0). \tag{4·5, 69}$$

We may now calculate the field of flow due to an isolated line vortex C. Replacing the volume element in (4·5, 66) by $A\,dl$, where dl is the element of length of C, and A its infinitesimal cross-section, and writing

$$\operatorname{curl} \mathbf{q} = (\xi, \eta, \zeta) = \omega \left(\frac{dx_0}{dl}, \frac{dy_0}{dl}, \frac{dz_0}{dl} \right),$$

we obtain

$$\Psi = {}^*\!\!\int_{C'} \omega \left(\frac{dx_0}{dl}, \frac{dy_0}{dl}, \frac{dz_0}{dl} \right) \frac{A\,dl}{2\pi \sqrt{\{(x - x_0)^2 - \beta^2((y - y_0)^2 + (z - z_0)^2)\}}},$$

or

$$\Psi = \frac{\gamma}{2\pi} {}^*\!\!\int_{C'} \frac{1}{\sqrt{\{(x - x_0)^2 - \beta^2((y - y_0)^2 + (z - z_0)^2)\}}} \, d\mathbf{l}, \tag{4·5, 70}$$

where γ is the strength of the vortex and $d\mathbf{l} = (dx_0, dy_0, dz_0)$ is the directed element of length. C' consists of the segments of C which satisfy (4·5, 69). The integral in (4·5, 70) like the more general integral in (4·5, 66) is actually an ordinary improper integral.

Suppose, in particular, that C consists of straight segments which are parallel either to the x-axis or to the y-axis. Then the integrals which occur in (4·5, 70) are one of the two following types:

$$\left. \begin{aligned} \int \frac{dx_0}{\sqrt{\{(x - x_0)^2 - \beta^2((y - y_0)^2 + (z - z_0)^2)\}}} \\ = -\cosh^{-1} \frac{x - x_0}{\beta \sqrt{\{(y - y_0)^2 + (z - z_0)^2\}}} + \text{constant}, \\ \int \frac{dy_0}{\sqrt{\{(x - x_0)^2 - \beta^2((y - y_0)^2 + (z - z_0)^2)\}}} \\ = -\frac{1}{\beta} \sin^{-1} \frac{\beta(y - y_0)}{\sqrt{\{(x - x_0)^2 - \beta^2(z - z_0)^2\}}} + \text{constant}. \end{aligned} \right\} \tag{4·5, 71}$$

Assume, for example, that C is a 'horseshoe-vortex' of strength γ, which consists of the straight segments

$$(x_1 \leqslant x_0 < \infty, y_0 = -y_1, z_0 = 0), \quad (x_0 = x_1, -y_1 \leqslant y_0 \leqslant y_1, z_0 = 0),$$
$$(x_1 \leqslant x_0 < \infty, y_0 = y_1, z_0 = 0),$$

where x_1 and y_1 are given constants. Using (4·5, 71) we find that the third component of Ψ, H, vanishes everywhere, and $F = G = 0$ for $x < x_1$, while for $x > x_1$, F and G are given by the following formulae:

$$F = \frac{\gamma}{2\pi}\left(\cosh^{-1}\frac{x - x_1}{\beta\sqrt{\{(y - y_1)^2 + z^2\}}} - \cosh^{-1}\frac{x - x_1}{\beta\sqrt{\{(y + y_1)^2 + z^2\}}}\right),$$

where the $\cosh^{-1}$ are replaced by 0 when their arguments are smaller than 1; and

$$G = -\frac{\gamma}{2\pi\beta}\left(\sin^{-1}\frac{\beta(y - y_1)}{\sqrt{\{(x - x_1)^2 - \beta^2 z^2\}}} - \sin^{-1}\frac{\beta(y + y_1)}{\sqrt{\{(x - x_1)^2 - \beta^2 z^2\}}}\right),$$

where the $\sin^{-1}$ are replaced by $\frac{1}{2}\pi$ or $-\frac{1}{2}\pi$ when their arguments are greater than 1, or smaller than -1, respectively.

We now obtain $\mathbf{q}_2 = (u, v, w)$ by taking the hyperbolic curl of $\Psi = (F, G, 0)$. Then $u = v = w = 0$ for $x < x_1$, while for $x > x_1$

$$
\left.
\begin{aligned}
u &= \frac{\gamma\beta^2}{2\pi}\mathscr{R}\left\{\frac{(y - y_1)z}{((x - x_1)^2 - \beta^2 z^2)\{(x - x_1)^2 - \beta^2((y - y_1)^2 + z^2)\}^{\frac{1}{2}}} \right. \\
&\qquad\left. - \frac{(y + y_1)z}{((x - x_1)^2 - \beta^2 z^2)\{(x - x_1)^2 - \beta^2((y + y_1)^2 + z^2)\}^{\frac{1}{2}}}\right\}, \\
v &= \frac{\gamma}{2\pi}\mathscr{R}\left\{\frac{(x - x_1)z}{((y - y_1)^2 + z^2)\{(x - x_1)^2 - \beta^2((y - y_1)^2 + z^2)\}^{\frac{1}{2}}} \right. \\
&\qquad\left. - \frac{(x - x_1)z}{((y + y_1)^2 + z^2)\{(x - x_1)^2 - \beta^2((y + y_1)^2 + z^2)\}^{\frac{1}{2}}}\right\}, \\
w &= \frac{\gamma}{2\pi}\mathscr{R}\left\{\frac{(x - x_1)(y - y_1)\{(x - x_1)^2 - \beta^2((y - y_1)^2 + 2z^2)\}}{((x - x_1)^2 - \beta^2 z^2)((y - y_1)^2 + z^2)\{(x - x_1)^2 - \beta^2((y - y_1)^2 + z^2)\}^{\frac{1}{2}}} \right. \\
&\qquad\left. - \frac{(x - x_1)(y + y_1)\{(x - x_1)^2 - \beta^2((y + y_1)^2 + 2z^2)\}}{((x - x_1)^2 - \beta^2 z^2)((y + y_1)^2 + z^2)\{(x - x_1)^2 - \beta^2((y + y_1)^2 + z^2)\}^{\frac{1}{2}}}\right\}.
\end{aligned}
\right\}
$$

$$(4\cdot5, 72)$$

In these formulae $\mathscr{R}$ indicates the real part of the expression in brackets, as usual. Except for the notation, (4·5, 72) agrees with the field of flow round a horseshoe-vortex calculated by Schlichting in the above-mentioned paper by an entirely different method.

Some care is required when it is desired to represent a volume of surface distribution of vorticity as a combination of line vortices. According to (4·5, 72) the velocity components u, v, w all vanish when the point (x, y, z) is outside both the after-cones $(x - x_1)^2 - \beta^2((y \pm y_1)^2 + z^2) = 0$ which emanate from the tips. But it can be shown that this is no longer the case if the vorticity along the span is distributed over a finite width Δx_0. This phenomenon is due to the fact that Ψ becomes discontinuous for a concentrated line vortex. It can occur only at points which belong to the envelope of the after-cones emanating from the vortex lines which are supposed to generate the surface or volume of the distribution.

4·6   Two-dimensional problems in linearized theory of supersonic flow

Consider the flow round an aerofoil which is situated approximately in the x, z-plane and extends from the z-axis, $x = 0$, to the line $x = c$. Suppose that the wing is generated by straight lines parallel to the z-axis, so that the aerofoil sections (wing profiles) are the same in all planes parallel to the x, y-plane. The main flow is supposed to be directed along the x-axis as in the preceding sections, so that the resulting flow is two-dimensional, $w = 0$, throughout, and we may confine our investigation to the x, y-plane.

According to the general theory, the disturbance produced in the flow by the presence of the wing, at any particular point P of the wing, is confined to the after-cone emanating from P. Thus the entire field of flow induced by the wing is bounded by the envelope of the after-cones which emanate from its leading edge. This is the family of cones

$$x^2 - \beta^2(y^2 + (z - z_0)^2) = 0 \quad (x > 0), \tag{4.6, 1}$$

where the parameter z_0 varies from $-\infty$ to $+\infty$. It is not difficult to see that the envelope of (4·6, 1) is the pair of half-planes

$$x \pm \beta y = 0 \quad (x > 0), \tag{4.6, 2}$$

which is represented by a pair of straight lines in the x, y-plane. These are the 'Mach lines' through the leading edge. Their slopes are given by

$$\pm \frac{1}{\beta} = \pm \frac{1}{\sqrt{(M^2 - 1)}} = \pm \tan \mu, \tag{4.6, 3}$$

where μ is the Mach angle as defined in § 4·1.

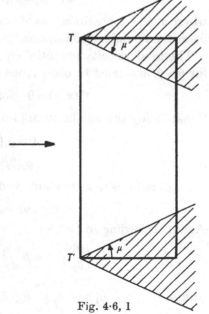

Even if the wing is of finite aspect ratio, conditions at those points of the wing which are outside the after-cones which emanate from the leading edge at the tips (Fig. 4·6, 1) will still be given by two-dimensional theory, so long as the geometry of the wing profiles along the span is two-dimensional in the sense defined above. This follows again from the fact that any disturbance in supersonic flow affects only the interior of the after-cone which emanates from the origin of the disturbance. We may formulate this in a slightly different fashion by saying that in steady supersonic

Fig. 4·6, 1

flow, conditions at any point P depend only on the disturbances which arise in the fore-cone of P.

Let $y = f_u(x)$ be the equation of the upper surface of the aerofoil, and $y = f_l(x)$ the equation of its lower surface, $0 \leqslant x \leqslant c$. Then $s_u(x) = \dfrac{df_u}{dx}$ or $s_l(x) = \dfrac{df_l}{dx}$ is the local slope of the aerofoil at the upper or lower surface, respectively, and since the induced velocity in the direction of the x-axis is in general small compared with U, the boundary condition at the aerofoil becomes

$$v = -\frac{\partial \Phi}{\partial y} = Us(x), \qquad (4·6, 4)$$

where $s(x) = s_u(x)$ and $s(x) = s_l(x)$ on the two surfaces of the wing respectively. This is the same condition as in the thin aerofoil theory of incompressible flow. As before, it will be satisfied for $y = 0$ rather than as the actual surface of the aerofoil, in keeping with the previous approximations.

The partial differential equation for Φ is

$$\beta^2 \frac{\partial^2 \Phi}{\partial x^2} - \frac{\partial^2 \Phi}{\partial y^2} = 0, \qquad (4·6, 5)$$

and we shall assume that Φ denotes the induced velocity potential, i.e. the potential of the flow which is due to the presence of the wing. The total velocity potential is then given by $-Ux + \Phi$.

We consider first conditions above the aerofoil, $y > 0$. The general solution of (4·6, 5) is

$$\Phi(x, y) = g_1(x - \beta y) + g_2(x + \beta y), \qquad (4·6, 6)$$

where g_1 and g_2 are functions of one argument which have to be determined from the boundary conditions.

The total velocity potential equals $-Ux$ upstream of the line $x - \beta y = 0$ for $y > 0$, and must be continuous across that line. Hence

$$\Phi(x, y) = 0 \quad \text{for} \quad x - \beta y = 0, \quad y > 0. \qquad (4·6, 7)$$

Substituting this condition in (4·6, 6), we obtain

$$0 = g_1(0) + g_2(2\beta y),$$

or

$$g_2(2\beta y) = -g_1(0).$$

Thus g_2 reduces to a constant, and (4·6, 6) becomes

$$\Phi(x, y) = g_1(x - \beta y) - g_1(0). \qquad (4·6, 8)$$

Again, according to (4·6, 4),

$$-\left(\frac{\partial \Phi}{\partial y}\right)_{y=0} = \beta \frac{dg_1}{dx} = Us(x) \quad \text{for} \quad 0 \leqslant x \leqslant c,$$

and so

$$g_1(x) = \frac{U}{\beta} \int_0^x s(\xi) \, d\xi + g_1(0) \quad \text{for} \quad 0 \leqslant x \leqslant c. \qquad (4·6, 9)$$

Hence

$$\Phi(x, y) = \frac{U}{\beta} \int_0^{x - \beta y} s_u(\xi) \, d\xi \quad \text{for} \quad y > 0, \ 0 \leqslant x - \beta y \leqslant c. \qquad (4·6, 10)$$

A similar argument shows that (for points below the aerofoil)

$$\Phi(x, y) = -\frac{U}{\beta} \int_0^{x+\beta y} s_l(\xi)\, d\xi \quad \text{for} \quad y < 0,\ 0 \leqslant x + \beta y \leqslant c. \quad (4\cdot6, 11)$$

The induced velocity in the direction of the x-axis is then given by

$$u'(x, +0) = -\frac{U}{\beta} s_u(x)$$

on the upper surface of the aerofoil, and by

$$u'(x, -0) = \frac{U}{\beta} s_l(x)$$

at its lower surface. According to the linearized form of Bernoulli's equation $(4\cdot1, 27)$ the pressure at any point on the upper surface of the aerofoil is then given by

$$p_u = p_0 + \frac{\rho_0 U^2}{\beta} s_u(x), \quad\quad\quad (4\cdot6, 12)$$

where p_0 is the pressure upstream of the aerofoil. Similarly, at points on the lower surface of the aerofoil

$$p_l = p_0 - \frac{\rho_0 U^2}{\beta} s_l(x). \quad\quad\quad (4\cdot6, 13)$$

We consider first a very thin but possibly cambered aerofoil at incidence. Then $f_u(x) = f_l(x) = f(x)$ for $0 \leqslant x \leqslant c$ and so $s_u(x) = s_l(x) = s(x) = df/dx$. Hence

$$p_u - p_l = \frac{\rho_0 U^2}{\beta}(s_u + s_l) = \frac{2\rho_0 U^2}{\beta} s(x) = \frac{2\rho_0 U^2}{\beta} \frac{df}{dx} \quad (4\cdot6, 14)$$

at any point of the aerofoil. The pressure acts in a direction normal to the surface of the aerofoil; but for a small slope $s(x)$, this direction is approximately parallel to the direction of the y-axis. Hence the resultant force in the direction of the y-axis, or lift, is given by

$$L = -\int_0^c (p_u - p_l)\, dx = -\frac{2\rho_0 U^2}{\beta}\int_0^c s(x)\, dx = \tfrac{1}{2}\rho_0 U^2 \frac{4}{\beta}(f(c) - f(0)).$$

$$(4\cdot6, 15)$$

To obtain the resultant force in the direction of the x-axis, or drag, we have to multiply the pressure difference by the sine of the local incidence, and this again is approximately equal to the local slope. Hence

$$D_W = \int_0^c (p_u - p_l)\, s(x)\, dx = \frac{2\rho_0 U^2}{\beta}\int_0^c (s(x))^2\, dx = \frac{2\rho_0 U^2}{\beta}\int_0^c \left(\frac{df}{dx}\right)^2 dx.$$

$$(4\cdot6, 16)$$

The drag is here called the wave drag (denoted by D_W) because it is associated with the system of disturbances or waves set up by the wing. As in the subsonic flow, the profile drag must be calculated separately.

Assume in particular that the wing is a flat plate at incidence α. Then the equation of the wing is

$$y = f(x) = x \tan \alpha \doteqdot x\alpha,$$

and so $s(x) = df/dx = \alpha$. Hence

$$L = -\tfrac{1}{2}\rho_0 U^2 c \frac{4\alpha}{\beta}, \quad D_W = \tfrac{1}{2}\rho_0 U^2 c \frac{4\alpha^2}{\beta}. \tag{4·6, 17}$$

and the corresponding lift and wave-drag coefficients are

$$C_L = -L/\tfrac{1}{2}\rho_0 U^2 c = \frac{4\alpha}{\beta}, \quad C_{Dw} = D_W/\tfrac{1}{2}\rho_0 U^2 c = \frac{4\alpha^2}{\beta}. \tag{4·6, 18}$$

These are Ackeret's formulae.‡ Now $\beta = \sqrt{(M^2 - 1)}$, and so the formulae show that the lift and drag coefficients decrease with increasing Mach number. It also appears from these formulae that in supersonic flow, potential theory yields a drag force which is associated with the lift even under two-dimensional conditions. No such drag was found to exist in subsonic flow.

It follows from (4·6, 14) that for a flat aerofoil at incidence, the pressure difference is constant across the chord of the wing. Hence the centre of pressure is at the mid-chord ($x = \tfrac{1}{2}c$), and the moment about the leading edge is

$$M = \tfrac{1}{2}cL = -\tfrac{1}{2}\rho_0 U^2 c^2 \frac{2\alpha}{\beta}. \tag{4·6, 19}$$

The corresponding moment coefficient is

$$C_M = -M/\tfrac{1}{2}\rho_0 U^2 c^2 = \frac{2\alpha}{\beta}. \tag{4·6, 20}$$

Next, consider a wing with symmetrical cross-section at zero incidence. Then

$$f_u(x) = f(x), \quad f_l(x) = -f(x), \quad f(0) = f(c) = 0, \quad s_u(x) = -s_l(x) = s(x) = \frac{df}{dx}.$$

The lift now vanishes, by symmetry. The pressure is given by

$$p = p_0 + \frac{\rho_0 U^2}{\beta} s(x),$$

both above and below the wing. The contribution of the constant part p_0 to the resultant drag vanishes, since the contour of the wing section is a closed curve. Hence

$$D_W = \frac{2\rho_0 U^2}{\beta} \int_0^c (s(x))^2 \, dx = \frac{2\rho_0 U^2}{\beta} \int_0^c \left(\frac{df}{dx}\right)^2 dx, \tag{4·6, 21}$$

which is formally identical with (4·6, 16).

For a double-wedge section with maximum thickness at the mid-chord (Fig. 4·6, 2a), we have

$$f(x) = \begin{cases} x \tan \alpha & \text{for} \quad 0 \leqslant x \leqslant \tfrac{1}{2}c, \\ (c - x) \tan \alpha & \text{for} \quad \tfrac{1}{2}c \leqslant x \leqslant c, \end{cases}$$

‡ J. Ackeret, 'Über Luftkräfte auf Flügel, die mit grösserer als Schallgeschwindigkeit bewegt werden', *Z. Flugtech.* vol. 16 (1925), pp. 72–4. Translated as *Tech. Memor. Nat. Adv. Comm. Aero., Wash.,* no. 317.

and so $(df/dx)^2 = \tan^2\alpha$ $(0 \leqslant x \leqslant c)$. Now $\tan\alpha$ is also equal to the maximum thickness-chord ratio $(t/c)_0$, and so

$$D_W = \tfrac{1}{2}\rho_0 U^2 c \frac{4}{\beta}(t/c)_0^2, \qquad (4\cdot6, 22)$$

and the corresponding drag coefficient is

$$C_{Dw} = \frac{4}{\beta}(t/c)_0^2. \qquad (4\cdot6, 23)$$

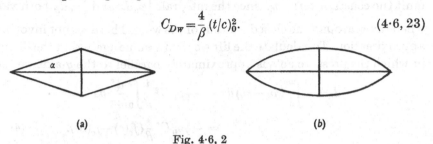

(a)   (b)

Fig. 4·6, 2

Again, for a wing with biparabolic cross-section and maximum thickness at the mid-chord (Fig. 4·6, 2b), we have

$$f(x) = \frac{2}{c}(t/c)_0 x(c-x) \quad (0 \leqslant x \leqslant c).$$

With this value of $f(x)$, (4·6, 21) becomes

$$D_W = \frac{2\rho_0 U^2}{\beta}\frac{4}{c^2}(t/c)_0^2 \int_0^c (c-2x)^2 \, dx$$

$$= \tfrac{1}{2}\rho_0 U^2 c \frac{16}{3\beta}(t/c)_0^2. \qquad (4\cdot6, 24)$$

The drag coefficient now is

$$C_{Dw} = \frac{16}{3\beta}(t/c)_0^2. \qquad (4\cdot6, 25)$$

Thus, for a given maximum thickness-chord ratio the wave-drag coefficient of a biparabolic wing is related to the wave-drag coefficient of a double-wedge wing as $4:3$.

Finally, let us consider the general case of a cambered wing of finite thickness at incidence. Thus $y = f_u(x)$ and $y = f_l(x)$ are different functions which describe the upper and lower surfaces, respectively. Then the equation of the mean camber line is

$$y = \tfrac{1}{2}(f_u(x) + f_l(x)) = f_c(x), \quad \text{say,}$$

and the thickness of the aerofoil at any given point is

$$f_u(x) - f_l(x) = 2f_t(x).$$

Then $f_u(x) = f_c(x) + f_t(x)$, $f_l(x) = f_c(x) - f_t(x)$, and according to (4·6, 12) and (4·6, 13) the pressure is

$$p_u = p_0 + \frac{\rho_0 U^2}{\beta}(f_t'(x) + f_c'(x))$$

at the top surface of the aerofoil, and

$$p_l = p_0 + \frac{\rho_0 U^2}{\beta} (f_t'(x) - f_c'(x))$$

at its bottom surface. To calculate the resultant force on the wing, we may omit the constant term p_0, since the integrals $\int p_0 dx$ and $\int p_0 dy$ both vanish when taken around the closed contour of the wing. This does not involve any approximation. To calculate the lift we then assume again that the direction in which the pressure acts is approximately parallel to the y-axis, so that

$$L = -\int_0^c (p_u - p_l) \, dx = -\tfrac{1}{2}\rho_0 U^2 \frac{4}{\beta} \int_0^c \frac{df_c}{dx} dx$$

$$= -\tfrac{1}{2}\rho_0 U^2 \frac{4}{\beta} (f_c(c) - f_c(0)). \qquad (4\cdot6, 26)$$

We note that this is precisely the lift on a flat aerofoil which coincides with the mean camber line of the given wing.

Again, the wave drag is now given by

$$D_W = \int_0^c (p_u s_u + p_l s_l) \, dx = \frac{\rho_0 U^2}{\beta} \int_0^c \{(f_t' + f_c')^2 + (f_t' - f_c')^2\} \, dx$$

$$= \tfrac{1}{2}\rho_0 U^2 \frac{4}{\beta} \left(\int_0^c \left(\frac{df_c}{dx}\right)^2 dx + \int_0^c \left(\frac{df_t}{dx}\right)^2 dx \right). \qquad (4\cdot6, 27)$$

Thus the wave drag is equal to the sum of the wave drags of a flat wing which coincides with the mean camber line of the given wing and of a wing with symmetrical section whose thickness coincides everywhere with the thickness of the given aerofoil and which is at zero incidence. It also follows from (4·6, 26) and (4·6, 27) that for a given wing at a given incidence the lift and drag vary as $U^2/\beta = U^2/\sqrt{(M^2 - 1)} = a_0^2 M^2/\sqrt{(M^2 - 1)}$.

So far we have determined the velocity potential only ahead of the Mach lines
$$x \pm \beta y = c,$$
which emanate from the trailing edge in the downstream direction. The argument which showed that $\Phi(x, y)$ must be of the form (4·6, 8) for $y > 0$ still applies aft of the Mach line $x - \beta y = c$. There are no sources or vortices along the x-axis aft of the trailing edge of the aerofoil, so that the velocity must be continuous across the x-axis aft of the trailing edge. Now the velocity components of the induced flow just above the x-axis are

$$u(x, +0) = \left(-\frac{\partial \Phi}{\partial x}\right)_{y=0} = -\frac{dg_1}{dx}$$

and

$$v(x, +0) = \left(-\frac{\partial \Phi}{\partial y}\right)_{y=0} = \beta \frac{dg_1}{dx}.$$

Similarly, on approaching the x-axis from the side of the negative y, we obtain

$$u(x, -0) = -\frac{dg_2}{dx}, \quad v(x, -0) = -\beta \frac{dg_2}{dx}.$$

Now we may assume that u and v are continuous across the x-axis behind the aerofoil, otherwise there would be sources or free vortices present in that region. Hence

$$\frac{dg_1}{dx} = \frac{dg_2}{dx}, \quad \beta\frac{dg_1}{dx} = -\beta\frac{dg_2}{dx},$$

and these conditions can only be satisfied simultaneously if

$$\frac{dg_1}{dx} = \frac{dg_2}{dx} = 0 \quad \text{for} \quad x > c.$$

Hence $g_1(x) = \text{constant}$, $g_2(x) = \text{constant}$ for $x > c$, although in general the two constants will be different. It follows that Φ is constant in the regions $y > 0$, $x - \beta y > c$, and $y < 0$, $x + \beta y > c$. This in turn implies that the induced velocities vanish in these two regions, and so within the limits of linearized two-dimensional theory, the aerofoil does not induce any downflow or upflow aft of its trailing edge.

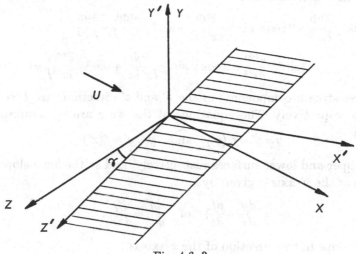

Fig. 4·6, 3

These results can be extended to the case of a wing with two-dimensional characteristics, as above, which is yawed by an angle γ, $0 \leqslant \gamma < \frac{1}{2}\pi - \mu$, relative to the z-axis (Fig. 4·6, 3). Thus the equation of the wing leading edge now is

$$x\cos\gamma - z\sin\gamma = 0,$$

while the free stream is still directed along the x-axis.

We introduce new coordinates along and normal to the leading edge of the wing, so

$$x' = x\cos\gamma - z\sin\gamma,$$

$$y' = y,$$

$$z' = x\sin\gamma + z\cos\gamma.$$

Then
$$\frac{\partial}{\partial x} = \cos \gamma \frac{\partial}{\partial x'} + \sin \gamma \frac{\partial}{\partial z'}, \quad \frac{\partial}{\partial z} = -\sin \gamma \frac{\partial}{\partial x'} + \cos \gamma \frac{\partial}{\partial z'}$$

and so
$$\frac{\partial^2 \Phi}{\partial x^2} = \left(\cos \gamma \frac{\partial}{\partial x'} + \sin \gamma \frac{\partial}{\partial z'} \right) \left(\cos \gamma \frac{\partial \Phi}{\partial x'} + \sin \gamma \frac{\partial \Phi}{\partial z'} \right)$$

$$= \cos^2 \gamma \frac{\partial^2 \Phi}{\partial x'^2} + 2 \cos \gamma \sin \gamma \frac{\partial^2 \Phi}{\partial x' \partial z'} + \sin^2 \gamma \frac{\partial \Phi}{\partial z'^2},$$

$$\frac{\partial^2 \Phi}{\partial z^2} = \left(-\sin \gamma \frac{\partial}{\partial x'} + \cos \gamma \frac{\partial}{\partial z'} \right) \left(-\sin \gamma \frac{\partial \Phi}{\partial x'} + \cos \gamma \frac{\partial \Phi}{\partial z'} \right)$$

$$= \sin^2 \gamma \frac{\partial^2 \Phi}{\partial x'^2} - 2 \sin \gamma \cos \gamma \frac{\partial^2 \Phi}{\partial x' \partial z'} + \cos^2 \gamma \frac{\partial^2 \Phi}{\partial z'^2},$$

and
$$\frac{\partial^2 \Phi}{\partial y^2} = \frac{\partial^2 \Phi}{\partial y'^2}.$$

Substituting these expressions in (4·5, 1), and replacing β by $\cot \mu$, where μ is the Mach angle, we obtain

$$\cot^2 \mu \left(\cos^2 \gamma \frac{\partial^2 \Phi}{\partial x'^2} + 2 \cos \gamma \sin \gamma \frac{\partial^2 \Phi}{\partial x' \partial z'} + \sin^2 \gamma \frac{\partial^2 \Phi}{\partial z'^2} \right) - \frac{\partial^2 \Phi}{\partial y'^2}$$

$$- \left(\sin^2 \gamma \frac{\partial^2 \Phi}{\partial x'^2} - 2 \cos \gamma \sin \gamma \frac{\partial^2 \Phi}{\partial x' \partial z'} + \cos^2 \gamma \frac{\partial^2 \Phi}{\partial z'^2} \right) = 0. \quad (4·6, 28)$$

The free-stream components in the x' and z' directions are $U \cos \gamma$ and $-U \sin \gamma$ respectively. The equations of the wing are by assumption of the form
$$y = y' = f_u(x') \quad \text{and} \quad y = y' = f_l(x')$$

for the upper and lower surfaces respectively, and so the local slope in the direction of the x'-axis is given by
$$\frac{dy'}{dx'} = \frac{df_u}{dx'} \quad \text{or} \quad \frac{dy'}{dx'} = \frac{df_l}{dx'},$$

while the slope in the direction of the x-axis is
$$\frac{dy}{dx} = \cos \gamma \frac{df_u}{dx'} \quad \text{or} \quad \frac{dy}{dx} = \cos \gamma \frac{df_l}{dx'}.$$

Assume for a moment that the free-stream velocity component in the direction of the x'-axis is kept constant, while a velocity $U \sin \gamma$ is superimposed on the given flow in the direction of the z'-axis. Then the resultant free-stream flow is parallel to the x'-axis. At the same time the boundary conditions at the wing are still satisfied. Now we know that the induced velocity potential of the modified flow is independent of z'. It follows that the induced velocity potential of the original flow also is independent of z'. Thus, the equation for the induced velocity potential is, by (4·6, 28),

$$(\cot^2 \mu \cos^2 \gamma - \sin^2 \gamma) \frac{\partial^2 \Phi}{\partial x'^2} - \frac{\partial^2 \Phi}{\partial y'^2} = 0. \quad (4·6, 29)$$

Now it was assumed earlier that $0 \leqslant \gamma < \frac{1}{2}\pi - \mu$. Hence $\cot \mu > \tan \gamma$, and so

$$\cot^2 \mu \cos^2 \gamma - \sin^2 \gamma = \beta_0^2 > 0.$$

(4·6, 29) may then be written as

$$\beta_0^2 \frac{\partial^2 \Phi}{\partial x'^2} - \frac{\partial^2 \Phi}{\partial y'^2} = 0, \qquad (4·6, 30)$$

whilst β_0^2 may be written as $\beta_0^2 = M_0^2 - 1$ in terms of the modified Mach number M_0, which is given by

$$M_0^2 = \beta_0^2 + 1 = \cot^2 \mu \cos^2 \gamma - \sin^2 \gamma + 1 = (\cot^2 \mu + 1) \cos^2 \gamma = M^2 \cos^2 \gamma,$$

or

$$M_0 = M \cos \gamma = \frac{U \cos \gamma}{a_0}.$$

Thus M_0 is simply the Mach number which is based on the velocity component in the direction of the x'-axis, normal to the leading edge of the wing. The boundary condition at the wing is

$$v = -\frac{\partial \Phi}{\partial y} = U \frac{dy}{dx}$$

at the upper or lower surface, and this may also be written as

$$v = -\frac{\partial \Phi}{\partial y'} = U \cos \gamma \frac{dy}{dx} \Big/ \cos \gamma = U \cos \gamma \frac{dy'}{dx'}. \qquad (4·6, 31)$$

Thus the problem of calculating the flow round a yawed two-dimensional wing can be reduced completely to the corresponding problem for the same wing, when unyawed, for a different free-stream velocity.

The pressure at any point of the aerofoil is given by

$$p - p_0 = -\rho_0 U u' = \rho_0 U \frac{\partial \Phi}{\partial x} = \rho_0 U \cos \gamma \frac{\partial \Phi}{\partial x'}.$$

We conclude that for a given wing whose incidence is fixed with respect to the primed system of coordinates, the pressure difference $p - p_0$ at any given point varies as $U^2 \cos^2 \gamma / \sqrt{(M^2 \cos^2 \gamma - 1)}$, if both the forward velocity and angle of yaw are variable. It follows that the same law applies to the lift, and also to the resultant force in the plane of the wing and normal to the leading edge, D' say. However, it must be understood that in this connexion both the lift and the force D' are taken per unit span, the span being measured along the leading edge of the wing. The wave drag, which is always taken parallel to the direction of flow, is then given by $D' \cos \gamma$, and so its law of variation is $U^2 \cos^3 \gamma / \sqrt{(M^2 \cos^2 \gamma - 1)}$. There is also a force component $-D' \sin \gamma$ along the z-axis.

Different laws of variation are obtained if we keep the section of the aerofoil constant in a plane which contains the direction of flow, while still being normal to the plane through the leading edge which contains the direction of flow (briefly, the x, y-plane). In that case, the slope in the direction of the x-axis varies as $\sec \gamma$ at corresponding points, and so the pressure difference varies as $U^2 \cos \gamma / \sqrt{(M^2 \cos^2 \gamma - 1)}$, and this is also the law of variation for lift and drag, taken per unit span, when the latter is measured normal to the direction of flow. Now

$$\frac{\partial}{\partial \gamma} \left(\frac{U^2 \cos \gamma}{\sqrt{(M^2 \cos^2 \gamma - 1)}} \right) = \frac{U^2 \sin \gamma}{(M^2 \cos^2 \gamma - 1)^{\frac{3}{2}}}.$$

This shows that according to this law of variation, the aerodynamic forces which act on the wing increase with increasing angle of yaw.

Consider, for example, a flat wing at incidence α. It is customary to measure the incidence relative to the free-stream direction, so that the second law of variation applies. Hence the lift is now given by

$$L = -\tfrac{1}{2} \rho_0 U^2 c \frac{4\alpha \cos \gamma}{\sqrt{(M^2 \cos^2 \gamma - 1)}}, \tag{4·6, 32}$$

and the corresponding wave drag is

$$D_W = \tfrac{1}{2} \rho_0 U^2 c \frac{4\alpha^2 \cos \gamma}{\sqrt{(M^2 \cos^2 \gamma - 1)}}. \tag{4·6, 33}$$

This yields the lift and wave-drag coefficients

$$C_L = \frac{4\alpha \cos \gamma}{\sqrt{(M^2 \cos^2 \gamma - 1)}}, \quad C_{D_W} = \frac{4\alpha^2 \cos \gamma}{\sqrt{(M^2 \cos^2 \gamma - 1)}}. \tag{4·6, 34}$$

Let us consider this case in more detail. The induced velocity potential satisfies the equation (4·6, 30), where $\beta_0^2 = M^2 \cos^2 \gamma - 1$. The general solution of (4·6, 30) is
$$\Phi = g_1(x' - \beta_0 y') + g_2(x' + \beta_0 y'),$$

or, in terms of the original system of coordinates,

$$\Phi = g_1(x \cos \gamma - z \sin \gamma - \beta_0 y) + g_2(x \cos \gamma - z \sin \gamma + \beta_0 y). \tag{4·6, 35}$$

The envelope of the Mach cones through the leading edge consists of the pair of planes
$$x \cos \gamma - z \sin \gamma \pm \beta_0 y = 0. \tag{4·6, 36}$$

For $y > 0$, Φ must again vanish at the plane $x \cos \gamma - z \sin \gamma - \beta_0 y = 0$, and this shows that g_2 is a constant and equal to $-g_1(0)$. The equation of the wing is
$$y = (x - z \tan \gamma) \tan \alpha \doteqdot (x - z \tan \gamma) \alpha,$$

and so its slope is $\dfrac{dy}{dx} \doteqdot \alpha$ for $0 \leqslant x - z \tan \gamma \leqslant c$.

The boundary condition is

$$v = -\frac{\partial \Phi}{\partial y} = \beta_0 g_1'(x \cos \gamma - z \sin \gamma) = U\alpha, \tag{4·6, 37}$$

where the prime indicates differentiation of g_1 with respect to its single argument. Then

$$
\left.
\begin{aligned}
u' &= -\frac{\partial \Phi}{\partial x} = -\cos \gamma \, g_1'(x \cos \gamma - z \sin \gamma) = -\frac{U\alpha \cos \gamma}{\beta_0} \\
\text{and} \qquad w &= -\frac{\partial \Phi}{\partial y} = +\sin \gamma \, g_1'(x \cos \gamma - z \sin \gamma) = \frac{U\alpha \sin \gamma}{\beta_0}.
\end{aligned}
\right\} \quad (4\cdot6, 38)
$$

These formulae are valid for $y > 0$ at the aerofoil and elsewhere in the region between the half-planes which are the envelopes to the after-cones above the x, z-plane, i.e. for

$$
0 \leqslant x \cos \gamma - z \sin \gamma - \beta_0 y \leqslant c \cos \gamma.
$$

The pressure at the wing is given by

$$
p_u - p_0 = -\rho_0 U u' = \rho_0 \frac{U^2 \alpha \cos \gamma}{\beta_0},
$$

and so the pressure difference between the upper and lower surfaces is

$$
p_u - p_l = 2\rho_0 \frac{U^2 \alpha \cos \gamma}{\beta_0} = \tfrac{1}{2}\rho_0 U^2 \frac{4\alpha \cos \gamma}{\sqrt{(M^2 \cos^2 \gamma - 1)}}.
$$

Hence, the lift is given by

$$
L = -\int_0^c (p_u - p_l)\, dx = -\tfrac{1}{2}\rho_0 U^2 \frac{4\alpha \cos \gamma}{\sqrt{(M^2 \cos^2 \gamma - 1)}}
$$

per unit span measured in a direction normal to the free-stream direction. This is in agreement with $(4\cdot6, 32)$ and provides a check on the general argument by which that formula was derived.

So long as $\gamma < \tfrac{1}{2}\pi - \mu$, the theory of a yawed two-dimensional wing possesses a concrete significance, and applies again exactly to the centre portion of a finite wing with two-dimensional characteristics, i.e. to the part of the wing which is not affected by the tips (Fig. 4·6, 4). However, as soon as γ exceeds $\tfrac{1}{2}\pi - \mu$, there is no region of the wing which remains unaffected by the tip conditions. The transformation of the equation $(4\cdot5, 1)$ to the primed system of coordinates can still be carried out formally, and the coefficient of $\partial^2 \Phi / \partial x'^2$ in $(4\cdot6, 29)$ is negative, so that we obtain an elliptic equation, as in the linearized theory of subsonic flow. This is in keeping with the fact that the Mach number $M_0 = M \cos \gamma$ is now less than unity, i.e. the free-stream velocity component normal to the leading edge is smaller than the velocity of sound a_0. This suggests that in cases where the component of the free-stream velocity normal to the leading edge is less than a_0, the characteristics of the flow will be, at least to some extent, subsonic. However, in this form the statement lacks precision, and it is more convenient to discuss this question in connexion with a concrete example in three-dimensional flow (§ 4·9).

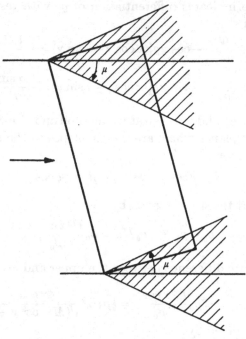

Fig. 4·6, 4

4·7 Methods based on distributions of supersonic sources: the symmetric case

It has been shown (§ 4·5) that the general theory of source and vortex distributions in incompressible flow, which was described in Chapter 1, possess its precise counterpart in steady supersonic flow. Accordingly, we may expect that the various methods for the solution of three-dimensional problems in incompressible flow which depend on source, doublet or vortex distributions to be available also in supersonic flow. However, it will appear that owing to the fact that in supersonic flow any disturbance is propagated only within the after-cone of its origin, various methods are available in supersonic flow, which do not exist, or which are of no practical importance under subsonic conditions. On the other hand, and for the same reason, lifting-line theory, which is so successful in dealing with wings of moderate aspect ratio at low speeds (§ 3·3), possess no counterpart at all in supersonic flow.

In this and the following sections we shall describe various methods which have proved useful in connexion with a large number of problems. Although there are problems which can be solved by more than one of these methods, it is, generally speaking, possible to select the most suitable in each particular case.

We assume, as in § 3·1, that the wing is situated approximately in the x, y-plane and that the main-stream velocity U is directed along the x-axis. Let $z = f_u(x, y)$ and $z = f_l(x, y)$ be the equations of the upper and lower surfaces of the aerofoil, then the boundary condition at the aerofoil is again

$$-\frac{\partial \Phi}{\partial z} = w = U s(x, y), \tag{4·7, 1}$$

where $s(x, y)$ is the local slope, $s(x, y) = \partial f_u / \partial x$ or $s(x, y) = \partial f_l / \partial x$.

We consider first the case of a symmetrical aerofoil at zero incidence. Then the upper and lower surfaces of the aerofoil are given by $z = \pm f(x, y)$, and the corresponding slope functions may be written as $\pm s(x, y)$. The boundary condition (4·7, 1) then shows that the induced velocity potential, which will be denoted simply by $\Phi(x, y, z)$, is symmetrical with respect to the x, y-plane, or

$$\Phi(x, y, z) = \Phi(x, y, -z), \tag{4·7, 2}$$

and so

$$\begin{aligned} u(x, y, z) &= U - \frac{\partial \Phi}{\partial x} = u(x, y, -z), \\ v(x, y, z) &= v(x, y, -z), \\ w(x, y, z) &= -w(x, y, -z). \end{aligned} \tag{4·7, 3}$$

and

This suggests that Φ can be represented by distribution of supersonic sources over the area S occupied by the wing in the x, y-plane. Thus, we write $\Phi(x, y, z)$ as

$$\Phi(x, y, z) = \frac{1}{2\pi} \int_{S'} \frac{\sigma(x_0, y_0) \, dx_0 dy_0}{\sqrt{\{(x - x_0)^2 - \beta^2((y - y_0)^2 + z^2)\}}}, \tag{4·7, 4}$$

where, for given (x, y, z), S' consists of the points $(x_0, y_0, 0)$ of S which satisfy the condition

$$(x - x_0)^2 - \beta^2((y - y_0)^2 + z^2) \geqslant 0 \quad (x > x_0), \tag{4·7, 5}$$

and $\sigma(x_0, y_0)$ is the source density, which remains to be determined. Applying (4·5, 54) for $\lambda = \mu = 0$, $\nu = 1$, we see that, at any point of S,

$$\left(\frac{\partial \Phi}{\partial z}\right)_{z = +0} - \left(\frac{\partial \Phi}{\partial z}\right)_{z = -0} = -\Delta q = -\sigma(x, y).$$

But

$$\left(\frac{\partial \Phi}{\partial z}\right)_{z = -0} = -w(x, y, -0) = w(x, y, +0) = -\left(\frac{\partial \Phi}{\partial z}\right)_{z = +0},$$

and so we may write

$$\sigma(x, y) = -2 \left(\frac{\partial \Phi}{\partial z}\right)_{z = +0}. \tag{4·7, 6}$$

Hence

$$\sigma(x, y) = 2 U s(x, y), \tag{4·7, 7}$$

by (4·7, 1), where $s(x, y)$ is the local slope of the upper surface. Substituting this expression in (4·7, 4) we obtain finally

$$\Phi(x, y, z) = \frac{U}{\pi} \int_{S'} \frac{s(x_0, y_0) \, dx_0 dy_0}{\sqrt{\{(x - x_0)^2 - \beta^2((y - y_0)^2 + z^2)\}}}. \tag{4·7, 8}$$

It remains to carry out the integration in each particular case. The procedure can be quite laborious even for relatively simple wing shapes.

It is sometimes convenient to integrate first along straight lines through a given vertex. For that reason it will be instructive to consider the following family of line distributions of supersonic sources.

Let
$$x = \lambda\tau, \quad y = \mu\tau, \quad z = \nu\tau \quad (\tau \geqslant 0) \tag{4·7, 9}$$

be the equation of a straight line l through the origin, where (λ, μ, ν) is a unit vector, and $\lambda > 0$. We assume in the first instance that l is in the interior of the after-cone emanating from the origin, $\lambda^2 - \beta^2(\mu^2 + \nu^2) > 0$.

Let $P(x, y, z)$ be any point in the after-cone of the origin. The hyperbolic distance of P from the origin (§ 4·5) is given by

$$r_h^2 = x^2 - \beta^2(y^2 + z^2) \quad (r_h > 0). \tag{4·7, 10}$$

From now on the subscript h will be omitted.

Define λ', μ', ν' by
$$\lambda' = x/r, \quad \mu' = \beta y/r, \quad \nu' = \beta z/r. \tag{4·7, 11}$$

Then $\lambda'^2 - (\mu'^2 + \nu'^2) = 1$. All the points along the straight line l correspond to the same set (λ', μ', ν') which can be expressed in terms of λ, μ, ν as follows:

$$\lambda' = \lambda/(\lambda^2 - \beta^2(\mu^2 + \nu^2)), \quad \mu' = \beta\mu/(\lambda^2 - \beta^2(\mu^2 + \nu^2)),$$
$$\nu' = \beta\nu/(\lambda^2 - \beta^2(\mu^2 + \nu^2)). \tag{4·7, 12}$$

Now let $\Phi(x, y, z)$ be the velocity potential due to a line distribution of sources along l, such that the source strength σ varies as the nth power of the distance from the origin $(n = 0, 1, 2, ...)$. Then σ may be written in the form

$$\sigma(x_0, y_0, z_0) = ar_0^n, \tag{4·7, 13}$$

where a is a constant and r_0 is the hyperbolic distance of (x_0, y_0, z_0) from the origin. Then

$$\Phi(x, y, z) = \frac{a}{2\pi} \int_{l'} \frac{r_0^n\, dl}{\sqrt{\{(x - x_0)^2 - \beta^2((y - y_0)^2 + (z - z_0)^2)\}}}, \tag{4·7, 14}$$

where the range of integration l' is restricted to the points (x_0, y_0, z_0) which satisfy the conditions

$$(x - x_0)^2 - \beta^2((y - y_0)^2 + (z - z_0)^2) > 0 \quad (x > x_0 > 0). \tag{4·7, 15}$$

Writing $\qquad \lambda_0' = x_0/r_0, \quad \mu_0' = \beta y_0/r_0, \quad \nu_0' = \beta z_0/r_0,$

we then have $\quad dl^2 = dx_0^2 + dy_0^2 + dz_0^2 = \dfrac{1}{\beta^2}(\beta^2\lambda_0'^2 + \mu_0'^2 + \nu_0'^2)\, dr_0^2.$

For any point $P(x, y, z)$ within the after-cone of the origin (4·7, 14) may therefore be written also as

$$\Phi(x, y, z) = \frac{a\sqrt{(\beta^2\lambda_0'^2 + \mu_0'^2 + \nu_0'^2)}}{2\pi\beta} \int_0 \frac{r_0^n\, dr_0}{\sqrt{\{(\lambda'r - \lambda_0'r_0)^2 - (\mu'r - \mu_0'r_0)^2 - (\nu'r - \nu_0'r_0)^2\}}}$$
$$= \frac{a\sqrt{(\beta^2\lambda_0'^2 + \mu_0'^2 + \nu_0'^2)}}{2\pi\beta} \int_0 \frac{r_0^n\, dr_0}{\sqrt{(r^2 + r_0^2 - 2rr_0h)}}, \tag{4·7, 16}$$

where
$$h = \lambda' \lambda_0' - \mu' \mu_0' - \nu' \nu_0' = \frac{xx_0 - \beta^2(yy_0 - zz_0)}{rr_0}. \tag{4·7, 17}$$

Note that h is independent of the variable of integration r_0. The upper limit of the range of integration in (4·7, 16) is given by the condition that (x_0, y_0, z_0) must be within the fore-cone of P. This certainly is the case for $r_0 = 0$. The roots of
$$r^2 + r_0^2 - 2rr_0 h = 0 \tag{4·7, 18}$$
for given r and h are
$$r_0 = r(h \pm \sqrt{(h^2 - 1)}).$$

These roots determine the points where l intersects the Mach cone through P. Now it is clear that the points of l cannot all belong to the fore-cone of P, and so the roots of (4·7, 18) must be real. Then r_0 leaves the fore-cone of P for the smaller roots of (4·7, 18),
$$r_0 = r(h - \sqrt{(h^2 - 1)}), \tag{4·7, 19}$$
and this must be the upper limit in the integration (4·7, 16). Incidentally, this shows that without further calculation that $h \geqslant 1$.

Substituting $s = r_0/r$ in (4·7, 16) we obtain
$$\Phi(x, y, z) = \frac{a\sqrt{(\beta^2 \lambda_0'^2 + \mu_0'^2 + \nu_0'^2)}}{2\pi\beta} r^n \int_0^{h - \sqrt{(h^2 - 1)}} \frac{s^n ds}{\sqrt{(1 + s^2 - 2hs)}}. \tag{4·7, 20}$$

But‡
$$\int_0^{h - \sqrt{(h^2 - 1)}} \frac{s^n ds}{\sqrt{(1 + s^2 - 2hs)}} = Q_n(h),$$

where $Q_n(h)$ is the nth Legendre function of the second kind. Hence Φ can be written as
$$\Phi(x, y, z) = \frac{a\sqrt{(\beta^2 \lambda_0'^2 + \mu_0'^2 + \nu_0'^2)}}{2\pi\beta} r^n Q_n(h). \tag{4·7, 21}$$

For $h > 1$ the function $Q_n(h)$ is of the form
$$Q_n(h) = \tfrac{1}{2} P_n(h) \log \frac{h+1}{h-1} - W_{n-1}(h), \tag{4·7, 22}$$

where $P_n(h)$ is the nth Legendre polynomial and $W_{n-1}(h)$ is a polynomial of degree $n - 1$. In particular
$$Q_0(h) = \tfrac{1}{2} \log \frac{h+1}{h-1}, \quad Q_1(h) = \tfrac{1}{2} h \log \frac{h+1}{h-1} - 1. \tag{4·7, 23}$$

Thus the Q_n are of an elementary type, and the integration (4·7, 16) can be carried out in each particular case without recourse to the theory of Legendre functions. Nevertheless, the vast amount of information, both analytical and numerical, which has been compiled concerning these functions warrants their explicit introduction.

‡ Compare E. W. Hobson, *The Theory of Spherical and Ellipsoidal Harmonics* (Cambridge University Press, 1931), p. 66.

Let $P(x, y, z)$ be any point within the after-cone of the origin. Put

$$x = r \cosh \psi, \quad y = \frac{1}{\beta} r \sinh \psi \cos \phi, \quad z = \frac{1}{\beta} r \sinh \psi \sin \phi, \quad (4·7, 24)$$

where r is the hyperbolic distance of P from the origin, $0 < r < \infty$ and $0 \leqslant \phi < 2\pi$, $0 \leqslant \psi < \infty$. Then ψ and ϕ are given uniquely in terms of x, y, z by the formulae

$$\tanh \psi = \beta \frac{\sqrt{(y^2 + z^2)}}{x}, \quad \tan \phi = \frac{z}{y},$$

provided ϕ is taken between 0 and π for positive z and between π and 2π for negative z. Also, for all values of r, ψ, ϕ within the specified interval of variation, the point (x, y, z), as given by (4·7, 24), is within the after-cone of the origin, since $r \cosh \psi > 0$ and

$$r \cosh^2 \psi - \beta^2 \frac{1}{\beta^2} r^2 \sinh^2 \psi \cos^2 \phi - \beta^2 \frac{1}{\beta^2} r^2 \sinh^2 \psi \sin^2 \phi = r^2 > 0.$$

The quantities λ', μ', ν' defined earlier are, in terms of ψ and ϕ,

$$\lambda' = \cosh \psi, \quad \mu' = \sinh \psi \cos \phi, \quad \nu' = \sinh \psi \sin \phi.$$

Hence

$$\begin{aligned} h = \lambda' \lambda_0' - \mu' \mu_0' - \nu' \nu_0' &= \cosh \psi \cosh \psi_0 - \sinh \psi \sinh \psi_0 (\cos \phi \cos \phi_0 \\ &\qquad + \sin \phi \sin \phi_0) \\ &= \cosh \psi \cosh \psi_0 - \sinh \psi \sinh \psi_0 \cos (\phi - \phi_0), \quad (4·7, 25) \end{aligned}$$

where ψ_0, ϕ_0 correspond to (x_0, y_0, z_0). It appears from (4·7, 25) that for given ψ and ψ_0, h attains its minimum value when $\cos (\phi - \phi_0) = 1$, $\phi = \phi_0$. In that case

$$h = \cosh \psi \cosh \psi_0 - \sinh \psi \sinh \psi_0 = \cosh (\psi - \psi_0).$$

Thus h obtains its minimum value (unity) only if $\psi = \psi_0$ and at the same time $\phi = \phi_0$.

Reverting to (4·7, 21) and (4·7, 22), we observe that $Q_n(h)$ has a logarithmic singularity for $h = 1$. In that case, the preceding argument shows that $\psi_0 = \psi$, $\phi_0 = \phi$, and so $\lambda_0' = \lambda'$, $\mu_0' = \mu'$, $\nu_0' = \nu'$ for all points of l, i.e. P is on l, and l meets the fore-cone of P at P, $r_0 = r$ in (4·7, 19).

If the point P is outside the after-cone of the origin, then $\Phi(x, y, z) = 0$.

Assume now that the line l is outside the after-cone of the origin, so that $\lambda^2 - \beta^2(\mu^2 + \nu^2) < 0$, where l is given by (4·7, 9), and that $\lambda \geqslant 0$. We consider again the velocity potential due to a distribution of sources along l which begins at the origin and varies as the nth power of the distance from the origin. We now define r_0 by

$$r_0^2 = -x_0^2 + \beta^2(y_0^2 + z_0^2) \quad (r_0 > 0), \quad (4·7, 26)$$

so that r_0 is again real along l.

The region of influence of l, i.e. the region which is affected by the presence of the sources along l, is bounded by the envelope of the after-cones emanating from l, beginning at the origin. The envelope of the after-cones of the points of l extending to infinity in both directions, consists of a pair of planes (compare (4·6, 36) for the case where the line l is in the x, z-plane). In the present case, we consider only the points of l for $\tau \geqslant 0$, i.e. beginning at the origin, and so the envelope consists partly of the planes just mentioned, and partly of the after-cone of the origin. It is not difficult to provide a geometrical construction of the curve at the intersection of the envelope with a plane m normal to the direction of flow, $x = x_1 > 0$, say (Fig. 4·7, 1). Let O' be the point where the x-axis intersects m, and let C be the circle of intersection of the after-cone of the origin with m. This is a circle of radius x_1/β about O'. Furthermore, let A be the point of intersection of l with m. The y- and z-coordinates of A are $x_1\mu/\lambda$ and $x_1\nu/\lambda$.

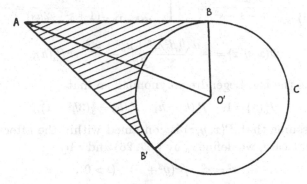

Fig. 4·7, 1

From A draw tangents AB and AB' to the circle C. Then the interior of the circle C, together with the interior of the curved triangle $AB'B$, constitutes the region of m which is affected by the source distribution. In the limiting case when l is in the y, z-plane ($\lambda = 0$), the point A moves to infinity. The two tangents to the circle are then parallel to l, and the shaded region is unlimited in one direction. Outside the region of influence determined above, the induced velocity potential $\Phi(x, y, z)$ vanishes. Consider now a point $P(x, y, z)$ within the region of influence of the source distribution, but outside the after-cone of the origin. Thus P belongs to the shaded area in Fig. 4·7, 1. For such a point, we define r by

$$r^2 = -x^2 + \beta^2(y^2 + z^2) \quad (r > 0). \tag{4·7, 27}$$

$\Phi(x, y, z)$ is given by (4·7, 14), where the range of integration is still defined by (4·7, 15). The source density σ may again be written as in (4·7, 13), where r_0 is now defined by (4·7, 26). Putting

$$\lambda' = x/r, \quad \mu' = \beta y/r, \quad \nu' = \beta z/r,$$

with similar definitions for $\lambda_0', \mu_0', \nu_0'$, and substituting in (4·7, 14), we now obtain

$$\Phi(x, y, z) = \frac{a\sqrt{(\beta^2\lambda_0'^2 + \mu_0'^2 + \nu_0'^2)}}{2\pi\beta} \int \frac{r_0^n \, dr_0}{\sqrt{(2rr_0 h - r^2 - r_0^2)}}, \tag{4·7, 28}$$

where $$h = -\lambda'\lambda_0' + \mu'\mu_0' + \nu'\nu_0' = \frac{-xx_0 + \beta^2(yy_0 + zz_0)}{rr_0}. \qquad (4\cdot7, 29)$$

The fore-cone of P intersects l at two points, corresponding to the roots of the equation $(4\cdot7, 18)$, $$r_0 = r(h \pm \sqrt{(h^2-1)}). \qquad (4\cdot7, 30)$$

These are the limits of integration in $(4\cdot7, 28)$. We see, incidentally, that $h \geqslant 1$, as before. The transformation $s = r_0/r$ reduces $(4\cdot7, 28)$ to

$$\Phi(x, y, z) = \frac{a\sqrt{(\beta^2\lambda_0'^2 + \mu_0'^2 + \nu_0'^2)}}{2\pi\beta} r^n \int_{h-\sqrt{(h^2-1)}}^{h+\sqrt{(h^2-1)}} \frac{s^n\, ds}{\sqrt{(2hs - 1 - s^2)}}.$$

But‡
$$\int_{h-\sqrt{(h^2+1)}}^{h+\sqrt{(h^2+1)}} \frac{s^n\, ds}{\sqrt{(2hs - 1 - s^2)}} = -i\int_{h-\sqrt{(h^2-1)}}^{h+\sqrt{(h^2-1)}} \frac{s^n\, ds}{\sqrt{(1 + s^2 - 2hs)}} = -\pi P_n(h).$$
$$(4\cdot7, 31)$$

Hence $$\Phi(x, y, z) = -\frac{a\sqrt{(\beta^2\lambda_0'^2 + \mu_0'^2 + \nu_0'^2)}}{2\beta} r^n P_n(h), \qquad (4\cdot7, 32)$$

where $P_n(h)$ is the nth Legendre polynomial, so that

$$P_0(h) = 1, \quad P_1(h) = h, \quad P_2(h) = \tfrac{1}{3}(3h^2 - 1). \qquad (4\cdot7, 33)$$

Finally, assume that $P(x, y, z)$ is contained within the after-cone of the origin. In that case, we define r_0 by $(4\cdot7, 26)$ and r by

$$r^2 = x^2 - \beta^2(y^2 + z^2) \quad (r > 0). \qquad (4\cdot7, 34)$$

The integral $(4\cdot7, 14)$ now becomes

$$\Phi(x, y, z) = \frac{a\sqrt{(\beta^2\lambda_0'^2 + \mu_0'^2 + \nu_0'^2)}}{2\pi\beta} \int \frac{r_0^n\, dr_0}{\sqrt{(r^2 - r_0^2 + 2rr_0 h)}}, \qquad (4\cdot7, 35)$$

where h is given by $(4\cdot7, 29)$. The lower limit of integration in $(4\cdot7, 35)$ is zero, corresponding to the origin, and the upper limit is one of the roots of the equation $$r^2 - r_0^2 + 2rr_0 h = 0. \qquad (4\cdot7, 36)$$

These are $$r_0 = rh \pm \sqrt{(r^2 h^2 + r^2)} = r(h \pm \sqrt{(h^2+1)}).$$

Since r_0 is positive, the appropriate root is

$$r_0 = r(h + \sqrt{(h^2+1)}).$$

Substitution of $s = r_0/r$ in $(4\cdot7, 35)$ leads to

$$\Phi(x, y, z) = \frac{a\sqrt{(\beta^2\lambda_0'^2 + \mu_0'^2 + \nu_0'^2)}}{2\pi\beta} r^n \int_0^{h+\sqrt{(h^2+1)}} \frac{s^n\, ds}{\sqrt{(1 - s^2 + 2hs)}}. \qquad (4\cdot7, 37)$$

Put $$k = ih, \quad t = is,$$

then $$\int_0^{h+\sqrt{(h^2+1)}} \frac{s^n\, ds}{\sqrt{(1 - s^2 + 2hs)}} = (-i)^{n+1} \int_0^{k-\sqrt{(k^2-1)}} \frac{t^n\, dt}{\sqrt{(1 + t^2 - 2kt)}},$$
$$(4\cdot7, 38)$$

where $$\arg(\sqrt{(k^2 - 1)}) = -\tfrac{1}{2}\pi.$$

‡ See E. T. Whittaker and G. N. Watson, *A Course in Modern Analysis* (4th ed., Cambridge University Press, 1927), p. 315.

For $h < 0$, $\arg k = -\tfrac{1}{2}\pi$, this choice of $\arg(\sqrt{(k^2-1)})$ is in agreement with the formula

$$Q_n(k) = \int_0^{k-\sqrt{(k^2-1)}} \frac{t^n dt}{\sqrt{(1+t^2-2kt)}}, \tag{4·7, 39}$$

where the Legendre function of the second kind, $Q_n(k)$, is, for complex k, equal to

$$Q_n(k) = \tfrac{1}{2}P_n(k)\log\frac{k+1}{k-1} - W_{n-1}(k). \tag{4·7, 40}$$

In this formula, the polynomials $P_n(k)$ and $W_{n-1}(k)$ are defined as in (4·7, 22), and

$$\mathscr{I}\left(\log\frac{k+1}{k-1}\right) = \arg(k+1) - \arg(k-1),$$

where both arguments on the right-hand side are taken to lie in the interval $(-\pi, \pi)$.‡ In particular, if $k = ih$ ($h < 0$), we have

$$\arg(k+1) = \tan^{-1}h, \quad \arg(k-1) = -\pi - \tan^{-1}h,$$

and so

$$\log\frac{k+1}{k-1} = 2\tan^{-1}h + \pi,$$

where

$$-\tfrac{1}{2}\pi < \tan^{-1}h < 0.$$

Hence, for $h < 0$,

$$Q_n(ih) = \frac{i}{2}P_n(ih)(2\tan^{-1}h + \pi) - W_{n-1}(ih), \tag{4·7, 41}$$

and so, by (4·7, 38) and (4·7, 39)

$$\int_0^{h+\sqrt{(h^2+1)}} \frac{s^n ds}{\sqrt{(1-s^2+2hs)}} = \frac{(-i)^n}{2}P_n(ih)(2\tan^{-1}h + \pi) - (-i)^{n+1}W_{n-1}(ih). \tag{4·7, 42}$$

Analytic continuation then shows that this relation still holds for $h \geqslant 0$, if we confine $\tan^{-1}h$ to the interval

$$-\tfrac{1}{2}\pi < \tan^{-1}h < \tfrac{1}{2}\pi. \tag{4·7, 43}$$

Let us denote the right-hand side of (4·7, 41) by $Q_n^*(ih)$, then

$$Q_n^*(ih) = Q_n(ih) \tag{4·7, 44}$$

for $h < 0$. However, if we use the standard definition of $Q_n(k)$, then (4·7, 44) is no longer true for $h > 0$. For such values of h we have§

$$Q_n(ih) = \frac{i}{2}P_n(ih)(2\tan^{-1}h - \pi) - W_{n-1}(ih), \tag{4·7, 45}$$

where $\tan^{-1}h$ is confined to the interval (4·7, 43). Hence, for $h > 0$,

$$Q_n^*(ih) = Q_n(ih) + i\pi P_n(ih). \tag{4·7, 46}$$

Clearly $Q_n^*(k)$ still is a solution of Legendre's equation of order n.

‡ E. W. Hobson, *The Theory of Spherical and Ellipsoidal Harmonics* (Cambridge University Press, 1931), pp. 51, 64–6.

§ *Ibid.* p. 51.

Finally, for the case $h=0$, the Legendre function of the second kind is, by the standard definition,

$$Q_n(0) = -W_{n-1}(0), \qquad (4\cdot7, 47)$$

and so

$$Q_n^*(0) = Q_n(0) + \frac{i\pi}{2} P_n(0). \qquad (4\cdot7, 48)$$

In particular

$$Q_0^*(ih) = \frac{i}{2}(2\tan^{-1}h + \pi), \quad Q_1^*(ih) = -\frac{h}{2}(2\tan^{-1}h + \pi) - 1. \qquad (4\cdot7, 49)$$

We now have $\displaystyle\int_0^{h+\surd(h^2+1)} \frac{s^n ds}{\sqrt{(1-s^2+2hs)}} = (-i)^{n+1} Q_n^*(ih),$

and so (4·7, 37) becomes

$$\Phi(x,y,z) = \frac{a\sqrt{(\beta^2\lambda_0'^2 + \mu_0'^2 + \nu_0'^2)}}{2\pi\beta} r^n(-i)^{n+1} Q_n^*(ih). \qquad (4\cdot7, 50)$$

To give a simple application of the formulae derived above we consider a source distribution of constant strength over the semi-infinite wedge S in the x, y-plane which is given by the conditions

$$(y - x\tan\gamma_1)(y - x\tan\gamma_2) < 0 \quad (x > 0), \qquad (4\cdot7, 51)$$

$$\frac{\pi}{2} > \tan\gamma_1 > \tan\gamma_2 > \frac{1}{\beta}. \qquad (4\cdot7, 52)$$

This area is bounded by the straight lines

$$y - x\tan\gamma_1 = 0, \quad y - x\tan\gamma_2 = 0, \qquad (4\cdot7, 53)$$

which are outside the Mach cone emanating from the origin, by (4·7, 52).

The source distribution is supposed constant over S, $\sigma(x_0, y_0) = \sigma_0$, say, so that (4·7, 4) becomes

$$\Phi(x,y,z) = \frac{\sigma_0}{2\pi} \int_{S'} \frac{dx_0 dy_0}{\sqrt{\{(x-x_0)^2 - \beta^2((y-y_0)^2 + z^2)\}}}. \qquad (4\cdot7, 54)$$

We shall be interested chiefly in points $(x, y, 0)$ on S. For such points $-x^2 + \beta^2 y^2 > 0$. Introduce the variables of integration $r_0 > 0$, $\psi_0 > 0$,

$$x_0 = r_0 \sinh\psi_0, \quad y_0 = \frac{1}{\beta} r_0 \cosh\psi_0. \qquad (4\cdot7, 55)$$

Then

$$r^2 = -x_0^2 + \beta^2 y_0^2,$$

in agreement with the definition of r_0 by (4·7, 26). The Jacobian of the transformation is given by

$$\frac{D(x_0, y_0)}{D(r_0, \psi_0)} = \begin{vmatrix} \sinh\psi_0 & \frac{1}{\beta}\cosh\psi_0 \\ r_0\cosh\psi_0 & \frac{1}{\beta}r_0\sinh\psi_0 \end{vmatrix} = -\frac{1}{\beta}r_0, \qquad (4\cdot7, 56)$$

and so (4·7, 54) becomes

$$\Phi(x,y,0) = -\frac{\sigma_0}{2\pi\beta} \int d\psi_0 \int \frac{r_0 dr_0}{\sqrt{(2rr_0 h - r^2 - r_0^2)}}. \qquad (4\cdot7, 57)$$

In this formula, r and h are given by (4·7, 27) and (4·7, 29), respectively, and the quantities $\lambda', \mu', \nu', \lambda_0', \mu_0', \nu_0'$ are defined as before. Thus, we now have

$$\lambda_0' = \sinh \psi_0, \quad \mu_0' = \cosh \psi_0, \quad \nu_0' - 0, \tag{4·7, 58}$$

and so
$$h = -\lambda' \sinh \psi_0 + \mu' \cosh \psi_0. \tag{4·7, 59}$$

Referring to (4·7, 28) and (4·7, 32), we then obtain from (4·7, 57)

$$\Phi(x, y, 0) = \frac{\sigma_0}{2\beta} r \int P_1(h)\, d\psi_0 = \frac{\sigma_0}{2\beta} r \int (-\lambda' \sinh \psi_0 + \mu' \cosh \psi_0)\, d\psi_0. \tag{4·7, 60}$$

The region of integration S' is a triangular area bounded by the two Mach lines through $(x, y, 0)$ and by

$$y_0 - x_0 \tan \gamma_1 = 0.$$

The value of ψ_0 which corresponds to this line is, by (4·7, 55), given by the condition

$$\frac{1}{\beta} \cosh \psi_0 - \sinh \psi_0 \tan \gamma_1 = 0,$$

or
$$\coth \psi_0 = \beta \tan \gamma_1. \tag{4·7, 61}$$

This value of ψ_0 is the lower limit of integration in (4·7, 60). The upper limit of integration corresponds to the straight line which passes through the origin and through $(x, y, 0)$. It is therefore given by

$$\sinh \psi_0 = \lambda_0' = \lambda', \quad \cosh \psi_0 = \mu_0' = \mu'. \tag{4·7, 62}$$

Now
$$\int (-\lambda' \sinh \psi_0 + \mu' \cosh \psi_0)\, d\psi_0 = -\lambda' \cosh \psi_0 + \mu' \sinh \psi_0 + \text{constant},$$

and so
$$\tag{4·7, 63}$$

$$\Phi(x, y, 0) = \frac{\sigma_0}{2\beta} r(-\lambda' \cosh \psi_0 + \mu' \sinh \psi_0) = \frac{\sigma_0}{2\beta}(-x \cosh \psi_0 + y\beta \sinh \psi_0), \tag{4·7, 64}$$

where ψ_0 is given by (4·7, 61). But

$$\sinh \psi_0 = \frac{1}{\sqrt{(\coth^2 \psi_0 - 1)}},$$

and so
$$\Phi(x, y, 0) = \frac{\sigma_0}{2\beta} \sinh \psi_0 (-x \coth \psi_0 + y\beta) = \frac{\sigma_0}{2} \frac{y - x \tan \gamma_1}{\sqrt{(\beta^2 \tan^2 \gamma_1 - 1)}}. \tag{4·7, 65}$$

Similar formulae can be derived if the wedge S is contained either partly or entirely within the after-cone of the origin. In general, these formulae will involve also the Legendre function of the second kind $Q_1(ih)$, and hence the inverse trigonometric function $\tan^{-1} h$. By superimposing the solutions obtained in this way, we may obtain the velocity potential due to any given wing at zero incidence which is bounded by plane surfaces.

We consider the example of an isosceles triangular wing (Delta wing) with similar double-wedge aerofoil sections which attain maximum thickness at mid-chord. Let AB be the base of the triangle and suppose that the direction of flow is parallel to the height CD on AB (Fig. 4·7, 2). Then $AB = 2s_0$ is the span of the wing and $CD = c$ its maximum chord. Let E be the mid-point of CD, then the wing attains its maximum thickness along the straight segments AE and BE.

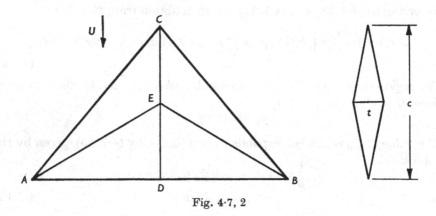

Fig. 4·7, 2

Let t be the thickness of the wing at E. Then the slope of the wing at the upper surface is

$$s(x_0, y_0) = \begin{cases} t/c = \tan\alpha & \text{on} \quad AEBCA, \\ -t/c = -\tan\alpha & \text{on} \quad ABEA. \end{cases} \qquad (4,7 \cdot 66)$$

Similarly, on the lower surface

$$s(x_0, y_0) = \begin{cases} -t/c = -\tan\alpha & \text{on} \quad AEBCA, \\ t/c = \tan\alpha & \text{on} \quad ABEA. \end{cases} \qquad (4\cdot7, 67)$$

Thus the induced velocity potential due to the wing, $\Phi(x, y, z)$, is given by (4·7, 4), where

$$\sigma(x_0, y_0) = \begin{cases} 2Ut/c & \text{on} \quad AEBCA, \\ -2Ut/c & \text{on} \quad ABEA. \end{cases} \qquad (4\cdot7, 68)$$

It follows that at the wing $\Phi(x, y, z)$ can also be obtained by the super-position of two velocity potentials $\Phi_1(x, y, z)$ and $\Phi_2(x, y, z)$

$$\Phi = \Phi_1 + \Phi_2,$$

where Φ_1 is due to a source distribution of constant strength $2Ut/c$ over the wedge enclosed between AC and BC, and Φ_2 is due to a source distribution of strength $-4Ut/c$ over the wedge enclosed between AE and BE. We note that the strength of the source distribution downstream of AB does not enter into the formula for the velocity potential at the wing.

Integration along straight lines through a fixed point is not the only, nor indeed always the quickest, way for the evaluation of (4·7, 4) or (4·7, 8).

An alternative procedure depends on the introduction of the variables of integration r_1, ψ_1, into (4·7, 4) or (4·7, 8), by

$$x_0 = x + r_1 \cosh \psi_1, \quad y_0 = y + \frac{1}{\beta} r_1 \sinh \psi_1. \qquad (4·7, 69)$$

By the use of this substitution, it has been shown‡ that if the wing is bounded by plane surfaces, then the velocity potential at the wing, as well as the resultant wave drag, can always be expressed in terms of elementary (algebraic, inverse trigonometric and logarithmic) functions.

Having determined the induced velocity potential Φ at the wing, we use the linearized Bernoulli equation (4·1, 27) to calculate the resultant wave drag. According to that formula, the pressure at any point $(x, y, 0)$ of the wing is

$$p = p_0 - \rho_0 U u' = p_0 + \rho_0 U \frac{\partial \Phi}{\partial x}. \qquad (4·7, 70)$$

Hence the resultant force on the wing is

$$\mathbf{R} = -\int p \, d\mathbf{S} = -\int \left(p_0 + \rho_0 U \frac{\partial \Phi}{\partial x} \right) d\mathbf{S},$$

where the integral is taken over both the upper and lower surface of the wing. Now

$$\int p_0 \, d\mathbf{S} = p_0 \int d\mathbf{S} = 0,$$

and so

$$\mathbf{R} = -\rho_0 U \int \frac{\partial \Phi}{\partial x} \, d\mathbf{S}. \qquad (4·7, 71)$$

The wave drag D_W is the component of $\mathbf{R}$ in the direction of the x-axis. Provided the local slope $s(x, y)$ of the wing is small everywhere, we may approximate the x component of a directed element of the upper surface $d\mathbf{S}$ by $-s(x, y) \, dx \, dy$. Hence

$$D_W = 2\rho_0 U \int_S \frac{\partial \Phi}{\partial x} s(x, y) \, dx \, dy, \qquad (4·7, 72)$$

where the integral is taken only once over the wing plan-form and $s(x, y)$ denotes the slope of the upper surface. (4·7, 72) does not apply if the slope of the wing becomes infinite locally, as is the case at a rounded leading edge.

Substituting (4·7, 8) in (4·7, 72) we obtain

$$D_W = \frac{2}{\pi} \rho_0 U^2 \int_S s(x, y) \, dx \, dy \, \frac{\partial}{\partial x} \int_{S'} \frac{s(x_0, y_0) \, dx_0 \, dy_0}{\sqrt{\{(x - x_0)^2 - \beta^2 (y - y_0)^2\}}},$$

or

$$D_W = \frac{2}{\pi} \rho_0 U^2 \int_S s(x, y) \, dx \, dy \,{}^* \!\!\int_{S'} \frac{s(x_0, y_0) (x_0 - x) \, dx_0 \, dy_0}{[(x - x_0)^2 - \beta^2 (y - y_0)^2]^{\frac{3}{2}}}, \qquad (4·7, 73)$$

where the asterisk indicates that the finite part is to be taken in the divergent second integral. For concrete applications it is advisable first to calculate Φ and then to evaluate (4·7, 72), but (4·7, 73) is useful for certain general considerations.

‡ A Robinson, 'Wave drag of an aerofoil at zero incidence', *Royal Aircraft Establishment Rep.* no. Aero. 2159 (1946).

For the case of a Delta wing with double-wedge section, which was considered above, the wave drag has been calculated by A. E. Puckett.‡ In Fig. 4·7, 3 the wave-drag coefficient $C_{D_W} = \dfrac{D_W}{\frac{1}{2}\rho_0 U^2 S}$ is plotted against Mach number for a wing with thickness-chord ratio t/c of 0·1 and various values

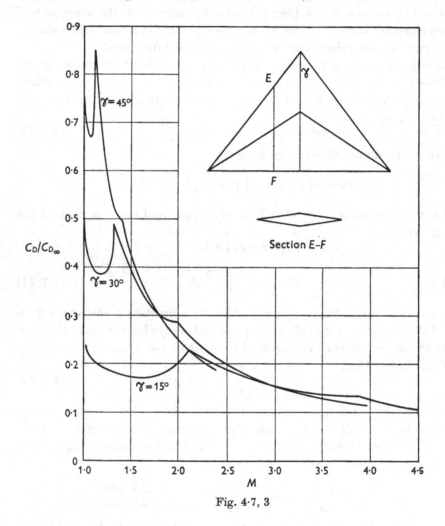

Fig. 4·7, 3

of the semi-apex angle γ. Results for other chordwise positions of the maximum thickness are included in Puckett's paper.

In Fig. 4·7, 4 the ratio $C_{D_W}/C_{D_{W_\infty}}$ is plotted against the parameter

$$\lambda = \beta \tan \gamma = \frac{\tan \gamma}{\tan \mu}, \qquad (4\cdot7, 74)$$

‡ A. E. Puckett, 'Supersonic wave drag of thin airfoils', *J. Aero. Sci.* vol. 13 (1946), pp. 475–84.

where μ is the Mach angle, and $C_{D_{W\infty}}$ is the two-dimensional value of the wave-drag coefficient for a double-wedge aerofoil with maximum thickness at the mid-chord, as given by (4·6, 23). Thus, in our present notation

$$C_{D_{W\infty}} = \frac{4}{\beta}(t/c)^2. \qquad (4\cdot7, 75)$$

We note that the different curves in Fig. 4·7, 3 all correspond to the same curve in Fig. 4·7, 4. The latter diagram may be interpreted in various ways. We may keep the free-stream velocity, and hence the Mach number, constant and regard γ as variable. As γ approaches $\frac{1}{2}\pi$, λ tends to infinity. Fig. 4·7, 4 then shows that the ratio $C_{D_W}/C_{D_{W\infty}}$ approaches unity, i.e. C_{D_W} approaches the two-dimensional wave-drag coefficient $C_{D_{W\infty}}$, as may be expected. The ratio $C_{D_W}/C_{D_{W\infty}}$ reaches its maximum at $\gamma = \mu$, that is to say, when the Mach lines through the apex coincide with the leading edges.

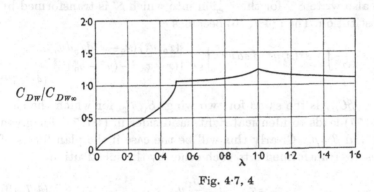

Fig. 4·7, 4

The slope of the curve is discontinuous for $\lambda = 1$, $\gamma = \mu$, and again for a lower value of λ, which corresponds to the case when the lines of maximum thickness (AE and BE in Fig. 4·7, 2) are Mach lines. We observe that in these cases the components of the free-stream velocity in directions normal to the leading edge, or to the lines of maximum thickness, in the x, y-plane, are equal to the velocity of sound.

It is equally instructive to assume that the wing is given while the Mach number is variable. Fig. 4·7, 4 then shows that the wave-drag coefficient tends to its two-dimensional value, $C_{D_W}/C_{D_{W\infty}} \to 1$, as the Mach number increases indefinitely.

We have noted that the different curves in Fig. 4·7, 3 all correspond to the same curve in Fig. 4·7, 4. For a given ratio t/c this amounts to saying that the quantity βC_{D_W} depends only on the ratio $\lambda = \beta \tan \gamma$. In other words, if we modify both the Mach number and the apex angle of the origin in such a way that the quantity λ stays constant, then βC_{D_W} also remains the same. This fact can be deduced without calculation from a more general principle which will now be established.

Referring to (4·7, 73), we see that

$$\beta C_{D_W} = \beta D_W / \tfrac{1}{2}\rho_0 U^2 S = \frac{4\beta}{\pi S} \int_S s(x,y)\,dx\,dy \,{}^*\!\!\int_{S'} \frac{s(x_0,y_0)\,(x_0-x)\,dx_0\,dy_0}{[(x-x_0)^2 - \beta^2(y-y_0)^2]^{\frac{3}{2}}}.$$

$$(4\cdot7, 76)$$

In this formula the symbol S has been used to denote both a region of integration and its area. We introduce the new variables

$$y^* = \beta y, \quad y_0^* = \beta y_0, \qquad\qquad (4\cdot7, 77)$$

and we put

$$s^*(x, y^*) = s\!\left(x, \frac{y^*}{\beta}\right) = s(x, y),$$

$$S^* = \beta S.$$

Then S^* is again the area of the region into which S is transformed by (4·7, 77). We also write $S^{*\prime}$ for the region into which S' is transformed by the same substitution. Then (4·7, 76) becomes

$$\beta C_{D_W} = \frac{4}{\pi S^*} \int_{S^*} s^*(x, y^*)\,dx\,dy^* \,{}^*\!\!\int_{S^{*\prime}} \frac{s(x_0, y_0^*)\,(x_0-x)\,dx_0\,dy_0^*}{[(x-x_0)^2 - (y^*-y_0^*)^2]^{\frac{3}{2}}}.$$

$$(4\cdot7, 78)$$

It follows that βC_{D_W} is the same for two wings S_1, S_2, for which the substitution (4·7, 77) leads to identical right-hand sides in (4·7, 78) for given values $\beta = \beta_1$ and $\beta = \beta_2$. Clearly this will be the case if the plan-forms of the two wings are transformed into each other by the substitution

$$x_2 = x_1, \quad y_2 = \frac{\beta_1}{\beta_2} y_1, \qquad\qquad (4\cdot7, 79)$$

and if the local slopes are the same at points which correspond according to this substitution. The condition is satisfied, inter alia, for two Delta wings with apexes at the origin, which have equal maximum chords and thickness-chord ratios, and whose semi-apex angles are related by the formula

$$\frac{\tan \gamma_1}{\tan \gamma_2} = \frac{\beta_2}{\beta_1}.$$

Another argument which leads to the same conclusion is as follows.

Let $\Phi_1(x, y, z)$ be the induced velocity potential which is appropriate to the flow around a specified wing S_1 at zero incidence, for a given Mach number M_1, and let $\beta_1 = \sqrt{(M_1^2 - 1)}$. Then

$$\Phi_2(x, y, z) = \Phi_1\!\left(\frac{\beta_1}{\beta_2} x, y, z\right)$$

satisfies the equation for the velocity potential (4·5, 1) for another

$$\beta = \beta_2 = \sqrt{(M_2^2 - 1)}.$$

Also, Φ_2 satisfies the boundary condition (4·7, 1) for the wing S_2 which is obtained from S_1 by stretching the x-coordinate of each point of the wing in the ratio

$$x_2 : x_1 = \beta_2 : \beta_1,$$

while leaving in the free-stream velocity U and the slope (at corresponding points) unaffected. Assuming that the free-stream pressure is the same in both cases, we find by the use of the linearized Bernoulli equation that the pressure increments at corresponding points are related by the formulae

$$(p_2 - p_0) : (p_1 - p_0) = \frac{\partial \Phi_2}{\partial x_2} : \frac{\partial \Phi_1}{\partial x_1} = \beta_1 : \beta_2.$$

It is not difficult to see that this must also be the ratio of the corresponding drag coefficients

$$(C_{Dw})_2 : (C_{Dw})_1 = \beta_1 : \beta_2,$$

or

$$\beta_2 (C_{Dw})_2 = \beta_1 (C_{Dw})_1. \qquad (4·7, 80)$$

The wings S_1 and S_2 are related to each other by (4·7, 79), except for a scalar factor which is irrelevant here. Thus (4·7, 80) is equivalent to the result obtained above by the other method.

We may draw another conclusion of a general nature from (4·7, 73), by rewriting it in the form

$$D_W = \frac{2}{\pi} \rho_0 U^2 \overset{*}{\int_T} s(x, y) s_0(x_0, y_0) \frac{(x_0 - x) \, dx \, dy \, dx_0 \, dy_0}{[(x - x_0)^2 - \beta^2 (y - y_0)^2]^{\frac{3}{2}}}. \qquad (4·7, 81)$$

The domain of integration T of the quadruple integral on the right-hand side is given by the condition that the points (x, y) and (x_0, y_0) belong to S such that

$$(x - x_0)^2 - \beta^2 (y - y_0)^2 > 0 \quad (x > x_0).$$

Now assume that the main flow is reversed in direction but kept constant in magnitude so that the free-stream velocity vector now is $(-U, 0, 0)$. Then the wave drag D'_W on the wing under the new conditions (regarded as positive when acting in the negative direction of the x-axis) is

$$D'_W = \frac{2}{\pi} \rho_0 U^2 \overset{*}{\int_{T'}} s(x, y) s(x_0, y_0) \frac{(x - x_0) \, dx \, dy \, dx_0 \, dy_0}{[(x - x_0)^2 - \beta^2 (y - y_0)^2]^{\frac{3}{2}}}. \qquad (4·7, 82)$$

The domain of integration T' is now given by the condition

$$(x - x_0)^2 - \beta^2 (y - y_0)^2 > 0 \quad (x_0 > x). \qquad (4·7, 83)$$

Exchanging the variables of integration x and y with x_0 and y_0 respectively in (4·7, 82), we obtain

$$D'_W = \frac{2}{\pi} \rho_0 U^2 \overset{*}{\int_T} s(x_0, y_0) s(x, y) \frac{(x_0 - x) \, dx_0 \, dy_0 \, dx \, dy}{[(x - x_0)^2 - \beta^2 (y - y_0)^2]^{\frac{3}{2}}}.$$

Hence

$$D'_W = D_W. \qquad (4·7, 84)$$

This is an important example of a group of results known as flow reversal theorems. It was first stated by Hayes‡ and von Kármán.§ Other theorems of the same type, some of which also refer to the lift, will be found in papers by Ursell and Ward,‖ and Heaslet and Spreiter.¶

4·8   Linearized theory of wings at incidence in supersonic flow

We saw in Chapter 3 that for incompressible conditions, the source distribution method applied only to the calculation of the pressure distribution on wings with symmetrical section at zero incidence. For the complementary case of a very thin wing at incidence we had to use instead a doublet distribution method (§ 3·6). This method has the drawback that there is no direct relation between the slope of the wing at a given point and the intensity of the doublet distribution at that point. It will now be shown that owing to the particular physical nature of supersonic flow, the source distribution method detailed above can also be applied to an important class of wings at incidence.

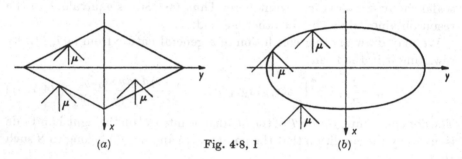

(a)   Fig. 4·8, 1   (b)

Consider a wing whose plan-form is convex (i.e. without re-entrant corners), and such that the slope of both leading and trailing edges satisfies the condition

$$\left|\frac{dx}{dy}\right| < \beta = \cot\mu \quad \text{everywhere.} \tag{4·8, 1}$$

It will be seen that the rhombic plan-form shown in Fig. 4·8, 1 a satisfies (4·8, 1), while for the plan-form of Fig. 4·8, 1 b the condition holds only at points which are sufficiently near to the centre line.

If (4·8, 1) is satisfied, then at any point at the edge of the wing the component of the free-stream velocity in a direction normal to the edge in the

‡ W. D. Hayes, 'Reversed flow theorems in supersonic aerodynamics', *Proc. 7th Int. Congr. Appl. Mech.* 1948, vol. 2, part 2, pp. 412–24.

§ Th. von Kármán, 'Supersonic aerodynamics—principles and applications', *J. Aero. Sci.* vol. 14 (1947), pp. 373–409.

‖ F. Ursell and G. N. Ward, 'On some general theorems in the linearised theory of compressible flow', *Quart. J. Mech. Appl. Math.* vol. 3 (1950), pp. 326–48.

¶ M. A. Heaslet and J. R. Spreiter, 'Reciprocity relations in aerodynamics', *Tech. Notes Nat. Adv. Comm. Aero., Wash.*, no. 2700 (1952).

x, y-plane is greater than the velocity of sound. The flow is then called 'definitely supersonic' or simply 'supersonic', although the latter terminology can be misleading when used out of context.

Now assume that we can deflect the lower surface locally in the neighbourhood of a point A while leaving the upper surface in its original position. As a result, the flow will be disturbed in the neighbourhood of A, and this disturbance will be propagated within the after-cone of A below the wing as far as the trailing edge (Fig. 4·8, 1a). Let B and C be two points at which the after-cone of A intersects the trailing edge. On reaching the trailing edge the disturbance will be propagated farther downstream within the envelope of the after-cones of the segment BC of the trailing edge. It follows that the deflexion of part or whole of the lower surface does not in any way affect the flow over the upper surface. In other words, the flow and pressure distribution over the upper surface of the wing depends only on the shape of the upper surface, and similarly, the flow and pressure distribution over the lower surface of the wing depends only on the shape of the lower surface. This is the 'principle of independence'.

Thus, given a wing at incidence which satisfies (4·8, 1), let $z = f(x, y)$ be the shape of the upper surface, and $s(x, y) = \partial f / \partial x$ the local slope. Irrespective of the *actual* shape of the lower surface, we may then imagine that the slope of the lower surface is given by $-s(x, y)$, so that there results a hypothetical wing which is symmetrical with respect to the x, y-plane. It follows that the induced velocity potential at the upper surface of the hypothetical wing is given by (4·7, 8), and, by the principle of independence, this is also the induced velocity potential at the upper surface of the given wing. Thus

$$\Phi(x, y, z) = \frac{U}{\pi} \int_{S'} \frac{s(x_0, y_0) \, dx_0 \, dy_0}{\sqrt{\{(x - x_0)^2 - \beta^2 ((y - y_0)^2 + z^2)\}}}, \qquad (4·8, 2)$$

where S' is the part of the wing which is enclosed within the fore-cone of (x, y, z). (4·8, 2) holds at the upper surface of the wing and also for all other points for which $z > 0$ and which are outside the after-cone of the trailing edge. If the wing is very thin, so that the slope of the actual wing at the lower surface equals the slope of the upper surface, then

$$\Phi(x, y, -z) = -\Phi(x, y, z) \qquad (4·8, 3)$$

by the usual conditions of symmetry.

Having written down (4·8, 2) we may verify directly, and without recourse to the earlier argument, that $\Phi(x, y, z)$ satisfies the correct boundary conditions at the upper surface of the wing, and, moreover, that Φ vanishes upstream of the after-cones emanating from the leading edge. However, it would appear that the formulation of the principle of independence makes the method more physically intuitive.

As an important example of the definitely supersonic case we may mention the flow round a flat unyawed Delta wing at incidence α, whose leading edges are outside the after-cone of the apex. This condition is expressed by the relation

$$\left|\frac{dx}{dy}\right| = \cot \gamma < \beta = \cot \mu, \qquad (4·8, 4)$$

so that (4·8, 1) is satisfied by the leading edges. It is clearly satisfied also by the trailing edge which is parallel to the y-axis, by assumption (Fig. 4·8, 2). We now have

$$s(x, y) = \tan \alpha \doteqdot \alpha, \qquad (4·8, 5)$$

from which we may derive the induced velocity potential by carrying out the integration in (4·8, 2). For the regions outside the after-cone of the origin (OAA' and $OB'B$ in Fig. 4·8, 2) the velocity potential and hence the

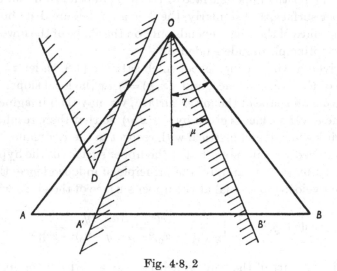

Fig. 4·8, 2

pressure distribution are exactly the same as for a flat two-dimensional wing which is yawed by an angle $\frac{1}{2}\pi - \gamma$ (see § 4·6). Indeed, the field of flow at a point D on OAA' depends only on the part of the wing which is within the fore-cone of D, and the conditions within the fore-cone remain unaffected if we modify the leading edge by producing the straight line OA indefinitely beyond O. By the analysis of § 4·6 (or by direct integration) we may then conclude that the pressure is constant over OAA' and $OB'B$.

Having determined the velocity potential for a case of definitely supersonic flow, we may then determine the resultant force on the wing, as given by (4·7, 71). Thus we obtain for the lift on a thin wing at incidence (see (4·8, 3))

$$L = -2\rho_0 U \int_S \frac{\partial \Phi}{\partial x} \, dx \, dy, \qquad (4·8, 6)$$

where the integral is taken over the upper surface of the wing only. As usual, the formula depends on the assumption that the angle between a directed surface element of the wing and the z-axis is small, and hence that the cosine of the angle may be set equal to 1. Using (4·8, 2) we may rewrite the expression for the lift as

$$L = \frac{2}{\pi} \rho_0 U^2 \int_T^* s(x_0, y_0) \frac{(x - x_0)\, dx\, dy\, dx_0\, dy_0}{[(x - x_0)^2 - \beta^2 (y - y_0)^2]^{\frac{3}{2}}}, \qquad (4·8, 7)$$

where T is defined as in (4·7, 81). The corresponding drag is still given by the formulae (4·7, 72) and (4·7, 81). In particular, if the wing is flat

$$s(x, y) = \tan \alpha \doteqdot \alpha,$$

then $$D = L\alpha. \qquad (4·8, 8)$$

This is evident also by virtue of the fact that the resultant force $\mathbf{R}$ is normal to the wing. The drag force (4·8, 8) is associated with the lift and, as in subsonic flow, it may be called the induced drag. The energy delivered by the wing to the air in working against this force is expended partly on the creation of a trailing vortex field, as in subsonic flow, and partly on the creation of compressibility waves (in the linearized approximation—acoustic waves), as in supersonic flow round a wing at zero incidence.

For a given flat wing at incidence α under definitely supersonic conditions, let us reverse the direction of the main flow while keeping its magnitude constant. At the same time, we change the incidence of the wing from α to $-\alpha$. Using the notation of (4·7, 84) for the drag, and a similar notation for the lift, we then have

$$D'_W = L'\alpha,$$

and at the same time $$D'_W = D_W. \qquad (4·8, 9)$$

Combining these equations with (4·8, 8) we obtain

$$L' = L, \qquad (4·8, 10)$$

which expresses the flow reversal theorem for the lift. The same relation may be derived directly from (4·8, 7).

Carrying out the calculation of lift and drag for the definitely supersonic case of a Delta wing, which was considered above, we find that the lift and wave-drag coefficients are given by precisely the same formulae as for two-dimensional conditions (4·6, 18)‡

$$C_L = \frac{4\alpha}{\beta}, \qquad C_D = \frac{4\alpha^2}{\beta}. \qquad (4·8, 11)$$

It will now be shown that these results can also be derived without any calculation by the use of (4·8, 9) and (4·8, 10). More generally, consider any

‡ See A. E. Puckett, 'Supersonic wave drag of thin airfoils', *J. Aero. Sci.* vol. 13 (1946), pp. 475–84; or A. Robinson, 'The wave drag of diamond-shaped aerofoils at zero incidence', *Rep. Memor. Aero. Res. Comm., Lond.*, no. 2394 (1946).

flat wing at incidence α, under definitely supersonic conditions, whose trailing edge is straight and parallel to the y-axis. To show that the formulae of (4·8, 11) hold for this type of wing, it is sufficient to establish the same formulae for the reversed flow case. Thus, we consider a flat wing S at incidence α under supersonic conditions, whose *leading* edge is straight and parallel to the y-axis. Then the induced velocity at a point A of S can be calculated by means of (4·8, 2). Thus, it depends only on the conditions within the pre-cone of A. But these are exactly the same as if the wing extended indefinitely to port and starboard. It follows that the pressure difference at A is given by (4·6, 14), where in the present case

$$\frac{df}{dx} = \tan\alpha \doteq \alpha.$$

Thus
$$p_u - p_l = \tfrac{1}{2}\rho_0 U^2 \frac{4\alpha}{\beta}. \tag{4·8, 12}$$

Multiplying this formula by the area S of the wing, we obtain the lift, and multiplying further by α, we obtain the wave drag, so

$$L = \tfrac{1}{2}\rho_0 U^2 S \frac{4\alpha}{\beta}, \quad D = \tfrac{1}{2}\rho_0 U^2 S \frac{4\alpha^2}{\beta}. \tag{4·8, 13}$$

The formulae of (4·8, 11) follow immediately from (4·8, 13). However, it should be noted that while equations (4·8, 11) and (4·8, 13) apply also to the reversed flow, the detailed pressure distribution as given by (4·8, 12) does not hold after reversal. In particular, while the pressure is constant over the areas OAA', OBB' in Fig. 4·8, 2, it is variable over $OA'B'$, i.e. within the after-cone of the apex.

Returning to the particular case of the Delta wing under definitely supersonic conditions, we shall show next that the pressure on the wing is constant along radial lines through the apex. Since the situation at the wing is not affected by conditions after the trailing edge, we may simplify matters by assuming that the wing extends downstream to infinity. By (4·8, 2) and (4·8, 5) the x-component of the induced velocity is given by

$$u'(x, y, z) = -\frac{\partial\Phi}{\partial x} = \frac{U\alpha}{\pi}* \int_{S'} \frac{(x - x_0)\, dx_0\, dy_0}{[(x - x_0)^2 - \beta^2((y - y_0)^2 + z^2)]^{\frac{3}{2}}} \tag{4·8, 14}$$

for points above the x, y-plane, i.e. for $z > 0$. Let (x_1, y_1, z_1), (x_2, y_2, z_2) $(z_1 > 0,\ z_2 > 0)$ be two points on the same radial line through the origin, so that

$$x_2 = mx_1, \quad y_2 = my_1, \quad z_2 = mz_1. \tag{4·8, 15}$$

The x-components of the induced velocity at these points are then given by

$$u_1' = u'(x_1, y_1, z_1) = \frac{U\alpha}{\pi}* \int_{S_1'} \frac{(x_1 - x_{01})\, dx_{01}\, dy_{01}}{[(x_1 - x_{01})^2 - \beta^2((y_1 - y_{01})^2 + z_1^2)]^{\frac{3}{2}}}$$

and
$$u_2' = u'(x_2, y_2, z_2) = \frac{U\alpha}{\pi}* \int_{S_2'} \frac{(x_2 - x_{02})\, dx_{02}\, dy_{02}}{[(x_2 - x_{02})^2 - \beta^2((y_2 - y_{02})^2 + z_2^2)]^{\frac{3}{2}}},$$

$$\tag{4·8, 16}$$

where the domains of integration S_1' and S_2' are transformed into each other by the substitution

$$x_{02} = mx_{01}, \quad y_{02} = my_{01}. \tag{4·8, 17}$$

Hence

$$*\!\!\int_{S_2'} \frac{(x_2 - x_{02})\, dx_{02} dy_{02}}{[(x_2 - x_{02})^2 - \beta^2((y_2 - y_{02})^2 + z_2^2)]^{\frac{3}{2}}}$$

$$= *\!\!\int_{S_1'} \frac{(mx_1 - mx_{01})\, m^2 dx_{01} dy_{01}}{[(mx_1 - mx_{01})^2 - \beta^2((my_1 - my_{01})^2 + m^2 z_1^2)]^{\frac{3}{2}}}$$

$$= *\!\!\int_{S_1'} \frac{(x_1 - x_{01})\, dx_{01} dy_{01}}{[(x_1 - x_{01})^2 - \beta^2((y_1 - y_{01})^2 + z_1^2)]^{\frac{3}{2}}}$$

or

$$u'(x_2, y_2, z_2) = u'(x_1, y_1, z_1),$$

and similarly

$$v(x_2, y_2, z_2) = v(x_1, y_1, z_1),$$

$$w(x_2, y_2, z_2) = w(x_1, y_1, z_1).$$

It then follows from Bernoulli's equation that the pressure also is the same at the two points, as asserted.

The same conclusion may be reached by means of dimensional arguments. The quantities which determine the induced velocity at a point P on a radial line l through the apex are, the free-stream velocity U, the velocity of sound a, the free-stream density ρ_0, the free-stream pressure p_0, the distance d of P from the origin, and a number of non-dimensional quantities, viz. the ratio of specific heats (see equation (4·1, 2)), the semi-apex angle of the wing and the incidence, and two angles which determine the direction of l. Now p_0 may be omitted from this list, since it is determined completely by ρ_0, and ρ_0 may then be discarded since it is the only remaining variable which involves mass. The other variables are all either non-dimensional, or they possess the dimensions of a velocity, with the exception of the distance d. It follows that the velocity components at P are of the form

$$u = Uf_1, \quad v = Uf_2, \quad w = Uf_3, \tag{4·8, 18}$$

where the functions f_1, f_2, f_3 depend on the non-dimensional variables enumerated above, and on the Mach number $M = U/a$, but are independent of d. This shows that the velocity (u, v, w) is constant along l, as asserted. We observe that this argument applies with equal force to a Delta wing whose leading edges are within the after-cone of the apex, $\beta \tan \gamma < 1$.

Since the pressure is also constant along radial lines through the origin, it follows that for small δx the total normal force on a segment of the wing between two lines x and $x + \delta x$, is proportional to x. If we denote this force, for the purpose of the present argument only, by $\Lambda(x)$, then $\Lambda(x) = \Lambda_0 x$, where Λ_0 is some constant. With this notation, the total lift is

$$L = \int_0^c \Lambda(x)\, dx = \Lambda_0 \frac{c^2}{2},$$

and the moment about the y-axis is

$$M = \int_0^c \Lambda(x)\, x\, dx = \Lambda_0 \frac{c^3}{3}.$$

It follows that the position of the centre of pressure in the chordwise direction is given by

$$x_0 = \frac{M}{L} = \tfrac{2}{3}c.$$

But

$$L = \tfrac{1}{2}\rho_0 U^2 S C_L = \tfrac{1}{2}\rho_0 U^2 S \frac{4\alpha}{\beta},$$

and so

$$M = Lx_0 = \tfrac{1}{2}\rho_0 U^2 S c \frac{8\alpha}{3\beta}.$$

Hence

$$C_M = \frac{8\alpha}{3\beta},$$

where it is understood that the pitching moment coefficient C_M has been defined in terms of the maximum chord, c, i.e. $C_M = M/\tfrac{1}{2}\rho_0 U^2 S c$.

It was shown by Evvard‡ that the source distribution method can be extended even to cases in which the condition of definitely supersonic flow (4·8, 1) is not satisfied everywhere along the edges. Evvard's ideas were developed further by Ward.§

For a given wing and Mach number, we shall call the (leading or trailing) edges of the wing supersonic at points where

$$\left| \frac{dx}{dy} \right| < \beta = \cot\mu, \tag{4·8, 19}$$

and subsonic if, on the contrary,

$$\left| \frac{dx}{dy} \right| > \beta = \cot\mu. \tag{4·8, 20}$$

To introduce the class of wings with which we shall be concerned, we consider the flow past a thin elliptic wing at incidence, such that the direction of the free stream is parallel to one of the axes of the projection of the wing on the x, y-plane. The starboard half of such a wing is shown in Fig. 4·8, 3.

Let A be the point of intersection at the centre line—in the present case the minor axis of the ellipse—with the leading edge. Then the leading edge is supersonic in the neighbourhood of A, as far as the point B which is given by the condition that the slope dx/dy of the leading edge at B is equal to β. The two Mach lines $OB : x - \beta y = $ constant, and $BB' : x + \beta y = $ constant, are shown on the diagram. The former is tangent to the leading edge at B.

‡ J. C. Evvard, 'Distribution of wave drag and lift in the vicinity of wing tips at supersonic speeds', *Tech. Notes Nat. Adv. Comm. Aero.*, Wash., no. 1382 (1947).

§ G. N. Ward, 'Supersonic flow past thin wings. I. General theory', *Quart J. Mech. Appl. Math.* vol. 2 (1949), pp. 136–52.

Moving farther along the edge in a clockwise direction, we come to the tip C, where $dx/dy = \infty$ (i.e. the tangent to the edge at C is parallel to the x-axis) and then to the point D where the Mach line $DD' : x + \beta y = \text{constant}$ is tangent to the edge. Then the edge is subsonic from B to D, and again supersonic from B to A', where A' is the point of intersection of the centre line with the trailing edge. Let the Mach line $x + \beta y = \text{constant}$ through C meet the trailing edge in a point C'. We shall assume that the point of intersection of BB' with the corresponding Mach line off the port wing is situated downstream of the trailing edge. In that case, the tips are said to be independent of each other.

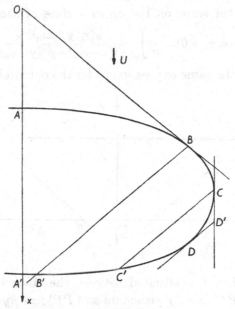

Fig. 4·8, 3

More generally, the method to be described will apply to any thin wing with convex plan-form whose leading edge is at least partly supersonic and whose tips are independent of each other. In the case of a rectangular wing (Fig. 4·8, 4) the assumption of independent wing tips implies for the aspect ratio $\mathcal{R}$,

$$\mathcal{R} > 2/\beta. \qquad (4·8, 21)$$

Reverting to the general case, let $P(x, y, +0)$ be any point on the upper surface of the wing. By (4·7, 6) and (4·7, 8) the induced velocity potential at P can be expressed by the formula

$$\Phi(x, y, +0) = -\frac{1}{\pi} \int_{S'} \left[\frac{\partial}{\partial z_0} \Phi(x_0, y_0, z_0) \right]_{z_0 = +0} \frac{dx_0 \, dy_0}{\sqrt{\{(x - x_0)^2 - \beta^2 (y - y_0)^2\}}},$$

$$(4·8, 22)$$

provided we take the double integral on the right-hand side over the region S' of the x, y-plane within the fore-cone of P in which $\left[\dfrac{\partial}{\partial z_0} \Phi(x_0, y_0, z_0)\right]_{z_0 = +0}$ is different from zero. Assume first that P is ahead of the Mach line BB' and also ahead of the corresponding Mach line from the port wing. This means, in the notation of Fig. 4·3, 3, that if P is on the starboard half of the wing then it belongs to the curvilinear quadrilateral $ABB'A'A$. Then $\partial \Phi / \partial z_0$ vanishes ahead of the leading edge, and so S' consists of a portion of the wing S. But on S,

$$\left[\frac{\partial}{\partial z_0} \Phi(x_0, y_0, z_0)\right]_{z_0 = +0} = -Us(x_0, y_0), \qquad (4\cdot8, 23)$$

where $s(x_0, y_0)$ is the slope on the upper surface of the wing. Hence, in that case

$$\Phi(x, y, +0) = \frac{U}{\pi} \int_{S'} \frac{s(x_0, y_0)\, dx_0\, dy_0}{\sqrt{\{(x-x_0)^2 - \beta^2 (y-y_0)^2\}}}, \qquad (4\cdot8, 24)$$

i.e. Φ is given by the same expression as for the definitely supersonic case.

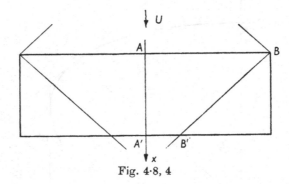

Fig. 4·8, 4

Assume next that P is situated between the Mach lines BB' and CC' (Fig. 4·8, 5). Let $PP' : x - \beta y = $ constant and $PP'' : x + \beta y = $ constant, be the two Mach lines through P, where P', P'' are the points of intersection of these lines with the leading edge, and let P''' be the point of intersection of the Mach lines PP'' and OB. Then $\partial \Phi / \partial z_0$ vanishes ahead of OBP''', since that region is unaffected by the presence of the wing, but it will in general be different from zero in the triangle $BP'''P''B$. It follows that S' now consists of the regions S_1 and S_2 which are bounded by the two curved triangles $P'BP''PP'$ and $BP'''P''B$, respectively. On S_1, $\partial \Phi / \partial z_0$ is given by (4·8, 23), while on S_2 we define $t(x_0, y_0)$ by

$$\left[\frac{\partial}{\partial z_0} \Phi(x_0, y_0, z_0)\right]_{z_0 = 0} = -Ut(x_0, y_0). \qquad (4\cdot8, 25)$$

Then (4·8, 22) becomes

$$\Phi(x, y, +0) = \frac{U}{\pi} \left\{ \int_{S_1} \frac{s(x_0, y_0)\, dx_0\, dy_0}{\sqrt{\{(x-x_0)^2 - \beta^2 (y-y_0)^2\}}} + \int_{S_2} \frac{t(x_0, y_0)\, dx_0\, dy_0}{\sqrt{\{(x-x_0)^2 - \beta^2 (y-y_0)^2\}}} \right\}.$$
$$(4\cdot8, 26)$$

However, while $s(x_0, y_0)$ is given by the geometry of the wing, $t(x_0, y_0)$ is a priori unknown, and so (4·8, 26) cannot be evaluated directly. The difficulty can be overcome by means of a remarkable principle due to Evvard.

Let Q be any point within the curved triangle $CBC''C$, where C'' is the point of intersection of two Mach lines OB and CC'. Through Q draw the two Mach lines $QQ'Q'' : x - \beta y = $ constant and $QQ''' : x + \beta y = $ constant, where Q', Q'' are the points of intersection of the first Mach line with the leading

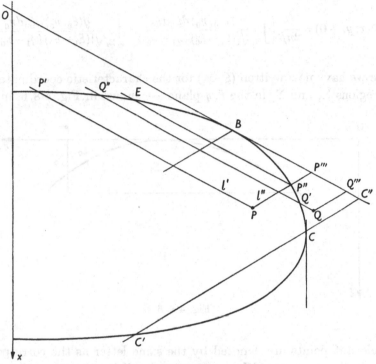

Fig. 4·8, 5

edge, and Q''' is the point of intersection of the second Mach line through Q with the Mach line OB. Then (4·8, 26) still applies at Q, if by S_1 we now mean the region bounded by the straight segment $Q'Q''$ and by the arc $Q''BQ'$, while S_2 is the curved quadrilateral $QQ'BQ'''Q$. But Φ is continuous everywhere except across the wing and the wake, and this, together with the relation $\Phi(x, y, -z) = -\Phi(x, y, z)$, implies that $\Phi(x, y, 0) = 0$ at Q. Hence, at Q

$$0 = \int_{S_1} \frac{s(x_0, y_0)\, dx_0\, dy_0}{\sqrt{\{(x-x_0)^2 - \beta^2(y-y_0)^2\}}} + \int_{S_2} \frac{t(x_0, y_0)\, dx_0\, dy_0}{\sqrt{\{(x-x_0)^2 - \beta^2(y-y_0)^2\}}}.$$

$$(4\cdot8, 27)$$

We introduce the 'characteristic coordinates' ξ, η by means of the formulae

$$\xi = x - \beta y, \quad \eta = x + \beta y, \qquad (4\cdot8, 28)$$

and we set further $\quad \xi_0 = x_0 - \beta y_0, \quad \eta_0 = x_0 + \beta y_0, \qquad (4\cdot8, 29)$

and
$$s(x_0, y_0) = f(\xi_0, \eta_0), \quad t(x_0, y_0) = g(\xi_0, \eta_0). \tag{4·8, 30}$$

It will now be seen that the Mach lines in the x, y-plane are then given by the two families $\xi = $ constant and $\eta = $ constant.

But
$$\frac{D(x_0, y_0)}{D(\xi_0, \eta_0)} = \frac{1}{2\beta}, \tag{4·8, 31}$$

and so we may replace (4·8, 26) by

$$\Phi(x, y, +0) = \frac{U}{2\pi\beta} \left\{ \int_{\Sigma_1} \frac{f(\xi_0, \eta_0) \, d\xi_0 d\eta_0}{\sqrt{\{(\xi_1 - \xi_0)(\eta_1 - \eta_0)\}}} + \int_{\Sigma_2} \frac{g(\xi_0, \eta_0) \, d\xi_0 d\eta_0}{\sqrt{\{(\xi_1 - \xi_0)(\eta_1 - \eta_0)\}}} \right\}, \tag{4·8, 32}$$

where we have now written (ξ_1, η_1) for the characteristic coordinates of P. The regions Σ_1 and Σ_2 in the ξ, η-plane are shown in Fig. 4·8, 6, in which

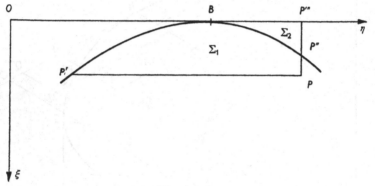

Fig. 4·8, 6

the relevant points are denoted by the same letter as the corresponding points in the x, y-plane. We assume for simplicity that the origin in the x, y-plane is at a point O on the Mach line OB. Then Σ_1 and Σ_2 are bounded by the curved triangles $PP'BP''P$ and $P''BP'''P''$ respectively. The co-ordinates of P are (ξ_1, η_1) and, in agreement with this, we may use the following notation for the coordinates of the remaining points:

$$B : (0, \eta_2), \quad P' : (\xi_1, \eta'), \quad P'' : (\xi'', \eta_1), \quad P''' : (0, \eta_1). \tag{4·8, 33}$$

Moreover, we may describe the arcs $P'B$ and BP'' by equations

$$\eta = F_1(\xi) \quad (0 \leqslant \xi \leqslant \xi_1), \tag{4·8, 34}$$

and
$$\eta = F_2(\xi) \quad (0 \leqslant \xi \leqslant \xi''), \tag{4·8, 35}$$

respectively, or by the single equation

$$\xi = G(\eta) \quad (\eta' \leqslant \eta \leqslant \eta_1). \tag{4·8, 36}$$

Then (4·8, 32) may be rewritten in the form

$$\Phi(x, y, +0) = \frac{U}{2\pi\beta} \left[\int_0^{\xi''} \frac{d\xi_0}{\sqrt{(\xi_1 - \xi_0)}} \left\{ \int_{F_1(\xi_0)}^{F_2(\xi_0)} \frac{f(\xi_0, \eta_0)}{\sqrt{(\eta_1 - \eta_0)}} d\eta_0 \right. \right.$$

$$\left. \left. + \int_{F_2(\xi_0)}^{\eta_1} \frac{g(\xi_0, \eta_0)}{\sqrt{(\eta_1 - \eta_0)}} d\eta_0 \right\} + \int_{\xi''}^{\xi_1} \frac{d\xi_0}{\sqrt{(\xi_1 - \xi_0)}} \int_{F_1(\xi_0)}^{\eta_1} \frac{f(\xi_0, \eta_0)}{\sqrt{(\eta_1 - \eta_0)}} d\eta_0 \right].$$

$$(4\cdot8, 37)$$

Now let Q be any point on the straight segment $P''P'''$, and let ξ_2 be the ξ-coordinate of Q. The η-coordinate of Q is η_1. Substituting characteristic coordinates in (4·8, 27) we obtain

$$0 = \int_0^{\xi_2} \frac{d\xi_0}{\sqrt{(\xi_2 - \xi_0)}} \left\{ \int_{F_1(\xi_2)}^{F_2(\xi_2)} \frac{f(\xi_0, \eta_0)}{\sqrt{(\eta_1 - \eta_0)}} d\eta_0 + \int_{F_2(\xi_0)}^{\eta_1} \frac{g(\xi_0, \eta_0)}{\sqrt{(\eta_1 - \eta_0)}} d\eta_0 \right\}. \quad (4\cdot8, 38)$$

This equation holds for all ξ_2 between 0 and ξ. We conclude that

$$\int_{F_1(\xi_0)}^{F_2(\xi_0)} \frac{f(\xi_0, \eta_0)}{\sqrt{\eta_1 - \eta_0}} d\eta_0 + \int_{F_2(\xi_0)}^{\eta_1} \frac{g(\xi_0, \eta_0)}{\sqrt{\eta_1 - \eta_0}} d\eta_0$$

vanishes. But this is precisely the expression in the curly brackets in (4·8, 37), and so (4·8, 37) reduces to

$$\Phi(x, y, +0) = \frac{U}{2\pi\beta} \int_{\xi''}^{\xi_1} \frac{d\xi_0}{\sqrt{(\xi_1 - \xi_0)}} \int_{F_1(\xi_0)}^{\eta_1} \frac{f(\xi_0, \eta_0)}{\sqrt{(\eta_1 - \eta_0)}} d\eta_0. \quad (4\cdot8, 39)$$

Transforming back to the physical plane, we obtain for the induced velocity potential at P,

$$\Phi(x, y, +0) = \frac{U}{\pi} \int_{S^*} \frac{s(x_0, y_0) \, dx_0 dy_0}{\sqrt{\{(x - x_0)^2 - \beta^2(y - y_0)^2\}}}. \quad (4\cdot8, 40)$$

In this formula, the domain of integration S^* is determined by the following procedure (Fig. 4·8, 5). Through P draw the two Mach lines l', l'', meeting the leading edge in P', P''. Let E be the second point of intersection of the other Mach line through P''. Then S^* is the region enclosed by the curved quadrilateral $PP'EP''P$. This rule for the calculation of the induced velocity potential embodies Evvard's principle.

To complete the calculation of the velocity potential at the wing, we have to determine Φ in the area bounded by the straight segment CC' and by the arc CDC' of the trailing edge (see Fig. 4·8, 3). We observe that in that region, and in the corresponding region on the port side, the velocity distribution is affected by conditions at the trailing edge and in the wake of the wing, whereas elsewhere on the wing the velocity is independent of these conditions. The general arguments of Chapters 1 and 3 lead us to conclude that the Joukowski condition as formulated in § 1·15 is satisfied along the arc CDC'. Moreover, we may assume again that the wake is confined to the region downstream of the trailing edge which is bounded laterally by the two straight lines $y = $ constant emanating from the tips. So long as the incidence remains finite everywhere on the wing as implicit in the general theory, the Joukowski condition is in fact satisfied automatically along an arc of the trailing edge whose end-points are C' and the corresponding point on port. Thus, along that arc, the Joukowski condition need not be introduced as a special hypothesis.

Put
$$\left[\frac{\partial}{\partial z_0}\Phi(x_0, y_0, z_0)\right]_{z_0=0} = -U\tau(x_0, y_0) \qquad (4·8, 41)$$

in the wake of the wing. Let P be a point in the region $CDC'C$, and draw the two Mach lines PP', PP'', where P' and P'' are the points of intersection of these lines with the edge (Fig. 4·8, 7). Let PP'' meet CC'' at E', where CC'' is the straight line $y=\text{constant}$ through the tip C. Finally, assume that the second Mach line through E' meets the leading edge at E and the trailing edge at E''.

We now apply Evvard's principle to the combined regions of wing and wake. This yields for the induced velocity potential at P

$$\Phi(x, y, +0) = \frac{U}{\pi}\left\{\int_{S^*}\frac{s(x_0, y_0)\,dx_0\,dy_0}{\sqrt{\{(x-x_0)^2-\beta^2(y-y_0)^2\}}} + \int_W\frac{\tau(x_0, y_0)\,dx_0\,dy_0}{\sqrt{\{(x-x_0)^2-\beta^2(y-y_0)^2\}}}\right\},$$
$$(4·8, 42)$$

where S^* is the region bounded by $PP'EE''P''P$, and W^* is bounded by $P''E'CE''P''$. However, we are not yet in a position to evaluate this formula since the function $\tau(x_0, y_0)$ is not known *a priori*.

It remains to evaluate $\tau(x_0, y_0)$ within the curved triangle $CDD'C$. For this purpose, we shall require the solution of 'Abel's integral equation'
$$\int_0^x\frac{h(\xi)\,d\xi}{\sqrt{(x-\xi)}} = k(x). \qquad (4·8, 43)$$

In this equation, $h(x)$ is the unknown function, while $k(x)$ is specified and is such that $k'(x)$ is continuous and $k(0)=0$.

The solution of this equation is‡
$$h(x) = \frac{1}{\pi}\frac{d}{dx}\int_0^x\frac{k(\xi)\,d\xi}{\sqrt{(x-\xi)}}. \qquad (4·8, 44)$$

Another form of the solution, which is obtained by integrating the right-hand side of (4·8, 44) by parts, is
$$h(x) = \frac{1}{\pi}\int_0^x\frac{k'(\xi)\,d\xi}{\sqrt{(x-\xi)}}. \qquad (4·8, 45)$$

The equation
$$\int_a^y\frac{h(\eta)\,d\eta}{\sqrt{(y-\eta)}} = k(y), \qquad (4·8, 46)$$

where $k(y)$ is continuous and satisfies the condition $k(a)=0$, is reduced to (4·8, 43) by the substitution
$$x=y-a, \quad \xi=\eta-a.$$

‡ E. T. Whittaker and G. N. Watson, *A Course in Modern Analysis* (4th ed., Cambridge University Press, 1927), p. 229.

This leads to the following equivalent formulae for the solution

$$h(y) = \frac{1}{\pi} \frac{d}{dy} \int_a^y \frac{k(\eta)d\eta}{\sqrt{(y-\eta)}}$$

(4·8, 47)

and

$$h(y) = \frac{1}{\pi} \int_a^y \frac{k'(\eta)\,d\eta}{\sqrt{(y-\eta)}}.$$

(4·8, 48)

It can readily be verified by substitution that this is indeed a solution of (4·8, 46).

Let Q be a point within CDD' (see Fig. 4·8, 7 for the notation). Applying Evvard's principle to the combined regions of wing and wake we obtain

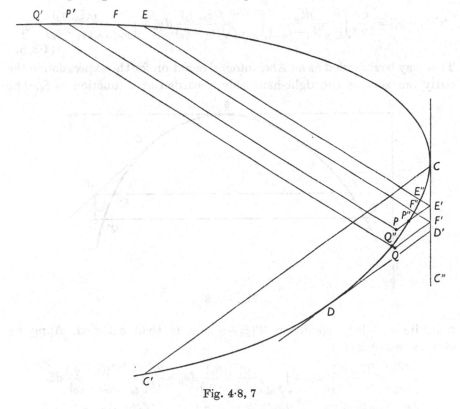

Fig. 4·8, 7

for the induced velocity potential at Q equation (4·8, 42), where S^* is now the region bounded by $Q''Q'FF''Q''$ and W^* is bounded by $QQ''F''F'Q$.

We introduce characteristic coordinates by (4·8, 28) and (4·8, 29), and put $s(x_0, y_0) = f(\xi_0, \eta_0)$ on the wing and $\tau(x_0, y_0) = g(\xi_0, \eta_0)$ on the wake. The points of the ξ, η-plane, which are shown in Fig. 4·8, 8, are denoted by the same letters as the corresponding points in the x, y-plane. Let (ξ_1, η_1) and (ξ_2, η_2) be the coordinates of Q and F in the ξ, η-plane, and let

$$\eta = F_1(\xi) \quad \text{and} \quad \eta = F_2(\xi)$$

(4·8, 49)

be the equations of the arcs which correspond to the leading and trailing edges respectively.

We note that the straight lines $y=$ constant correspond to lines $\xi-\eta=$ constant in the ξ, η-plane, and we may write

$$\Phi(x, y, +0) = \Psi(\xi - \eta) \tag{4·8, 50}$$

in the wake. Since $\Phi(x, y, 0) = 0$ at points of the x, y-plane which do not belong to the wing or the wake, it follows from considerations of continuity that $\Phi(x, y, 0) = 0$ also along the line $y =$ constant through C, and in particular at the point $F'(\xi_2, \eta_1)$. Thus

$$\Psi(\xi_2 - \eta_1) = 0 \tag{4·8, 51}$$

$(4·8, 42)$ now becomes

$$\Psi(\xi_1 - \eta_1) = \frac{U}{2\pi\beta} \int_{\xi_2}^{\xi_1} \frac{d\xi_0}{\sqrt{(\xi_1 - \xi_0)}} \left\{ \int_{F_1(\xi_0)}^{F_2(\xi_0)} \frac{f(\xi_0, \eta_0)}{\sqrt{(\eta_1 - \eta_0)}} d\eta_0 + \int_{F_2(\xi_0)}^{\eta_1} \frac{g(\xi_0, \eta_0)}{\sqrt{(\eta_1 - \eta_0)}} d\eta_0 \right\}. \tag{4·8, 52}$$

This may be regarded as an Abel integral equation for the expression in the curly brackets on the right-hand side, regarded as a function of ξ_1. The

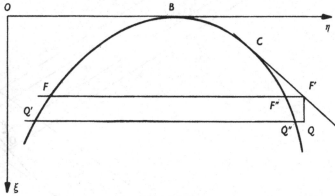

Fig. 4·8, 8

requisite ancillary condition $\Psi(\xi_2 - \eta_1) = 0$ is then satisfied. Applying $(4·8, 48)$ we obtain

$$\int_{F_1(\xi_1)}^{F_2(\xi_1)} \frac{f(\xi_1, \eta_0)}{\sqrt{(\eta_1 - \eta_0)}} d\eta_0 + \int_{F_2(\xi_1)}^{\eta_1} \frac{g(\xi_1, \eta_0)}{\sqrt{(\eta_1 - \eta_0)}} d\eta_0 = \frac{2\beta}{U} \int_{\xi_2}^{\xi_1} \frac{\Psi'(\xi_0 - \eta_1)}{\sqrt{(\xi_1 - \xi_0)}} d\xi_0$$

or $$\int_{F_2(\xi_1)}^{\eta_1} \frac{g(\xi_1, \eta_0)}{\sqrt{(\eta_1 - \eta_0)}} d\eta_0 = \frac{2\beta}{U} \int_{\xi_2}^{\xi_1} \frac{\Psi'(\xi_0 - \eta_1)}{\sqrt{(\xi_1 - \xi_0)}} d\xi_0 - \int_{F_1(\xi_1)}^{F_2(\xi_1)} \frac{f(\xi_1, \eta_0)}{\sqrt{(\eta_1 - \eta_0)}} d\eta_0. \tag{4·8, 53}$$

This in turn is an Abel integral equation for $g(\xi_1, \eta_1)$ when regarded as a function of η_1, for given ξ_1. Assuming that $g(\xi_1, \eta_1)$ is bounded in the neighbourhood of $\eta_1 = F_2(\xi_1)$, $|g(\xi_1, \eta_1)| < M$, say, we have

$$\left| \int_{F_2(\xi_1)}^{\eta_1} \frac{g(\xi_1, \eta_0)}{\sqrt{(\eta_1 - \eta_0)}} d\eta_0 \right| \leqslant M \int_{F_2(\xi_1)}^{\eta_1} \frac{d\eta_0}{\sqrt{(\eta_1 - \eta_0)}} = 2M \sqrt{\{\eta_1 - F_2(\xi_1)\}},$$

and so $$\lim_{\eta_1 \to F_2(\xi_1)} \int_{F_2(\xi_1)}^{\eta_1} \frac{g(\xi_1, \eta_0)}{\sqrt{(\eta_1 - \eta_0)}} d\eta_0 = 0. \tag{4·8, 54}$$

This shows that the right-hand side of (4·8, 53) must vanish for $\eta_1 = F_2(\xi_1)$, which is a required ancillary condition. Using (4·8, 47) we then obtain

$$g(\xi_1, \eta_1) = \frac{1}{\pi} \left\{ \frac{2\beta}{U} \frac{d}{d\eta_1} \int_{F_2(\xi_1)}^{\eta_1} \frac{d\eta_0}{\sqrt{(\eta_1 - \eta_0)}} \int_{\xi_2}^{\xi_1} \frac{\Psi'(\xi_0 - \eta_0)\, d\xi_0}{\sqrt{(\xi_1 - \xi_0)}} \right.$$
$$\left. - \frac{d}{d\eta_1} \int_{F_2(\xi_1)}^{\eta_1} \frac{d\eta_0}{\sqrt{(\eta_1 - \eta_0)}} \int_{F_1(\xi_1)}^{F_2(\xi_1)} \frac{f(\xi_1, \eta')}{\sqrt{(\eta_0 - \eta')}}\, d\eta' \right\}. \quad (4·8, 55)$$

We may carry out the integration with respect to η_0 in the second repeated integral on the right-hand side. Then

$$g(\xi_1, \eta_1) = \frac{1}{\pi} \left\{ \frac{2\beta}{U} \frac{d}{d\eta_1} \int_{F_2(\xi_1)}^{\eta_1} \frac{d\eta_0}{\sqrt{(\eta_1 - \eta_0)}} \int_{\xi_2}^{\xi_1} \frac{\Psi'(\xi_0 - \eta_0)}{\sqrt{(\xi_1 - \xi_0)}}\, d\xi_0 \right.$$
$$\left. - \frac{1}{\sqrt{\{\eta_1 - F_2(\xi_1)\}}} \int_{F_1(\xi_1)}^{F_2(\xi_1)} \frac{f(\xi_1, \eta')\sqrt{\{F_2(\xi_1) - \eta'\}}}{\eta_1 - \eta'}\, d\eta' \right\}. \quad (4·8, 56)$$

Let the function $\Omega(\xi)$ be defined by

$$\Omega(\xi) = \frac{2\beta}{U} \int_{\gamma_1}^{\xi} \frac{\Psi'(\xi_0)\, d\xi_0}{\sqrt{(\xi - \xi_0)}}, \quad (4·8, 57)$$

where $\gamma_1 = \xi_2 - \eta_1$. Then (4·8, 56) becomes

$$g(\xi_1, \eta_1) = \frac{1}{\pi} \left\{ \frac{d}{d\eta_1} \int_{F_2(\xi_1)}^{\eta_1} \frac{\Omega(\xi_1 - \eta_0)}{\sqrt{(\eta_1 - \eta_0)}}\, d\eta_0 \right.$$
$$\left. - \frac{1}{\sqrt{\{\eta_1 - F_2(\xi_1)\}}} \int_{F_1(\xi_1)}^{F_2(\xi_1)} \frac{f(\xi_1, \eta')\sqrt{\{F_2(\xi_1) - \eta'\}}}{\eta_1 - \eta'}\, d\eta' \right\}. \quad (4·8, 58)$$

Now
$$\int_{F_2(\xi_1)}^{\eta_1} \frac{\Omega(\xi_1 - \eta_0)}{\sqrt{(\eta_1 - \eta_0)}}\, d\eta_0 = 2\Omega(\xi_1 - F_2(\xi_1))\sqrt{\{\eta_1 - F_2(\xi_1)\}}$$
$$- 2\int_{F_2(\xi_1)}^{\eta_1} \Omega'(\xi_1 - \eta_0)\sqrt{(\eta_1 - \eta_0)}\, d\eta_0.$$

Hence, carrying out the differentiation in the first term of (4·8, 58), we obtain

$$g(\xi_1, \eta_1) = \frac{1}{\pi} \left\{ -\int_{F_2(\xi_1)}^{\eta_1} \frac{\Omega'(\xi_1 - \eta_0)}{\sqrt{(\eta_1 - \eta_0)}}\, d\eta_0 + \frac{1}{\sqrt{\{\eta_1 - F_2(\xi_1)\}}} \right.$$
$$\left. \times \left[\Omega(\xi_1 - F_2(\xi_1)) - \int_{F_1(\xi_1)}^{F_2(\xi_1)} \frac{f(\xi_1, \eta')\sqrt{\{F_2(\xi_1) - \eta'\}}}{\eta_1 - \eta'}\, d\eta' \right] \right\}. \quad (4·8, 59)$$

As the point $Q(\xi_1, \eta_1)$ approaches a point of the arc which corresponds to the trailing edge, η_1 approaches $F_2(\xi_1)$, and so $\dfrac{1}{\sqrt{\{\eta_1 - F_2(\xi_1)\}}}$ tends to infinity. But $g(\xi_1, \eta_1)$ is proportional to the z-component of the velocity which must remain finite at the trailing edge, by virtue of the Joukowski condition. It follows that the expression in square brackets on the right-hand side of (4·8, 59) vanishes for $\eta_1 = F_2(\xi_1)$. Hence

$$\Omega(\xi_1 - F_2(\xi_1)) = \int_{F_1(\xi_1)}^{F_2(\xi_1)} \frac{f(\xi_1, \eta')}{\sqrt{\{F_2(\xi_1) - \eta'\}}}\, d\eta'. \quad (4·8, 60)$$

In order to evaluate (4·8, 59) for given $Q:(\xi_1, \eta_1)$, we require $Q(\xi)$ for the interval which is intercepted on the ξ-axis by the lines QQ_1 and $Q''Q_1''$ which are parallel to CD'. All these intervals are contained in $C_1 D_1$, where C_1 and D_1 are the points of intersection of the ξ-axis with CD' and with the parallel line through D (Fig. 4·8, 9). Letting $Q(\xi_1, \eta_1)$ vary along the arc CD in (4·8, 60) we obtain the values of $\Omega(\xi)$ for all values of ξ in the interval $C_1 D_1$. Accordingly, we are now in a position to evaluate $g(\xi_1, \eta_1)$ and hence $\tau(x, y)$ for all points within $CDD'C$. Substitution in (4·8, 42) then yields the

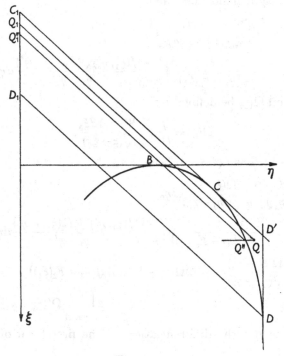

Fig. 4·8, 9

required values of the velocity potential for points within $CDC'C$. The reader is referred to Ward's paper[‡] for a proof that the solution obtained in this way actually satisfies the Joukowski condition under sufficiently wide assumptions regarding the regularity of the specified data. In particular, it can be shown that in that case the x-component of the induced velocity and hence the pressure difference vanish along the subsonic arc CD of the trailing edge.

The present analysis can be extended to include cases where the tips are no longer independent. Complete analytical solutions are available so long as the point C' in which the Mach line CC' (Fig. 4·8, 3) intersects the trailing

‡ G. N. Ward, 'Supersonic flow past thin wings. I. General theory', *Quart. J. Mech. Appl. Math.* vol. 2 (1949), pp. 136–52.

edge, belongs to the supersonic arc of the trailing edge, with a similar condition for the corresponding Mach line through the port tip.

For many important cases, the application of Evvard's principle alone is sufficient to calculate the velocity potential over the entire wing. Thus for a rectangular wing with independent wing tips, we have to distinguish two regions. In both regions, the velocity potential may be calculated

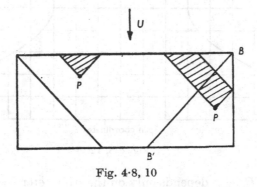

Fig. 4·8, 10

by means of (4·8, 40), where the areas of integration are indicated in Fig. 4·8, 10. It is not difficult to show that Evvard's method is still applicable if the two Mach lines BB' and EE' intersect on the wing, provided

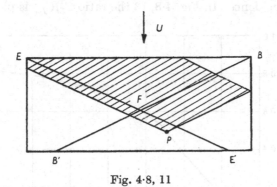

Fig. 4·8, 11

these lines do not meet the opposite tips (Fig. 4·8, 11). The condition is equivalent to

$$\mathcal{R} > \frac{1}{\beta} = \tan\mu. \tag{4·8, 61}$$

For a point P within the triangle $B'E'F$ Evvard's principle must now be applied with respect to both tips as shown on the diagram. Having determined the velocity potential on a rectangular wing which satisfies (4·8, 61), we may then determine the pressure distribution, and the lift and wave drag in the usual way. Fig. 4·8, 12 shows the spanwise lift distribution for a rectangular wing with independent wing tips. We note that the lift density drops to zero at the tips. For a flat rectangular wing which satisfies (4·8, 61), the lift and drag coefficients are given by

$$C_L = \frac{4\alpha}{\beta}\left(1 - \frac{1}{2R\beta}\right), \quad C_D = \frac{4\alpha^2}{\beta}\left(1 - \frac{1}{2R\beta}\right). \qquad (4·8, 62)$$

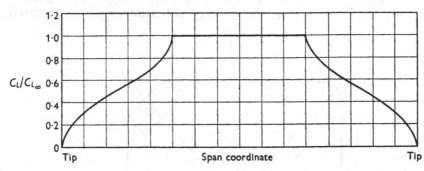

Fig. 4·8, 12

Thus the ratio C_L/C_{L_∞} depends only on the parameter $\lambda = R\beta$,

$$C_L/C_{L_\infty} = 1 - \frac{1}{2\lambda}, \qquad (4·8, 63)$$

where $C_{L_\infty} = 4\alpha/\beta$ is the two-dimensional value of the lift coefficient of a flat wing at incidence. In Fig. 4·8, 13 the ratio C_L/C_{L_∞} is plotted against λ.

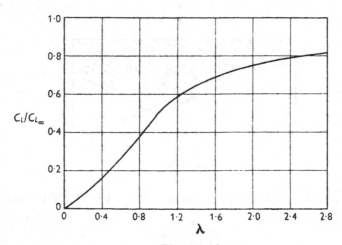

Fig. 4·8, 13

The theory detailed above breaks down for $\lambda < 1$, and in that region the curve was obtained by a different method (see §4·10). We have observed that βC_L depends only on the parameter λ. This fact can also be established independently of any detailed analysis by the general argument by means of which we established (4·7, 80).

It seems that no detailed solutions have been worked out by the above methods for wings with curved subsonic leading edges. However, it appears from other solutions (see § 4·9 below) that in the neighbourhood of the subsonic leading edges of a wing at incidence, the pressure becomes infinite. This gives rise to a suction force at the leading edge, which must be taken into account in the calculation of the overall lift and drag on the wing. The detailed analysis of this force will be found at the end of the next section.

4·9   The method of special coordinates

The methods of the previous section are not suitable for the calculation of the flow around a Delta wing with subsonic leading edges, i.e. for

$$\gamma < \mu, \qquad (4·9, 1)$$

where γ is the semi-apex angle (Fig. 4·9, 1). This problem is of considerable importance for both practical and theoretical reasons. To solve it, we adapt the classical method of orthogonal coordinates for the solution of Laplace's equation to supersonic conditions.‡

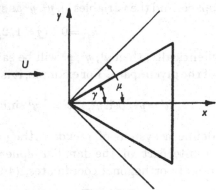

Fig. 4·9, 1

Let x^1, x^2, x^3 and y^1, y^2, y^3 be two sets of variables which are connected by the relations

$$\begin{aligned} x^j &= f_j(y^1, y^2, y^3) \quad (j = 1, 2, 3), \\ y^j &= F_j(x^1, x^2, x^3) \quad (j = 1, 2, 3). \end{aligned} \qquad (4·9, 2)$$

We suppose that the transformation given by (4·9, 2) is non-singular in a given region, so that the determinants of the transformation satisfy

$$\left| \frac{\partial f_j}{\partial y^k} \right| \neq 0, \quad \left| \frac{\partial F_j}{\partial x^k} \right| \neq 0.$$

Now
$$dx^j = \frac{\partial f_j}{\partial y^k} dy^k, \qquad (4·9, 3)$$

where we have used the double suffix notation, i.e. it is understood that the right-hand side must be summed over k. Then, for given $\beta > 0$,

$$(dx^1)^2 - \beta^2(dx^2)^2 - \beta^2(dx^3)^2 = h_1^2(dy^1)^2 - h_2^2(dy^2)^2 - h_3^2(dy^3)^2$$
$$+ 2h_{12}dy^1 dy^2 + 2h_{23}dy^2 dy^3 + 2h_{13}dy^1 dy^3, \quad (4·9, 4)$$

‡ See A. Robinson, 'Aerofoil theory of a flat delta wing at supersonic speeds', *Rep. Memor. Aero. Res. Comm., Lond.*, no. 2548 (1946). The problem has also been solved by H. J. Stewart, 'The lift of a delta wing at supersonic speeds', *Quart. Appl. Math.* vol. 4 (1946), pp. 246–54; Stewart used the cone field method of § 4·10.

where

$$h_1^2 = \left(\frac{\partial f_1}{\partial y^1}\right)^2 - \beta^2\left(\frac{\partial f_2}{\partial y^1}\right)^2 - \beta^2\left(\frac{\partial f_3}{\partial y^1}\right)^2,$$

$$h_2^2 = -\left(\frac{\partial f_1}{\partial y^2}\right)^2 + \beta^2\left(\frac{\partial f_2}{\partial y^2}\right)^2 + \beta^2\left(\frac{\partial f_3}{\partial y^2}\right)^2, \qquad (4·9, 5)$$

$$h_3^2 = -\left(\frac{\partial f_1}{\partial y^3}\right)^2 + \beta^2\left(\frac{\partial f_2}{\partial y^3}\right)^2 + \beta^2\left(\frac{\partial f_3}{\partial y^3}\right)^2,$$

and

$$h_{jk} = \left(\frac{\partial f_1}{\partial y^j}\right)\left(\frac{\partial f_1}{\partial y^k}\right) - \beta^2\left(\frac{\partial f_2}{\partial y^j}\right)\left(\frac{\partial f_2}{\partial y^k}\right) - \beta^2\left(\frac{\partial f_3}{\partial y^j}\right)\left(\frac{\partial f_3}{\partial y^k}\right) \quad (j=1,2,3; k=1,2,3; j<k).$$
$$(4·9, 6)$$

If x^1, x^2, x^3 are rectangular Cartesian coordinates in three-dimensional space, and the variables y^1, y^2, y^3 as specified by (4·9, 2) satisfy the conditions

$$h_{jk} = 0 \quad (j=1,2,3; k=1,2,3; j<k), \qquad (4·9, 7)$$

identically, then y^1, y^2, y^3 will be said to be pseudo-orthogonal coordinates of the given space. For example, it is not difficult to verify that the equations

$$x^1 = y^1 \cosh y^2, \quad x^2 = \frac{1}{\beta} y^1 \sinh y^2 \cos y^3, \quad x^3 = \frac{1}{\beta} y^1 \sinh y^2 \sin y^3$$

define a system of pseudo-orthogonal coordinates. These constitute the counterpart of the familiar spherical coordinates. For any system of pseudo-orthogonal coordinates, (4·9, 4) reduces to

$$(dx^1)^2 - \beta^2(dx^2)^2 - \beta^2(dx^3)^2 = h_1^2(dy^1)^2 - h_2^2(dy^2)^2 - h_3^2(dy^3)^2. \qquad (4·9, 8)$$

It will be necessary presently to transform the differential expression

$$\beta^2 \frac{\partial^2 \Phi}{\partial x^1 \partial x^1} - \frac{\partial^2 \Phi}{\partial x^2 \partial x^2} - \frac{\partial^2 \Phi}{\partial x^3 \partial x^3} \qquad (4·9, 9)$$

into pseudo-orthogonal coordinates, where Φ is an arbitrary scalar function. As in the case of Laplace's equation, the explicit transformation of this expression would be extremely tedius, but the work can be shortened greatly through the use of the tensor calculus.‡

Let g_{jk} be the fundamental covariant tensor of the quadratic differential form $(dx^1)^2 - \beta^2(dx^2)^2 - \beta^2(dx^3)^2$, i.e.

$$[g_{jk}] = \begin{bmatrix} 1 & 0 & 0 \\ 0 & -\beta^2 & 0 \\ 0 & 0 & -\beta^2 \end{bmatrix}. \qquad (4·9, 10)$$

The corresponding determinant is

$$g = |g_{jk}| = \beta^4.$$

‡ See, for example, O. Veblen, *Invariants of Quadratic Differential Forms*, Cambridge Tracts in Mathematics and Physics, no. 24 (Cambridge University Press, 1933).

Similarly, let G_{jk} be the fundamental covariant tensor of the quadratic differential form $h_1^2(dy^1)^2 - h_2^2(dy^2)^2 - h_3^2(dy^3)^2$,

$$[G_{jk}] = \begin{bmatrix} h_1^2 & 0 & 0 \\ 0 & -h_2^2 & 0 \\ 0 & 0 & -h_3^2 \end{bmatrix}, \qquad (4·9, 11)$$

so that

$$G = |G_{jk}| = h_1^2 h_2^2 h_3^2.$$

The corresponding contravariant tensors are

$$[g^{jk}] = \begin{bmatrix} 1 & 0 & 0 \\ 0 & -\dfrac{1}{\beta^2} & 0 \\ 0 & 0 & -\dfrac{1}{\beta^2} \end{bmatrix} \quad \text{and} \quad [G^{jk}] = \begin{bmatrix} \dfrac{1}{h_1^2} & 0 & 0 \\ 0 & -\dfrac{1}{h_2^2} & 0 \\ 0 & 0 & -\dfrac{1}{h_3^2} \end{bmatrix}. \quad (4·9, 12)$$

The expression $\dfrac{1}{\sqrt{g}} \dfrac{\partial}{\partial x^j}\left(\sqrt{g}\, g^{jk} \dfrac{\partial \Phi}{\partial x^k}\right)$ is called the generalized Laplacian (or the second differential parameter) of Φ. It is known to be an absolute scalar, so that

$$\frac{1}{\sqrt{g}} \frac{\partial}{\partial x^j}\left(\sqrt{g}\, g^{jk} \frac{\partial \Phi}{\partial x^k}\right) = \frac{1}{\sqrt{G}} \frac{\partial}{\partial y^j}\left(\sqrt{G}\, G^{jk} \frac{\partial \Phi}{\partial y^k}\right). \qquad (4·9, 13)$$

Writing down this equation explicitly, we obtain

$$\beta^2 \frac{\partial^2 \Phi}{\partial x^1 \partial x^1} - \frac{\partial^2 \Phi}{\partial x^2 \partial x^2} - \frac{\partial^2 \Phi}{\partial x^3 \partial x^3}$$

$$= \frac{\beta^2}{h_1 h_2 h_3}\left\{\frac{\partial}{\partial y^2}\left(\frac{h_2 h_3}{h_1} \frac{\partial \Phi}{\partial y^1}\right) - \frac{\partial}{\partial y^2}\left(\frac{h_3 h_1}{h_2} \frac{\partial \Phi}{\partial y^2}\right) - \frac{\partial}{\partial y^3}\left(\frac{h_1 h_2}{h_3} \frac{\partial \Phi}{\partial y^3}\right)\right\}, \quad (4·9, 14)$$

which is the required expression.

Next, we wish to find formulae for $\partial G_j/\partial x^k$ ($j = 1, 2, 3$; $k = 1, 2, 3$), in terms of y^1, y^2, y^3. We have $dx^j = a_{jk} dy^k$,

where $a_{jk} = \partial f_j/\partial y^k$. Solving for dy^k, we obtain

$$dy^k = A_{kj} dx^j, \qquad (4·9, 15)$$

where $[A_{kj}]$ is the inverse of the matrix $[a_{jk}]$. On the other hand, evidently $A_{kj} = \partial G_k/\partial x^j$.

Now let y^k be a system of pseudo-orthogonal coordinates for the Cartesian coordinates x^j. Using the relations of pseudo-orthogonality, we find by direct matrix multiplication

$$\begin{bmatrix} \dfrac{\partial f_1}{\partial y^1}\Big/h_1^2 & -\beta^2\dfrac{\partial f_2}{\partial y^1}\Big/h_1^2 & -\beta^2\dfrac{\partial f_3}{\partial y^1}\Big/h_1^2 \\ -\dfrac{\partial f_1}{\partial y^2}\Big/h_2^2 & \beta^2\dfrac{\partial f_2}{\partial y^2}\Big/h_2^2 & \beta^2\dfrac{\partial f_3}{\partial y^2}\Big/h_2^2 \\ -\dfrac{\partial f_1}{\partial y^3}\Big/h_3^2 & \beta^2\dfrac{\partial f_2}{\partial y^3}\Big/h_3^2 & \beta^2\dfrac{\partial f_3}{\partial y^3}\Big/h_3^2 \end{bmatrix} \begin{bmatrix} \dfrac{\partial f_1}{\partial y^1} & \dfrac{\partial f_1}{\partial y^2} & \dfrac{\partial f_1}{\partial y^3} \\ \dfrac{\partial f_2}{\partial y^1} & \dfrac{\partial f_2}{\partial y^2} & \dfrac{\partial f_2}{\partial y^3} \\ \dfrac{\partial f_3}{\partial y^1} & \dfrac{\partial f_3}{\partial y^2} & \dfrac{\partial f_3}{\partial y^3} \end{bmatrix} = \begin{bmatrix} 1 & 0 & 0 \\ 0 & 1 & 0 \\ 0 & 0 & 1 \end{bmatrix}.$$

$$(4·9, 16)$$

The second matrix on the left-hand side of (4·9, 16) is identical with $[a_{jk}]$, so that the first must be $[A_{kj}]$. Hence

$$
\left.
\begin{aligned}
\frac{\partial G_1}{\partial x^1} &= \ \frac{\partial f_1}{\partial y^1}\bigg/h_1^2, & \frac{\partial G_1}{\partial x^2} &= -\beta^2\frac{\partial f_2}{\partial y^1}\bigg/h_1^2, & \frac{\partial G_1}{\partial x^3} &= -\beta^2\frac{\partial f_3}{\partial y^1}\bigg/h_1^2, \\[2mm]
\frac{\partial G_2}{\partial x^1} &= -\frac{\partial f_1}{\partial y^2}\bigg/h_2^2, & \frac{\partial G_2}{\partial x^2} &= \ \beta^2\frac{\partial f_2}{\partial y^2}\bigg/h_2^2, & \frac{\partial G_2}{\partial x^3} &= \ \beta^2\frac{\partial f_3}{\partial y^2}\bigg/h_2^2, \\[2mm]
\frac{\partial G_3}{\partial x^1} &= -\frac{\partial f_1}{\partial y^3}\bigg/h_3^2, & \frac{\partial G_3}{\partial x^2} &= \ \beta^2\frac{\partial f_2}{\partial y^3}\bigg/h_3^2, & \frac{\partial G_3}{\partial x^3} &= \ \beta^2\frac{\partial f_3}{\partial y^3}\bigg/h_3^2.
\end{aligned}
\right\} \quad (4·9, 17)
$$

A particular system of pseudo-orthogonal coordinates, which is suitable for the solution of problems relating to Delta wings with subsonic leading edges, is defined by the following equations:

$$
\left.
\begin{aligned}
x &= \frac{r\sigma\nu}{k}, \\[3mm]
y &= \frac{r\sqrt{\{(\sigma^2 - k^2)(\nu^2 - k^2)\}}}{\beta k k'}, \\[3mm]
z &= \frac{r\sqrt{\{(\sigma^2 - 1)(1 - \nu^2)\}}}{\beta k'}.
\end{aligned}
\right\} \quad (4·9, 18)
$$

In these formulae we have reintroduced the usual notation, x, y, z, for the Cartesian coordinates, while the curvilinear coordinates are r, σ, ν. k and k' are real constants between 0 and 1 which are related to the geometry of the Delta wing by the equation

$$k'^2 = 1 - k^2 = \beta^2\tan^2\gamma. \quad (4·9, 19)$$

The ranges of r, μ, ν will be defined by

$$0 \leqslant r < \infty, \quad 1 \leqslant \sigma < \infty, \quad k \leqslant \nu < 1. \quad (4·9, 20)$$

By eliminating first σ and ν, then r and ν, and finally r and σ from (4·9, 18), we obtain the following coordinate surfaces, $r = $ constant, $\sigma = $ constant, $\nu = $ constant:

$$
\left.
\begin{aligned}
x^2 - \beta^2(y^2 + z^2) &= r^2, \\[2mm]
\frac{x^2}{\sigma^2} - \frac{\beta^2 y^2}{\sigma^2 - k^2} - \frac{\beta^2 z^2}{\sigma^2 - 1} &= 0, \\[2mm]
\frac{x^2}{\nu^2} - \frac{\beta^2 y^2}{\nu^2 - k^2} + \frac{\beta^2 z^2}{1 - \nu^2} &= 0.
\end{aligned}
\right\} \quad (4·9, 21)
$$

Of these equations, the first represents a family of hyperboloids of two sheets while the other two represent families of cones. For the specified ranges, the coordinates r, σ, ν can represent only points inside the after-cone of the apex, i.e.

$$x^2 - \beta^2(y^2 + z^2) \geqslant 0 \quad (x \geqslant 0). \quad (4·9, 22)$$

By solving (4·9, 18) for r, σ, ν, we find that to every point (x, y, z) which satisfies (4·9, 22), there corresponds exactly one triplet (r, σ, ν) in the specified range. On the other hand, to each triplet (r, σ, ν) there corresponds in general four points (x, y, z), according to the determination of the square roots in (4·9, 18). To avoid this ambiguity we may replace σ, ν by new variables σ', ν', where

$$\sigma' = \int_\sigma^\infty \frac{dt}{\sqrt{\{(t^2 - k^2)(t^2 - 1)\}}}, \quad \nu' = \int_\nu^k \frac{dt}{\sqrt{\{(t^2 - k^2)(1 - t^2)\}}}, \quad (4\cdot9, 23)$$

i.e. in terms of elliptic functions,

$$\sigma = \operatorname{ns}(\sigma', k), \quad \nu = k \operatorname{nd}(\nu', k'). \quad (4\cdot9, 24)$$

Then

$$
\begin{aligned}
x &= r \operatorname{ns}(\sigma', k) \operatorname{nd}(\nu', k'), \\
y &= \frac{r}{\beta} \operatorname{ds}(\sigma', k) \operatorname{sd}(\nu', k'), \\
z &= \frac{r}{\beta} \operatorname{cs}(\sigma', k) \operatorname{cd}(\nu', k').
\end{aligned}
\right\} \quad (4\cdot9, 25)
$$

As σ' varies from 0 to $K(k)$, σ varies from ∞ to 1. As ν' varies from $-2K'(k)$ to 0, ν varies from k to 1 and back to k, and this process is repeated as ν' increases to $2K'(k)$. Different points (x, y, z) now correspond to different triplets (r, σ', ν'), since the functions on the right-hand side of (4·9, 25) are one-valued. However, for the analysis which is to follow we may as well use the coordinates (r, σ, ν).

As σ tends to infinity, the cones of the second family in (4·9, 21) approach the after-cone of the origin,

$$x^2 - \beta^2(y^2 + z^2) = 0,$$

while, as $\sigma \to 1$, these cones tend to the (two-sided) angular region in the x, y-plane which is given by

$$x^2 - \frac{\beta^2 y^2}{k'^2} \geqslant 0 \quad (x \geqslant 0),$$

i.e. by

$$x^2 \tan^2 \gamma - y^2 \geqslant 0 \quad (x \geqslant 0). \quad (4\cdot9, 26)$$

Let $x = c$ be the equation of the trailing edge of the given Delta wing, so that c is the maximum chord. Then the wing coincides with (4·9, 26) for $x \leqslant c$, while, for $x > c$, we may modify the specified boundary conditions at will without affecting conditions at the wing. In particular, we may therefore assume that the trailing edge of the wing has been moved to infinity so that the modified wing actually coincides with (4·9, 26).

As $\nu \to 1$, the cones of the third family of (4·9, 21) approach the angular region

$$x^2 \tan^2 \gamma - y^2 \leqslant 0.$$

The intersections of the cones of the second family with the plane $x = 1$ are ellipses, varying between the circle

$$y^2 + z^2 = 1/\beta^2,$$

and the slit

$$z = 0, \quad y^2 \leqslant \tan^2 \gamma.$$

The intersections of the cones of the third family with the plane $x = 1$ are hyperbolae.

The formulae (4·9, 18) yield, by partial differentiation,

$$
\left.
\begin{aligned}
&\frac{\partial x}{\partial r} = \frac{\sigma \nu}{k}, && \frac{\partial x}{\partial \sigma} = \frac{\nu r}{k}, && \frac{\partial x}{\partial \nu} = \frac{r \sigma}{k}, \\
&\frac{\partial y}{\partial r} = \frac{\sqrt{\{(\sigma^2 - k^2)(\nu^2 - k^2)\}}}{\beta k k'}, && \frac{\partial y}{\partial \sigma} = r \frac{\sigma}{\beta k k'} \frac{\sqrt{(\nu^2 - k^2)}}{\sqrt{(\sigma^2 - k^2)}}, && \frac{\partial y}{\partial \nu} = r \frac{\nu}{\beta k k'} \frac{\sqrt{(\sigma^2 - k^2)}}{\sqrt{(\nu^2 - k^2)}}, \\
&\frac{\partial z}{\partial r} = \frac{\sqrt{\{(\sigma^2 - 1)(1 - \nu^2)\}}}{\beta k'}, && \frac{\partial z}{\partial \sigma} = r \frac{\sigma}{\beta k'} \frac{\sqrt{(1 - \nu^2)}}{\sqrt{(\sigma^2 - 1)}}, && \frac{\partial z}{\partial \nu} = r \frac{\nu}{\beta k'} \frac{\sqrt{(\sigma^2 - 1)}}{\sqrt{(1 - \nu^2)}}.
\end{aligned}
\right\}
$$

(4·9, 27)

Hence

$$h_{12} = h_{23} = h_{13} = 0,$$

so that the system of coordinates (r, σ, ν) is indeed pseudo-orthogonal, and

$$
\left.
\begin{aligned}
h_1^2 &= 1, \\
h_2^2 &= \frac{r^2(\sigma^2 - \nu^2)}{(\sigma^2 - k^2)(\sigma^2 - 1)}, \\
h_3^2 &= \frac{r^2(\sigma^2 - \nu^2)}{(\nu^2 - k^2)(1 - \nu^2)}.
\end{aligned}
\right\}
$$

(4·9, 28)

Using (4·9, 15), we then find that the equation of supersonic flow (4·5, 1) becomes

$$
(\sigma^2 - \nu^2)\frac{\partial}{\partial r}\left(r^2 \frac{\partial \Phi}{\partial r}\right) - \sqrt{\{(\sigma^2 - k^2)(\sigma^2 - 1)\}}\frac{\partial}{\partial \sigma}\left(\sqrt{\{(\sigma^2 - k^2)(\sigma^2 - 1)\}}\frac{\partial \Phi}{\partial \sigma}\right)
$$
$$
- \sqrt{\{(\nu^2 - k^2)(1 - \nu^2)\}}\frac{\partial}{\partial \nu}\left(\sqrt{\{(\nu^2 - k^2)(1 - \nu^2)\}}\frac{\partial \Phi}{\partial \nu}\right) = 0. \quad (4\cdot9, 29)
$$

We try to find particular solutions of (4·9, 29) which are of the form

$$\Phi(r, \sigma, \nu) = r^n \Psi(\sigma, \nu). \quad (4\cdot9, 30)$$

Substitution in (4·9, 29) leads to the following equation for Ψ:

$$
n(n+1)(\sigma^2 - \nu^2)\Psi - \sqrt{\{(\sigma^2 - k^2)(\sigma^2 - 1)\}}\frac{\partial}{\partial \sigma}\left(\sqrt{\{(\sigma^2 - k^2)(\sigma^2 - 1)\}}\frac{\partial \Psi}{\partial \sigma}\right)
$$
$$
- \sqrt{\{(\nu^2 - k^2)(1 - \nu^2)\}}\frac{\partial}{\partial \nu}\left(\sqrt{\{(\nu^2 - k^2)(1 - \nu^2)\}}\frac{\partial \Psi}{\partial \nu}\right) = 0. \quad (4\cdot9, 31)
$$

Next we separate the variables in (4·9, 31) by putting

$$\Psi = G(\sigma) H(\nu). \quad (4\cdot9, 32)$$

This leads to

$$
\frac{1}{G(\sigma)}\left[n(n+1)\sigma^2 G(\sigma) - \sigma(2\sigma^2 - k^2 - 1)\frac{dG}{d\sigma} - (\sigma^2 - k^2)(\sigma^2 - 1)\frac{d^2 G}{d\sigma^2}\right]
$$
$$
= \frac{1}{H(\nu)}\left[n(n+1)\nu^2 H(\nu) - \nu(2\nu^2 - k^2 - 1)\frac{dH}{d\nu} - (\nu^2 - k^2)(\nu^2 - 1)\frac{d^2 H}{d\nu^2}\right] H.
$$

(4·9, 33)

The left-hand side of this equation is independent of ν, the right-hand side is independent of σ. Hence both sides equal a constant, which will be denoted by h. It follows that $G(\sigma)$ has to satisfy the differential equation

$$(n(n+1)\sigma^2 - h)G(\sigma) - \sigma(2\sigma^2 - k^2 - 1)\frac{dG}{d\sigma} - (\sigma^2 - k^2)(\sigma^2 - 1)\frac{d^2G}{d\sigma^2} = 0,$$

(4·9, 34)

with an exactly similar equation for $H(\nu)$.

(4·9, 34) is Lamé's equation.‡ For a given non-negative integer n, h can be determined in $(2n+1)$ different ways, so that $G(\sigma)$ is of one of the following four forms:

$$\left.\begin{aligned}
K(\sigma) &= a_0\sigma^n + a_1\sigma^{n-2} + \cdots, \\
L(\sigma) &= \sqrt{(\sigma^2 - k^2)}\,(a_0\sigma^{n-1} + a_1\sigma^{n-3} + \cdots), \\
M(\sigma) &= \sqrt{(\sigma^2 - 1)}\,(a_0\sigma^{n-1} + a_1\sigma^{n-3} + \cdots), \\
N(\sigma) &= \sqrt{\{(\sigma^2 - k^2)(\sigma^2 - 1)\}}\,(a_0\sigma^{n-2} + a_1\sigma^{n-4} + \cdots).
\end{aligned}\right\}$$

(4·9, 35)

In these formulae, the expressions $a_0\sigma^n + a_1\sigma^{n-2} + \cdots$ are all polynomials in σ. Thus, for $n = 0$ the only solution of the above-mentioned type, except for a constant factor, is

$$E_0^1(\sigma) = 1.$$

(4·9, 36)

For $n = 1$ there are three independent solutions:

$$E_1^1(\sigma) = \sigma, \quad E_1^2(\sigma) = \sqrt{(\sigma^2 - k^2)}, \quad E_1^3(\sigma) = \sqrt{(\sigma^2 - 1)}. \qquad (4\cdot9, 37)$$

Solutions of this type are known as Lamé functions of the first kind. Except for $E_0^1(\sigma)$, they all tend to infinity with σ. Assume that a solution of this type, $E_n^m(\sigma)$, has already been determined for given n and for an appropriate h. Then a second solution of Lamé's equation can be found in the usual way by setting

$$G(\sigma) = E_n^m(\sigma)\,G^*(\sigma).$$

Imposing the condition that the new solutions vanish at infinity, we obtain

$$F_n^m(\sigma) = E_n^m(\sigma)\int_\sigma^\infty \frac{dt}{[E_n^m(t)]^2\,\sqrt{\{(t^2 - k^2)(t^2 - 1)\}}}. \qquad (4\cdot9, 38)$$

The functions $F_n^m(\sigma)$ are known as Lamé functions of the second kind. In particular

$$F_0^1(\sigma) = \int_\sigma^\infty \frac{dt}{\sqrt{\{(t^2 - k^2)(t^2 - 1)\}}}, \qquad (4\cdot9, 39)$$

and

$$\left.\begin{aligned}
F_1^1(\sigma) &= \sigma\int_\sigma^\infty \frac{dt}{t^2\,\sqrt{\{(t^2 - k^2)(t^2 - 1)\}}}, \\
F_1^2(\sigma) &= \sqrt{(\sigma^2 - k^2)}\int_\sigma^\infty \frac{dt}{(t^2 - k^2)^{\frac{3}{2}}\,\sqrt{(t^2 - 1)}}, \\
F_1^3(\sigma) &= \sqrt{(\sigma^2 - 1)}\int_\sigma^\infty \frac{dt}{(t^2 - 1)^{\frac{3}{2}}\,\sqrt{(t^2 - k^2)}}.
\end{aligned}\right\}$$

(4·9, 40)

‡ E. W. Hobson, *The Theory of Spherical and Ellipsoidal Harmonics* (Cambridge University Press, 1931), pp. 459 et seq.

It can be shown that the functions $E_n^m(\sigma)$ which belong to the first or second of the four types specified in (4·9, 35) do not vanish for $\sigma \geqslant 1$, and the same applies to $E_n^m(\sigma)/\sqrt{(\sigma^2-1)}$, if $E_n^m(\sigma)$ belongs to the third or fourth type. In the former case, the expression (4·9, 38) for $F_n^m(\sigma)$ remains determinate as σ approaches 1 from above. In the latter case, we put

$$P(\sigma) = E_n^m(\sigma)/\sqrt{(\sigma^2-1)},$$

for given $E_n^m(\sigma)$. Then

$$F_n^m(\sigma) = \sqrt{(\sigma^2-1)}\, P(\sigma) \int_\sigma^\infty \frac{dt}{[P(t)]^2\,(t^2-1)^{\frac{3}{2}}\sqrt{(t^2-k^2)}}$$

$$= P(\sigma)\sqrt{(\sigma^2-1)}\left\{-\left[\frac{1}{t[P(t)]^2\sqrt{\{(t^2-1)(t^2-k^2)\}}}\right]_\sigma^\infty\right.$$

$$\left. + \int_\sigma^\infty \frac{d}{dt}\left[\frac{1}{t[P(t)]^2\sqrt{(t^2-k^2)}}\right]\frac{dt}{\sqrt{(t^2-1)}}\right\}.$$

As σ tends to 1, the last integral tends to a finite limit. Hence

$$\lim_{\sigma\to 1} F_n^m(\sigma) = \lim_{\sigma\to 1} \frac{P(\sigma)\sqrt{(\sigma^2-1)}}{\sigma[P(\sigma)]^2\sqrt{\{(\sigma^2-1)(\sigma^2-k^2)\}}} = \frac{1}{k'P(1)}. \qquad (4\cdot9, 41)$$

We shall also require an expression for the limit of $\sqrt{(\sigma^2-1)}\dfrac{d}{d\sigma}F_n^m(\sigma)$ as σ approaches 1. For functions $E_n^m(\sigma)$ which belong to the first or second type detailed in (4·9, 35),

$$\sqrt{(\sigma^2-1)}\frac{d}{d\sigma}F_n^m(\sigma) = \sqrt{(\sigma^2-1)}\left\{\frac{d}{d\sigma}E_n^m(\sigma)\int_\sigma^\infty \frac{dt}{[E_n^m(t)]^2\sqrt{\{(t^2-k^2)(t^2-1)\}}}\right.$$

$$\left. - \frac{1}{E_n^m(\sigma)\sqrt{\{(\sigma^2-k^2)(\sigma^2-1)\}}}\right\},$$

and so

$$H_n^m = \lim_{\sigma\to 1}\sqrt{(\sigma^2-1)}\frac{d}{d\sigma}F_n^m(\sigma) = -\frac{1}{k'E_n^m(1)}. \qquad (4\cdot9, 42)$$

On the other hand, for functions of the third or fourth type,

$$\sqrt{(\sigma^2-1)}\frac{d}{d\sigma}F_n^m(\sigma) = \sqrt{(\sigma^2-1)}\frac{d}{d\sigma}\left[\frac{1}{\sigma P(\sigma)\sqrt{(\sigma^2-k^2)}}\right.$$

$$\left. + P(\sigma)\sqrt{(\sigma^2-1)}\int_\sigma^\infty \frac{d}{dt}\left[\frac{1}{t[P(t)]^2\sqrt{(t^2-k^2)}}\right]\frac{dt}{\sqrt{(t^2-1)}}\right],$$

and so

$$\lim_{\sigma\to 1}\sqrt{(\sigma^2-1)}\frac{d}{d\sigma}F_n^m(\sigma) = \lim_{\sigma\to 1}\sqrt{(\sigma^2-1)}\frac{d}{d\sigma}\left[P(\sigma)\sqrt{(\sigma^2-1)}\right.$$

$$\left. \times \int_\sigma^\infty \frac{d}{dt}\left[\frac{1}{t[P(t)]^2\sqrt{(t^2-k^2)}}\right]\frac{dt}{\sqrt{(t^2-1)}}\right]$$

$$= \lim_{\sigma\to 1}\left\{\sigma P(\sigma)\int_\sigma^\infty \frac{d}{dt}\left[\frac{1}{t[P(t)]^2\sqrt{(t^2-k^2)}}\right]\frac{dt}{\sqrt{(t^2-1)}}\right.$$

$$\left. - P(\sigma)(\sigma^2-1)\frac{d}{d\sigma}\left[\frac{1}{\sigma[P(\sigma)]^2\sqrt{(\sigma^2-k^2)}}\right]\frac{1}{\sqrt{(\sigma^2-1)}}\right\}.$$

or $\quad H_n^m = \lim\limits_{\sigma \to 1} \sqrt{(\sigma^2 - 1)} \dfrac{d}{d\sigma} F_n^m(\sigma) = P(1) \displaystyle\int_1^\infty \dfrac{d}{dt}\left[\dfrac{1}{t[P(t)]^2\sqrt{(t^2 - k^2)}}\right]\dfrac{dt}{\sqrt{(t^2 - 1)}}.$

$$(4\cdot 0, 43)$$

Particular solutions of (4·9, 29) may now be written down as products of r^n and of the appropriate Lamé functions of the first or second kind. In the physical problem, the continuity of the velocity potential implies that Φ vanishes on the after-cone of the origin $x^2 - \beta^2(y^2 + z^2) = 0$, and this condition is satisfied if we choose for $G(\sigma)$ a function of the second kind. Since Φ must be finite and one-valued within the after-cone of the origin, it can be shown that only functions of the first kind need be considered for $H(\nu)$. Thus, we shall confine ourselves to particular solutions which are of the form

$$\Phi = r^n F_n^m(\sigma)\, E_n^m(\nu) \quad (n = 0, 1, 2, \ldots; \; m = 1, \ldots, 2n+1). \qquad (4\cdot 9, 44)$$

We shall require the partial derivatives $\partial\Phi/\partial x$, $\partial\Phi/\partial z$, for points on the wing, i.e. for $\sigma = 1$. For this purpose, we first calculate $\partial r/\partial x$, $\partial\sigma/\partial x$, $\partial\nu/\partial x$, $\partial r/\partial z$, $\partial\sigma/\partial z$, $\partial\nu/\partial z$, as given by (4·9, 17), and taking into account (4·9, 27) and (4·9, 28). We find

$$\frac{\partial r}{\partial x} = \frac{\sigma\nu}{k}, \quad \frac{\partial\sigma}{\partial x} = -\frac{\nu(\sigma^2 - k^2)(\sigma^2 - 1)}{kr(\sigma^2 - \nu^2)}, \quad \frac{\partial\nu}{\partial x} = -\frac{\sigma(\nu^2 - k^2)(1 - \nu^2)}{kr(\sigma^2 - \nu^2)},$$

and $\quad \dfrac{\partial r}{\partial z} = -\dfrac{\beta^2\sqrt{\{(\sigma^2 - 1)(1 - \nu^2)\}}}{k'}, \quad \dfrac{\partial\sigma}{\partial z} = \dfrac{\beta\sigma\sqrt{\{(1 - \nu^2)(\sigma^2 - 1)\}(\sigma^2 - k^2)}}{rk'(\sigma^2 - \nu^2)},$

$$\frac{\partial\nu}{\partial z} = \frac{\beta\nu\sqrt{\{(1 - \nu^2)(\sigma^2 - 1)\}(\nu^2 - k^2)}}{rk'(\sigma^2 - \nu^2)}.$$

$$(4\cdot 9, 45)$$

Hence $\quad \lim\limits_{\sigma \to 1} \dfrac{\partial\Phi}{\partial x} = \lim\limits_{\sigma \to 1}\left(\dfrac{\partial\Phi}{\partial r}\dfrac{\partial r}{\partial x}\right) + \lim\limits_{\sigma \to 1}\left(\dfrac{\partial\Phi}{\partial\sigma}\dfrac{\partial\sigma}{\partial x}\right) + \lim\limits_{\sigma \to 1}\left(\dfrac{\partial\Phi}{\partial\nu}\dfrac{\partial\nu}{\partial x}\right)$

$$= \frac{F_n^m(1)}{k} r^{n-1}\left\{n\nu E_n^m(\nu) - (\nu^2 - k^2)\frac{dE_n^m}{d\nu}\right\}. \qquad (4\cdot 9, 46)$$

Similarly

$$\lim\limits_{\sigma \to 1} \frac{\partial\Phi}{\partial z} = \lim\limits_{\sigma \to 1}\left(\frac{\partial\Phi}{\partial\sigma}\frac{\partial\sigma}{\partial z}\right) = \frac{\beta k' r^{n-1}}{\sqrt{(1 - \nu^2)}}\left(\lim\limits_{\sigma \to 1}\sqrt{(\sigma^2 - 1)}\frac{dF_n^m}{d\sigma}\right)E_n^m(\nu),$$

or $\quad \lim\limits_{\sigma \to 1} \dfrac{\partial\Phi}{\partial z} = \beta k' r^{n-1}\dfrac{E_n^m(\nu)}{\sqrt{(1 - \nu^2)}}\lim\limits_{\sigma \to 1}\sqrt{(\sigma^2 - 1)}\dfrac{dF_n^m}{d\sigma}, \qquad (4\cdot 9, 47)$

where the limit on the right-hand side is given either by (4·9, 42) or (4·9, 43). Thus

$$\lim\limits_{\sigma \to 1} \frac{\partial\Phi}{\partial z} = \beta k' H_n^m r^{n-1}\frac{E_n^m(\nu)}{\sqrt{(1 - \nu^2)}}. \qquad (4\cdot 9, 48)$$

Hence, if $\qquad\qquad \Phi = C r^n F_n^m(\sigma)\, E_n^m(\nu) \qquad\qquad (4\cdot 9, 49)$

is the induced velocity potential for a particular flow, where C is a constant, then the corresponding slope (see (4·7, 1)) is given by

$$s = -\frac{1}{U}\left(\frac{\partial\Phi}{\partial z}\right)_{z=0} = -C\frac{\beta k'}{U} H_n^m r^{n-1}\frac{E_n^m(\nu)}{\sqrt{(1 - \nu^2)}}. \qquad (4\cdot 9, 50)$$

Conversely, for a given wing, we may try to satisfy the correct boundary condition at the wing by a linear combination of functions of the type (4·9, 49). For some important cases, a single function is already sufficient.

Consider the flow around a flat Delta wing at incidence α. The slope is now given by
$$s(x, y) = \tan \alpha \doteqdot \alpha. \tag{4·9, 51}$$

Accordingly, we try to find a particular Φ for which the right-hand side of (4·9, 50) reduces to a constant. Referring to (4·9, 37), we see immediately that this will be true for $n = 1$, $m = 3$. In that case,
$$\left.\begin{aligned} E_1^3(\nu) &= \sqrt{\nu^2 - 1} = i\sqrt{1 - \nu^2}, \\ F_1^3(\sigma) &= \sqrt{(\sigma^2 - 1)} \int_\sigma^\infty \frac{dt}{(t^2 - 1)^{\frac{3}{2}} \sqrt{(t^2 - k^2)}}, \end{aligned}\right\} \tag{4·9, 52}$$

and so, by (4·9, 43),
$$H_1^3 = \int_1^\infty \frac{d}{dt}\left(\frac{1}{t\sqrt{(t^2 - k^2)}}\right) \frac{dt}{\sqrt{(t^2 - 1)}}.$$

This integral can be evaluated by standard methods,‡ a suitable substitution in the present case being
$$t = \text{ns}\,(n, k).$$

The result is
$$H_1^3 = -\frac{1}{k'^2} E(k), \tag{4·9, 53}$$

where $E(k)$ is the complete elliptic integral of the second kind,
$$E(k) = \int_0^{\frac{1}{2}\pi} \sqrt{(1 - k^2 \sin^2 \psi)}\, d\psi. \tag{4·9, 54}$$

Then (4·9, 50) becomes
$$s = -iC\frac{\beta}{Uk'} E(k),$$

and this must be equal to $\tan \alpha \doteqdot \alpha$ by (4·9, 51). Hence
$$C = i\frac{U\alpha}{\beta}\frac{k'}{E(k)}, \tag{4·9, 55}$$

and
$$\begin{aligned} \Phi &= -\frac{U\alpha}{\beta}\frac{k'}{E(k)} r\sqrt{\{(1 - \nu^2)(\sigma^2 - 1)\}} \int_\sigma^\infty \frac{dt}{(t^2 - 1)^{\frac{3}{2}}\sqrt{(t^2 - k^2)}} \\ &= -U\alpha \frac{k'^2}{E(k)} z \int_\sigma^\infty \frac{dt}{(t^2 - 1)^{\frac{3}{2}}\sqrt{(t^2 - k^2)}}. \end{aligned} \tag{4·9, 56}$$

These expressions for Φ become indeterminate for $\sigma = 1$, i.e. on the wing. In that case, we may use (4·9, 41), and obtain
$$\Phi = -\frac{U\alpha}{\beta E(k)} r\sqrt{(1 - \nu^2)}. \tag{4·9, 57}$$

Now, for $\sigma = 1$,
$$x = \frac{r\nu}{y}, \quad y = \frac{r\sqrt{(\nu^2 - k^2)}}{\beta k}, \quad z = 0. \tag{4·9, 58}$$

‡ E. T. Whittaker and G. N. Watson, *A Course in Modern Analysis* (4th ed., Cambridge University Press, 1927), ch. 22.

Hence $\qquad r^2(1 - \nu^2) = x^2 - \beta^2 y^2 - k^2 x^2 = \beta^2(x^2 \tan^2 \gamma - y^2).$

Substituting in (4·9, 57), we therefore obtain for the velocity potential on the wing,

$$\Phi = -\frac{U\alpha}{E(k)} \sqrt{(x^2 \tan^2 \gamma - y^2)} = -\frac{U\alpha}{E'(\beta \tan \gamma)} \sqrt{(x^2 \tan^2 \gamma - y^2)}, \quad (4\cdot9, 59)$$

where E' is the complementary complete elliptic integral of the second kind,

$$E'(\beta \tan \gamma) = \int_0^{\frac{1}{2}\pi} \sqrt{\{1 - (1 - \beta^2 \tan^2 \gamma) \sin^2 \psi\}} \, d\psi. \quad (4\cdot9, 60)$$

The appropriate sign of the square root in (4·9, 59) is $+$ on the upper surface and $-$ on the lower surface. Hence, by the linearized Bernoulli equation (4·1, 27) the pressure difference between the upper and lower surfaces at any point of the wing is

$$p_u - p_l = 2\rho_0 U \left(\frac{\partial \Phi}{\partial x}\right)_{z = +0} = -\frac{2\rho_0 U^2 x \tan^2 \gamma}{E'(\beta \tan \gamma) \sqrt{(x^2 \tan^2 \gamma - y^2)}}. \quad (4\cdot9, 61)$$

The spanwise lift distribution is given by

$$l(y) = -\int_0^c (p_u - p_l) \, dx = -2\rho_0 U \Phi(c, y, +0)$$

$$= 2\rho_0 U^2 \alpha \frac{\sqrt{(c^2 \tan^2 \gamma - y^2)}}{E'(\beta \tan \gamma)}, \quad (4\cdot9, 62)$$

where c is the maximum chord on the wing. The total lift is then obtained by integrating $l(y)$ along the span from $y = -c \tan \gamma$ to $y = c \tan \gamma$. The result is

$$L = \pi \rho_0 U^2 \alpha c^2 \frac{\tan^2 \gamma}{E'(\beta \tan \gamma)}. \quad (4\cdot9, 63)$$

The lift coefficient, C_L, based on the surface area $S = c^2 \tan \gamma$, is then given by

$$C_L = L / \tfrac{1}{2} \rho_0 U^2 S = \frac{2\pi \alpha \tan \gamma}{E'(\beta \tan \gamma)}. \quad (4\cdot9, 64)$$

The ratio $\qquad \dfrac{C_L}{C_{L_\infty}} = \dfrac{2\pi \alpha \tan \gamma}{E'(\beta \tan \gamma)} \Big/ \dfrac{4\alpha}{\beta} = \pi \dfrac{\lambda}{E'(\lambda)}, \qquad (4\cdot9, 65)$

depends only on the parameter

$$\lambda = \beta \tan \gamma = \frac{\tan \gamma}{\tan \mu}, \quad (4\cdot9, 66)$$

in agreement with the general result of the preceding section. This ratio is plotted in Fig. 4·9, 2.

It will be seen that the pressure becomes infinite at the leading edge, as under subsonic conditions, and this has to be taken into account in the calculation of the drag. As in the subsonic case, the total force on the wing can be divided into a pressure integral which is calculated in the usual way, and a suction force which acts at the leading edge. The longitudinal component of the pressure integral is given by

$$D_p = L\alpha = \pi\rho_0 U^2 \alpha^2 c^2 \frac{\tan^2\gamma}{E'(\beta\tan\gamma)}. \qquad (4\cdot9, 67)$$

For subsonic conditions, the magnitude of the suction force per unit length of the leading edge was calculated in §4·2 (see equation (4·2, 31)). A survey of the arguments used in the derivation of (4·2, 31) shows that the formulae apply also at the subsonic leading edges of a wing in supersonic flow. In particular, for an infinitely thin wing ($r_0 = 0$), (4·2, 31) yields

$$X = -\pi\rho_0 \sqrt{(1 - M^2 \cos^2\beta_0)}\, C_1^2 = -\pi\rho_0 \sqrt{(1 - M^2 \sin^2\gamma)}\, C_1^2. \quad (4\cdot9, 68)$$

To determine C_1 we note that by (4·9, 57)

$$u' = -\frac{\partial\Phi}{\partial x} = \frac{U\alpha}{E'(\beta\tan\gamma)} \frac{x\tan^2\gamma}{\sqrt{(x^2\tan^2\gamma - y^2)}}.$$

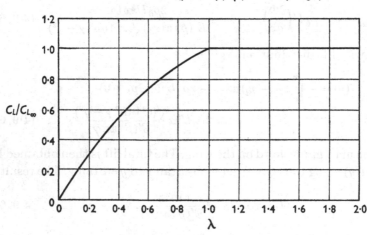

Fig. 4·9, 2

Let $(x_0, y_0, 0)$ be a point on the port leading edge, so that $y_0 = -x_0 \tan\gamma$. By (4·2, 25), C_1 may be obtained as the limit

$$C_1 = \lim_{x \to x_0} \sqrt{(x - x_0)}\, u = \lim_{x \to x_0} \sqrt{(x - x_0)}\, u'$$

$$= \frac{U\alpha}{E'(\beta\tan\gamma)} \lim_{x \to x_0} \frac{x\tan^2\gamma}{\sqrt{(x\tan\gamma - y_0)}} \lim_{x \to x_0} \sqrt{\frac{x - x_0}{x\tan\gamma + y_0}},$$

or

$$C_1 = \frac{U\alpha \sqrt{(|y_0|\tan\gamma)}}{\sqrt{2}\, E'(\beta\tan\gamma)}, \qquad (4\cdot9, 69)$$

and this expression is valid also on the starboard side.

To obtain the x-component of the suction force per unit span, we have to divide (4·9, 68) by $|\sin\gamma|$. The result is, taking into account (4·9, 69),

$$-\pi\rho_0 \sqrt{(\operatorname{cosec}^2\gamma - M^2)} \frac{U^2\alpha^2 |y_0|\tan\gamma}{2[E'(\beta\tan\gamma)]^2} = -\pi\rho_0 U^2\alpha^2 \frac{\sqrt{(\cot^2\gamma - \beta^2)}\,|y_0|\tan\gamma}{2[E'(\beta\tan\gamma)]^2}.$$

Integrating this expression across the span, we obtain the total contribution of the suction force to the drag,

$$D_s = -\pi\rho_0 U^2 \alpha^2 c^2 \frac{\tan^2\gamma \sqrt{(1-\beta^2\tan^2\gamma)}}{2[E'(\beta\tan\gamma)]^2}. \qquad (4\cdot9,70)$$

As the minus sign on the right-hand side of (4·9, 70) shows, the suction force actually reduces the drag. Adding (4·9, 67) and (4·9, 70), we obtain for the total induced drag

$$D = D_p + D_s = \pi\rho_0 U^2 \alpha^2 c^2 \frac{\tan^2\gamma}{E'(\beta\tan\gamma)}\left(1 - \frac{\sqrt{(1-\beta^2\tan^2\gamma)}}{2E'(\beta\tan\gamma)}\right). \quad (4\cdot9,71)$$

The corresponding drag coefficient is

$$C_D = D/\tfrac{1}{2}\rho_0 U^2 S = \frac{2\pi\alpha^2\tan\gamma}{E'(\beta\tan\gamma)}\left(1 - \frac{\sqrt{(1-\beta^2\tan^2\gamma)}}{2E'(\beta\tan\gamma)}\right). \qquad (4\cdot9,72)$$

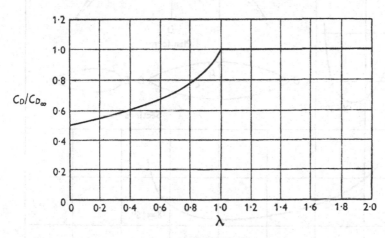

Fig. 4·9, 3

Let C_{D_∞} be the two-dimensional drag coefficient for a flat wing at incidence, $C_{D_\infty} = 4\alpha^2/\beta$. This is also the drag coefficient for a flat Delta wing with supersonic leading edges. Then

$$\frac{C_D}{C_{D_\infty}} = \frac{\pi\lambda}{2E'(\lambda)}\left(1 - \frac{\sqrt{(1-\lambda^2)}}{2E'(\lambda)}\right), \qquad (4\cdot9,73)$$

where λ is given by (4·9, 66). C_D/C_{D_∞} is plotted against λ in Fig. 4·9, 3.

Finally, the chordwise position of the centre of pressure is given by $x = \tfrac{2}{3}c$, as can be shown by the dimensional argument used in the discussion of the Delta wing with supersonic leading edges.

The functions (4·9, 44) can be employed for the solution of a variety of other problems. Thus, Squire has used a linear combination of one function of the family (4·9, 44) with $n = 1$ and two functions with $n = 2$ to calculate the flow over a particular Delta wing with rounded leading edges.‡ The drag at zero incidence is shown in Fig. 4·9, 4. The shape of the aerofoil sections is as indicated in the figure, the value of the maximum thickness to chord ratio, t_0/c, in this case being 0·01. Calculation of the drag involves use of

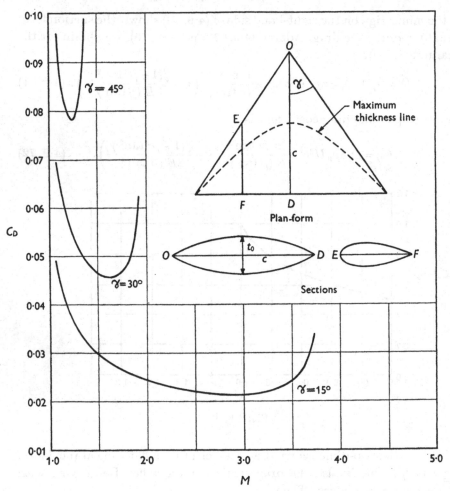

Fig. 4·9, 4. (*Courtesy H.M.S.O.*)

(4·2, 31) because of the presence of a suction force at the rounded leading edges. An extensive survey of various solutions corresponding partly to thick symmetrical wings and partly to thin cambered and twisted wings has been made by G. M. Roper.§

‡ H. B. Squire, 'An example of wing theory at supersonic speeds', *Rep. Memor. Aero. Res. Comm., Lond.*, no. 2549 (1947).

§ G. M. Roper, 'Some applications of the Lamé function solutions of the linearised supersonic flow equations. Part I.—Finite swept-back wings with symmetrical sections and rounded leading edges. Part II.—Cambered and twisted wings', *Rep. Memor. Aero. Res. Comm., Lond.*, no. 2865 (1951).

So far, we have confined ourselves to the assumption that the wing plan-form is symmetrical with respect to the x-axis. We shall now indicate how the results obtained can be made of use for the solution of similar problems for yawed Delta wings.

For given β, consider the transformation

$$\left.\begin{aligned} x &= x' \cosh \psi + y' \beta \sinh \psi, \\ y &= x' \frac{1}{\beta} \sinh \psi + y' \cosh \psi, \\ z &= z', \end{aligned}\right\} \tag{4·9, 74}$$

where ψ is a constant. This transformation is due to R. T. Jones.‡ The inverse transformation to (4·9, 74) is

$$\left.\begin{aligned} x' &= x \cosh x - y\beta \sinh \psi, \\ y' &= -x \frac{1}{\beta} \sinh \psi + y \cosh \psi, \\ z' &= z. \end{aligned}\right\} \tag{4·9, 75}$$

It is evident that (4·9, 74) carries the x, y-plane into the x', y'-plane. Also, let (x_0, y_0, z_0) be any point in the x, y, z-space and let (x'_0, y'_0, z'_0) be the corresponding point in the x', y', z'-space. Then

$$\begin{aligned} (x - x_0)^2 - \beta^2 \{(y - y_0)^2 + (z - z_0)^2\} \\ = [(x' - x'_0) \cosh \psi + (y' - y'_0) \beta \sinh \psi]^2 \\ - \beta^2 \left\{ \left[(x' - x'_0) \frac{1}{\beta} \sinh \psi + (y' - y'_0) \cosh \psi \right]^2 + (z' - z'_0)^2 \right\} \\ = (x' - x'_0)^2 - \beta^2 \{(y' - y'_0)^2 + (z' - z'_0)^2\}. \end{aligned} \tag{4·9, 76}$$

It follows that the Mach cones of the x, y, z-space are transformed into the Mach cones of the x', y', z'-space. More particularly, it is not difficult to see that fore-cones will be transformed into fore-cones and after-cones into after-cones.

We may regard x', y', z' as coordinates in x, y, z-space. Then the evaluation of (4·9, 6) shows that x', y', z' represent a pseudo-orthogonal system. Calculating h_1, h_2, h_3 by means of (4·9, 5), we obtain

$$h_1 = 1, \quad h_2 = \beta, \quad h_3 = \beta. \tag{4·9, 77}$$

Applying (4·9, 15) we find that for any scalar function Φ,

$$\beta^2 \frac{\partial^2 \Phi}{\partial x^2} - \frac{\partial^2 \Phi}{\partial y^2} - \frac{\partial^2 \Phi}{\partial z^2} = \beta^2 \frac{\partial^2 \Phi}{\partial x'^2} - \frac{\partial^2 \Phi}{\partial y'^2} - \frac{\partial^2 \Phi}{\partial z'^2}. \tag{4·9, 78}$$

It follows that any function Φ which is a solution of the equation of compressible flow in the x, y, z-space is a solution of the corresponding equation in the x', y', z'-space and vice versa.

‡ R. T. Jones, 'Thin oblique airfoils at supersonic speed', *N.A.C.A. Rep.* no. 851 (1946).

Now let $\Phi(x, y, z)$ be the induced velocity potential for a particular case of supersonic flow round a wing S which, as usual, is situated approximately in the x, y-plane. Φ is characterized by the following properties: it is a solution of the linearized equation of compressible flow (4·5, 1); it satisfies the boundary condition (4·7, 1) at the wing; it vanishes upstream of the envelope of the after-cones which emanate from the leading edges; it satisfies the Joukowski condition at the subsonic trailing edges; and it is continuous away from the wing, except possibly across the wake. At the same time, $\partial\Phi/\partial x$ must be continuous even across the wake.

For given ψ, the transformation (4·9, 74) carries the plan-form R of the wing S into a certain region R' in the x', y'-plane. We define a wing S' with plan-form R' by the rule that at any point of R', the slopes of the upper and lower surfaces of S' are the same as at the corresponding point of S.

Since Mach cones are carried into Mach cones by the transformation, it follows that supersonic edges are carried into supersonic edges, and subsonic edges into subsonic edges. At the same time, it is quite possible that (4·9, 74) transforms part of the leading edge of S into part of the trailing edge of S' and vice versa. However, we shall exclude this possibility and shall deal only with cases in which leading edges are carried into leading edges and trailing edges into trailing edges.

We now regard Φ as a function of x', y', z'. As such, Φ has the following properties: it is the solution of the linearized equation of compressible flow; it satisfies the correct boundary condition at the wing S', for the same main-stream velocity U as in the x, y, z-space; and it vanishes upstream of the envelope of the after-cones which emanate from the leading edge of S', since (4·9, 74) carries after-cones into after-cones, and, moreover, carries leading edges into leading edges, owing to the special assumption made above. For the same reason, the Joukowski condition holds at the subsonic trailing edges of S', but, generally speaking, the necessary conditions will not be satisfied at the wake of the wing. We conclude that Φ represents the induced velocity potential in the region of the x', y', z'-plane which is unaffected by the wake, i.e. upstream of the envelope of the after-cones which emanate from the trailing edge. In particular, Φ is equal to the induced velocity potential everywhere on the wing S' if the trailing edge is supersonic throughout.

We consider the example of a yawed Delta wing S' with subsonic leading edges in the x', y', z'-space. The incidence α' of the wing measures a rotation round an axis parallel to the trailing edge. Let γ' be the semi-apex angle of the wing and β_0 the angle of yaw (Fig. 4·9, 5). In order to formulate the boundary condition (4·7, 1), we have to measure the slope of the wing in the z', x'-plane. Thus

$$s(x', y') = \text{constant} = \tan\alpha,$$

where α is given by

$$\sin\alpha = \sin\alpha' \cos\beta_0.$$

Or, since α' is supposed small,

$$s(x', y') = \alpha' \cos \beta_0. \qquad (4·9, 79)$$

Assume that $\beta_0 > 0$ as in Fig. 4·9, 5. In order that UA and OB be subsonic leading edges, and AB a supersonic trailing edge, we must have

$$\beta_0 < \gamma' < \mu - \beta_0 \quad \text{and} \quad \beta_0 + \mu < \tfrac{1}{2}\pi.$$

We now use the transformation (4·9, 74) in order to reduce the case in hand to the corresponding problem for an unyawed Delta wing S. The semi-apex angle γ of that wing, and the appropriate ψ in (4·9, 74), may be determined in the following way.

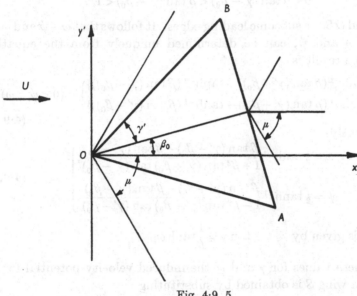

Fig. 4·9, 5

The parametric equations of the starboard leading edge of S are

$$x = t, \quad y = t \tan \gamma.$$

Hence, and by (4·9, 75), the parametric equations of the starboard leading edge of S', OB, are

$$x' = t(\cosh \psi - \beta \tan \gamma \sinh \psi),$$

$$y' = t\frac{1}{\beta}(-\sinh \psi + \beta \tan \gamma \cosh \psi).$$

It follows that the slope of OB, $\tan(\beta_0 + \gamma')$, can also be written as

$$\tan(\beta_0 + \gamma') = \frac{1}{\beta}\frac{-\sinh \psi + \beta \tan \gamma \cosh \psi}{\cosh \psi - \beta \tan \gamma \sinh \psi} = \frac{1}{\beta}\frac{\lambda - \mu}{1 - \lambda \mu}, \qquad (4·9, 80)$$

where

$$\lambda = \beta \tan \gamma, \quad \mu = \tanh \mu.$$

Similarly, we obtain for the slope of OA,

$$\tan (\beta_0 - \gamma') = -\frac{1}{\beta} \frac{\lambda + \mu}{1 + \lambda \mu}. \tag{4·9, 81}$$

λ is a real number between zero and 1, and so we may introduce a quantity by

$$\lambda = \tanh \omega \quad (\omega > 0). \tag{4·9, 82}$$

Then by (4·9, 80) and (4·9, 81),

$$\left. \begin{aligned} \tanh (\omega + \psi) &= \beta \tan (\gamma' + \beta_0), \\ \tanh (\omega - \psi) &= \beta \tan (\gamma' - \beta_0). \end{aligned} \right\} \tag{4·9, 83}$$

But

$$0 < \beta \tan (\gamma' - \beta_0) < \beta \tan (\gamma' + \beta_0) < 1,$$

since OA and OB are subsonic leading edges. It follows that $\omega + \psi$ and $\omega - \psi$ and hence ω and ψ, can be determined uniquely from the equations (4·9, 83). The result is

$$\left. \begin{aligned} \omega &= \tfrac{1}{2} \{ \tanh^{-1} (\beta \tan (\gamma' + \beta_0)) + \tanh^{-1} (\beta \tan (\gamma' - \beta_0)) \} \\ \psi &= \tfrac{1}{2} \{ \tanh^{-1} (\beta \tan (\gamma' + \beta_0)) - \tanh^{-1} (\beta \tan (\gamma' - \beta_0)) \} \end{aligned} \right\} \quad (0 < \psi < \omega),$$

$$\tag{4·9, 84}$$

or, equivalently,

$$\left. \begin{aligned} \omega &= \tfrac{1}{2} \tanh^{-1} \frac{\beta \tan (\gamma' + \beta_0) + \beta \tan (\gamma' - \beta_0)}{1 + \beta^2 \tan (\gamma' + \beta_0) \tan (\gamma' - \beta_0)}, \\ \psi &= \tfrac{1}{2} \tanh^{-1} \frac{\beta \tan (\gamma' + \beta_0) - \beta \tan (\gamma' - \beta_0)}{1 - \beta^2 \tan (\gamma' + \beta_0) \tan (\gamma' - \beta_0)}. \end{aligned} \right\} \tag{4·9, 85}$$

Then $\tan \gamma$ is given by

$$\tan \gamma = \frac{1}{\beta} \tanh \omega,$$

and with these values for γ and ψ the induced velocity potential for the yawed Delta wing S is obtained by substituting

$$x = x' \cosh \psi + y' \beta \sinh \psi, \quad y = x' \frac{1}{\beta} \sinh \psi + y' \cosh \psi$$

in the formula

$$\Phi = -\frac{U\alpha}{E'(\beta \tan \gamma)} \sqrt{(x^2 \tan^2 \gamma - y^2)}$$

(see (4·9, 74) and (4·9, 59)). From this expression for the induced velocity potential, the pressure and aerodynamic forces may then be calculated in the usual way.

4·10 Cone fields

A field of flow in which the velocity components along any line emanating from a fixed point O are proportional to the nth power of the distance from O, is called a cone field of order n, where n is a given constant. O will be called the apex of the field. Examples of cone fields occurred in earlier sections. Thus, a distribution of sources along a straight line emanating from a point O, whose intensity is proportional to the nth power of the distance from O, generates a cone field of order $n-1$ (see §4·7). Similarly, the solutions (4·9, 44) are cone fields of order $n-1$. The field of flow round a flat Delta wing at incidence, with either subsonic of supersonic leading edges, is a cone field of order zero, neglecting conditions downstream of the trailing edge. The importance of this concept was first recognized by Busemann.‡ It will now be shown that problems relating to cone fields of order zero can be reduced either to complex variable theory, or to the two-dimensional wave equation or to a combination of these two.§

It will be assumed that the apex of the cone field coincides with the origin. We shall use different transformations to reduce the equation of compressible flow inside or outside the after-cone of the apex, respectively. Inside the after-cone of the apex, we introduce a system of pseudo-orthogonal coordinates, r, ψ, ϕ, by

$$x = r \cosh \psi, \quad y = \frac{1}{\beta} r \sinh \psi \cos \phi, \quad z = \frac{1}{\beta} r \sinh \psi \sin \phi, \quad (4·10, 1)$$

$$(0 \leqslant r < \infty) \qquad (0 \leqslant \psi < \infty) \qquad (0 \leqslant \phi < 2\pi)$$

We verify by direct calculation that the conditions (4·9, 7) are satisfied. Evaluating (4·9, 5), we obtain

$$h_1 = 1, \quad h_2 = r, \quad h_3 = r \sinh \psi. \qquad (4·10, 2)$$

The coordinates (r, ψ, ϕ) constitute the pseudo-orthogonal counterpart of the familiar spherical polar coordinates.

Using (4·9, 15), we find that the linearized equation of compressible flow (4·5, 1), becomes

$$\sinh^2 \psi \frac{\partial}{\partial r} \left(r^2 \frac{\partial \Phi}{\partial r} \right) - \sinh \psi \frac{\partial}{\partial \psi} \left(\sinh \psi \frac{\partial \Phi}{\partial \psi} \right) - \frac{\partial^2 \Phi}{\partial \phi^2} = 0. \qquad (4·10, 3)$$

Equation (4·10, 3) is satisfied also by the velocity components. In particular, in a cone field of order zero, these are independent of r, and so

$$\sinh \psi \frac{\partial}{\partial \psi} \left(\sinh \psi \frac{\partial u'}{\partial \psi} \right) + \frac{\partial^2 u'}{\partial \phi^2} = 0, \qquad (4·10, 4)$$

‡ A. Busemann, 'Aerodynamischer Auftrieb bei Überschallgeschwindigkeit', *Luftfahrtforsch.* vol. 12 (1935), pp. 210–20.

§ The following reports relating to the general theory of cone fields may be mentioned: P. Germain, 'Quelques applications de la théorie des mouvements coniques à l'aérodynamique supersonique', *Proc. 7th Int. Congr. Appl. Mech.* (London, 1948); S. Goldstein and G. N. Ward, 'The linearised theory of conical fields in supersonic flow, with applications to plane aerofoils', *Aero. Quart.* vol. 2 (1950), pp. 39–84; W. D. Hayes, 'Linearised conical supersonic flow', *Proc. 6th Int. Congr. Appl. Mech.* (Paris, 1946); P. A. Lagerstrom, 'Linearised supersonic theory of conical wings', *Tech. Notes Nat. Adv. Comm. Aero.*, *Wash.*, no. 1685 (1948); A. Robinson and J. H. Hunter-Tod, 'The aerodynamic derivatives with respect to sideslip for a delta wing with small dihedral at zero incidence at supersonic speeds', *Rep. Memor. Aero. Res. Comm., Lond.*, no. 2410 (1948); H. J. Stewart, 'The lift of a delta wing at supersonic speeds', *Quart. Appl. Math.* vol. 4 (1946), pp. 246–54.

with similar equations for v and w. To simplify (4·10, 4), we put

$$\sinh\psi\,\frac{\partial}{\partial\psi}=\frac{\partial}{\partial\sigma}, \quad d\sigma=\frac{d\psi}{\sinh\psi}, \quad \sigma=\log\tanh\tfrac{1}{2}\psi. \qquad (4\cdot10,5)$$

Then (4·10, 4) is transformed into Laplace's equation

$$\frac{\partial^2 u'}{\partial\sigma^2}+\frac{\partial^2 u'}{\partial\phi^2}=0. \qquad (4\cdot10,6)$$

Thus u' may be represented as the real part of an analytic function of the complex variable $\sigma+i\phi$. Introducing the complex variable ξ by

$$\xi=e^{\sigma+i\phi}, \qquad (4\cdot10,7)$$

we may then write also

$$u'=\mathscr{R}f_1(\xi),$$

and similarly

$$\left.\begin{array}{l}v=\mathscr{R}f_2(\xi),\\[4pt]w=\mathscr{R}f_3(\xi),\end{array}\right\} \qquad (4\cdot10,8)$$

where f_1, f_2, f_3 are analytic functions of ξ. To express ξ directly in terms of x, y, z, we observe that

$$e^\sigma=\tanh\tfrac{1}{2}\psi=\frac{\sinh\psi}{1+\cosh\psi}=\frac{r\sinh\psi}{r+x}. \qquad (4\cdot10,9)$$

Hence

$$\left.\begin{array}{l}\eta=e^\sigma\cos\phi=\dfrac{r\sinh\psi\cos\phi}{r+x}=\dfrac{\beta y}{\sqrt{\{x^2-\beta^2(y^2+z^2)\}}+x},\\[12pt]\zeta=e^\sigma\sin\phi=\dfrac{r\sinh\psi\sin\phi}{r+x}=\dfrac{\beta z}{\sqrt{\{x^2-\beta^2(y^2+z^2)\}}+x},\end{array}\right\} \qquad (4\cdot10,10)$$

and so

$$\xi=\eta+i\zeta=\frac{\beta(y+iz)}{\sqrt{\{x^2-\beta^2(y^2+z^2)\}}+x}. \qquad (4\cdot10,11)$$

The partial derivatives of ξ with respect to x, y and z are, by straightforward calculation,

$$\frac{\partial\xi}{\partial x}=-\frac{1}{r}\frac{\beta(y+iz)}{r+x}=-\frac{1}{r}\xi,$$

$$\frac{\partial\xi}{\partial y}=\frac{\beta[(r+x)r+\beta^2 y(y+iz)]}{(r+x)^2 r}=\frac{\beta(y+iz)}{2r(r+x)}\frac{\beta^2(y+iz)^2+(r+x)^2}{(r+x)(y+iz)}$$

$$=\frac{\beta}{2r}(\xi^2+1),$$

$$\frac{\partial\xi}{\partial z}=\frac{\beta[i(r+x)r+\beta^2 z(y+iz)]}{(r+x)^2 r}=-\frac{i\beta(y+iz)}{2r(r+x)}\frac{\beta^2(y+iz)-(r+x)^2}{(r+x)(y+iz)}$$

$$=-\frac{i\beta}{2r}(\xi^2-1).$$

Hence

$$\frac{\partial\xi}{\partial x}:\frac{\partial\xi}{\partial y}:\frac{\partial\xi}{\partial z}=\frac{2}{\beta}\xi:-(\xi^2+1):i(\xi^2-1). \qquad (4\cdot10,12)$$

The condition of irrotational flow is

$$\frac{\partial w}{\partial y}=\frac{\partial v}{\partial z}, \quad \frac{\partial u'}{\partial z}=\frac{\partial w}{\partial x}, \quad \frac{\partial v}{\partial x}=\frac{\partial u'}{\partial y}, \tag{4·10, 13}$$

and this condition will certainly be satisfied if

$$\frac{\partial f_3}{\partial y}=\frac{\partial f_2}{\partial z}, \quad \frac{\partial f_1}{\partial z}=\frac{\partial f_3}{\partial x}, \quad \frac{\partial f_2}{\partial x}=\frac{\partial f_1}{\partial y},$$

i.e. if $\quad f_3'(\xi)\dfrac{\partial \xi}{\partial y}=f_2'(\xi)\dfrac{\partial \xi}{\partial z}, \quad f_1'(\xi)\dfrac{\partial \xi}{\partial z}=f_3'(\xi)\dfrac{\partial \xi}{\partial x}, \quad f_2'(\xi)\dfrac{\partial \xi}{\partial x}=f_1'(\xi)\dfrac{\partial \xi}{\partial y}.$ (4·10, 14)

(4·10, 14) is equivalent to

$$f_1'(\xi):f_2'(\xi):f_3'(\xi)=\frac{\partial \xi}{\partial x}:\frac{\partial \xi}{\partial y}:\frac{\partial \xi}{\partial z}, \tag{4·10, 15}$$

and combining this relation with (4·10, 12) we obtain

$$\frac{f_1'(\xi)}{\dfrac{2}{\beta}\xi}=\frac{f_2'(\xi)}{-(\xi^2+1)}=\frac{f_3'(\xi)}{i(\xi^2-1)}=f(\xi), \quad \text{say.} \tag{4·10, 16}$$

We may therefore express u', v, w, in terms of the single analytic function $f(\xi)$ as follows:

$$\left.\begin{aligned}
u' &= \frac{2}{\beta}\mathscr{R}\int^{\xi}\omega f(\omega)\,d\omega, \\[2mm]
v &= -\mathscr{R}\int^{\xi}(\omega^2+1)f(\omega)\,d\omega, \\[2mm]
w &= \mathscr{R}\int^{\xi}i\,(\omega^2-1)f(\omega)\,d\omega.
\end{aligned}\right\} \tag{4·10, 17}$$

For a given analytic function $f(\xi)$, the velocity components u', v, w, as defined by (4·10, 17), all satisfy the linearized equation of compressible flow (4·5, 1), and together they constitute an irrotational field of flow. Let Φ be the corresponding velocity potential. Since u', v, w satisfy (4·5, 1), it follows that

$$\left(\frac{\partial}{\partial x}, \frac{\partial}{\partial y}, \frac{\partial}{\partial z}\right)\left(\beta^2\frac{\partial^2\Phi}{\partial x^2}-\frac{\partial^2\Phi}{\partial y^2}-\frac{\partial^2\Phi}{\partial z^2}\right)=0$$

identically, and so $\quad \beta^2\dfrac{\partial^2\Phi}{\partial x^2}-\dfrac{\partial^2\Phi}{\partial y^2}-\dfrac{\partial^2\Phi}{\partial z^2}=\text{constant.}$ (4·10, 18)

Now the quantities u', v, w are constant along straight lines emanating from the origin, and so Φ is proportional to the distance from the origin along such lines, and the left-hand side of (4·10, 18) is inversely proportional to the distance from the origin. But this is possible only if the constant of the right-hand side of (4·10, 18) is equal to zero, i.e. Φ also satisfies (4·5, 1).

Since Φ is arbitrary to the extent of the addition of a constant, we may take $\Phi = 0$ at the origin. Then Φ is a homogeneous function of degree 1 in x, y, z, and so, by Euler's theorem on homogeneous functions,

$$\Phi = x\frac{\partial\Phi}{\partial x} + y\frac{\partial\Phi}{\partial y} + z\frac{\partial\Phi}{\partial z},$$

or
$$\Phi = -(xu' + yv + zw). \tag{4·10, 19}$$

For points outside the Mach cone of the origin we introduce the variables r, ψ, ϕ by

$$x = r\sinh\psi, \quad y = \frac{1}{\beta}r\cosh\psi\cos\phi, \quad z = \frac{1}{\beta}r\cosh\psi\sin\phi. \tag{4·10, 20}$$

This is again a system of pseudo-orthogonal coordinates, but we now have

$$h_1^2 = -1, \quad h_2^2 = -r^2, \quad h_3^2 = r^2\cosh^2\psi. \tag{4·10, 21}$$

It follows that h_1 and h_2 are imaginary, but we may nevertheless apply (4·9, 15) formally. The choice of signs for the h_j ($h_1 = i$ or $h_1 = -i$, etc.) is irrelevant. We obtain in place of (4·5, 1)

$$\cosh^2\psi\frac{\partial}{\partial r}\left(r^2\frac{\partial\Phi}{\partial r}\right) - \cosh\psi\frac{\partial}{\partial\psi}\left(\cosh\psi\frac{\partial\Phi}{\partial\psi}\right) + \frac{\partial^2\Phi}{\partial\phi^2} = 0. \tag{4·10, 22}$$

Thus, in a cone-field of order zero, u' satisfies the equation

$$\cosh\psi\frac{\partial}{\partial\psi}\left(\cosh\psi\frac{\partial u'}{\partial\psi}\right) - \frac{\partial^2 u'}{\partial\phi^2} = 0, \tag{4·10, 23}$$

and the same equation is satisfied by v and w.

Put
$$\sigma = \int\frac{d\psi}{\cosh\psi} = 2\tan^{-1}e^\psi,$$

so that
$$\cosh\psi = \tfrac{1}{2}(\tan\tfrac{1}{2}\sigma + \cot\tfrac{1}{2}\sigma) = \frac{1}{\sin\sigma},$$

$$\sinh\psi = \tfrac{1}{2}(\tan\tfrac{1}{2}\sigma - \cot\tfrac{1}{2}\sigma) = \cot\sigma,$$

$$x = r\cot\sigma, \quad y = r\frac{\cos\phi}{\beta\sin\sigma}, \quad z = r\frac{\sin\phi}{\beta\sin\sigma}, \tag{4·10, 24}$$

and (4·10, 23) is transformed into the wave equation in two variables

$$\frac{\partial^2 u'}{\partial\sigma^2} - \frac{\partial^2 u'}{\partial\phi^2} = 0. \tag{4·10, 25}$$

The general solution of (4·10, 25) is

$$u' = g_1(\sigma + \phi) + G_1(\sigma - \phi). \tag{4·10, 26}$$

v and w satisfy the same partial differential equation, and may therefore be written as
$$\left.\begin{aligned} v &= g_2(\sigma + \phi) + G_2(\sigma - \phi), \\ w &= g_3(\sigma + \phi) + G_3(\sigma - \phi). \end{aligned}\right\} \tag{4·10, 27}$$

The condition of irrotational flow (4·10, 13) is therefore satisfied by

$$
\left.
\begin{aligned}
g_3'(\sigma+\phi)\frac{\partial}{\partial y}(\sigma+\phi) &= g_2'(\sigma+\phi)\frac{\partial}{\partial z}(\sigma+\phi), \\[2mm]
g_1'(\sigma+\phi)\frac{\partial}{\partial z}(\sigma+\phi) &= g_3'(\sigma+\phi)\frac{\partial}{\partial x}(\sigma+\phi), \\[2mm]
g_2'(\sigma+\phi)\frac{\partial}{\partial x}(\sigma+\phi) &= g_1'(\sigma+\phi)\frac{\partial}{\partial y}(\sigma+\phi),
\end{aligned}
\right\}
\tag{4·10, 28}
$$

with similar relations for G_1, G_2, G_3. They are equivalent to

$$
\left.
\begin{aligned}
g_1'(\sigma+\phi):g_2'(\sigma+\phi):g_3'(\sigma+\phi) &= \frac{\partial}{\partial x}(\sigma+\phi):\frac{\partial}{\partial y}(\sigma+\phi):\frac{\partial}{\partial z}(\sigma+\phi), \\[2mm]
G_1'(\sigma-\phi):G_2'(\sigma-\phi):G_3'(\sigma-\phi) &= \frac{\partial}{\partial x}(\sigma-\phi):\frac{\partial}{\partial y}(\sigma-\phi):\frac{\partial}{\partial z}(\sigma-\phi).
\end{aligned}
\right\}
\tag{4·10, 29}
$$

Put
$$
\eta = e^{i(\sigma+\phi)} = \frac{\beta}{r(\cot\sigma - i)}\left(\frac{r\cos\phi}{\beta\sin\sigma} + i\,\frac{r\sin\phi}{\beta\sin\sigma}\right)
$$

$$
= \frac{\beta(y+iz)}{x-ir},
$$

then
$$
\frac{\partial\eta}{\partial x} = -\frac{\beta(y+iz)}{(x-ir)^2}\left(1+i\frac{x}{r}\right) = -i\frac{\eta}{r},
$$

$$
\frac{\partial\eta}{\partial y} = \frac{\beta[(x-ir)r + i\beta^2 y(y+iz)]}{(x-ir)^2 r} = \frac{\beta i}{2r}\left(\eta+\frac{1}{\eta}\right)
$$

$$
= \beta i\frac{\eta}{r}\cos(\sigma+\phi),
$$

$$
\frac{\partial\eta}{\partial z} = \frac{\beta[i(x-ir)r + i\beta^2 z(y+iz)]}{(x-ir)^2 r} = \frac{\beta\eta}{2r}\left(\eta-\frac{1}{\eta}\right)
$$

$$
= \beta i\frac{\eta}{r}\sin(\sigma+\phi).
$$

Hence
$$
\frac{\partial}{\partial x}(\sigma+\phi):\frac{\partial}{\partial y}(\sigma+\phi):\frac{\partial}{\partial z}(\sigma+\phi) = \frac{\partial\eta}{\partial x}:\frac{\partial\eta}{\partial y}:\frac{\partial\eta}{\partial z} = -\frac{1}{\beta}:\cos(\sigma+\phi):\sin(\sigma+\phi),
$$

and similarly
$$
\frac{\partial}{\partial x}(\sigma-\phi):\frac{\partial}{\partial y}(\sigma-\phi):\frac{\partial}{\partial z}(\sigma-\phi) = -\frac{1}{\beta}:\cos(\sigma-\phi):-\sin(\sigma-\phi).
$$

Using these relations, we obtain from (4·10, 29)

$$
\left.
\begin{aligned}
g_1'(\sigma+\phi):g_2'(\sigma+\phi):g_3'(\sigma+\phi) &= -\frac{1}{\beta}:\cos(\sigma+\phi):\sin(\sigma+\phi), \\[2mm]
G_1'(\sigma-\phi):G_2'(\sigma-\phi):G_3'(\sigma-\phi) &= -\frac{1}{\beta}:\cos(\sigma-\phi):-\sin(\sigma-\phi),
\end{aligned}
\right\}
\tag{4·10, 30}
$$

or
$$\left.\begin{aligned}\frac{g_1'(\sigma+\phi)}{-\dfrac{1}{\beta}}=\frac{g_2'(\sigma+\phi)}{\cos{(\sigma+\phi)}}=\frac{g_3'(\sigma+\phi)}{\sin{(\sigma+\phi)}}=g(\sigma+\phi),\\[2mm]\end{aligned}\right.$$

and
$$\left.\frac{G_1'(\sigma-\phi)}{-\dfrac{1}{\beta}}=\frac{G_2'(\sigma-\phi)}{\cos{(\sigma-\phi)}}=\frac{G_3'(\sigma-\phi)}{-\sin{(\sigma-\phi)}}=G(\sigma-\phi).\right\} \quad (4\cdot10,31)$$

Hence the general solution for u', v, w is

$$\left.\begin{aligned}u'&=-\frac{1}{\beta}\int^{\sigma+\phi}g(\omega)\,d\omega-\frac{1}{\beta}\int^{\sigma-\phi}G(\omega)\,d\omega,\\[2mm]v&=\int^{\sigma+\phi}g(\omega)\cos\omega\,d\omega+\int^{\sigma-\phi}G(\omega)\cos\omega\,d\omega,\\[2mm]w&=\int^{\sigma+\phi}g(\omega)\sin\omega\,d\omega-\int^{\sigma-\phi}G(\omega)\sin\omega\,d\omega.\end{aligned}\right\} \quad (4\cdot10,32)$$

It can be shown as before that the field of flow given by (4·10, 32) possesses a velocity potential Φ which is a solution of (4·5, 1) and which may be expressed as

$$\Phi=-(xu'+yv+zw).$$

In all applications of cone-field theory, the induced flow vanishes for $x<0$, since otherwise it would have to be finite non-vanishing also at points infinitely far upstream, by the cone-field property. We may therefore confine the variables in (4·10, 20) to the ranges

$$0\leqslant r<\infty,\quad 0\leqslant\psi<\infty,\quad 0\leqslant\phi<2\pi. \quad (4\cdot10,33)$$

This covers the region outside the Mach cone of the apex downstream of the z, x-plane ($x\geqslant0$). The corresponding range for σ is

$$0<\sigma\leqslant\tfrac{1}{2}\pi. \quad (4\cdot10,34)$$

Since u', v, w are one-valued, it follows from (4·10, 32) that $g(\omega)$, $G(\omega)$ must be periodic functions of period 2π such that

$$\left.\begin{aligned}\int_0^{2\pi}g(\omega)\,d\omega&=\int_0^{2\pi}g(\omega)\cos\omega\,d\omega=\int_0^{2\pi}g(\omega)\sin\omega\,d\omega\\[2mm]\end{aligned}\right.$$

and
$$\left.\int_0^{2\pi}G(\omega)\,d\omega=\int_0^{2\pi}G(\omega)\cos\omega\,d\omega=\int_0^{2\pi}G(\omega)\sin\omega\,d\omega.\right\} \quad (4\cdot10,35)$$

The functions $f(\omega)$, $g(\omega)$, $G(\omega)$ remain to be determined from the boundary condition. If the wing is situated approximately in the x, y-plane then the usual procedure leads to the boundary condition

$$w=U\alpha, \quad (4\cdot10,36)$$

where the incidence must be constant along straight lines through the origin. However, the special assumptions of cone-field theory still apply if the wing is not approximately plane (o.g. a Delta wing with finite dihedral) provided the surface of the wing is generated by straight lines through the origin.

Some skill is required for the determination of the appropriate functions f, g, G in each particular case. We shall consider the example of a flat Delta wing at incidence which is entirely within the after-cone of the apex, and which is yawed by a positive angle β_0 greater than its semi-apex angle γ (Fig. 4·10, 1). Under these conditions, the edge OA becomes a subsonic trailing edge, along which the Joukowski condition must be satisfied. The

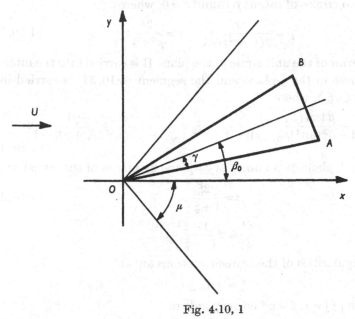

Fig. 4·10, 1

problem cannot be dealt with by the methods developed so far, but a solution can be obtained by cone-field theory on the entire wing when the trailing edge AB is supersonic. In that case, the wing may be produced to infinity beyond AB, without affecting the situation on the wing. If the trailing edge AB is subsonic then cone-field theory fails to provide the correct solution over the rear portion of the wing which is influenced by the presence of AB (in Fig. 4·10, 1 this is the part of the wing which is within the after-cone of B). Since the induced velocities vanish outside the after-cone of O, we only have to determine the flow within that cone.

Consider the plane Π which is given by the equation

$$x = \text{constant} = \beta.$$

In this plane, the interior of the after-cone of O is the interior of the unit circle

$$y^2 + z^2 = 1,$$

and the trace of the wing is the straight segment

$$z = 0, \quad \beta \tan(\beta_0 - \gamma) \leqslant y \leqslant \beta \tan(\beta_0 + \gamma). \tag{4·10, 37}$$

The value of ξ corresponding to any point (β, y, z) in the plane Π is, by (4·10, 11),

$$\xi = \eta + i\zeta = \frac{y + iz}{\sqrt{(1 - y^2 - z^2)} + 1}. \tag{4·10, 38}$$

(4·10, 38) transforms the origin $(y = z = 0)$ of Π into the origin $\xi = 0$ of the complex ξ-plane. Straight lines through $y = z = 0$ are carried into straight lines of the same slope through $\xi = 0$, and circles of radius r round $y = z = 0$ are carried into circles of radius ρ round $\xi = 0$, where

$$\rho = \frac{r}{\sqrt{(1 - r^2)} + 1}, \quad r = \frac{2\rho}{\rho^2 + 1}. \tag{4·10, 39}$$

Thus, the interior of the unit circle in the plane Π is carried into the interior of the unit circle in the ξ-plane and the segment (4·10, 37) is carried into the segment $\langle \xi_1, \xi_2 \rangle$, where

$$\xi_1 = \frac{\beta \tan(\beta_0 - \gamma)}{\sqrt{\{1 - \beta^2 \tan^2(\beta_0 - \gamma)\}} + 1}, \quad \xi_2 = \frac{\beta \tan(\beta_0 + \gamma)}{\sqrt{\{1 - \beta^2 \tan^2(\beta_0 + \gamma)\}} + 1}. \tag{4·10, 40}$$

We map the ξ-plane on an auxiliary t-plane by means of the transformation

$$t = \frac{2\xi}{1 + \xi^2}. \tag{4·10, 41}$$

Then

$$\frac{dt}{d\xi} = 2\frac{1 - \xi^2}{(1 + \xi^2)^2},$$

and so the singularities of the transformations are at

$$\xi = \pm 1, \pm i.$$

The unit circle $|\xi| = 1$, $\xi = e^{i\phi}$ corresponds to

$$t = \frac{2e^{i\phi}}{e^{2i\phi} + 1} = \sec \phi.$$

As ϕ varies from 0 to $\frac{1}{2}\pi$, t varies from 1 to $+\infty$ along the real axis; as ϕ varies from $-\frac{1}{2}\pi$ to π, t varies from $-\infty$ to -1; as ϕ varies from π to $\frac{3}{2}\pi$, t varies back from $t = -1$ to $t = -\infty$; and finally, as ϕ varies from $\frac{3}{2}\pi$ to 2π, t varies from $t = \infty$ down to $t = 1$. Thus the unit circle in the ξ-plane corresponds to the entire t-plane cut along the real axis from $t = 1$ to $+\infty$ and from $t = -1$ to $-\infty$. The relation between the two planes is indicated in Fig. 4·10, 2. We may express t directly in terms of the coordinates, y, z of the plane Π (see (4·10, 38)),

$$t = \frac{2}{\xi + 1/\xi} = \frac{2}{\dfrac{y + iz}{1 + \sqrt{(1 - y^2 - z^2)}} + \dfrac{y - iz}{1 - \sqrt{(1 - y^2 - z^2)}}} = \frac{y^2 + z^2}{y - iz\sqrt{(1 - y^2 - z^2)}}$$

or

$$t = \frac{y + iz\sqrt{(1 - y^2 - z^2)}}{1 - z^2}. \tag{4·10, 42}$$

In particular, for points on the y-axis of II,

$$t = y,$$

and so the edges of the wing correspond to

$$t_1 = \beta \tan(\beta_0 - \gamma), \quad t_2 = \beta \tan(\beta_0 + \gamma). \tag{4·10, 43}$$

By (4·10, 8) $\quad u' = \mathscr{R}f_1(\xi) = \mathscr{R}F_1(t), \quad$ say,

and similarly $\quad v = \mathscr{R}f_2(\xi) = \mathscr{R}F_2(t),$

$$w = \mathscr{R}f_3(\xi) = \mathscr{R}F_3(t), \tag{4·10, 44}$$

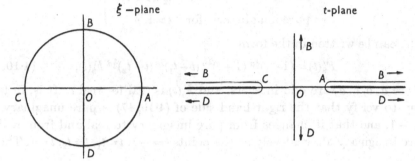

Fig. 4·10, 2

where the functions F_1, F_2, F_3 are connected by the relations

$$\frac{F_1'(t)}{\dfrac{2}{\beta}\xi} = \frac{F_2'(t)}{-(\xi^2+1)} = \frac{F_3'(t)}{i(\xi^2-1)},$$

or

$$\frac{F_1'(t)}{\dfrac{it}{\beta\sqrt{(1-t^2)}}} = \frac{F_2'(t)}{\dfrac{-i}{\sqrt{(1-t^2)}}} = F_3'(t), \tag{4·10, 45}$$

by virtue of (4·10, 16). The sign of the square root in (4·10, 45) is to be taken as positive when t is real and numerically smaller than 1.

We now try to determine F_1, F_2, F_3 subject to the following conditions for u', v, w, which are related to F_1, F_2, F_3, by (4·10, 44):

$u' = v = w = 0$ at the after-cone of the apex, i.e. for $|t| > 1$, t real;

$w = U\alpha$ at the wing, i.e. for $t_1 < t < t_2$;

u' and w are continuous across the wake, i.e. for $0 < t < t_1$, but v may be discontinuous across the wake. Also, u' must be an odd function of z, and so $u' = 0$ for $-1 < t < t_1$ and $t_2 < t < 1$.

These conditions imply

$$\mathscr{R}F_3(t) = 0 \quad \text{for} \quad |t| > 1, \; t \text{ real},$$

$$\mathscr{R}F_3(t) = U\alpha \quad \text{for} \quad t_1 < t < t_2,$$

$$\mathscr{R}F_1(t) = 0 \quad \text{for} \quad -1 < t < t_1 \quad \text{and} \quad t_2 < t < 1,$$

and so
$$\mathscr{R}F_3'(t) = 0 \quad \text{for} \quad |t| > 1 \quad \text{and} \quad t_1 < t < t_2,$$
$$\mathscr{R}F_1'(t) = 0 \quad \text{for} \quad -1 < t < t_2 \quad \text{and} \quad t_2 < t < 1.$$

Now (4·10, 45) shows that for $|t| < 1$, t real, $\mathscr{R}F_1'(t) = 0$, implies $\mathscr{I}F_3'(t) = 0$. Hence the function $F_3'(t)$ is

$$
\left.
\begin{array}{lll}
\text{pure imaginary} & \text{for} & t < -1, \\
\text{real} & \text{for} & -1 < t < t_1, \\
\text{pure imaginary} & \text{for} & t_1 < t < t_2, \\
\text{real} & \text{for} & t_2 < t < 1, \\
\text{pure imaginary} & \text{for} & t > 1.
\end{array}
\right\}
\tag{4·10, 46}
$$

$F_3'(t)$ can be written in the form

$$F_3'(t) = (1-t)^{\frac{1}{2}a}(1+t)^{\frac{1}{2}b}(t-t_1)^{\frac{1}{2}c}(t-t_2)^{\frac{1}{2}d}F_0(t), \tag{4·10, 47}$$

where a, b, c, d, are odd integers, and $F_0(t)$ is real for real t. It is, in fact, easy to verify that the right-hand side of (4·10, 47) is pure imaginary for $t < -1$, and that it changes from pure imaginary to real and from real to pure imaginary alternatively at the points $t = -1$, $t = t_1$, $t = t_2$, $t = 1$. Then

$$F_1'(t) = \frac{i}{\beta}(1-t)^{\frac{1}{2}(a-1)}(1+t)^{\frac{1}{2}(b-1)}(t-t_1)^{\frac{1}{2}c}(t-t_2)^{\frac{1}{2}d}tF_0(t) \tag{4·10, 48}$$

by (4·10, 46), from which $F_1(t)$ is determined by integration. Now the evidence of earlier examples suggests that $u' = \mathscr{R}F_1(t)$ becomes infinite at the leading edge, for $y = t = t_2$, while it vanishes as the square root of the distance from the trailing edge, $y = t = t_1$. Accordingly, we take $d = -3$, $c = -1$. Moreover, we may assume that all factors of the form $(1-t)^n$, $(1+t)^n$, where n is a non-vanishing integer, are included in $(1-t)^{\frac{1}{2}(a-1)}$, $(1+t)^{\frac{1}{2}(b-1)}$. Since w remains finite (actually it is zero) for $t = \pm 1$, it follows from (4·10, 47) that we must have $a \geqslant 1$, $b \geqslant 1$. And since u' remains finite for $t = \pm i\infty$, $F_1'(t)$ must tend to zero more strongly than t^{-1} as t tends to infinity, and so $tF_0(t)$ must be bounded at infinity. Also, the singularities of $F_1'(t)$ are supposed absorbed in the first four variable factors on the right-hand side of (4·10, 48), and so $tF_0(t)$ must be regular for all finite t. This shows that $tF_0(t)$ reduces to a constant, and, furthermore, we must have $a = b = 1$. Hence, finally

$$F_3'(t) = \frac{C}{t}(1-t^2)^{\frac{1}{2}}(t-t_1)^{-\frac{1}{2}}(t-t_2)^{-\frac{3}{2}}, \tag{4·10, 49}$$

and, by (4·10, 45),
$$
\left.
\begin{array}{l}
F_1'(t) = \dfrac{iC}{\beta}(t-t_1)^{-\frac{1}{2}}(t-t_2)^{-\frac{3}{2}}, \\[3mm]
F_2'(t) = -\dfrac{iC}{t}(t-t_1)^{-\frac{1}{2}}(t-t_2)^{-\frac{3}{2}}.
\end{array}
\right\}
\tag{4·10, 50}
$$

F_1 and F_2 (and hence u' and v) may now be obtained from (4·10, 50) by integration, and the result can be expressed in terms of elementary functions.

We may take the lower limit of these integrals at any point on the after-cone of the origin, e.g. $t = -1$, since the induced velocity components vanish at the cone. In particular

$$F_1(t) = \frac{iC}{\beta} \int_{-1}^{t} \frac{dt}{(t-t_1)^{\frac{1}{2}}(t-t_2)^{\frac{3}{2}}} = \frac{2iC}{\beta(t_2-t_1)} \left[\sqrt{\frac{t-t_1}{t-t_2}} \right]_{-1}^{t}. \quad (4\cdot10, 51)$$

But this formula shows that $F_1(t_1)$ is pure imaginary, and so

$$u' = \mathscr{R}F_1(t) = \mathscr{R}\{F_1(t) - F_1(t_1)\},$$

or
$$u' = \frac{2C}{\beta(t_2-t_1)} \mathscr{I} \sqrt{\frac{t-t_1}{t-t_2}}. \quad (4\cdot10, 52)$$

The constant C is obtained from the condition

$$U\alpha = w = \mathscr{R} \int_{-1}^{t_0} \frac{C}{t}(1-t^2)^{\frac{1}{2}}(t-t_1)^{-\frac{1}{2}}(t-t_2)^{-\frac{3}{2}}dt, \quad (4\cdot10, 53)$$

where t_0 may be any point in the interval $\langle t_1, t_2 \rangle$. Since w is constant on the wing, the integrand on the right-hand side of $(4\cdot10, 53)$ must be imaginary in the interval $\langle t_1, t_2 \rangle$. Hence C must be real, as was assumed already in the formula $(4\cdot10, 52)$. The path of integration in $(4\cdot9, 53)$ must be chosen so as to avoid the origin, since the integrand has a pole at that point. This can be done, for example, by taking the integral along the real axis, except for a small semicircle round the origin. But the residue of the integrand at the origin is real and so that contribution of the semicircle to the integral is pure imaginary. It follows that we may take the integral in $(4\cdot10, 53)$ entirely along the real axis, provided it is understood that the principal value of the integral is taken at the origin. Putting $t_0 = t_1$, we then obtain

$$C = U\alpha \bigg/ \int_{-1}^{t_1} \frac{1}{t}(1-t^2)^{\frac{1}{2}}(t-t_1)^{-\frac{1}{2}}(t-t_2)^{-\frac{3}{2}}dt. \quad (4\cdot10, 54)$$

The denominator on the right-hand side can be evaluated in terms of complete elliptic integrals of all three kinds.‡

Having determined u', we may calculate the aerodynamic forces on the wing in the usual way, taking into account the suction force at the leading edge. The pressure difference across the wing is proportional to u', and so to $\mathscr{I} \sqrt{\dfrac{t-t_1}{t-t_2}} = \sqrt{\dfrac{t-t_1}{t_2-t}}$. Put

$$s_0 = \tfrac{1}{2}(t_2+t_1), \quad s = \tfrac{1}{2}(t_2-t_1), \quad t = s_0 + s\cos\theta,$$

then
$$\sqrt{\frac{t-t_1}{t_2-t}} = \sqrt{\frac{1+\cos\theta}{1-\cos\theta}} = \cot\tfrac{1}{2}\theta. \quad (4\cdot10, 55)$$

Comparing $(4\cdot10, 55)$ with $(2\cdot4, 16)$ we see that the pressure distribution across the span of the wing, for a given chord section, is of the same type

‡ See S. Goldstein and G. N. Ward, 'The linearised theory of conical fields in supersonic flow, with applications to plane aerofoils', *Aero. Quart.* vol. 2 (1950), pp. 39–84.

as the chordwise pressure distribution on a two-dimensional flat plate at incidence in incompressible flow.

For certain plan-forms, the velocity potential can be obtained by the superposition of a number of cone fields. Consider, for example, a flat rectangular wing at incidence α, as shown in Fig. 4·10, 3. According to § 4·8, this problem can also be solved by Evvard's method. It is assumed that the Mach lines which pass over the wing from the end-points of the

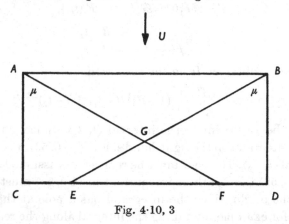

Fig. 4·10, 3

leading edge intersect on the wing, but that they do not meet the opposite tips. Then the velocity potential over $AGBA$ is given by Ackeret's two-dimensional theory and the velocity potentials over $ACEGA$ and $BGFDB$ are given by the cone fields due to two infinite wedges at incidence, as shown in Fig. 4·10, 4. In order to obtain the induced velocity potential over EFG,

Fig. 4·10, 4

we only have to add the induced velocity potentials Φ_A and Φ_B which are due to the cone fields with apexes at A and B respectively, and to subtract the two-dimensional velocity potential Φ_{AB} which would exist if the leading edge were produced to infinity beyond A and B, so

$$\Phi = \Phi_A + \Phi_B - \Phi_{AB}. \tag{4·10, 56}$$

Indeed, the induced velocity potential is given by Φ_A and Φ_B within the after-cones of A and B, respectively. It follows that if we define Φ by

(4·10, 56) within the region R common to the two cones, then the correct boundary condition (4·10, 36) is satisfied at the wing, and Φ is continuous on the boundary of R with the adjacent regions of the after-cones of A and B.

The above method of superposition fails if the Mach lines AG and BG meet the opposite tips (the segments DB and AC).

Germain has indicated‡ how the investigation of cone fields of higher order can be reduced to the study of cone fields of order zero, thus becoming amenable to the treatment by a complex variable method of the type detailed above.

4·11 Other methods and problems in the linearized theory of supersonic flow

The following additional methods and problems of supersonic flow will be dealt with briefly.

The supersonic counterparts of doublets and vortices in incompressible flow were developed in §4·5. The importance of these concepts for supersonic aerofoil theory is somewhat smaller than for incompressible or subsonic flow, in view of the existence of the methods described in the preceding sections, which have no subsonic counterparts. Nevertheless, doublet and vortex distribution methods have been used also in supersonic theory. The velocity potential due to a surface distribution of doublets over an area S of the x, y-plane is, by (4·5, 46),

$$\Phi(x,y,z) = -\beta^2 \int_{S'}^{*} \frac{\sigma(x_0, y_0)\, z\, dx_0\, dy_0}{[(x-x_0)^2 - \beta^2((y-y_0)^2 + z^2)]^{\frac{3}{2}}}, \qquad (4·11, 1)$$

where the region S' is defined by (4·7, 5). Φ may also be written as

$$\Phi(x,y,z) = -\frac{\partial}{\partial z} \int_{S'}^{*} \frac{\sigma(x_0, y_0)\, dx_0\, dy_0}{\sqrt{\{(x-x_0)^2 - \beta^2((y-y_0)^2 + z^2)\}}}, \qquad (4·11, 2)$$

i.e. it may be regarded as the z-component of the velocity field due to a distribution of sources of strength σ over S'. It then follows, as in the incompressible case, that

$$\Phi(x,y,+0) - \Phi(x,y,-0) = 2\pi\sigma. \qquad (4·11, 3)$$

Φ is an odd function of z, and so

$$\Phi(x,y,-0) = -\Phi(x,y,+0) \quad \text{and} \quad \Phi(x,y,+0) = \pi\sigma. \qquad (4·11, 4)$$

The pressure difference across a thin wing is proportional to the difference

‡ P. Germain, 'La théorie des mouvements homogènes et son application au calcul de certaines ailes delta en régime supersonique', *Rech. aéro.* no. 7 (Jan.-Fév. 1949); 'Hypothèses linéarisées', *Actes du colloque international de mécanique* (Poitiers, 1950); vol. 2.

between the x-components of the velocity. Hence, if the pressure difference across the wing is specified, we may obtain from it the difference

$$\Phi(x, y, +0) - \Phi(x, y, -0)$$

by integrating with respect to x. The doublet density is then determined by (4·11, 3), and the velocity potential can be calculated from (4·11, 2). We see therefore that the doublet distribution method provides a direct means for calculating the velocity potential, and hence the incidence distribution, on a thin wing for a given pressure distribution. On the other hand, since there is no direct relation between local doublet strength and local slope or incidence, the method is less suitable for the calculation of the pressure distribution on a given wing. Nevertheless, Brown[‡] has treated the problem of the flat Delta wing with subsonic leading edges (see § 4·8) by means of a doublet distribution method. Again, historically, the first paper on finite span wings in supersonic flow[§] was based on a distribution of supersonic horseshoe vortices (see (4·5, 72)). Although the paper, which deals with a case of a flat rectangular wing at incidence, contains an analytical mistake which invalidates the results, the method used is basically sound. Goldsworthy[||] has used the technique of Evvard and Ward (see § 4·8) in connexion with a doublet distribution method in order to determine the slope of a wing at zero incidence for a specified pressure distribution on a given plan-form.

Doublet and vortex distribution methods have also been used for the calculation of the flow in the wake of a given wing. They are, in fact, rather more important in this connexion, since, in general, the methods of the preceding sections describe the field of flow correctly only at the wing.[¶] We observe that the field of flow due to a surface distribution of vorticity on the x, y-plane can be obtained by a passage to the limit in (4·5, 67). The vorticity of the infinitesimally thick layer is given by

$$\xi\, dz_0 = -\{v(x_0, y_0, +0) - v(x_0, y_0, -0)\} = -\Delta v, \quad \text{say,}$$
$$\eta\, dz_0 = \quad u(x_0, y_0, +0) - u(x_0, y_0, -0) = \quad \Delta u,$$
$$\zeta\, dz_0 = 0,$$

‡ C. E. Brown, 'Theoretical lift and drag of thin triangular wings at supersonic speeds', *N.A.C.A. Rep.* no. 839 (1946). See also ch. 15 of A. Ferri, *Elements of Aerodynamics of Supersonic Flows* (Macmillan, 1949).

§ H. Schlichting, 'Tragflügeltheorie bei Überschallgeschwindigkeit', *Luftfahrtforsch.* vol. 13 (1936), pp. 320–35.

|| F. A. Goldsworthy, 'Supersonic flow over thin symmetrical wings with given surface pressure distribution', *Aero. Quart.* vol. 3 (1952), pp. 263–79.

¶ See: M. A Heaslet and H. Lomax, 'The use of source-sink and doublet distributions extended to the solution of arbitrary boundary problems in supersonic flow', *Tech. Notes Nat. Adv. Comm. Aero., Wash.,* no. 1515 (1948); H. Mirels and R. C. Haefeli, 'Line vortex theory for calculation of supersonic downwash', *Tech. Notes Nat. Adv. Comm. Aero., Wash.,* no. 1925 (1949); A. Robinson and J. H. Hunter-Tod, 'Bound and trailing vortices in the theory of supersonic flow, and the downwash in the wake of a delta wing', *Rep. Memor. Aero. Res. Comm., Lond.,* no. 2409 (1947).

and inserting these expressions in (4·5, 67) we obtain

$$u = -\frac{\beta^2}{2\pi} \overset{*}{\int_{S'}} (z - z_0)\, \Delta u\, \frac{dx_0\, dy_0}{[(x - x_0)^2 - \beta^2((y - y_0)^2 + z^2)]^{\frac{3}{2}}},$$

$$v = -\frac{\beta^2}{2\pi} \overset{*}{\int_{S'}} (z - z_0)\, \Delta v\, \frac{dx_0\, dy_0}{[(x - x_0)^2 - \beta^2((y - y_0)^2 + z^2)]^{\frac{3}{2}}},$$

$$w = \frac{\beta^2}{2\pi} \overset{*}{\int_{S'}} ((y - y_0)\, \Delta v + (x - x_0)\, \Delta u)\, \frac{dx_0\, dy_0}{[(x - x_0)^2 - \beta^2((y - y_0)^2 + z^2)]^{\frac{3}{2}}}.$$

$$(4\cdot11, 5)$$

With the aid of these formulae, the flow round a thin wing at incidence can be represented by a vorticity distribution over the wing and the wake.

The following example may illustrate the idea which is common to the so-called lift cancellation methods.

Consider a flat trapezoidal wing S at incidence α, as shown in Fig. 4·11, 1. We assume that the aspect ratio of the wing is sufficiently large so that the Mach lines through the end-points of the leading edge do not intersect on the wing. It is then sufficient to show how to calculate the induced velocity potential Φ in the starboard tip region only. Let Φ_0 be the induced velocity potential which is appropriate to the wing S_0 obtained by producing the leading and trailing edges of S to infinity in both directions. Φ_0 coincides with Φ in the region I, inboard of the Mach line BF, but is different from the unknown Φ in the region II (the triangle BFC). Instead of calculating Φ directly, we may first determine the

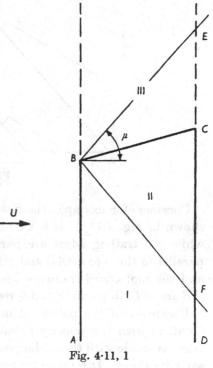

Fig. 4·11, 1

difference $\Phi - \Phi_0 = \Phi_1$. Φ_1 vanishes ahead of the leading edge everywhere. It also vanishes in the region I, while in the region II we have

$$\frac{\partial \Phi_1}{\partial z} = \frac{\partial \Phi}{\partial z} - \frac{\partial \Phi_0}{\partial z} = -U\alpha - (-U\alpha) = 0. \qquad (4\cdot11, 6)$$

Let III be the region between the leading and trailing edges outboard of the starboard tip BC. Then Φ is continuous across III and, being an odd function of z, it therefore vanishes on III. Hence, on III, Φ_1 satisfies the condition

$$\Phi_1 = -\Phi_0, \qquad (4\cdot11, 7)$$

where Φ_0 is known. Regions II and III together may be regarded as a new wing, the 'cancellation wing' S'. The transformed problem is then to determine Φ_1 on the cancellation wing subject to the conditions (4·11, 6) and (4·11, 7). The term 'lift cancellation' for this procedure is due to the fact that the superposition of Φ_1 on Φ_0 cancels the lift, or discontinuity of pressure, across the wing III.

The above problem can be solved more easily by Evvard's method (§ 4·8), or, alternatively, by a cone-field method (§ 4·10). However, there are other cases which so far have been dealt with mainly by methods for which the lift cancellation idea is essential. This is true, in particular, for swept-back wings with subsonic trailing edges.

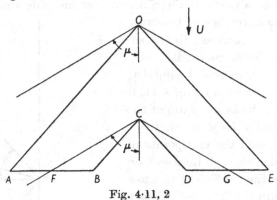

Fig. 4·11, 2

Consider, for example, the flat swept-back wing S whose plan-form is shown in Fig. 4·11, 2. It is assumed that the leading edges are subsonic, while the trailing edges are partly subsonic (BC and CD) and partly parallel to the y-axis (AB and DE). The Mach lines through the point C, and the root-chord trailing edge are supposed to intersect the segments AB and DE, in points F and G respectively.

Upstream of the polygonal line $AFCGE$ the velocity potential Φ is identical with the velocity potential Φ_0 for a Delta wing whose leading edges coincide with the leading edges of the given wing and which has the same incidence. It remains to determine Φ on the cancellation wing CFG. Put $\Phi_1 = \Phi - \Phi_0$, then Φ_1 vanishes upstream of the line FCG, while satisfying the conditions

$$\frac{\partial \Phi_1}{\partial z} = 0 \quad \text{on} \quad CFB \text{ and } CDG,$$

and

$$\frac{\partial \Phi_1}{\partial x} = -\frac{\partial \Phi_0}{\partial x} \quad \text{on} \quad CBD.$$

The last condition follows from the fact that

$$u' = -\frac{\partial \Phi}{\partial x} = 0$$

in the wake of the wing. In addition, it is necessary to ensure that the Joukowski condition is satisfied by Φ as the trailing edge is approached from upstream. A more complicated situation arises when the tips of the wing are parallel to the direction of flow (Figs. 4·11, 3 a and b). In these cases there are cancellation regions associated with the tips of the wing which

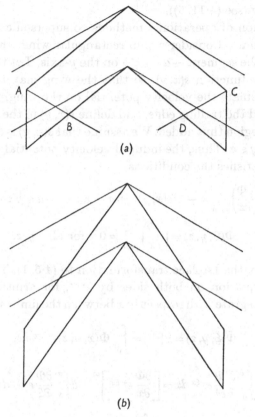

(a)

(b)

Fig. 4·11, 3

overlap the central cancellation region considered earlier. These complications necessitate the adoption of approximate methods. The problem has been dealt with by Cohen‡ who employed a method of superposition of infinitesimal cone fields to effect the lift cancellation. She found that the pressure is discontinuous across the Mach lines AB and CD in the case of Fig. 4·11, 3 a, such that the residual pressure difference aft of these lines is almost negligible. Interpreting this result in terms of the physical

‡ D. Cohen, 'Theoretical loading at supersonic speeds of flat swept-back wings with interacting trailing and leading edges', *Tech. Notes Nat. Adv. Comm. Aero., Wash.*, no. 1991 (1949); 'The theoretical lift of flat swept-back wings at supersonic speeds', *Tech. Notes Nat. Adv. Comm. Aero., Wash.*, no. 1555 (1948).

conditions, we would expect at least a rapid variation of the pressure in place of the discontinuity predicted by linearized theory.

Amongst other investigations which make use of the lift cancellation technique, we may mention a paper by Goodman‡ who employs doublet distributions (see (4·11, 1)), and a report by Mirels,§ whose work is based in the first instance on the representation of the flow by a surface distribution of vorticity (see (4·11, 5)).

The application of operational methods to supersonic wing theory was initiated by Gunn.‖ Consider a thin rectangular wing whose leading edge coincides with the segment $-a \leqslant y \leqslant a$ on the y-axis. Let the slope be given by an arbitrary function $s(x, y)$, so that the wing may be cambered and twisted. To calculate the velocity potential on the wing, we may produce the wing beyond the trailing edge, and define $s(x, y)$ in the most convenient manner in the region thus added. We assume that $s(x, y)$ is a known function for $x \geqslant 0$, $-a \leqslant y \leqslant a$. Thus, the induced velocity potential Φ is a solution of (4·5, 1) which satisfies the conditions

$$\left(\frac{\partial \Phi}{\partial z}\right)_{z=0} = -Us(x, y) \quad \text{for} \quad x \geqslant 0, \ -a \leqslant y \leqslant a, \qquad (4\cdot11, 8)$$

and

$$\Phi(0, y, z) = \left(\frac{\partial \Phi}{\partial x}\right)_{x=0} = 0 \quad \text{for all} \quad y, z.$$

We now apply the Laplace transformation to (4·5, 1). That is to say, we multiply the equation on both sides by $e^{-\xi x}$, for arbitrary ξ such that $\mathcal{R}(\xi) > 0$, and integrate with respect to x between the limits 0 and ∞. Putting

$$\bar{\Phi}(\xi, y, z) = \mathfrak{L}[\Phi] = \int_0^\infty \Phi(x, y, z) e^{-\xi x} dx,$$

we then have

$$\int_0^\infty \frac{\partial^2 \Phi}{\partial x^2} e^{-\xi x} dx = \left[\frac{\partial \Phi}{\partial x} e^{-\xi x}\right]_0^\infty + \xi \int_0^\infty \frac{\partial \Phi}{\partial \xi} e^{-\xi x} dx$$

$$= \xi^2 \int_0^\infty \Phi e^{-\xi x} dx$$

$$= \xi^2 \bar{\Phi},$$

while

$$\int_0^\infty \frac{\partial^2 \Phi}{\partial y^2} e^{-\xi x} dx = \frac{\partial^2}{\partial y^2} \int_0^\infty \Phi e^{-\xi x} dx$$

$$= \frac{\partial^2 \bar{\Phi}}{\partial y^2},$$

---

‡ T. R. Goodman, 'The lift distribution on conical and non-conical flow regions of thin finite wings in a supersonic stream', *J. Aero. Sci.* vol. 16 (1949), pp. 365–74.

§ H. Mirels, 'Lift cancellation technique in linearized supersonic wing theory', *N.A.C.A. Rep.* no. 1004 (1951).

‖ J. C. Gunn, 'Linearized supersonic aerofoil theory', Parts I and II. *Phil. Trans.* A. vol. 240, no. 820 (1947), pp. 327–73.

and
$$\int_0^\infty \frac{\partial^2 \Phi}{\partial z^2} e^{-\xi x} \, dx = \frac{\partial^2 \bar{\Phi}}{\partial z^2}.$$

Hence (4·5, 1) is replaced by

$$\beta^2 \xi^2 \bar{\Phi} - \frac{\partial^2 \bar{\Phi}}{\partial y^2} - \frac{\partial^2 \bar{\Phi}}{\partial z^2} = 0. \qquad (4\cdot 11, 9)$$

Similarly, multiplying (4·11, 8) by $e^{-\xi x}$ and integrating between 0 and ∞, we obtain

$$\left(\frac{\partial \bar{\Phi}}{\partial z} \right)_{z=0} = \bar{s}(\xi, y) \quad (-a \leqslant y \leqslant a), \qquad (4\cdot 11, 10)$$

where
$$\bar{s}(\xi, y) = \mathfrak{L}[s] = \int_0^\infty s(x, y) \, e^{-\xi x} \, dx.$$

(4·11, 9) and (4·11, 10) together constitute a boundary-value problem involving a partial differential equation of two variables only. The conditions at infinity have to be taken into account in order to make the problem determinate. Assuming that this problem has been solved, and $\bar{\Phi}$ determined, we may then obtain Φ by means of the inversion formula for the Laplace transform:[‡]

$$\Phi(x, y, z) = \frac{1}{2\pi i} \int_{\gamma - i\infty}^{\gamma + i\infty} \bar{\Phi}(\xi, y, z) \, e^{\xi x} \, d\xi. \qquad (4\cdot 11, 11)$$

For given y and z, the constant γ in (4·11, 11) must be chosen in such a way that $\bar{\Phi}(\xi, y, z)$ is regular for all ξ such that $\mathscr{R}(\xi) \geqslant \gamma$, i.e. for all values of ξ which are on or to the right of the straight line $\mathscr{R}(\xi) = \gamma$ in the complex ξ-plane.

The individual steps which are required to carry out this analysis are by no means simple. Thus, while the method possesses a certain theoretical interest, it must be regarded as inferior, both in generality and in ease of handling, to the methods developed in the preceding sections. On the other hand, it may be mentioned here that operational methods have acquired greater practical importance in connexion with the problems of unsteady supersonic wing theory.[§]

Another method which is chiefly of theoretical and historical interest is von Kármán's 'acoustic analogy'.[||]

The connexion between acoustics and linearized supersonic flow is more than an analogy, since both subjects are concerned largely with the same medium, air, are based on the same approximations, and depend on the same differential equation (4·1, 10). Thus, to a certain extent, the two theories are not merely analogous but coinciding. The theories diverge

[‡] See, for example, H. S. Carslaw and J. C. Jaeger, *Operational Methods in Applied Mathematics* (Oxford University Press, 2nd ed., 1948).

[§] J. W. Miles, 'The oscillating rectangular airfoil at supersonic speeds', *Quart. J. Appl. Math.* vol. 9 (1951), pp. 47–65.

[||] Th. von Kármán, 'Supersonic aerodynamics—principles and applications', *J. Aero. Sci.* vol. 14 (1947), pp. 373–409.

because they are concerned with different particular cases, or, mathematically speaking, because the boundary conditions appropriate to the two subjects are of a different type. Moreover, even when identical problems are considered, the quantities calculated by the aerodynamicist as a matter of course may be of small interest to the acoustics expert, and vice versa.

The problems of steady supersonic flow do not possess direct interpretations in acoustics. However, we may compare the partial differential equation which governs the flow ((4·5, 1) or (4·11, 9)) with the equation to which (4·1, 10) reduces when the motion is independent of x, viz.

$$\frac{1}{a_0^2}\frac{\partial^2\Phi}{\partial t^2} - \frac{\partial^2\Phi}{\partial x^2} - \frac{\partial^2\Phi}{\partial y^2} = 0. \tag{4·11, 12}$$

If we identify t formally with x, and a_0^{-1} with β, then (4·11, 12) is reduced to (4·11, 9). Thus, we identify an unsteady two-dimensional phenomenon with a steady phenomenon in three dimensions. This is the basis of von Kármán's acoustic analogy. The aerofoil in x, y, z-space corresponds to a distribution of variable acoustic impulses along segments of the y-axis. Thus, for a symmetrical wing at zero incidence, the acoustic impulses are represented by sources which are proportional to the normal velocity at any given time. In acoustics, it is natural to Fourier-analyse the variable strength of a source, i.e. to represent it for given y, by an integral

$$\int_{-\infty}^{\infty} (f(k) - ig(k))\, e^{ikt}\, dk.$$

In three dimensions, a diverging acoustic source is given by the expression

$$\Phi = \frac{C}{r} e^{ik(t-r/a_0)},$$

where r denotes the distance from the origin of the source and C and k are arbitrary constants. From this, a two-dimensional source is obtained, by integration, in terms of Bessel functions,‡ and the potential can then be written down as a Fourier integral. Finally, in the acoustic analogy, the wave drag is found to correspond to the rate at which energy is emitted by the system of acoustic sources.

The analysis has been developed in more detail by Chang.§ It is applied to the problem of determining the shape of a wing of given plan-form for a preassigned distribution of the pressure.

Problems for which linearization leads to partial differential equations other than (4·5, 1) arise when an aerofoil is placed in a non-uniform steady supersonic air stream. Non-uniformity of the main stream may occur in

‡ See H. Lamb, *Hydrodynamics* (6th ed., Cambridge Univ. Press, 1932), ch. 10.
§ C. C. Chang, 'Applications of Von Kármán's integral method in supersonic wing theory', *Tech. Notes Nat. Adv. Comm. Aero., Wash.*, no. 2317 (1951).

a wind tunnel or in an open jet. It is of some importance to be able to estimate the corresponding corrections which have to be made to the results obtained on the assumption that the main stream is uniform.

Suppose, for example, that the flow is two-dimensional and that $U(x, y)$, $V(x, y)$ are the velocity components of the main stream such that the flow is symmetrical with respect to the y-axis (i.e. $U(x, y) = U(x, -y)$, $V(x, y) = -V(x, -y)$). An aerofoil is placed, approximately, along the y-axis. It is assumed that U is of the order of magnitude of a_0, while V is small compared with U. Further suitable assumptions are made on the relative orders of magnitude of the quantities involved. With these assumptions, the linearization of (4·1, 6) for two-dimensional steady flow leads to the equation

$$\left(\frac{U^2}{a_0^2} - 1\right)\frac{\partial^2 \Phi}{\partial x^2} + \frac{2U}{a_0^2}\frac{\partial U}{\partial x}\frac{\partial \Phi}{\partial x} - \frac{\partial^2 \Phi}{\partial y^2} = 0.$$

In the particular case when the pressure along the x-axis in the main flow is a linear function of x, the problem can be reduced to one in three-dimensional steady supersonic flow by an artifice.‡

4·12   Higher order theories of supersonic flow

The linearized equation (4·1, 20), which for the supersonic case has been considered at some length in §§ 4·5–4·12, provides an adequate approximation to the complete non-linear equation (4·1, 21) for Mach numbers that are not too close to 1. Roughly speaking, the minimum supersonic Mach number for which it may be used is of the order 1·15, but the figure depends on the aerofoil shape, and, in particular, on whether the leading edge is sharp or rounded. For aerofoils with sharp leading edges the higher order effects may be regarded as corrections to the linearized solution, but if the leading edge is blunt, strong shock waves can occur which change completely the nature of the flow.

For two-dimensional aerofoils it is possible to find exact solutions of the compressible flow equations; to do this we require to introduce the concepts of characteristic lines (the Mach lines of § 4·6) and of shock waves. We first return to the full non-linearized irrotational flow equation (4·1, 21),

$$\left(1 - \frac{\Phi_x^2}{a^2}\right)\Phi_{xx} + \left(1 - \frac{\Phi_y^2}{a^2}\right)\Phi_{yy} + \left(1 - \frac{\Phi_z^2}{a^2}\right)\Phi_{zz} - 2\frac{\Phi_x \Phi_y}{a^2}\Phi_{xy}$$

$$- 2\frac{\Phi_z \Phi_x}{a^2}\Phi_{zx} - 2\frac{\Phi_y \Phi_z}{a^2}\Phi_{yz} = 0. \quad (4·12, 1)$$

‡ See A. Robinson, 'Non-uniform supersonic flows', *Quart. Appl. Math.* vol. 10 (1953), pp. 307–19. Compare also H. F. Ludloff and M. B. Friedman, 'Corrections for lift, drag, and moment of an airfoil in a supersonic tunnel having a given pressure gradient', *Tech. Notes Nat. Adv. Comm. Aero., Wash.*, no. 2849 (1952).

If Φ_x^2, Φ_y^2, $\Phi_z^2 < a^2$, this is a hyperbolic quasi-linear differential equation of the second order. It is sufficient for our purpose to define the characteristic surfaces (or 'characteristics') for such an equation as surfaces across which there can be discontinuities in one or more of the *second-order* derivatives of Φ. In the two-dimensional case, the characteristic surfaces become characteristic lines; (4·12, 1) then reduces to

$$\left(1 - \frac{\Phi_x^2}{a^2}\right)\Phi_{xx} + \left(1 - \frac{\Phi_y^2}{a^2}\right)\Phi_{yy} - 2\frac{\Phi_x\Phi_y}{a^2}\Phi_{xy} = 0. \qquad (4·12, 2)$$

Let us suppose a line $C : x = x(s)$, $y = y(s)$ is a characteristic in the sense just defined, and let us write

$$\Delta\Phi_{xx} = (\Phi_{xx})_1 - (\Phi_{xx})_2,$$

$$\Delta\Phi_{yy} = (\Phi_{yy})_1 - (\Phi_{yy})_2,$$

$$\Delta\Phi_{xy} = \Delta\Phi_{yx} = (\Phi_{xy})_1 - (\Phi_{xy})_2 = (\Phi_{yx})_1 - (\Phi_{yx})_2,$$

where the suffices 1 and 2 refer to adjacent points on either side of C. Since

$$\frac{\partial\Phi}{\partial s} = \Phi_x x'(s) + \Phi_y y'(s), \qquad (4·12, 3)$$

and since $\partial^2\Phi/\partial x\,\partial s$, $\partial^2\Phi/\partial y\,\partial s$ will be continuous across C, we obtain, differentiating (4·12, 3) with respect to x and y at 1 and 2,

$$\left.\begin{aligned} x'(s)\,\Delta\Phi_{xx} + y'(s)\,\Delta\Phi_{xy} = 0, \\ x'(s)\,\Delta\Phi_{xy} + y'(s)\,\Delta\Phi_{yy} = 0. \end{aligned}\right\} \qquad (4·12, 4)$$

Also we have

$$\left(1 - \frac{\Phi_x^2}{a^2}\right)\Delta\Phi_{xx} + \left(1 - \frac{\Phi_y^2}{a^2}\right)\Delta\Phi_{yy} - 2\frac{\Phi_x\Phi_y}{a^2}\Delta\Phi_{xy} = 0, \qquad (4·12, 5)$$

(4·12, 4) and (4·12, 5) will possess a non-trivial solution for $\Delta\Phi_{xx}$, $\Delta\Phi_{xy}$, $\Delta\Phi_{yy}$, provided

$$\begin{vmatrix} 1 - \dfrac{\Phi_x^2}{a^2} & -2\dfrac{\Phi_x\Phi_y}{a^2} & 1 - \dfrac{\Phi_y^2}{a^2} \\[2mm] x'(s) & y'(s) & 0 \\[2mm] 0 & x'(s) & y'(s) \end{vmatrix} = 0, \qquad (4·12, 6)$$

and this equation defines the characteristics in terms of the velocity components, $u = -\Phi_x$, $v = -\Phi_y$. Expanding (4·12, 6) we obtain

$$(a^2 - u^2)\,(y'(s))^2 + 2uv\,x'(s)\,y'(s) + (a^2 - v^2)\,(x'(s))^2 = 0, \qquad (4·12, 7)$$

a quadratic equation for the slope of the characteristics. Its solution is

$$\zeta = \frac{dy}{dx} = \frac{-uv \pm a\sqrt{(u^2 + v^2 - a^2)}}{a^2 - u^2}, \qquad (4·12, 8)$$

so that the two characteristic directions at a point are determined by the local values of the velocity and the speed of sound. We observe that they are real only if $u^2 + v^2 \geqslant a^2$, i.e. in sonic and supersonic flow. If we refer (4·12, 8) to axes (x', y') along and normal to the local velocity direction, such that the x'-axis has the direction of the velocity, then (4·12, 8) becomes

$$\frac{dy'}{dx'} = \mp \frac{1}{\sqrt{(M^2 - 1)}} = \mp \frac{1}{\beta}, \qquad (4·12, 9)$$

where $M = \dfrac{q}{a} = \dfrac{\sqrt{(u^2 + v^2)}}{a^2}$. Hence the characteristic directions are the same as for the Mach lines (§ 4·6) at the point in question. The Mach lines as defined in § 4·6 for linearized flow are straight lines, but we may now extend their definition to cover non-linearized flow by identifying them with the characteristics (4·12, 8).

In general, the solutions of (4·12, 8) depend upon the boundary conditions of the problem considered, but in the hodograph plane integrals of the differential equation corresponding to (4·12, 8) can be obtained independently of the boundary conditions. Thus, we have

$$\frac{du}{dx} = \frac{\partial u}{\partial x} + \zeta \frac{\partial u}{\partial y},$$

$$\frac{dv}{dx} = \frac{\partial v}{\partial x} + \zeta \frac{\partial v}{\partial y},$$

where ζ is the slope of one of the characteristics as defined by (4·12, 8). Using the irrotational flow condition $\partial u / \partial y = \partial v / \partial x$, we obtain

$$\frac{\partial u}{\partial x} = \frac{du}{dx} - \zeta \frac{dv}{dx} + \zeta^2 \frac{\partial v}{\partial y},$$

and substituting in (4·12, 2) we get

$$\left(1 - \frac{u^2}{a^2}\right) \frac{du}{dx} - \left\{\left(1 - \frac{u^2}{a^2}\right) \zeta + \frac{2uv}{a^2}\right\} \frac{dv}{dx}$$

$$+ \left\{\left(1 - \frac{u^2}{a^2}\right) \zeta^2 + \frac{2uv}{a^2} \zeta + 1 - \frac{v^2}{a^2}\right\} \frac{\partial v}{\partial y} = 0. \qquad (4·12, 10)$$

The last term in (4·12, 10) is zero, through (4·12, 7), and hence

$$\left(1 - \frac{u^2}{a^2}\right) \frac{du}{dx} - \left\{\left(1 - \frac{u^2}{a^2}\right) \zeta + \frac{2uv}{a^2}\right\} \frac{dv}{dx} = 0. \qquad (4·12, 11)$$

Corresponding to the two solutions (4·12, 8) we thus have

$$\omega = \frac{du}{dv} = \frac{uv \pm a\sqrt{(u^2 + v^2 - a^2)}}{a^2 - u^2}. \qquad (4·12, 12)$$

If we denote by ζ_+ and ζ_- the slopes of the characteristics in the x, y-plane corresponding to the positive and negative signs in (4·12, 8) respectively, and ω_+ and ω_- for the slopes in the hodograph plane (4·12, 12), equations (4·12, 8) and (4·12, 12) show that

$$\left.\begin{array}{l} \zeta_+ = -\omega_-, \\ \zeta_- = -\omega_+. \end{array}\right\} \qquad (4·12, 13)$$

Equation (4·12, 12) can be integrated by quadrature without the use of boundary conditions. We put

$$q^2 = u^2 + v^2 \quad \text{and} \quad \tan\theta = \frac{v}{u}, \tag{4·12, 14}$$

so that q and θ are polar coordinates in the hodograph plane. Differentiation yields

$$q\frac{dq}{dv} = v + u\frac{du}{dv},$$

and

$$\frac{d\theta}{dv} = \frac{u - v\dfrac{du}{dv}}{u^2 + v^2}.$$

Using (4·12, 12), we find readily that

$$\frac{1}{q}\frac{dq}{d\theta} = \frac{a(av \pm u\sqrt{(u^2 + v^2 - a^2)})}{\mp\sqrt{(u^2 + v^2 - a^2)}\,(\pm\sqrt{(u^2 + v^2 - a^2)} + av)},$$

or

$$\frac{1}{q}\frac{dq}{d\theta} = \mp\frac{1}{\sqrt{(M^2 - 1)}} = \tan\mu, \tag{4·12, 15}$$

where M is the local Mach number q/a.

Combined with the differential form of Bernoulli's equation

$$q\,dq + \frac{1}{\rho}\,dp = 0,$$

(4·12, 15) gives the important relation

$$\frac{1}{\rho}\frac{dp}{d\theta} = \pm\frac{q^2}{\sqrt{(M^2 - 1)}}. \tag{4·12, 16}$$

Use of the alternative form

$$\tfrac{1}{2}q^2 + \frac{1}{\gamma - 1}a^2 = \text{constant},$$

for Bernoulli's equation (cf. (4·1, 4)), shows that

$$\frac{dq}{q} = \frac{dM}{M(1 + \tfrac{1}{2}(\gamma - 1)M^2)},$$

and substituting in (4·12, 15), we get

$$d\theta = \mp\frac{\sqrt{(M^2 - 1)}}{M(1 + \tfrac{1}{2}(\gamma - 1)M^2)}\,dM, \tag{4·12, 17}$$

which can be integrated to give finally

$$\mp\theta = \mu - \sqrt{\frac{\gamma + 1}{\gamma - 1}}\tan^{-1}\left(\sqrt{\frac{\gamma + 1}{\gamma - 1}}\tan\mu\right) + \text{constant}, \tag{4·12, 18}$$

where we have again written $\mu = \sin^{-1}1/M$ for the Mach angle. This relationship holds along any Mach line or characteristic of the flow. We note that in (4·12, 18) the positive sign holds for the ζ_+ characteristic of equation

(4·12, 8) (i.e. the characteristic with the positive sign on the right-hand side) and the negative sign for the ζ_- characteristic. For a given value of the constant on the right-hand side, (4·12, 18) defines an epicycloid. Tables of θ and μ, which are useful for calculations, can be found from many sources.

We shall now consider the possibility of finding a solution for a two-dimensional supersonic flow field in which all of one set of characteristics in the x, y-plane are straight lines. The type of flow we envisage is one in which there is a region I of uniform flow (for which all the characteristics are straight lines) next to a region II of non-uniform flow. The boundary between the two regions is supposed to consist of a single straight characteristic c, and we wish to prove that at least one set of characteristics in the region II consists also of straight lines. The type of flow in region II is then called a region of simple wave flow. Let us suppose that c is, say, a ζ_+ characteristic; it corresponds in the u, v-hodograph plane to a ω_+ characteristic, which in this case is a single point C. Since in the u, v-plane there can be only one member of each set of characteristics passing through each point with slopes given by (4·12, 15), it follows that the ω_- characteristics consist of a single curve c' through the point C. Every member of the ω_+ characteristics is a point lying on this curve, and therefore all the corresponding ζ_+ characteristics are straight lines. Moreover, not only the ζ_- characteristics, but all curves which cut the ζ_+ characteristics, correspond to the same curve c', and thus the relations (4·12, 15) and (4·12, 18) (with the negative signs) hold for all these curves. If, in particular, we take such a curve to be a solid boundary to which the flow is tangential, we may suppose θ to be the slope of this surface. (4·12, 18) then gives us immediately the Mach number distribution on the surface in terms of a single constant which has to be determined from initial conditions.

Let us apply these conclusions to flow round a convex bend (Fig. 4·12, 1), in which we suppose the initial flow to be uniform in the region forward of the Mach line emanating from the start of the bend. The Mach line bounding the region of uniform flow may consist of either a ζ_+ or a ζ_- characteristic, as shown in the figure, though we shall demonstrate presently that the ζ_- line is the only one that can occur in real physical flow. Assuming for the moment, however, that it is a ζ_+ characteristic, the change in velocity round a corner is given by (4·12, 18) with the positive sign, where θ denotes the slope of the bend and the constant is determined by the initial uniform flow Mach number. From (4·12, 17) we see that the Mach number decreases round the bend, i.e. we get flow compression. Similarly for the case shown in (b) in Fig. 4·12, 1 with the ζ_- characteristics bounding the initial flow, we obtain an expansion of the flow.

The flow again becomes uniform in the region beyond the Mach line emanating from the end of the corner. However, it may be shown that

(4·12, 18) has a maximum value for a given incident flow Mach number, so that there is a limiting value of the angle through which supersonic flow can be turned in this manner. For corners greater than this maximum, the theoretical solution is as shown in Fig. 4·12, 2. The flow separates from the surface when it has been deflected through the maximum possible angle, leaving a region of cavitation (zero pressure) between it and the surface. The maximum angle increases as the Mach number decreases, having its

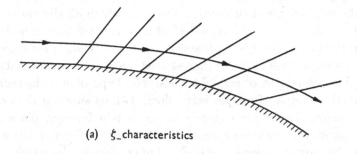

(a) ζ_- characteristics

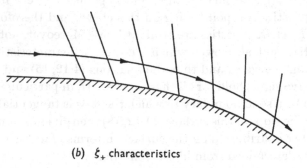

(b) ζ_+ characteristics

Fig. 4·12, 1

greatest value with an incident Mach number of unity for which the maximum deflexion is approximately 130°. Flow deflexions of such an order of magnitude are very unlikely to occur for normal aerofoils.

So far in the treatment there has been no indication as to which of the two solutions represented by (a) and (b) in Fig. 4·12, 1 is physically correct. But we can show with the aid of the results of linearized theory that it is the backward-facing ζ_- characteristics that are the physically possible ones. Thus, we may consider that in any small region of supersonic flow in which there are no discontinuities of velocity, the flow is linearized, in the sense that the changes in velocity occurring in the region are small compared with the magnitude of the velocity. It follows from the considerations of §4·1 that locally the direction in which a small disturbance would be propagated in such a region is along the Mach line making an angle less than 90° with the direction of flow, and the area in that region that can be influenced by the conditions at a point within it lies inside a wedge, the sides of which make an angle $\mu = \sin^{-1} 1/M$ with the flow direction. Therefore, in the case considered here, the initial direction of the bounding Mach line between the uniform and the non-uniform flow, emanating from the start of the bend, must be downstream, viz. the correct solution is the one in which the ζ_- characteristics are straight lines.

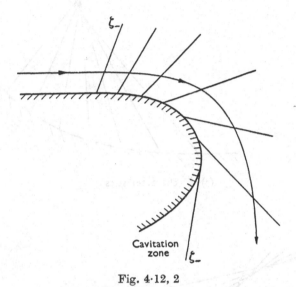

Fig. 4·12, 2

The analysis given above for flow round a bend is equally applicable to flow round a sharp corner. All the characteristics originate from the sharp corner forming an expansion fan, or a 'centred' simple wave. This type of flow is known as 'Prandtl–Meyer flow'.‡

In both solutions for a flow round a convex bend, as can be seen from (4·12, 9), the straight characteristics are divergent away from the surface and do not meet each other in the flow field. However, when the corresponding case of flow round a concave bend is considered, with either the ζ_+ or the ζ_- characteristics straight lines, the characteristics are convergent and intersect at a finite distance from the body (Fig. 4·12, 3). Since the flow velocity is constant along every straight Mach line it follows that the velocity at the points of intersection of the characteristics is not single-

‡ The original papers are: L. Prandtl, 'Neue Untersuchungen über die strömende Bewegung der Gase und Dämpfe', *Phys. Z.* vol. 8 (1907), pp. 23–31; Th. Meyer, 'Über zweidimensionale Bewegungsvorgänge in einem Gas, das mit Überschallgeschwindigkeit strömt', *Forschungsh. Ver. dtsch. Ing.* vol. 62 (1908), pp. 31–67.

valued, and the method of solution breaks down. In particular, Prandtl–Meyer flow in a sharp concave corner, which is governed by equation (4·12, 18), can be valid nowhere. To deal with a case of flow turning through a concave bend we therefore have recourse to the concept of a shock wave, the appearance of which in this case is suggested by the convergence of the characteristics, indicating that the changes of velocity, pressure, etc., take

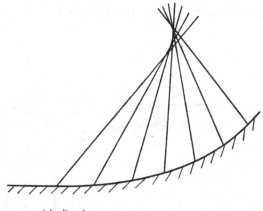

(a) ζ_- characteristics

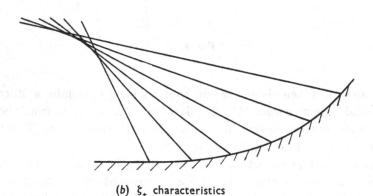

(b) ζ_+ characteristics

Fig. 4·12, 3

place more rapidly. To cover the properties of shock waves necessary in dealing with aerofoil theory, it will be sufficient to outline some of their chief properties.

We consider the possibility in two-dimensional flow of the occurrence of lines of discontinuity in the velocity, pressure and density within the body of the fluid (plane shock waves). To relate values on either side of a shock, which we signify by the suffices 1 and 2, we use the condition of continuity of mass flow, of momentum and of total energy across the shock. The resulting equations are

$$\rho_1 q_{1n} = \rho_2 q_{2n},$$
$$p_1 + \rho_1 q_{1n}^2 = p_2 + \rho_2 q_{2n}^2,$$
$$\rho_1 q_{1n} q_{1t} = \rho_2 q_{2n} q_{2t},$$

and

$$\tfrac{1}{2} q_1^2 + \frac{\gamma}{\gamma - 1} \frac{p_1}{\rho_1} = \tfrac{1}{2} q_2^2 + \frac{\gamma}{\gamma - 1} \frac{p_2}{\rho_2} = \frac{1}{2} \frac{\gamma + 1}{\gamma - 1} q^{*2},$$

$$(4·12, 19)$$

where the suffices n and t refer to components of the velocity normal and tangential to the shock wave (Fig. 4·12, 4). In the last equation q^* is the critical velocity, defined as the flow velocity which equals the local velocity of sound. From Bernoulli's equation, written as (cf. equation (4·1, 4))

$$\tfrac{1}{2} q^2 + \frac{\gamma}{\gamma - 1} \frac{p}{\rho} = \tfrac{1}{2} q^2 + \frac{a^2}{a^2 - 1} = \text{constant},$$

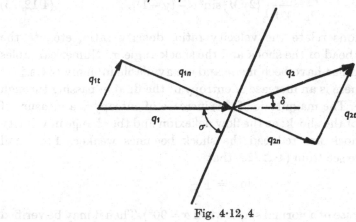

Fig. 4·12, 4

we see that

$$\tfrac{1}{2} q^2 + \frac{\gamma}{\gamma - 1} \frac{p}{\rho} = \tfrac{1}{2} q^{*2} + \frac{q^{*2}}{\gamma - 1} = \frac{1}{2} \frac{\gamma + 1}{\gamma - 1} q^{*2},$$

which verifies the last equation of (4·12, 19).

Equations (4·12, 19) can be used to establish various relations between the values of the variables before and after the shock. We shall merely state the most important of these:[‡]

$$q_{1t} = q_{2t}, \qquad\qquad (4·12, 20)$$

$$q_n^{*2} = q_{1n} q_{2n}; \qquad\qquad (4·12, 21)$$

the Rankine–Hugonoit equation, relating the pressure and density across the shock:

$$\frac{p_2}{p_1} = \frac{\dfrac{\gamma + 1}{\gamma - 1} \dfrac{\rho_2}{\rho_1} - 1}{\dfrac{\gamma + 1}{\gamma - 1} - \dfrac{\rho_2}{\rho_1}}. \qquad\qquad (4·12, 22)$$

‡ Details may be found in R. Courant and K. O. Friedrichs, *Supersonic Flow and Shock Waves* (Interscience, New York, 1948).

In many applications the flow deflexion δ is known, and relations are therefore desired which express the velocities, pressures, etc., behind the shock in terms of the values ahead of the shock and δ. First, the relation

$$\tan \delta = \frac{2(M_1^2 \sin^2 \sigma - 1)}{\tan \sigma \{2 + M_1^2(\gamma + \cos 2\sigma)\}}, \qquad (4·12, 23)$$

enables the shock wave angle σ (see Fig. 4·12, 4) to be found from the incident Mach number and the flow deflexion. The Mach number following the shock can be found from

$$M_2^2 = \frac{1 + \frac{1}{2}(\gamma - 1) M_1^2}{\gamma M_1^2 \sin^2 \sigma - \frac{1}{2}(\gamma - 1)} + \frac{M_1^2 \cos^2 \sigma}{1 + \frac{1}{2}(\gamma - 1) M_1^2 \sin^2 \sigma}, \qquad (4·12, 24)$$

and the pressure ratio from

$$\frac{p_2}{p_1} = \frac{1}{\gamma + 1} \{2\gamma M_1^2 \sin^2 \sigma - (\gamma - 1)\}. \qquad (4·12, 25)$$

Further equations relate the velocity ratio, density ratio, etc., to the Mach number ahead of the shock and the shock angle σ. Numerical tables from these formulae have been made and are available in many texts.‡

In general, there is an increase of entropy of the fluid in passing through the shock wave. The magnitude of this increase of entropy is a measure of the 'strength' of the shock; as the flow deflexion and the change in velocity through the shock are reduced the shock becomes weaker. For small deflexions, δ, we see from (4·12, 23) that

$$\sin \sigma \doteqdot \pm \frac{1}{M}$$

(excluding the case of a normal shock when $\sigma = 90°$). Then it may be verified that (4·12, 24) and (4·12, 25) yield $M_1 \doteqdot M_2$ and $p_1 \doteqdot p_2$. Hence, in this case the shock wave reduces approximately to a Mach wave and the flow through it is isentropic.

An extremely useful representation of transition through shock waves is the hodograph curve of q_2/q^*, for given values of $u_1/q^* = q_1/q^*$. Here q^* is the critical velocity, and we take (u, v) to be the components of q, with u parallel to the incident velocity q_1. The governing equation, obtained from (4·12, 19), is

$$\frac{v_2^2}{q^{*2}} = \left(\frac{q_1}{q^*} - \frac{q_1^2}{q^{*2}} \right)^2 \frac{\dfrac{q_1}{q^*} \dfrac{u_2}{q^*} - 1}{\dfrac{2}{\gamma + 1} \dfrac{q_1^2}{q^{*2}} - \dfrac{q_1}{q^*} \dfrac{u_2}{q^*} + 1}. \qquad (4·12, 26)$$

Plotting v_2/q^* against u_2/q^*, we obtain curves of the type shown in Fig. 4·12, 5 for each value of the incident Mach number. Such curves are known as

‡ For example, H. W. Emmons, *Gas Dynamics Tables for Air* (Dover, New York, 1947); or *Compressible Airflow Tables* (Oxford University Press, 1952), sect. III.

shock polar curves. From them we can find many of the important properties of shock waves. Thus, the points P and Q on the u-axis correspond respectively to $q_1=q_2$ (i.e. a vanishingly weak shock wave), and to $q_1q_2=q^{*2}$, a strong normal shock from supersonic velocity represented by the vector $\overrightarrow{OP}$ to subsonic velocity $\overrightarrow{OQ}$. The line of demarcation between subsonic and supersonic flow behind the shock is shown in the figure by the dotted circle. Secondly, we can see that for a given flow deflexion, δ, and incident Mach number there are in general three possible solutions to (4·12, 26), as shown by the points A, B, C in Fig. 4·12, 5. Point C corresponds to transition

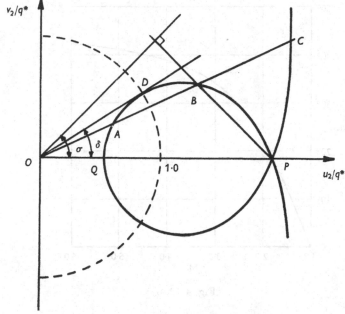

Fig. 4·12, 5

to a velocity after the shock greater than the incident velocity, and it may be shown that this involves a decrease in entropy of the flow through the shock. This transition is therefore ruled out. The other two points A and B represent a strong shock, with subsonic flow after it, and a relatively weak supersonic shock. Since the shock angle is given by the normal to the line joining the points q_2/q^* and P, we see that the strong shock approximates to a normal shock, whilst the weak shock is an oblique shock with an angle approximately equal to the Mach angle, which is given by the direction of the normal to the tangent to the shock polar at the point P. There is no theoretical basis for preferring one of these two possible solutions to the other, but it appears from experiment that, if compatible with the flow conditions, the weaker shock is more likely to occur. Finally, from the shock polar curve we see that there is a maximum deflexion of the flow that is possible through a shock wave, where both the solutions represented by the points A and B meet at the point D. Fig. 4·12, 6 shows the variation of this maximum possible value $\delta_{\max}$ with Mach number ahead of the shock.

Having summarized the properties of Mach and shock waves we are now in a position to discuss their application to the calculation of flow about two-dimensional aerofoils.

We consider first a double-wedge aerofoil (Fig. 4·12, 7), the flow deflexions at the leading edge being assumed less than the maximum allowed in plane-shock theory for the given incident flow Mach number (Fig. 4·12, 6).

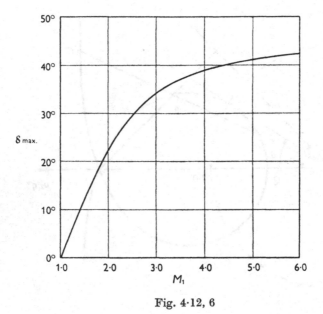

Fig. 4·12, 6

For large enough incidence, we may assume a Prandtl–Meyer expansion on the upper surface at the leading edge. The flow is uniform in front of the Mach wave emanating from the leading edge, and is again uniform and parallel to the aerofoil surface after the last Mach wave of the Prandtl–Meyer expansion fan. The governing equation is (4·12, 18), with the constant found from the main-stream velocity. A similar flow expansion can be calculated at the vertex B of the double wedge. At the trailing edge C the flow must be compressed to join the flow from the lower surface. Since an abrupt deflexion of flow at C is required, there is no possibility of a compression wave accomplishing this, and a plane shock wave is therefore introduced, with strength determined from the uniform flow Mach number in front of it and equations (4·12, 22)–(4·12, 25). The shock is assumed to be weak (corresponding to the point B in the shock polar diagram, Fig. 4·12, 5) with supersonic flow behind it. Along the lower surface of the aerofoil similar considerations apply; a plane shock is attached to the lower surface of the leading edge, and expansion fans form at D and C (see Fig. 4·12, 7).

The forces acting on the aerofoil can then be found by integration of the pressure over the aerofoil surface, but no explicit analytic expressions are obtainable, in contrast to the results of linearized theory. Details of calculation procedures may be found in several publications.‡

This simple picture of the nature of the flow about a wedge aerofoil, although verified satisfactorily by experiment, in fact is not theoretically quite correct. In the first instance, the flows in the section ECG and FCG of the wake, originating from the upper and lower surfaces respectively, cannot be matched along CG if we assume uniform flow throughout the

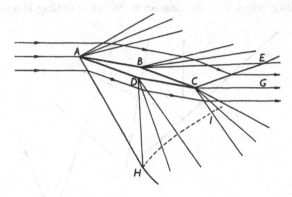

Fig. 4·12, 7

wake. For, in the general case, there are two independent physical quantities (the Mach numbers and the pressures on each side of CG) to be matched, and only one adjustable degree of freedom (the flow direction $\overrightarrow{CG}$); in practice we would expect a vortex wake to start at C, which would separate the two flow regions. Alternatively, the situation may be regarded as arising from the fact that the entropy increase occurring in the compression shocks, will, in general, be different on the upper and lower surfaces; thus, the discrepancy in the simple representation of Fig. 4·12, 7 will become smaller as the shock strengths decrease, and we might expect to obtain a theoretically consistent solution if we were to assume isentropic flow through the shocks. But in this case we find a difficulty of a different nature, and the solution in the wake is not unique, since now the Prandtl–Meyer flow equation (4·12, 18) is applicable to both expansions and

‡ For example: H. R. Ivey, G. W. Stickle and A. Schuettler, 'Charts for determining the characteristics of sharp-nose airfoils in two-dimensional flow at supersonic speeds', *Tech. Notes Nat. Adv. Comm. Aero., Wash.,* no. 1143 (1947); N. Edmonson, F. D. Murnaghan and R. M. Snow, 'The theory and practice of two-dimensional supersonic pressure calculations', *Johns Hopk. Univ. Appl. Phys. Lab. Bumblebee Rep.* no. 26 (1945).

compressions, and the two wake regions match whatever angle is chosen for the final flow direction. However, the problem thus presented is not important, since the difference between the velocities in the regions ECG and FCG obtained from the exact compression shock equations is never large if we assume the final flow direction to be the same as the incident flow direction. And, moreover (excluding the effect of viscosity), the nature of the wake and the shocks and expansions at the trailing edge cannot affect the pressure on the aerofoil itself or the forces acting on it.

A second effect omitted in the simple representation of Fig. 4·12, 7, and one which may in certain circumstances affect the aerofoil pressure distribution, is the interaction of the expansion fan at D with the shock wave from the leading edge. The expansion wave on meeting the shock wave at

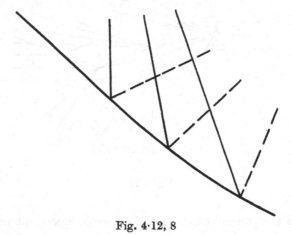

Fig. 4·12, 8

H reduces the deflexion of flow through it, and thus reduces its intensity and bends it in the manner shown in Fig. 4·12, 8. The change in entropy through the shock varies along the curved part of the shock and produces a region of rotational flow behind it. In fact, a more detailed study shows that weak reflexion compression waves also appear at the shock line, the first of which is shown as HI in Figs. 4·8, 7 and 4·8, 8. It will be seen that in the case illustrated in Fig. 4·12, 7 the reflected wave does not meet the aerofoil, although in other cases it is possible for the pressure distribution on the aerofoil to be altered by the presence of such reflected waves. In a similar manner, the expansion fan from C meets the shock and further reduces its intensity, so that with increasing distance from the aerofoil the shock intensity tends to zero, and its direction approaches the direction of the main-flow Mach waves. On the upper surface of the aerofoil there is a corresponding interaction of the trailing edge shock with the expansion waves from the upper surface, with the result that at infinity the shock waves are altogether eliminated. In this way the shocks are limited in extent, and the entropy increase for the complete flow system is finite. The calculation of the rotational flow behind the curved shock waves is a step-by-step process, and, although possible,‡ it is extremely arduous to work out the exact flow taking into account the interactions of shocks and Mach waves. The effects on the aerodynamic forces are negligible in all practical cases.

At smaller incidences the double-wedge aerofoil of Fig. 4·12, 7 can have shock wave on both upper and lower surfaces of the leading edge. The flow pattern is then similar to that on the lower surface at the higher angle of incidence, except that a shock wave can occur on both surfaces of the trailing

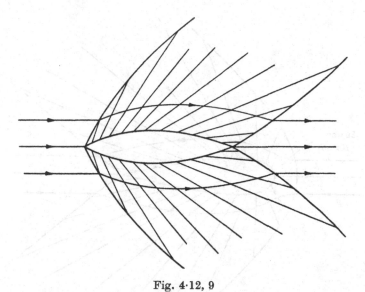

Fig. 4·12, 9

edge. As an example we have shown in Fig. 4·12, 9 the flow at zero incidence of a sharp-nosed aerofoil with continuously curved surfaces. This case illustrates, more clearly than the previous one, the Mach shock wave interaction effect in producing curved shocks at both leading and trailing edges, and shows that the entire flow over the aerofoil surfaces can be rotational, thus contradicting the basic assumptions of the supersonic flow field theory developed in this section. In spite of this, the results of calculations made for such aerofoils, using the method described above for the double-wedged aerofoil, are in good agreement with experiment, and, indeed, it may be shown that for all practical types of supersonic aerofoil (i.e. thin aerofoils with 'sharp' leading edges) the strengths of the compression waves reflected from the shocks are sufficiently small to be neglected.

‡ See A. Ferri, *Elements of Aerodynamics of Supersonic Flows* (Macmillan, 1949), chs. 5 and 7.

So far we have explicitly excluded cases in which the incidence of the aerofoil was large, or the leading edge rounded. We have seen that there is a maximum-flow deflexion through a shock for a given incident Mach number, as illustrated in Fig. 4·12, 6. This means that the flow configurations of Figs. 4·12, 7 and 4·12, 9, with an attached edge shock wave which deflects the flow from the incident flow direction to the direction of the aerofoil leading edge tangent, are impossible if the incidence is too high, or if the leading edge angle is too large. However, for an aerofoil with a non-zero leading edge radius, the maximum flow deflexion is exceeded at

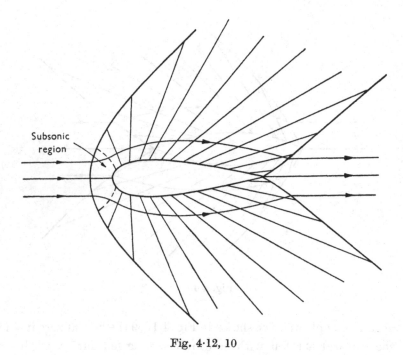

Subsonic region

Fig. 4·12, 10

the leading edge for all Mach numbers. Experiment shows that in such cases the shock at the leading edge is detached, the flow then being similar to that illustrated in Fig. 4·12, 10. The shock wave in front of the leading edge (or 'bow wave') is of the strong type; the velocity behind it is subsonic, and is reduced to zero at the leading edge stagnation point of the aerofoil. The shock is strongest at the centre, where it is normal; on either side the flow deflexion increases, and (as can be seen from the shock polar diagram, Fig. 4·12, 5) the shock becomes weaker until the point of maximum flow deflexion is reached. Here the shock changes its character, and becomes a member of the weak family of shocks, behind which the flow is supersonic. Moving out along the shock wave, the shock gradually loses its strength and eventually becomes a Mach line with no deflexion of the flow through it.

It is clear from this description that the calculation of the flow in such a case is not easy. For conventional supersonic aerofoils with sharp leading edges the situation will not occur except in the transonic régime. Numerical methods of calculation are possible, but it is suspected that the influence of boundary layers on the shock waves is large enough to change radically the nature of the solutions obtained from inviscid theory.

The general method described above, which employs both shock compression and Prandtl–Meyer expansion formulae, can be simplified and calculation facilitated by using equation (4·12, 18) for simple wave flow, in place of the shock wave equations (4·12, 23) and (4·12, 24). The former equation, as we have seen, is strictly applicable only to *expansion* round a corner. However, it will be shown that for thin aerofoils at small incidence, where shock strengths are small, this procedure is a legitimate approximation. It has been proved that in the limit when the flow deflexion through a shock becomes zero, the shock angle σ is equal to the Mach angle μ. For small shock strengths we may therefore write an expansion for the pressure coefficient as

$$C_p = \frac{p_2 - p_1}{\frac{1}{2}\rho_1 q_1^2} = C_1\delta + C_2\delta^2 + C_3\delta^3 + \ldots, \qquad (4\cdot12, 27)$$

where δ is the flow deflexion through the shock. The coefficients in (4·12, 27) can be found from (4·12, 23) and (4·12, 25). They are

$$\left.\begin{aligned}
C_1 &= \frac{2}{\surd(M_1^2 - 1)}, \\[2mm]
C_2 &= \frac{1}{M_1^2 - 1}\left\{\frac{(\gamma + 1)M_1^4}{2(M_1^2 - 1)} - 1\right\}, \\[2mm]
C_3 &= \frac{0\cdot36M_1^8 - 1\cdot493M_1^6 + 3\cdot6M_1^4 - 2M_1^2 + 1\cdot333}{(M_1^2 - 1)^{3\cdot5}}, \\[2mm]
&\ldots\ldots\ldots\ldots\ldots\ldots\ldots\ldots\ldots\ldots\ldots\ldots\ldots\ldots\ldots,
\end{aligned}\right\} \qquad (4\cdot12, 28)$$

where, in the last equation, for brevity, we have introduced the usually accepted value of γ for air (1·40).

If we make a similar expansion for Prandtl–Meyer flow in terms of the deflexion θ (with the negative sign in (4·12, 18)), we find that C_1 and C_2 are the same as in (4·12, 28), and C_3 is changed to the value

$$C_3' = \frac{0\cdot4M_1^8 - 1\cdot813M_1^6 + 4M_1^4 - 2M_1^2 + 1\cdot333}{(M_1^2 - 1)^{3\cdot5}}. \qquad (4\cdot12, 29)$$

Thus, up to the second order in the flow deflexion, shock and Prandtl–Meyer theory give the same results for the change in pressure, and comparison of (4·12, 28) and (4·12, 29) shows that even the third-order terms are of similar magnitude if the Mach number is not too high. When the first two terms of (4·12, 27) only are retained, the resulting equation is known as

Busemann's parabolic formula,‡ and replaces the use of shock-wave relations in calculations. The first term alone yields Ackeret's linearized formula (§ 4·6). The effect of the higher order terms on the aerodynamic forces can be found by integration of (4·12, 27) over the chord. A summary of the results obtained is given in Hilton's book on *High Speed Aerodynamics*.§ Fig. 4·12, 11 gives a comparison of the exact shock equation results and Busemann's approximate formula when applied to flow past a wedge. Differences become significant for large wedge angles at low Mach angles.

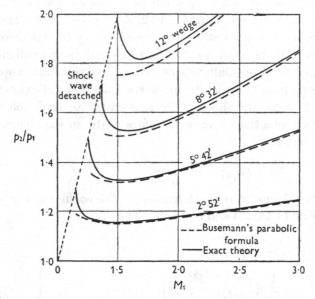

Fig. 4·12, 11. (*Courtesy H.M.S.O.*)

The present section has dealt thus far exclusively with two-dimensional theory. In three dimensions no general procedure for solution of the nonlinearized potential equation has been found possible. The main difficulty in formulation of the problem is with the boundary condition associated with the shock waves. Thus, although generalized three-dimensional characteristics methods have been devised which are applicable to supersonic regions, and relaxation methods are capable of dealing with subsonic or mixed flow regions, it is always necessary to know the boundary conditions or initial values in front of the wing either at or behind the shock. Since the shock position and strength depends on the flow behind it, these boundary conditions are, in general, functions of the solution itself, and thus an iteration process is the only method of dealing with the problem. Exceptions

---

‡ A. Busemann and O. Walchner, 'Profileigenschaften bei Überschallgeschwindigkeit', *Forsch. IngWes.* vol. 4 (1933), pp. 87–92.

§ W. F. Hilton, *High Speed Aerodynamics* (Longmans, Green, 1951), pp. 226–41.

occur only for special geometric and shock configurations, and we shall discuss a few of these.

The classic Taylor–Macoll exact results‡ hold for an infinite unyawed circular cone. They can be extended to cover the case of a slightly yawed cone on the assumption that the flow is irrotational. This ignores the non-uniformity of the shock which results from the asymmetry of the configuration. Exact analysis therefore involves the use of rotational flow equations, and recently Ferrari§ has introduced a method of treating rotational conical flows. The Taylor–Macoll solution may be used also as the starting point of a three-dimensional characteristics method (or a relaxation method in quasi-subsonic flows) for axisymmetric flow.‖ The extension to slightly yawed symmetric bodies follows as for the infinite cone. Such cases do not, however, fall within the province of aerofoil theory, and we shall not discuss them here. The slightly more general case of flow about conical bodies (of which a flat Delta wing is an extreme example) is of more interest in our study.

It can be shown, and it is also apparent intuitively, that in the exact flow about a body with conical symmetry, the flow variables are constant along rays emanating from the cone apex, i.e. the body generates a cone field of order zero, in terms of the definition of § 4·8. It is necessary, however, that the shock wave be straight, and this rules out the cases of subsonic flow regions after the shock and of a finite length body with shock. If, on the other hand, there is no shock wave associated with the flow in a certain region (e.g. on the upper surface of a Delta wing at incidence), the body can be of finite length and the flow field still conical. The problem is now considerably simpler than in the general case, for it is two-dimensional in the sense that the flow patterns are the same in all parallel sections of the cone. The difficulties are again centred around determination of the external boundary conditions. Moore¶ has developed a second-order theory of conical flow for aerofoils with subsonic leading edges, the case in which the wing lies entirely within the envelope of the Mach lines from the tip of the aerofoil (the tip Mach cone). A solution of the differential equations is presented in integral form in a complex plane. Boundary conditions are satisfied approximately at a mean plane.

A general numerical procedure that can be used for conical flows has been

‡ G. I. Taylor and J. W. Macoll, 'The air pressure over a cone moving at high speeds', *Proc. Roy. Soc.* A, vol. 139 (1933), pp. 278–311.

§ C. Ferrari, 'Sui moti Conici', published in *Onore di Modesto Panetti*. Publ. L'Aerotecnica, Associazione Tecnica Automobile and La Termotecnica, Turin, 1950. (Translated in *Tech. Memor. Nat. Adv. Comm. Aero., Wash.*, no. 1333.)

‖ A. Ferri, *Elements of Aerodynamics of Supersonic Flows* (Macmillan, 1949), ch. 13.

¶ F. K. Moore, 'Second approximation to supersonic conical flows', *J. Aero. Sci.* vol. 17 (1950), p. 328.

described by Maslen‡ and by Fowell.§ Solution is by means of characteristics in the supersonic régime and by relaxation in the mixed flow
régimes. Boundary conditions to be applied when using these techniques
are obtained either by an iteration procedure involving the shock-wave
equations, or, in such cases as the supersonic expansion at the leading
edge of a wing at incidence, by use of the Prandtl–Meyer equation. In the
subsonic régime the process is particularly lengthy, since for flow with
shocks it involves complete solution of the flow field for each degree of
approximation. The initial shock position in the iteration method may be

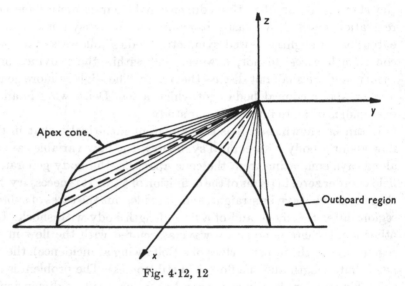

Fig. 4·12, 12

assumed to correspond to two-dimensional flow. If the flow is supersonic
after the shock the initial direction only of the shock at the leading edge is
required to start the characteristics method, and the analysis is easier.
In such a technique all results are obtained by numerical calculation.
However, in the case of a thin wing at incidence with supersonic leading
edges, the flow on the upper surface is irrotational (shocks are only formed
on the lower surface and there is no interaction between the upper and lower
surfaces), and results of interest can be obtained without numerical evaluation of the flow field.

Fig. 4·12, 12 illustrates the general nature of the flow on the upper surface
of a flat Delta wing at incidence. Outboard of the after cone of the apex
the flow is effectively two-dimensional in planes normal to the leading edge,

‡ S. H. Maslen, 'Supersonic conical flows', *Tech. Notes Nat. Adv. Comm. Aero.*,
Wash., no. 2651 (1952).

§ L. R. Fowell, 'An exact solution for the delta wing', *Symposium on High Speed
Aerodynamics*, National Aeronautical Establishment, Ottawa (1953), pp. 47–52. Also
'An exact theory of supersonic flow around a delta wing', *Rep. Univ. of Toronto
Inst. of Aerophysics*, no. 30 (1955).

since the influence of the apex can be felt only within the cone, and outside this region the wing is thus equivalent to an infinite yawed wing. However, the after cone of the apex shown corresponds to the free-stream Mach number M_0, and due to the expansion at the leading edge, the Mach number in some regions will be larger than M_0; hence, corresponding to these higher velocities, the cone angle will be smaller and the region of influence of the wing apex reduced. It may be shown‡ that for the complete compressible potential equation (4·1, 21) the equation of the surface bounding the region of influence of the apex (the characteristic conoid) is

$$(x^2 + y^2 + z^2)(q^2 - a^2) = (ux + vy + wz)^2, \qquad (4·12, 30)$$

where q is the magnitude of the velocity. Within this surface the flow is no longer two-dimensional, and other methods of solution have to be used. In our case the problem can be reduced to the solution of an equation of the elliptic type within the conoid by the transformation $Y = y/x$, $Z = z/x$. The boundary then reduces to the parabolic line of the flow ($q = a$). Boundary conditions are obtained from the results of the Prandtl–Meyer flow analysis in the outboard sections of the wing and from the slope of the wing.

4·13   Transonic flow

When the free-stream velocity is equal to the velocity of sound ($M = U/a_0 = 1$), the linearized equation of compressible flow (4·1, 20) is reduced to

$$\frac{\partial^2 \Phi}{\partial y^2} + \frac{\partial^2 \Phi}{\partial z^2} = 0. \qquad (4·13, 1)$$

(4·13, 1) is identical with the governing equation of the theory of wings of small aspect ratio (§ 3·10), which leads to concrete expressions for the aerodynamic forces. Moreover, it appears that the results obtained in this way for the lift and the induced drag of a wing of finite aspect ratio are the limits of the values of these quantities as M *tends* to one from above or from below (compare § 4·2); in other words, according to linearized theory these quantities are continuous across the speed of sound.§ In Fig. 4·13, 1 the lift-curve slope $dC_L/d\alpha$ is plotted against Mach number for flat Delta wings of various aspect ratios $\mathcal{R}$ and semi-apex angles γ. For $M > 1$, the value of $dC_L/d\alpha$, according to linearized theory, is known exactly (see § 4·8), while for $M < 1$ the available evidence is not quite adequate, and some extrapolation is necessary. Note that for $M = 1$, $dC_L/d\alpha = \tfrac{1}{2}\pi\mathcal{R}$, as in the theory of wings of small aspect ratio.

It must be stated that for transonic speeds, the results provided by

‡ R. Courant and D. Hilbert, *Methoden der Mathematischer Physik* (Julius Springer, Berlin, 1937), ch. 5.

§ See A. Robinson and A. D. Young, 'A note on the application of the linearised theory for compressible flow to transonic speed', *Rep. Memor. Aero. Res. Comm., Lond.*, no. 2399 (1947).

linearized theory are definitely not reliable. Shock waves, and more particularly problems of shock-wave–boundary-layer interaction are beyond the scope of linearized theory, although these phenomena may be very important in determining the lift and drag in the transonic range.

As the aspect ratio becomes very large, the lift-curve slope also becomes very large for $M = 1$. In the limit, i.e. for two-dimensional conditions, the lift-curve slope becomes infinite as $M = 1$ is approached either from below or from above. Indeed, the lift-curve slope for a two-dimensional flat plate is, according to linearized theory,

$$\frac{dC_L}{d\alpha} = \begin{cases} \dfrac{2\pi}{\sqrt{(1 - M^2)}} & \text{for} \quad M < 1, \\[2ex] \dfrac{4}{\sqrt{(M^2 - 1)}} & \text{for} \quad M > 1, \end{cases}$$

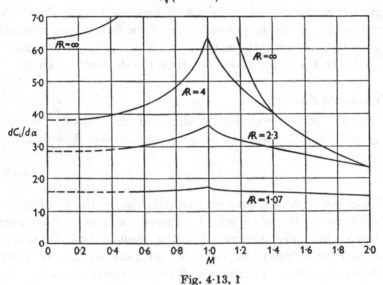

Fig. 4·13, 1

(see equations (4·2, 37) and (4·6, 18)). Thus, the theory breaks down completely at $M = 1$. For this case, a non-linear approximate theory has been developed by von Kármán.‡ From (4·1, 6) we obtain, for steady compressible flow in two-dimensions,

$$\left(1 - \frac{\Phi_x^2}{a^2}\right)\Phi_{xx} + \left(1 - \frac{\Phi_y^2}{a^2}\right)\Phi_{yy} - 2\frac{\Phi_x\Phi_y}{a^2}\Phi_{xy} = 0. \tag{4·13, 2}$$

Under the same conditions, gravity forces being neglected, we obtain from (4·1, 4)

$$a^2 + \frac{\gamma - 1}{2}q^2 = \text{constant} = (\gamma - 1)H.$$

‡ Th. von Kármán, 'The similarity law of transonic flow', *J. Math. Phys.* vol. 26 (1947), pp. 182–90.

Hence, if we denote by a^* the value of a at points where the Mach number q/a equals 1, the so-called 'critical velocity of sound' (note that a^* is exactly the same as the critical velocity q^*, used in §4·12),

$$a^2 + \frac{\gamma-1}{2}(\Phi_x^2 + \Phi_y^2) = \frac{\gamma+1}{2}a^{*2}. \tag{4·13, 3}$$

We assume that the local Mach number q/a differs only by a small quantity from 1 throughout the field of flow. Since the y-component of the velocity is small compared with the free-stream velocity for flow past an aerofoil at small incidence, we may write

$$u = -\frac{\partial \Phi}{\partial x} = a^* - \frac{\partial \Phi^*}{\partial x}, \quad v = -\frac{\partial \Phi}{\partial y} = -\frac{\partial \Phi^*}{\partial y}, \tag{4·13, 4}$$

where Φ^* is a perturbation velocity potential. Substituting (4·13, 4) in (4·13, 2), and neglecting the terms

$$\frac{1}{a^2}\left(\frac{\partial \Phi^*}{\partial x}\right)^2, \quad \frac{1}{a^2}\left(\frac{\partial \Phi^*}{\partial y}\right)^2, \quad \frac{1}{a^2}\frac{\partial \Phi^*}{\partial x}\frac{\partial \Phi^*}{\partial y},$$

which are of the second order of smallness, we then obtain

$$\left(1 - \frac{1}{a^2}\left(a^{*2} - 2a^*\frac{\partial \Phi^*}{\partial x}\right)\right)\frac{\partial^2 \Phi^*}{\partial x^2} + \frac{\partial^2 \Phi^*}{\partial y^2} = 0. \tag{4·13, 5}$$

Similarly, substituting (4·13, 4) in (4·13, 3) and neglecting terms which are of the second order of smallness, we find

$$a^2 + \frac{\gamma-1}{2}\left(a^{*2} - 2a^*\frac{\partial \Phi^*}{\partial x}\right) = \frac{\gamma+1}{2}a^{*2},$$

or

$$a^2 - a^{*2} = (\gamma-1)a^*\frac{\partial \Phi^*}{\partial x}.$$

It follows that the coefficient of $\partial^2 \Phi^*/\partial x^2$ may also be written as

$$\frac{1}{a^2}\left(a^2 - a^{*2} + 2a^*\frac{\partial \Phi^*}{\partial x}\right) = (\gamma+1)\frac{a^*}{a^2}\frac{\partial \Phi^*}{\partial x} \doteqdot \frac{\gamma+1}{a^*}\frac{\partial \Phi^*}{\partial x}.$$

Then (4·13, 5) becomes

$$\frac{\gamma+1}{a^*}\frac{\partial \Phi^*}{\partial x}\frac{\partial^2 \Phi^*}{\partial x^2} + \frac{\partial^2 \Phi^*}{\partial y^2} = 0. \tag{4·13, 6}$$

This equation forms the basis of von Kármán's analysis of transonic flow.

The free-stream velocity is supposed to be of magnitude U and directed along the x-axis. Hence, at infinity

$$\left.\begin{aligned}\frac{\partial \Phi^*}{\partial x} &= a^* - U, \\[2mm] \frac{\partial \Phi^*}{\partial y} &= 0.\end{aligned}\right\} \tag{4·13, 7}$$

Now, by (4·13, 3), we have, at infinity,

$$a^2 + \frac{\gamma-1}{2} U^2 = \frac{\gamma+1}{2} a^{*2},$$

or

$$\frac{\gamma+1}{2} (a^{*2} - U^2) = a^2 - U^2,$$

$$a^* - U = \frac{2}{\gamma+1} (1-M) a^* \frac{a(a+U)}{a^*(a^*+U)},$$

where $M = U/a$ is the free-stream Mach number. The factor $\dfrac{a(a+U)}{a^*(a^*+U)}$ is, in the first approximation, equal to 1, and so

$$a^* - U \doteqdot \frac{2}{\gamma+1} (1-M) a^*, \qquad (4·13, 8)$$

and (4·13, 7) may be replaced by

$$\left.\begin{aligned} \frac{\partial \Phi^*}{\partial x} &= \frac{2a^*}{\gamma+1} (1-M), \\ \frac{\partial \Phi^*}{\partial y} &= 0. \end{aligned}\right\} \qquad (4·13, 9)$$

Let the equation of the aerofoil be

$$y = f_u(x), \quad y = f_l(x)$$

at the upper and lower surfaces respectively. Then the slope is given by

$$s_u(x) = \frac{dy}{dx} = f_u'(x), \quad s_l(x) = \frac{dy}{dx} = f_l'(x). \qquad (4·13, 10)$$

Since the x-component of the velocity at the aerofoil is supposed to differ but little from a^*, we may write the boundary condition at the aerofoil in the form

$$\left.\begin{aligned} -\left(\frac{\partial \Phi^*}{\partial y}\right)_{y=+0} &= a^* s_u(x), \\ -\left(\frac{\partial \Phi^*}{\partial y}\right)_{y=-0} &= a^* s_l(x). \end{aligned}\right\} \qquad (4·13, 11)$$

Let τ be the maximum thickness-chord ratio of the aerofoil, except if the latter is infinitely thin. In that case we take τ as the incidence. Alternatively, we may take τ as the ratio of the normal projections of the aerofoil on the y- and x-axes. Then the slope may also be represented by

$$s_u(x) = \tau h_u\!\left(\frac{x}{c}\right), \quad s_l(x) = \tau h_l\!\left(\frac{x}{c}\right), \qquad (4·13, 12)$$

where c is the chord of the aerofoil and h_u and h_l are functions of order of magnitude 1. Then (4·13, 11) becomes

$$-\left(\frac{\partial \Phi^*}{\partial y}\right)_{y=+0} = a^* \tau h_u\!\left(\frac{x}{c}\right), \quad -\left(\frac{\partial \Phi^*}{\partial y}\right)_{y=-0} = a^* \tau h_l\!\left(\frac{x}{c}\right). \qquad (4·13, 13)$$

Φ^* is a solution of the partial differential equation (4·13, 6) subject to the ancillary conditions (4·13, 9) and (4·13, 13). We introduce non-dimensional variables ξ, η, Ψ^*, by

$$x = c\xi, \quad y = c\eta, \quad \Phi^* = ca^*\Psi. \tag{4·13, 14}$$

Then (4·13, 6), (4·13, 9) and (4·13, 13) become

$$(\gamma+1)\frac{\partial\Psi^*}{\partial\xi}\frac{\partial^2\Psi^*}{\partial\xi^2} + \frac{\partial^2\Psi^*}{\partial\eta^2} = 0, \tag{4·13, 15}$$

$$\left.\begin{aligned} \frac{\partial\Psi^*}{\partial\xi} &= \frac{2}{\gamma+1}(1-M) \\[2mm] \frac{\partial\Psi^*}{\partial\eta} &= 0 \end{aligned}\right\} \quad \text{at infinity} \tag{4·13, 16}$$

and

$$\left.\begin{aligned} -\left(\frac{\partial\Psi^*}{\partial\eta}\right)_{\eta=+0} &= \tau h_u(\xi) \\[2mm] -\left(\frac{\partial\Psi^*}{\partial\eta}\right)_{\eta=-0} &= \tau h_l(\xi) \end{aligned}\right\} \quad \text{at the aerofoil.} \tag{4·13, 17}$$

For given functions h_u and h_l, the problem now depends only on the parameters $(\gamma+1)$, $(1-M)$ and τ. We now try to find a law of similarity for this problem, i.e. a rule which correlates the results for different values of the parameters. For this purpose we set

$$\xi = \xi', \quad \eta = k_1\eta', \quad \Psi^* = k_2\Psi', \tag{4·13, 18}$$

where the constants k_1 and k_2 will be determined presently. Substituting in (4·13, 15), (4·13, 16) and (4·13, 17), we obtain

$$(\gamma+1)k_1^2 k_2 \frac{\partial\Psi'}{\partial\xi'}\frac{\partial^2\Psi'}{\partial\xi'^2} + \frac{\partial^2\Psi'}{\partial\eta'^2} = 0, \tag{4·13, 19}$$

$$\frac{\partial\Psi'}{\partial\xi'} = \frac{2(1-M)}{k_2(\gamma+1)}, \quad \frac{\partial\Psi'}{d\eta'} = 0 \quad \text{at infinity,} \tag{4·13, 20}$$

$$-\left(\frac{\partial\Psi'}{\partial\eta'}\right)_{\eta'=\pm0} = \frac{k_1\tau}{k_2}h_u(\xi), \quad -\left(\frac{\partial\Psi'}{\partial\eta'}\right)_{\eta'=\mp0} = \frac{k_1\tau}{k_2}h_l(\xi) \quad \text{at the aerofoil.}$$
$$\tag{4·13, 21}$$

Now assume that k_1, k_2 satisfy the conditions

$$\frac{2(1-M)}{k_2(\gamma+1)} = 1, \quad \frac{k_1\tau}{k_2} = 1,$$

i.e. that

$$k_1 = \frac{2(1-M)}{\tau(\gamma+1)}, \quad k_2 = \frac{2(1-M)}{\gamma+1}. \tag{4·13, 22}$$

Then the equations (4·13, 20) and (4·13, 21) become

$$\frac{\partial\Psi'}{\partial\xi'} = 1, \quad \frac{\partial\Psi'}{\partial\eta'} = 0, \tag{4·13, 23}$$

and

$$-\left(\frac{\partial\Psi'}{\partial\eta'}\right)_{\eta'=\pm0} = h_u(\xi), \quad -\left(\frac{\partial\Psi'}{\partial\eta'}\right)_{\eta'=\mp0} = h_l(\xi). \tag{4·13, 24}$$

The ambiguous expressions $\eta' = \pm 0$, $\eta' = \mp 0$, in (4·13, 21) and (4·13, 24), are due to the fact that k_1 is positive or negative according as M is smaller or greater than 1. By using the values k_1, k_2 which are given by (4·13, 22), we have eliminated the parameters $(\gamma + 1)$, $(1 - M)$ and τ from the ancillary conditions. At the same time, the partial differential equation (4·13, 19) becomes

$$\frac{8(1-M)^3}{\tau^2(\gamma+1)^2} \frac{\partial \Psi''}{\partial \xi'} \frac{\partial^2 \Psi''}{\partial \xi'^2} + \frac{\partial^2 \Psi''}{\partial \eta'^2} = 0, \tag{4·13, 25}$$

or

$$2k \frac{\partial \Psi''}{\partial \xi'} \frac{\partial^2 \Psi''}{\partial \xi'^2} + \frac{\partial^2 \Psi''}{\partial \eta'^2} = 0, \tag{4·13, 26}$$

where

$$k = \frac{4(1-M)^3}{\tau^2(\gamma+1)^2}. \tag{4·13, 27}$$

Thus the two cases for which the functions $h_u(x/c)$, $h_l(x/c)$ and the constant k are the same, are reduced to the same system (4·13, 23), (4·13, 24) and (4·13, 26). Such cases may be called similar. The solutions for similar cases can therefore be obtained from each other by scaling up the variables in accordance with (4·13, 14), (4·13, 18) and (4·13, 22). We note that these equations can be combined into

$$x = c\xi', \quad y = \frac{2(1-M)}{\tau(\gamma+1)} c\eta', \quad \Phi^* = \frac{2(1-M)}{\gamma+1} ca^* \Psi''. \tag{4·13, 28}$$

To calculate the pressure at any point of the wing we may use the linearized form of Bernoulli's equation (4·1, 27). Thus

$$p - p_0 = -\rho_0 U(u - U),$$

where p_0 and ρ_0 are the pressure and density at infinity, respectively. It follows that the pressure coefficient C_p is given by

$$C_p = \frac{p - p_0}{\frac{1}{2}\rho_0 U^2} = -\frac{2(u - U)}{U}.$$

Now

$$u = -\frac{\partial \Phi}{\partial x} = a^* - \frac{\partial \Phi^*}{\partial x} = a^* \left(1 - \frac{2(1-M)}{\gamma+1} \frac{\partial \Psi''}{\partial \xi'}\right),$$

by (4·13, 4) and (4·13, 28), and

$$U = a^* - \frac{2}{\gamma+1}(1-M)a^*,$$

by (4·13, 8). Hence

$$C_p = \frac{2a^*}{U} \left(\frac{2(1-M)}{\gamma+1} \frac{\partial \Psi''}{\partial \xi'} - \frac{2(1-M)}{\gamma+1}\right) \div \frac{4(1-M)}{\gamma+1} \left(\frac{\partial \Psi''}{\partial \xi'} - 1\right). \tag{4·13, 29}$$

Now

$$\frac{1-M}{\gamma+1} = \sqrt[3]{\frac{4(1-M)^3}{\tau^2(\gamma+1)^2}} \sqrt[3]{\frac{\tau^2}{4(\gamma+1)}} = \frac{k^{\frac{1}{3}}\tau^{\frac{2}{3}}}{4^{\frac{1}{3}}(\gamma+1)^{\frac{1}{3}}}.$$

Hence C_p may be written in the form

$$C_p = \frac{\tau^{\frac{2}{3}}}{(\gamma+1)^{\frac{1}{3}}} f(k, \xi'). \tag{4·13, 30}$$

We may replace k by another parameter, λ, which is defined by

$$\lambda = (2k)^{-\frac{1}{3}} = \frac{((\gamma+1)\tau)^{\frac{1}{3}}}{\sqrt{\{2(1-M)\}}} \doteqdot \frac{((\gamma+1)\tau)^{\frac{1}{3}}}{\sqrt{(1-M^2)}}. \qquad (4·13, 31)$$

Note that the last expression in $(4·13, 31)$ depends on the approximation $1 + M \doteqdot 2$. Then $(4·13, 30)$ is equivalent to

$$C_p = \frac{\tau^{\frac{2}{3}}}{(\gamma+1)^{\frac{1}{3}}} F\left(\frac{((\gamma+1)\tau)^{\frac{1}{3}}}{\sqrt{(1-M^2)}}, \frac{x}{c}\right). \qquad (4·13, 32)$$

The excess pressure is obtained by multiplying $(4·12, 32)$ by $\frac{1}{2}\rho_0 U^2$, and the lift on the aerofoil is then found by integration with respect to x. It follows that the lift coefficient can be written in the form

$$C_L = \frac{\tau^{\frac{2}{3}}}{(\gamma+1)^{\frac{1}{3}}} \mathfrak{L}\left(\frac{((\gamma+1)\tau)^{\frac{1}{3}}}{\sqrt{(1-M^2)}}\right). \qquad (4·13, 33)$$

Similarly, the drag is obtained by multiplying the excess pressure by the slope before integration. The slope is proportional to τ, and so the drag coefficient may be written as

$$C_D = \frac{\tau^{\frac{5}{3}}}{(\gamma+1)^{\frac{1}{3}}} \mathfrak{D}\left(\frac{((\gamma+1)\tau)^{\frac{1}{3}}}{\sqrt{(1-M^2)}}\right). \qquad (4·13, 34)$$

Finally, the moment coefficient is given by

$$C_M = \frac{\tau^{\frac{2}{3}}}{(\gamma+1)^{\frac{1}{3}}} \mathfrak{M}\left(\frac{((\gamma+1)\tau)^{\frac{1}{3}}}{\sqrt{(1-M^2)}}\right), \qquad (4·13, 35)$$

which has the same form as $(4·13, 33)$.

The above theory is based on the assumption of potential flow. It has been pointed out by von Kármán‡ that the results may still be expected to apply if the shock waves which occur in the flow are approximately normal to the direction of the free stream, and if the vorticity outside the boundary layer remains small. For $M > 1$ we may replace $\sqrt{(1-M^2)}$ by $\sqrt{(M^2-1)}$ in $(4·13, 33)$, $(4·13, 34)$ and $(4·13, 35)$ in order to avoid imaginary values of the argument. Both for $M < 1$ and for $M > 1$ these formulae are compatible with the results provided by linearized theory, as can be seen by assuming that $\mathfrak{L}$, $\mathfrak{D}$ and $\mathfrak{M}$ are constant multiples of their arguments.

Accepting the idea that equations $(4·13, 33)$–$(4·13, 35)$ remain applicable to an actual physical case, we conclude that $\mathfrak{L}$, $\mathfrak{D}$ and $\mathfrak{M}$ tend to certain finite limits $\mathfrak{L}(\infty)$, $\mathfrak{D}(\infty)$, $\mathfrak{M}(\infty)$ for $M = 1$. Hence, for $M = 1$,

$$C_L = \frac{\tau^{\frac{2}{3}}}{(\gamma+1)^{\frac{1}{3}}} \mathfrak{L}(\infty), \quad C_D = \frac{\tau^{\frac{5}{3}}}{(\gamma+1)^{\frac{1}{3}}} \mathfrak{D}(\infty), \quad C_M = \frac{\tau^{\frac{2}{3}}}{(\gamma+1)^{\frac{1}{3}}} \mathfrak{M}(\infty).$$
$$(4·13, 36)$$

‡ Th. von Kármán, 'The similarity law of transonic flow', *J. Math. Phys.* vol. 26 (1947), pp. 182–90.

Thus we obtain the rule that at sonic speeds the lift and moment coefficients are proportional to the $\frac{2}{3}$ power of the thickness-chord ratio (or the incidence) and inversely proportional to the $\frac{1}{3}$ power of $(\gamma + 1)$; while the drag coefficient is proportional to the $\frac{5}{3}$ power of the thickness-chord ratio and inversely proportional to the $\frac{1}{3}$ power of $(\gamma + 1)$. We observe that the skin-friction drag of the wing is not included in these considerations, but the importance of this term at transonic speeds is small compared with the drag due to compressibility. The above laws have been found to be in reasonably good agreement with experiment.

The theory does not enable us to calculate the detailed pressure distribution round a given aerofoil under transonic conditions by theoretical means. Only the beginnings of such a theory have been developed so far, and even these include some of the most advanced analytical methods to be employed in the field of aerodynamics.‡

‡ See *Modern Developments in Fluid Dynamics. High Speed Flow*, ed. Howarth (Oxford University Press, 1953), ch. 7, by M. J. Lighthill, where reference is also made to the work of T. M. Cherry.

CHAPTER 5

THEORY OF AEROFOILS IN
UNSTEADY MOTION

5·1 Introduction

It is shown in classical hydrodynamics‡ that in the absence of free vortices the forces and moments which act on a rigid body in unsteady motion in an incompressible fluid depend only on the instantaneous linear and angular components of velocity and acceleration of the body. However, this result has little bearing on aerofoil theory, since in general there is a sheet of vortices trailing behind a moving wing. Owing to the presence of this sheet the forces which act on the wing at any given moment depend on the entire history of the motion.

We shall confine the analysis to a number of particular cases which are of practical importance. These may be divided into two categories, according as the medium through which the wing moves is, or is not, otherwise undisturbed. To the first category belongs (i) the accelerated rectilinear motion of a wing at a fixed altitude; this problem may be of some importance in connexion with take-off or landing considerations and, possibly, for the conditions which obtain when a wing passes through the speed of sound; (ii) the oscillatory motion of a wing relative to a mean position which is itself displaced with constant speed along a straight line, or the divergent motion from an initial position with the same properties. In both cases mentioned under (ii) the wing may be either rigid or deformable, the second alternative being particularly important for aero-elastic investigations, the first for problems of stability and control. Also to the first category belongs (iii) the motion of a wing in steady pitch or roll or yaw; as in the steady case, the theory is least reliable for yaw.

Under the second heading, we shall consider the effect of a gust on a wing in uniform rectilinear motion. The determination of the corresponding aerodynamic loads is necessary for structural purposes, and the occurrence of gusts also leads to special problems in stability and response.

As in the theory of steady motion, the above cases can be analysed only on the basis of a number of simplifying assumptions. Thus, we shall suppose throughout that the fluid is inviscid, while using the argument of §1·14 to show that the Joukowski condition must still be satisfied. The trailing vortex sheet, which may now exist even in two-dimensional flow, is taken to occupy approximately the area covered by the wing in its previous

‡ See H. Lamb, *Hydrodynamics* (6th revised ed., Cambridge University Press, 1932), ch. 6.

motion, so that the distortion of the sheet through mutual interference is neglected, as in the steady case. Again, most of the work which deals with compressible flow conditions is based on linearized theory, and hence on the wave equation. There are other simplifications which have no counterpart in the theory of steady flow. For example, one usually neglects the curvature of the vortex sheet in the cases of roll and pitch mentioned under (iii) above.

Additional errors may be introduced by the indiscriminate use of aerodynamic derivatives, and a preliminary discussion may help to clear up some of the difficulties which have arisen in connexion with this concept.‡

Consider an aerodynamic force acting on a wing in unsteady motion, e.g. the lift L. As stated above, the value of L depends on the position and attitude of the wing at all times preceding the moment under consideration. To simplify matters let us assume that the wing is two-dimensional and that its mid-point moves along a straight line with constant velocity U. Then the lift at time $t = 0$ depends on the incidence $\alpha(t)$ which must be specified for all $t < 0$. We write $L = L[\alpha(t)]$, where the square brackets indicate the fact that L is not an ordinary function of a numerical variable α, but a numerical variable which depends on a function, a so-called functional. However, if $\alpha(t)$ possesses a MacLaurin series which converges for all t (i.e. if $\alpha(t)$ is an integral function)

$$\alpha(t) = \alpha(0) + \alpha'(0)\,t + \frac{\alpha''(0)}{2!}\,t^2 + \dots + \frac{\alpha^{(n)}(0)}{n!}\,t^n + \dots, \qquad (5·1,1)$$

then we may regard L also as a function of α and of its derivatives for $t = 0$. Thus, the arguments are now numerical variables, but their number is infinite

$$L = L(\alpha_0, \alpha_1, \dots, \alpha_n, \dots), \quad \text{where} \quad \alpha_n = \left(\frac{d^n\alpha}{dt^n}\right)_{t=0} \quad (n = 0, 1, 2, \dots).$$
$$(5·1,2)$$

Now let us give a small increment $\Delta\alpha_n$ to the nth derivative. Then the corresponding increment of L may be denoted by ΔL,

$$\Delta L = L(\alpha_0, \alpha_1, \dots, \alpha_n + \Delta\alpha_n, \dots) - L(\alpha_0, \alpha_1, \dots, \alpha_n, \dots), \qquad (5·1,3)$$

provided L exists for both sets of arguments $(\alpha_0, \alpha_1, \dots, \alpha_n, \dots)$ and $(\alpha_0, \alpha_1, \dots, \alpha_n + \Delta\alpha_n, \dots)$. The aerodynamic derivative of L with respect to $\alpha_n = \dfrac{d^n\alpha}{dt^n}$ at $(\alpha_0, \alpha_1, \dots, \alpha_n, \dots)$ is now defined by

$$L_\alpha(n) = \lim_{\Delta\alpha_n \to 0} \frac{\Delta L}{\Delta\alpha_n}, \qquad (5·1,4)$$

where it is assumed that the limit on the right-hand side exists.

‡ See B. M. Jones's article in vol. 5 of *Aerodynamic Theory* (ed. W. F. Durand, Julius Springer, Berlin, 1935). Also W. J. Duncan, *The Principles of Control and Stability of Aircraft* (Cambridge Aeronautical Series, 1952), §§ 3·5 and 3·6.

(5·1, 4) is typical of the accepted definition of an aerodynamic derivative. Unfortunately, the assumption that the limit on the right-hand side of (5·1, 4) exists is untenable, as is shown by the following considerations.

As will appear later, L is a linear function of $\alpha(t)$, i.e.

$$L[c_1\alpha_1(t) + c_2\alpha_2(t)] = c_1 L[\alpha_1(t)] + c_2 L[\alpha_2(t)] \qquad (5·1, 5)$$

for any functions $\alpha_1(t)$, $\alpha_2(t)$ and constants c_1, c_2. Hence, questions of convergence apart,

$$L[\alpha(t)] = \alpha(0)\, L_\alpha + \alpha'(0)\, L_{\alpha'} + \ldots + \alpha^{(n)}(0)\, L_{\alpha^{(n)}} + \ldots, \qquad (5·1, 6)$$

where $L_{\alpha^{(n)}}$ $(n = 0, 1, 2, \ldots)$ is the limit defined in (5·1, 4).

We shall denote the value of L at time $t = t_0$ by $L = L[\alpha(t), t_0]$, so that in particular $L[\alpha(t)] = L[\alpha(t), 0]$. Defining the function $\alpha^*(t)$ by

$$\alpha^*(t) = \alpha(t + t_0).$$

We then have

$$L[\alpha(t), t_0] = L[\alpha^*(t), 0] = L[\alpha^*(t)]$$
$$= \alpha^*(0)\, L_\alpha + \alpha^{*\prime}(0)\, L_{\alpha'} + \ldots + \alpha^{*(n)}(0)\, L_{\alpha^{(n)}} + \ldots. \qquad (5·1, 7)$$

In this formula, $\alpha^{*(n)}(0)$ is the nth derivative of $\alpha^*(t)$ for $t = 0$, i.e. it is the nth derivative of $\alpha(t)$ for $t = t_0$. Thus

$$L[\alpha(t), t_0] = \alpha(t_0)\, L_\alpha + \alpha'(t_0)\, L_{\alpha'} + \ldots + \alpha^{(n)}(t_0)\, L_{\alpha^{(n)}} + \ldots. \qquad (5·1, 8)$$

If the operations involved in the above analysis are permissible, it must therefore be possible to write the lift for any given $\alpha(t)$ in the form of (5·1, 8) where the coefficients $L_{\alpha^{(n)}}$ are independent of the particular $\alpha(t)$ under consideration.

Consider in particular the case of exponentially divergent motion in pitch, for which

$$\alpha(t) = A\, e^{at} = A \exp\left(\frac{U}{c}\mu t\right),$$

A real, a real positive, $\mu = ac/U$. Then

$$\alpha^{(n)}(t_0) = Aa^n e^{at_0} = A\left(\frac{U}{c}\right)^n \mu^n e^{at_0}, \qquad (5·1, 9)$$

and so, if the representation (5·1, 8) were applicable,

$$L(A\, e^{at}, t_0) = A\left(L_\alpha + L_{\alpha'}\frac{U}{c}\mu + \ldots + L_{\alpha^{(n)}}\left(\frac{U}{c}\right)^n \mu^n + \ldots\right) e^{at_0}. \qquad (5·1, 10)$$

According to (5·1, 10), $L[A\, e^{at}, t_0]\, e^{-at_0}$ is a power series of μ. On the other hand, as will be shown in the next section, $L[A\, e^{at}, t_0]\, e^{-at_0}$, while remaining finite for $\mu = 0$, has a logarithmic singularity at that point, and so the MacLaurin series for this function does not exist. To investigate this phenomenon further, let us go back to the definition of L_{α_n} by (5·1, 4).

Using the fact that L is a linear functional of $\alpha(t)$, we obtain from (5·1, 3), for specified n,

$$\Delta L = L[\alpha_0 + a_1 t + \ldots + (\alpha_n + \Delta \alpha_n)\, t^n + \ldots] - L[\alpha_0 + \alpha_1 t + \ldots + \alpha_n t^n + \ldots]$$
$$= \Delta \alpha_n L[t^n].$$

Hence

$$L_{\alpha^{(n)}} = \lim_{\Delta \alpha_n \to 0} \frac{\Delta L}{\Delta \alpha_n} = \lim_{\Delta \alpha_n \to 0} \frac{\Delta \alpha_n L[t^n]}{\Delta \alpha_n} = L[t^n]. \tag{5·1, 11}$$

But for a time variation of the incidence $\alpha = t^n$ $(n \geqslant 1)$, α tends to $\pm \infty$ as t approaches $-\infty$ and so the problem is physically meaningless and does not possess a solution. Thus the original definition of the aerodynamic derivatives fails in practice. The quantities which are accepted instead as aerodynamic derivatives are obtained by representing $L[\alpha(t), t_0]$, for the case of divergent motion or of oscillatory motion, in the form

$$L[\alpha(t), t_0] = \alpha(t_0)\, \lambda_\alpha + \alpha'(t_0)\, \lambda_{\alpha'} + \alpha''(t_0)\, \lambda_{\alpha''}. \tag{5·1, 12}$$

In this equation λ_α, $\lambda_{\alpha'}$, $\lambda_{\alpha''}$, are independent of time but dependent on the characteristics of $\alpha(t)$, e.g. the constant a introduced above for divergent motion, or the frequency for oscillatory motion. If we put

$$\lambda_\alpha = L_\alpha, \quad \lambda_{\alpha'} = L_{\alpha'}, \quad \lambda_{\alpha''} = L_{\alpha''}, \tag{5·1, 13}$$

in (5·1, 12), then the result is formally a special case of (5·1, 8). Accordingly, we regard λ_α, $\lambda_{\alpha'}$, $\lambda_{\alpha''}$ as the aerodynamic derivatives of the lift with respect to α, α', α''. These are not derivatives in the sense of the original definition, since the latter—if they existed—would have to be independent of the characteristics of $\alpha(t)$. Nevertheless, the procedure is acceptable so long as we apply the derivatives only to the particular case for which they were derived in the first instance. On the other hand, it is not surprising that the corresponding derivatives obtained in this way for different types of motion may differ from each other considerably, and may even have different signs. Such is the case for the supersonic damping derivatives in pitch (§ 5·7), a phenomenon which has caused a certain amount of confusion.

We conclude that aerodynamic derivatives as defined by (5·1, 12) should be applied only in circumstances which are at least similar to the circumstances for which the derivatives were determined.

5·2 Aerofoil theory for unsteady incompressible flow

We consider a two-dimensional very thin aerofoil which moves (approximately) along a straight line in an incompressible fluid and at the same time executes a motion of small amplitude in the transverse direction. We admit the possibility that the wing is deformable, so that its elements may be displaced relative to each other.

We choose a system of coordinates which is fixed relative to the surrounding air, such that the aerofoil moves along the x'-axis in the negative direction. Then the position of the aerofoil at any time is described by the coordinate of the normal projection of its leading edge on the x'-axis. $x' = x_L = x_L(t)$, say, and by the equation of the aerofoil

$$y' = f(x', t), \quad x_L \leqslant x' \leqslant x_T = x_L + c, \qquad (5 \cdot 2, 1)$$

where c is the chord length and x_T is the coordinate of the trailing edge, or, more precisely, of the normal projection of the trailing edge on the x'-axis.

It is supposed that the aerofoil started from rest in the undisturbed atmosphere at some previous moment of time (possibly remote), so that the field of flow is irrotational at any subsequent time except in the wake of the aerofoil, i.e. in the region through which the aerofoil travelled previously. It is in keeping with the approximations of earlier chapters to assume that this 'vortex wake' is infinitely thin and that it coincides with a segment of the x'-axis. At all points other than the wing or wake the velocity potential exists and is continuous and satisfies Laplace's equation

$$\frac{\partial^2 \Phi}{\partial x'^2} + \frac{\partial^2 \Phi}{\partial y'^2} = 0. \qquad (5 \cdot 2, 2)$$

The boundary condition at the wing is (see $(1 \cdot 2, 11)$)

$$v' = u' \frac{\partial f}{\partial x'} - \frac{\partial f}{\partial t} = 0 \quad (x_L \leqslant x' \leqslant x_T, \ y = 0), \qquad (5 \cdot 2, 3)$$

where u', v' are the velocity components in the primed system of coordinates. In addition, we have Bernoulli's equation for unsteady flow in an incompressible fluid (§ 1·6) which may be written in the form

$$\frac{p}{\rho} + \frac{1}{2} \left(\left(\frac{\partial \Phi}{\partial x'} \right)^2 + \left(\frac{\partial \Phi}{\partial y'} \right)^2 \right) = \frac{\partial \Phi}{\partial t} + h(t). \qquad (5 \cdot 2, 4)$$

The function $h(t)$ on the right-hand side may be omitted or retained, as convenient. The pressure p is continuous everywhere except across the aerofoil. In particular, it must be continuous across the wake.

We say that the aerofoil occupies its 'mean position' when it coincides with its normal projection on the x'-axis, $\langle x_L, x_T \rangle$. Let (x, y) be a second system of coordinates which is fixed relative to the mean position of the wing and such that the x'-axis coincides with the x-axis. More particularly, take the origin O of the new system of the mid-point of the aerofoil. Then the abscissa of O is in the primed system

$$x_0' = \tfrac{1}{2}(x_L + x_T) = g(t), \quad \text{say}. \qquad (5 \cdot 2, 5)$$

$g(t)$ must be a decreasing function of the time, since the aerofoil is supposed to move along the x'-axis in the negative direction. Then

$$x = x' - g(t), \quad y = y'. \qquad (5 \cdot 2, 6)$$

Let $U(t)$ be the (positive) forward velocity of the aerofoil, $U(t) = -g'(t)$. Then

$$\frac{\partial \Phi}{\partial x} = \frac{\partial \Phi}{\partial x'}, \quad \frac{\partial \Phi}{\partial y} = \frac{\partial \Phi}{\partial y'},$$

$$\left(\frac{\partial \Phi}{\partial t}\right)_x = \left(\frac{\partial \Phi}{\partial t}\right)_{x'} + g'(t)\frac{\partial \Phi}{\partial x'} = \left(\frac{\partial \Phi}{\partial t}\right)_{x'} - U(t)\frac{\partial \Phi}{\partial x'},$$

$$(5\cdot2, 7)$$

where the subscripts x and x' indicate that x or x' is kept constant in the differentiation, as the case may be.

We note that $-\partial\Phi/\partial x'$ is only the induced x-component of velocity. The total velocity component in that direction relative to the mean position of the aerofoil is $U - \partial\Phi/\partial x$. $(5\cdot2, 7)$ shows that for the 'steady case' when $(\partial\Phi/\partial t)_x = 0$, we have

$$\left(\frac{\partial \Phi}{\partial t}\right)_{x'} = U\frac{\partial \Phi}{\partial x'}. \qquad (5\cdot2, 8)$$

Also, for the steady case we may assume as usual that the induced velocities $-\partial\Phi/\partial x = -\partial\Phi/\partial x'$ and $-\partial\Phi/\partial y = -\partial\Phi/\partial y'$, are small compared with U. It then follows from $(5\cdot2, 8)$ that the term $\frac{1}{2}((\partial\Phi/\partial x')^2 + (\partial\Phi/\partial y')^2)$ is small compared with $(\partial\Phi/\partial t)_{x'}$. Accordingly, we may replace $(5\cdot2, 4)$ by

$$-\frac{p}{\rho} = \left(\frac{\partial \Phi}{\partial t}\right)_{x'} + h(t). \qquad (5\cdot2, 9)$$

We shall assume that the relative orders of magnitude are the same in the unsteady case now under consideration, so that $(5\cdot2, 9)$ holds for this case also. Since the pressure is continuous everywhere except across the aerofoil, it follows from $(5\cdot2, 9)$ that, to the order of accuracy of that formula, the function

$$\Omega(x', y', t) = \left(\frac{\partial \Phi}{\partial t}\right)_{x'} \qquad (5\cdot2, 10)$$

also is continuous everywhere except across the aerofoil. Ω in Prandtl's acceleration potential for unsteady flow. Except for a constant in space, $\rho\Omega$ is an approximation to the pressure, but it is advantageous to define Ω in the first instance by $(5\cdot2, 10)$. Using that definition we see immediately that Ω is a solution of Laplace's equation

$$\frac{\partial^2 \Omega}{\partial x'^2} + \frac{\partial^2 \Omega}{\partial y'^2} = 0, \qquad (5\cdot2, 11)$$

and this relation is exact. On the other hand, if we identify $\rho\Omega$ with $p - \rho h(t)$, then Ω is strictly continuous everywhere except across the wing, but Laplace's equation is satisfied only approximately, and in fact it is not easy to ascribe a precise meaning to such an approximation. In the unprimed system of coordinates, the acceleration potential is given by

$$\Omega = \left(\frac{\partial \Phi}{\partial t}\right)_x + U(t)\frac{\partial \Phi}{\partial x}, \qquad (5\cdot2, 12)$$

and this reduces to the definition of § 3·9 when Φ is independent of time.

It follows from (5·2, 6) and (5·2, 11) that Ω is a solution of Laplace's equation also in the unprimed system of coordinates

$$\frac{\partial^2 \Omega}{\partial x^2} + \frac{\partial^2 \Omega}{\partial y^2} = 0. \tag{5·2, 13}$$

Let
$$y = F(x, t) \quad (-\tfrac{1}{2}c \leqslant x \leqslant \tfrac{1}{2}c, \, y = 0) \tag{5·2, 14}$$

be the equation of the wing in the unprimed system

$$F(x, t) = F(x' - g(t), t) = f(x', t) = f(x + g(t), t). \tag{5·2, 15}$$

The corresponding boundary condition is

$$-\frac{\partial \Phi}{\partial y} - \left(U - \frac{\partial \Phi}{\partial x}\right)\frac{\partial F}{\partial x} - \left(\frac{\partial F}{\partial t}\right)_x = 0. \tag{5·2, 16}$$

Neglecting $\partial \Phi / \partial x$, which is small compared with U, we obtain

$$-\frac{\partial \Phi}{\partial y} = U \frac{\partial F}{\partial x} + \left(\frac{\partial F}{\partial t}\right)_x. \tag{5·2, 17}$$

But
$$\left(\frac{\partial}{\partial t}\right)_{x'} = \left(\frac{\partial}{\partial t}\right)_x - g'(t)\frac{\partial}{\partial x} = U\frac{\partial}{\partial x} + \left(\frac{\partial}{\partial t}\right)_x, \tag{5·2, 18}$$

and so the same approximation yields for the primed system of coordinates

$$-\frac{\partial \Phi}{\partial y'} = \left(\frac{\partial f}{\partial t}\right)_{x'} \quad (x_L \leqslant x' \leqslant x_T, \, y' = 0). \tag{5·2, 19}$$

The function Φ is arbitrary to the extent of the addition or subtraction of a function of the time only. Thus we may assume that $\Omega = (\partial \Phi / \partial t)_{x'}$ vanishes at infinity at all times. (5·2, 9) shows that in that case $h(t)$ is the value of p/ρ at infinity. By (5·2, 13) we may therefore regard Ω as the real part of a variable ω which depends on $z = x + iy$ and on the time t, and which, moreover, is an analytic function of z and vanishes for $z = \infty$ at all times. ω will be called the 'complex acceleration potential'.

We map the z-plane, cut along the x-axis from $x = -\tfrac{1}{2}c$ to $x = \tfrac{1}{2}c$, on the region outside the unit circle in an auxiliary ζ-plane by means of the transformation

$$z = \frac{c}{4}\left(\zeta + \frac{1}{\zeta}\right). \tag{5·2, 20}$$

Then ω is an analytic function of ζ which is regular outside the unit circle and vanishes at infinity. We represent ω by the expansion

$$\omega = i\left(-a_0 \frac{\zeta}{1+\zeta} + \sum_{n=1}^{\infty} \frac{a_n}{\zeta^n}\right), \tag{5·2, 21}$$

where $a_0, a_1, a_2, \ldots$ are functions of time.

The infinite series in this expression does not require any explanation. The presence of the first term is due to the fact that ω may be infinite at the leading edge. This is also true for steady conditions which are included in the present analysis as a special case (see § 2·8). It will be shown presently that the first term in (5·2, 21) indeed represents such an infinity.

We assumed earlier that the wing is very thin, so that $F(x, t)$ is the same on both surfaces of the wing. A familiar argument (compare § 3·1) then shows that Ω is an odd function of y and hence that Ω vanishes along the x-axis, except at the wing. It follows that $\Omega = \mathscr{R}(\omega)$ vanishes also along the real axis of the ζ-plane, and this will be the case if the a_n are real.

Put $\zeta = e^{i\theta}$ at the unit circle. Then

$$\frac{\zeta}{\zeta + 1} = \frac{e^{i\theta}}{e^{i\theta} + 1} = \frac{1}{2} \frac{\cos \frac{1}{2}\theta + i \sin \frac{1}{2}\theta}{\cos \frac{1}{2}\theta},$$

and so at the aerofoil

$$\omega = i \left(-\frac{a_0}{2} \frac{\cos \frac{1}{2}\theta + i \sin \frac{1}{2}\theta}{\cos \frac{1}{2}\theta} + \sum_{n=1}^{\infty} a_n (\cos n\theta - i \sin n\theta) \right),$$

and
$$\Omega = \frac{1}{2} a_0 \tan \frac{1}{2}\theta + \sum_{n=1}^{\infty} a_n \sin n\theta. \tag{5·2, 22}$$

We have to determine the coefficients a_n $(n = 0, 1, 2, \ldots)$ by means of the boundary conditions (5·2, 17). Let

$$G(x, t) = G(\tfrac{1}{2}c \cos \theta, t) = U \frac{\partial F}{\partial x} + \left(\frac{\partial F}{\partial t} \right)_x = \frac{b_0}{2} + \sum_{n=1}^{\infty} b_n \cos n\theta, \tag{5·2, 23}$$

where b_0, b_1, b_2, ... are functions of time, and θ is defined as in (5·2, 22). Only the cosine terms appear in this Fourier series for $G(x, t)$ because G is the same on both sides of the aerofoil. Then

$$-\frac{\partial \Phi}{\partial y} = \frac{b_0}{2} + \sum_{n=1}^{\infty} b_n \cos n\theta. \tag{5·2, 24}$$

We wish to express $\partial \Phi / \partial y$ in terms of ω. By (5·2, 10)

$$\Phi(x_1', y_1', t_1) = \int_{t_0}^{t_1} \Omega \, dt = \mathscr{R} \int_{t_0}^{t_1} \omega \, dt \tag{5·2, 25}$$

at a time t_1, at any given point (x_1', y_1'), where t_0 is an earlier moment of time when Φ is supposed to have vanished at and in the neighbourhood of (x_1', y_1').

We now introduce in place of the time t the new variable

$$\tau = -x_0' = -g(t) \tag{5·2, 26}$$

(see (5·2, 5)), so that, except for the sign, τ measures the displacement of the aerofoil along the x'-axis. It follows from our earlier assumptions that τ increases with increasing t, and

$$\frac{d\tau}{dt} = -g'(t) = U(t) = U(g^{-1}(-\tau)) = \chi(\tau), \quad \text{say.}$$

We write ω as a function of the two variables z and τ, $\omega = \omega(z, \tau)$, and we regard $a_0, a_1, \ldots, b_0, b_1, \ldots$ also as functions of τ. Substituting τ as variable of integration in (5·2, 25) and putting $\tau_0 = -g(t_0)$, $\tau_1 = -g(t_1)$, we then obtain

$$\Phi(x_1', y_1', t_1) = \mathcal{R} \int_{\tau_0}^{\tau_1} \omega(x_1' + \tau + iy_1', \tau) \frac{d\tau}{\chi(\tau)}, \qquad (5\cdot2, 27)$$

since the value of z which corresponds to (x_1', y_1') at any time t is

$$x_1' - g(t) + iy_1' = x_1' + \tau + iy_1',$$

by (5·2, 6). Also

$$\frac{\partial \omega}{\partial y'} = \frac{\partial \omega}{\partial y} = i\omega'(z, \tau),$$

where $\omega'(z, \tau)$ is the derivative of ω with respect to the complex variable z. Hence

$$\left(\frac{\partial \Phi}{\partial y'}\right)_{x'=x_1', y'=y_1', t=t_1} = -\mathcal{I} \int_{\tau_0}^{\tau_1} \omega'(x_1' + \tau + iy_1', \tau) \frac{d\tau}{\chi(\tau)}. \qquad (5\cdot2, 28)$$

We now introduce

$$z = x_1' + \tau + iy_1$$

as variable of integration in (5·2, 28). Then

$$\left(\frac{\partial \Phi}{\partial y'}\right)_{x'=x_1', y'=y_1', t=t_1} = -\mathcal{I} \int_{z_0}^{z_1} \omega'(z, \tau_1 + z - z_1) \frac{dz}{\chi(\tau_1 + z - z_1)}, \qquad (5\cdot2, 29)$$

where

$$z_1 = x_1 + iy_1 = x_1' + \tau_1 + iy_1' = x_1' - g(t_1) + iy_1', \qquad (5\cdot3, 30)$$

and (x_1, y_1) are the coordinates of the point (x, y) at time $t = t_1$. The lower limit z_0 in (5·2, 29) only has to satisfy the condition that $\partial \Phi / \partial y' = \partial \Phi / \partial y$ vanishes when $z_1 = z_0$, i.e. when $x_1' - g(t_1) = \mathcal{R}(z_0)$. Since the wing moves in the negative direction of the x'-axis, the choice $\mathcal{R}(z_0) = -\infty$ is appropriate. The path of integration in (5·2, 29) is defined initially as a straight line parallel to the x-axis. We now assume that $g(t)$ is an analytic function. Then U, g^{-1} and χ also are analytic functions of their arguments, and we shall assume that the same applies to ω when regarded as a function of its second argument. (Since ω is for the moment undetermined, the last statement must be verified a posteriori.) Accordingly, we may deform the path of integration in (5·2, 29) in the z-plane as the need arises, provided we do not leave the domain of regularity of the integrand.

Combining (5·2, 24) and (5·2, 27) we obtain for a point of the wing which is given by

$$z_1 = x_1 = \tfrac{1}{2}c \cos \theta \quad (-\tfrac{1}{2}c < x_1 < \tfrac{1}{2}c), \qquad (5\cdot2, 31)$$

$$\mathcal{I} \int_{-\infty}^{x_1} \omega'(z, \tau_1 + z - x_1) \frac{dz}{\chi(\tau_1 + z - x_1)} = \frac{b_0}{2} + \sum_{n=1}^{\infty} b_n \cos n\theta. \qquad (5\cdot2, 32)$$

The function ω, and hence also ω', must be expected to possess a singularity at the leading edge. Accordingly, we take the path of integration on the left-hand side of (5·2, 32) along the real axis except for a small curve round the leading edge (Fig. 5·2, 1). (5·2, 32) is the condition which determines the coefficients of (5·2, 21). If we wish to express the condition in terms of t_1, then we obtain the more complicated equation

$$\mathscr{I}\int_{-\infty}^{x_1} \omega'(z, g^{-1}(g(t_1)+x_1-z)) \frac{dz}{U(g^{-1}(g(t_1)+x_1-z))} = \frac{b_0}{2} + \sum_{n=1}^{\infty} b_n \cos n\theta.$$

(5·2, 33)

In this formula it has now been assumed that the second argument of ω' is the time.

The integral on the left-hand side of (5·2, 32) may be written in the form

$$\int_{-\infty}^{x_1} \psi(z, \xi)\, dz,$$

where

$$\xi = \tau_1 - x_1.$$

Applying the operation

$$\frac{\partial}{\partial x_1} + \frac{\partial}{\partial \tau_1}$$

(5·2, 34)

z-plane

Fig. 5·2, 1

to the integrand, we obtain

$$\frac{\partial \psi}{\partial x_1} + \frac{\partial \psi}{\partial \tau_1} = \frac{\partial \psi}{\partial \xi}\left(\frac{\partial \xi}{\partial x_1} + \frac{\partial \xi}{\partial \tau_1}\right),$$

whilst applying it to the upper limit of the integral in (5·2, 32) we obtain the integrand. Hence

$$\mathscr{I}\omega'(x_1, \tau_1) = \chi(\tau_1)\left(\frac{\partial}{\partial x_1} + \frac{\partial}{\partial \tau_1}\right)\left(\frac{b_0}{2} + \sum_{n=1}^{\infty} b_n \cos n\theta\right).$$

(5·2, 35)

To carry out the differentiation on the right-hand side, we note that

$$\frac{\partial}{\partial x_1} = \frac{2}{c}\frac{\partial}{\partial \cos \theta} = -\frac{2}{c \sin \theta}\frac{\partial}{\partial \theta},$$

and

$$\chi(\tau_1) = U(t_1), \quad \chi(\tau_1)\frac{\partial}{\partial \tau_1} = \frac{\partial}{\partial t_1}.$$

Hence the right-hand side of (5·2, 35) is equal to

$$\frac{2U(t_1)}{c \sin \theta}\sum_{n=1}^{\infty} nb_n \sin n\theta + \frac{b_0'}{2} + \sum_{n=1}^{\infty} b_n' \cos n\theta,$$

(5·2, 36)

where

$$b_n' = \left(\frac{db_n}{dt}\right)_{t=t_1} \quad (n = 0, 1, 2, ...).$$

Also

$$\omega' = \frac{\partial \omega}{\partial \xi}\bigg/\frac{\partial z}{\partial \xi} = -i\left(\frac{a_0}{(\zeta+1)^2} + \sum_{n=1}^{\infty} \frac{na_n}{\zeta^{n+1}}\right)\bigg/\frac{c}{4}\left(1 - \frac{1}{\zeta^2}\right),$$

(5·2, 37)

and so at the aerofoil where $\zeta = e^{i\theta}$

$$\omega' = -\frac{4i}{c}\left(\frac{a_0 e^{i\theta}}{(e^{i\theta}+1)^2} + \sum_{n=1}^{\infty} n a_n e^{-in\theta}\right)\bigg/(e^{i\theta}-e^{-i\theta})$$

$$= -\frac{2}{c\sin\theta}\left(\frac{u_0}{2(\cos\theta+1)} + \sum_{n=1}^{\infty} n a_n(\cos n\theta - i\sin n\theta)\right).$$

(5·2, 35) now becomes

$$\frac{2}{c\sin\theta}\sum_{n=1}^{\infty} n a_n \sin n\theta = \frac{2U}{c\sin\theta}\sum_{n=1}^{\infty} n b_n \sin n\theta + \frac{b_0'}{2} + \sum_{n=1}^{\infty} b_n' \cos n\theta. \quad (5\cdot2, 38)$$

Multiplying (5·2, 38) by $\frac{1}{2}(c\sin\theta)$ and making use of the formula

$$\cos n\theta \sin\theta = \tfrac{1}{2}(\sin(n+1)\theta - \sin(n-1)\theta),$$

we find
$$\sum_{n=1}^{\infty} n a_n \sin n\theta = \sum_{n=1}^{\infty}\left(\frac{c}{4}b_{n-1}' + n U b_n - \frac{c}{4}b_{n+1}'\right)\sin n\theta. \quad (5\cdot2, 39)$$

Comparing coefficients, we obtain finally

$$a_n(t) = \frac{c}{4n}b_{n-1}'(t) + U(t)\,b_n(t) - \frac{c}{4n}b_{n+1}'(t) \quad (n=1,2,\dots), \quad (5\cdot2, 40)$$

where we have replaced t_1 by t.

Instead of expressing $a_n\ (n\geqslant 1)$ by (5·2, 40), we may also represent these coefficients in terms of derivatives with respect to $\tau = \tau_1$,

$$a_n = \chi(\tau)\left(\frac{c}{4n}\frac{db_{n-1}}{d\tau} + b_n - \frac{c}{4n}\frac{db_{n+1}}{d\tau}\right) \quad (n=1,2,\dots). \quad (5\cdot2, 41)$$

We have yet to determine a_0. Substituting (5·2, 41) in (5·2, 37), we obtain

$$\omega'(z,\tau) = -\frac{4i\zeta^2}{c(\zeta^2-1)}\left\{\frac{a_0}{(\zeta+1)^2} + \chi(\tau)\sum_{n=1}^{\infty}\left(\frac{c}{4}\frac{db_{n-1}}{d\tau} + n b_n - \frac{c}{4}\frac{db_{n+1}}{d\tau}\right)\zeta^{-n-1}\right\}$$

$$= -i\left\{\frac{4a_0\zeta^2}{c(\zeta^2-1)(\zeta+1)^2} + \chi(\tau)\left(\frac{1}{\zeta^2-1}\frac{db_0}{d\tau} + \frac{\zeta}{\zeta^2-1}\frac{db_1}{d\tau}\right.\right.$$

$$\left.\left. + \sum_{n=1}^{\infty}\left\{\frac{db_n}{d\tau}\left(\frac{1}{\zeta^{n+2}} - \frac{1}{\zeta^n}\right) + \frac{4n b_n}{c\zeta^{n+1}}\frac{\zeta^2}{\zeta^2-1}\right\}\right)\right\}$$

$$= -i\left\{\frac{4a_0\zeta^2}{c(\zeta^2-1)(\zeta+1)^2} + \chi(\tau)\left(\frac{1}{\zeta^2-1}\frac{db_0}{d\tau} + \frac{\zeta}{\zeta^2-1}\frac{db_1}{d\tau}\right.\right.$$

$$\left.\left. + \sum_{n=1}^{\infty}\left\{\frac{4n b_n}{c\zeta^{n-1}(\zeta^2-1)} - \frac{1}{\zeta^n}\frac{db_n}{d\tau}\right\}\right)\right\}.$$

Now, for $n = 1, 2, \dots$

$$\frac{4n b_n}{c\zeta^{n-1}(\zeta^2-1)} - \frac{1}{\zeta^n}\frac{db_n}{d\tau} = -\left(\frac{\partial}{\partial\zeta}\left(\frac{b_n}{\zeta^n}\right)\bigg/\frac{c}{4}\left(1-\frac{1}{\zeta^2}\right) + \frac{d}{d\tau}\left(\frac{b_n}{\zeta^n}\right)\right),$$

and so
$$\omega'(z,\tau) = -\frac{i}{\zeta^2-1}\left\{4a_0\frac{\zeta^2}{c(\zeta+1)^2} + \chi(\tau)\left(\frac{db_0}{d\tau} + \zeta\frac{db_1}{d\tau}\right)\right\} + \left[i\chi(\tau)\left(\frac{\partial}{\partial z} + \frac{\partial}{\partial\tau}\right)b(z,\tau)\right],$$

$$(5\cdot2, 42)$$

where
$$b(z,\tau)=\sum_{n=1}^{\infty}\frac{b_n}{\zeta^n},\quad z=\frac{c}{4}\left(\zeta+\frac{1}{\zeta}\right). \tag{5·2, 43}$$

We substitute (5·2, 42) on the left-hand side of (5·2, 32) and consider first the contribution from the expression in square brackets,

$$\mathscr{I}\int_{-\infty}^{x_1}\left\{i\left[\chi(\tau)\left(\frac{\partial}{\partial z}+\frac{\partial}{\partial \tau}\right)b(z,\tau)\frac{1}{\chi(\tau)}\right]_{\tau=\tau_1+z-x_1}\right\}dz$$
$$=\mathscr{R}\int_{-\infty}^{x_1}\left\{\left[\left(\frac{\partial}{\partial z}+\frac{\partial}{\partial \tau}\right)b(z,\tau)\right]_{\tau=\tau_1+z-x_1}\right\}dz. \tag{5·2, 44}$$

The value of τ in the integral of (5·2, 44) is connected with z by the relation
$$\tau=z+\tau_1-x_1. \tag{5·2, 45}$$

Thus, for given x_1 and τ_1, the integration in (5·2, 44) is carried out with respect to z along a straight line in the z,τ-plane whose equation is given by (5·2, 45). The inverse operation, i.e. differentiation with respect to z along the straight line (5·2, 45), is
$$\frac{d}{dz}=\frac{\partial}{\partial z}+\frac{d\tau}{dz}\frac{\partial}{\partial \tau}=\frac{\partial}{\partial z}+\frac{\partial}{\partial \tau}. \tag{5·2, 46}$$

Hence (5·2, 44) may also be written as
$$\mathscr{R}\int_{-\infty}^{x_1}\frac{db}{dz}dz=\mathscr{R}(b(x_1,\tau_1))+\text{constant}. \tag{5·2, 47}$$

Letting x_1 tend to $-\infty$ we see that the constant on the right-hand side of (5·2, 47) must be zero. Inserting
$$\mathscr{R}(b(x,\tau_1))=\mathscr{R}\left(\sum_{n=1}^{\infty}\frac{b_n}{\zeta^n}\right)=\sum_{n=1}^{\infty}b_n\cos n\theta$$

on the left-hand side of (5·2, 32), we see that this expression is cancelled by the infinite series on the right-hand side. The equation then becomes

$$-\mathscr{R}\int_{-\infty}^{x_1}\left(4a_0\frac{\zeta^2}{c(\zeta+1)^2\chi(\tau)}+\frac{db_0}{d\tau}+\zeta\frac{db_1}{d\tau}\right)\frac{dz}{\zeta^2-1}=\left(\frac{b_0}{2}\right)_{\tau=\tau_1}, \tag{5·2, 48}$$

where
$$z=\frac{c}{4}\left(\zeta+\frac{1}{\zeta}\right),$$

and where the value of τ on the left-hand side is given by (5·2, 45).

Introducing ζ as variable of integration in (5·2, 48), and putting $x_1=\frac{1}{2}c\cos\theta_1$, we obtain

$$\mathscr{R}\int_{-\infty}^{e^{i\theta_1}}\left(\frac{4a_0}{c\chi(\tau)(\zeta+1)^2}+\frac{1}{\zeta^2}\frac{db_0}{d\tau}+\frac{1}{\zeta}\frac{db_1}{d\tau}\right)d\zeta=-\frac{2}{c}(b_0)_{\tau=\tau_1},$$

or
$$\mathscr{R}\int_{-\infty}^{e^{i\theta_1}}\frac{4a_0\,d\zeta}{c(\zeta+1)^2\chi(\tau)}=-\frac{2}{c}(b_0)_{\tau=\tau_1}-\mathscr{R}\int_{-\infty}^{e^{i\theta_1}}\left(\frac{1}{\zeta^2}\frac{db_0}{d\tau}+\frac{1}{\zeta}\frac{db_1}{d\tau}\right)d\zeta. \tag{5·2, 49}$$

To continue, we require the following integral

$$\int_{-\infty}^{e^{i\theta_1}} \frac{d\zeta}{(\zeta+1)^2} = \left[-\frac{1}{\zeta+1} \right]_{-\infty}^{e^{i\theta_1}} = -\frac{1}{2} \frac{\cos \frac{1}{2}\theta_1 + i \sin \frac{1}{2}\theta_1}{\cos \frac{1}{2}\theta_1}. \tag{5·2, 50}$$

Then

$$\mathscr{R}\int_{-\infty}^{e^{i\theta_1}} \frac{d\zeta}{(\zeta+1)^2} = -\frac{1}{2}. \tag{5·2, 51}$$

Now consider the integral

$$\int_{-\infty}^{e^{i\theta_1}} \frac{4a_0}{c(\zeta+1)^2 \chi(\tau)} d\zeta \tag{5·2, 52}$$

which occurs in (5·2, 49). The argument τ on the left-hand side is given by (5·2, 45), which may also be written as

$$\tau = \tau_1 + \frac{c}{4}\left(\zeta + \frac{1}{\zeta} - 2\cos\theta_1 \right). \tag{5·2, 53}$$

Note that τ occurs also as the argument of a_0.

The integral (5·2, 52) diverges for $\theta_1 = \pi$, $e^{i\theta_1} = -1$. However, for $0 < \theta_1 < \pi$,

$$\int_{-\infty}^{e^{i\theta_1}} \frac{4a_0}{c(\zeta+1)^2 \chi(\tau)} d\zeta$$

$$= \int_{-\infty}^{e^{i\theta_1}} \left(\frac{4a_0}{c(\zeta+1)^2 \chi(\tau)} - \frac{4(a_0)_{\tau=\tau_1}}{c(\zeta+1)^2 \chi(\tau_1)} \right) d\zeta + \frac{4(a_0)_{\tau=\tau_1}}{c\chi(\tau_1)} \int_{-\infty}^{e^{i\theta_1}} \frac{d\zeta}{(\zeta+1)^2},$$

and so by, (5·2, 52),

$$\mathscr{R}\int_{-\infty}^{e^{i\theta_1}} \frac{4a_0}{c(\zeta+1)^2 \chi(\tau)} d\zeta = \mathscr{R}\int_{-\infty}^{e^{i\theta_1}} \left(\frac{4a_0}{c(\zeta+1)^2 \chi(\tau)} - \frac{4(a_0)_{\tau=\tau_1}}{c(\zeta+1)^2 \chi(\tau_1)} \right) d\zeta - \frac{2(a_0)_{\tau=\tau_1}}{c\chi(\tau_1)}. \tag{5·2, 54}$$

Note that the integral on the right-hand side remains finite as θ_1 tends to π, since τ as given by (5·2, 53) becomes equal to τ, for $\theta_1 = \pi$ and $\zeta = -1$. Hence, substituting the right-hand side of (5·2, 54) on the left-hand side of (5·2, 49) and setting θ_1 equal to π, we obtain

$$\int_{-\infty}^{-1} \left(\frac{a_0}{\chi(\tau)} - \frac{(a_0)_{\tau=\tau_1}}{\chi(\tau)} \right) \frac{d\zeta}{(\zeta+1)^2} - \frac{(a_0)_{\tau=\tau_1}}{2\chi(\tau_1)}$$

$$= -\frac{1}{2}(b_0)_{\tau=\tau_1} - \frac{c}{4}\int_{-\infty}^{-1} \left(\frac{1}{\zeta^2} \frac{db_0}{d\tau} + \frac{1}{\zeta} \frac{db_1}{d\tau} \right) d\zeta. \tag{5·2, 55}$$

Except where it is indicated that $\tau = \tau_1$, the argument τ in this equation is obtained by setting $\theta_1 = \pi$ in (5·2, 53), so

$$\tau = \tau_1 + \frac{c}{4\zeta}(\zeta+1)^2. \tag{5·2, 56}$$

We observe that equation (5·2, 56) involves real quantities only. It may be regarded as an integral equation for the determination of the coefficient a_0 as a function of τ.

If τ is expressed in terms of t by means of (5·2, 26), then (5·2, 55) is transformed into the following integral equation for a_0 as a function of t,

$$\int_{-\infty}^{-1} \left(\frac{a_0(t)}{U(t)} - \frac{a_0(t_1)}{U(t_1)} \right) \frac{d\zeta}{(\zeta+1)^2} - \frac{a_0(t_1)}{2U(t_1)}$$

$$= - \tfrac{1}{2} b_0(t_1) - \frac{c}{4} \int_{-\infty}^{-1} \left(\frac{b_0'(t)}{\zeta^2} + \frac{b_1'(t)}{\zeta} \right) \frac{d\zeta}{U(t)}, \quad (5·2, 57)$$

where
$$t = g^{-1} \left(g(t_1) - \frac{c}{4\zeta} (\zeta+1)^2 \right). \quad (5·2, 58)$$

In the succeeding sections we shall study particular solutions of equation (5·2, 57).

5·3 Two-dimensional motion with constant forward velocity

Suppose in the first instance that the aerofoil has constant velocity U. We may assume that the mid-point of the aerofoil coincides with the origin of the fixed system of coordinates at time $t = 0$, and so

$$g(t) = - Ut.$$

Then (5·2, 58) becomes
$$t = t_1 + \frac{c}{4U\zeta} (\zeta+1)^2, \quad (5·3, 1)$$

and the integral equation for $a_0(t)$ may be written in the form

$$\int_{-\infty}^{-1} (a_0(t) - a_0(t_1)) \frac{d\zeta}{(\zeta+1)^2} - \tfrac{1}{2} a_0(t_1) = - \tfrac{1}{2} U b_0(t_1) - \frac{c}{4} \int_{-\infty}^{-1} \left(\frac{b_0'(t)}{\zeta^2} + \frac{b_1'(t)}{\zeta} \right) d\zeta. \quad (5·3, 2)$$

Now suppose more particularly that the displacement function of the aerofoil (see (5·2, 15)) is of the form

$$F(x, t) = F_0(x) e^{at} \quad (a > 0). \quad (5·3, 3)$$

This is the case of exponentially divergent motion. It is understood that the formula refers to the system of coordinates which is fixed relative to the mean position of the aerofoil. Then by (5·2, 23),

$$G(x, t) = U \frac{\partial F}{\partial x} + \frac{\partial F}{\partial t} = (U F_0'(x) + a F_0(x)) e^{at} = G_0(x) e^{at}. \quad (5·3, 4)$$

We expand $G_0(x) = G_0(\tfrac{1}{2} c \cos \theta)$ into a Fourier series,

$$G_0(x) = \frac{B_0}{2} + \sum_{n=1}^{\infty} B_n \cos n\theta, \quad (5·3, 5)$$

where $B_0, B_1, B_2, \ldots$ are constants. Comparison with (5·2, 23) shows
$$b_n = B_n e^{at} \quad (n = 0, 1, 2, \ldots). \quad (5·3, 6)$$

By (5·2, 40), the coefficients $a_n(t)$ of the complex acceleration potential are then given by

$$a_n(t) = U \left(\frac{ca}{4nU} B_{n-1} + B_n - \frac{ca}{4nU} B_{n+1} \right) e^{at}, \quad (5·3, 7)$$

or $$a_n(t) = U\left(\frac{\mu}{4n}B_{n-1} + B_n - \frac{\mu}{4n}B_{n+1}\right)e^{at} \tag{5·3, 8}$$

for $n = 1, 2, \ldots$, where μ is the non-dimensional parameter $\mu = ca/U$.

In order to determine $a_0(t)$ it is natural to assume that it also depends on the time only through an exponential factor e^{at},

$$a_0(t) = A\,e^{at}, \quad \text{say.} \tag{5·3, 9}$$

Substituting (5·3, 8) and (5·3, 9) in (5·3, 2), we obtain

$$A\left(\int_{-\infty}^{-1}(e^{at} - e^{at_1})\frac{d\zeta}{(\zeta+1)^2} - \tfrac{1}{2}e^{at_1}\right) = -\frac{U}{2}B_0 e^{at_1} - \frac{ca}{4}\int_{-\infty}^{-1}\left(\frac{B_0}{\zeta^2} + \frac{B_1}{\zeta}\right)e^{at}\,d\zeta, \tag{5·3, 10}$$

where $$t = t_1 + \frac{c}{4U\zeta}(\zeta+1)^2.$$

Hence, dividing (5·3, 10) by e^{at_1},

$$A\left(\int_{-\infty}^{-1}\left\{\exp\left(\frac{\mu}{4\zeta}(\zeta+1)^2\right) - 1\right\}\frac{d\zeta}{(\zeta+1)^2} - \frac{1}{2}\right)$$

$$= -\frac{ca}{4}\left(\frac{2}{\mu}B_0 + B_0\int_{-\infty}^{-1}\frac{1}{\zeta^2}\exp\left(\frac{\mu}{4\zeta}(\zeta+1)^2\right)d\zeta + B_1\int_{-\infty}^{-1}\frac{1}{\zeta}\exp\left(\frac{\mu}{4\zeta}(\zeta+1)^2\right)d\zeta\right). \tag{5·3, 11}$$

It remains to calculate the three integrals in (5·3, 11). Putting $\zeta = -e^{\eta}$, we obtain for the integral on the left-hand side,

$$\int_{-\infty}^{-1}\left\{\exp\left(\frac{\mu}{4\zeta}(\zeta+1)^2\right) - 1\right\}\frac{d\zeta}{(\zeta+1)^2}$$

$$= -\frac{1}{2}\int_0^{\infty}\{\exp\left(\tfrac{1}{2}\mu(1-\cosh\eta)\right) - 1\}\frac{d\eta}{1-\cosh\eta}. \tag{5·3, 12}$$

Now $$(\coth\tfrac{1}{2}\eta)' = \frac{1}{1-\cosh\eta}.$$

Hence, integrating by parts in (5·3, 12),

$$-\frac{1}{2}\left\{[(\exp\left(\tfrac{1}{2}\mu(1-\cosh\eta)\right) - 1)\coth\tfrac{1}{2}\eta]_0^{\infty}\right.$$

$$\left. + \tfrac{1}{2}\mu\int_0^{\infty}\exp\left(\tfrac{1}{2}\mu(1-\cosh\eta)\right)\sinh\eta\,\coth\tfrac{1}{2}\eta\,d\eta\right\}. \tag{5·3, 13}$$

The integrated part in (5·3, 13) is equal to

$$\lim_{\eta\to\infty}(\exp\left(\tfrac{1}{2}\mu(1-\cosh\eta)\right) - 1)\coth\tfrac{1}{2}\eta = -1$$

in the upper limit. It becomes indeterminate in the lower limit and we apply de l'Hôpital's rule:

$$\lim_{\eta\to 0}\{\exp\left(\tfrac{1}{2}\mu(1-\cosh\eta)\right) - 1\}\coth\tfrac{1}{2}\eta$$

$$= \lim_{\eta\to 0}\cosh\tfrac{1}{2}\eta\lim_{\eta\to 0}\frac{(\exp\left(\tfrac{1}{2}\mu(1-\cosh\eta)\right) - 1)'}{(\sinh\tfrac{1}{2}\eta)'}$$

$$= \lim_{\eta\to 0} -\frac{\tfrac{1}{2}\mu\sinh\eta\exp\left(\tfrac{1}{2}\mu(1-\cosh\eta)\right)}{\tfrac{1}{2}\cosh\tfrac{1}{2}\eta}$$

$$= 0.$$

The integral in (5·3, 13) can be evaluated in terms of modified Bessel functions of the second kind by means of the formula‡

$$K_n(x) = \int_0^\infty e^{-x \cosh \eta} \cosh n\eta \, d\eta \quad (x > 0, \ n = 0, 1, 2, \ldots). \quad (5·3, 14)$$

Then

$$\int_0^\infty \exp\left(\tfrac{1}{2}\mu(1 - \cosh \eta)\right) \sinh \eta \coth \tfrac{1}{2}\eta \, d\eta = e^{\frac{1}{2}\mu} \int_0^\infty e^{-\frac{1}{2}\mu \cosh \eta} (1 + \cosh \eta) \, d\eta$$

$$= e^{\frac{1}{2}\mu} (K_0(\tfrac{1}{2}\mu) + K_1(\tfrac{1}{2}\mu)). \quad (5·3, 15)$$

Hence (5·3, 13) becomes

$$\tfrac{1}{2} - \tfrac{1}{4}\mu \, e^{\frac{1}{2}\mu}(K_0(\tfrac{1}{2}\mu) + K_1(\tfrac{1}{2}\mu)).$$

Similarly, substituting $-e^\eta$ for ζ in the integrals on the right-hand side of (5·3, 11), we obtain

$$\int_{-\infty}^{-1} \frac{1}{\zeta} \exp\left(\frac{\mu}{4\zeta}(\zeta + 1)^2\right) d\zeta = e^{\frac{1}{2}\mu} \int_0^\infty e^{-\frac{1}{2}\mu \cosh \eta} \, d\eta = -e^{\frac{1}{2}\mu} K_0(\tfrac{1}{2}\mu), \quad (5·3, 16)$$

and

$$\int_{-\infty}^{-1} \frac{1}{\zeta^2} \exp\left(\frac{\mu}{4\zeta}(\zeta + 1)^2\right) d\zeta = e^{\frac{1}{2}\mu} \int_0^\infty e^{-\frac{1}{2}\mu \cosh \eta} e^{-\eta} \, d\eta$$

$$= e^{\frac{1}{2}\mu}\left[\int_0^\infty e^{-\frac{1}{2}\mu \cosh \eta} \cosh \eta \, d\eta - \int_0^\infty e^{-\frac{1}{2}\mu \cosh \eta} \sinh \eta \, d\eta\right]$$

$$= e^{\frac{1}{2}\mu} K_1(\tfrac{1}{2}\mu) - \frac{2}{\mu}. \quad (5·3, 17)$$

Substituting in (5·3, 11) from (5·3, 15), (5·3, 16) and (5·3, 17)

$$-A\frac{\mu}{4} e^{\frac{1}{2}\mu} (K_0(\tfrac{1}{2}\mu) + K_1(\tfrac{1}{2}\mu)) = \frac{ca}{4} e^{\frac{1}{2}\mu} (B_1 K_0(\tfrac{1}{2}\mu) - B_0 K_1(\tfrac{1}{2}\mu)),$$

or

$$A = U \frac{B_0 K_1(\tfrac{1}{2}\mu) - B_1 K_0(\tfrac{1}{2}\mu)}{K_1(\tfrac{1}{2}\mu) + K_0(\tfrac{1}{2}\mu)}. \quad (5·3, 18)$$

Hence

$$a_0(t) = A \, e^{at} = U \frac{B_0 K_1(\tfrac{1}{2}\mu) - B_1 K_0(\tfrac{1}{2}\mu)}{K_1(\tfrac{1}{2}\mu) + K_0(\tfrac{1}{2}\mu)} e^{at}. \quad (5·3, 19)$$

Introducing the function

$$J(\mu) = \frac{2K_0(\tfrac{1}{2}\mu)}{K_1(\tfrac{1}{2}\mu) + K_0(\tfrac{1}{2}\mu)}, \quad (5·3, 20)$$

we may write (5·3, 19) also as

$$a_0(t) = \tfrac{1}{2}U\{(2 - J(\mu)) B_0 - J(\mu) B_1\} e^{at}. \quad (5·3, 21)$$

Values of $J(\mu)$ are given in Table 5·3, 1.§

‡ See G. N. Watson, *A Treatise on the Theory of Bessel Functions* (2nd ed., Cambridge University Press, 1944), p. 181.

§ These values are selected from a more extensive tabulation made by W. P. Jones, 'Summary of formulae and notations used in two-dimensional derivative theory', *Rep. Memor. Aero. Res. Comm., Lond.*, no. 1958 (1941).

We shall now determine the aerodynamic forces acting on the aerofoil, first for the general case considered in § 5·2.

By (5·2, 9) and (5·2, 10), $\dfrac{p}{\rho} = \Omega + h(t)$,   (5·3, 22)

where $h(t)$ must be equal to the value of p/ρ at infinity since Ω vanishes there. Thus $\rho\Omega$ is the pressure increment, and the discontinuity of the pressure across any point of the aerofoil is

$$\Delta p = \rho(\Omega(x', +0, t) - \Omega(x', -0, t)) = \rho\left(a_0 \tan \tfrac{1}{2}\theta + 2 \sum_{n=1}^{\infty} a_n \sin n\theta\right),$$
(5·3, 23)

from (5·2, 22), where $x' = \tfrac{1}{2}c \cos \theta$ $(0 \leqslant \theta \leqslant \pi)$.

Table 5·3, 1

μ	0	0·10	0·20	0·30	0·40	0·60	0·80	1·00	2·00
J	0	0·27052	0·39526	0·47723	0·53692	0·61984	0·67570	0·71637	0·82317

Integrating (5·3, 23), we obtain the total lift on the aerofoil,

$$L = -\int_{-\frac{1}{2}c}^{\frac{1}{2}c} \Delta p \, dx' = -\tfrac{1}{2}\rho c \int_0^\pi \left(a_0 \tan \tfrac{1}{2}\theta + 2 \sum_{n=1}^{\infty} a_n \sin n\theta\right) \sin \theta \, d\theta,$$

or $L = -\tfrac{1}{2}\pi\rho c(a_0(t) + a_1(t))$.   (5·3, 24)

This formula for the lift neglects the slope of the wing in assuming that the pressure acts in a direction normal to the x'-axis. It also neglects the presence of the suction force at the leading edge. This is in keeping with the remaining simplifying assumptions of the theory.

Similarly, we obtain for the moment round the mid-chord of the (mean position of the) aerofoil

$$M = -\rho\int_{-\frac{1}{2}c}^{\frac{1}{2}c} \Delta p x' \, dx' = -\tfrac{1}{4}\rho c^2 \int_0^\pi \left(a_0 \tan \tfrac{1}{2}\theta + 2 \sum_{n=1}^{\infty} a_n \sin n\theta\right) \cos \theta \sin \theta \, d\theta$$

or $M = \tfrac{1}{8}\pi\rho c^2(a_0(t) - a_2(t))$.   (5·3, 25)

For unsteady conditions the resultant aerodynamic force on the wing will also possess an x-component which may be either positive (drag) or negative (thrust). To calculate this force, we multiply the pressure difference by the local slope of the aerofoil and integrate between the leading and trailing edges. In addition, it is now necessary to take into account the suction force at the leading edge. One determines this force by applying the momentum principle to a small region S bounded by a circle which surrounds the leading edge. It is not difficult to show by means of (5·2, 25) that Φ and $(\partial\Phi/\partial t)_x$ remain bounded near the leading edge. It then follows from (5·2, 12) that $\partial\Phi/\partial x$ becomes infinite at that point as $U^{-1}\Omega$, i.e. as the square root of the distance from the leading edge. Hence the momentum within S vanishes in the limit and the equation is the same as for the steady case. As a result the formula (2·8, 30) for the suction force is still valid. Moreover, in the definition of the constant C which occurs in that formula, we may replace the velocity component u (or u') by $U^{-1}\Omega$ for which an explicit expression is more readily available.

Another method for the determination of the x-component of the aerodynamic force, for the particular case of a rigid wing, is given by Nekrasov.‡ Although the problem is of considerable interest in connexion with the flight of birds and its imitations, it is of small importance for normal aircraft. Even the drag which arises in rectilinear accelerated flight is usually neglected, being small compared with the ordinary inertia forces which came into play at the same time. For these reasons, we shall not consider the determination of the force in more detail.

Returning to the case of exponentially divergent motion at constant forward speed, we obtain for the lift and moment, by means of (5·3, 7), (5·3, 21), (5·3, 23) and (5·3, 25),

$$L = - \tfrac{1}{2}\pi\rho Uc\{(1 - \tfrac{1}{2}J(\mu))\,B_0 - \tfrac{1}{2}J(\mu)\,B_1 + (\tfrac{1}{4}\mu B_0 + B_1 - \tfrac{1}{4}\mu B_2)\}\,e^{at},$$

or
$$L = - \tfrac{1}{2}\pi\rho Uc\{(1 - \tfrac{1}{2}J(\mu) + \tfrac{1}{4}\mu)\,B_0 + (1 - \tfrac{1}{2}J(\mu))\,B_1 - \tfrac{1}{4}\mu B_2\}\,e^{at},$$
$$\tag{5·3, 26}$$

and
$$M = \tfrac{1}{8}\pi\rho Uc^2\{(1 - \tfrac{1}{2}J(\mu))\,B_0 - (\tfrac{1}{8}\mu + \tfrac{1}{2}J(\mu))\,B_1 - B_2 + \tfrac{1}{8}\mu B_3\}\,e^{at}. \tag{5·3, 27}$$

Let us assume in particular that the aerofoil diverges vertically, so that (see (5·3, 3) and (5·3, 4))

$$\left.\begin{aligned}F_0(x) &= \text{constant} = y_0,\\[4pt]G_0(x) &= ay_0,\\[4pt]B_0 = 2ay_0, \quad B_1 &= B_2 = \ldots = 0.\end{aligned}\right\} \tag{5·3, 28}$$

Then
$$L = - \pi\rho Ucay_0(1 - \tfrac{1}{2}J(\mu) + \tfrac{1}{4}\mu)\,e^{at}, \tag{5·3, 29}$$

$$M = \tfrac{1}{4}\pi\rho Uc^2 ay_0(1 - \tfrac{1}{2}J(\mu))\,e^{at}. \tag{5·3, 30}$$

The functions $K_0(\tfrac{1}{2}\mu)$ and $K_1(\tfrac{1}{2}\mu)$ possess logarithmic singularities at the origin,§ and the same then applies to $J(\mu)$. This confirms a statement made in the introduction (§ 5·1). Now

$$y = F(x, t) = y_0\,e^{at},$$

and so
$$y'(t) = \frac{\partial y}{\partial t} = ay_0\,e^{at}, \quad y''(t) = \frac{\partial^2 y}{\partial t^2} = a^2 y_0\,e^{at}.$$

---

‡ A. I. Nekrasov, 'Wing theory for unsteady flow', *A.R.C. Paper*, no. 11792 (translated from the Russian by S. W. Skan), ch. 6.

§ G. N. Watson, *A Treatise on the Theory of Bessel Functions* (2nd ed., Cambridge University Press, 1941), p. 80.

We may therefore write the lift in the form

$$L = -\pi\rho Ucy'(t)(1 - \tfrac{1}{2}J(\mu)) - \pi\rho c^2 y''(t)\tfrac{1}{4},$$

or
$$L = y(t)L_y + y'(t)L_{y'} + y''(t)L_{y''}, \qquad (5\cdot3, 31)$$

where
$$L_y = 0,$$
$$\left.\begin{aligned} L_{y'} &= -\pi\rho Uc(1 - \tfrac{1}{2}J(\mu)), \\ L_{y''} &= -\pi\rho c^2\tfrac{1}{4}. \end{aligned}\right\} \qquad (5\cdot3, 32)$$

In accordance with the introduction (see (5·1, 12) and (5·1, 13)), we regard L_y, $L_{y'}$, $L_{y''}$ as the aerodynamic derivatives of the lift with respect to displacement, velocity and acceleration.

Similarly we may rewrite (5·3, 30) in the form

$$M = \tfrac{1}{4}\pi\rho Uc^2 y'(t)(1 - \tfrac{1}{2}J(\mu)), \qquad (5\cdot3, 33)$$

or
$$M = y(t)M_y + y'(t)M_{y'} + y''(t)M_{y''}, $$

where
$$M_y = 0,$$
$$\left.\begin{aligned} M_{y'} &= \tfrac{1}{4}\pi\rho Uc^2(1 - \tfrac{1}{2}J(\mu)), \\ M_{y''} &= 0. \end{aligned}\right\} \qquad (5\cdot3, 34)$$

Next we consider divergent motion in pitch round the origin. That is to say, at every chord station x,

$$y = F(x, t) = -\alpha_0 x e^{at}, \quad F_0(x) = -\alpha_0 x,$$

where α_0 is a constant. Hence

$$G_0(x) = -\alpha_0(U + ax) = -\alpha_0 U - \frac{\alpha_0 ac}{2}\cos\theta, \qquad (5\cdot3, 35)$$

and so
$$B_0 = -2\alpha_0 U, \quad B_1 = -\frac{\alpha_0 ac}{2}, \quad B_2 = B_3 = \ldots = 0. \qquad (5\cdot3, 36)$$

Substituting in (5·3, 26), we obtain

$$L = \pi\rho U^2 c\alpha_0\{(1 - \tfrac{1}{2}J(\mu) + \tfrac{1}{4}\mu) + \mu(\tfrac{1}{4} - \tfrac{1}{8}J(\mu))\}e^{at}, \qquad (5\cdot3, 37)$$

or
$$L = \alpha(t)L_\alpha + \alpha'(t)L_{\alpha'} + \alpha''(t)L_{\alpha''}, \qquad (5\cdot3, 38)$$

where
$$\alpha(t) = \alpha_0 e^{at}, \quad \alpha'(t) = a\alpha_0 e^{at}, \quad \alpha''(t) = a^2\alpha_0 e^{at},$$
$$\left.\begin{aligned} L_\alpha &= \pi\rho U^2 c(1 - \tfrac{1}{2}J(\mu)), \\ L_{\alpha'} &= \pi\rho Uc^2(\tfrac{1}{2} - \tfrac{1}{8}J(\mu)), \\ L_{\alpha''} &= 0. \end{aligned}\right\} \qquad (5\cdot3, 39)$$

Similarly, substituting in (5·3, 27),

$$M = -\tfrac{1}{8}\pi\rho U^2 c^2\alpha_0\{(2 - J(\mu)) - \tfrac{1}{2}\mu(\tfrac{1}{8}\mu + \tfrac{1}{2}J(\mu))\}e^{at}, \qquad (5\cdot3, 40)$$

or
$$M = \alpha(t)M_\alpha + \alpha'(t)M_{\alpha'} + \alpha''(t)M_{\alpha''}, \qquad (5\cdot3, 41)$$

where
$$M_\alpha = -\pi\rho U^2 c^2(\tfrac{1}{4} - \tfrac{1}{8}J(\mu)),$$
$$M_{\alpha'} = \pi\rho U c^3 \tfrac{1}{32}J(\mu),$$
$$M_{\alpha''} = \pi\rho c^4 \tfrac{1}{128}.$$
(5·3, 42)

Some simple modifications are required if a point other than the mid-chord is taken both as point of reference for the moment and as pivot for the pitching motion. Assume that the new point of reference P is at a distance hc aft of the leading edge. Apart from referring the moment to this point, we also have to take into account that for small α a pure rotation α round P is equal to a rotation α round the mid-chord together with a displacement $(h-\tfrac{1}{2})\alpha c$ in the direction of the y-axis. The results are stated below in standard notation which is as follows. The new x-axis is taken in the direction of motion of the (mean position of the) wing, i.e. in the opposite direction to that used above, the new z-axis is taken in the direction opposite to that of the y-axis used above, and the new y-axis is defined so that the three new axes form a right-handed system. Thus, the new y-axis points to starboard. The lift now acts in the negative direction of the z-axis and we define the force component Z by $Z = -L$. The incidence is denoted by α, as before, but the primes are replaced by dots to indicate differentiation with respect to time. The pitching moment is denoted by M, as before, but it is now regarded as positive when it tends to raise the leading edge (nose-up moment). With this notation

$$Z_z = 0,$$
$$\frac{Z_{\dot z}}{\pi\rho Uc} = -(1 - \tfrac{1}{2}J(\mu)),$$
$$\frac{Z_{\ddot z}}{\pi\rho c^2} = -\tfrac{1}{4},$$
$$M_z = 0,$$
$$\frac{M_{\dot z}}{\pi\rho Uc^2} = (h - \tfrac{1}{4})(1 - \tfrac{1}{2}J(\mu)),$$
$$\frac{M_{\ddot z}}{\pi\rho c^3} = \tfrac{1}{4}h - \tfrac{1}{8},$$
$$\frac{Z_\alpha}{\pi\rho U^2 c} = -(1 - \tfrac{1}{2}J(\mu)),$$
$$\frac{Z_{\dot\alpha}}{\pi\rho Uc^2} = h - 1 + (\tfrac{3}{8} - \tfrac{1}{2}h)J(\mu),$$
$$\frac{Z_{\ddot\alpha}}{\pi\rho c^3} = \tfrac{1}{4}h - \tfrac{1}{8},$$
$$\frac{M_\alpha}{\pi\rho U^2 c^2} = (h - \tfrac{1}{4})(1 - \tfrac{1}{2}J(\mu)),$$
$$\frac{M_{\dot\alpha}}{\pi\rho Uc^3} = (\tfrac{3}{4} - h)\{(h - \tfrac{1}{2}) + (\tfrac{1}{8} - \tfrac{1}{2}h)J(\mu)\},$$
$$\frac{M_{\ddot\alpha}}{\pi\rho c^4} = -\tfrac{3}{128} - \tfrac{1}{4}(h - \tfrac{1}{4})(h - \tfrac{3}{4}).$$
(5·3, 43)

The derivatives with respect to the displacements, linear or angular, e.g. M_z, Z_α, are called stiffness derivatives, by analogy with the theory of elasticity, the derivatives with respect to the velocities, $Z_{\dot z}$, etc., are called damping derivatives, and the derivatives with respect to the accelerations are called acceleration derivatives (or, with reversed sign, aerodynamic inertias). When applied to aero-elastic problems the derivatives with respect to α, $\dot\alpha$, $\ddot\alpha$ are called torsional derivatives, while the derivatives with respect to $\dot z$, $\ddot z$ are called flexural derivatives. The latter name is due to that fact that when the wing flexes in an upward or downward direction, the individual sections are displaced in the direction of the z-axis, and the formulae (5·3, 42) may then be used in a 'strip-theory' approximation (compare § 3·8).

Next we consider the case where the displacement function $F(x, t)$ represents a divergent oscillation

$$F_c(x, t) = F_0(x)\, e^{bt} \cos \nu t \quad (b > 0), \qquad (5\cdot3, 44)$$

or

$$F_s(x, t) = F_0(x)\, e^{bt} \sin \nu t \quad (b > 0). \qquad (5\cdot3, 45)$$

It is found expedient to consider (5·3, 44) and (5·3, 45) simultaneously by putting $\quad y = F(x, t) = F_c(x, t) + iF_s(x, t) = F_0(x)\, e^{bt}(\cos \nu t + i \sin \nu t)$,

or

$$y = F(x, t) = F_0(x)\, e^{(b+i\nu)t}. \qquad (5\cdot3, 46)$$

By setting $b + i\nu = a$, we reduce the problem formally to the case of exponentially divergent motion considered above (see equation (5·3, 3)). We may therefore expect to get the correct results for (5·3, 46) and hence for (5·3, 44) and (5·3, 45) by setting $a = b + i\nu$ in (5·3, 43), in which a intervenes through the intermediary of μ. However, since the coincidence of (5·3, 46) with (5·3, 3) is only formal, it is necessary to check that every step in the analysis of divergent motion can be interpreted in terms of the simultaneous treatment of the cases given by (5·3, 44) and (5·3, 45).

The first step in the analysis is the determination of the function $G(x, t)$:

$$G(x, t) = U \frac{\partial F}{\partial x} + \frac{\partial F}{\partial t} = (UF_0'(x) + i\nu F_0(x))\, e^{(b+i\nu)t}$$

$$= \frac{b_0}{2} + \sum_{n=1}^{\infty} b_n \cos n\theta$$

$$= \left(\frac{B_0}{2} + \sum_{n=1}^{\infty} B_n \cos n\theta\right) e^{(b+i\nu)t},$$

where the coefficients b_n, B_n are now complex. The real part of this expression corresponds to (5·3, 44) and the imaginary part to (5·3, 45). The coefficients of the acceleration potential, excepting the first, are then obtained by means of the linear operation with real coefficients, which is given by (5·2, 40). Hence the real and imaginary parts of $a_n(t)$ as given by (5·3, 8) for $a = b + i\nu$, $n = 1, 2, 3, \ldots$ again correspond to (5·3, 44) and (5·3, 45) respectively.

Next we have to determine $a_0(t)$ as the solution of the integral equation (5·2, 57). We note that both the range of integration in (5·2, 57) and the coefficients of a_0, b_0, b_0', b_1', are real. Hence the real and imaginary parts of $a_0(t)$ as given by (5·3, 19) correspond to (5·3, 44) and (5·3, 45), where $x = b + i\nu$ and $\mu = \dfrac{(b + i\nu)c}{U}$, and where the functions $K_0(x)$, $K_1(x)$ are still defined by (5·3, 14). We note that (5·3, 14) is valid for $\mathscr{R}(x) > 0$, a condition which is satisfied in the present case by virtue of the assumption $b > 0$.

It follows immediately that the lift, and the moment about the origin, for displacements

$$y = y_0 e^{bt} \cos \nu t,$$

and

$$y = y_0 e^{bt} \sin \nu t,$$

are given by the real and imaginary parts of (5·3, 29) and (5·3, 30), for $a = b + i\nu$. In a similar way, the lift and moment for pitching motion about the mid-chord are given by (5·3, 37) and (5·3, 40).

The case of harmonic oscillations may be obtained from the above by passing to the limit as b approaches zero. We cannot actually put $b = 0$ in (5·3, 14), since the integral on the right-hand side diverges for pure imaginary x for $n = 1$.

To carry out the passage to the limit we observe‡ that

$$K_n(x) = \tfrac{1}{2}\pi i^{-n-1} H_n^{(2)}(-ix),$$

where $H_n^{(2)}(-ix)$ is a Hankel function

$$H_n^{(2)}(-ix) = J_n(-ix) - iY_n(-ix), \qquad (5·3, 47)$$

J_n and Y_n being Bessel functions of the first and second kind respectively. Hence, for $b > 0$,

$$K_n(b + i\nu) = \tfrac{1}{2}\pi i^{-n-1}(J_n(\nu - ib) - iY_n(\nu - ib)),$$

and so in the limit

$$\lim_{b \to 0} \int_0^\infty e^{-(b+i\nu)\cosh \eta} \cosh n\eta \, d\eta = \tfrac{1}{2}\pi i^{-n-1}(J_n(\nu) - iY_n(\nu)). \qquad (5·3, 48)$$

Hence, putting $\lambda = \nu c/U$, $\mu = i\lambda$,

$$J(\mu) = J(i\lambda) = \frac{2K_0(\tfrac{1}{2}i\lambda)}{K_0(\tfrac{1}{2}i\lambda) + K_1(\tfrac{1}{2}i\lambda)}$$

$$= 2\frac{Y_0(\tfrac{1}{2}\lambda) + iJ_0(\tfrac{1}{2}\lambda)}{Y_0(\tfrac{1}{2}\lambda) + J_1(\tfrac{1}{2}\lambda) + i(J_0(\tfrac{1}{2}\lambda) - Y_1(\tfrac{1}{2}\lambda))}$$

$$= 2\frac{(Y_0(\tfrac{1}{2}\lambda))^2 + (J_0(\tfrac{1}{2}\lambda))^2 + Y_0(\tfrac{1}{2}\lambda)\,J_0(\tfrac{1}{2}\lambda) - J_0(\tfrac{1}{2}\lambda)\,Y_1(\tfrac{1}{2}\lambda) + i(J_0(\tfrac{1}{2}\lambda)\,J_1(\tfrac{1}{2}\lambda) + Y_0(\tfrac{1}{2}\lambda)\,Y_1(\tfrac{1}{2}\lambda))}{(J_0(\tfrac{1}{2}\lambda) - Y_1(\tfrac{1}{2}\lambda))^2 + (J_1(\tfrac{1}{2}\lambda) + Y_0(\tfrac{1}{2}\lambda))^2},$$

‡ G. N. Watson, *A Treatise on the Theory of Bessel Functions* (2nd ed., Cambridge University Press, 1944), p. 78.

or
$$J(i\lambda) = H(\lambda) + i\lambda G(\lambda), \tag{5·3, 49}$$

where
$$\left. \begin{aligned} H(\lambda) &= 2\frac{J_0^2 + Y_0^2 + Y_0 J_1 - J_0 Y_1}{(J_0 - Y_1)^2 + (J_1 + Y_0)^2}, \\ G(\lambda) &= \frac{2}{\lambda}\frac{J_0 J_1 + Y_0 Y_1}{(J_0 - Y_1)^2 + (J_1 + Y_0)^2}. \end{aligned} \right\} \tag{5·3, 50}$$

In these formulae the argument of the functions on the right-hand side is $\frac{1}{2}\lambda$. A brief tabulation of $H(\lambda)$ and $G(\lambda)$ is given in Table 5·3, 2.

Table 5·3, 2

λ	H	G
0·00	0·00000	∞
0·02	0·03516	4·56521
0·04	0·07255	3·76040
0·10	0·18198	2·61280
0·20	0·33615	1·72302
0·40	0·54484	0·94312
0·60	0·67006	0·59773
1·00	0·80413	0·30142
2·00	0·92113	0·10027
4·00	0·97409	0·02885
6·00	0·98744	0·01334
10·00	0·99521	0·00492
∞	1·00000	0·00000

We now have
$$y = F(x, t) = F_0(x)\, e^{ivt}, \tag{5·3, 51}$$

and so for oscillation parallel to the y-axis (vertical oscillation)
$$F_0(x) = y_0, \quad F(x, t) = y_0\, e^{ivt}.$$
Then the lift is given by
$$\begin{aligned} L &= -\pi\rho U c v y_0 i (1 - \tfrac{1}{2}(H(\lambda) + i\lambda G(\lambda)) + \tfrac{1}{4}i\lambda)\, e^{ivt} \\ &= -\pi\rho U^2 y_0 (i\lambda(1 - \tfrac{1}{2}H(\lambda)) - \lambda^2(G(\lambda) + \tfrac{1}{4}))\, e^{ivt} \\ &= -\pi\rho U c (1 - \tfrac{1}{2}H(\lambda))\, y'(t) - \pi\rho c^2 (G(\lambda) + \tfrac{1}{4})\, y''(t), \end{aligned} \tag{5·3, 52}$$

or
$$L = y(t)\, L_y + y'(t)\, L_{y'} + y''(t)\, L_{y''},$$

where
$$\left. \begin{aligned} L_y &= 0, \\ L_{y'} &= -\pi\rho U c (1 - \tfrac{1}{2}H(\lambda)), \\ L_{y''} &= -\pi\rho c^2 (\tfrac{1}{4} + G(\lambda)). \end{aligned} \right\} \tag{5·3, 53}$$

Similarly, the moment is given by
$$M = y(t)\, M_y + y'(t)\, M_{y'} + y''(t)\, M_{y''}, \tag{5·3, 54}$$

where
$$M_y = 0,$$
$$M_{y'} = \tfrac{1}{4}\pi\rho Uc^2(1 - \tfrac{1}{2}H(\lambda)),$$
$$M_{y''} = -\tfrac{1}{8}\pi\rho c^3 G(\lambda).$$
(5·3, 55)

The aerodynamic derivatives for pitching oscillations can be determined in the same way. A list of the oscillatory derivatives is given below in the standard notation used in (5·3, 43). Results are quoted again with reference to a point which is at a distance hc aft of the leading edge

$$Z_z = 0, \quad \frac{Z_{\dot{z}}}{\pi\rho Uc} = -(1 - \tfrac{1}{2}H(\lambda)),$$

$$\frac{Z_{\ddot{z}}}{\pi\rho c^2} = \tfrac{1}{2}G(\lambda) - \tfrac{1}{4},$$

$$M_z = 0, \quad \frac{M_{\dot{z}}}{\pi\rho Uc^2} = (h - \tfrac{1}{4})(1 - \tfrac{1}{2}H(\lambda)),$$

$$\frac{M_{\ddot{z}}}{\pi\rho c^3} = \tfrac{1}{4}h - \tfrac{1}{8} + (\tfrac{1}{8} - \tfrac{1}{2}h)G(\lambda),$$

$$\frac{Z_\alpha}{\pi\rho U^2 c} = -(1 - \tfrac{1}{2}H(\lambda)),$$

$$\frac{Z_{\dot{\alpha}}}{\pi\rho Uc^2} = h - 1 + (\tfrac{3}{8} - \tfrac{1}{2}h)H(\lambda) + \tfrac{1}{2}G(\lambda),$$

$$\frac{Z_{\ddot{\alpha}}}{\pi\rho c^3} = \tfrac{1}{4}h - \tfrac{1}{8} + (\tfrac{3}{8} - \tfrac{1}{2}h)G(\lambda),$$

$$\frac{M_\alpha}{\pi\rho U^2 c^2} = (h - \tfrac{1}{4})(1 - \tfrac{1}{2}H(\lambda)),$$

$$\frac{M_{\dot{\alpha}}}{\pi\rho Uc^3} = (\tfrac{3}{4} - h)\{(h - \tfrac{1}{2}) + (\tfrac{1}{8} - \tfrac{1}{2}h)H(\lambda)\} + (\tfrac{1}{8} - \tfrac{1}{2}h)G(\lambda),$$

$$\frac{M_{\ddot{\alpha}}}{\pi\rho c^4} = -\tfrac{3}{128} - (\tfrac{3}{4} - h)(\tfrac{1}{4} - h)(\tfrac{1}{4} - \tfrac{1}{2}G(\lambda)).$$
(5·3, 56)

Separating the forces and moments into real and imaginary parts, we see that the aerodynamic damping derivatives ($L_{\dot{z}}$, $M_{\dot{z}}$, $L_{\dot{\alpha}}$, $M_{\dot{\alpha}}$) give rise to the terms which are in quadrature with the motion (i.e. $\pm 90°$ out of phase).

By superimposing solutions of the harmonic type as in (5·3, 51) or of the more general type (5·3, 46), we may solve the problem for a general variation of $F(x, t)$ subject only to certain conditions of regularity. Thus, assume that $F(x, t)$ can be expressed as a Fourier integral with respect to the time,

$$F(x, t) = \int_{-\infty}^{\infty} K(x, s)\, e^{ils}\, ds.$$
(5·3, 57)

The function $K(x, s)$ is then given by

$$K(x, s) = \frac{1}{2\pi} \int_{-\infty}^{\infty} F(x, t)\, e^{-ist}\, dt,$$
(5·3, 58)

by Fourier's integral theorem. For any particular s, define $F_0(x)$ in (5·3, 51) by
$$F_0(x) = K(x, s).$$

The corresponding lift and pitching moment about the mid-chord are given by (5·3, 26) and (5·3, 27), where $a = is$, $\mu = ics/U$. For given ρ, U, c and $F(x, t)$, we regard these as functions of s and write

$$L = l(s), \quad M = m(s). \tag{5·3, 59}$$

Superimposing the solutions for varying s, we then obtain for the total lift

$$L = \int_{-\infty}^{\infty} l(s)\, e^{its}\, ds, \tag{5·3, 60}$$

and for the pitching moment

$$M = \int_{-\infty}^{\infty} m(s)\, e^{its}\, ds. \tag{5·3, 61}$$

If the Fourier integral (5·3, 58) does not exist, it may still be possible to apply the principle of superposition by using a complex path of integration. As an important example of the procedure we consider a flat aerofoil which advances with a constant velocity U, and has incidence zero initially, for $t < 0$. At time $t = 0$ the aerofoil is suddenly pulled up to an incidence α_0 which remains constant thereafter. Since a flat aerofoil in rectilinear motion at zero incidence does not produce any motion in the surrounding fluid, we may assume alternatively that the aerofoil is initially at rest and only begins to move at time $t = 0$ with constant velocity U and constant incidence α_0. So far as the motion of the surrounding medium and the dynamic reactions on the wing are concerned, the two problems are completely equivalent, but only the first problem satisfies the assumption $U = $ constant, on which the analysis of this section is based.

Our assumption is that the position of the aerofoil is given by

$$F(x, t) = \begin{cases} 0 & (t < 0), \\ -\alpha_0 x & (t > 0), \end{cases} \tag{5·3, 62}$$

Hence

$$G(x, t) = \begin{cases} 0 & (t < 0), \\ -U\alpha_0 & (t > 0), \end{cases} \tag{5·3, 63}$$

and so

$$\begin{aligned} b_0 = b_1 = b_2 = \ldots = 0 \quad (t < 0), \\ b_0 = -2U\alpha_0, \quad b_1 = b_2 = \ldots = 0 \quad (t > 0). \end{aligned} \tag{5·3, 64}$$

It must be pointed out that it was assumed at one point in the analysis that the velocity U and the coefficients b_0, b_1, b_2, ... are analytic functions of the time. However, once having found the solution for the general type of motion (5·3, 46), we may obtain other solutions by superposition even if the displacement function $F(x, t)$ which results from the superposition is not an analytic function of the time.

To find the coefficients $a_1(t)$, $a_2(t)$, ... of the acceleration potential, we employ (5·2, 40). This yields

$$a_1 = a_2 = a_3 = \ldots = 0 \tag{5·3, 65}$$

for all times except $t = 0$.

In order to determine a_0, we represent b_0 by the following integral:

$$b_0 = -\frac{U\alpha_0}{\pi i} \int_{\gamma-i\infty}^{\gamma+i\infty} e^{ts} \frac{ds}{s}. \qquad (5\cdot3, 66)$$

In this formula γ is any real positive constant and the integration is along a straight line parallel to the imaginary axis. To establish that $(5\cdot3, 66)$ is equal to b_0 as given by $(5\cdot3, 64)$, we have to show that

$$\frac{1}{2\pi i} \int_{\gamma-i\infty}^{\gamma+i\infty} e^{ts} \frac{ds}{s} = \begin{cases} 0 & (t<0), \\ 1 & (t>0). \end{cases} \qquad (5\cdot3, 67)$$

The left-hand side of $(5\cdot3, 67)$ is

$$\lim_{R\to\infty} \frac{1}{2\pi i} \int_{\gamma-iR}^{\gamma+iR} e^{ts} \frac{ds}{s}.$$

Let C be a circle of radius R round the point $s = (\gamma, 0)$, and let C_1 and C_2 be the two halves of C to the left and to the right of the straight line $s = \gamma$, respectively. For $t < 0$, the integral

$$\int_{C_2} e^{ts} \frac{ds}{s}$$

vanishes in the limit, as R tends to infinity, by Jordan's lemma‡ and for $t > 0$, the integral

$$\int_{C_1} e^{ts} \frac{ds}{s}$$

vanishes in the limit, as R tends to infinity, by the same lemma. Hence, for $t < 0$,

$$\lim_{R\to\infty} \frac{1}{2\pi i} \int_{\gamma-iR}^{\gamma+iR} e^{ts} \frac{ds}{s} = \lim_{R\to\infty} \int_{C_2^*} e^{ts} \frac{ds}{s}, \qquad (5\cdot3, 68)$$

where C_2^* is the closed contour made up of the segment $(\gamma - iR, \gamma + iR)$ and of the semicircle C_2 with the same radius R, taken into mathematically negative direction. But the integrand is regular within the region bounded by C_2^*, and so the integrand on the right-hand side of $(5\cdot3, 68)$ vanishes identically for all R. Hence

$$\lim_{R\to\infty} \frac{1}{2\pi i} \int_{\gamma-iR}^{\gamma+iR} e^{ts} \frac{ds}{s} = 0,$$

proving one part of our assertion. Similarly, for $t > 0$,

$$\lim_{R\to\infty} \frac{1}{2\pi i} \int_{\gamma-iR}^{\gamma+iR} e^{ts} \frac{ds}{s} = \lim_{R\to\infty} \frac{1}{2\pi i} \int_{C_1^*} e^{ts} \frac{ds}{s},$$

where C_1^* is the closed contour made up of the segment $(\gamma - iR, \gamma + iR)$ and of the semicircle C_1 with the same radius R, taken in the mathematically

‡ See E. T. Whittaker and G. N. Watson, *A Course of Modern Analysis* (4th ed., Cambridge University Press, 1927), p. 115.

positive direction. The integrand is negative on and in the interior of C_1, except at the origin where it has a simple pole with residue 1, and so

$$\lim_{R\to\infty} \frac{1}{2\pi i} \int_{\gamma-iR}^{\gamma+iR} e^{ts}\frac{ds}{s} = \lim_{R\to\infty} \frac{1}{2\pi i}\cdot 2\pi i = 1,$$

proving the second part of our assertion.

Now the solution of the integral equation (5·3, 2) for $b_1(t) = 0$, $b_0(t) = B_0\,e^{at}$, $\mathcal{R}(a) > 0$, is, by (5·3, 21),

$$a_0(t) = \tfrac{1}{2}U(2 - J(\mu))\,B_0\,e^{at},$$

where $\mu = ca/U$. It follows by superposition that when b_0 is given by (5·3, 66), while b_1 vanishes identically, the corresponding $a_0(t)$ is

$$a_0(t) = -\frac{U^2\alpha_0}{\pi i}\int_{\gamma-i\infty}^{\gamma+i\infty}\left(1 - \tfrac{1}{2}J\left(\frac{cs}{U}\right)\right)e^{ts}\frac{ds}{s}. \tag{5·3, 69}$$

Write $ts = w$, $t\gamma = \kappa$, $Ut/c = \sigma$, then (5·3, 69) becomes

$$a_0(t) = -\frac{U^2\alpha_0}{\pi i}\int_{\kappa-i\infty}^{\kappa+i\infty}\left(1 - \tfrac{1}{2}J\left(\frac{w}{\sigma}\right)\right)e^{w}\frac{dw}{w}. \tag{5·3, 70}$$

σ is a non-dimensional parameter, the distance of advance of the aerofoil from the time $t = 0$ measured in terms of the chord length. Let

$$k(\sigma) = \frac{1}{\pi i}\int_{\kappa-i\infty}^{\kappa+i\infty}\left(1 - \tfrac{1}{2}J\left(\frac{w}{\sigma}\right)\right)e^{w}\frac{dw}{w}, \tag{5·3, 71}$$

then

$$a_0(t) = -U^2\alpha_0 k(\sigma),$$

and so, for $t > 0$,

$$L = \tfrac{1}{2}\rho U^2 c\pi\alpha_0 k(\sigma),$$

$$C_L = \pi\alpha_0 k(\sigma). \tag{5·3, 72}$$

H. G. Küssner‡ gives the following expansion of $k(\sigma)$ as a series of rational functions:

$$k(\sigma) = 1 + \frac{\sigma}{2(\sigma+2)} + \frac{\sigma^2}{4(\sigma+2)^2} + \frac{\sigma^3}{3!}\left(\frac{1}{(\sigma+2)^3} - \frac{1}{4(\sigma+2)^2(\sigma+4)}\right) + \dots.$$

$$\tag{5·3, 73}$$

Values of $k(\sigma)$ are given in Table 5·3, 3.

Table 5·3, 3

σ	0	2	4	6	8	10	∞
$k(\sigma)$	1·000	1·339	1·516	1·625	1·698	1·750	2·000

It will be seen that just after the aerofoil has been pulled up to incidence α_0, or just after the beginning of the motion in the second interpretation,

‡ H. G. Küssner, 'Das zweidimensionale Problem der beliebig bewegten Tragfläche unter Berücksichtigung von Partial-bewegung der Flüssigkeit', *Luftfahrtforsch.* vol. 17 (1940), pp. 355–61.

the lift coefficient is $\pi\alpha_0$, i.e. one-half the steady two-dimensional value $2\pi\alpha_0$. As might be expected, the lift coefficient tends to the steady value asymptotically. The phenomenon of lift deficiency just after the start of the motion, or just after the incidence of the wing has been changed abruptly, is known as the Wagner effect.‡

The first paper on the problem of the oscillating aerofoil appears to be due to Birnbaum.§ Birnbaum's analysis is applicable to small values of λ. For general λ the aerodynamic derivatives for oscillating rigid aerofoils were first calculated by Glauert‖ and by Duncan and Collar.¶ The latter calculated also the derivatives for the case of exponential divergence. Later work on the subject is due to Theodorsen,‡‡ Kassner and Fingado,§§ von Kármán and Sears‖‖ and Küssner.¶¶ A description of work by I. N. Sedov has been given by Nekrasov.‡‡‡ The concept of the acceleration potential is employed by Possio§§§ and Küssner‖‖‖ in investigations which include two-dimensional problems in an incompressible fluid as special cases.

In unsteady flow, unlike in steady flow, there exists a vortex wake, i.e. a surface across which the tangential velocity component is discontinuous, even under two-dimensional conditions. For methods which do not employ the concept of the acceleration potential, the determination of the strength of the vortex wake, and of the field of flow due to it, forms an integral part of the analysis. The concept of the acceleration potential is convenient precisely because it eliminates the need for considering the discontinuity across the wake. This entails a considerable simplification in the analysis, but at the same time it implies that an important physical feature of the

‡ H. Wagner, 'Über die Entstehung des dynamischen Auftriebes von Trag-flügeln', *Z. angew. Math. Mech.* vol. 5 (1925), pp. 17–35.

§ W. Birnbaum, 'Das ebene Problem des schlagenden Flügels', *Z. angew. Math. Mech.* vol. 4 (1924), pp. 277–92.

‖ H. Glauert, 'The force and moment on an oscillating aerofoil', *Rep. Memor. Aero. Res. Comm., Lond.*, no. 1242 (1929).

¶ W. J. Duncan and A. R. Collar, 'Resistance derivatives of flutter theory', Part I, *Rep. Memor. Aero. Res. Comm., Lond.*, no. 1500 (1932).

‡‡ Th. Theodorsen, 'General theory of aerodynamic instability and the mechanism of flutter', *Tech. Rep. Nat. Adv. Comm. Aero., Wash.*, no. 496 (1935).

§§ R. Kassner and H. Fingado, 'Das ebene Problem der Flügelschwingung', *Luftfahrtforsch.* vol. 13 (1936), pp. 374–87.

‖‖ Th. von Kármán and W. R. Sears, 'Airfoil theory for non-uniform motion', *J. Aero. Sci.* vol. 5 (1938), pp. 379–90.

¶¶ H. G. Küssner, 'Das zweidimensionale Problem der beliebig bewegten Trag-fläche unter Berücksichtigung von Partial-bewegung der Flüssigkeit', *Luftfahrtforsch.* vol. 17 (1940), pp. 355–61.

‡‡‡ A. J. Nekrasov, 'Wing theory for unsteady flow', *A.R.C. Paper*, no. 11792 (translated from the Russian by S. W. Skan), ch. 6.

§§§ C. Possio, 'Aerodinamica sul profilo oscillanto in un fluido compressibile a velocita iposonora', *Aerotecnica*, vol. 18 (1938), pp. 441–58.

‖‖‖ H. G. Küssner, 'Allgemeine Tragflächentheorie', *Luftfahrtforsch.* vol. 17 (1940), pp. 370–78 . (Translated in *Tech. Memor. Nat. Adv. Comm. Aero., Wash.*, no. 979.) This is discussed further in § 5·6.

phenomenon remains out of sight. However, once the acceleration potential has been found, the strength of the vortex wake can be determined in the following way.

It is assumed that the vortex wake coincides with the x-axis downstream of the trailing edge $x = \frac{1}{2}c$. The strength of the vortex sheet at a point $P(x, 0)$ of the wake (Fig. 5·3, 1), referred to the system of coordinates which moves with the aerofoil, is denoted by $\gamma(x_1, t_1)$. Consider the circulation round a small rectangular circuit $ABCDA$, whose sides AB, CD are parallel to the x-axis and close to it on either side, while DA passes through P. Then the circulation round the circuit at any time is, approximately,

$$-\left(\Phi(x_1, +0, t) - \Phi(x_1 + \delta x_1, +0, t) + \Phi(x_1 + \delta x_1, -0, t) - \Phi(x_1, -0, t)\right)$$

$$\doteq \left(\left(\frac{\partial \Phi}{\partial x}\right)_{y=+0} - \left(\frac{\partial \Phi}{\partial x}\right)_{y=-0}\right) \delta x, \quad (5\cdot3, 74)$$

Fig. 5·3, 1

where δx is the length of AB and of CD. (5·3, 74) must be equal to the total vorticity within the circuit, $\gamma \, \delta x$. Also,

$$\left(\frac{\partial \Phi}{\partial x}\right)_{y=-0} = -\left(\frac{\partial \Phi}{\partial x}\right)_{y=+0},$$

and so

$$\gamma(x, t) = 2\left(\frac{\partial \Phi}{\partial x}\right)_{y=+0} = -2\left(\frac{\partial \Phi}{\partial x}\right)_{y=-0}. \quad (5\cdot3, 75)$$

Expressing $\partial \Phi/\partial x$ in terms of the complex acceleration potential we then obtain (compare (5·2, 27))

$$\gamma(x_1, t_1) = \lim_{y_1 \to +0} \left\{-2\mathscr{R} \int_{-\infty + iy_1}^{z_1} \omega'(z, g^{-1}(g(t_1) + z_1 - z)) \frac{dz}{U(g^{-1}(g(t_1) + z_1 - z))}\right\},$$
$$(5\cdot3, 76)$$

where $z_1 = x_1 + iy_1$, and the expression $\lim\limits_{y_1 \to +0}$ indicates that y_1 tends to zero from above.

Even without calculating $\gamma(x, t)$ explicitly we may draw an important general conclusion from (5·3, 75). According to that equation, we have at any point of the wake

$$U\frac{\partial \gamma}{\partial x} + \frac{\partial \gamma}{\partial t} = 2\left(U\frac{\partial}{\partial x} + \frac{\partial}{\partial t}\right)\left(\frac{\partial \Phi}{\partial x}\right)_{y=+0}$$

$$= 2\frac{\partial}{\partial x}\left(U\frac{\partial}{\partial x} + \frac{\partial}{\partial t}\right)\Phi(x, +0) = 2\frac{\partial}{\partial x}\Omega(x, +0). \quad (5\cdot3, 77)$$

But Ω is an odd function of y which is continuous across the wake, and so $\Omega(x, +0) = 0$. Hence

$$U\frac{\partial\gamma}{\partial x} + \frac{\partial\gamma}{\partial t} = 0. \tag{5·3, 78}$$

In the primed system of coordinates, which is fixed relative to the air at infinity, the equation becomes

$$\frac{\partial\gamma}{\partial t'} = 0. \tag{5·3, 79}$$

We may express this verbally by saying that after being shed by the wing at its trailing edge, the vortices of the wake remain at rest in the surrounding fluid. On the other hand, if we place ourselves in a system which is at rest relative to the aerofoil, then our result states that the vortices in the wake are swept along with the velocity of the main stream. This is the linear approximation to the statement that the vortices move with the fluid. In fact, it can be shown quite generally that in the two-dimensional motion of an incompressible fluid the vorticity of any fluid particle is constant or, which is the same, that the vortices retain constant strength while moving with the fluid.

Indeed, Euler's equations for two-dimensional flow without body forces are (see (1·4, 9))

$$\frac{\partial u}{\partial t} + u\frac{\partial u}{\partial x} + v\frac{\partial u}{\partial y} = -\frac{1}{\rho}\frac{\partial p}{\partial x},$$

$$\frac{\partial v}{\partial t} + u\frac{\partial v}{\partial x} + v\frac{\partial v}{\partial y} = -\frac{1}{\rho}\frac{\partial p}{\partial y}.$$

Differentiating the second equation with respect to x and the first with respect to y and subtracting, we obtain

$$\frac{\partial\zeta}{\partial t} + u\frac{\partial\zeta}{\partial x} + v\frac{\partial\zeta}{\partial y} + \frac{\partial u}{\partial x}\frac{\partial v}{\partial x} + \frac{\partial v}{\partial x}\frac{\partial v}{\partial y} - \frac{\partial u}{\partial y}\frac{\partial u}{\partial x} - \frac{\partial v}{\partial y}\frac{\partial u}{\partial y} = 0,$$

where $\zeta = \partial v/\partial x - \partial u/\partial y$ is the vorticity. The sum of the last four terms on the left-hand side vanishes and so

$$\frac{D\zeta}{Dt} = \frac{\partial\zeta}{\partial t} + u\frac{\partial\zeta}{\partial x} + v\frac{\partial\zeta}{\partial y} = 0. \tag{5·3, 80}$$

This proves the assertion. Since the free-stream velocity is large compared with the induced velocities we may linearize, so

$$U\frac{\partial\zeta}{\partial x} + \frac{\partial\zeta}{\partial t} = 0. \tag{5·3, 81}$$

Equation (5·3, 78) is obtained from (5·3, 81) by a passage to the limit for the case of an infinitely thin layer of vorticity. Alternatively, (5·3, 78) can be derived from Kelvin's theorem on the constancy of the circulation round a closed circuit which moves with the fluid (§ 1·5).

If it is desired to determine $\gamma(x, t)$ explicitly, then the following procedure may be adopted in preference to the direct evaluation of (5·3, 76).

Consider first the circulation $\Gamma(t)$ round a single closed curve which surrounds the aerofoil and crosses the x-axis at the trailing edge. $\Gamma(t)$ is given by

$$\Gamma(t) = \Phi(\tfrac{1}{2}c, +0, t) - \Phi(\tfrac{1}{2}c, -0, t)$$

$$= -\mathscr{R}\int_C \omega(z, g^{-1}(g(t) + \tfrac{1}{2}c - z)) \frac{dz}{U(g^{-1}(g(t) + \tfrac{1}{2}c - z))} \qquad (5·3, 82)$$

(compare (5·2, 25)–(5·2, 27)). We may assume in particular that the integral on the right-hand side is taken along both sides of the mean position of the aerofoil, so that the path of integration C is real.

Suppose now that the aerofoil is in exponentially divergent motion relative to its mean position while the forward speed U is constant. In that case the complex acceleration potential is of the form

$$\omega = i\left(-A_0 \frac{\zeta}{\zeta+1} + \sum_{n=1}^{\infty} \frac{A_n}{\zeta^n}\right) e^{at} \qquad (a > 0). \qquad (5·3, 83)$$

where

$$z = \frac{c}{4}\left(\zeta + \frac{1}{\zeta}\right), \qquad dz = \frac{c}{4}\left(1 - \frac{1}{\zeta^2}\right).$$

Also

$$g^{-1}(g(t) + \tfrac{1}{2}c - z) = t - \frac{c}{2U}\left(1 - \frac{1}{2}\left(\zeta + \frac{1}{\zeta}\right)\right),$$

and so $\Gamma(t)$ becomes

$$\Gamma(t) = -\mathscr{R}\left\{\frac{ic}{4U} e^{at-\frac{1}{2}\mu} \int_{C'} \left(-A_0 \frac{\zeta}{\zeta+1} + \sum_{n=1}^{\infty} \frac{A_n}{\zeta^n}\right)\left(1 - \frac{1}{\zeta^2}\right) \exp\left(\frac{\mu}{4}\left(\zeta + \frac{1}{\zeta}\right)\right) d\zeta\right\},$$

$$(5·3, 84)$$

where $\mu = ac/U$ and C' is the unit circle in the ζ-plane. Exchanging the order of summation and integration in (5·3, 84)

$$\Gamma(t) = -\mathscr{R}\left\{\frac{ic}{4U} e^{at-\frac{1}{2}\mu} \left[-A_0 \int_{C'} \left(1 - \frac{1}{\zeta}\right) \exp\left(\frac{\mu}{4}\left(\zeta + \frac{1}{\zeta}\right)\right) d\zeta\right.\right.$$

$$\left.\left. + \sum_{n=1}^{\infty} A_n \int_{C'} \left(\frac{1}{\zeta^n} - \frac{1}{\zeta^{n+2}}\right) \exp\left(\frac{\mu}{4}\left(\zeta + \frac{1}{\zeta}\right)\right) d\zeta\right]\right\}. \qquad (5·3, 85)$$

But‡

$$\int_{C'} \frac{1}{\zeta^n} \exp\left(\frac{\mu}{4}\left(\zeta + \frac{1}{\zeta}\right)\right) d\zeta = 2\pi i I_{n-1}(\tfrac{1}{2}\mu) \qquad (n = 0, \pm 1, \pm 2, \ldots),$$

$$(5·3, 86)$$

where I_n is the modified Bessel function of the first kind, so that in particular

$$\int_{C'} \exp\left(\frac{\mu}{4}\left(\zeta + \frac{1}{\zeta}\right)\right) d\zeta = 2\pi i I_{-1}(\tfrac{1}{2}\mu) = 2\pi i I_1(\tfrac{1}{2}\mu).$$

‡ G. N. Watson, *A Treatise on the Theory of Bessel Functions* (2nd ed., Cambridge University Press, 1944), pp. 20, 77.

Hence (5·3, 85) becomes

$$\Gamma(t) = -\frac{\pi c}{2U} e^{at-\frac{1}{2}\mu}\left[A_0(I_1(\tfrac{1}{2}\mu) - I_0(\tfrac{1}{2}\mu)) + \sum_{n=1}^{\infty} A_n(I_{n+1}(\tfrac{1}{2}\mu) - I_{n-1}(\tfrac{1}{2}\mu)) \right].$$
$$(5·3, 87)$$

We may now modify (5·3, 87) by using the following recurrence formula:‡

$$I_{n-1}(z) - I_{n+1}(z) = \frac{2n}{z} I_n(z). \tag{5·3, 88}$$

Substituting the right-hand side of (5·3, 88) in the infinite series on the right-hand side of (5·3, 87) we find

$$\Gamma(t) = -\frac{\pi c}{2U} e^{at-\frac{1}{2}\mu}\left[A_0(I_1(\tfrac{1}{2}\mu) - I_0(\tfrac{1}{2}\mu)) - \sum_{n=1}^{\infty} \frac{4n}{\mu} A_n I_n(\tfrac{1}{2}\mu) \right]. \tag{5·3, 89}$$

The coefficients A_n in this formula are given by

$$\left. \begin{aligned} A_0 &= U \frac{B_0 K_1(\tfrac{1}{2}\mu) - B_1 K_0(\tfrac{1}{2}\mu)}{K_1(\tfrac{1}{2}\mu) + K_0(\tfrac{1}{2}\mu)}, \\ A_n &= U\left(\frac{\mu}{4n} B_{n-1} + B_n - \frac{\mu}{4n} B_{n+1}\right) \quad (n = 1, 2, \dots) \end{aligned} \right\} \tag{5·3, 90}$$

(see (5·3, 8) and (5·3, 19)), and so

$$\Gamma(t) = -\frac{\pi c}{2U} e^{at-\frac{1}{2}\mu}\left[A_0(I_1(\tfrac{1}{2}\mu) - I_0(\tfrac{1}{2}\mu)) - U\sum_{n=1}^{\infty}\left(B_{n-1} + \frac{4n}{\mu} B_n - B_{n+1}\right) I_n(\tfrac{1}{2}\mu) \right]. \tag{5·3, 91}$$

The infinite series on the right-hand side of (5·3, 91) becomes, after rearrangement and subsequent application of (5·3, 88),

$$B_0 I_1(\tfrac{1}{2}\mu) + B_1\left(\frac{4}{\mu} I_1(\tfrac{1}{2}\mu) + I_2(\tfrac{1}{2}\mu)\right) + \sum_{n=2}^{\infty} B_n\left(I_{n+1}(\tfrac{1}{2}\mu) + \frac{4n}{\mu} I_n(\tfrac{1}{2}\mu) - I_{n-1}(\tfrac{1}{2}\mu)\right)$$
$$= B_0 I_1(\tfrac{1}{2}\mu) + B_1 I_0(\tfrac{1}{2}\mu). \tag{5·3, 92}$$

Hence

$$\begin{aligned} \Gamma(t) &= -\frac{\pi c}{2} e^{at-\frac{1}{2}\mu}\left[\frac{B_0 K_1(\tfrac{1}{2}\mu) - B_1 K_0(\tfrac{1}{2}\mu)}{K_1(\tfrac{1}{2}\mu) + K_0(\tfrac{1}{2}\mu)} (I_1(\tfrac{1}{2}\mu) - I_0(\tfrac{1}{2}\mu)) \right. \\ &\qquad\qquad\qquad \left. - (B_0 I_1(\tfrac{1}{2}\mu) + B_1 I_0(\tfrac{1}{2}\mu)) \right] \\ &= \frac{\pi c}{2} e^{at-\frac{1}{2}\mu} \frac{K_1(\tfrac{1}{2}\mu) I_0(\tfrac{1}{2}\mu) + K_0(\tfrac{1}{2}\mu) I_1(\tfrac{1}{2}\mu)}{K_1(\tfrac{1}{2}\mu) + K_0(\tfrac{1}{2}\mu)} (B_0 + B_1). \end{aligned} \tag{5·3, 93}$$

Now§

$$K_1(\tfrac{1}{2}\mu) I_0(\tfrac{1}{2}\mu) + K_0(\tfrac{1}{2}\mu) I_1(\tfrac{1}{2}\mu) = \frac{2}{\mu}, \tag{5·3, 94}$$

and so

$$\Gamma(t) = \pi c\, e^{at-\frac{1}{2}\mu} \frac{B_0 + B_1}{\mu(K_1(\tfrac{1}{2}\mu) + K_0(\tfrac{1}{2}\mu))}. \tag{5·3, 95}$$

---

‡ G. N. Watson, *A Treatise on the Theory of Bessel Functions* (2nd ed., Cambridge University Press, 1944), p. 79.

§ *Ibid.* p. 80.

We note that as a, and hence μ, tends to 0, $\mu(K_1(\tfrac{1}{2}\mu) + K_0(\tfrac{1}{2}\mu))$ tends to 2, and so $\Gamma(t)$ becomes in the limit

$$\Gamma = \frac{\pi}{2}c(B_0 + B_1). \tag{5·3, 96}$$

Also, in this case the coefficients $a_0(t)$ and $a_1(t)$ in (5·3, 24) become equal to UB_0 and UB_1 respectively (see (5·3, 19) and (5·3, 8)), and so the lift becomes

$$L = -\frac{\pi}{2}\rho Uc(B_0 + B_1) = -\rho U\Gamma, \tag{5·3, 97}$$

in agreement with the theorem of Kutta and Joukowski (see (2·3, 5)).

The strength of the vorticity shed into the fluid at the trailing edge must be equal, except for the sign, to the rate of change of the circulation round the aerofoil per unit distance of the forward motion,

$$\gamma(\tfrac{1}{2}c, t) = -\frac{1}{U}\frac{d\Gamma}{dt}. \tag{5·3, 98}$$

Hence, from (5·3, 95),

$$\gamma(\tfrac{1}{2}c, t) = -\pi\, e^{at - \tfrac{1}{2}\mu}\frac{B_0 + B_1}{K_1(\tfrac{1}{2}\mu) + K_0(\tfrac{1}{2}\mu)}.$$

It was pointed out earlier that γ remains constant at any point of the wake which is fixed in the primed system of coordinates. Thus, in the system of coordinates which moves with the aerofoil,

$$\gamma(x, t) = -\pi\, e^{a(t - x/U)}\frac{B_0 + B_1}{K_1(\tfrac{1}{2}\mu) + K_0(\tfrac{1}{2}\mu)}. \tag{5·3, 99}$$

It can be shown that, in the sense explained above ((5·3, 44) et seq.), equations (5·3, 95) and (5·3, 99) still hold if a is complex ($\mathcal{R}(a) \geqslant 0$). For more general cases, the circulation and vorticity can then be found by superposition. Thus we obtain for the case of a flat plate at incidence α_0 which started from rest at time $t = 0$ and whose forward velocity U is constant (see (5·3, 62) et seq.),

$$\Gamma(t) = 2iU^2\alpha_0\int_{\gamma - i\infty}^{\gamma + i\infty}\frac{\exp\{(t - c/2U)s\}}{K_1(cs/2U) + K_0(cs/2U)}\frac{ds}{s^2} \quad (t > 0), \tag{5·3, 100}$$

$$\gamma(x, t) = -iU\alpha_0\int_{\gamma - i\infty}^{\gamma + i\infty}\frac{\exp\{(t - x/U)s\}}{K_1(cs/2U) + K_0(cs/2U)}\frac{ds}{s} \quad (t > 0, x > \tfrac{1}{2}c).$$

$$\tag{5·3, 101}$$

Write

$$ts = w, \quad t\gamma = \kappa, \quad Ut/c = \sigma, \quad x/c = \xi.$$

Then equations (5·3, 100) and (5·3, 101) become

$$\Gamma(t) = 2iU^2\alpha_0 t\int_{\kappa - i\infty}^{\kappa + i\infty}\frac{\exp\{(1 - 1/2\sigma)w\}}{K_1(w/2\sigma) + K_0(w/2\sigma)}\frac{dw}{w^2}, \tag{5·3, 102}$$

and

$$\gamma(\xi, t) = -iU\alpha_0\int_{\kappa - i\infty}^{\kappa + i\infty}\frac{\exp\{(1 - \xi/\sigma)w\}}{K_1(w/2\sigma) + K_0(w/2\sigma)}\frac{dw}{w}. \tag{5·3, 103}$$

Farren and Walker have measured experimentally the growth of circulation about an aerofoil started from rest, and have obtained confirmation of the theoretical predictions.‡

5·4   Motion with non-uniform average velocity

In this section we shall consider a few examples in which the mean velocity of the aerofoil relative to the surrounding medium is no longer constant. Thus, first, we shall calculate the aerodynamic forces produced by a gust. This problem belongs to the second category mentioned in the introduction (§ 5·1), since there exists a disturbance in the atmosphere independently of the passage of the wing through it. We assume that the gust is confined to a bounded region, so that it has a meaning to introduce a system of co-ordinates (x', y') which is at rest relative to the atmosphere at infinity. Let $U_g(x', y')$, $V_g(x', y')$ be the velocity components in the gust, supposed independent of time, and small compared with the constant forward velocity U of the aerofoil. We assume further that the motion induced by the presence of the aerofoil possess a velocity potential $\Phi(x', y', t')$, so that the total velocity components are given by

$$u' = U_g - \frac{\partial \Phi}{\partial x'}, \quad v' = V_g - \frac{\partial \Phi}{\partial y'}. \tag{5·4, 1}$$

The assumption that the motion induced by the aerofoil is irrotational, and hence that it possesses a velocity potential, is strictly justifiable only if the gust motion is irrotational. In actual fact, a certain amount of vorticity may well be present in the gust, but it is supposed here that its distribution is not affected by the presence of the wing.

Let the motion of the aerofoil be defined by (5·2, 1). Then the boundary condition at the aerofoil is (see (1·2, 11))

$$V_g - \frac{\partial \Phi}{\partial y'} - \left(U_g - \frac{\partial \Phi}{\partial x'}\right) \frac{\partial f}{\partial x'} - \left(\frac{\partial f}{\partial t}\right)_{x'} = 0 \quad (x_L \leqslant x' \leqslant x_T, \, y = 0). \tag{5·4, 2}$$

Or, introducing the equation of the aerofoil referred to a system of co-ordinates which is fixed relative to the aerofoil (compare (5·2, 15), (5·2, 16) and (5·2, 17)),

$$V_g - \frac{\partial \Phi}{\partial y} - \left(U + U_g - \frac{\partial \Phi}{\partial x}\right) \frac{\partial F}{\partial x} - \frac{\partial F}{\partial t} = 0. \tag{5·4, 3}$$

Since both U_g and $\partial \Phi / \partial x$ are supposed small compared with U we may replace (5·4, 3) by

$$V_g - \frac{\partial \Phi}{\partial y} - \left(U \frac{\partial F}{\partial x} + \frac{\partial F}{\partial t}\right) = 0. \tag{5·4, 4}$$

‡ W. S. Farren, 'An apparatus for the measurement of two-dimensional flows at high Reynolds numbers with an application to the growth of circulation round a wing started impulsively from rest', *Proc. 3rd Int. Congr. Appl. Mech.* (Stockholm, 1930), pp. 323–30. P. B. Walker, 'Growth of circulation about a wing and an apparatus for measuring fluid motion'. *Rep. Mem. Aero. Res. Comm., Lond.*, no. 1402 (1931).

This boundary condition for Φ is linear and we may therefore split Φ into two parts, Φ_1 and Φ_2, which are due, respectively, to the geometry and attitude of the aerofoil and to the presence of the gust. Thus Φ_1 satisfies

$$-\frac{\partial \Phi_1}{\partial y} - \left(U\frac{\partial F}{\partial x} + \frac{\partial F}{\partial t}\right) = 0, \qquad (5\cdot4,5)$$

and Φ_2 satisfies

$$V_g - \frac{\partial \Phi_2}{\partial y} = 0. \qquad (5\cdot4,6)$$

$(5\cdot4,5)$ is independent of the presence of the gust and is included in the general analysis of §5·3. $(5\cdot4,6)$ is the condition to which $(5\cdot4,4)$ reduces if the aerofoil is flat and at zero incidence. Assuming this to be the case, we may omit the suffix in $(5\cdot4,6)$, and so the condition becomes

$$-\frac{\partial \Phi}{\partial y'} = -V_g(x', y') \quad (x_L \leqslant x' \leqslant x_T, \; y = 0). \qquad (5\cdot4,7)$$

Or, in the unprimed system of coordinates,

$$-\frac{\partial \Phi}{\partial y} = -V_g(x - Ut, y) \quad (-\tfrac{1}{2}c \leqslant x \leqslant \tfrac{1}{2}c, \; y = 0). \qquad (5\cdot4,8)$$

Thus, we may again determine Φ by the general method of §5·3 by equating $-V_g(x - Ut, 0)$ to

$$G(x, t) = \frac{b_0}{2} + \sum_{n=1}^{\infty} b_n \cos n\theta. \qquad (5\cdot4,9)$$

This shows that an equivalent problem is obtained for an aerofoil which moves in an otherwise undisturbed atmosphere, provided the equation of the aerofoil $y = F(x, t)$ satisfies the condition

$$U\frac{\partial F}{\partial x} + \frac{\partial F}{\partial t} = -V_g(x - Ut, 0). \qquad (5\cdot4,10)$$

It will be seen that the problem is completely determined by the variation of V_g along the x-axis. Thus, so long as U_g is small compared with U, as assumed in the present analysis, its effect on the aerodynamic forces is negligible.‡

Consider the example of a 'sharp-edged gust'

$$V_g(x', 0) = \begin{cases} V_0 & (x' < 0), \\ 0 & (x' > 0), \end{cases} \qquad (5\cdot4,11)$$

where V_0 is constant. Then

$$G(x, t) = -V_g(x - Ut, 0) = \begin{cases} 0, & -\tfrac{1}{2}c \leqslant x \leqslant \tfrac{1}{2}c, & t < -c/2U, \\ 0, & Ut < x \leqslant \tfrac{1}{2}c, & -c/2U < t < c/2U, \\ -V_0, & -\tfrac{1}{2}c \leqslant x \leqslant Ut, & -c/2U < t < c/2U, \\ -V_0, & -\tfrac{1}{2}c \leqslant x \leqslant \tfrac{1}{2}c, & t > c/2U. \end{cases}$$

$$(5\cdot4,12)$$

‡ Compare W. J. Duncan, *The Principles of Control and Stability of Aircraft* (Cambridge Aeronautical Series, 1952), p. 57.

The same $G(x, t)$ is obtained for a deformable aerofoil moving in an undisturbed atmosphere, if the aerofoil is at zero incidence before the time $t = -c/2U$ and at incidence $\alpha = V_0/U$ after $t = c/2U$, while during the intervening period it is at incidence V_0/U over its front portion, for $x < Ut$, and at zero incidence over the remaining rear portion.

The above case is taken as fundamental by von Kármán and Sears,[‡] but is solved by them by an essentially different method. Küssner,[§] to whom the solution for the complex exponential case (5·3, 46) and its generalization by superposition is due, obtains his results in the first instance for a linearly graded gust. Such a gust is given by

$$V_g(x', 0) = \begin{cases} v_0 x' & (x' < 0), \\ 0 & (x' > 0). \end{cases} \tag{5·4, 13}$$

Superposition of two cases of this type and a passage to the limit as the gust gradient v_0 becomes very large again yields a sharp-edged gust.

The lift due to the sharp-edged gust at any time t must be proportional to v_0 and can therefore be written in the form

$$L = \tfrac{1}{2} C_L \rho U v_0 c,$$

where C_L depends only on the time t and on the constants U and c. By dimensional reasoning alone, C_L must therefore be a function of

$$\sigma = \frac{Ut}{c} + \frac{1}{2} = \frac{2Ut + c}{2c},$$

$$C_L = k_g(\sigma), \quad \text{say}. \tag{5·4, 14}$$

σ is the distance that the leading edge of the aerofoil has advanced into the gust region at time t, measured in multiples of the chord length. Thus

$$k_g(\sigma) = 0 \quad \text{for} \quad \sigma < 0.$$

The variation of $k_g(\sigma)$ for $\sigma > 0$ is shown in Fig. 5·4, 1.

Consider now a smoothly graded gust given by

$$V_g(x', 0) = \begin{cases} V_0(x') & (x' \leqslant 0), \\ 0 & (x' > 0), \end{cases} \tag{5·4, 15}$$

where $V_0(x')$ is a differentiable function which vanishes for $x' = 0$ as well as for large negative x'. We may then regard (5·4, 15) as the result of the superposition of infinitesimal sharp-edged gusts, $(dV_0/dx) \delta x$,

$$V_0(x') = \int_{-\infty}^{x'} V_0'(\xi) \, d\xi \quad \text{for} \quad x' \leqslant 0,$$

‡ Th. von Kármán and W. R. Sears, 'Airfoil theory for non-uniform motion', *J. Aero. Sci.* vol. 5 (1938), pp. 379–390.

§ H. G. Küssner, 'Das zweidimensionale Problem der beliebig bewegten Tragfläche unter Berücksichtigung von Partial-bewegung der Flüssigkeit', *Luftfahrtforsch.* vol. 17 (1940), pp. 355–61.

where $V_0' = dV_0/dx'$. The corresponding lift coefficient therefore is

$$C_L = C_L(t) = \int_{-\frac{1}{2}(2Ut+c)}^{0} k_g\left(\sigma + \frac{\xi}{c}\right) V_0'(\xi)\, d\xi$$

$$= \int_{-\frac{1}{2}(2Ut+c)}^{0} k_g\left(\frac{2(Ut+\xi)+c}{2c}\right) V_0'(\xi)\, d\xi \quad \text{for} \quad t > -c/2U$$

and $\quad C_L(t) = 0 \quad$ for $\quad t < -c/2U.$

$$(5·4, 16)$$

We now return to the more general case of variable forward velocity $U(t) = \chi(\tau) > 0$. The analysis of § 5·2 is applicable to this case.

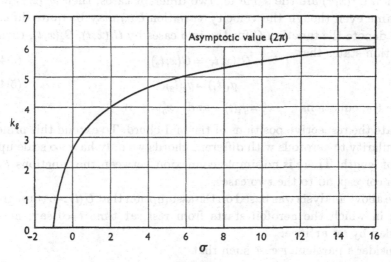

Fig. 5·4, 1

Put $\qquad \dfrac{\alpha_n(t)}{U(t)} = \dfrac{a_n(g^{-1}(-\tau))}{\chi(\tau)} = a_n^*(\tau) \quad (n = 0, 1, 2, \ldots)$ (5·4, 17)

and $\qquad b_n(t) = b_n(g^{-1}(-\tau)) = b_n^*(\tau) \quad (n = 0, 1, 2, \ldots).$ (5·4, 18)

Then (5·2, 41) and (5·2, 55) can also be written as

$$a_n^*(\tau) = \frac{c}{4n} b_{n-1}^{*'}(\tau) + b_n^*(\tau) - \frac{c}{4n} b_{n+1}^{*'}(\tau) \quad (n = 1, 2, \ldots) \quad (5·4, 19)$$

and $\qquad \displaystyle\int_{-\infty}^{-1} (a_0^*(\tau) - a_0^*(\tau_1)) \frac{d\zeta}{(\zeta+1)^2} - \tfrac{1}{2} a_0^*(\tau_1)$

$$= -\tfrac{1}{2} b_0^*(\tau_1) - \frac{c}{4} \int_{-\infty}^{-1} \left(\frac{b_0^{*'}(\tau)}{\zeta^2} + \frac{b_1^{*'}(\tau)}{\zeta}\right) d\zeta. \quad (5·4, 20)$$

(5·4, 19) and (5·4, 20) can be obtained formally from (5·2, 40) and (5·2, 57), respectively, by setting $U(t) = 1$, replacing t by τ, and affixing asterisks to the a_n and b_n in the latter formulae.

Now the $b_n(t)$ are the Fourier coefficients of the function

$$G(x, t) = G(\tfrac{1}{2}c\cos\theta, t) = G(x, g^{-1}(-\tau)) = G^*(x, \tau), \quad \text{say,}$$

(see (5·2, 23)) and the $a_n(t)$ are the coefficients of the acceleration potentials ω and Ω (see (5·2, 21) and (5·2, 22)). It follows that if the function $G^*(x, \tau)$ is specified, for a given chord length c, then the quantities

$$w^*(z, \tau) = \frac{w}{U} = i\left(-a_0^* \frac{\zeta}{\zeta + 1} + \sum_{n=1}^{\infty} \frac{a_n^*}{\zeta^n}\right) \tag{5·4, 21}$$

and

$$\Omega^*(\tfrac{1}{2}c \cos \theta, \tau) = \frac{\Omega}{U} = \tfrac{1}{2}a_0^* \tan \tfrac{1}{2}\theta + \sum_{n=1}^{\infty} a_n^* \sin n\theta \tag{5·4, 22}$$

are determined completely as functions of z or θ, and of the position of the mid-chord of the aerofoil. That is to say, if the chord length c and the function $G^*(x, \tau)$ are the same for two different cases, then $w^*(z, \tau)$ also is the same even though the velocity variation $U(t)$ may be quite different. If we denote the function G for the two cases by $G_1(x, t)$, $G_2(x, t)$, then the condition states that

$$G_1(x, t_1) = G(x, t_2), \tag{5·4, 23}$$

if

$$g_1(t_1) = g_2(t_2), \tag{5·4, 24}$$

where the equations $\quad x_0' = g_1(t) \quad$ and $\quad x_0' = g_2(t)$

indicate the respective positions of the mid-chord. To extend this principle of similarity to aerofoils with different chords we only have to scale up the unit of length. There is no simple connexion between the functions $F(x, t)$ which correspond to the two cases.

The above analysis was based on the assumption that $U(t) > 0$ throughout. Cases in which the aerofoil starts from rest, at time $t = 0$ say, may be included by an artifice.

Consider a particular case such that

$$U(t)\begin{cases} = 0 & (t < 0), \\ > 0 & (t > 0), \end{cases} \tag{5·4, 25}$$

while $F(x, t)$ is arbitrary for $t > 0$ and satisfies the condition $\partial F/\partial t = 0$ for $t < 0$.

Let t_0 be a small positive quantity, then $U_0 = U(t_0)$ is positive by assumption. We consider a new case, for which $F(x, t)$ is the same function as before while the velocity $U_1(t)$ is given by

$$U_1(t) = \begin{cases} U_0 & (t < t_0), \\ U(t) & (t > t_0). \end{cases} \tag{5·4, 26}$$

The function $U_1(t)$ defined in this way is positive everywhere and so the general theory is applicable. Also, the function $G(x, t)$ which determines the boundary condition is the same in both cases for $t < 0$ and $t > t_0$. It follows that as t_0 tends to zero in (5·4, 26), we obtain the desired solution of (5·4, 25).

It will appear that for a concrete case the passage to the limit is unnecessary. Thus, we may set $t_0 = 0$ immediately, reducing (5·4, 26) to (5·4, 25).

Assume, for example, that the aerofoil is a flat plate at constant incidence α_0, so that

$$F(x, t) = -\alpha_0 x, \quad -\tfrac{1}{2}c \leqslant x \leqslant \tfrac{1}{2}c. \tag{5·4, 27}$$

Suppose that at time $t = 0$ the aerofoil starts from rest with constant acceleration $\beta > 0$. Then, for $t > 0$,

$$\left.\begin{aligned} U(t) &= \beta t, \\ g(t) &= -\tfrac{1}{2}\beta t^2. \end{aligned}\right\} \tag{5·4, 28}$$

Hence, for $t > 0$, $\qquad G(x, t) = U\dfrac{\partial F}{\partial x} + \dfrac{\partial F}{\partial t} = -\alpha_0 \beta t.$

But $\qquad\qquad\qquad\qquad \tau = -g(t) = \tfrac{1}{2}\beta t^2,$

$$t = \sqrt{\frac{2\tau}{\beta}},$$

and so

$$\left.\begin{aligned} G^*(x, \tau) &= -\alpha_0 \sqrt{(2\beta\tau)}, \\ b_0^*(\tau) &= -2\alpha_0 \sqrt{(2\beta\tau)} = -\alpha_0 \sqrt{(8\beta\tau)}, \\ b_1^*(\tau) &= b_2^*(\tau) = \dots = 0. \end{aligned}\right\} \tag{5·4, 29}$$

Then, by (5·4, 19), $\qquad \left.\begin{aligned} a_1^*(\tau) &= \frac{c}{4}\frac{db_0^*}{d\tau} = -c\alpha_0\sqrt{\frac{\beta}{8\tau}}, \\ a_2^*(\tau) &= a_3^*(\tau) = \dots = 0. \end{aligned}\right\} \tag{5·4, 30}$

$a_0^*(\tau)$ is given by the integral equation (5·4, 20). To solve that equation, we try to represent b_0^* by a complex integral of the form

$$b_0^*(\tau) = \frac{1}{2\pi i}\int_{\gamma-i\infty}^{\gamma+i\infty} e^{\tau s} f(s)\, ds, \tag{5·4, 31}$$

where γ is a real constant and the integration is carried out along a straight line parallel to the imaginary axis. In order to determine $f(s)$ we observe that the right-hand side of (5·4, 31) is the Fourier–Mellin formula which represents the inverse of the Laplace transform.‡ Accordingly, we may expect $f(s)$ to be given by the Laplace transform

$$f(s) = \int_0^\infty e^{-s\tau} b_0^*(\tau)\, d\tau = -\alpha_0 \sqrt{(8\beta)}\int_0^\infty e^{-s\tau}\sqrt{\tau}\, d\tau. \tag{5·4, 32}$$

Substituting $\theta = \sqrt{\tau}$ in the integral on the right-hand side, we obtain

$$\begin{aligned} \int_0^\infty e^{-s\tau}\sqrt{\tau}\, d\tau &= 2\int_0^\infty e^{-s\theta^2}\theta^2\, d\theta \\ &= -2\frac{d}{ds}\int_0^\infty e^{-s\theta^2}\, d\theta \\ &= -2\frac{d}{ds}\frac{1}{\sqrt{s}}\int_0^\infty e^{-\eta^2}\, d\eta \\ &= \frac{1}{2}\sqrt{\frac{\pi}{s^3}}, \end{aligned}$$

‡ See, for example, H. S. Carslaw and J. C. Jaeger, *Operational Methods in Applied Mathematics* (2nd ed., Oxford University Press, 1949), ch. 4.

since
$$\int_0^\infty e^{-\eta^2}d\eta = \tfrac{1}{2}\sqrt{\pi}.$$

Hence
$$f(s) = -\alpha_0 \sqrt{\frac{2\pi\beta}{s^3}}. \tag{5·4, 33}$$

$f(s)$ is regular for $\mathscr{R}(s) > 0$, and so (5·4, 31) holds for $\tau > 0$, for all $\gamma > 0$. For $\tau < 0$, we may apply Jordan's lemma, as in (5·3, 68), in order to show that the right-hand side of (5·4, 31) vanishes. Hence

$$b_0^*(\tau) = i\alpha_0 \sqrt{\frac{\beta}{2\pi}} \int_{\gamma-i\infty}^{\gamma+i\infty} e^{\tau s} \frac{ds}{s^{\frac{3}{2}}}. \tag{5·4, 34}$$

Setting $U = 1$ in (5·3, 21), we see that the solution of the integral equation (5·4, 20) is, for $b_1^*(\tau) = 0$, $b_0^*(\tau) = B_0 e^{a\tau}$, $\mathscr{R}(a) > 0$,

$$a_0^*(\tau) = \tfrac{1}{2}(2 - J(\mu)) B_0 e^{a\tau}, \tag{5·4, 35}$$

where $\mu = ca$. Using the principle of superposition we then find that in the present case

$$a_0^*(\tau) = i\alpha_0 \sqrt{\frac{\beta}{2\pi}} \int_{\gamma-i\infty}^{\gamma+i\infty} (1 - \tfrac{1}{2}J(cs)) e^{\tau s} \frac{ds}{s^{\frac{3}{2}}}. \tag{5·4, 36}$$

Put $\tau s = w$, $\tau\gamma = \kappa$, $\tau/c = \sigma$, then (5·4, 36) becomes

$$a_0^*(\tau) = i\alpha_0 \sqrt{\frac{\beta c\sigma}{2\pi}} \int_{\kappa-i\infty}^{\kappa+i\infty} \left(1 - \tfrac{1}{2}J\left(\frac{w}{\sigma}\right)\right) e^w \frac{dw}{w^{\frac{3}{2}}}. \tag{5·4, 37}$$

Now if we take κ small positive, then $|e^w| \doteqdot e^\kappa$ is uniformly close to 1 along the path of integration. Also, it can be shown that J is bounded along the path of integration and that it vanishes at the origin (compare (5·3, 49) and the Table 5·3, 2 for $H(\lambda)$ and $G(\lambda)$). Since (5·4, 37) contains the factor $w^{-\frac{3}{2}}$ it then follows that for sufficiently large values of τ or σ, J affects (5·4, 37) only for small values of the argument and may therefore be omitted. Hence (5·4, 37) becomes approximately

$$a_0^*(\tau) \doteqdot i\alpha_0 \sqrt{\frac{\beta c\sigma}{2\pi}} \int_{\kappa-i\infty}^{\kappa+i\infty} e^w \frac{dw}{w^{\frac{3}{2}}}. \tag{5·4, 38}$$

But the right-hand side of (5·4, 38) is (except for a factor $\sqrt{(c\sigma)}$) identical with the right-hand side of (5·4, 34), if we set $\tau = 1$ in the latter. Hence (for large values of τ or σ)

$$a_0^*(\tau) = -\alpha_0 \sqrt{(8\beta c\sigma)} = -\alpha_0 \sqrt{(8\beta\tau)}. \tag{5·4, 39}$$

Now
$$\beta\tau = \tfrac{1}{2}\beta^2 t^2 = \tfrac{1}{2}U^2,$$

and so (5·4, 39) may also be written as

$$\alpha_0^*(\tau) = -2U\alpha_0. \tag{5·4, 40}$$

Similarly, substituting first $c\sigma$ and then $U^2/2\beta$ for τ in (5·4, 30), we obtain

$$a_1^*(\tau) = -\alpha_0 \sqrt{\frac{\beta c}{\text{\small o}\_}} = -\frac{c\beta}{\text{\small o}\textit{U}}\alpha_0. \tag{5·4, 41}$$

Hence, by (5·4, 19),

$$a_0(t) = -2U^2\alpha_0, \quad a_1(t) = -\frac{c\beta}{2}\alpha_0, \quad a_2(t) = a_3(t) = \ldots = 0. \quad (5\cdot4, 42)$$

The lift on the aerofoil is given by (5·3, 24). For large τ or σ, a_1 is small compared with a_0 as given by the approximate formula in (5·4, 42) and should be omitted, in keeping with the quality of the approximation. Hence

$$L = \pi\rho U^2 c\alpha_0, \quad (5\cdot4, 43)$$

which is the same as if the wing were advancing with constant speed U. A similar conclusion applies to the moment.

No numerical evaluation of (5·4, 37) has been carried out so far. The same case has been treated by Wagner‡ by a different method. However, Wagner's results are only approximate and are not reliable for small σ.

According to the general principle of similarity stated above, it must be possible to find an equivalent case for which the forward velocity of the wing is a constant, U_1 say. In the present case,

$$G^*(x, \tau) = \begin{cases} 0 & (\tau < 0), \\ -\alpha_0 \sqrt{(2\beta\tau)} & (\tau > 0), \end{cases}$$

by (5·4, 29), and this must be the function G^* also for the case of constant $U = U_1$. In that case, $\tau = U_1 t$ and so

$$G(x, t) = U_1 \frac{\partial F}{\partial x} + \frac{\partial F}{\partial t} = \begin{cases} 0 & (t < 0), \\ -\alpha_0 \sqrt{(2\beta U_1 t)} & (t > 0). \end{cases} \quad (5\cdot4, 44)$$

Integrating, we obtain

$$F(x, t) = \begin{cases} 0 & (t < 0), \\ -\frac{2}{3}\alpha_0 \sqrt{(2\beta U_1 t^3)} & (t > 0). \end{cases} \quad (5\cdot4, 45)$$

This procedure can be applied quite generally. For any function $G^*(x, \tau)$ which is given by a case of variable forward velocity, and for arbitrary constant $U_1 > 0$, the function $F(x, t)$ which corresponds to the constant velocity U_1 satisfies

$$U_1 \frac{\partial F}{\partial x} + \frac{\partial F}{\partial t} = G^*(x, \tau) = G^*(x, U_1 t). \quad (5\cdot4, 46)$$

Solving (5·4, 46) for $F(x, t)$, we obtain

$$F(x, t) = \int G^*(x + U_1(t' - t), U_1 t') \, dt'. \quad (5\cdot4, 47)$$

We shall now consider the motion of an aerofoil along an approximately circular path. Let Q be a predetermined point on the aerofoil. We assume that Q is confined throughout to the neighbourhood of a fixed circle C with centre O and radius R. Let Q_0 be the normal projection of Q on C, i.e. the intersection of C with the radial line OQ. We choose a fixed coordinate

‡ H. Wagner, 'Über die Entstehung des dynamischen Auftriebes von Tragflügeln', *Z. angew. Math. Mech.* vol. 5 (1925), pp. 17–35.

system (x', y') with origin at O, and a moving system (x, y) with origin at Q_0 and such that the x-axis is tangential to C at Q_0, pointing along C in the negative sense, while the y-axis coincides with OQ and points away from O. It will be convenient to denote the time by t or t' according as to whether it occurs in connexion with the primed or unprimed system of coordinates, $t' = t$.

Let Q_0 be given by the equation

$$x_0' = R\cos\theta, \quad y_0' = R\sin\theta, \quad \theta = \theta(t'), \qquad (5\cdot4, 48)$$

so that R, θ are the polar coordinates of Q_0. It will be assumed that θ is an increasing function of time. Then, for any given point, the coordinates in the fixed and moving systems are related by the equations

$$\left. \begin{array}{l} x' = x_0' + (\quad x\sin\theta + y\cos\theta) = \quad x\sin\theta + (R+y)\cos\theta, \\ y' = y_0' + (-x\cos\theta + y\sin\theta) = -x\cos\theta + (R+y)\sin\theta, \end{array} \right\} \quad (5\cdot4, 49)$$

and conversely,

$$\left. \begin{array}{l} x = x'\sin\theta - y'\cos\theta, \\ y = x'\cos\theta + y'\sin\theta - R. \end{array} \right\} \qquad (5\cdot4, 50)$$

Let the velocity components of the fluid in the fixed system be u', v', and at any time t let u, v be the components of the absolute velocity of the fluid referred to the moving system. Then

$$\left. \begin{array}{l} u' = \quad u\sin\theta + v\sin\theta, \\ v' = -u\cos\theta + v\sin\theta. \end{array} \right\} \qquad (5\cdot4, 51)$$

Also, let the equation of the aerofoil in the moving system be

$$F(x, y, t) = 0. \qquad (5\cdot4, 52)$$

Then the equation of the aerofoil in the fixed system becomes

$$f(x', y', t') \equiv F(x'\sin\theta - y'\cos\theta, x'\cos\theta + y'\sin\theta - R, t) = 0. \quad (5\cdot4, 53)$$

Also, the boundary condition at the aerofoil is (see $(1\cdot2, 11)$)

$$u'\frac{\partial f}{\partial x'} + v'\frac{\partial f}{\partial y'} + \frac{\partial f}{\partial t'} = 0. \qquad (5\cdot4, 54)$$

Or, using $(5\cdot4, 51)$ and $(5\cdot4, 53)$,

$$(u\sin\theta + v\cos\theta)\left(\frac{\partial F}{\partial x}\sin\theta + \frac{\partial F}{\partial y}\cos\theta\right)$$
$$+ (-u\cos\theta + v\sin\theta)\left(-\frac{\partial F}{\partial x}\cos\theta + \frac{\partial F}{\partial y}\sin\theta\right)$$
$$+ \frac{\partial F}{\partial x}(x'\cos\theta + y'\sin\theta)\frac{d\theta}{dt} - \frac{\partial F}{\partial y}(x'\sin\theta - y'\cos\theta)\frac{d\theta}{dt} + \frac{\partial F}{\partial t} = 0.$$

Simplifying, and putting $R\dfrac{d\theta}{dt} = U(t)$, we obtain

$$u\frac{\partial F}{\partial x} + v\frac{\partial F}{\partial y} + \left\{\left(\frac{y}{R}+1\right)\frac{\partial F}{\partial x} - \frac{x}{R}\frac{\partial F}{\partial y}\right\}U(t) + \frac{\partial F}{\partial t} = 0. \qquad (5\cdot4, 55)$$

We assume that y is small compared with R at all points of the aerofoil. Then the term y/R in (5·4, 55) may be neglected. The term is strictly equal to zero if the aerofoil is a flat plate which coincides with the x-axis at all times.

Suppose that $F(x, y, t)$ is given in the form

$$F(x, y, t) = y - k(x, t). \tag{5·4, 56}$$

Then (5·4, 55) becomes

$$v - \frac{x}{R} U(t) - (u + U(t)) \frac{\partial k}{\partial x} - \frac{\partial k}{\partial t} = 0. \tag{5·4, 57}$$

The velocity component u may be expected to be small compared with the forward velocity of the aerofoil $U(t)$ and will therefore be neglected in (5·4, 57). This leads to the condition

$$-\frac{\partial \Phi}{\partial y} = v = U(t) \left(\frac{\partial k}{\partial x} + \frac{x}{R} \right) + \frac{\partial k}{\partial t}. \tag{5·4, 58}$$

Let Φ be the velocity potential of the motion relative to the fixed system of coordinates. We assume that the chord length c is small compared with the radius R, and we may then expect that the relative orders of magnitude of $u'^2 = (\partial \Phi/\partial x')^2$ and $v'^2 = (\partial \Phi/\partial y')^2$ on one hand, and of $\partial \Phi/\partial t'$ on the other, will be the same as in the case of unsteady rectilinear motion. Hence Bernoulli's equation (5·2, 4) may again be replaced by (5·2, 9), showing that the acceleration potential

$$\Omega(x', y', t') = \frac{\partial \Phi}{\partial t'} \tag{5·4, 59}$$

is continuous everywhere except across the aerofoil. Ω is a solution of Laplace's equation.

We now assume that Q_0 is the projection of the mid-chord. Then the projection of the aerofoil on the x-axis (its 'mean position') extends from $x = -\frac{1}{2}c$ to $x = \frac{1}{2}c$. Expressing Φ and Ω at any time as functions of x, y and t (see (5·4, 50)), we have the relation

$$\Omega = \frac{\partial \Phi}{\partial t'} = \frac{\partial \Phi}{\partial x} \frac{\partial x}{\partial t'} + \frac{\partial \Phi}{\partial y} \frac{\partial y}{\partial t'} + \frac{\partial \Phi}{\partial t} \frac{\partial t}{\partial t'},$$

or

$$\Omega = \left\{ (y + R) \frac{\partial \Phi}{\partial x} - x \frac{\partial \Phi}{\partial y} \right\} \theta'(t) + \frac{\partial \Phi}{\partial t}. \tag{5·4, 60}$$

Putting

$$x + iy = z = \frac{c}{4} \left(\zeta + \frac{1}{\zeta} \right), \tag{5·4, 61}$$

we may again regard Ω as the real part of a complex acceleration potential

$$\omega = i \left(-a_0 \frac{\zeta}{\zeta + 1} + \sum_{n=1}^{\infty} \frac{a_n}{\zeta^n} \right) \tag{5·4, 62}$$

(compare (5·2, 20) and (5·2, 21)).

In particular, if U is constant, and the function $k(x, t)$ (see (5·4, 56)) is independent of the time, then (5·4, 58) becomes

$$-\frac{\partial \Phi}{\partial y} = U\left(\frac{dk}{dx} + \frac{x}{R}\right). \tag{5·4, 63}$$

Then Φ and Ω are also independent of t when expressed in terms of the moving coordinates, and the coefficients $a_0, a_1, a_2, \ldots$ are constant. At the projection of the aerofoil on the x-axis, x is small compared with R, while $y = 0$. Bearing in mind that $R\theta'(t) = U$, we may therefore replace (5·4, 60) in the neighbourhood of the aerofoil by

$$\Omega = U\frac{\partial \Phi}{\partial x}. \tag{5·4, 64}$$

Thus the problem is reduced formally to one of steady motion of a thin aerofoil (§ 2·8). The boundary condition (5·4, 63) shows that the curvature of the flight path is equivalent to a chordwise variation of the slope

$$s(x) = \frac{x}{R} = \frac{c}{2R}\cos\theta. \tag{5·4, 65}$$

Assuming that $k(x, t) = 0$ identically, as will be the case for a flat plate, we then obtain from (2·8, 14) and (2·8, 15),

$$\Omega = U\frac{\partial \Phi}{\partial x} = U^2\frac{c}{2R}\sin\theta. \tag{5·4, 66}$$

Thus $$a_0 = a_2 = \ldots = 0, \quad a_1 = U^2\frac{c}{2R}. \tag{5·4, 67}$$

The same result is obtained from (5·2, 40) and (5·2, 57) by making the coefficients a_n, b_n, which are involved in these equations, independent of the time. The application of (5·3, 24) and (5·3, 25) then yields

$$\left. \begin{aligned} L &= -\tfrac{1}{4}\pi\rho U^2\frac{c^2}{R}, \\ M &= 0, \end{aligned} \right\} \tag{5·4, 68}$$

where M is the moment about the mid-chord Q_0.

The period of revolution of the aerofoil round the centre of the circle C is $2\pi R/U$. During the same time the aerofoil revolves just once round its own mid-chord and so its angular velocity in this motion is

$$q = 2\pi : \frac{2\pi R}{U} = \frac{U}{R}. \tag{5·4, 69}$$

In terms of this quantity, (5·4, 67) becomes

$$L = -\tfrac{1}{2}\pi\rho Uc^2 q = L_q q, \tag{5·4, 70}$$

where $$\frac{L_q}{\pi\rho Uc^2} = -\frac{1}{4}. \tag{5·4, 71}$$

L_q is an aerodynamic derivative in the sense explained in the introduction.

Assume now that Q_0 is, more generally, at a distance hc aft of the leading edge. Then the aerofoil extends from $x = -hc$ to $x = (1-h)c$, and the mid-chord corresponds to $x_m = (\frac{1}{2} - h)c$. The appropriate transformation replacing (5·4, 61) is

$$x - x_m + iy = z - x_m = \frac{c}{4}\left(\zeta + \frac{1}{\zeta}\right), \qquad (5\cdot4, 72)$$

and the slope for the equivalent steady case is

$$s(x) = \frac{x}{R} = \frac{x_m}{R} + \frac{x - x_m}{R} = (\tfrac{1}{2} - h)\frac{c}{R} + \frac{c}{2R}\cos\theta. \qquad (5\cdot4, 73)$$

Hence Ω becomes, again by (2·8, 14) and (2·8, 15), or alternatively by (5·2, 40) and (5·2, 57),

$$\Omega = U^2\left\{(\tfrac{1}{2} - h)\frac{c}{R}\tan\tfrac{1}{2}\theta + \frac{c}{2R}\sin\theta\right\}, \qquad (5\cdot4, 74)$$

i.e. $\qquad a_0 = U^2(1-2h)\dfrac{c}{R}, \quad a_1 = U^2\dfrac{c}{2R}, \quad a_2 = a_3 = \ldots = 0. \qquad (5\cdot4, 75)$

Then the lift is $\qquad L = -\pi\rho U^2\dfrac{c^2}{R}(\tfrac{3}{4} - h) = L_q q, \qquad (5\cdot4, 76)$

where $\qquad \dfrac{L_q}{\pi\rho U c^2} = -(\tfrac{3}{4} - h), \qquad (5\cdot4, 77)$

while the moment about the mid-chord $(x_m, 0)$ is

$$M_m = \pi\rho U^2\frac{c^3}{R}(\tfrac{1}{8} - \tfrac{1}{4}h). $$

It follows that the moment about Q_0 is given by

$$M = M_m + x_m L = -\pi\rho U^2\frac{c^3}{R}(\tfrac{1}{2} - h)^2 = M_q q, \qquad (5\cdot4, 78)$$

where $\qquad \dfrac{M_q}{\pi\rho U c^3} = -(\tfrac{1}{2} - h)^2. \qquad (5\cdot4, 79)$

L_q and M_q are called the rotary pitching derivatives. They can be determined experimentally on a whirling arm. The agreement between theory and experiment is reasonable.[‡] It must be explained that these derivatives are not applicable to the oscillatory case considered earlier.

It will be seen that the procedure developed above depends on the use of the theory of steady motion, together with the correct boundary condition for unsteady flow. This is the so-called quasi-static method.

The motion of the wing relative to its mean position on the x-axis is given by the function $k(x, t)$ in equation (5·4, 58). Its effect is additive to that of the rotary motion considered above. One can show that the pressure distribution and aerodynamic forces due to $k(x, t)$ are approximately the same as for a rectilinear flight path, provided one neglects the fact that the

‡ E. F. Relf, T. Lavender and E. Ower, 'The determination of rotary derivatives', *Rep. Memor. Aero. Res. Comm.*, *Lond.*, no. 809 (1921).

wing may fly into a region which was occupied by it during a previous revolution. This will happen, for example, if U is constant. In any case, however, the general assumptions of aerofoil theory for unsteady motion are not sufficiently accurate to permit the prediction of the evolution of the field of flow in the wake of an aerofoil, over the length of time which would normally be required for a complete revolution. Although it would be possible to realize this case by means of an oscillating aerofoil which is mounted on a steadily rotating whirling arm, the problem is not likely to arise in full-scale flight.

5·5   Three-dimensional incompressible unsteady flow

The concept of the acceleration potential is applicable also to three-dimensional flow conditions. Suppose that a wing executes an unsteady motion of small magnitude relative to a 'mean position' M which is itself displaced without rotation in a fixed direction.

Let the displacement of any point of the mean position M (e.g. of the centre-line leading edge) measured from the time $t = 0$, be $d = f(t)$. Then the forward velocity of the wing is $U(t) = f'(t)$. We choose a system of co-ordinates (x', y', z') which is fixed in the surrounding medium such that the positive direction of the x'-axis is opposite to the direction of motion of M and such that M is situated in the x', y'-plane. Then the system of co-ordinates (x, y, z) which is given by

$$x = x' + f(t), \quad y = y', \quad z = z' \tag{5·5, 1}$$

is fixed relative to M.

In order to determine the position of the wing completely we require in addition an equation

$$z = F(x, y, t), \tag{5·5, 2}$$

which describes the deflexion of the wing in a direction normal to its mean position. Then the boundary condition (1·2, 11) is linearized to

$$-\frac{\partial \Phi}{\partial z} = U \frac{\partial F}{\partial x} + \frac{\partial F}{\partial t} = G(x, y, t). \tag{5·5, 3}$$

Φ is the velocity potential of the flow referred to the primed system of coordinates. It is also the induced velocity potential of the flow relative to the wing, i.e. the field of flow obtained by the vectorial subtraction of the apparent velocity at infinity, $U(t)$.

We define the acceleration potential Ω as in (5·2, 10). Then Ω is a solution of Laplace's equation in three dimensions both in the primed and in the unprimed system of coordinates, and we may assume that Ω vanishes at infinity. Moreover, linearizing Bernoulli's equation we find that within the order of accuracy of that approximation, Ω is continuous everywhere except across the wing. To express the boundary condition (5·5, 3) in terms of Ω, we integrate (5·2, 10). Expressing the result in terms of the system of coordinates which moves with the wing, and differentiating with respect to z, we obtain for a point $(x_1, y_1, 0)$ on the (mean position of the) wing

$$\lim_{z_1 \to 0} \int_{-\infty}^{x_1} \Omega_z(x, y_1, z_1, \tau_1 + x - x_1) \frac{dx}{\chi(\tau_1 + x - x_1)} = G(x_1, y_1, t_1). \quad (5·5, 4)$$

In this equation, G is defined by $(5·5, 3)$ and Ω is regarded as a function of the variables x, y, z and $\tau = f(t)$. $\chi(\tau)$ is the forward velocity of the wing expressed as a function of τ, $\chi(\tau) = U(f^{-1}(\tau))$, and τ_1 is given by $\tau_1 = f(t_1)$. Ω_z denotes the partial derivative of Ω with respect to z. The detailed procedure for the derivation of $(5·5, 4)$ is similar to that used for the two-dimensional case ($\S 5·2$), but the convenient concept of the complex acceleration potential is no longer at our disposal.

The method of the acceleration potential has been applied to the oscillatory motion of a circular aerofoil by Schade‡ and Kočin.§ Using the acceleration potential, and making certain approximations, Küssner∥ has obtained a lifting-line theory which may be expected to be valid for wings of sufficiently high aspect ratio. It is a special solution of equation $(5·6, 21)$, derived by Küssner for compressible three-dimensional flow. This general equation is discussed in the next section. Küssner's analysis amends an earlier lifting-line theory due to Cicala.¶ Other methods for the determination of the oscillatory derivatives have been developed by W. P. Jones in a series of papers whose value is enhanced by the inclusion of detailed numerical results.‡‡ Jones's approach is closely related to the method of $\S 3·6$.

All the above-mentioned papers assume constant forward velocity. For that case $\tau = Ut$, and so $(5·5, 4)$ becomes

$$\lim_{z_1 \to 0} \frac{1}{U} \int_{-\infty}^{x_1} \Omega_z(x, y_1, z_1, Ut_1 + x - x_1) \, dx = G(x_1, y_1, t_1). \quad (5·5, 5)$$

For variable forward velocity, put

$$\Omega^*(x, y, z, t) = \frac{\Omega(x, y, z, t)}{\chi(\tau)}, \quad G^*(x, y, \tau) = G(x, y, f^{-1}(\tau)). \quad (5·5, 6)$$

Then $(5·5, 4)$ becomes

$$\lim_{z_1 \to 0} \int_{-\infty}^{x_1} \Omega_z^*(x, y_1, z_1, \tau_1 + x - x_1) \, dx = G^*(x_1, y_1, \tau_1). \quad (5·5, 7)$$

‡ Th. Schade, 'Theorie der schwingenden kreisförmigen Tragfläche auf Potentialtheoretischer Grundlage', *Luftfahrtforsch.* vol. 17 (1940), pp. 387–400.

§ N. E. Kočin, 'On the steady oscillations of a wing of circular planform' (in Russian). *Appl. Math. Mech.* vol. 6 (1942), p. 287. Compare Nekrasov, 'Wing theory for unsteady flow', *A.R.C. Paper*, no. 11792, ch. 10.

∥ H. G. Küssner, 'Allgemeine Tragflächentheorie', *Luftfahrtforsch.* vol. 17 (1940), pp. 370–78. (Translated in *Tech. Memor. Nat. Adv. Comm. Aero., Wash.*, no. 979.)

¶ P. Cicala, 'Comparison of theory with experiment in the phenomenon of wing flutter', *Tech. Memor. Nat. Adv. Comm. Aero., Wash.*, no. 887 (1939) (translated from *Aerotecnica*, vol. 18 (1938), pp. 412–33).

‡‡ Two of these are: 'Theoretical air-load and derivative coefficients for rectangular wings', *Rep. Memor. Aero. Res. Comm., Lond.*, no. 2142 (1943), and 'Aerodynamic forces on wings in non-uniform motion', *Rep. Memor. Aero. Res. Comm., Lond.*, no. 2117 (1945).

Now compare the cases of two wings of the same plan-form for which the functions f and G are given by f_1, G_1 and f_2, G_2 respectively. Assume that for any two moments of time t_1, t_2 such that

$$f_1(t_1) = f_2(t_2), \tag{5·5, 8}$$

the condition

$$G_1(x, y, t_1) = G_2(x, y, t_2) \tag{5·5, 9}$$

is satisfied. Then the function $G^*(x, y, \tau)$ is the same for the two cases and it follows that they both lead to identical equations (5·5, 7). Hence Ω^* also is the same for the two cases, and so the solution of one of them immediately provides the solution for the other. In particular, given $G_1(x, y, t)$ and $f_1(t)$, we may construct a second case for which G_2 and f_2 satisfy the condition laid down above, and such that the forward velocity in the second case is equal to unity, in the given scale of measurement. For this purpose, we put

$$f_2(t) = t,$$

$$G_2(x, y, t) = G_1(x, y, f^{-1}(t)).$$

(5·5, 8) now becomes

$$f_1(t_1) = t_2,$$

and if this condition is satisfied, then

$$G_2(x, y, t_2) = G_1(x, y, f_1^{-1}(t_2)) = G_1(x, y, f_1^{-1}(t_1)) = G_1(x, y, t_1), \tag{5·5, 10}$$

in agreement with (5·5, 9).

We observe that f_2 and G_2 have been defined in such a way that $t = \tau$, implying that equations (5·5, 5) and (5·5, 7) coincide for this case.

Consider next a flat wing which rotates about an axis l in its own plane and at the same time is displaced with constant linear velocity in a direction parallel to l. We choose l as the x-axis of a system of coordinates which moves with the wing, such that the positive direction of the x-axis is opposed to the direction of motion of the wing; and we then choose the y- and z-axes in the plane of the wing and normal to it.

Let p be the angular velocity of the wing, and let Φ be the velocity potential of the motion referred to a system of coordinates which is at rest relative to the medium at infinity, but which coincides instantaneously with the system of coordinates defined above. Then it is not difficult to see directly that the boundary condition at the wing is

$$-w = \frac{\partial \Phi}{\partial z} = py. \tag{5·5, 11}$$

Equation (5·5, 11) can also be derived formally from the general boundary condition for unsteady flow (1·2, 11).

We now assume that the product pb is small compared with the forward velocity U, where b is the span of the wing. The following remarks may be sufficient in order to explain the procedure which it is usual to adopt in this case.

As the wing advances, a sheet of trailing vortices is formed in its wake. The velocity distribution induced by the trailing vortices at the wing is determined by the law of Biot–Savart (equation (1·9, 21)). It will be seen that according to that law the effect of the trailing vortex elements which are sufficiently far downstream (say more than three span lengths) is unimportant. On the other hand, if the angular velocity p is sufficiently small, as assumed above, then the remaining portion of the vortex wake, just downstream of the wing, is approximately plane. So far as the calculation of the flow round the wing is concerned, we therefore assume that the

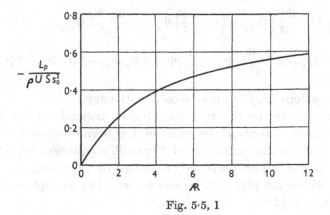

Fig. 5·5, 1

vortex wake is situated in the x, y-plane. Thus the entire problem is reduced to one of steady flow in three dimensions, and this simplified problem can be solved by any one of the methods described in Chapter 3. It appears from (5·5, 11) that the boundary condition can now be interpreted as that due to a linearly varying twist in the spanwise direction for an equivalent steady case. This is another case of the use of the quasi-static method.

For example, if the aspect ratio of the wing is moderately high, we may use lifting-line theory (§ 3·3). The resulting moment about the x-axis, or rolling moment, can be written in the form $L = L_p p$. (The symbols L and p are employed elsewhere with a different meaning, but no confusion is likely.) The variation of L_p with aspect ratio is shown in Fig. 5·5, 1 for a wing of elliptic plan-form.

The effects of camber or incidence may be calculated by the theory of steady motion and superimposed on the results for the rolling motion of a flat wing just considered. This again involves the simplifying assumption that the vortex wake is plane.

Curvilinear flight of wings of finite span has been considered by Wieselsberger‡ and Glauert.§ Wieselsberger treats the case of steady yaw when the wing moves along a circular path in its own plane. The analysis involves the calculation of the downwash due to the curved trailing vortices. Similarly, Glauert‖ has calculated the effect of the curvature of the trailing vortices on a wing which moves along a circular path in a plane normal to its span. This is the three-dimensional version of a problem which has been considered previously (§ 5·4).

5·6 Aerofoil theory for unsteady subsonic compressible flow

The general equations for the irrotational flow of a compressible fluid were obtained in § 4·1. Equation (4·1, 6) is the differential equation for the velocity potential Φ,

$$
\left(1 - \frac{\Phi_{x'}^2}{a^2}\right)\Phi_{x'x'} + \left(1 - \frac{\Phi_{y'}^2}{a^2}\right)\Phi_{y'y'} + \left(1 - \frac{\Phi_{z'}^2}{a^2}\right)\Phi_{z'z'} - 2\frac{\Phi_{x'}\Phi_{y'}}{a^2}\Phi_{x'y'}
$$

$$
- 2\frac{\Phi_{z'}\Phi_{x'}}{a^2}\Phi_{z'x'} - 2\frac{\Phi_{y'}\Phi_{z'}}{a^2}\Phi_{y'z'} = \frac{1}{a^2}(\Phi_{tt} - 2\Phi_{x'}\Phi_{x't} - 2\Phi_{y'}\Phi_{y't} - 2\Phi_{z'}\Phi_{z't}).
$$

$$(5·6, 1)$$

The coordinate system (x', y', z') is supposed stationary.

In all applications to the theory of aerofoils in unsteady compressible flow only the linearized form of the equation has been employed. This is in conformity with the linearization of Bernoulli's equation introduced already for the incompressible unsteady case (see (5·2, 9)), and the same approximation will be adopted in this section also. Thus we replace (5·6, 1) by (cf. equation (4·1, 12))

$$
\frac{\partial^2 \Phi}{\partial x'^2} + \frac{\partial^2 \Phi}{\partial y'^2} + \frac{\partial^2 \Phi}{\partial z'^2} = \frac{1}{a_0^2}\frac{\partial^2 \Phi}{\partial t^2},
$$

$$(5·6, 2)$$

on the assumption that the induced velocities are small compared with a_0, the velocity of sound in the undisturbed medium. Again, we have the linearized Bernoulli equation (5·2, 9)

$$
\int \frac{dp}{\rho} \doteq \frac{p}{\rho} = \left(\frac{\partial \Phi}{\partial t}\right)_{x'} + h(t),
$$

$$(5·6, 3)$$

where the function $h(t)$ may be omitted if desired by incorporation into Φ. As for incompressible flow, the acceleration potential is defined by

$$
\Omega(x', y', t) = \left(\frac{\partial \Phi}{\partial t}\right)_{x'} = \frac{p}{\rho} - h(t),
$$

$$(5·6, 4)$$

‡ C. Wieselsberger, 'Zur Theorie des Tragflügels bei gekrümmter Flugbahn', *Z. angew. Math. Mech.* vol. 2 (1922), pp. 323–40.

§ H. Glauert, 'Calculation of the rotary derivatives due to yawing for a monoplane wing', *Rep. Memor. Aero. Res. Comm., Lond.*, no. 866 (1923).

‖ H. Glauert, 'The lift and pitching moment of an aerofoil due to uniform angular velocity of pitch', *Rep. Memor. Aero. Res. Comm., Lond.*, no. 1216 (1928).

so that knowledge of Ω enables the calculation of the pressure in the linear approximation.

Let us take the equation of the aerofoil to be

$$z' = f(x', y', t) \quad (x_L \leqslant x' \leqslant x_T), \tag{5·6, 5}$$

where x_L and x_T are the coordinates of the leading and trailing edges respectively. The aerofoil is supposed to lie approximately along the x-axis, moving with a mean velocity $U(t)$ in the negative x direction. f may take different values f_u and f_l on the upper and lower surfaces of the wing, though in the present section we shall be concerned only with thin wings for which f_u is taken to be equal to f_l.

The boundary condition at the wing is (cf. (1·2, 11))

$$w' = u' \frac{\partial f}{\partial x'} + \frac{\partial f}{\partial t}, \tag{5·6, 6}$$

where (u', v', w') is the velocity of flow in the (x', y', z') system. (5·6, 6) assumes that the spanwise curvature of the aerofoil is small, so that $\partial f / \partial y$ may be neglected (though clearly this breaks down at the tips).

If we introduce coordinates (x, y, z) moving with the mean velocity $U(t)$ of the aerofoil, such that

$$x = x' - x_L, \quad y = y', \quad z = z', \tag{5·6, 7}$$

and putting $\qquad x_L = g(t), \quad U(t) = -g'(t) > 0, \tag{5·6, 8}$

we find that the boundary condition (5·6, 6) becomes

$$w - (u + U) \frac{\partial F}{\partial x} - \frac{\partial F}{\partial t} = 0, \tag{5·6, 9}$$

where $\qquad z = F(x, y, t), \quad (-\tfrac{1}{2}c \leqslant x \leqslant \tfrac{1}{2}c),$

is the equation of the aerofoil surface in the moving coordinates. For small values of the induced velocity u we obtain the linearized boundary condition

$$w = G(x, y, t) = U \frac{\partial F}{\partial x} + \frac{\partial F}{\partial t}. \tag{5·6, 10}$$

For a prescribed motion of the aerofoil w is determined by (5·6, 10). From $w = -\partial \Phi / \partial z$ and (5·6, 4) an equation can then be found for the acceleration potential, and this in turn enables the pressure at the aerofoil to be calculated. The relationship between Ω and w is completely analogous to the equation for two-dimensional incompressible flow (5·2, 29). To establish the equation for compressible flow we proceed as follows. For given (x', y', z') we have from (5·6, 4)

$$\Phi(x_1', y_1', z_1', t_1) = \int_{t_0}^{t_1} \Omega(x_1', y_1', z_1', t) \, dt, \tag{5·6, 11}$$

it being understood that in the integration x_1', y_1', z_1' are to be kept constant.

The initial value of $t = t_0$ is chosen such that $\Phi(x_1', y_1', z_1', t_0) = 0$. If in the integrand we rewrite Ω as a function of (x, y, z, t), we get

$$\Phi(x_1', y_1', z_1', t_1) = \int_{t_0}^{t_1} \Omega(x_1' - g(t), y_1', z_1', t)\, dt, \qquad (5·6, 12)$$

since $x_1 = x_1' - g(t)$, $y_1 = y_1'$, $z_1 = z_1'$.

Introduce as variable of integration

$$\tau = -g(t),$$

so that (cf. equation (5·2, 27))

$$\Phi(x_1', y_1', z_1', t_1) = \int_{\tau_0}^{\tau_1} \Omega(x_1' + \tau, y_1', z_1', \tau) \frac{d\tau}{\chi(\tau)}, \qquad (5·6, 13)$$

where $\tau_1 = -g(t_1)$, $\tau_0 = -g(t_0)$, and $\chi(\tau)$ is defined as in equation (5·2, 27) by

$$\chi(\tau) = \frac{d\tau}{dt}.$$

Again putting $\qquad\qquad\qquad \xi = x_1' + \tau, \qquad\qquad\qquad\qquad\qquad (5·6, 14)$

(5·6, 13) becomes

$$\Phi(x_1', y_1', z_1', t_1) = \int_{x_0}^{x_1} \Omega(\xi, y_1', z_1', \tau_1 + \xi - x_1) \frac{d\xi}{\chi(\tau_1 + \xi - x_1)}, \qquad (5·6, 15)$$

where $x_1 = \xi_1 = x_1' + \tau_1$, $x_0 = x_1' + \tau_0$. The lower limit is chosen so that $\Phi_{x=x_0} = 0$, and thus it is reasonable to take $x_0 = -\infty$ (i.e. x_0 far ahead of the aerofoil), as in the incompressible case (§ 5·2).

Suppose that Φ on the left-hand side of (5·6, 15) is written as a function of (x, y, z, t), then, observing that $y' = y$, $z' = z$, we obtain finally

$$\Phi(x_1, y_1, z_1, t_1) = \int_{-\infty}^{x_1} \Omega(\xi, y_1, z_1, \tau_1 + \xi - x_1) \frac{d\xi}{\chi(\tau_1 + \xi - x_1)}, \qquad (5·6, 16)$$

which is the equation corresponding to (5·2, 29).

Differentiation of (5·6, 16) with respect to z yields the velocity w, which is supposed known from the boundary condition (5·6, 10). The resulting integral equation has then to be solved for Ω as a function of the aerofoil position. This can be approached in a number of ways. Küssner‡ uses the fundamental solution of the wave equation (5·6, 2) that corresponds to a moving acoustic source. The basic solution of (5·6, 2) corresponding to a source located at the origin, was found in § 4·1 to be

$$\Phi(x', y', z', t) = \frac{1}{r'} f\!\left(t - \frac{r'}{a_0}\right), \qquad (5·6, 17)$$

(cf. (4·1, 31)), where $r'^2 = x'^2 + y'^2 + z'^2$.

To obtain the potential of a source moving with a constant velocity U in the negative direction we may either apply the Lorentz transformation, as done by Küssner, or we may look for the solution of the wave equation when

‡ H. G. Küssner, 'Allgemeine Tragflächentheorie', *Luftfahrtforsch.* vol. 17 (1940), pp. 370–8 (translated in *Tech. Memor. Nat. Adv. Comm. Aero., Wash.*, no. 979).

expressed in terms of the moving coordinate system (x, y, z) that corresponds to $(5·6, 17)$. Thus $(5·6, 2)$ in the moving system is

$$\left(1 - \frac{U^2}{a_0^2}\right)\frac{\partial^2 \Phi}{\partial x^2} - \frac{2U}{a_0^2}\frac{\partial^2 \Phi}{\partial x \partial t} + \frac{\partial^2 \Phi}{\partial y^2} + \frac{\partial^2 \Phi}{\partial z^2} = \frac{1}{a_0^2}\frac{\partial^2 \Phi}{\partial t^2}. \qquad (5·6, 18)$$

According to the Lorentz transformation the potential at a point (x', y', z', t) due to a moving source is obtained from the potential for a stationary source by introducing the quantities

$$\frac{x' + Ut}{\sqrt{(1 - U^2/a_0^2)}}, \quad y', \quad z', \quad \frac{t + Ux'/a_0^2}{\sqrt{(1 - U^2/a_0^2)}} \quad (U < a_0),$$

in the expression $(5·6, 17)$ on the right-hand side.‡ If we also transform to coordinates moving with the source, i.e. we put

$$x = x' + Ut, \quad y = y', \quad z = z',$$

$(5·6, 17)$ becomes

$$\Phi(x, y, z, t) = \sqrt{(1 - M^2)}\frac{f\left(\sqrt{(1 - M^2)}\left\{t + \dfrac{xM}{a_0(1 - M^2)} - \dfrac{\sqrt{\{x^2 + (1 - M^2)(y^2 + z^2)\}}}{a_0(1 - M^2)}\right\}\right)}{\sqrt{\{x^2 + (1 - M^2)(y^2 + z^2)\}}}, \qquad (5·6, 19)$$

where $M = U/a_0$. It may be verified by direct substitution that this is a solution of $(5·6, 18)$. When the strength f of the source is constant $(5·6, 19)$ reduces to the steady-state solution discussed in § 4·1.

If the source is located not at the origin but at a point (x_0, y_0, z_0), the potential is

$$\Phi(x, y, z, t) = \sqrt{(1 - M^2)}$$
$$\times \frac{f\left(\sqrt{(1 - M^2)}\left\{t + \dfrac{(x - x_0)M}{a_0(1 - M^2)} - \dfrac{\sqrt{[(x - x_0)^2 + (1 - M^2)\{(y - y_0)^2 + (z - z_0)^2\}]}}{a_0(1 - M^2)}\right\}\right)}{\sqrt{[(x - x_0)^2 + (1 - M^2)\{(y - y_0) + (z - z_0)^2\}]}}. \qquad (5·6, 20)$$

As in § 5·3 we shall be concerned in this section only with the theory for thin aerofoils. In this case a doublet distribution is the type of singularity required to represent the wing and wake (cf. § 3·7). For an aerofoil in which the surface lies approximately in the plane $z = 0$, the potential due to a doublet at the wing is found by differentiation of $(5·6, 20)$ with respect to z_0 at the point (z_0, y_0, z_0) $(z_0 = 0)$. It follows from $(5·6, 4)$ that the acceleration potential due to such a doublet at (x_0, y_0, z_0) is

$$\Omega(x, y, z, t) = \frac{\partial}{\partial t}\left[\frac{\partial}{\partial z_0}\frac{Kf}{\sqrt{[(x - x_0)^2 + (1 - M^2)\{(y - y_0)^2 + (z - z_0)^2\}]}}\right]_{z_0 = 0}$$

$$= -\frac{\partial}{\partial z}\left[\frac{K(\partial f/\partial t)}{\sqrt{[(x - x_0)^2 + (1 - M^2)\{(y - y_0)^2 + (z - z_0)^2\}]}}\right]_{z_0 = 0},$$

‡ See, for example, G. Joos, *Theoretical Physics* (Blackie, 1943), p. 230.

where K is a constant multiplier with dimensions of length. We may define an acceleration potential doublet strength σ by

$$\frac{1}{4\pi}\sigma = K\sqrt{(1-M^2)}\frac{\partial f}{\partial t},$$

in which case

$$\Omega(x,y,z,t) = -\frac{1}{4\pi}\frac{\partial}{\partial z}\left[\frac{\sigma\left(t+\dfrac{(x-x_0)M}{a_0(1-M^2)} - \dfrac{\sqrt{[(x-x_0)^2+(1-M^2)\{(y-y_0)^2+z^2\}]}}{a_0(1-M^2)}\right)}{\sqrt{[(x-x_0)^2+(1-M^2)\{(y-y_0)^2+z^2\}]}}\right].$$
$$(5\cdot6,21)$$

Introducing this expression for Ω into the integral $(5\cdot6,16)$ we obtain

$$\Phi(x,y,z,t) = -\frac{1}{4\pi U}\int_{-\infty}^{x-x_0}\frac{\partial}{\partial z}\left[\frac{\sigma\left(t+\dfrac{\xi-x}{U}+\dfrac{\xi M}{a_0(1-M^2)} - \dfrac{\sqrt{[\xi^2+(1-M^2)\{(y-y_0)^2+z^2\}]}}{a_0(1-M^2)}\right)}{\sqrt{[\xi^2+(1-M^2)\{(y-y_0)^2+z^2\}]}}\right]d\xi.$$
$$(5\cdot6,22)$$

In using $(5\cdot6,16)$ we have employed the fact that the velocity $U(t)$ is a constant, in which case

$$\tau = Ut + \text{constant} \quad \text{and} \quad \chi(\tau) = U.$$

Hence the normal velocity at a point (x,y,z), due to a doublet at $(x_0,y_0,0)$, is

$$w(x,y,z,t) = -\frac{\partial\Phi}{\partial z} = \frac{1}{4\pi U}\int_{-\infty}^{x-x_0}\frac{\partial^2}{\partial z^2}\left[\frac{\sigma\left(t+\dfrac{\xi-x}{U}+\dfrac{\xi M}{a_0(1-M^2)} - \dfrac{\sqrt{[\xi^2+(1-M^2)\{(y-y_0)^2+z^2\}]}}{a_0(1-M^2)}\right)}{\sqrt{[\xi^2+(1-M^2)\{(y-y_0)^2+z^2\}]}}\right]d\xi.$$
$$(5\cdot6,23)$$

We next replace σ in $(5\cdot6,23)$ by $\sigma'\,dx_0\,dy_0$, and integrate over the wing surface S, so that σ' represents a surface distribution of doublets on the wing. Thus

$$w(x,y,z,t)$$
$$= \frac{1}{4\pi U}\int_S\int_{-\infty}^{x-x_0}\frac{\partial^2}{\partial z^2}\left[\frac{\sigma'\left(t+\dfrac{\xi-x}{U}+\dfrac{\xi M}{a_0(1-M^2)} - \dfrac{\sqrt{[\xi^2+(1-M^2)\{(y-y_0)^2+z^2\}]}}{a_0(1-M^2)}\right)}{\sqrt{[\xi^2+(1-M^2)\{(y-y_0)^2+z^2\}]}}\right]dx_0\,dy_0\,d\xi.$$
$$(5\cdot6,24)$$

is the velocity at (x,y,z) due to a surface density σ' of doublets.

The three-dimensional form of $(5\cdot6,24)$ has been of limited use in problems of compressible flow. In two dimensions, introducing the new coordinate system by replacing w by v and z by y, as customary, it becomes

$$
\begin{aligned}
v(x,y,t) \\
= \frac{U}{4\pi} \int_{-\frac{1}{2}c}^{\frac{1}{2}c} \int_{-\infty}^{\infty} \int_{-\infty}^{x-x_0} A(x_0) \frac{\partial^2}{\partial y^2} \left[\frac{\exp\left\{ i\nu\left(t + \frac{\xi-x}{U} + \frac{\xi M}{a_0(1-M^2)} \right. \right.}{\left. \left. - \frac{\sqrt{\{\xi^2+(1-M^2)(y^2+y_0^2)\}}}{a_0(1-M^2)} \right) \right\}}{\sqrt{\{\xi^2+(1-M^2)(y^2+y_0^2)\}}} \right] dx_0\, dy_0\, d\xi.
\end{aligned}
$$

$$(5\cdot6, 25)$$

Here we have specified a complex harmonic time variation of σ'

$$\sigma'(x,t) = U^2 A(x)\, e^{i\nu t}. \tag{5·6, 26}$$

A general time dependence for σ' (with a rigid aerofoil) may be obtained by superposition of such expressions with different values of A and ν.

The integration with respect to y_0 in (5·6, 25) can be expressed as a Bessel function. Thus, the second Bessel function of the third kind (or Hankel function) of zero order has the integral representation‡

$$H_0^{(2)}(u) = \frac{2i}{\pi} \int_1^\infty \exp(-iup) \frac{dp}{\sqrt{(p^2-1)}}.$$

Putting

$$u = \frac{\nu\sqrt{\{\xi^2+(1-M^2)y^2\}}}{a_0(1-M^2)},$$

and

$$p = \sqrt{\left(\frac{(1-M^2)y_0^2}{\xi^2+(1-M^2)y^2} + 1 \right)},$$

we get

$$\int_{-\infty}^{\infty} \frac{\exp\left\{ -i\nu \frac{\sqrt{\{\xi^2+(1-M^2)(y^2+y_0^2)\}}}{a_0(1-M^2)} \right\}}{\sqrt{\{\xi^2-(1-M^2)(y^2+y_0^2)\}}} dy_0 = 2\int_1^\infty \frac{\exp(-iup)}{\sqrt{\{(1-M^2)(p^2-1)\}}} dp$$

$$= -\frac{\pi i}{\sqrt{(1-M^2)}} H_0^{(2)}(u),$$

so that (5·6, 25) can be written

$$
v(x,y,t) = \frac{-iU}{4\sqrt{(1-M^2)}} \int_{-\frac{1}{2}c}^{\frac{1}{2}c} \int_{-\infty}^{\infty} A(x_0) \exp\left[i\nu\left(t + \frac{\xi-x}{U} + \frac{\xi M}{a_0(1-M^2)} \right) \right]
$$
$$
\times \frac{\partial^2}{\partial y^2} H_0^{(2)}\left(\frac{\nu\sqrt{\{\xi^2+(1-M^2)y^2\}}}{a_0(1-M^2)} \right) dx_0\, d\xi. \tag{5·6, 27}
$$

Equation (5·6, 27) has also been derived by Possio in a paper which gives a method for solving it for a specified motion of the wing.§ Possio derives the equation directly as follows. The two-dimensional form of the wave equation (5·6, 2) is

$$\frac{\partial^2 \Phi}{\partial x^2} + \frac{\partial^2 \Phi}{\partial y^2} = \frac{1}{a_0^2} \frac{\partial^2 \Phi}{\partial t^2}, \tag{5·6, 28}$$

‡ See G. N. Watson, *A Treatise on the Theory of Bessel Functions* (2nd ed., Cambridge University Press, 1944), p. 170.

§ C. Possio, 'L'Azione Aerodinamica sul profilo, oscillanto in un fluido compressibile a velocita iposonora', *Aerotecnica*, *Roma*, vol. 18 (1938), pp. 441–58 (translated as *A.R.C. Paper*, no. 3799).

where we again take the y-axis to be normal to the plane of the aerofoil and normal to the main-flow velocity U, which is assumed to be constant and directed along the negative x-axis.

Because of $(5\cdot6, 4)$ it is evident that the acceleration potential also satisfies $(5\cdot6, 25)$, and Possio takes as a fundamental source solution

$$\Omega = \frac{iAU^2}{4\sqrt{(1-M^2)}} \exp\left[i\nu\left(t+\frac{Mx}{a_0(1-M^2)}\right)\right] H_0^{(2)}\left(\frac{\nu\sqrt{\{x^2+(1-M^2)\,y^2\}}}{a_0(1-M^2)}\right).$$

$$(5\cdot6, 29)$$

That this is indeed a solution may be verified by letting

$$\Omega = F(u)\exp\left[i\nu\left(t+\frac{Mx}{a_0(1-M^2)}\right)\right],$$

where

$$u = \nu\sqrt{\{\xi^2+(1-M^2)\,y^2\}}/a_0(1-M^2),$$

and substituting into the wave equation. It is then found that F has to satisfy the ordinary differential equation

$$\frac{d^2F}{du^2} + \frac{1}{u}\frac{dF}{du} + F = 0,$$

which is Bessel's equation of zero order, with $H_0^{(2)} = J_0 - iY_0$ as a particular (complex) solution.‡

The doublet corresponding to $(5\cdot6, 29)$ is

$$\frac{iAU^2}{4\sqrt{(1-M^2)}} \exp\left[i\nu\left(t+\frac{Mx}{a_0(1-M^2)}\right)\right] \frac{\partial}{\partial y} H_0^{(2)}\left(\frac{\nu\sqrt{\{x^2+(1-M^2)\,y^2\}}}{a_0(1-M^2)}\right).$$

Hence for a doublet situated at the point $(x_0, 0)$

$$\Omega(x,y,z,t) = \frac{iA(x_0)\,U^2}{4\sqrt{(1-M^2)}} \exp\left[i\nu\left(t+\frac{M(x-x_0)}{a_0(1-M^2)}\right)\right]\frac{\partial}{\partial y}$$

$$\times H_0^{(2)}\left(\frac{\nu\sqrt{\{(x-x_0)^2+(1-M^2)\,y^2\}}}{a_0(1-M^2)}\right)$$

$$= \frac{-i\nu A(x_0)\,U^2 y\exp\left[i\nu\left(t+\frac{M(x-x_0)}{a_0(1-M^2)}\right)\right]}{4a_0\sqrt{(1-M^2)}((x-x_0)^2+(1-M^2)y^2)^{\frac{1}{2}}}$$

$$\times H_1^{(2)}\left(\frac{\nu\sqrt{\{(x-x_0)^2+(1-M^2)\,y^2\}}}{a_0(1-M^2)}\right),\quad (5\cdot6, 30)$$

since $-(H_0^{(2)})' = H_1^{(2)}$, the Hankel function of the first order.

Using $(5\cdot6, 16)$, we find the velocity corresponding to this solution to be

$$v = -\frac{\partial\Phi}{\partial y} = \frac{-iU}{4\sqrt{(1-M^2)}}\int_{-\infty}^{x-x_0} A(x_0)\exp\left[i\nu\left(t+\frac{\xi-x}{U}+\frac{\xi M}{a_0(1-M^2)}\right)\right]$$

$$\times \frac{\partial^2}{\partial y^2} H_0^{(2)}\left(\frac{\nu\sqrt{\{\xi^2+(1-M^2)\,y^2\}}}{a_0(1-M^2)}\right) d\xi.\quad (5\cdot6, 31)$$

‡ G. N. Watson, *A Treatise on the Theory of Bessel Functions* (2nd ed., Cambridge University Press, 1944), ch. 3.

For a distribution of the doublet strengths $A(x_0)$ over the chord, $(5·6, 31)$ leads directly to $(5·6, 27)$, which is the equation derived from Küssner's general equation $(5·6, 23)$.

Numerical solutions of $(5·6, 27)$ are obtained by choosing a suitable series expansion for $A(x_0)$. Possio adopts a series of the form (compare $(2·10, 13)$)

$$A(x_0) = \frac{A_0}{2} \tan \tfrac{1}{2}\theta + \sum_{n=1}^{\infty} A_n \sin n\theta, \tag{5·6, 32}$$

where $x_0 = \tfrac{1}{2}c \cos \theta$. The forces acting on the aerofoil can be represented directly in terms of the coefficients $A_0, A_1, \ldots$, precisely as in the steady case. Thus, taking $A(x_0)$ to be the doublet *density* distribution, we find from $(5·6, 4)$ and $(5·6, 30)$ that the discontinuity in pressure across the aerofoil at a point $(x, 0)$ is

$$\Delta p = \rho(\Omega(x, +0, t) - \Omega(x, -0, t))$$

$$= \frac{-2i\rho\nu U^2 e^{i\nu t}}{4a_0 \sqrt{(1 - M^2)}} \left[\int_{-\frac{1}{2}c}^{\frac{1}{2}c} \frac{A(x_0)\, y \exp\left[i\nu \dfrac{M(x - x_0)}{a_0(1 - M^2)} \right]}{\sqrt{\{(x - x_0)^2 + (1 - M^2)y^2\}}} \right.$$

$$\left. \times H_1^{(2)}\!\left(\frac{\nu \sqrt{\{(x - x_0)^2 + (1 - M^2)y^2\}}}{a_0(1 - M^2)} \right) dx_0 \right]_{y=0}.$$

It can be shown that the only contribution made to the integral occurs for the value $x_0 = x$ (where the integrand has a singularity). Its value can be found by using the asymptotic value of $H_1^{(2)}$ for $u \sim 0$,‡

$$H_1^{(2)}(u) \sim \frac{2i}{\pi u}.$$

Then $\quad \Delta p = \dfrac{-2i\rho\nu U^2 e^{i\nu t} A(x)}{4a_0 \sqrt{(1 - M^2)}} \left[\displaystyle\int_{-\frac{1}{2}c}^{\frac{1}{2}c} \frac{2ia_0(1 - M^2)y}{\nu\pi((x - x_0)^2 + (1 - M^2)y^2)} dx_0 \right]_{y=0}$

$$= \frac{\rho U^2 e^{i\nu t} A(x)}{\pi \sqrt{(1 - M^2)}} \left[\left[\sqrt{(1 - M^2)} \tan^{-1}\!\left(\frac{x - x_0}{\sqrt{(1 - M^2)}\, y} \right) \right]_{-\frac{1}{2}c}^{\frac{1}{2}c} \right]_{y=0}$$

$$= \rho U^2 e^{i\nu t} A(x). \tag{5·6, 33}$$

This equation is equivalent to $(5·3, 23)$ with

$$a_n(t) = U^2 e^{i\nu t} A_n.$$

Hence we can derive by integration the forces acting on the aerofoil exactly as in § 5·3 for the incompressible case. $(5·3, 24)$ and $(5·3, 25)$ thus give for the lift and the moment about the mid-chord

$$L = -\tfrac{1}{2}\pi\rho U^2 c(A_0 + A_1) e^{i\nu t},$$
$$M = \tfrac{1}{8}\pi\rho U^2 c^2 (A_0 - A_2) e^{i\nu t}. \tag{5·6, 34}$$

and

‡ G. N. Watson, *A Treatise on the Theory of Bessel Functions* (2nd ed., Cambridge University Press, 1944), pp. 40, 62.

A_0, A_1, ... can be determined from (5·6, 27) and the boundary condition (5·6, 10). The latter, written in the currently used coordinate system, is

$$v = U\frac{\partial F}{\partial x} + \frac{\partial F}{\partial t}. \tag{5·6, 35}$$

We can now write (5·6, 27) as

$$v(x,y,t) = U\frac{\partial F}{\partial x} + \frac{\partial F}{\partial t} = \nu e^{i\nu t}\int_{-\frac{1}{2}c}^{\frac{1}{2}c} A(x_0)\,h(x-x_0)\,dx_0, \tag{5·6, 36}$$

where

$$h(x-x_0) = -\frac{i\lambda}{4c\,\sqrt{(1-M^2)}}\int_{-\infty}^{x-x_0}\exp\left[i\nu\left(\frac{\xi-x}{U} + \frac{\xi M}{a_0(1-M^2)}\right)\right]$$
$$\times \frac{\partial^2}{\partial y^2}H_0^{(2)}\left(\frac{\nu\,\sqrt{\{\xi^2+(1-M^2)\,y^2\}}}{a_0(1-M^2)}\right)d\xi$$

and

$$\lambda = \nu c/U.$$

$F(x,t)$ having been specified, it is in general possible to solve (5·6, 36) for

$$A(x_0) = \tfrac{1}{2}A_0\tan\tfrac{1}{2}\theta + \sum_{n=1}^{\infty}A_n\sin n\theta$$

only by numerical methods. A standard procedure is to calculate v at a number of chord stations x at each of which (for given values of λ and M) the integral in (5·6, 36) can be evaluated in terms of the constants A_0, A_1, ... A number of simultaneous algebraic equations is thus obtained for a limited number of the A_n. As an aid to such a calculation $h = h(\lambda, M)$ can be evaluated without a knowledge of the particular motion of the aerofoil (see a paper by Frazer‡ for tables of h for a value of $M = 0·7$). Since only numerical values of the coefficients A_0, A_1, ... can be found and not their functional dependence on the frequency ν or the frequency parameter λ, it is not possible to separate all the aerodynamic derivatives L_y, $L_{y'}$, ..., M_y, $M_{y'}$, ... Thus, considering the case of vertical oscillation $y = y_0 e^{i\nu t}$,

$$L = y_0 e^{i\nu t}(L_y + i\nu L_{y'} - \nu^2 L_{y''}), \tag{5·6, 37}$$

where $L_{y^{(k)}}$ is the aerodynamic derivative with respect to the kth derivative of the displacement. Equating (5·6, 37)–(5·6, 34) we find

$$\frac{y_0(L_y + i\nu L_{y'} - \nu^2 L_{y''})}{\pi\rho U^2 c} = -\tfrac{1}{2}(A_0 + A_1).$$

A_0 and A_1 in general are complex, so that this equation can only yield values of $L_y - \nu^2 L_{y''}$ and $\nu L_{y'}$. To find the individual values of L_y and $L_{y''}$ the frequency dependence of A_0 and A_1 must be determined. However, it will be noticed that it is the damping derivative L_y which is known uniquely, and in many respects this derivative is the most interesting for practical applications. Similar conclusions can be drawn for the derivatives in pitch and for the moment derivatives. In all cases it is the damping

‡ R. A. Frazer, 'Possio's subsonic derivative theory and its application to flexural-torsional wing flutter', *Rep. Memor. Aero. Res. Comm., Lond.*, no. 2553 (1942).

derivative which is given uniquely in terms of A_0 and A_1 or A_2, whilst the derivatives with respect to displacement and acceleration always appear in combination (see Frazer).

To date the most extensive calculations of subsonic aerodynamic derivatives have been made by Schade,‡ who assumes an expansion for the acceleration potential in a series of Legendre functions. A different approach, utilizing the theory of Fourier transforms, is given by Temple§ in *Modern Developments in Fluid Dynamics*. Timman‖ has shown how to solve the two-dimensional case of an oscillating aerofoil by representing the acceleration potential as an expansion involving Mathieu functions, the terms of the expansion being obtained as particular solutions of the wave equation in terms of elliptic coordinates. Further papers are due to Dietze¶ and Schwartz.‡‡

5·7   Aerofoil theory for unsteady supersonic flow

The existing theory of aerofoils in unsteady supersonic flow is—with some few exceptions—based on the assumption that the velocities induced in the medium by the motion of the wing are small, and hence that the equations governing the flow may be linearized (see §4·1). This leads, as in §5·6, to the wave equation

$$\frac{\partial^2 \Phi}{\partial x^2} + \frac{\partial^2 \Phi}{\partial y^2} + \frac{\partial^2 \Phi}{\partial z^2} - \frac{1}{a_0^2} \frac{\partial^2 \Phi}{\partial t^2} = 0, \tag{5·7, 1}$$

where we now take the (x, y, z) system of coordinates to be at rest relative to the medium far away from the wing; Φ is the velocity potential and a_0 is the velocity of sound. For two-dimensional flow, when the velocity potential is independent of z, (5·7, 1) becomes

$$\frac{\partial^2 \Phi}{\partial x^2} + \frac{\partial^2 \Phi}{\partial y^2} - \frac{1}{a_0^2} \frac{\partial^2 \Phi}{\partial t^2} = 0. \tag{5·7, 2}$$

Under three-dimensional conditions, a disturbance which originated at a point (x_0, y_0, z_0) at a time t_0 is propagated within the region

$$a_0^2(t - t_0)^2 - (x - x_0)^2 - (y - y_0)^2 - (z - z_0)^2 \geqslant 0 \quad (t \geqslant t_0) \tag{5·7, 3}$$

‡ Th. Schade, 'Beitrag zu Zahlentafeln zur Luftkraftberechnung der schwingenden Tragfläche in ebener Unterschallströmung—Teil II', ZWB UM 3211 (1944) (translated as *A.R.C. Paper*, no. 10,108).

§ *Modern Developments in Fluid Dynamics. High Speed Flow*, ed. Howarth (Oxford University Press, 1953), pp. 350–6.

‖ R. Timman, 'Beschouwingen over de luchtkrachten op trillende vliegtuigvleugels' (Thesis, Delft, 1946).

¶ F. Dietze, 'Untersuchungen über die Genauigkeit des Verfahrens von Possio zur angenäherten Ermittlung der Luftkräfte am schwingenden Flügel im kompressiblen Medium bei Unterschallgeschwindigkeit', ZWB FB 1554 (1942).

‡‡ L. Schwartz, 'Zahlentafeln zur Luftkraftberechnung der schwingenden Tragfläche in kompressibler ebener Unterschallströmung', ZWB FB 1838 (1943).

(see §4·1). It is not difficult to see that the condition becomes for two-dimensional conditions, as above,

$$a_0^2(t-t_0)^2 - (x-x_0)^2 - (y-y_0)^2 \geqslant 0 \quad (t \geqslant t_0). \tag{5·7, 4}$$

Indeed, the three-dimensional interpretation of a two-dimensional disturbance which originates at a point (x_0, y_0) at a time t_0, is a system of disturbances which originate simultaneously at time t_0 at the points of the straight line l which is given by

$$x = x_0, \quad y = y_0.$$

A point $P(x, y, z)$ can be affected by the system of disturbances at a time t only if there exists a point (x_0, y_0, z_0) on l and such that (5·7, 3) is satisfied. But this will be the case if (5·7, 4) holds for given x, y, z, t, x_0, y_0, since in that case we only have to make z_0 equal to z. On the other hand, if (5·7, 4) does not hold, for given x, y, z, t, x_0, y_0, then (5·7, 3) clearly does not hold for any z_0.

We now consider two-dimensional conditions such that aerofoil moves approximately along the x-axis. Then we may simplify the analysis as usual by satisfying the boundary condition at the normal projection Π of the wing on the x-axis. Suppose that Π is given by

$$x_L \leqslant x \leqslant x_T, \quad x_L = g(t), \quad x_T = g(t) + c, \tag{5·7, 5}$$

where $g(t)$ is a function of the time and c is the chord. We assume that $g(t)$ is a decreasing function of the time so that the aerofoil moves in the negative direction of the x-axis. Then $U(t) = -g'(t)$ is the (positive) forward velocity of the aerofoil. Let x' be the distance of any point of the aerofoil from the leading edge, so that

$$x' = x - x_L = x - g(t). \tag{5·7, 6}$$

The y-coordinates of the aerofoil at any given moment may then be specified either in the form

$$y = f(x, t) \quad (x_L \leqslant x \leqslant x_T), \tag{5·7, 7}$$

or, in terms of the distance x' from the leading edge, by an equation

$$y = F(x', t) \quad (0 \leqslant x' \leqslant c). \tag{5·7, 8}$$

The functions f and F may take different values f_u, F_u and f_l, F_l, respectively, on the upper and lower surfaces of the wing. By (5·7, 6), the functions f and F are related by the equation

$$f(x, t) = F(x', t) = F(x - g(t), t). \tag{5·7, 9}$$

The boundary condition at the wing is (see (1·2, 11))

$$v - u\frac{\partial f}{\partial x} - \frac{\partial f}{\partial t} = 0, \tag{5·7, 10}$$

or

$$v - u\frac{\partial F}{\partial x'} + g'(t)\frac{\partial F}{\partial x'} - \frac{\partial F}{\partial t} = 0,$$

$$v - (u + U)\frac{\partial F}{\partial x'} - \frac{\partial F}{\partial t} = 0.$$

Since the induced velocities are supposed small compared with the forward velocities we may neglect u compared with U and obtain

$$\left(-\frac{\partial \Phi}{\partial y}\right)_{y=0} = v = U\frac{\partial F}{\partial x'} + \frac{\partial F}{\partial t} = G(x', t), \quad \text{say} \quad (0 \leqslant x' \leqslant c). \quad (5\cdot7, 11)$$

G may again take different values G_u, G_l on the upper and lower surfaces of the aerofoil. Assume now that the Mach number of the forward motion of the aerofoil is always greater than 1,

$$M = U/a_0 > 1. \quad (5\cdot7, 12)$$

Since the motion of the fluid is entirely due to the presence of the aerofoil, it follows that the velocity of the fluid at a point (x, y) at a time t can be

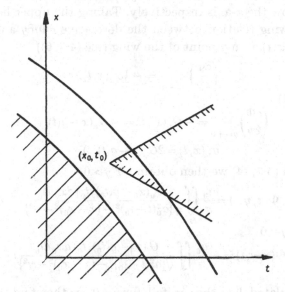

Fig. 5·7, 1

different from zero only if there exists a point $(x_0, 0)$ which belongs to the aerofoil at a time t_0, such that condition $(5\cdot7, 4)$ is satisfied. In other words, the effect of the presence of the aerofoil at a point $(x_0, 0)$ at a time t_0 is felt only within the cone $(5\cdot7, 4)$, and any point which is not within such a cone at a given moment of time, cannot be affected, at that moment, by the presence of the aerofoil. It follows that in the region of the x, t-plane which satisfies the condition

$$x < g(t) \quad (5\cdot7, 13)$$

—the shaded area in Fig. 5·7, 1, the fluid is not disturbed by the presence of the aerofoil. Accordingly, we may apply the analysis developed in §4·5 for steady supersonic flow in three dimensions to the present case, provided we replace the symbols x, y, z, β employed there, by t, x, y, a_0^{-1} respectively. This substitution transforms the equation of steady supersonic flow $(4\cdot5, 1)$

into equation (5·7, 2). The steady counterpart of our present aerofoil in the x, t-plane then possesses supersonic leading and trailing edges throughout. This is the definitely supersonic case, and the analysis given for that case in §§ 4·7 and 4·8 is applicable. Accordingly, we may represent the velocity potential Φ at a point (x, y) at a time t by the following integral:

$$\Phi(x, y, t) = \frac{a_0}{2\pi} \iint\limits_R \frac{\sigma(x_0, t_0)\, dx_0\, dt_0}{\sqrt{\{a_0^2(t - t_0)^2 - (x - x_0)^2 - y^2\}}}. \qquad (5·7, 14)$$

The region of integration R is given by

$$a_0^2(t - t_0)^2 - (x - x_0)^2 - y^2 > 0 \quad (t_0 < t),$$

and the definition of the source density σ is different for the half-planes above and below the x-axis respectively. Taking the upper half-plane, we have the following relation between the derivative $\partial\Phi/\partial y$ and the source density $\sigma = \sigma_u(x, t)$ at any point of the wing (see (4·7, 6))

$$\left(\frac{\partial\Phi}{\partial y}\right)_{y=+0} = -\tfrac{1}{2}\sigma_u(x, t). \qquad (5·7, 15)$$

But by (5·7, 11),

$$\left(\frac{\partial\Phi}{\partial y}\right)_{y=+0} = -G_u(x', t) = -G_u(x - g(t), t), \qquad (5·7, 16)$$

and so

$$\sigma_u(x, t) = 2G_u(x - g(t), t). \qquad (5·7, 17)$$

Substituting in (5·7, 14) we then obtain for $y > 0$

$$\Phi(x, y, t) = \frac{a_0}{\pi} \iint\limits_R \frac{G_u(x_0 - g(t_0), t_0)\, dx_0\, dt_0}{\sqrt{\{a_0^2(t - t_0)^2 - (x - x_0)^2 - y^2\}}}. \qquad (5·7, 18)$$

Similarly, for $y < 0$

$$\Phi(x, y, t) = -\frac{a_0}{\pi} \iint\limits_R \frac{G_l(x_0 - g(t_0), t_0)\, dx_0\, dt_0}{\sqrt{\{a_0^2(t - t_0)^2 - (x - x_0)^2 - y^2\}}}. \qquad (5·7, 19)$$

Having calculated Φ at the aerofoil, for $y = 0$, we then find the pressure p and hence the aerodynamic forces from the linearized Bernoulli equation (equation (5·2, 9))

$$p = p_0 + \rho \frac{\partial\Phi}{\partial t}, \qquad (5·7, 20)$$

where p_0 is the pressure at infinity.

The case of a rigid aerofoil which is accelerated in the direction of its forward motion (i.e. in the direction of the x-axis) is included in the above analysis. For that case $F(x', t)$ is independent of time and so

$$G(x', t) = U \frac{\partial F}{\partial x'} = -g'(t) \frac{\partial F}{\partial x'}. \qquad (5·7, 21)$$

It can be shown‡ that the resultant pressure on the wing can be analysed into three parts. Of these, the first is the value of the pressure predicted for

‡ A Robinson, 'On some problems of unsteady supersonic aerofoil theory', *Rep. Coll. Aero. Cranfield*, no. 16 (1948). Also in *Proc. Int. Cong. App. Mech.*, London, 1948, pp. 500–14.

steady motion on the basis of the instantaneous forward velocity, while the second term depends on the square of the velocity during a limited period preceding the moment of time under consideration, and the third term depends directly on the forward acceleration. It appears that under normal conditions the second and third terms are small compared with the first, so that in this case, the use of steady-flow theory is adequate.

It may be mentioned that, provided the forward velocity of the aerofoil was supersonic initially, the methods of Chapter 4 will yield formal results even if the aerofoil is decelerated to subsonic forward speeds. The practical value of the results obtained for this case is questionable.‡

We consider next the oscillatory motion of an aerofoil which is displaced with constant forward velocity, U. Using complex notation as explained in § 5·2, we assume that $F(x', t)$ is given by

$$F(x', t) = F_0(x') e^{i\nu t}. \qquad (5\cdot7, 22)$$

We shall confine our attention to the upper half of the x, y-plane, where the velocity distribution depends only on the boundary condition at the top surface of the wing. The subscript u will be omitted. Then

$$G(x', t) = G_0(x') e^{i\nu t},$$

where
$$G_0(x') = U F_0'(x') + i\nu F_0(x'). \qquad (5\cdot7, 23)$$

We assume $g(t) = -Ut$ and define

$$x' = x - g(t) = x + Ut \qquad (5\cdot7, 24)$$

for general x. Then the system of coordinates (x', y) moves with the normal projection of the wing on the x-axis, Π. Introducing $x_0' = x_0 + Ut_0$ as variable of integration in $(5\cdot7, 18)$, we obtain

$$\Phi = \frac{a_0}{\pi} \int_{R'} \frac{G_0(x_0') e^{i\nu t_0}}{\sqrt{[a_0^2(t - t_0)^2 - \{(x' - x_0') - U(t - t_0)\}^2 - y^2]}} dx_0' dt_0, \qquad (5\cdot7, 25)$$

where R' is given by

$$a_0^2(t - t_0)^2 - ((x' - x_0') - U(t - t_0))^2 - y^2 > 0 \quad (t_0 < t).$$

We write the expression under the sign of the square root in $(5\cdot7, 25)$ as

$$\frac{a_0^2(x' - x_0')^2}{U^2 - a_0^2} - y^2 - (U^2 - a_0^2) \left\{ (t - t_0)^2 - \frac{2U(x' - x_0')(t - t_0)}{U^2 - a_0^2} + \frac{U^2(x' - x_0')^2}{(U^2 - a_0^2)^2} \right\}$$

$$= \frac{1}{\beta^2} (x' - x_0')^2 - y^2 - \beta^2 a_0^2 \left(t - t_0 - \frac{M(x' - x_0')}{\beta^2 a_0} \right)^2,$$

where
$$\beta^2 = M^2 - 1 = \left(\frac{U}{a_0} \right)^2 - 1$$

Put
$$b^2 = \frac{1}{\beta^2} (x' - x_0')^2 - y^2,$$

---

‡ Compare with a paper by C. S. Gardner and H. F. Ludloff, 'Influence of acceleration on aerodynamic characteristics of thin airfoils in supersonic and transonic flight', *J. Aero. Sci.* vol. 17 (1950), pp. 47–59.

and introduce the variable of integration

$$t' = \frac{\beta a_0}{b}\left(t - t_0 - \frac{M(x' - x_0')}{\beta^2 a_0}\right)$$

in place of t_0 in (5·7, 25). Then the integral becomes

$$\Phi = \frac{1}{\pi\beta}\int_0^{x'-\beta y} dx_0' \int_{-1}^1 G_0(x_0')\exp\left\{iv\left(t - \frac{M(x'-x_0')}{\beta^2 a_0}\right)\right\} e^{i\mu t'}\frac{dt'}{\sqrt{(1-t'^2)}}, \quad (5·7, 26)$$

where

$$\mu = \frac{vb}{\beta a_0} = \frac{v}{\beta^2 a_0}\sqrt{\{(x'-x_0')^2 - \beta^2 y^2\}}. \quad (5·7, 27)$$

(5·7, 26) is applicable for $x' > \beta y$; if $x' \leqslant \beta y$, then $\Phi = 0$. As stated, we consider only the upper half-plane.

Now‡

$$\int_{-1}^1 e^{-i\mu t'}\frac{dt'}{\sqrt{(1-t'^2)}} = \int_{-\frac12\pi}^{\frac12\pi} e^{i\mu\sin\theta}\, d\theta$$

$$= \pi J_0(\mu)$$

$$= \pi J_0\left(\frac{v}{\beta^2 a_0}\sqrt{\{(x'-x_0')^2 - \beta^2 y^2\}}\right), \quad (5·7, 28)$$

where $J_0(\mu)$ is the Bessel function of order zero of the first kind. Substituting in (5·7, 26) we obtain finally

$$\Phi(x', y, t) = \frac{1}{\beta}\exp\left\{iv\left(t - \frac{Mx'}{\beta^2 a_0}\right)\right\}$$

$$\times \int_0^{x'-\beta y} G_0(x_0')\exp\left(\frac{ivMx_0'}{\beta^2 a_0}\right) J_0\left(\frac{v}{\beta^2 a_0}\sqrt{\{(x'-x_0')^2 - y^2\}}\right) dx_0'. \quad (5·7, 29)$$

It is understood that in this formula all the variables refer now to a system of coordinates which moves with the (normal projection of the) wing.

In particular, on the upper surface of the wing,

$$\Phi(x', +0, t) = \frac{1}{\beta}\exp\left\{iv\left(t - \frac{Mx'}{\beta^2 a_0}\right)\right\}\int_0^{x'} G_0(x_0')\exp\left(\frac{ivMx_0'}{\beta^2 a_0}\right) J_0\left(\frac{v}{\beta^2 a_0}(x'-x_0')\right) dx_0'. \quad (5·7, 30)$$

Let c be the chord of the aerofoil. We introduce the non-dimensional quantities

$$\lambda = \frac{vc}{U}, \quad \tau = \frac{Ut}{c}, \quad \xi = \frac{x'}{c}, \quad \xi_0 = \frac{x'-x_0'}{c};$$

and take into account that

$$\frac{vc}{\beta^2 a_0} = \lambda\frac{M}{M^2 - 1} = \lambda\sec^2\mu_0\sin\mu_0,$$

and

$$\frac{vMc}{\beta^2 a_0} = \lambda\frac{M^2}{M^2 - 1} = \lambda\sec^2\mu_0,$$

where μ_0 is the Mach angle, $\sin\mu_0 = M^{-1}$. Then (5·7, 30) becomes

$$\Phi\left(c\xi, +0, \frac{c\tau}{U}\right) = c\tan\mu_0\, e^{i\lambda\tau}\int_0^\xi G_0(c(\xi-\xi_0))\, e^{-i\lambda\sec^2\mu_0\xi_0} J_0(\lambda\sec^2\mu_0\sin\mu_0\xi_0)\, d\xi_0. \quad (5·7, 31)$$

‡ G. N. Watson, *A Treatise on the Theory of Bessel Functions* (2nd ed., Cambridge University Press, 1944), p. 21.

(5·7, 31) is in agreement with the results obtained by Possio,‡ von Borbely § and Temple and Jahn.|| Compare the careful study of the mathematical aspects of the problem by Todd,¶ also the work of Miles.‡‡ Temple and Jahn employ a variant of the classical method of Riemann–Green. They derive a formula which coincides with (5·7, 31) except for a slightly different notation.

Having determined Φ we may calculate the pressure at any point by means of (5·7, 20). The time derivative in that formula refers to a system which is at rest in the surrounding medium, i.e. it is $(\partial\Phi/\partial t)_x$ in the notation of § 5·2. This becomes in the system of coordinates which moves with the aerofoil

$$\left(\frac{\partial\Phi}{\partial t}\right)_x = \left(\frac{\partial\Phi}{\partial t}\right)_{x'} + \frac{\partial x'}{\partial t}\frac{\partial\Phi}{\partial x'} = \left(\frac{\partial\Phi}{\partial t}\right)_{x'} + U\frac{\partial\Phi}{\partial x'}.$$

Hence (5·7, 20) becomes

$$p = p_0 + \rho\left(\left(\frac{\partial\Phi}{\partial t}\right)_{x'} + U\frac{\partial\Phi}{\partial x'}\right). \tag{5·7, 32}$$

For a very thin aerofoil, the slope of the wing is the same at corresponding points on the upper and lower surfaces. In that case the value of Φ and hence of $p - p_0$ at a point of the lower surface is numerically equal but opposite in sign to the value at the corresponding point of the upper surface. Hence the pressure difference between the two surfaces is at any point of the wing

$$\Delta p = 2\rho\left(\left(\frac{\partial\Phi}{\partial t}\right)_{x'} + U\frac{\partial\Phi}{\partial x'}\right) = \frac{2\rho U}{c}\left(\frac{\partial\Phi}{\partial\tau} + \frac{\partial\Phi}{\partial\xi}\right), \tag{5·7, 33}$$

where Φ is given by (5·7, 31). This yields

$$L = -\int_0^c \Delta p\, dx' = -c\int_0^1 \Delta p\, d\xi = -2\rho U\left(\int_0^1 \frac{\partial\Phi}{\partial\tau}\, d\xi + \Phi\left(c, +0, \frac{c\tau}{U}\right)\right) \tag{5·7, 34}$$

for the lift, and

$$M = -\int_0^c \Delta p\, x'\, dx' = -c^2\int_0^1 \Delta p\, \xi\, d\xi = -2\rho U c\int_0^1 \left(\frac{\partial\Phi}{\partial\tau} + \frac{\partial\Phi}{\partial\xi}\right)\xi\, d\xi \tag{5·7, 35}$$

for the pitching moment.

‡ C. Possio, 'L'Azione Aerodinamica sul profilo oscillanto alla velocita ultrasonora', *Acta pontif. Acad. Sci.* vol. 1 (1937), p. 93 (translation *A.R.C. Paper*, no. 7668).

§ S. von Borbely, 'Über die Luftkräfte, die auf einen harmonisch schwingenden zweidimensionalen Flügel bei Überschallgeschwindigkeit wirken', *Z. angew. Math. Mech.* vol. 22 (1942), pp. 190–205.

|| G. Temple and H. A. Jahn, 'Flutter at supersonic speeds. Part I. Mid-chord derivative coefficients for a thin aerofoil at zero incidence', *Rep. Memor. Aero. Res. Comm., Lond.*, no. 2140 (1945).

¶ O. Todd, 'On some boundary value problems in the theory of the non-uniform supersonic motion of an aerofoil', *Rep. Memor. Aero. Res. Comm., Lond.*, no. 2141 (1945).

‡‡ J. W. Miles, 'The aerodynamic forces on an oscillating airfoil at supersonic speeds', *J. Aero. Sci.* vol. 14 (1942), pp. 351–8.

In particular, for a flat plate at zero incidence which oscillates in a direction parallel to the y-axis,

$$F_0(x) = \text{constant} = y_0, \left.\right\} \tag{5·7, 36}$$
$$G_0(x) = i\nu y_0,$$

and so (5·7, 31) becomes

$$\Phi\left(c\xi, +0, \frac{c\tau}{U}\right) = i\nu y_0 c \tan \mu_0 e^{i\lambda\tau} \int_0^\xi f(\xi_0)\, d\xi_0, \tag{5·7, 37}$$

where

$$f(\xi) = e^{i\lambda \sec^2\mu_0 \xi} J_0(\lambda \sec^2 \mu_0 \sin \mu_0 \xi). \tag{5·7, 38}$$

Hence $\quad L = 2\rho U^2 \lambda y_0 \tan \mu_0 e^{i\lambda\tau}\left\{\lambda \int_0^1 d\xi \int_0^\xi f(\xi_0)\, d\xi_0 - i\int_0^1 f(\xi)\, d\xi\right\},$

or, integrating by parts in the first integral,

$$L = 2\rho U^2 \lambda y_0 \tan \mu_0 e^{i\lambda\tau} \int_0^1 (\lambda(1-\xi) - i) f(\xi)\, d\xi. \tag{5·7, 39}$$

Let $f_r(\xi)$ and $f_i(\xi)$ be the real and imaginary part of $f(\xi)$, respectively, and let

$$r_n = \int_0^1 \xi^n f_r(\xi)\, d\xi = \int_0^1 \xi^n \cos(\lambda \sec^2 \mu_0 \xi) J_0(\lambda \sec^2 \mu_0 \sin \mu_0 \xi)\, d\xi$$

and $\quad s_n = -\int_0^1 \xi^n f_i(\xi)\, d\xi = \int_0^1 \xi^n \sin(\lambda \sec^2 \mu_0 \xi) J_0(\lambda \sec^2 \mu_0 \sin \mu_0 \xi)\, d\xi.$

$$\tag{5·7, 40}$$

Then (5·7, 39) may also be written as

$$L = 2\rho U^2 \lambda y_0 \tan \mu_0 e^{i\lambda\tau}\{(\lambda - i)(r_0 - is_0) - \lambda(r_1 - is_1)\}$$
$$= 2\rho U^2 \lambda y_0 \tan \mu_0\{(\lambda(r_0 - r_1) - s_0)\cos \lambda\tau + (\lambda(s_0 - s_1) + r_0)\sin \lambda\tau$$
$$+ i[-(\lambda(s_0 - s_1) + r_0)\cos \lambda\tau + (\lambda(r_0 - r_1) - s_0)\sin \lambda\tau]\}. \tag{5·7, 41}$$

Taking real parts in (5·7, 22), (5·7, 36) and (5·7, 41), we find that the lift corresponding to a deflexion

$$y = F(x', t) = y_0 \cos \nu t = y_0 \cos \lambda\tau \tag{5·7, 42}$$

is $\quad L = 2\rho U^2 \lambda y_0 \tan \mu_0\{(\lambda(r_0 - r_1) - s_0)\cos \lambda\tau + (\lambda(s_0 - s_1) + r_0)\sin \lambda\tau\}.$

$$\tag{5·7, 43}$$

Since $\quad y'(t) = \dfrac{\partial y}{\partial t} = -\nu y_0 \sin \nu t = -\dfrac{U y_0}{c}\lambda \sin \lambda t,$

we may write (5·7, 43) also in the form

$$L = y(t)\, L_y + y'(t)\, L_{y'}, \tag{5·7, 44}$$

where $\quad L_y = 2\rho U^2 \tan \mu_0(\lambda^2(r_0 - r_1) - \lambda s_0)$

and $\quad L_{y'} = -2\rho U c \tan \mu_0(\lambda(s_0 - s_1) + r_0).$

$$\left.\right\} \tag{5·7, 45}$$

This is the form in which the aerodynamic derivatives are given by Temple and Jahn.[‡] They obtain their results with opposite sign by taking their

‡ G. Temple and H. A. Jahn, 'Flutter at supersonic speeds. Part I. Mid-chord derivative coefficients for a thin aerofoil at zero incidence', *Rep. Memor. Aero. Res. Comm., Lond.*, no. 2140 (1945).

z-coordinate (which corresponds to the present y-coordinate) in the opposite direction. Similarly, in the formulae given below, the pitching moment is positive when it tends to turn the wing in a mathematically positive direction (nose down moment), whereas in the paper by Temple and Jahn a nose-up moment is taken as positive.

The above is not the only possible representation of the lift in terms of aerodynamic derivatives. Indeed, it would be more in keeping with the notation used for incompressible flow (§§ 5·2–5·5) to write

$$L = yL_y^* + y'L_{y'}^* + y''L_{y''}^*,$$

where

$$L_y^* = 0,$$

$$L_{y'}^* = L_{y'} = -2\rho Uc \tan\mu_0(\lambda(s_0 - s_1) + r_0),$$

$$L_{y''}^* = \qquad -2\rho c^2 \tan\mu_0\left(r_0 - r_1 - \frac{s_0}{\lambda}\right).$$

Note that s_0/λ remains finite as λ tends to 0.

Both representations are legitimate in the sense explained in §5·1. Adopting the first alternative (equation (5·7, 44)) also for the other derivatives, we may write the pitching moment in the form

$$M = yM_y + y'M_{y'}. \tag{5·7, 46}$$

Similarly, for pitching motion,

$$\left.\begin{aligned} L &= \alpha L_\alpha + \alpha' L_{\alpha'}, \\ M &= \alpha M_\alpha + \alpha' M_{\alpha'}. \end{aligned}\right\} \tag{5·7, 47}$$

By the use of simple statics and kinematics, we may work out the corresponding results for a point of reference which is at a distance hc aft of the leading edge (see equation (5·3, 43)). Then L_y and $L_{y'}$ are still given by (5·7, 45), while the remaining six derivatives are as follows:

$$\left.\begin{aligned} L_\alpha &= 2\rho U^2 c \tan\mu_0(\tfrac{1}{2}\lambda^2(-r_0 + 2r_1 - r_2) + 2\lambda(s_0 - s_1) + r_0 \\ &\quad + h(\lambda^2(r_0 - r_1) - \lambda s_0)), \\ L_{\alpha'} &= 2\rho Uc^2 \tan\mu_0(\tfrac{1}{2}\lambda(2s_0 - 2s_1 + s_2) + 2(r_0 - r_1) - \frac{s_0}{\lambda} \\ &\quad - h(\lambda(s_0 - s_1) + r_0)), \\ M_y &= 2\rho U^2 c \tan\mu_0(\tfrac{1}{2}\lambda^2(r_0 - r_2) - \lambda s_1 - h(\lambda^2(r_0 - r_1) + \lambda s_0)), \\ M_{y'} &= 2\rho Uc^2 \tan\mu_0(-\tfrac{1}{2}\lambda(s_0 - s_2) - r_1 + h(\lambda(s_0 - s_1) + r_0)), \\ M_\alpha &= 2\rho U^2 c^2 \tan\mu_0(\tfrac{1}{6}\lambda^2(-2r_0 + 3r_1 - r_3) + \lambda(s_0 - s_2) + r_1 \\ &\quad + h(\lambda^2(r_0 - r_1) + \lambda(-2s_0 + s_1) - r_0) - h^2(\lambda^2(r_0 - r_1) - \lambda s_0)), \\ M_{\alpha'} &= 2\rho Uc^3 \tan\mu_0(\tfrac{1}{6}\lambda(2s_0 - 3s_1 + s_3) + r_0 - r_2 - \frac{1}{\lambda}s_1 \\ &\quad + h(-\lambda(s_0 - s_1) - 2r_0 + r_1 + \frac{1}{\lambda}s_0) + h^2(\lambda(s_0 - s_1) + r_0)). \end{aligned}\right\}$$

$$\tag{5·7, 48}$$

Let us consider the pitching moment derivatives M_α and $M_{\alpha'}$ in more detail. For small values of the frequency parameter λ, it suffices to retain only the terms which are linear in λ or independent of it. Now, from (5·7, 40),

$$\left.\begin{aligned} r_n &= \frac{1}{n+1} + O(\lambda^2), \\[2mm] s_n &= \frac{1}{n+2}\lambda\sec^2\mu_0 + O(\lambda^3), \end{aligned}\right\} \tag{5·7, 49}$$

where $O(\lambda^2)$, $O(\lambda^3)$ are functions which tend to zero with λ as λ^2 or as λ^3, respectively. It follows that M_α and $M_{\alpha'}$ are, for small λ,

$$\left.\begin{aligned} M_\alpha &= 2\rho U^2 c^2 \tan\mu_0(\tfrac{1}{2}-h), \\ M_{\alpha'} &= 2\rho U c^3 \tan\mu_0(\tfrac{2}{3}-\tfrac{1}{3}\sec^2\mu_0 + h(-\tfrac{3}{2}+\tfrac{1}{2}\sec^2\mu_0) + h^2). \end{aligned}\right\} \tag{5·7, 50}$$

In particular, if the reference point is at mid-chord, $h=\tfrac{1}{2}$, and so

$$\left.\begin{aligned} M_\alpha &= 0, \\ M_{\alpha'} &= \tfrac{1}{6}\rho U c^3 \tan\mu_0(1-\tan^2\mu_0). \end{aligned}\right\} \tag{5·7, 51}$$

It will be seen that $M_{\alpha'}$ is negative for $\tan^2\mu_0 - 1 > 0$. This condition is, in terms of the Mach number, $M_0 = \operatorname{cosec}\mu_0$,

$$(M_0^2 - 1)^{-1} - 1 > 0,$$

i.e. $$M_0 < \sqrt{2}. \tag{5·7, 52}$$

Since $M_\alpha = 0$, the pitching moment is given by

$$M = \alpha' M_{\alpha'}.$$

But according to our sign convention, a negative pitching moment tends to increase the incidence, and simple dynamical consideration then shows that the pitching motion in this case is unstable. However, it appears that the phenomenon just described may be masked by the influence of the wing thickness.‡ The experimental evidence available on this point is inconclusive.

To calculate the rotary pitching derivatives we may use the quasi-static method described in §5·4 (equations (5·4, 48)–(5·4, 79)). The method is legitimate provided the chord of the aerofoil, c, is small compared with the radius of curvature of the flight path, and the non-dimensional parameter qc/U is small compared with unity.§ Collar‖ has applied the quasi-static

‡ See W. P. Jones, 'Negative torsional aerodynamic damping at supersonic speeds', *Rep. Memor. Aero. Res. Comm., Lond.*, no. 2194 (1946), and M. J. Lighthill, 'Oscillating airfoils at high Mach numbers', *J. Aero. Sci.* vol. 20 (1953), pp. 402–6.

§ R. Westley, 'The potential due to a source moving through a compressible fluid, and applications to some rotary derivatives of an aerofoil', *Rep. Coll. Aero. Cranfield*, no. 54 (1952).

‖ A. R. Collar, 'Resistance derivatives of flutter theory. Part II. Results for supersonic speds', *Rep. Memor. Aero. Res. Comm., Lond.*, no. 2139 (1944).

method also to the calculation of the oscillatory derivatives considered above. For the limiting case of small λ this leads to results which are in agreement with $(5\cdot7, 48)$, excepting the angular velocity derivatives $L_{\alpha'}$, $M_{\alpha'}$, for which Collar's results differ from the limiting values obtained from $(5\cdot7, 48)$ by the absence of a factor $1 - \tan^2\mu_0$. Other problems in aerofoil theory for unsteady supersonic flow have been considered by Strang‡ and Chang.§

For a wing of finite span, the character of a problem of unsteady supersonic motion depends on the inclination of the edges of the wing relative to the Mach cones. Let us assume that the wing executes a motion of small magnitude relative to a mean position which advances at constant speed in a fixed direction, such that the edges of the wing are supersonic throughout. In that case we may use a distribution of diverging acoustic sources (see § 4·1 and compare with § 5·6) in order to express the flow in the neighbourhood of the wing. In general, different source distributions will represent the flow above and below the wing.||

Problems of unsteady supersonic motion when some of the wing edges are subsonic have been considered by Miles,¶ who used a Fourier transform procedure. The method has been developed further by Temple.‡‡ The problem of an oscillating Delta wing with subsonic leading edges has been considered by Robinson.§§ This and other problems of oscillatory supersonic flow have been treated also by Germain and Bader |||| by their 'method of homogeneous fields' (compare § 4·10). Other papers on the subject are due to Gardner, Chang, Stewart and Li, and Stewartson.¶¶ Rotary derivatives

---

‡ W. J. Strang, 'A physical theory of supersonic aerofoils in unsteady flow', *Proc. Roy. Soc.* A, vol. 195 (1948), pp. 245–64.

§ C. C. Chang, 'The transient reaction of an airfoil due to change in angle of attack at supersonic speed', *J. Aero. Sci.* vol. 15 (1948), pp. 635–55.

|| See I. E. Garrick and S. I. Rubinow, 'Theoretical study of air forces on an oscillating or steady thin wing in a supersonic main stream', *Tech. Notes Nat. Adv. Comm. Aero., Wash.*, no. 1383 (1947).

¶ J. W. Miles, 'The oscillating rectangular airfoil at supersonic speeds', *Quart. J. Appl. Math.* vol. 9 (1951), pp. 47–65; 'Harmonic and transient motion of a swept wing in supersonic flow', *J. Aero. Sci.* vol. 15 (1948), pp. 343–6.

‡‡ Consult G. Temple, article in ch. 9 of *Modern Developments in Fluid Dynamics. High Speed Flow*, ed. Howarth (Oxford University Press, 1953).

§§ A. Robinson, 'On some problems of unsteady supersonic aerofoil theory', *Rep. Coll. Aero. Cranfield*, no. 16 (1948).

|||| P. Germain and R. Bader, 'Quelques remarques sur le mouvement vibratoire d'une aile en régime supersonique', *Rech. Aéro.* no. 11 (1949), pp. 3–13.

¶¶ C. Gardner, 'Time dependent linearized supersonic flow past planar wings', *Commun. Pure Appl. Math.* vol. 3 (1950), pp. 33–8; C. C. Chang, 'The aerodynamic behaviour of a hamonically oscillating finite sweptback wing in supersonic flow', *Tech. Notes Nat. Adv. Comm. Aero., Wash.*, no. 2467 (1951); H. J. Stewart and Ting-Yi Li, 'Source-superposition method of solution of a periodically oscillating wing at supersonic speeds', *Quart. Appl. Math.* vol. 9 (1951), pp. 31–45; K. Stewartson, 'On the linearised potential theory of unsteady supersonic motion', *Quart. J. Mech. Appl. Math.* vol. 5 (1952), pp. 137–54.

were calculated by Robinson using a quasi-static procedure.‡ The rotary derivatives should not be confused with the corresponding oscillatory derivatives. However, Miles has shown§ that the results provided by the quasi-static methods can be applied also to the calculation of certain oscillatory derivatives for small values of the frequency parameter λ. Other investigations of the problem for small values of λ are due to Miles‖ and Berndt.¶ For pointed wings of small aspect ratio Ribner and Malve-stuto‡‡ obtained a number of derivatives by the method of Munk and Jones (see § 3·10). The advantage of these simplified methods is that they provide compact results without undue labour. The numerical evaluation of the more exact procedures mentioned above is usually so tedious that the provision of concrete numerical data has tended to lag far behind the development of new methods.

‡ A. Robinson, 'Rotary derivatives of a delta wing at supersonic speeds', *J. R. Aero. Soc.* vol. 52 (1948), pp. 735–52.

§ J. W. Miles, 'On damping in pitch for delta wings', *J. Aero. Sci.* vol. 16 (1949), pp. 574–5.

‖ J. W. Miles, 'Quasi-stationary airfoil theory in subsonic compressible flow', *Quart. Appl. Math.* vol. 8 (1951), pp. 351–8.

¶ S. B. Berndt, 'On the theory of slowly oscillating delta wings at supersonic speeds', *Rep. Aero. Res. Inst. Sweden* (FFA), no. 43 (1952).

‡‡ H. S. Ribner and F. S. Malvestuto Jr., 'Stability derivatives of triangular wings at supersonic speeds', *Tech. Rep. Nat. Adv. Comm. Aero., Wash.*, no. 908 (1948).

LIST OF PRINCIPAL SYMBOLS

There are inevitably several different usages of a given symbol in this text; the following is a list of the most important ones that are used consistently throughout the book.

$a = \left(\dfrac{\partial p}{\partial \rho}\right)^{\frac{1}{2}}$ velocity of sound

a defined by $F = F_0 e^{at}$ in exponentially divergent motion (Ch. 5)

$a_0 = \dfrac{\partial C_L}{\partial \alpha}$ lift curve slope

R aspect ratio = (span)^2/area

c chord of aerofoil

$C_L = \dfrac{L}{\frac{1}{2}\rho U^2 S}$ drag coefficient

$C_D = \dfrac{D}{\frac{1}{2}\rho U^2 S}$ lift coefficient

$C_M = \dfrac{M}{\rho U^2 S s_0}$ pitching moment coefficient

$C_p = \dfrac{p}{\frac{1}{2}\rho U^2}$ pressure coefficient

D/Dt differentiation following the motion of the fluid

$\dfrac{\partial}{\partial n}$ differentiation normal to a surface

D drag

D_i induced drag

D_0 profile drag

D_W wave drag

$f(x, y) = z$ equation of aerofoil (steady motion)

$F(x, y, t) = z$ equation of aerofoil (unsteady motion)

f_c mean camber line

f_t half thickness of aerofoil

l lift per unit span

L lift

M Mach number

M pitching moment

M_0 pitching moment about the quarter chord

p pressure

p angular velocity about x-axis (Ch. 5)

q angular velocity about y-axis (Ch. 5)

$\mathbf{q} = (u, v, w)$ velocity vector

$\mathbf{q}' = (u', v', w')$	induced velocity vector
$r = \{x^2 + \beta^2(y^2 + z^2)\}^{\frac{1}{2}}$	hyperbolic distance from the origin
r	angular velocity about z-axis (Ch. 5)
s	local slope of aerofoil
s_0	semi-span of wing
S	wing area
t	time
t	thickness of aerofoil
u	velocity component in x-direction
u'	induced velocity component in x-direction
U	free stream velocity in x-direction
v	velocity component in y-direction
V	free stream velocity in y-direction
w	velocity component in z-direction
$w = u - iv$	complex velocity in the two-dimensional case (Ch. 2)
y_c	mean camber line
$z = x + iy$	complex variable
α	incidence
β	angle of yaw
$\beta = \|M^2 - 1\|^{\frac{1}{2}}$	
γ	ratio of specific heats
γ	semi-apex angle of Delta wing
γ, Γ	circulation, and strength of line vortex
Γ'	strength of horseshoe vortex
δ	boundary layer thickness (Chs. 1, 2)
δ_1	boundary layer displacement thickness
ϵ	trailing edge angle
ζ	z-component of vorticity
$\zeta = \xi + i\eta$	complex variable
η	y-component of vorticity
θ_1	boundary layer momentum thickness
$\lambda = \beta \tan \gamma$	(Ch. 4)
$\lambda = vc/U$	frequency parameter (Ch. 5)
μ	viscosity
$\mu = \sin^{-1} 1/M$	Mach angle
$\mu = ca/U$	(Ch. 5)
ν	frequency of oscillation
ξ	x component of vorticity
$\Pi = \Phi + i\Psi$	complex potential
ρ	density
σ	source strength or density
τ	thickness to chord ratio
τ	doublet strength or density

Φ   velocity potential
Φ′   induced velocity potential
$\chi = y + iz$ complex variable (Ch. 3)
Ψ   stream function
$\boldsymbol{\omega} = (\xi, \eta, \zeta)$ vorticity vector
ω   complex acceleration potential
Ω   acceleration potential

Subscripts

L   leading edge value
T   trailing edge value
+   value above aerofoil
−   value below aerofoil
u   value on upper surface of aerofoil
l   value on lower surface of aerofoil
x   aerodynamic derivative with respect to variable x (Ch. 5)
0   undisturbed (main stream) value
∞   value far downstream in the Trefftz plane, also two di-
         mensional value

Coordinates

In three dimensions the coordinate system used is as follows. The x-, y-
and z-axes (or x'-, y'- and z'-axes) form a right-handed system with the
x-axis pointing in the direction of the flow, i.e. in a direction opposite to
the direction of motion of the wing. The y-axis is taken to point to starboard, and the z-axis upwards. Thus, in this system a positive angle of
attack means a positive angular displacement α in the x, z-plane; similarly,
a pitching moment is positive when it induces a 'nose-up' attitude and
an anticlockwise circulation about the wing (viewed from starboard) is
positive. A positive roll leads to downward movement of the starboard
wing and a positive yaw means that the starboard wing is in front of the port
wing. In Chapter 5 the aerodynamic derivatives are in general referred to
the same system of coordinates, but mention is also made of an alternative
formulation that is used sometimes in that context. Here the directions of
all three axes are reversed, and in place of the lift force L a force $Z = -L$
is introduced.

In two-dimensional theory the x-axis is again taken to be parallel to
flow direction, but, in order to accord with standard practice, the other
axis is now chosen to be the y-axis. Thus, in this case a positive pitching
moment leads to a 'nose-down' attitude, and a positive circulation is anticlockwise. Note, however, that the angle of incidence α is always defined to
be positive when the angle of attack is positive.

ADDITIONAL RECENT REFERENCES

CHAPTER 2

H. K. Cheng and N. Rott. Generalizations of the inversion formula of thin airfoil theory. *J. Rat. Mech. Anal.* vol. 3 (1954), pp. 357–82.

W. Frank. Zur Berechnung von Potentialströmungsfeldern. *Öst. IngArch.* vol. 8 (1954), pp. 97–107.

H. C. Garner. Simple evaluation of the theoretical lift slope and aerodynamic centre of symmetrical aerofoils. *Rep. Memor. Aero. Res. Coun., Lond.*, no. 2847 (1951).

M. B. Glauert. The application of the exact method of aerofoil design. *Rep. Memor. Aero. Res. Coun., Lond.*, no. 2683 (1947).

D. S. Jones. Note on the steady flow of a fluid past a thin aerofoil. *Quart. J. Math.* vol. 6 (1955), pp. 4–8.

D. A. Spence. Prediction of the characteristics of two-dimensional airfoils. *J. Aero. Sci.* vol. 21 (1954), pp. 577–87.

D. A. Spence and N. A. Routledge. Velocity calculations by conformal mapping for two-dimensional aerofoils. *Roy. Aircraft Est. Rep. Aero.* no. 2539 (1955).

CHAPTER 3

G. G. Brebner. The application of camber and twist to swept wings in incompressible flow. *Curr. Paper Aero. Res. Coun.* no. 171 (1954).

R. Brescia. Studio dell'interferenza delle gallerie aerodinamiche con pareti a fessure. Cf. *Atti Accad. Torino*, Cl. Fis. Mat. Nat., vol. 87 (1953), pp. 225–44.

H. Fujikawa. The lift on the symmetrical Joukowski aerofoil in a stream bounded by a wall. *J. Phys. Soc. Japan*, vol. 9 (1954), pp. 233–9, 240–3.

H. C. Garner. Theoretical calculation of the distribution of aerodynamic loading on a delta wing. *Rep. Memor. Aero. Res. Coun., Lond.*, no. 2819 (1949).

G. J. Hancock. The design of thin finite wings in incompressible flow. *Aero. Quart.* vol. 5 (1954), pp. 119–30.

O. Holme and F. Hjelte. On the calculation of the pressure distribution on three dimensional wings at zero incidence in incompressible flow. *K. tekn. Högsk., Inst. Flygtekn., Tech. Note*, no. 23 (1953), 28s.

D. Küchemann. The distribution of lift over the surface of swept wings. *Aero. Quart.* vol. 4 (1953), pp. 261–78.

P. F. Maeder. Some aspects of the behaviour of perforated transonic tunnel walls. *Tech. Rep. Div. Engng, Brown Univ.* WT-15 (1954).

A. Muggia. Sulla teoria delle superfici portanti. *Atti Accad. Torino*, Cl. Sci. Mat. Nat., vol. 87 (1953), pp. 193–9.

K. Nickel. Der höchstmögliche Auftrieb von Tragflügeln. *Z. angew. Math. Mech.* vol. 34 (1954), pp. 374–85.

E. Truckenbrodt. Das Geschwindigkeitspotential der tragenden Fläche bei inkompressibler Strömung. *Z. angew. Math. Mech.* vol. 33 (1953), pp. 165–73.

J. Weber. The effect of aspect ratio on the chordwise load distribution of flat and cambered rectangular wings. *Roy. Aircraft Est. Rep. Aero.* no. 2525 (1954).

L. C. Woods. Second order terms in two-dimensional tunnel blockage. *Aero. Quart.* vol. 4 (1953), pp. 361–72.

CHAPTER 4

S. Asaka. On the velocity distribution on the surface of a symmetric aerofoil at high speeds. I. *Nat. Sci. Rep. Ochanomizu Univ.* vol. 4 (1954), pp. 213–26.

S. Bergmann and M. Schiffer. *Kernel Functions and Elliptic Differential Equations in Mathematical Physics*, ch. 2 (Academic Press Inc., New York, 1953).

L. Bers. Existence and uniqueness of a subsonic flow past a given profile. *Commun. Pure Appl. Math.* vol. 7 (1954), pp. 441–504.

W. Chester. Supersonic flow past wing-body combinations. *Aero. Quart.* vol. 4 (1953), pp. 287–314.

M. H. Clarkson. A second order theory for three dimensional wings in supersonic flow. *Quart. J. Mech. and Appl. Math.* vol. 7 (1954), pp. 203–21.

J. D. Cole. Transonic limits of linearized theory. *Guggenheim Aero. Lab., California Inst. Tech., O.S.R. Tech. Note*, no. 228 (1954).

M. Cooper and F. C. Grant. Minimum-wave-drag airfoil sections for arrow wings. *Tech. Notes Nat. Adv. Comm. Aero., Wash.*, no. 3183 (1954).

A. J. Eggers Jr., C. A. Syvertson and S. Kraus. A study of inviscid flow about airfoils at high supersonic speeds. *Rep. Nat. Adv. Comm. Aero., Wash.*, no. 1123 (1953).

F. I. Frankl and E. A. Karpovich. *Gas Dynamics of Thin Bodies.* Trans. M. D. Friedman. (Interscience, London and New York, 1953.)

P. Germain. Sur l'écoulement subsonique au voisinage de la pointe avant d'une aile delta. *Rech. aéro.* no. 44 (1955), pp. 3–8.

J. H. Giese and H. Cohn. Canonical equations for non-linearized steady irrotational conical flow. *Quart. Appl. Math.* vol. 13 (1955), pp. 351–60.

J. H. Giese and H. Cohn. Two new non-linearized conical flows. *Quart. Appl. Math.* vol. 11 (1953), pp. 101–8.

E. W. Graham, B. J. Beane and R. M. Licher. The drag on non-planar thickness distributions. *Aero. Quart.* vol. 6 (1955), pp. 99–113.

F. C. Grant. The proper combination of lift loading for least drag on a supersonic wing. *Tech. Notes Nat. Adv. Comm. Aero., Wash.*, no. 3533 (1955).

F. M. Hamaker, S. E. Neice and T. J. Wong. The similarity law for hypersonic flow and requirements for dynamic similarity of related bodies in free flight. *Rep. Nat. Adv. Comm. Aero., Wash.*, no. 1147 (1953).

K. C. Harder and E. B. Klunker. On slender-body theory at transonic speeds. *Res. Memor. Nat. Adv. Comm. Aero., Wash.*, no. L 54 A 29a (1954).

M. A. Heaslet and H. Lomax. The calculation of pressure on slender airplanes in subsonic and supersonic flow. *Rep. Nat. Adv. Comm. Aero., Wash.*, no. 1185 (1954).

C. Kaplan. On the small-disturbance iteration method for the flow of a compressible fluid with application to a parabolic cylinder. *Tech. Notes Nat. Adv. Comm. Aero., Wash.*, no. 3318 (1955).

F. Keune. On the subsonic, transonic and supersonic flow around low aspect ratio wings with incidence and thickness. *K. tekn. Högsk., Inst. Flygtekn., Tech. Notes*, no. 28 (1953), 32s.

F. Keune. The influence of camber and geometrical twist on low aspect ratio wings of finite thickness in subsonic, transonic and supersonic flow. *K. tekn. Högsk., Inst. Flygtekn., Tech. Notes*, no. 29 (1953), 13s.

A. F. Kryučin. On the problem of transonic flow about a profile. *Appl. Math. Mech., Leningr.*, vol. 18 (1954), pp. 547–60.

G. N. Lance. The delta wing in a non-uniform supersonic stream. *Aero. Quart.* vol. 5 (1954), pp. 55–72.

G. N. Lance. The lift of twisted and cambered wings in supersonic flow. *Aero. Quart.* vol. 6 (1955), pp. 149–63.

H. R. Lawrence and A. H. Flax. Wing-body interference at subsonic and supersonic speeds—survey and new developments. *J. Aero. Sci.* vol. 21 (1954), pp. 289–324.

J. Legras. La seconde approximation de l'aile élancée en écoulement subsonique. *Rech. aéro.* no. 42 (1954), pp. 17–21.

D. C. M. Leslie and J. D. Perry. Wave drag of wings at supersonic speeds. *Proc. Roy. Soc. A*, vol. 225 (1954), pp. 213–35.

H. C. Levy. Exact solutions for transonic flow past cusped aerofoils. *Rep. A.R.L. Aust.* A 87 (1954).

J. C. Martin. A vector study of linearized supersonic flow applications to non-planar problems. *Rep. Nat. Adv. Comm. Aero., Wash.*, no. 1143 (1953).

J. W. Miles. A general solution for the rectangular airfoil in supersonic flow. *Quart. Appl. Math.* vol. 11 (1953), pp. 1–8.

H. Mirels. Aerodynamics of slender wings and wing-body combinations having swept trailing edges. *Tech. Notes Nat. Adv. Comm. Aero., Wash.*, no. 3105 (1954).

G. Morikawa. Conical potential solutions in linearized supersonic flow. *J. Aero. Sci.* vol. 20 (1953), pp. 283–5.

T. Nonweiler. The theoretical wave drag at zero lift of fully tapered swept wings of arbitrary section. *Rep. Coll. Aero. Cranfield*, no. 76 (1953).

D. C. Pack and S. I. Pai. Similarity laws for supersonic flows. *Q. Appl. Math.* vol. 11 (1954), pp. 377–84.

H. Richter and W. Müller. Zur Tschaplyginschen Hodographenmethode bei Unter-schallströmungen mit Zirkulation. *Z. angew. Math. Mech.* vol. 35 (1955), pp. 1–11.

A. M. Rodriguez, P. A. Lagerstrom and E. W. Graham. Theorems concerning the drag reduction of wings of fixed planform. *J. Aero. Sci.* vol. 21 (1954), pp. 1–7.

W. R. Sears (ed.). *General Theory of High Speed Aerodynamics. High Speed Aero-dynamics and Jet Propulsion* (Princeton University Press, N.J. 1954).

J. B. Serrin. Comparison theorems for subsonic flows. *J. Math. Phys.* vol. 33 (1954), pp. 27–45.

J. R. Spreiter. On the application of transonic similarity rules to wings of finite span. *Rep. Nat. Adv. Comm. Aero., Wash.*, no. 1153 (1954).

J. R. Spreiter and A. Alksne. Theoretical prediction of pressure distributions on non-lifting airfoils at high subsonic speeds. *Tech. Notes Nat. Adv. Comm. Aero., Wash.*, no. 3096 (1954).

T. Theodorsen. Limits and classification of supersonic flows. *Tech. Note Univ. Maryland Inst. Fluid Dynamics Appl. Math.* BN-23.

F. G. Tricomi. Beispiel einer Strömung mit Durchgang durch die Schallgeschwindig-keit. *Mh. Math.* vol. 58 (1954), pp. 160–71.

M. D. Van Dyke. A study of hypersonic small-disturbance theory. *Rep. Nat. Adv. Comm. Aero., Wash.*, no. 1194 (1954).

M. D. Van Dyke. Second order subsonic airfoil-section theory and its practical application. *Tech. Notes Nat. Adv. Comm. Aero., Wash.*, no. 3390 (1955).

A. I. van de Vooren. An approach to lifting surface theory. *Rep. Nat. Lucht-vaartlaboratorium, Amst.*, no. F 129 (1953).

W. G. Vincenti and N. H. Fisher. Calculation of the supersonic pressure distribution on a single-curved tapered wing in regions not influenced by the root or tip. *Tech. Notes Nat. Adv. Comm. Aero., Wash.*, no. 3499 (1955).

G. N. Ward. *Linearized Theory of Steady High Speed Flow* (Cambridge University Press, 1955).

G. N. Ward. The drag of source distributions in linearised supersonic flow. *Rep. Coll. Aero. Cranfield*, no. 88 (1955).

J. Weber. The calculation of the pressure distribution on thick wings of small aspect ratio at zero lift in subsonic flow. *Roy. Aircraft Est. Rep. Aero.* no. 2519 (1954).

L. C. Woods. The application of the polygon method to the calculation of the com-pressible subsonic flow round some two-dimensional profiles. *Curr. Paper, Aero. Res. Coun. Lond.*, no. 115 (1953).

L. C. Woods. The design of two-dimensional aerofoils with mixed boundary con-ditions. *Quart. Appl. Math.* vol. 13 (1955), pp. 139–46.

L. C. Woods and A. Thom. A new relaxation treatment of the compressible two-dimensional flow about an aerofoil with circulation. *Rep. Memor. Aero. Res. Coun., Lond.*, no. 2727 (1950).

CHAPTER 5

W. E. A. Acum. Aerodynamic forces on rectangular wings oscillating in a supersonic air stream. *Rep. Memor. Aero. Res. Coun., Lond.,* no. 2763 (1950).

S. Bergmann and M. Schiffer. *Kernel Funtions and Elliptic Differential Equations in Mathematical Physics,* Ch. 2 (Academic Press Inc., New York, 1953).

S. B. Berndt. On the theory of slowly oscillating delta wings at supersonic speeds. *Rep. Flygtekn. Försöksanstalt.* no. 43 (1952).

A. H. Flax. Reverse-flow and variational theorems for lifting surfaces in nonstationary compressible flow. *J. Aero. Sci.* vol. 20 (1953), pp. 120–6.

L. E. Fraenkel. On the unsteady motion of a slender body through a compressible fluid. *Aero. Quart.* vol. 6 (1955), pp. 59–80.

F. I. Frankl and E. A. Karpovich. *Gas Dynamics of Thin Bodies* (Interscience, London and New York, 1953).

W. P. Jones. Supersonic theory for oscillating wings of any plan form. *Rep. Memor. Aero. Res. Coun. Lond.,* no. 2655 (1948).

W. P. Jones. The influence of thickness/chord ratio on supersonic derivatives for oscillating aerofoils. *Rep. Memor. Aero. Res. Coun., Lond.,* no. 2679 (1948).

H. G. Küssner. A general method for solving problems of the unsteady lifting surface theory in the subsonic range. *J. Aero. Sci.* vol. 21 (1954), pp. 17–26.

H. G. Küssner. *A Review of the Two-dimensional Problem of Unsteady Lifting Surface Theory during the last Thirty Years* (Inst. Fluid Dynamics Appl. Math., Univ. Maryland, Lect. Ser. no. 23).

H. G. Küssner. The difference property of the kernel of the unsteady lifting surface theory. *J. Aero. Sci.* vol. 22 (1955), pp. 227–30.

E. A. Krasil'ščikova. Unsteady motion of a profile in a compressible fluid. *Dokl. Akad. Nauk. SSSR,* vol. 94 (1954), pp. 397–400.

E. A. Krasil'ščikova. Unsteady motions of a wing of infinite span. *Izvestiya Akad. Nauk. SSSR. Otd. Techn. Nauk,* no. 2 (1954), pp. 25–41.

H. Lomax. Lift developed on unrestrained rectangular wings entering gusts at subsonic and supersonic speeds. *Rep. Nat. Adv. Comm. Aero., Wash.,* no. 1162 (1954).

H. Lomax, M. A. Heaslet, F. B. Fuller and L. Sluder. Two- and three-dimensional unsteady lift problems in high-speed flight. *Rep. Nat. Adv. Comm. Aero.* no. 1077 (1952).

M. Marini. Le azioni aerodinamiche su ali a freccia in moto oscillatorio. *Aerotecnica,* vol. 33 (1953), pp. 275–87.

J. C. Martin and N. Gerber. The effect of thickness on airfoils with constant vertical acceleration at supersonic speeds. *J. Aero. Sci.* vol. 22 (1955), pp. 179–88.

J. C. Martin and F. S. Malvestuto. Aerodynamics of a rectangular wing of infinite aspect ratio at high angles of attack and supersonic speeds. *Tech. Notes Nat. Adv. Comm. Aero., Wash.,* no. 3421 (1955).

H. Merbt and M. Landahl. Aerodynamic forces on oscillating low aspect ratio wings in compressible flow. *K. tekn. Högsk.,* Inst. Flygtekn., Tech. Notes, no. 30, 20s (1953).

J. W. Miles. On the low aspect ratio oscillating rectangular wing in supersonic flows. *Aero. Quart.* vol. 4 (1954), pp. 231–44.

J. W. Miles. Linearization of the equations of non-steady flow in a compressible fluid. *J. Math. Phys.* vol. 33 (1954), pp. 135–43.

J. W. Miles. On the transformation of the linearized equations of unsteady supersonic flow. *Quart. Appl. Math.* vol. 12 (1954), pp. 1–12.

A. Robinson. *On some Problems of Unsteady Wing Theory.* Second Canadian Symposium on Aerodynamics, Inst. of Aerophysics, Univ. of Toronto, 1954.

W. R. Sears (ed.). *General Theory of High Speed Aerodynamics. High Speed Aerodynamics and Jet Propulsion* (Princeton Univ. Press, N.J., 1954).

R. Timman. La théorie des profils minces en écoulement non stationnaire en fluide incompressible ou compressible. *Journées de Mécanique des Fluides, Marseille,* (1952) pp. 285–327.

R. Timman. Linearized theory of the oscillating airfoil in compressible subsonic flow. *J. Aero. Sci.* vol. 21 (1954), pp. 230–50.

M. Tobak. On the use of the indicial function concept in the analysis of unsteady motions of wings and wing-tail combinations. *Rep. Nat. Adv. Comm. Aero., Wash.,* no. 1188 (1954).

M. D. Van Dyke. Supersonic flow past oscillating airfoils including nonlinear thickness effects. *Rep. Nat. Adv. Comm. Aero., Wash.,* no. 1183 (1955).

C. E. Watkins and J. H. Berman. On the kernel function of the integral equation relating lift and downwash distributions of oscillating wings in supersonic flow. *Tech. Notes Nat. Adv. Comm. Aero., Wash.,* no. 3438 (1955).

L. C. Woods. The lift and moment acting on a thick aerofoil in unsteady motion. *Phil. Trans.* A., vol. 247 (1954), pp. 131–62.

INDEX OF AUTHORS

INDEX OF SUBJECTS

Printed in the United States
By Bookmasters